2.8 ABSOLUTE VALUE

$$|x| = \begin{cases} x & \text{when } x \geq 0 \\ -x & \text{when } x < 0 \end{cases}$$

If $a > 0$, then

$$|a| = \sqrt{a^2}$$

$|x| = a$ is equivalent to $x = a$ or $x = -a$.

$|x| < a$ is equivalent to $-a < x < a$.

$|x| > a$ is equivalent to $x > a$ or $x < -a$.

$$|ab| = |a||b| \qquad \left|\frac{a}{b}\right| = \frac{|a|}{|b|} \qquad |a + b| \leq |a| + |b|$$

3.1 THE RECTANGULAR COORDINATE SYSTEM

Distance formula: $d(PQ) = \sqrt{(x_2 - x_1)^2 + (y_2 - y_1)^2}$

Midpoint formula: $\left(\dfrac{x_1 + x_2}{2}, \dfrac{y_1 + y_2}{2}\right)$

3.2 THE SLOPE OF A NONVERTICAL LINE

Formula for slope: $m = \dfrac{y_2 - y_1}{x_2 - x_1} \quad (x_2 \neq x_1)$

Lines with equal slopes are parallel.

Lines with slopes that are negative reciprocals are perpendicular.

3.3 WRITING EQUATIONS OF LINES

Point–slope form: $y - y_1 = m(x - x_1)$

Slope–intercept form: $y = mx + b$

General form: $Ax + By = C$

3.4 GRAPHS OF EQUATIONS

Tests for symmetry:

If $(-x, y)$ lies on a graph whenever (x, y) does, the graph is symmetric about the y-axis.

If $(x, -y)$ lies on a graph whenever (x, y) does, the graph is symmetric about the x-axis.

If $(-x, -y)$ lies on a graph whenever (x, y) does, the graph is symmetric about the origin.

Equations of a circle with radius r:

$(x - h)^2 + (y - k)^2 = r^2$; center at (h, k)

$x^2 + y^2 = r^2$; center at origin

3.5 PROPORTION AND VARIATION

If k is a constant,

$y = kx \quad y$ varies directly with x

$y = \dfrac{k}{x} \quad y$ varies inversely with x

$y = kxz \quad y$ varies jointly with x and z

4.2 QUADRATIC FUNCTIONS

The graph of

$$y - k = a(x - h)^2 \quad (a \neq 0)$$

is a parabola with vertex at (h, k).

The graph of

$$y = ax^2 + bx + c \quad (a \neq 0)$$

is a parabola with vertex at

$$\left(-\frac{b}{2a}, c - \frac{b^2}{4a}\right).$$

4.4 TRANSLATING AND STRETCHING GRAPHS

If $k > 0$, the graph of $\begin{cases} y = f(x) + k \\ y = f(x) - k \end{cases}$ is identical to the graph

of $y = f(x)$, except it is translated k units $\begin{cases} \text{upward} \\ \text{downward} \end{cases}$.

If $k > 0$, the graph of $\begin{cases} y = f(x - k) \\ y = f(x + k) \end{cases}$ is identical to the graph

of $y = f(x)$, except it is translated k units to the $\begin{cases} \text{right} \\ \text{left} \end{cases}$.

4.6 OPERATIONS ON FUNCTIONS

If the ranges of functions f and g are subsets of the real numbers, then

$$(f + g)(x) = f(x) + g(x)$$
$$(f - g)(x) = f(x) - g(x)$$
$$(f \cdot g)(x) = f(x) \cdot g(x)$$
$$(f/g)(x) = \frac{f(x)}{g(x)} \quad (g(x) \neq 0)$$
$$(f \circ g)(x) = f(g(x))$$

Turn the graphing calculator into a powerful tool for your success!

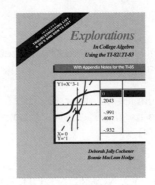

Explorations in College Algebra Using the TI-82/TI-83: With Appendix Notes for the TI-85
by Deborah J. Cochener and Bonnie M. Hodge,
both of Austin Peay State University

$19.00 Single Copy Price. 384 pages. Spiral bound. 8 1/2 x 11.
ISBN: 0-534-34228-0. © 1997. Published by Brooks/Cole.

You can quickly learn to use the graphing calculator to develop problem-soving and critical-thinking skills that will help improve your performance in your college algebra course

Designed to help you succeed in your college algebra course, this unique and student-friendly workbook improves both your understanding and retention of college algebra concepts—using the graphing calculator. By integrating technology into mathematics, the authors help you develop problem-solving and critical-thinking skills.

To guide you in your explorations, you'll find:
- hands-on applications with solutions
- correlation charts that relate course topics to the workbook units
- key charts (specific to the TI-82, TI-83, and TI-85) that show which units introduce keys on the calculator
- a "Troubleshooting Section" to help you avoid common errors

This is the best graphing calculator lab manual supplement that I have seen. After reviewing it, I feel that is time for a change ...to a more subjective, comprehensive, exercise packed lab manual ...the lab manual being Explorations in College Algebra Using the TI-82/TI-83
Chris Bendixen, Sue Bennett College

I did not find anything that I did not like about this manuscript. I thought it was well written and organized. The authors did a good job of describing the use of specific functions of the graphing calculators and following through with good examples that illustrated their use.
David Ray, University of Tennessee-Martin

Topics

This text contains 50 units divided into the following subsections:
Textbook Correlation Charts
    Basic Calculator Operations
    Graphically Solving Equations and Inequalities
    Graphing and Applications of Equations in Two Variables
    Conic Sections
    Miscellaneous (Matrices, Combinatorics and Probability, and Sequences and Series)
    Stat Plots
    Programming
Troubleshooting
Calculator Menus

Order your copy today!

To receive your copy of *Explorations in College Algebra Using the TI-82/TI-83: With Appendix Notes for the TI-85,* simply mail in the order form attached.

ORDER FORM

To order, simply fill out this coupon and return it to Brooks/Cole along with your check, money order, or credit card information.

_____Yes! I would like to order *Explorations in College Algebra Using the TI-82/TI-83: With Appendix Notes for the TI-85*, by Deborah J. Cochener and Bonnie M. Hodge, ISBN: 0-534-34228-0 for $19.00.

Residents of: AL, AZ, CA, CT, CO, FL, GA, IL, IN, KS, KY, LA, MA, MD, MI, MN, MO, NC, NJ, NY, OH, PA, RI, SC, TN, TX, UT, VA, WA, WI must add appropriate state sales tax.

Subtotal _____
Tax _____
Handling __$4.00__
Total _____

Payment Options

_____ Check or money order enclosed

or bill my ____VISA ____MasterCard ____American Express

Card Number: _____

Expiration Date: _____

Signature: _____

Note: Credit card billing and shipping addresses must be the same.

Please ship my order to: (Please print.)

Name _____

Street Address_____

City _____ State _____ Zip+4_____

Telephone (   )_____ e-mail _____

Mail to:

Brooks/Cole Publishing Company
Dept. 8BCMA130
511 Forest Lodge Road, Pacific Grove, California 93950-5098
Phone: (408) 373-0728; Fax: (408) 375-6414
Internet: http://www.brookscole.com

I(T)P **International Thomson Publishing Education Group**

8/97

P.S. If this book is appropriate for a course that you teach, please send your request for a complimentary review copy, on department letterhead, to the address listed above. Prices subject to change without notice.

ORDER FORM

_____Yes! Send me a copy of *Scientific Notebook™ for Windows®95 and Windows NT® 4.0* (ISBN: 0-534-34864-5)

_____Copies x $59.95* =_____

Residents of: AL, AZ, CA, CT, CO, FL, GA, IL, IN, KS, KY, LA, MA, MD, MI, MN, MO, NC, NJ, NY, OH, PA, RI, SC, TN, TX, UT, VA, WA, WI must add appropriate state sales tax.

Subtotal _____
Tax _____
Handling ___$4.00___
Total Due _____

Payment Options

_____ Check or money order enclosed

Bill my ____VISA ____MasterCard ____American Express

Card Number: _____

Expiration Date: _____

Signature: _____

Please ship my order to: *(Credit card billing and shipping addresses must be the same)*

Name _____

Institution _____

_____

Street Address_____

_____

City _____ State _____ Zip+4_____

Telephone (   )_____ e-mail _____

Your credit card will not be billed until your order is shipped. Prices subject to change without notice. We will refund payment for unshipped out-of-stock titles after 120 days and for not-yet-published titles after 180 days unless an earlier date is requested in writing from you.

Mail to:

Brooks/Cole Publishing Company
Source Code 8BCTC022
511 Forest Lodge Road
Pacific Grove, California 93950-5098
Phone: (408) 373-0728; Fax: (408) 375-6414
e-mail: info@brookscole.com

* Call after 12/1/97 for current prices: 1-800-487-3575

SIXTH EDITION

College Algebra

To our wives, Carol and Martha;
and our children:
Kristy and Steven;
Sarah, Heidi, and David

Books in the Gustafson/Frisk Series

SIXTH EDITION

College Algebra ■ ■ ■ ■ ■ ■ ■ ■ ■

R. David Gustafson
Rock Valley College

Peter D. Frisk
Rock Valley College

Brooks/Cole Publishing Company

I(T)P™ An International Thomson Publishing Company

Pacific Grove • Albany • Bonn • Boston • Cincinnati • Detroit • London • Madrid • Melbourne
Mexico City • New York • Paris • San Francisco • Singapore • Tokyo • Toronto • Washington

Publisher: *Robert W. Pirtle*
Project Development Editor: *Elizabeth Rammel*
Marketing Team: *Jennifer Huber, Christine Davis*
Editorial Assistant: *Melissa Duge*
Production Editor: *Ellen Brownstein*
Production Service: *Hoyt Publishing Services*
Manuscript Editor: *David Hoyt*
Permissions Editor: *Carline Haga*

Interior Design: *E. Kelly Shoemaker, Vernon Boes*
Interior Illustration: *Lori Heckelman*
Cover Design: *Roy Neuhaus*
Cover Photo: *M. Gibson/H. Armstrong Roberts*
Art Coordinator: *David Hoyt*
Typesetting: *The Clarinda Company*
Cover Printing: *Phoenix Color Corp.*
Printing and Binding: *World Color (Taunton)*

For more information, contact:

BROOKS/COLE PUBLISHING COMPANY
511 Forest Lodge Road
Pacific Grove, CA 93950
USA

International Thomson Publishing Europe
Berkshire House 168-173
High Holborn
London WC1V 7AA
England

Thomas Nelson Australia
102 Dodds Street
South Melbourne, 3205
Victoria, Australia

Nelson Canada
1120 Birchmount Road
Scarborough, Ontario
Canada M1K 5G4

International Thomson Editores
Seneca 53
Col. Polanco
11560 México, D. F., México

International Thomson Publishing GmbH
Königswinterer Strasse 418
53227 Bonn
Germany

International Thomson Publishing Asia
221 Henderson Road
#05-10 Henderson Building
Singapore 0315

International Thomson Publishing Japan
Hirakawacho Kyowa Building, 3F
2-2-1 Hirakawacho
Chiyoda-ku, Tokyo 102
Japan

Printed in the United States of America

10   9   8   7   6   5   4   3   2   1

Library of Congress Cataloging-in-Publication Data

Gustafson, R. David (Roy David), [date]
    College algebra / R. David Gustafson, Peter Frisk.—6th ed.
        p.   cm.
    Includes index.
    ISBN 0-534-35159-X (hardcover : alk. paper)
    1. Algebra.   I. Frisk, Peter D., [date]   .   II. Title.
QA154.2.G87   1997
512.9—dc21

To the Instructor

College Algebra has been extensively revised in this sixth edition. This revision was motivated by the need to prepare students better for trigonometry, precalculus, statistics, business mathematics, finite mathematics, liberal arts mathematics, or everyday life.

It has been our goal to write a text that

- Is relevant and easy to understand.
- Stresses the concept of function throughout.
- Uses real-life applications to motivate problem solving.
- Develops critical thinking skills in all students.
- Develops the necessary skills for students to continue in their study of mathematics.

We believe that we have accomplished this goal through a successful blending of content and pedagogy. We present comprehensive, in-depth, precise coverage of the topics of college algebra, incorporated into a framework of tested teaching strategies combined with carefully selected pedagogical features. In keeping with the spirit of the NCTM standards and the recommendations of the American Mathematical Association of Two-Year Colleges, this text emphasizes conceptual understanding, problem solving, and the use of technology.

■ CHANGES IN THE SIXTH EDITION

The overall effects of the changes made to the sixth edition are as follows:

- *To increase the emphasis on learning mathematics through graphing.* Although graphing calculators are incorporated throughout the book, their use is not required. All of the topics are fully discussed in traditional ways.

Of course, we recommend that instructors use the graphing-calculator material.

- *To increase the emphasis on problem solving through realistic applications.* The variety of application problems has been increased significantly, and all application problems are labeled with special titles.

- *To fine-tune the presentation of topics* for better flow of ideas and for clarity.

- *To increase the visual interest by the use of color.* We include color not just as a design feature, but to highlight terms that instructors would point to in a classroom discussion.

To make the book more useful to students, we have:

1. Added Self Checks following most examples.

2. Integrated more applications throughout the text.

3. Added a chapter on mathematics of finance.

4. Added Vocabulary and Concepts exercises to each exercise set.

5. Added Discovery and Writing exercises to each exercise set.

6. Added Review exercises to each exercise set.

7. Included cumulative review exercises after every two chapters.

8. Divided many sections into more subsections.

9. Included Problems and Projects sections at the end of each chapter, for group learning.

10. Produced a set of text-specific tutorial videotapes. Video symbols [⊙⊙] in the text mark the examples taught on tape.

11. Produced a completely new question bank for the computerized testing system.

Some of the specific changes made to chapters include:

Chapter 1 does more work with interval notation. It continues to be a thorough review of basic algebra.

Chapter 2 covers equations and inequalities. The solutions to inequalities are given in terms of graphs and intervals.

Chapter 3 covers graphing lines, slope, writing equations of lines, general graphing of other equations, and ratio and proportion. Since graphing-calculator material has been integrated into the other sections of the chapter, the old Section 3.5 has been eliminated. We now present slope as a rate of change.

Chapter 4 no longer discusses relations. More emphasis is given to defining functions by tables, graphs, and equations. There is more work on finding domains and ranges of functions. The material on translations and stretching has been rewritten to simplify it. Reflection about the *x*-axis is now taught as a separate topic rather than as a special case of a stretching.

We now include applications of rational functions.

Chapters 5 and 6 have been interchanged. Chapter 5 now covers Exponential and Logarithmic Functions. This material has been reorganized to integrate the applications into the chapter, rather than having a separate section on applications.

Chapter 6 now covers solving polynomial equations. Synthetic division has been integrated into the first section, and the material on nested intervals has been dropped. We believe this change smooths the flow of ideas and certainly improves the pace of the chapter.

Chapter 7 deals with systems of linear equations. Graphing calculators are used to enhance the discussion of solving systems by graphing. The application problems have been made more relevant. The material on graphing linear inequalities in two variables is covered in the section discussing systems of linear inequalities.

Chapter 8 deals with conic sections and quadratic systems.

Chapter 9 has been reorganized. The section on arithmetic and geometric sequences has been divided into separate sections. The applications of sequences have been integrated into these sections.

Chapter 10 includes effective rates of interest, present value, annuities and future value, sinking funds, and amortization.

All chapters include more work with **calculators and graphing calculators.**

■ FEATURES

We have kept the pedagogical features that made previous editions of the book so successful:

Solid Mathematics   The treatment of college algebra remains direct and straightforward. Although the treatment is mathematically sound, it is not so rigorous that it will confuse students. Every effort has been made to ensure the accuracy of the mathematics and of the answers to the exercises. The book has been critiqued by dozens of reviewers. Each author and a problem checker have worked every exercise. All exercise sets have Vocabulary and Concepts problems, Practice problems, Discovery and Writing problems, and Review problems. The book contains more than 4,000 exercises.

Accessibility to Students   The book is written for students to read and understand. The numerous problems within each exercise set are carefully keyed to over 400 worked examples, in which author's notes explain many of the steps used in the problem-solving process. Most examples are followed by a Self Check. The answers to the odd-numbered exercises appear in an appendix in the Student Edition.

Review is incorporated into the book in many ways: There are chapter summaries, review exercises at the end of each chapter, cumulative review exercises after every two chapters, and endpapers that list, in order of presentation, the important formulas developed in the book.

Emphasis on Applications   To show that mathematics is useful, we include a large number of word problems and applications throughout the book.

■ ORGANIZATION AND COVERAGE

The book can be used in a variety of ways. To maintain optimum flexibility, many chapters are sufficiently independent to allow you to pick and choose topics that are relevant to your students' needs. The following diagram shows how the chapters are related.

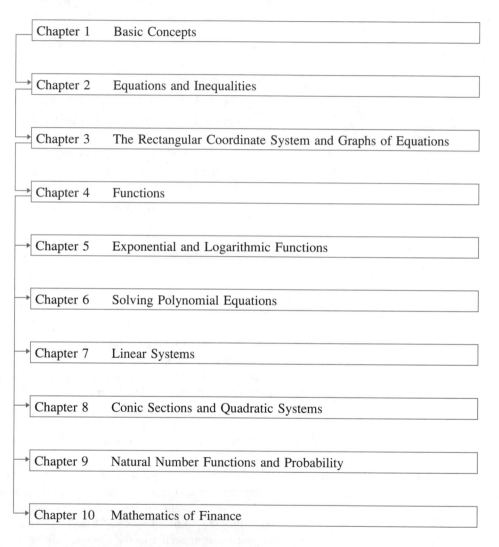

Chapter 1	Basic Concepts
Chapter 2	Equations and Inequalities
Chapter 3	The Rectangular Coordinate System and Graphs of Equations
Chapter 4	Functions
Chapter 5	Exponential and Logarithmic Functions
Chapter 6	Solving Polynomial Equations
Chapter 7	Linear Systems
Chapter 8	Conic Sections and Quadratic Systems
Chapter 9	Natural Number Functions and Probability
Chapter 10	Mathematics of Finance

■ CALCULATORS AND COMPUTERS

The use of calculators is assumed throughout the book. We believe that students should learn calculator skills in the mathematics classroom. They will then be prepared to use calculators in science and business classes and for nonacademic purposes. The directions within each exercise set indicate which exercises require calculators.

■ ANCILLARIES FOR THE INSTRUCTOR

Instructor's Edition In the Instructor's Edition, the answer is printed in blue next to each exercise.

Complete Solutions Manual The Complete Solutions Manual contains solutions to all even- and odd-numbered exercises from the text, including Chapter Summaries, Chapter Reviews, and Chapter Tests. (ISBN 0-534-35160-3)

ITP Tools *ITP Tools,* a fully integrated suite of programs that includes ITP Test/Tutorial, ITP Class Manager, and ITP Test On-Line, provides text-specific algorithmic testing and tutorial options designed to offer you more flexibility and your students greater continuity. Computer platforms it supports include Windows 3.1, Windows 95, Windows NT, Macintosh, and Power Macintosh.

- ITP Test/Tutorial features algorithmic test generation drawing from a bank of text-specific items as well as the ability to import and edit your own questions and graphics. The tutorial feature offers feedback for all objective questions delivered through ITP Test On-Line.
- ITP Test On-Line allows the students to complete an on-line test, practice test, or tutorial. Tests and tutorials can be delivered on a floppy disk, local hard disk, LAN (Local Area Network) volume, or via an Internet IP address.
- ITP Class Manager provides the ability to extract and track scores from on-line tests and practice tests created by ITP Test.

For a more complete description of each component of ITP Tools and to download a self-running demonstration, please visit Brooks/Cole's web site:
http://www.brookscole.com
(Win: ISBN 0-534-35164-6; Mac: ISBN 0-534-35163-8)

Test Manual This manual contains four ready-to-use forms of every chapter test—two multiple-choice and two free-response. (ISBN 0-534-35166-2)

Text-Specific Video Tutorial Series A set of book-specific videotapes is available without charge to adopters of this text (limited to one set per department, although copies may be made). The videos feature segments taught by the authors, worked-out solutions to many examples in the book, and animated graphics. (ISBN 0-534-35167-0)

■ ANCILLARY FOR THE STUDENT

Student Solutions Manual and Study Guide This manual features worked solutions to all the odd-numbered exercises in the text, as well as solutions to the Chapter Reviews, Chapter Summaries, and Chapter Tests. It also offers hints and additional problems for practice, similar to those in the text. (ISBN 0-534-35161-1)

■ EXAMPLES OF FEATURES IN THE TEXT

Videotape icons show which ▶
examples are taught on tape.

Most examples
have Self Checks. ▶

200   CHAPTER 3   THE RECTANGULAR COORDINATE SYSTEM AND GRAPHS OF EQUATIONS

EXAMPLE 1   Find the slope of the line passing through $P(-1, -2)$ and $Q(7, 8)$. (See Figure 3-17.)

Solution   We can let $P(x_1, y_1) = P(-1, -2)$ and $Q(x_2, y_2) = Q(7, 8)$. Then we substitute -1 for x_1, -2 for y_1, 7 for x_2, and 8 for y_2 to get

$$m = \frac{\text{change in } y}{\text{change in } x}$$

$$m = \frac{y_2 - y_1}{x_2 - x_1}$$

$$= \frac{8 - (-2)}{7 - (-1)}$$

$$= \frac{10}{8}$$

$$= \frac{5}{4}$$

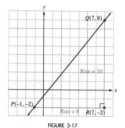

FIGURE 3-17

The slope of the line is $\frac{5}{4}$. We would have obtained the same result if we had let $P(x_1, y_1) = P(7, 8)$ and $Q(x_2, y_2) = Q(-1, -2)$.   ■

Self Check   Find the slope of the line passing through $P(-3, -4)$ and $Q(5, 9)$.
Answer $\frac{13}{8}$

WARNING!   When calculating slope, always subtract the y values and the x values in the same order.

286   CHAPTER 4   FUNCTIONS

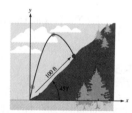

ILLUSTRATION 1

24. Maximizing area   The rectangular garden in Illustration 2 has a width of x and a perimeter of 100 feet. Find x such that the area of the rectangle is maximum.

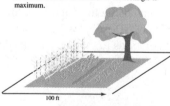

100 ft

ILLUSTRATION 2

25. Maximizing storage area   A farmer wants to partition a rectangular feed storage area in a corner of his barn. The barn walls form two sides of the stall, and the farmer has 50 feet of partition for the remaining two sides. What dimensions will maximize the area? (See Illustration 3.)

26. Maximizing grazing area   A rancher wishes to enclose a rectangular partitioned corral with 1800 feet of fencing. (See Illustration 4.) What dimensions of the corral would enclose the largest possible area? Find the maximum area.

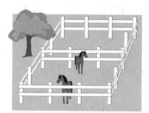

ILLUSTRATION 4

27. Sheet metal fabrication   A 24-inch-wide sheet of metal is to be bent into a rectangular trough with the cross section shown in Illustration 5. Find the dimensions that will maximize the amount of water the trough can hold. That is, find the dimensions that will maximize the cross-sectional area.

Depth
24 in.
Width

ILLUSTRATION 5

▲
Students are warned
about common errors.

◀ Graphics have
been improved.

◀ All applications
have titles.

$$y = -950x + 4750$$
$$0 = -950x + 4750$$
$$-4750 = -950x \qquad \text{Subtract 4750 from both sides.}$$
$$5 = x \qquad \text{Divide both sides by } -950.$$

The computer will have no value in 5 years. ∎

Depreciation

ACCENT ON TECHNOLOGY

To solve Example 5 with a graphing calculator, we graph $y = -950x + 4750$ in the window $X = [-10, 10]$ and $Y = [-10, 5000]$, as shown in Figure 3-12(a). We chose the maximum y-value of 5000 because y is almost 5000 when $x = 0$. We then trace to get Figure 3-12(b). We then zoom and trace again to get Figure 3-12(c), which shows that $y = 0$ when $x = 5$.

(a)

X=4.893617   Y=101.06383
(b)

X=5   Y=9.5E-7
(c)

FIGURE 3-12

Word equations always appear before mathematical equations.

Historical notes appear in the margins.

George Polya
(1888–1985)

Polya, a Hungarian, became a professor of mathematics at Stanford University. His approach to problem solving made him very popular with faculty and students. His book *How to Solve It* became a best-seller. His problem-solving approach involves four steps:
1. Understand the problem.
2. Devise a plan.
3. Carry out the plan.
4. Check back.

14% investment	0.14	$10,000 - x$	$0.14(10,000 - x)$

The total income from these two investments can be expressed in two ways: as $1275 and as the sum of the incomes of the two investments.

The income from the 9% investment	+	the income from the 14% investment	=	the total investment
$0.09x$	+	$0.14(10,000 - x)$	=	1275

We can solve this equation for x.

$$0.09x + 0.14(10,000 - x) = 1275$$
$$9x + 14(10,000 - x) = 127,500 \qquad \text{Multiply both sides by 100 to eliminate the decimal points.}$$
$$9x + 140,000 - 14x = 127,500 \qquad \text{Use the distributive property to remove parentheses.}$$
$$-5x + 140,000 = 127,500 \qquad \text{Combine like terms.}$$
$$-5x = -12,500 \qquad \text{Subtract 140,000 from both sides.}$$
$$x = 2500 \qquad \text{Divide both sides by } -5.$$

◄ Author's notes explain the steps in the problem-solving process.

◄ Accent on Technology features appear throughout the text.

94. **Stopping distance** How much farther does it take to stop at 60 mph than at 30 mph?

95. **CB radio** The CB radio of a trucker covers the circular area shown in Illustration 1. Find the equation of that circle, in general form.

ILLUSTRATION 2

DISCOVERY AND WRITING *The solution of the inequality $P(x) < 0$ consists of those numbers x for which the graph of $y = P(x)$ lies below the x-axis. To solve $P(x) < 0$, we graph $y = P(x)$ and trace to find numbers x that produce negative values of y. In Exercises 97–98, solve each inequality.*

97. $x^2 + x - 6 < 0$

98. $x^2 - 3x - 10 > 0$

REVIEW *Solve each equation.*

99. $3(x + 2) + x = 5x$

100. $12b + 6(3 - b) = b + 3$

◄ Graphing-calculator problems are marked by an icon.

◄ Exercise sets include Discovery and Writing exercises and Review problems.

170   CHAPTER 2   EQUATIONS AND INEQUALITIES

▮▮▮▮▮▮▮▮▮ PROBLEMS AND PROJECTS

1. A total of 736 people paid $7772 to attend a concert. Senior citizen tickets cost $7, and all others cost $12. For the rights to perform the music, a 5% royalty on the receipts from senior citizen tickets and a 9% royalty on the remaining receipts must be paid. How much is owed in royalties?

2. Find the distance x required to balance the lever shown in Illustration 1. Research beginning and intermediate algebra books to find the principles from physics that are necessary to solve the problem.

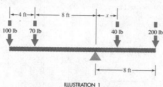

ILLUSTRATION 1

3. Present to a friend the proof of the quadratic formula given in Section 2.3. Then present the following proof.

$$ax^2 + bx + c = 0$$
$$ax^2 + bx = -c$$
$$4a(ax^2) + 4a(bx) = 4a(-c)$$
$$4a^2x^2 + 4abx = -4ac$$
$$4a^2x^2 + 4abx + b^2 = b^2 - 4ac$$
$$(2ax + b)^2 = b^2 - 4ac$$
$$2ax + b = \pm\sqrt{b^2 - 4ac}$$
$$2ax = -b \pm \sqrt{b^2 - 4ac}$$
$$x = \frac{-b \pm \sqrt{b^2 - 4ac}}{2a}$$

Which proof did your friend understand better? Why? Which proof did you prefer?

4. A researcher wants to estimate the mean real estate tax paid by homeowners living in Rockford, IL. To do so, he decides to select a random sample of homeowners and compute the mean tax paid by the homeowners in that sample. How large must the sample be for the researcher to be 95% certain that his computed sample mean will be within ... is, within ... homeow... deviation.

From ...

$$\frac{3.84...}{N}$$

where E ... sample si...

PROJECT 1   Use a geometry textbook to find the ...

1.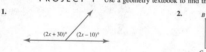

$(2x + 30)°$ $(2x - 10)°$

2.

◀ Each chapter has a Problems and Projects section.

---

356   CHAPTER 4   FUNCTIONS

In Questions 27–28, graph each function.

27. $y = (x - 3)^2 + 1$

28. $y = \sqrt{x - 1} + 5$

In Questions 29–30, find the range of f by finding the domain of f^{-1}.

29. $y = \dfrac{3}{x} - 2$ range: $(-\infty, -2) \cup (-2, \infty)$

30. $y = \dfrac{3x - 1}{x - 3}$ range: $(-\infty, 3) \cup (3, \infty)$

◀ Cumulative review exercises appear after every second chapter.

▮ Cumulative Review Exercises

In Exercises 1–2, use the x- and y-intercepts to graph each equation.

1. $5x - 3y = 15$

2. $3x + 2y = 12$

In Exercises 3–4, find the length, the midpoint, and the slope of the line segment PQ.

3. $P\left(-2, \dfrac{7}{2}\right)$; $Q\left(3, -\dfrac{1}{2}\right)$ $\sqrt{41}$; $\left(\frac{1}{2}, \frac{3}{2}\right)$; $-\frac{4}{5}$

4. $P(3, 7)$; $Q(-7, 3)$ $2\sqrt{29}$; $(-2, 5)$; $\frac{2}{5}$

In Exercises 5–8, write the equation of the line with the given properties. Give the answer in slope-intercept form.

5. The line passes through $(-3, 5)$ and $(3, -7)$. $y = -2x - 1$

6. The line passes through $\left(\frac{3}{2}, \frac{5}{2}\right)$ and has a slope of $\frac{7}{2}$. $y = \frac{7}{2}x - \frac{11}{4}$

7. The line is parallel to $3x - 5y = 7$ and passes through $(-5, 3)$. $y = \frac{3}{5}x + 6$

8. The line is perpendicular to $x - 4y = 12$ and passes through the origin. $y = -4x$

◀ In the Instructor's Edition, answers are printed in blue next to the problems.

To the Student

Congratulations. You now own a state-of-art textbook that has been written especially for you. We have tried to write a book that you can read and understand. The book features a carefully written narrative and an extensive number of worked examples with Self Checks.

To get the most of this course, you must read and study the textbook properly. We recommend that you work the examples on paper first and then work the Self Checks. Only after you thoroughly understand the concepts taught in the examples should you attempt to work the exercises. The *Student Solutions Manual and Study Guide* contains solutions to the odd-numbered exercises.

Since the material presented in *College Algebra,* 6th edition will be of value to you in later years, we suggest that you keep this book. It will be a good source of reference and will keep at your fingertips the material you have learned here.

We wish you well.

Acknowledgments

We are grateful to the following people who reviewed the manuscript in its various stages. They all had valuable suggestions that have been incorporated into the text.

The reviewers include Richard Andrews, University of Wisconsin; James Arnold, University of Wisconsin; Ronald Atkinson, Tennessee State University; Wilson Banks, Illinois State University; Jerry Bloomberg, Essex Community College; Elaine Bouldin, Middle Tennessee State University; Dale Boye, Schoolcraft College; Lee R. Clancy, Golden West College; Jan Collins, Embry Riddle College; Cecilia Cooper, William & Harper College; Romae J. Cormier, Northern Illinois University; John S. Cross, University of Northern Iowa; M. Hilary Davies, University of Alaska at Anchorage; Grace DeVelbiss, Sinclair Community College; Lena Dexter, Faulkner State Junior College; Emily Dickinson, University of Arkansas; Robert E. Eicken, Illinois Central College; Eric Ellis, Essex Community College; Eunice F. Everett, Seminole Community College; Dale Ewen, Parkland College; Harold Farmer, Wallace Community College–Hanceville; Ronald J. Fischer, Evergreen Valley College; Mary Jane Gates, University of Arkansas at Little Rock; Marvin Goodman, Monmouth College; Jerry Gustafson, Beloit College; Jerome Hahn, Bradley University; Douglas Hall, Michigan State University; Robert Hall, University of Wisconsin; David Hansen, Monterey Peninsula College; Kevin Hastings, University of Delaware; William Hinrichs, Rock Valley College; Arthur M. Hobbs, Texas A & M University; Jack E. Hofer, California Polytechnic State University; Ingrid Holzner, University of Wisconsin; Warren Jaech, Tacoma Community College; Nancy Johnson, Broward Community College; William B. Jones, University of Colorado; Barbara Juister, Elgin Community College; David Kinsey, University of Southern Indiana; Helen Kriegsman, Pittsburg State University; Marjorie O. Labhart, University of Southern Indiana; Jaclyn LeFebvre, Illinois Central College;

Judy McKinney, California Polytechnic Institute at Pomona; Sandra McLaurin, University of North Carolina; Marcus McWaters, University of Southern Florida; Donna Menard, University of Massachusetts, Dartmouth; James W. Mettler, Pennsylvania State University; Eldon L. Miller, University of Mississippi; Stuart E. Mills, Louisiana State University, Shreveport; Mila Mogilevskaya, Wichita State University; Gilbert W. Nelson, North Dakota State; Marie Neuberth, Catonsville City College; Anthony Peressini, University of Illinois; David L. Phillips, University of Southern Colorado; William H. Price, Middle Tennessee State University; Ronald Putthoff, University of Southern Mississippi; Janet P. Ray, Seattle Central Community College; Barbara Riggs, Tennessee Technological University; Paul Schaefer, SUNY, Geneseo; Vincent P. Schielack, Jr., Texas A & M University; Robert Sharpton, Miami Dade Community College; L. Thomas Shiflett, Southwest Missouri State University; Richard Slinkman, Bemidji State University; Merreline Smith, California Polytechnic Institute at Pomona; John Snyder, Sinclair Community College; Warren Strickland, Del Mar College; Ray Tebbetts, San Antonio College; Faye Thames, Lamar State University; Douglas Tharp, University of Houston–Downtown; Carol M. Walker, Hinds Community College; William Waller, University of Houston–Downtown; Carroll G. Wells, Western Kentucky University; William H. White, University of South Carolina at Spartanburg; Charles R. Williams, Midwestern State University; Harry Wolff, University of Wisconsin; Clifton Whyburn, University of Houston; Albert Zechmann, University of Nebraska.

We wish to thank the staff at Brooks/Cole, especially Bob Pirtle, Vernon Boes, Ellen Brownstein, and Melissa Duge, for their competent work and support. We give special thanks to David Hoyt, for his editing skills and good nature, and to Lori Heckelman for fine artwork.

R. David Gustafson
Peter D. Frisk

■ ■ ■ ■ ■ ■ ■ ■ ■ ■ CONTENTS

Page numbers for Examples are **boldface;** page numbers for exercises are lightface.

SIXTH EDITION

College Algebra

1 Basic Concepts

IN THIS FIRST CHAPTER, WE REVIEW MANY CONCEPTS AND SKILLS LEARNED IN PREVIOUS ALGEBRA COURSES. BE SURE TO MASTER THIS MATERIAL NOW BECAUSE IT IS THE BASIS FOR THE REST OF THIS COURSE.

1.1 Sets of Real Numbers

■ SETS OF REAL NUMBERS ■ PROPERTIES OF REAL NUMBERS ■ GRAPHING SUBSETS OF REAL NUMBERS ■ INEQUALITY SYMBOLS ■ INTERVALS ■ ABSOLUTE VALUE ■ DISTANCE ON A NUMBER LINE

A **set** is a collection of objects, such as a set of dishes or a set of golf clubs. The set of vowels of the English language can be denoted as {a, e, i, o, u}, where the braces { } are read as "the set of."

■ SETS OF REAL NUMBERS

Most of the numbers that we use in business, science, and industry are *real numbers.*

Real Numbers
A **real number** is any number that can be expressed as a decimal.

Some examples of real numbers are

9, because $9 = 9.0000000. . .$ Read = as "is equal to." The ellipsis . . . indicates that the 0's go on forever.

$\frac{3}{4}$, because $\frac{3}{4} = 0.75$

$-\frac{1}{3}$, because $-\frac{1}{3} = -0.\overline{3}$ $-0.\overline{3}$ is a shortcut way of writing $-0.3333. . . .$

$\sqrt{5} = 2.236067978. . .$

There are many sets of numbers that are included in the set of real numbers. These sets, called **subsets of the real numbers,** have special names. Recall that when a letter is used to represent a number, it is called a **variable.**

Subsets of the Real Numbers

Natural numbers
The set of numbers used for counting: {1, 2, 3, 4, 5, 6, . . . }

Whole numbers
The set of natural numbers and 0: {0, 1, 2, 3, 4, 5, 6, . . .}

Integers
The set of natural numbers along with their negatives and 0:

$$\{\ldots, -5, -4, -3, -2, -1, 0, 1, 2, 3, 4, 5, \ldots\}$$

Rational numbers
The rational numbers are the fractions. In **set-builder notation,** the rational numbers are $\{x \mid x$ can be written in the form $\frac{a}{b}$, where a and b are integers and $b \neq 0\}$. The symbol "$\mid$" is read as "such that." The rational numbers can also be described as the decimals that either terminate or repeat. Some examples of rational numbers are

$$5 = \frac{5}{1} \qquad \frac{3}{4} = 0.75 \qquad \frac{1}{3} = 0.\overline{3} \qquad \frac{5}{11} = 0.454545\ldots = 0.\overline{45}$$

Irrational numbers
All real numbers that are not rational. The irrational numbers are the decimals that don't terminate and don't repeat. Some examples are

$$\sqrt{2} = 1.414213\ldots \qquad \pi = 3.141592\ldots \qquad \sqrt{5} = 2.236067\ldots$$

Figure 1-1 shows how the above sets of numbers are related.

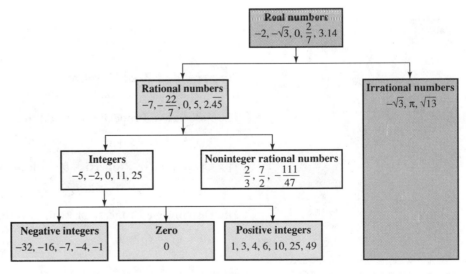

FIGURE 1-1

We will often mention two subsets of the natural numbers. A **prime number** is a natural number, greater than 1, that is divisible only by itself and 1. A **composite number** is a natural number, greater than 1, that is not prime.

The set of prime numbers: $\{2, 3, 5, 7, 11, 13, 17, 19, 23, 29, 31, \ldots\}$

The set of composite numbers: $\{4, 6, 8, 9, 10, 12, 14, 15, 16, 18, 20, 21, \ldots\}$

There are also two important subsets of the integers. The **even integers** are those integers that are divisible by 2. The **odd integers** are those integers that are not divisible by 2.

The set of even integers: $\{ \ldots, -10, -8, -6, -4, -2, 0, 2, 4, 6, 8, 10, \ldots \}$

The set of odd integers: $\{ \ldots, -9, -7, -5, -3, -1, 1, 3, 5, 7, 9, \ldots \}$

EXAMPLE 1   In the set $\left\{ -3, -2, 0, \frac{1}{2}, 1, \sqrt{5}, 2, 4, 5, 6 \right\}$, list all   **a.** even integers,   **b.** prime numbers,   **c.** rational numbers.

Solution   **a.** $-2, 0, 2, 4, 6$

b. $2, 5$

c. $-3, -2, 0, \frac{1}{2}, 1, 2, 4, 5, 6$ ∎

Self Check   In the set in Example 1, list all   **a.** odd integers,   **b.** composite numbers,   **c.** irrational numbers.

Answers   **a.** $-3, 1, 5,$ **b.** $4, 6,$ **c.** $\sqrt{5}$

■ PROPERTIES OF REAL NUMBERS

When we work with real numbers, we will use the following properties.

Properties of Real Numbers
If a, b, and c are real numbers, then

The Associative Properties for Addition and Multiplication:

$$(a + b) + c = a + (b + c) \qquad (ab)c = a(bc)$$

The Commutative Properties for Addition and Multiplication:

$$a + b = b + a \qquad ab = ba$$

The Distributive Property of Multiplication over Addition:

$$a(b + c) = ab + ac$$

EXAMPLE 2   Tell which property of real numbers justifies each statement.

a. $(9 + 2) + 3 = 9 + (2 + 3)$

b. $3(x + 2) = 3x + 3 \cdot 2$

Solution   **a.** Associative property of addition

b. Distributive property   ∎

Self Check   Tell which property of real numbers justifies each statement.

a. $mn = nm$ **b.** $(xy)z = x(yz)$ **c.** $p + q = q + p$

Answers   **a.** Commutative property of multiplication,   **b.** Associative property of multiplication,   **c.** Commutative property of addition

■ GRAPHING SUBSETS OF REAL NUMBERS

We can graph subsets of real numbers on the **number line.** The number line shown
in Figure 1-2 continues forever in both directions. The **positive numbers** are represented by the points to the right of 0, and the **negative numbers** are represented
by the points to the left of 0.

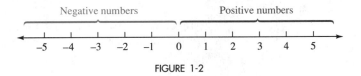

FIGURE 1-2

 WARNING!   Zero is neither positive nor negative.

Figure 1-3(a) shows the graph of the natural numbers from 1 to 5. The point associated with each number is called the **graph** of the number, and the number is
called the **coordinate** of its point. Figure 1-3(b) shows the graph of the prime numbers that are less than 10. Figure 1-3(c) shows the graph of the integers from -4 to
3. Figure 1-3(d) shows the graph of the real numbers $-\frac{3}{4}$, $\frac{5}{3}$, $0.\overline{3}$, and $-\sqrt{5}$.

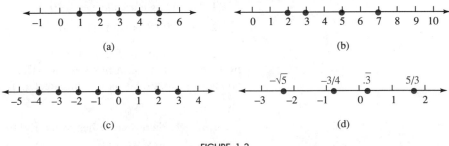

FIGURE 1-3

The graphs in Figure 1-3 suggest that there is a **one-to-one correspondence**
between the set of real numbers and the points on a number line. This means that
to each real number there corresponds exactly one point on the number line, and
that to each point on the number line there corresponds exactly one real-number coordinate.

EXAMPLE 3   Graph the set $\left\{-3, -\frac{4}{3}, 0, \sqrt{2}\right\}$.

Solution   Note that $\sqrt{2} \approx 1.414$.

Self Check   Graph the set $\left\{-2, \frac{3}{4}, \sqrt{3}\right\}$. (*Hint:* $\sqrt{3} \approx 1.7$.)

Answer

■ INEQUALITY SYMBOLS

To show that two quantities are equal, we use the $=$ sign. To show that two quantities are not equal, we use one of the following **inequality symbols.**

Symbol	Read as	Examples		
$\neq$	"is not equal to"	$5 \neq 8$	and	$0.25 \neq \frac{1}{3}$
$<$	"is less than"	$12 < 20$	and	$0.17 < 1.1$
$>$	"is greater than"	$15 > 9$	and	$\frac{1}{2} > 0.2$
$\leq$	"is less than or equal to"	$25 \leq 25$	and	$1.7 \leq 2.3$
$\geq$	"is greater than or equal to"	$19 \geq 19$	and	$15.2 \geq 13.7$

It is often possible to write an inequality with the inequality symbol pointing in the opposite direction. For example,

$12 < 20$ is equivalent to $20 > 12$

$2.3 \geq -1.7$ is equivalent to $-1.7 \leq 2.3$

In each graph in Figure 1-3, the coordinates of points get larger as we move from left to right on a number line. Thus, if a and b are the coordinates of two points, the one to the right is the greater. This suggests the following general principle:

If $a > b$, then point a lies to the right of point b on a number line.

If $a < b$, then point a lies to the left of point b on a number line.

■ INTERVALS

Figure 1-4(a) shows the graph of the **inequality** $x > -2$ (or $-2 < x$). This graph includes all real numbers x that are greater than -2. The parenthesis at -2 indicates that -2 is not included in the graph. Figure 1-4(b) shows the graph of the inequality $x \leq 5$ (or $5 \geq x$). The bracket at 5 indicates that 5 is included in the graph.

(a)   (b)

FIGURE 1-4

Two inequalities are often written as a single expression to form a **compound inequality.** For example, the compound inequality

$$5 < x < 12$$

FIGURE 1-5

is a combination of the inequalities $5 < x$ and $x < 12$. It is read as "5 is less than x, and x is less than 12" and means that x is between 5 and 12. Its graph is shown in Figure 1-5.

The graphs shown in Figures 1-4 and 1-5 are portions of a number line called **intervals.** The interval shown in Figure 1-6(a) is denoted by the inequality $-2 < x < 4$, or as $(-2, 4)$. The parentheses indicate that the endpoints are not included. The interval shown in Figure 1-6(b) is denoted by the inequality $x > 1$, or as $(1, \infty)$.

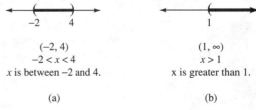

$(-2, 4)$ $(1, \infty)$
$-2 < x < 4$ $x > 1$
x is between -2 and 4. x is greater than 1.

(a)   (b)

FIGURE 1-6

WARNING!   The symbol ∞ (infinity) is not a real number. It is used to indicate that the graph in Figure 1-6(b) extends infinitely far to the right.

EXAMPLE 4   Write the inequality $-3 < x < 5$ in interval notation and graph it.

Solution   This is the interval $(-3, 5)$. Its graph includes all real numbers between -3 and 5, as shown in Figure 1-7.

$-3 \qquad 5$

FIGURE 1-7   ∎

 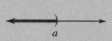

If an interval extends forever in one direction, it is called an **unbounded interval.**

Unbounded Intervals

The interval (a, ∞) includes all real numbers x such that $x > a$.

The interval $[a, \infty)$ includes all real numbers x such that $x \geq a$.

The interval $(-\infty, a)$ includes all real numbers x such that $x < a$.

The interval $(-\infty, a]$ includes all real numbers x such that $x \leq a$.

The interval $(-\infty, \infty)$ includes all real numbers x such that $-\infty < x < \infty$. The graph of this interval is the entire number line.

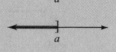

A bounded interval with no endpoints is called a **open interval.** A bounded interval with one endpoint is called a **half-open interval.** Figure 1-8(a) shows the open interval between -3 and 2. Figure 1-8(b) shows the half-open interval between -2 and 3, including -2.

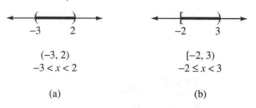

$(-3, 2)$ $\qquad\qquad$ $[-2, 3)$
$-3 < x < 2$ $\qquad\quad$ $-2 \leq x < 3$

(a) $\qquad\qquad\qquad$ (b)

FIGURE 1-8

Intervals that contain both endpoints are called **closed intervals.** Figure 1-9 shows the graph of a closed interval from -2 to 4.

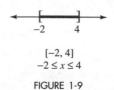

$[-2, 4]$
$-2 \leq x \leq 4$

FIGURE 1-9

**Amalie Noether
(1882–1935)**

Albert Einstein described Noether as the most creative female mathematical genius since the beginning of higher education for women. Her work was in the area of abstract algebra.

Open Intervals
The interval (a, b) includes all real numbers x such that $a < x < b$.

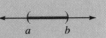

Half-Open Intervals
The interval $[a, b)$ includes all real numbers x such that $a \leq x < b$.

The interval $(a, b]$ includes all real numbers x such that $a < x \leq b$.

Closed Intervals
The interval $[a, b]$ includes all real numbers x such that $a \leq x \leq b$.

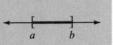

EXAMPLE 5   Write the inequality $3 \leq x$ in interval notation and graph it.

Solution   The inequality $3 \leq x$ can be written in the form $x \geq 3$. This is the interval $[3, \infty)$. Its graph includes all real numbers greater than or equal to 3, as shown in Figure 1-10.

FIGURE 1-10   ■

Self Check   Write the inequality $-2 < x \leq 5$ in interval notation and graph it.
Answer $(-2, 5]$,

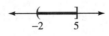

EXAMPLE 6   Write the inequality $5 \geq x \geq -1$ in interval notation and graph it.

Solution   The inequality $5 \geq x \geq -1$ can be written in the form

$$-1 \leq x \leq 5$$

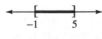

FIGURE 1-11

This is the interval $[-1, 5]$. Its graph includes all real numbers from -1 to 5. The graph is shown in Figure 1-11.   ■

Self Check   Write the inequality $0 \leq x \leq 3$ in interval notation and graph it.
Answer $[0, 3]$,

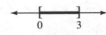

Another type of compound inequality is

$x < -2$ or $x \geq 3$ Read as "x is less than -2 or x is greater than or equal to 3."

This inequality is called the **union** of two intervals. In interval notation, it is written as

$(-\infty, -2) \cup [3, \infty)$ Read the symbol $\cup$ as "union."

Its graph is shown in Figure 1-12.

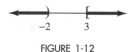

FIGURE 1-12

■ ABSOLUTE VALUE

The **absolute value** of a real number x (denoted as $|x|$) is the distance on a number line between 0 and the point with a coordinate of x. For example, points with coordinates of 4 and -4 both lie four units from 0, as shown in Figure 1-13. Therefore, it follows that

$|-4| = |4| = 4$

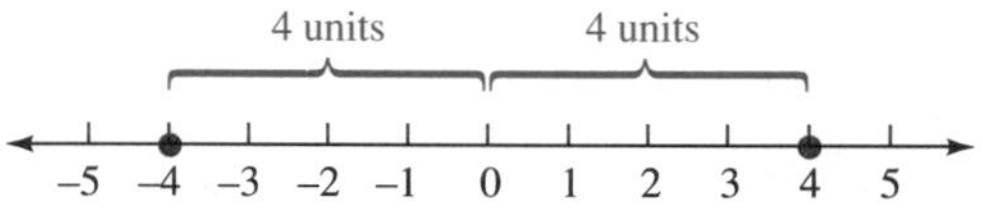

FIGURE 1-13

In general, for any real number x,

$|-x| = |x|$

We can define absolute value algebraically as follows.

Absolute Value
If x is a real number, then

$|x| = x$ when $x \geq 0$

$|x| = -x$ when $x < 0$

This definition indicates that when x is positive or 0, then x is its own absolute value. However, when x is negative, then $-x$ (which is positive) is its absolute value. Thus, $|x|$ is always nonnegative.

$|x| \geq 0$ **for all real numbers** x

WARNING!   Remember that x is not always positive and $-x$ is not always negative.

EXAMPLE 7 Write each number without using absolute value symbols: **a.** $|3|$, **b.** $|-4|$, **c.** $|0|$, **d.** $-|-8|$.

Solution **a.** $|3| = 3$ **b.** $|-4| = 4$ **c.** $|0| = 0$ **d.** $-|-8| = -(8) = -8$ ■

Self Check Write each number without using absolute value symbols: **a.** $|-10|$, **b.** $|12|$, **c.** $-|6|$.

Answers **a.** 10, **b.** 12, **c.** -6

EXAMPLE 8 Write each number without using absolute value symbols: **a.** $|\pi - 1|$, **b.** $|2 - \pi|$, **c.** $|2 - x|$ if $x \geq 5$.

Solution **a.** Since $\pi \approx 3.1416$, $\pi - 1$ is positive. Thus, $\pi - 1$ is its own absolute value.

$$|\pi - 1| = \pi - 1$$

b. Since $2 - \pi$ is negative, its absolute value is $-(2 - \pi)$.

$$|2 - \pi| = -(2 - \pi) = -2 - (-\pi) = -2 + \pi = \pi - 2$$

c. Since $x \geq 5$, the expression $2 - x$ is negative. Thus, its absolute value is $-(2 - x)$.

$$|2 - x| = -(2 - x) = -2 + x = x - 2$$ ■

Self Check Write each number without using absolute value symbols. $\left(\textit{Hint: } \sqrt{5} \approx 2.236.\right)$
  a. $\left|2 - \sqrt{5}\right|$ **b.** $|2 - x|$ if $x \leq 1$

Answers **a.** $\sqrt{5} - 2$, **b.** $2 - x$

■ DISTANCE ON A NUMBER LINE

On the number line shown in Figure 1-14, the distance between the points with co-ordinates of 1 and 4 is $4 - 1$, or 3 units. However, if the subtraction were done in the other order, the result would be $1 - 4$, or -3 units. To guarantee that the distance between two points is always positive, we can use absolute value symbols. Thus, the distance d between two points with coordinates of 1 and 4 is

$$d = |4 - 1| = |1 - 4| = 3$$

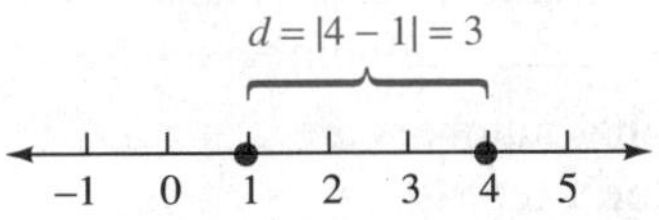

$$d = |4 - 1| = 3$$

FIGURE 1-14

In general, we have the following definition for the distance between two points on the number line.

> **Distance between Two Points**
>
> If a and b are the coordinates of two points on the number line, the distance between the points is given by the formulas
>
> $$d = |a - b| \qquad \text{or} \qquad d = |b - a|$$

EXAMPLE 9   Find the distance on a number line between points with coordinates of   **a.** 3 and 5,   **b.** -2 and 3,   and   **c.** -5 and -1.

Solution   **a.** $d = |5 - 3| = |2| = 2$

or

$$d = |3 - 5| = |-2| = 2$$

b. $d = |3 - (-2)| = |3 + 2| = |5| = 5$

or

$$d = |-2 - 3| = |-5| = 5$$

c. $d = |-1 - (-5)| = |-1 + 5| = |4| = 4$

or

$$d = |-5 - (-1)| = |-5 + 1| = |-4| = 4 \qquad \blacksquare$$

Self Check   Find the distance on a number line between points with coordinates of   **a.** 4 and 10   and   **b.** -2 and -7.

Answers   **a.** 6,   **b.** 5

E X E R C I S E   1.1

VOCABULARY AND CONCEPTS   *Fill in the blank to make a true statement.*

1. A real number is any number that can be expressed as a _____ .

2. A _____ is a letter that is used to represent a number.

3. The smallest prime number is __ .

4. All integers that are divisible by 2 are called _____ integers.

5. Natural numbers greater than 1 that are not prime are called _____ numbers.

6. Fractions such as $\frac{2}{3}$, $\frac{8}{2}$, and $-\frac{7}{9}$ are called _____ numbers.

7. Irrational numbers are _____ that don't terminate and don't repeat.

8. The symbol __ is read as "is less than or equal to."

9. On a number line, the _____ numbers are to the left of 0.

10. The only integer that is neither positive nor negative is __ .

11. The associative property of addition states that $(x + y) + z =$ _____ .

12. The commutative property of multiplication states that $xy =$ ___ .

13. Use the distributive property to complete the statement: $5(m + 2) =$ _____.

14. The statement $(m + n)p = p(m + n)$ illustrates the _____ property of _____.

15. The graph of an _____ is a portion of a number line.

16. The graph of an open interval has ___ endpoints.

17. The graph of a closed interval has ___ endpoints.

18. The graph of a _____ interval has one endpoint.

19. Except for 0, the absolute value of every number is _____.

20. The _____ between two points on a number line is always positive.

PRACTICE   *In Exercises 21–30, consider the following set:* $\left\{-5, -4, -\frac{2}{3}, 0, 1, \sqrt{2}, 2, 2.75, 6, \text{ and } 7\right\}$.

21. Which numbers are natural numbers?

22. Which numbers are whole numbers?

23. Which numbers are integers?

24. Which numbers are rational numbers?

25. Which numbers are irrational numbers?

26. Which numbers are prime numbers?

27. Which numbers are composite numbers?

28. Which numbers are even integers?

29. Which numbers are odd integers?

30. Which numbers are negative numbers?

In Exercises 31–38, graph each subset of the real numbers on a number line.

31. The natural numbers between 1 and 5

32. The composite numbers less than 10

33. The prime numbers between 10 and 20

34. The integers from -2 to 4

35. The integers between -5 and 0

36. The even integers between -9 and -1

37. The odd integers between -6 and 4

38. $-0.7, 1.75,$ and $3\dfrac{7}{8}$

In Exercises 39–54, write each inequality in interval notation and graph the interval.

39. $x > 2$

40. $x < 4$

41. $0 < x < 5$

42. $-2 < x < 3$

43. $x > -4$

44. $x < 3$

45. $-2 \le x < 2$

46. $-4 < x \le 1$

47. $x \le 5$

48. $x \geq -1$

49. $-5 < x \leq 0$

50. $-3 \leq x < 4$

51. $-2 \leq x \leq 3$

52. $-4 \leq x \leq 4$

53. $6 \geq x \geq 2$

54. $3 \geq x \geq -2$

In Exercises 55–58, write each compound inequality as the union of two intervals and graph the result.

55. $x < -2$ or $x > 2$

56. $x \leq -5$ or $x > 0$

57. $x \leq -1$ or $x \geq 3$

58. $x < -3$ or $x \geq 2$

In Exercises 59–74, write each expression without using absolute value symbols.

59. $|13|$

60. $|-17|$

61. $|0|$

62. $-|63|$

63. $-|-8|$

64. $|-25|$

65. $-|32|$

66. $-|-6|$

67. $|\pi - 5|$

68. $|8 - \pi|$

69. $|\pi - \pi|$

70. $|2\pi|$

71. $|x + 1|$ and $x \geq 2$

72. $|x + 1|$ and $x \leq -2$

73. $|x - 4|$ and $x < 0$

74. $|x - 7|$ and $x > 10$

In Exercises 75–78, find the distance between each pair of points on the number line.

75. 3 and 8

76. -5 and 12

77. -8 and -3

78. 6 and -20

APPLICATIONS

79. What subset of the real numbers would you use to describe the populations of several cities?

80. What subset of the real numbers would you use to describe the subdivisions of an inch on a ruler?

81. What subset of the real numbers would you use to report temperatures in several cities?

82. What subset of the real numbers would you use to describe the financial condition of a business?

DISCOVERY AND WRITING

83. Explain why $-x$ could be positive.

84. Explain why every integer is a rational number.

85. Is the statement $|ab| = |a| \cdot |b|$ always true? Explain.

86. Is the statement $\left|\dfrac{a}{b}\right| = \dfrac{|a|}{|b|}$ always true? Explain.

87. Is the statement $|a + b| = |a| + |b|$ always true? Explain.

88. Under what conditions will the statement given in Exercise 87 be true?

1.2 Integer Exponents and Scientific Notation

■ NATURAL-NUMBER EXPONENTS ■ RULES OF EXPONENTS ■ ORDER OF OPERATIONS
■ SCIENTIFIC NOTATION ■ USING SCIENTIFIC NOTATION TO SIMPLIFY COMPUTATIONS

■ NATURAL-NUMBER EXPONENTS

When two or more quantities are multiplied together, each quantity is called a **factor** of the product. For example, the expression x^4 indicates that x is to be used as a factor four times.

$$x^4 = x \cdot x \cdot x \cdot x$$

In general, the following is true.

> **Natural-Number Exponents**
> For any natural number n,
>
> $$x^n = \overbrace{x \cdot x \cdot x \cdot \cdots \cdot x}^{n \text{ factors of } x}$$

In the **exponential expression** x^n, x is called the **base**, and n is called the **exponent** or the **power** to which the base is raised. The expression x^n is called a **power of x.** From the definition, we see that a natural-number exponent tells how many times the base of an exponential expression is to be used as a factor in a product. If an exponent is 1, it is usually not written:

$$x^1 = x$$

EXAMPLE 1   Write each expression without using exponents:
a. 4^2,   **b.** $(-4)^2$,   **c.** 5^3,   **d.** $(-5)^3$,   **e.** $3x^4$, and   **f.** $(3x)^4$.

Solution   **a.** $4^2 = 4 \cdot 4 = 16$ Read 4^2 as "four squared."

b. $(-4)^2 = (-4)(-4) = 16$

c. $5^3 = 5 \cdot 5 \cdot 5 = 125$ Read 5^3 as "five cubed."

d. $(-5)^3 = (-5)(-5)(-5) = -125$

e. $3x^4 = 3 \cdot x \cdot x \cdot x \cdot x$ Read x^4 as "x to the fourth power."

f. $(3x)^4 = (3x)(3x)(3x)(3x) = 81x \cdot x \cdot x \cdot x$ ■

Self Check   Write each expression without using exponents: **a.** 7^3,   **b.** $(-3)^2$,   **c.** $5a^3$, and **d.** $(5a)^3$.

Answers   **a.** $7 \cdot 7 \cdot 7 = 343$,   **b.** $(-3)(-3) = 9$,   **c.** $5 \cdot a \cdot a \cdot a$,
d. $(5a)(5a)(5a) = 125a \cdot a \cdot a$

WARNING! It is important to note the distinction between ax^n and $(ax)^n$.

$$\overbrace{ax^n = a \cdot x \cdot x \cdot x \cdot \cdots \cdot x}^{n \text{ factors of } x} \qquad \overbrace{(ax)^n = (ax)(ax)(ax) \cdot \cdots \cdot (ax)}^{n \text{ factors of } ax}$$

It is also important to note the distinction between $-x^n$ and $(-x)^n$.

$$\overbrace{-x^n = -(x \cdot x \cdot x \cdot \cdots \cdot x)}^{n \text{ factors of } x} \qquad \overbrace{(-x)^n = (-x)(-x)(-x) \cdot \cdots \cdot (-x)}^{n \text{ factors of } -x}$$

■ ■ ■ ■ ■ ■ ■ ■ ■   ## Using a Calculator to Find Powers

**ACCENT ON
TECHNOLOGY**

We can use calculators to find powers of numbers. For example, to find 2.35^3 with a scientific calculator, we press these numbers and these keys:

2.35 $\boxed{y^x}$ 3 $\boxed{=}$

The display will read $\boxed{12.977875}$.

To find 2.35 with a graphing calculator, we press these numbers and these keys:

2.35 $\boxed{\wedge}$ 3 $\boxed{\text{ENTER}}$

The display will read 2.35^3

12.977875

■ ■ ■ ■ ■ ■ ■ ■ ■   In either case, we see that $2.35^3 = 12.977875$.

■ RULES OF EXPONENTS

We begin to review the rules of exponents by considering the product $x^m x^n$. Since x^m indicates that x is to be used as a factor m times, and since x^n indicates that x is to be used as a factor n times, there are $m + n$ factors of x in the product $x^m x^n$.

$$x^m x^n = \overbrace{\underbrace{x \cdot x \cdot x \cdot \cdots \cdot x}_{m \text{ factors of } x} \cdot \underbrace{x \cdot x \cdot x \cdot \cdots \cdot x}_{n \text{ factors of } x}}^{m + n \text{ factors of } x}$$

This suggests that to multiply exponential expressions with the same base, we *keep the base and add the exponents.*

> **Product Rule for Exponents**
> If m and n are natural numbers, then
> $$x^m x^n = x^{m+n}$$

WARNING! The product rule applies only to exponential expressions with the same base. A product of two powers with different bases, such as $x^4 y^3$, cannot be simplified.

To find another property of exponents, we consider the exponential expression $(x^m)^n$. The exponent n indicates that x^m is to be used as a factor n times. This implies that x is to be used as a factor mn times.

$$(x^m)^n = \overbrace{\overbrace{(x^m)(x^m)(x^m) \cdots \cdot (x^m)}^{n \text{ factors of } x^m}}^{mn \text{ factors of } x} = x^{mn}$$

This suggests that to raise an exponential expression to a power, we *keep the base and multiply the exponents*.

To raise a product to a power, we raise each factor to that power.

$$(xy)^n = \overbrace{(xy)(xy)(xy) \cdots \cdot (xy)}^{n \text{ factors of } xy} = \overbrace{(x \cdot x \cdot x \cdots \cdot x)}^{n \text{ factors of } x} \overbrace{(y \cdot y \cdot y \cdots \cdot y)}^{n \text{ factors of } y} = x^n y^n$$

To raise a fraction to a power, we raise both the numerator and the denominator to that power. If $y \neq 0$, then

$$\left(\frac{x}{y}\right)^n = \overbrace{\left(\frac{x}{y}\right)\left(\frac{x}{y}\right)\left(\frac{x}{y}\right) \cdots \cdot \left(\frac{x}{y}\right)}^{n \text{ factors of } \frac{x}{y}}$$

$$= \frac{\overbrace{x\,x\,x \cdots \cdot x}^{n \text{ factors of } x}}{\underbrace{y\,y\,y \cdots \cdot y}_{n \text{ factors of } y}}$$

$$= \frac{x^n}{y^n}$$

The previous three results are called the **power rules of exponents.**

Power Rules of Exponents

If m and n are natural numbers, then

$$(x^m)^n = x^{mn} \qquad (xy)^n = x^n y^n \qquad \left(\frac{x}{y}\right)^n = \frac{x^n}{y^n} \quad (y \neq 0)$$

EXAMPLE 2   Simplify **a.** $x^5 x^7$, **b.** $x^2 y^3 x^5 y$, **c.** $(x^4)^9$, **d.** $(x^2 x^5)^3$, **e.** $\left(\dfrac{x}{y^2}\right)^5$, and

f. $\left(\dfrac{5x^2 y}{z^3}\right)^2$.

Solution   **a.** $x^5 x^7 = x^{5+7} = x^{12}$

b. $x^2 y^3 x^5 y = x^{2+5} y^{3+1} = x^7 y^4$

c. $(x^4)^9 = x^{4 \cdot 9} = x^{36}$

d. $(x^2 x^5)^3 = (x^7)^3 = x^{21}$

e. $\left(\dfrac{x}{y^2}\right)^5 = \dfrac{x^5}{(y^2)^5} = \dfrac{x^5}{y^{10}}$ $(y \neq 0)$

f. $\left(\dfrac{5x^2y}{z^3}\right)^2 = \dfrac{5^2(x^2)^2 y^2}{(z^3)^2} = \dfrac{25x^4y^2}{z^6}$ $(z \neq 0)$ ∎

Self Check   Simplify each expression: **a.** $(y^3)^2$,   **b.** $(a^2a^4)^3$,   **c.** $(x^2)^3(x^3)^2$, and

d. $\left(\dfrac{3a^3b^2}{c^3}\right)^3$ $(c \neq 0)$.

Answers   **a.** y^6,   **b.** a^{18},   **c.** x^{12},   **d.** $\dfrac{27a^9b^6}{c^9}$

If we assume that the rules for natural-number exponents hold for exponents of 0, we can write

$$x^0 x^n = x^{0+n} = x^n = 1x^n$$

Since $x^0 x^n = 1x^n$, it follows that if $x \neq 0$, then $x^0 = 1$.

Zero Exponent
$$x^0 = 1 \quad (x \neq 0)$$

If we assume that the rules for natural-number exponents hold for exponents that are negative integers, we can write

$$x^{-n} x^n = x^{-n+n} = x^0 = 1 \quad (x \neq 0)$$

However, we know that

$$\dfrac{1}{x^n} \cdot x^n = 1 \quad (x \neq 0) \qquad \dfrac{1}{x^n} \cdot x^n = \dfrac{x^n}{x^n}, \text{ and any nonzero number divided by itself is 1.}$$

Since $x^{-n} x^n = \dfrac{1}{x^n} \cdot x^n$, it follows that $x^{-n} = \dfrac{1}{x^n}$ $(x \neq 0)$.

Negative Exponents
If n is an integer and $x \neq 0$, then
$$x^{-n} = \dfrac{1}{x^n} \qquad \text{and} \qquad \dfrac{1}{x^{-n}} = x^n$$

Because of the two previous definitions, all of the rules for natural-number exponents hold for integer exponents.

OO e,f

EXAMPLE 3 Simplify and write all answers without using negative exponents: **a.** $(3x)^0$, **b.** $3(x^0)$, **c.** x^{-4}, **d.** $\dfrac{1}{x^{-6}}$, **e.** $x^{-3}x$, and **f.** $(x^{-4}x^8)^{-5}$.

Solution **a.** $(3x)^0 = 1$ **b.** $3(x^0) = 3(1) = 3$ **c.** $x^{-4} = \dfrac{1}{x^4}$

d. $\dfrac{1}{x^{-6}} = x^6$ **e.** $x^{-3}x = x^{-3+1}$ **f.** $(x^{-4}x^8)^{-5} = (x^4)^{-5}$

$= x^{-2}$ $= x^{-20}$

$= \dfrac{1}{x^2}$ $= \dfrac{1}{x^{20}}$ ∎

Self Check Simplify and write all answers without using negative exponents: **a.** $7a^0$, **b.** $3a^{-2}$, **c.** $a^{-4}a^2$, and **d.** $(a^3a^{-7})^3$.

Answers **a.** 7, **b.** $\dfrac{3}{a^2}$, **c.** $\dfrac{1}{a^2}$, **d.** $\dfrac{1}{a^{12}}$

To develop the quotient rule for exponents, we proceed as follows:

$$\frac{x^m}{x^n} = x^m\left(\frac{1}{x^n}\right) = x^m x^{-n} = x^{m+(-n)} = x^{m-n} \quad (x \neq 0)$$

This suggests that to divide two exponential expressions with the same nonzero base, we *keep the base and subtract the exponent in the denominator from the exponent in the numerator.*

> **Quotient Rule for Exponents**
> If m and n are integers, then
>
> $$\frac{x^m}{x^n} = x^{m-n} \quad (x \neq 0)$$

EXAMPLE 4 Simplify and write all answers without using negative exponents: **a.** $\dfrac{x^8}{x^5}$ and **b.** $\dfrac{x^2x^4}{x^{-5}}$.

Solution **a.** $\dfrac{x^8}{x^5} = x^{8-5}$ **b.** $\dfrac{x^2x^4}{x^{-5}} = \dfrac{x^6}{x^{-5}}$

$= x^3$ $= x^{6-(-5)}$

$= x^{11}$ ∎

Self Check   Simplify and write all answers without using negative exponents: **a.** $\dfrac{x^{-6}}{x^2}$ and **b.** $\dfrac{x^4x^{-3}}{x^2}$.

Answers   **a.** $\dfrac{1}{x^8}$,   **b.** $\dfrac{1}{x}$

EXAMPLE 5   Simplify and write all answers without using negative exponents:

a. $\left(\dfrac{x^3y^{-2}}{x^{-2}y^3}\right)^{-2}$ and   **b.** $\left(\dfrac{x}{y}\right)^{-n}$.

Solution   **a.** $\left(\dfrac{x^3y^{-2}}{x^{-2}y^3}\right)^{-2} = (x^{3-(-2)}y^{-2-3})^{-2}$

$= (x^5y^{-5})^{-2}$

$= x^{-10}y^{10}$

$= \dfrac{y^{10}}{x^{10}}$

b. $\left(\dfrac{x}{y}\right)^{-n} = \dfrac{x^{-n}}{y^{-n}}$

$= \dfrac{x^{-n}x^ny^n}{y^{-n}x^ny^n}$ Multiply numerator and denominator by 1 in the form $\dfrac{x^ny^n}{x^ny^n}$.

$= \dfrac{x^0y^n}{y^0x^n}$ $x^{-n}x^n = x^0$ and $y^{-n}y^n = y^0$.

$= \dfrac{y^n}{x^n}$ $x^0 = 1$ and $y^0 = 1$.

$= \left(\dfrac{y}{x}\right)^n$

∎

Self Check   Simplify each expression. Write all answers without using negative exponents.

a. $\left(\dfrac{x^4y^{-3}}{x^{-3}y^2}\right)^2$ and   **b.** $\left(\dfrac{2a}{3b}\right)^{-3}$.

Answers   **a.** $\dfrac{x^{14}}{y^{10}}$,   **b.** $\dfrac{27b^3}{8a^3}$

Part **b** of Example 5 establishes the following rule.

A Fraction to a Negative Power
If n is a natural number, then

$$\left(\dfrac{x}{y}\right)^{-n} = \left(\dfrac{y}{x}\right)^n \quad (x \neq 0 \text{ and } y \neq 0)$$

■ ORDER OF OPERATIONS

When several operations occur in an expression, the operations should be done in the following order.

> **Order of Operations**
>
> If an expression does not contain grouping symbols such as parentheses or brackets, follow these steps:
>
> 1. Find the values of any exponential expressions.
> 2. Do all multiplications and/or divisions, working from left to right.
> 3. Do all additions and/or subtractions, working from left to right.
>
> If an expression contains grouping symbols, use the rules above to do the calculations within each pair of grouping symbols, working from the innermost pair to the outermost pair.
>
> In a fraction, simplify the numerator and the denominator separately. Then simplify the fraction, if possible.

EXAMPLE 6 If $x = -2$, $y = 3$, and $z = -4$, evaluate   **a.** $-x^2 + y^2z$ and   **b.** $\dfrac{2z^3 - 3y^2}{5x^2}$.

Solution **a.** $-x^2 + y^2z = -(-2)^2 + 3^2(-4)$

$$= -(4) + 9(-4) \qquad \text{Evaluate the powers.}$$

$$= -4 + (-36) \qquad \text{Do the multiplication.}$$

$$= -40 \qquad \text{Do the addition.}$$

b. $\dfrac{2z^3 - 3y^2}{5x^2} = \dfrac{2(-4)^3 - 3(3)^2}{5(-2)^2}$

$$= \dfrac{2(-64) - 3(9)}{5(4)} \qquad \text{Evaluate the powers.}$$

$$= \dfrac{-128 - 27}{20} \qquad \text{Do the multiplications.}$$

$$= \dfrac{-155}{20} \qquad \text{Do the subtraction.}$$

$$= -\dfrac{31}{4} \qquad \text{Simplify the fraction.}$$

■

Self Check If $x = 3$ and $y = -2$, evaluate $\dfrac{2x^2 - 3y^2}{x - y}$.

Answer: $\dfrac{6}{5}$

■ SCIENTIFIC NOTATION

Scientists often work with numbers that are either very large or very small. These numbers can be written compactly by expressing them in *scientific notation*.

> **Scientific Notation**
> A number is written in **scientific notation** when it is written in the form
>
> $$N \times 10^n$$
>
> where $1 \le N \le 10$ and n is an integer.

Light travels 29,980,000,000 centimeters per second. To express this number in scientific notation, we must express the number as the product of a number between 1 and 10 and some integer power of 10. The number 2.998 lies between 1 and 10. To get 29,980,000,000, the decimal point in 2.998 must be moved ten places to the right. This is accomplished by multiplying 2.998 by 10^{10}.

Standard notation ⟶ $29{,}980{,}000{,}000 = 2.998 \times 10^{10}$ ⟵ Scientific notation

One meter is approximately 0.0006214 mile. To express this number in scientific notation, we must express the number as the product of a number between 1 and 10 and some integer power of 10. The number 6.214 lies between 1 and 10. To get 0.0006214, the decimal point in 6.214 must be moved four places to the left. This is accomplished by multiplying 6.214 by $\frac{1}{10^4}$ or by multiplying 6.214 by 10^{-4}.

Standard notation ⟶ $0.0006214 = 6.214 \times 10^{-4}$ ⟵ Scientific notation

The following numbers are written in both standard and scientific notation. In each case, the exponent gives the number of places the decimal point moves, and the sign of the exponent indicates the direction in which it moves.

a. $3\,7\,2\,0\,0\,0 = 3.72 \times 10^5$ 5 places to the right.

b. $0.0\,0\,0\,5\,3\,7 = 5.37 \times 10^{-4}$ 4 places to the left.

c. $7.36 = 7.36 \times 10^0$ No movement of the decimal point.

EXAMPLE 7   Write   **a.** 62,000 and   **b.** −0.0027 in scientific notation.

Solution   **a.** We must express 62,000 as a product of a number between 1 and 10 and some integer power of 10. This is accomplished by multiplying 6.2 by 10^4.

$$62{,}000 = 6.2 \times 10^4$$

b. We must express −0.0027 as a product of a number whose absolute value is between 1 and 10 and some integer power of 10. This is accomplished by multiplying 2.7 by 10^{-3}.

$$-0.0027 = -2.7 \times 10^{-3}$$

■

Self Check   Write   **a.** 93,000,000 and   **b.** −0.0000087 in scientific notation.

Answers   **a.** 9.3×10^7,   **b.** -8.7×10^{-6}

EXAMPLE 8   Write   **a.** 7.35×10^2 and   **b.** 3.27×10^{-5} in standard notation.

Solution   **a.** The factor of 10^2 indicates that 7.35 must be multiplied by 2 factors of 10. Because each multiplication by 10 moves the decimal point one place to the right, we have

$$7.35 \times 10^2 = 735$$

b. The factor of 10^{-5} indicates that 3.27 must be divided by 5 factors of 10. Because each division by 10 moves the decimal point one place to the left, we have

$$3.27 \times 10^{-5} = 0.0000327$$ ∎

Self Check   Write   **a.** 6.3×10^3 and   **b.** 9.1×10^{-4} in standard notation.

Answers   **a.** 6300,   **b.** 0.00091

∎ USING SCIENTIFIC NOTATION TO SIMPLIFY COMPUTATIONS

Another advantage of scientific notation becomes evident when we must multiply and divide combinations of very large and very small numbers.

EXAMPLE 9   Use scientific notation to calculate

$$\frac{(3,400,000)(0.00002)}{170,000,000}$$

Solution   After changing each number into scientific notation, we can do the arithmetic on the numbers and the exponential expressions separately.

$$\frac{(3,400,000)(0.00002)}{170,000,000} = \frac{(3.4 \times 10^6)(2.0 \times 10^{-5})}{1.7 \times 10^8}$$

$$= \frac{6.8}{1.7} \times 10^{6+(-5)-8}$$

$$= 4.0 \times 10^{-7}$$

$$= 0.0000004$$ ∎

Self Check   Use scientific notation to simplify

$$\frac{(192,000)(0.0015)}{(0.0032)(4500)}$$

Answer   20

■ ■ ■ ■ ■ ■ ■ ■ ■

ACCENT ON TECHNOLOGY

Calculators often give answers in scientific notation. For example, if we use a calculator to find 21^8, the display will read

On a scientific calculator		*On a graphing calculator*
3.7822855936 ^{10}	or	21^ 8
		3.7822855936E10

In either case, the result is given in scientific notation and means $3.7822855936 \times 10^{10}$.

We can enter numbers into a calculator in scientific form. For example, to enter 0.000000000061 (which is 6.1×10^{-11}), we enter these numbers and press these keys:

6.1 **EXP** 11 **+/−**   On a scientific calculator

6.1 **EE** **(−)** 11   On a graphing calculator

To use a scientific calculator to simplify

$$\frac{21^8}{0.000000000061}$$

we must enter the denominator in scientific notation, because there are too many digits to enter it directly. To do the calculation with a scientific calculator, we press these numbers and these keys:

21 y^x 8 **=** ÷ 6.1 **EXP** **+/−** 11 **=**

The display will read 6.200468748 ^{20} . In standard notation, the answer is approximately 620,046,874,800,000,000,000.

The steps are similar on a graphing calculator.

■ ■ ■ ■ ■ ■ ■ ■ ■

EXERCISE 1.2

VOCABULARY AND CONCEPTS *Fill in the blank to make a true statement.*

1. Each quantity in a product is called a _____ of the product.

2. A _____ number exponent tells how many times a base is used as a factor.

3. In the expression $(2x)^3$, __ is the exponent and __ is the base.

4. The expression x^n is called an _____ expression.

5. A number is in _____ notation when it is written in the form $N \times 10^n$, where $1 \le N \le 10$ and n is an _____.

6. Unless parentheses indicate otherwise, _____ are done before additions.

In Exercises 7–12, complete each formula.

7. $x^m x^n = $ _____

8. $(x^m)^n = $ _____

9. $(xy)^n = $ _____

10. $\dfrac{x^m}{x^n} = $ _____

11. $x^0 = $ __

12. $x^{-n} = $ _____

PRACTICE   *In Exercises 13–20, write each number without using exponents.*

13. 13^2

14. 10^3

15. -5^2

16. $(-5)^2$

17. $4x^3$

18. $(4x)^3$

19. $(-5x)^4$

20. $-6x^2$

In Exercises 21–28, write each expression using exponents.

21. $7xxx$

22. $-8yyyy$

23. $(-x)(-x)$

24. $(2a)(2a)(2a)$

25. $(3t)(3t)(-3t)$

26. $-(2b)(2b)(2b)(2b)$

27. $xxxyy$

28. $aaabbbb$

In Exercises 29–32, use a calculator to simplify each expression.

29. 2.2^3

30. 7.1^4

31. -0.5^4

32. $(-0.2)^4$

In Exercises 33–78, simplify each expression. Express all answers without using negative exponents. Assume that all variables are restricted to those numbers for which the expression is defined.

33. $x^2 x^3$

34. $y^3 y^4$

35. $(z^2)^3$

36. $(t^6)^7$

37. $(y^5 y^2)^3$

38. $(a^3 a^6)a^4$

39. $(z^2)^3(z^4)^5$

40. $(t^3)^4(t^5)^2$

41. $(3r)^3$

42. $(-2y)^4$

43. $(x^2 y)^3$

44. $(x^3 z^4)^6$

45. $\left(\dfrac{a^2}{b}\right)^3$

46. $\left(\dfrac{x}{y^3}\right)^4$

47. $(-x)^0$

48. $4x^0$

49. $(4x)^0$

50. $-2x^0$

51. z^{-4}

52. $\dfrac{1}{t^{-2}}$

53. $y^{-2} y^{-3}$

54. $-m^{-2} m^3$

55. $(x^3 x^{-4})^{-2}$

56. $(y^{-2} y^3)^{-4}$

57. $\dfrac{x^7}{x^3}$

58. $\dfrac{r^5}{r^2}$

59. $\dfrac{a^{21}}{a^{17}}$

60. $\dfrac{t^{13}}{t^4}$

61. $\dfrac{(x^2)^2}{x^2 x}$

62. $\dfrac{s^9 s^3}{(s^2)^2}$

63. $\left(\dfrac{m^3}{n^2}\right)^3$

64. $\left(\dfrac{t^4}{t^3}\right)^3$

65. $\dfrac{(a^3)^{-2}}{aa^2}$

66. $\dfrac{r^9 r^{-3}}{(r^{-2})^3}$

67. $\left(\dfrac{a^{-3}}{b^{-1}}\right)^{-4}$

68. $\left(\dfrac{t^{-4}}{t^{-3}}\right)^{-2}$

69. $\left(\dfrac{r^4 r^{-6}}{r^3 r^{-3}}\right)^2$

70. $\dfrac{(x^{-3} x^2)^2}{(x^2 x^{-5})^{-3}}$

71. $\left(\dfrac{x^5 y^{-2}}{x^{-3} y^2}\right)^4$

72. $\left(\dfrac{x^{-7} y^5}{x^7 y^{-4}}\right)^3$

73. $\left(\dfrac{5x^{-3} y^{-2}}{3x^2 y^{-3}}\right)^{-2}$

74. $\left(\dfrac{3x^2 y^{-5}}{2x^{-2} y^{-6}}\right)^{-3}$

75. $\left(\dfrac{3x^5 y^{-3}}{6x^{-5} y^3}\right)^{-2}$

76. $\left(\dfrac{12x^{-4} y^3 z^{-5}}{4x^4 y^{-3} z^5}\right)^3$

77. $\dfrac{(8^{-2} z^{-3} y)^{-1}}{(5y^2 z^{-2})^3 (5yz^{-2})^{-1}}$

78. $\dfrac{(m^{-2} n^3 p^4)^{-2}(mn^{-2} p^3)^4}{(mn^{-2} p^3)^{-4}(mn^2 p)^{-1}}$

In Exercises 79–90, let $x = -2$, $y = 0$, and $z = 3$, and evaluate each expression.

79. x^2

80. $-x^2$

81. x^3

82. $-x^3$

83. $(-xz)^3$

84. $-xz^3$

85. $\dfrac{-(x^2z^3)}{z^2 - y^2}$

86. $\dfrac{z^2(x^2 - y^2)}{x^3z}$

87. $5x^2 - 3y^3z$

88. $3(x - z)^2 + 2(y - z)^3$

89. $\dfrac{-3x^{-3}z^{-2}}{6x^2z^{-3}}$

90. $\dfrac{(-5x^2z^{-3})^2}{5xz^{-2}}$

In Exercises 91–100, express each number in scientific notation.

91. 372,000

92. 89,500

93. $-177,000,000$

94. $-23,470,000,000$

95. 0.007

96. 0.00052

97. -0.000000693

98. -0.000000089

99. one trillion

100. one millionth

In Exercises 101–108, express each number in standard notation.

101. 9.37×10^5

102. 4.26×10^9

103. 2.21×10^{-5}

104. 2.774×10^{-2}

105. 0.00032×10^4

106. 9300.0×10^{-4}

107. -3.2×10^{-3}

108. -7.25×10^3

In Exercises 109–114, use the method of Example 9 to do each calculation. Write all answers in scientific notation.

109. $\dfrac{(65,000)(45,000)}{250,000}$

110. $\dfrac{(0.000000045)(0.00000012)}{45,000,000}$

111. $\dfrac{(0.00000035)(170,000)}{0.00000085}$

112. $\dfrac{(0.0000000144)(12,000)}{600,000}$

113. $\dfrac{(45,000,000,000)(212,000)}{0.00018}$

114. $\dfrac{(0.00000000275)(4750)}{500,000,000,000}$

APPLICATIONS *In Exercises 115–118, use scientific notation to compute each answer. Write all answers in scientific notation.*

115. The speed of sound in air is 3.31×10^4 centimeters per second. Compute the speed of sound in meters per minute.

116. Calculate the volume of a box that has dimensions of 6000 by 9700 by 4700 millimeters.

117. The mass of one proton is 0.0000000000000000000000000167248 gram. Find the mass of one billion protons.

118. The speed of light in a vacuum is approximately 30,000,000,000 centimeters per second. Find the speed of light in miles per hour. (160,934.4 cm = 1 mile.)

DISCOVERY AND WRITING   *In Exercises 119–124, write each expression with a single base.*

119. $x^n x^2$

120. $\dfrac{x^m}{x^3}$

121. $\dfrac{x^m x^2}{x^3}$

122. $\dfrac{x^{3m+5}}{x^2}$

123. $x^{m+1} x^3$

124. $a^{n-3} a^3$

125. Explain why $-x^4$ and $(-x)^4$ represent different numbers.

126. Explain why 32×10^2 is not in scientific notation.

REVIEW

127. Graph the interval $(-2, 4)$.

128. Evaluate $|\pi - 5|$.

1.3 Fractional Exponents and Radicals

■ FRACTIONAL EXPONENTS WITH NONNEGATIVE BASES ■ FRACTIONAL EXPONENTS WITH NEGATIVE BASES ■ FRACTIONAL EXPONENTS WITH NUMERATORS GREATER THAN 1 ■ RADICAL EXPRESSIONS ■ SIMPLIFYING AND COMBINING RADICALS ■ RATIONALIZING DENOMINATORS AND NUMERATORS

■ FRACTIONAL EXPONENTS WITH NONNEGATIVE BASES

If we apply the rule $(x^m)^n = x^{mn}$ to $(25^{1/2})^2$, we obtain

$$(25^{1/2})^2 = 25^{(1/2)2} \qquad \text{Keep the base and multiply the exponents.}$$
$$= 25^1 \qquad \tfrac{1}{2} \cdot 2 = 1.$$
$$= 25$$

Thus, $25^{1/2}$ is a real number whose square is 25. Although $5^2 = 25$ and $(-5)^2 = 25$, we define $25^{1/2}$ to be the positive real number whose square is 25:

$$25^{1/2} = 5 \qquad \text{Read } 25^{1/2} \text{ as "the square root of 25."}$$

Fractional Exponents

If $a > 0$ and n is a natural number, then $a^{1/n}$ is the positive real number such that

$$(a^{1/n})^n = a \qquad \text{Read } a^{1/n} \text{ as "the } n\text{th root of } a \text{."}$$

If $a = 0$, then $a^{1/n} = 0$.

If n is even in the expression $a^{1/n}$, we must often use absolute value symbols to guarantee that the root is nonnegative.

$(49x^2)^{1/2} = 7|x|$

Because $(7|x|)^2 = 49x^2$. Since x could be negative, we use absolute value symbols to guarantee that the square root is nonnegative.

$(16x^4)^{1/4} = 2|x|$

Because $(2|x|)^4 = 16x^4$. Since x could be negative, we use absolute value symbols to guarantee that the fourth root is nonnegative. Read $(16x^4)^{1/4}$ as "the fourth root of $16x^4$."

$(729x^{12})^{1/6} = 3x^2$

Because $(3x^2)^6 = 729x^{12}$. Since $3x^2$ is always positive or 0, absolute value symbols are unnecessary. Read $(729x^{12})^{1/6}$ as "the sixth root of $729x^{12}$."

EXAMPLE 1

a. $27^{1/3} = 3$

Because $3^3 = 27$. Read $27^{1/3}$ as "the cube root of 27."

b. $\left(\dfrac{1}{81}\right)^{1/4} = \dfrac{1}{3}$

Because $\left(\dfrac{1}{3}\right)^4 = \dfrac{1}{81}$.

c. $-32^{1/5} = -(32^{1/5})$

Read $32^{1/5}$ as "the fifth root of 32."

$\qquad\quad = -(2)$

$32^{1/5} = 2$, because $2^5 = 32$.

$\qquad\quad = -2$

d. $(64a^6)^{1/6} = 2|a|$

Because $(2|a|)^6 = 64a^6$. Since a could be negative, use absolute value symbols to guarantee that the sixth root is nonnegative.

e. $(81x^4)^{1/2} = 9x^2$

Because $(9x^2)^2 = 81x^4$. No absolute value symbols are needed, because $9x^2$ is always nonnegative. ∎

Self Check   Simplify each expression: **a.** $100^{1/2}$,   **b.** $243^{1/5}$,   **c.** $(16a^4)^{1/4}$, and **d.** $(16a^4)^{1/2}$.

Answers   **a.** 10,   **b.** 3,   **c.** $2|a|$,   **d.** $4a^2$

WARNING! $(-64)^{1/2}$ is not a real number, because no real number squared is -64. In the expression $a^{1/n}$, there is no real-number nth root of a when a is negative and n is even.

■ **FRACTIONAL EXPONENTS WITH NEGATIVE BASES**

If we apply the rule $(x^m)^n = x^{mn}$ to $(-64^{1/3})^3$, we obtain

$(-64^{1/3})^3 = (-64)^{1/3 \cdot 3}$

Keep the base and multiply the exponents.

$\qquad\qquad\; = (-64)^1$

$\frac{1}{3} \cdot 3 = 1$.

$\qquad\qquad\; = -64$

Thus, $(-64)^{1/3}$ is a number whose cube is -64. Since $(-4)^3 = -64$, we have

$(-64)^{1/3} = -4$

Read $(-64)^{1/3}$ as "the cube root of -64."

If n is odd in the expression $a^{1/n}$, we don't need to use absolute value symbols, because odd roots can be negative.

$(-27x^3)^{1/3} = -3x$ Because $(-3x)^3 = -27x^3$.

$(-128a^7)^{1/7} = -2a$ Because $(-2a)^7 = -128a^7$. Read $(-128a^7)^{1/7}$ as "the seventh root of $-128a^7$."

EXAMPLE 2

  a. $(-8)^{1/3} = -2$ Because $(-2)^3 = -8$.

  b. $(-3{,}125)^{1/5} = -5$ Because $(-5)^5 = -3125$.

  c. $(-1000x^3)^{1/3} = -10x$ Because $(-10x)^3 = -1000x^3$.

  d. $(-128x^{14})^{1/7} = -2x^2$ Because $(-2x^2)^7 = -128x^{14}$. ∎

Self Check   Simplify each expression: **a.** $(-125)^{1/3}$,   **b.** $(-8x^3)^{1/3}$, and   **c.** $(-100{,}000a^{10})^{1/5}$.

Answers   **a.** -5,   **b.** $-2x$,   **c.** $-10a^2$

We summarize the definitions concerning $a^{1/n}$ as follows.

> If n is a natural number and a is a real number in the expression $a^{1/n}$, then
>
> If $a > 0$, then $a^{1/n}$ is the positive number such that $(a^{1/n})^n = a$.
>
> If $a = 0$, then $a^{1/n} = 0$.
>
> If $a < 0$, $\begin{cases} \text{and } n \text{ is odd, then } a^{1/n} \text{ is the real number such that } (a^{1/n})^n = a. \\ \text{and } n \text{ is even, then } a^{1/n} \text{ is not a real number.} \end{cases}$

■ FRACTIONAL EXPONENTS WITH NUMERATORS GREATER THAN 1

The definition of $a^{1/n}$ can be extended to fractional exponents with numerators other than 1. For example, $4^{3/2}$ can be written as either

$(4^{1/2})^3$ or $(4^3)^{1/2}$ Because of the power rule, $(x^m)^n = x^{mn}$.

This suggests the following rule.

Rule for Fractional Exponents

If m and n are positive integers, the fraction $\frac{m}{n}$ is in lowest terms, and $a^{1/n}$ is a real number, then

$$a^{m/n} = (a^{1/n})^m = (a^m)^{1/n}$$

In the previous rule, we view the expression $a^{m/n}$ in two ways:

1. $(a^{1/n})^m$: the mth power of the nth root of a
2. $(a^m)^{1/n}$: the nth root of the mth power of a

For example, $(-27)^{2/3}$ can be simplified in two ways:

$$(-27)^{2/3} = [(-27)^{1/3}]^2 \qquad \text{or} \qquad (-27)^{2/3} = [(-27)^2]^{1/3}$$
$$= (-3)^2 \qquad\qquad\qquad\qquad = (729)^{1/3}$$
$$= 9 \qquad\qquad\qquad\qquad\qquad = 9$$

As this example suggests, it is usually easier to take the root of the base first in order to avoid large numbers.

Negative Fractional Exponents

If m and n are positive integers, the fraction $\frac{m}{n}$ is in lowest terms, and $a^{1/n}$ is a real number, then

$$a^{-m/n} = \frac{1}{a^{m/n}} \qquad \text{and} \qquad \frac{1}{a^{-m/n}} = a^{m/n} \quad (a \neq 0)$$

EXAMPLE 3

a. $25^{3/2} = (25^{1/2})^3$
$= 5^3$
$= 125$

b. $\left(-\dfrac{x^6}{1000}\right)^{2/3} = \left[\left(-\dfrac{x^6}{1000}\right)^{1/3}\right]^2$
$= \left(-\dfrac{x^2}{10}\right)^2$
$= \dfrac{x^4}{100}$

c. $32^{-2/5} = \dfrac{1}{32^{2/5}}$
$= \dfrac{1}{(32^{1/5})^2}$
$= \dfrac{1}{2^2}$
$= \dfrac{1}{4}$

d. $\dfrac{1}{81^{-3/4}} = 81^{3/4}$
$= (81^{1/4})^3$
$= 3^3$
$= 27$

$\blacksquare$

Self Check   Simplify each expression: **a.** $49^{3/2}$,   **b.** $16^{-3/4}$, and   **c.** $\dfrac{1}{(27x^3)^{-2/3}}$.

Answers   **a.** 343,   **b.** $\dfrac{1}{8}$,   **c.** $9x^2$

Because of the definition of fractional exponents, they follow the same rules as integer exponents.

EXAMPLE 4 Assume that all variables represent positive numbers. Write all answers without using negative exponents.

a. $(36x)^{1/2} = 36^{1/2}x^{1/2}$
$\qquad\quad = 6x^{1/2}$

b. $\dfrac{(a^{1/3}b^{2/3})^6}{(y^3)^2} = \dfrac{a^{6/3}b^{12/3}}{y^6}$
$\qquad\qquad\qquad = \dfrac{a^2b^4}{y^6}$

c. $\dfrac{a^{x/2}a^{x/4}}{a^{x/6}} = a^{x/2+x/4-x/6}$
$\qquad\qquad = a^{6x/12+3x/12-2x/12}$
$\qquad\qquad = a^{7x/12}$

d. $\left[\dfrac{-c^{-2/5}}{c^{4/5}}\right]^{5/3} = (-c^{-2/5-4/5})^{5/3}$
$\qquad\qquad\qquad = [(-1)(c^{-6/5})]^{5/3}$
$\qquad\qquad\qquad = (-1)^{5/3}(c^{-6/5})^{5/3}$
$\qquad\qquad\qquad = -1c^{-30/15}$
$\qquad\qquad\qquad = -c^{-2}$
$\qquad\qquad\qquad = -\dfrac{1}{c^2}$ ■

Self Check Use the directions for Example 4.

a. $\left(\dfrac{y^2}{49}\right)^{1/2}$, **b.** $\dfrac{b^{3/7}b^{2/7}}{b^{4/7}}$, and **c.** $\dfrac{(9r^4s)^{1/2}}{rs^{-3/2}}$.

Answers **a.** $\dfrac{y}{7}$, **b.** $b^{1/7}$, **c.** $3s^2$

■ RADICAL EXPRESSIONS

Radical signs can also be used to express roots of numbers.

> **Definition of $\sqrt[n]{a}$**
> If n is a natural number greater than 1 and if $a^{1/n}$ is a real number, then
> $$\sqrt[n]{a} = a^{1/n}$$

In the **radical expression** $\sqrt[n]{a}$, the symbol $\sqrt{}$ is the **radical sign,** a is the **radicand,** and n is the **index** (or the **order**) of the radical. If the order is 2, the expression is a **square root,** and we do not write the index.

$$\sqrt{a} = \sqrt[2]{a}$$

If the index of a radical is 3, the radical is a **cube root.**

---

nth Root of a Nonnegative Number

If n is a natural number greater than 1 and $a \geq 0$, then $\sqrt[n]{a}$ is the nonnegative number whose nth power is a.

$$\left(\sqrt[n]{a}\right)^n = a$$

The nonnegative number $\sqrt[n]{a}$ is called the **principal nth root** of a.

If 2 is substituted for n in the equation $\left(\sqrt[n]{a}\right)^n = a$, we have

$$\left(\sqrt[2]{a}\right)^2 = \left(\sqrt{a}\right)^2 = \sqrt{a}\,\sqrt{a} = a$$

This shows that if a number a can be factored into two equal factors, then either of those factors is a square root of a. Furthermore, if a can be factored into n equal factors, any of those factors is an nth root of a.

If n is an odd natural number greater than 1 in the expression $\sqrt[n]{a}$, we can remove the restriction that $a \geq 0$.

---

nth Root of a Number

If n is an odd natural number greater than 1 and a is any real number, then $\sqrt[n]{a}$ is the real number whose nth power is a.

$$\left(\sqrt[n]{a}\right)^n = a$$

EXAMPLE 5

a. $\sqrt{81} = 9$ Because $9^2 = 81$.

b. $\sqrt[3]{-125} = -5$ Because $(-5)^3 = -125$.

c. $-\sqrt[4]{256} = -\left(\sqrt[4]{256}\right)$
$$= -(4)$$
$$= -4$$

d. $-\sqrt[5]{-243} = -\left(\sqrt[5]{-243}\right)$
$$= -(-3)$$
$$= 3 \qquad \blacksquare$$

Self Check   Find each root: **a.** $\sqrt{169}$,   **b.** $\sqrt[3]{216}$, and   **c.** $\sqrt[5]{-\dfrac{1}{32}}$.

Answers   **a.** 13,   **b.** 6,   **c.** $-\dfrac{1}{2}$

We summarize the definitions concerning $\sqrt[n]{a}$ as follows.

If n is a natural number greater than 1 and a is a real number, then

If $a > 0$, then $\sqrt[n]{a}$ is the positive number such that $\left(\sqrt[n]{a}\right)^n = a$.

If $a = 0$, then $\sqrt[n]{a} = 0$.

If $a < 0$ $\begin{cases} \text{and } n \text{ is odd, then } \sqrt[n]{a} \text{ is the real number such that } \left(\sqrt[n]{a}\right)^n = a. \\ \text{and } n \text{ is even, then } \sqrt[n]{a} \text{ is not a real number.} \end{cases}$

We have seen that $a^{m/n} = (a^{1/n})^m = (a^m)^{1/n}$. This same fact can be stated in radical notation.

$$a^{m/n} = \left(\sqrt[n]{a}\right)^m = \sqrt[n]{a^m}$$

Thus, *the mth power of the nth root of a is the same as the nth root of the mth power of a.* For example, to find $\sqrt[3]{27^2}$, we can proceed in either of two ways:

$$\sqrt[3]{27^2} = \left(\sqrt[3]{27}\right)^2 = 3^2 = 9 \qquad \text{or} \qquad \sqrt[3]{27^2} = \sqrt[3]{729} = 9$$

By definition, $\sqrt{a^2}$ represents a nonnegative number. If a could be negative, we must use absolute value symbols to guarantee that $\sqrt{a^2}$ is nonnegative. Thus, if a is unrestricted,

$$\sqrt{a^2} = |a|$$

A similar argument holds when the index is any even natural number. The symbol $\sqrt[4]{a^4}$, for example, means the *positive* fourth root of a^4. Thus, if a is unrestricted,

$$\sqrt[4]{a^4} = |a|$$

EXAMPLE 6   If x could be any real number, simplify   **a.** $\sqrt[6]{64x^6}$,   **b.** $\sqrt[3]{x^3}$, and   **c.** $\sqrt{9x^8}$.

Solution   **a.** $\sqrt[6]{64x^6} = 2|x|$ Use absolute value symbols to guarantee that the result is nonnegative.

b. $\sqrt[3]{x^3} = x$ Because the index is odd, no absolute value symbols are needed.

c. $\sqrt{9x^8} = 3x^4$ Because $3x^4$ is always nonnegative, no absolute value symbols are needed.   ∎

Self Check   Use the directions for Example 6:   **a.** $\sqrt[4]{16x^4}$,   **b.** $\sqrt[3]{27y^3}$, and   **c.** $\sqrt[4]{x^8}$.

Answers   **a.** $2|x|$,   **b.** $3y$,   **c.** x^2

■ SIMPLIFYING AND COMBINING RADICALS

Many properties of exponents have counterparts in radical notation. For example, since $a^{1/n}b^{1/n} = (ab)^{1/n}$ and $\dfrac{a^{1/n}}{b^{1/n}} = \left(\dfrac{a}{b}\right)^{1/n}$, we have the following.

Multiplication and Division Properties of Radicals

$$\sqrt[n]{a}\sqrt[n]{b} = \sqrt[n]{ab}$$

$$\frac{\sqrt[n]{a}}{\sqrt[n]{b}} = \sqrt[n]{\frac{a}{b}} \quad (b \neq 0)$$

In words, we say

The product of two nth roots is equal to the nth root of their product.

The quotient of two nth roots is equal to the nth root of their quotient.

 WARNING!   These properties involve either the nth root of the product of two numbers or the nth root of the quotient of two numbers. There is no such property for sums or differences. For example, $\sqrt{9 + 4} \neq \sqrt{9} + \sqrt{4}$, because

$$\sqrt{9 + 4} = \sqrt{13} \qquad \text{but} \qquad \sqrt{9} + \sqrt{4} = 3 + 2 = 5$$

and $\sqrt{13} \neq 5$. In general,

$$\sqrt{a + b} \neq \sqrt{a} + \sqrt{b} \qquad \text{and} \qquad \sqrt{a - b} \neq \sqrt{a} - \sqrt{b}$$

Numbers such as 1, 4, 9, 16, 25, and 36 that are squares of positive integers are called **perfect squares.** Expressions such as x^2 and x^6 are also perfect squares, because each one is the square of another expression.

Numbers such as 1, 8, 27, 64, 125, and 216 that are cubes of positive integers are called **perfect cubes.** Expressions such as x^3 and x^9 are also perfect cubes, because each one is the cube of another expression. There are also perfect fourth powers, perfect fifth powers, and so on.

We can use these perfect powers and the multiplication property of radicals to simplify radical expressions. For example, to simplify $\sqrt{12x^5}$, we factor $12x^5$ so that one factor is the largest perfect square that divides $12x^5$. In this case, it is $4x^4$. We then rewrite $12x^5$ as $4x^4 \cdot 3x$ and simplify.

$$\sqrt{12x^5} = \sqrt{4x^4 \cdot 3x} \qquad \text{Factor } 12x^5 \text{ as } 4x^4 \cdot 3x.$$
$$= \sqrt{4x^4}\sqrt{3x} \qquad \text{Use the multiplication property of radicals: } \sqrt{ab} = \sqrt{a}\sqrt{b}.$$
$$= 2x^2\sqrt{3x} \qquad \sqrt{4x^4} = 2x^2.$$

To simplify $\sqrt[3]{432x^9y}$, we find the largest perfect cube factor of $432x^9y$ (which is $216x^9$) and proceed as follows:

$$\sqrt[3]{432x^9y} = \sqrt[3]{216x^9 \cdot 2y} \qquad \text{Factor } 432x^9y \text{ as } 216x^9 \cdot 2y.$$
$$= \sqrt[3]{216x^9}\sqrt[3]{2y} \qquad \text{Use the multiplication property of radicals: } \sqrt[3]{ab} = \sqrt[3]{a}\sqrt[3]{b}.$$
$$= 6x^3\sqrt[3]{2y} \qquad \sqrt[3]{216x^9} = 6x^3.$$

Radical expressions with the same index and the same radicand are called **like** or **similar radicals.** We can combine the like radicals in $3\sqrt{2} + 2\sqrt{2}$ by using the distributive property.

$$3\sqrt{2} + 2\sqrt{2} = (3+2)\sqrt{2}$$
$$= 5\sqrt{2}$$

This example suggests that to combine like radicals, we *add their numerical coefficients and keep the same radical.*

When radicals have the same index but different radicands, we can often change them to equivalent forms having the same radicand. We can then combine them. For example, to simplify $\sqrt{27} - \sqrt{12}$, we simplify both radicals and combine like radicals.

$$\sqrt{27} - \sqrt{12} = \sqrt{9 \cdot 3} - \sqrt{4 \cdot 3}$$
$$= \sqrt{9}\sqrt{3} - \sqrt{4}\sqrt{3}$$
$$= 3\sqrt{3} - 2\sqrt{3} \qquad \sqrt{9} = 3 \text{ and } \sqrt{4} = 2.$$
$$= \sqrt{3}$$

EXAMPLE 7 Simplify **a.** $\sqrt{50} + \sqrt{200}$ and **b.** $3z\sqrt[5]{64z} - 2\sqrt[5]{2z^6}$.

Solution **a.** $\sqrt{50} + \sqrt{200} = \sqrt{25 \cdot 2} + \sqrt{100 \cdot 2}$
$$= \sqrt{25}\sqrt{2} + \sqrt{100}\sqrt{2}$$
$$= 5\sqrt{2} + 10\sqrt{2}$$
$$= 15\sqrt{2}$$

b. $3z\sqrt[5]{64z} - 2\sqrt[5]{2z^6} = 3z\sqrt[5]{32 \cdot 2z} - 2\sqrt[5]{z^5 \cdot 2z}$
$$= 3z\sqrt[5]{32}\sqrt[5]{2z} - 2\sqrt[5]{z^5}\sqrt[5]{2z}$$
$$= 3z(2)\sqrt[5]{2z} - 2z\sqrt[5]{2z}$$
$$= 6z\sqrt[5]{2z} - 2z\sqrt[5]{2z}$$
$$= 4z\sqrt[5]{2z}$$

Self Check Simplify: **a.** $\sqrt{18} - \sqrt{8}$ and **b.** $2\sqrt[3]{81a^4} + a\sqrt[3]{24a}$.
Answers **a.** $\sqrt{2}$, **b.** $8a\sqrt[3]{3a}$

■ RATIONALIZING DENOMINATORS AND NUMERATORS

A process called **rationalizing the denominator** enables us to write a fraction such as

$$\frac{\sqrt{5}}{\sqrt{3}}$$

as a fraction with a rational number in the denominator. All that we must do is multiply both the numerator and the denominator by $\sqrt{3}$.

$$\frac{\sqrt{5}}{\sqrt{3}} = \frac{\sqrt{5}\sqrt{3}}{\sqrt{3}\sqrt{3}} = \frac{\sqrt{15}}{3}$$

To rationalize the numerator, we multiply the numerator and denominator by $\sqrt{5}$.

$$\frac{\sqrt{5}}{\sqrt{3}} = \frac{\sqrt{5}\sqrt{5}}{\sqrt{3}\sqrt{5}} = \frac{5}{\sqrt{15}}$$

EXAMPLE 8 c,d

Rationalize each denominator and simplify (assume that all variables represent positive numbers):

a. $\dfrac{1}{\sqrt{7}}$, **b.** $\sqrt[3]{\dfrac{3}{4}}$, **c.** $\sqrt{\dfrac{3}{x}}$, and **d.** $\sqrt{\dfrac{3a^3}{5x^5}}$.

Solution

a. $\dfrac{1}{\sqrt{7}} = \dfrac{1\sqrt{7}}{\sqrt{7}\sqrt{7}}$

$= \dfrac{\sqrt{7}}{7}$

b. $\sqrt[3]{\dfrac{3}{4}} = \dfrac{\sqrt[3]{3}}{\sqrt[3]{4}}$

$= \dfrac{\sqrt[3]{3}\sqrt[3]{2}}{\sqrt[3]{4}\sqrt[3]{2}}$

$= \dfrac{\sqrt[3]{6}}{\sqrt[3]{8}}$

$= \dfrac{\sqrt[3]{6}}{2}$

c. $\sqrt{\dfrac{3}{x}} = \dfrac{\sqrt{3}}{\sqrt{x}}$

$= \dfrac{\sqrt{3}\sqrt{x}}{\sqrt{x}\sqrt{x}}$

$= \dfrac{\sqrt{3x}}{x}$

d. $\sqrt{\dfrac{3a^3}{5x^5}} = \dfrac{\sqrt{3a^3}}{\sqrt{5x^5}}$

$= \dfrac{\sqrt{3a^3}\sqrt{5x}}{\sqrt{5x^5}\sqrt{5x}}$

$= \dfrac{\sqrt{15a^3x}}{\sqrt{25x^6}}$

$= \dfrac{\sqrt{a^2}\sqrt{15ax}}{5x^3}$

$= \dfrac{a\sqrt{15ax}}{5x^3}$

■

Self Check Use the directions for Example 8: **a.** $\dfrac{6}{\sqrt{6}}$ and **b.** $\sqrt[3]{\dfrac{2}{5x}}$.

Answers **a.** $\sqrt{6}$, **b.** $\dfrac{\sqrt[3]{50x^2}}{5x}$

EXAMPLE 9 Rationalize each numerator and simplify (assume that all variables represent positive numbers):
a. $\dfrac{\sqrt{x}}{7}$ and **b.** $\dfrac{2\sqrt[3]{9x}}{3}$.

Solution **a.** $\dfrac{\sqrt{x}}{7} = \dfrac{\sqrt{x}\cdot\sqrt{x}}{7\sqrt{x}}$

$= \dfrac{x}{7\sqrt{x}}$

b. $\dfrac{2\sqrt[3]{9x}}{3} = \dfrac{2\sqrt[3]{9x}\cdot\sqrt[3]{3x^2}}{3\sqrt[3]{3x^2}}$

$= \dfrac{2\sqrt[3]{27x^3}}{3\sqrt[3]{3x^2}}$

$= \dfrac{2(3x)}{3\sqrt[3]{3x^2}}$

$= \dfrac{2x}{\sqrt[3]{3x^2}}$ Divide out the 3's.

Self Check Use the directions for Example 9: **a.** $\dfrac{\sqrt{2x}}{5}$ and **b.** $\dfrac{3\sqrt[3]{2y^2}}{6}$.

Answers **a.** $\dfrac{2x}{5\sqrt{2x}}$, **b.** $\dfrac{y}{\sqrt[3]{4y}}$

After rationalizing denominators, we can often combine like terms.

EXAMPLE 10 Simplify $\sqrt{\dfrac{1}{2}} + \sqrt{\dfrac{1}{8}}$.

Solution $\sqrt{\dfrac{1}{2}} + \sqrt{\dfrac{1}{8}} = \dfrac{1}{\sqrt{2}} + \dfrac{1}{\sqrt{8}}$

$= \dfrac{1\sqrt{2}}{\sqrt{2}\sqrt{2}} + \dfrac{1\sqrt{2}}{\sqrt{8}\sqrt{2}}$

$= \dfrac{\sqrt{2}}{2} + \dfrac{\sqrt{2}}{\sqrt{16}}$

$= \dfrac{\sqrt{2}}{2} + \dfrac{\sqrt{2}}{4}$

$= \dfrac{3\sqrt{2}}{4}$

Self Check   Simplify: $\sqrt[3]{\dfrac{x}{2}} - \sqrt[3]{\dfrac{x}{16}}$.

Answer $\dfrac{\sqrt[3]{4x}}{4}$

Another property of radicals can be derived from the properties of exponents. If all of the expressions represent real numbers, then

$$\sqrt[n]{\sqrt[m]{x}} = \sqrt[n]{x^{1/m}} = (x^{1/m})^{1/n} = x^{1/(mn)} = \sqrt[mn]{x}$$

$$\sqrt[m]{\sqrt[n]{x}} = \sqrt[m]{x^{1/n}} = (x^{1/n})^{1/m} = x^{1/(nm)} = \sqrt[mn]{x}$$

These results are summarized in the following theorem.

Theorem
If all of the expressions involved represent real numbers, then

$$\sqrt[m]{\sqrt[n]{x}} = \sqrt[n]{\sqrt[m]{x}} = \sqrt[mn]{x}$$

We can use the previous theorem to simplify many radicals. For example,

$$\sqrt[3]{\sqrt{8}} = \sqrt{\sqrt[3]{8}} = \sqrt{2}$$

Fractional exponents can be used to simplify many radical expressions, as shown in the following example.

EXAMPLE 11   Simplify: **a.** $\sqrt[6]{4}$,   **b.** $\sqrt[12]{x^3}$, and   **c.** $\sqrt[9]{8y^3}$. (Assume that x and y are positive numbers.)

Solution   **a.** $\sqrt[6]{4} = 4^{1/6} = (2^2)^{1/6} = 2^{2/6} = 2^{1/3} = \sqrt[3]{2}$

b. $\sqrt[12]{x^3} = x^{3/12} = x^{1/4} = \sqrt[4]{x}$

c. $\sqrt[9]{8y^3} = (2^3 y^3)^{1/9} = (2y)^{3/9} = (2y)^{1/3} = \sqrt[3]{2y}$ ■

Self Check   Simplify: **a.** $\sqrt[4]{4}$ and   **b.** $\sqrt[9]{27x^3}$.
Answers   **a.** $\sqrt{2}$,   **b.** $\sqrt[3]{3x}$

EXERCISE 1.3

VOCABULARY AND CONCEPTS   *Fill in the blank to make a true statement.*

1. If $a = 0$, then $a^{1/n} =$ ___.

2. If $a > 0$ and n is a natural number, then $a^{1/n}$ is a _____ number.

3. If $a < 0$ and n is an even number, then $a^{1/n}$ is ___ a real number.

4. $6^{2/3}$ can be written as _____ or _____.

5. $\sqrt[n]{a} = $ ____.

6. $\sqrt{a^2} = $ ____.

7. $\sqrt[n]{a}\sqrt[n]{b} = $ ____.

8. $\sqrt[n]{\dfrac{a}{b}} = $ ____.

9. $\sqrt{x+y}$ ___ $\sqrt{x} + \sqrt{y}$

10. $\sqrt[m]{\sqrt[n]{x}}$ or $\sqrt[n]{\sqrt[m]{x}}$ can be written as ____.

PRACTICE   *In Exercises 11–22, simplify each expression.*

11. $9^{1/2}$

12. $8^{1/3}$

13. $\left(\dfrac{1}{25}\right)^{1/2}$

14. $\left(\dfrac{16}{625}\right)^{1/4}$

15. $-81^{1/4}$

16. $-\left(\dfrac{8}{27}\right)^{1/3}$

17. $(10,000)^{1/4}$

18. $(243)^{1/5}$

19. $(-64)^{1/3}$

20. $\left(-\dfrac{27}{8}\right)^{1/3}$

21. $(64)^{1/3}$

22. $(-125)^{1/3}$

In Exercises 23–34, simplify each expression. Use absolute symbols when necessary.

23. $(16a^2)^{1/2}$

24. $(25a^4)^{1/2}$

25. $(16a^4)^{1/4}$

26. $(-64a^3)^{1/3}$

27. $(-32a^5)^{1/5}$

28. $(64a^6)^{1/6}$

29. $(-216b^6)^{1/3}$

30. $(256t^8)^{1/4}$

31. $\left(\dfrac{16a^4}{25b^2}\right)^{1/2}$

32. $\left(-\dfrac{a^5}{32b^{10}}\right)^{1/5}$

33. $\left(-\dfrac{1000x^6}{27y^3}\right)^{1/3}$

34. $\left(\dfrac{49t^2}{100z^4}\right)^{1/2}$

In Exercises 35–50, simplify each expression. Write all answers without using negative exponents.

35. $4^{3/2}$

36. $8^{2/3}$

37. $-16^{3/2}$

38. $(-8)^{2/3}$

39. $-1000^{2/3}$

40. $100^{3/2}$

41. $64^{-1/2}$

42. $25^{-1/2}$

43. $64^{-3/2}$

44. $49^{-3/2}$

45. $-9^{-3/2}$

46. $(-27)^{-2/3}$

47. $\left(\dfrac{4}{9}\right)^{5/2}$

48. $\left(\dfrac{25}{81}\right)^{3/2}$

49. $\left(-\dfrac{27}{64}\right)^{-2/3}$

50. $\left(\dfrac{125}{8}\right)^{-4/3}$

In Exercises 51–64, simplify each expression. Write all answers without using negative exponents. Assume that all variables represent positive numbers.

51. $(100s^4)^{1/2}$

52. $(64u^6v^3)^{1/3}$

53. $(32y^{10}z^5)^{-1/5}$

54. $(625a^4b^8)^{-1/4}$

55. $(x^{10}y^5)^{3/5}$

56. $(64a^6b^{12})^{5/6}$

57. $(r^8s^{16})^{-3/4}$

58. $(-8x^9y^{12})^{-2/3}$

59. $\left(-\dfrac{8a^6}{125b^9}\right)^{2/3}$

60. $\left(\dfrac{16x^4}{625y^8}\right)^{3/4}$

61. $\left(\dfrac{27r^6}{1000s^{12}}\right)^{-2/3}$

62. $\left(-\dfrac{32m^{10}}{243n^{15}}\right)^{-2/5}$

63. $\dfrac{a^{2/5}a^{4/5}}{a^{1/5}}$

64. $\dfrac{x^{6/7}x^{3/7}}{x^{2/7}x^{5/7}}$

In Exercises 65–72, simplify each radical.

65. $\sqrt{49}$ **66.** $\sqrt{81}$ **67.** $\sqrt[3]{125}$ **68.** $\sqrt[3]{-64}$

69. $-\sqrt[4]{81}$ **70.** $\sqrt[5]{-243}$ **71.** $\sqrt[5]{-\dfrac{32}{100,000}}$ **72.** $\sqrt[4]{\dfrac{256}{625}}$

In Exercises 73–80, simplify each expression, using absolute value symbols when necessary. Write answers without using negative exponents.

73. $\sqrt{36x^2}$ **74.** $-\sqrt{25y^2}$ **75.** $\sqrt{9y^4}$ **76.** $\sqrt{a^4b^8}$

77. $\sqrt[3]{8y^3}$ **78.** $\sqrt[3]{-27z^9}$ **79.** $\sqrt[4]{\dfrac{x^4y^8}{z^{12}}}$ **80.** $\sqrt[5]{\dfrac{a^{10}b^5}{c^{15}}}$

In Exercises 81–94, simplify each expression. Assume that all variables represent positive numbers so that no absolute value symbols are needed.

81. $\sqrt{8} - \sqrt{2}$ **82.** $\sqrt{75} - 2\sqrt{27}$ **83.** $\sqrt{200x^2} + \sqrt{98x^2}$

84. $\sqrt{128a^3} - a\sqrt{162a}$ **85.** $2\sqrt{48y^5} - 3y\sqrt{12y^3}$ **86.** $y\sqrt{112y} + 4\sqrt{175y^3}$

87. $2\sqrt[3]{81} + 3\sqrt[3]{24}$ **88.** $3\sqrt[4]{32} - 2\sqrt[4]{162}$ **89.** $\sqrt[4]{768z^5} + \sqrt[4]{48z^5}$

90. $-2\sqrt[5]{64y^2} + 3\sqrt[5]{486y^2}$ **91.** $\sqrt{8x^2y} - x\sqrt{2y} + \sqrt{50x^2y}$ **92.** $3x\sqrt{18x} + 2\sqrt{2x^3} - \sqrt{72x^3}$

93. $\sqrt[3]{16xy^4} + y\sqrt[3]{2xy} - \sqrt[3]{54xy^4}$ **94.** $\sqrt[4]{512x^5} - \sqrt[4]{32x^5} + \sqrt[4]{1250x^5}$

In Exercises 95–102, assume that all variables are positive numbers. Rationalize each denominator and simplify.

95. $\dfrac{3}{\sqrt{3}}$ **96.** $\dfrac{2}{\sqrt{x}}$ **97.** $\dfrac{2}{\sqrt[3]{2}}$ **98.** $\dfrac{5a}{\sqrt[3]{25a}}$

99. $\dfrac{2b}{\sqrt[4]{3a^2}}$ **100.** $\sqrt{\dfrac{x}{2y}}$ **101.** $\sqrt[3]{\dfrac{2u^4}{9v}}$ **102.** $\sqrt[3]{-\dfrac{3s^5}{4r^2}}$

In Exercises 103–108, assume that all variables are positive numbers. Rationalize each numerator and simplify.

103. $\dfrac{\sqrt{5}}{10}$ **104.** $\dfrac{\sqrt{y}}{3}$ **105.** $\dfrac{\sqrt[3]{9}}{3}$ **106.** $\dfrac{\sqrt[3]{16b^2}}{16}$

107. $\dfrac{\sqrt[5]{16b^3}}{64a}$ **108.** $\sqrt{\dfrac{3x}{57}}$

In Exercises 109–112, rationalize each denominator and simplify.

109. $\sqrt{\dfrac{1}{3}} - \sqrt{\dfrac{1}{27}}$ **110.** $\sqrt[3]{\dfrac{1}{2}} + \sqrt[3]{\dfrac{1}{16}}$

111. $\sqrt{\dfrac{x}{8}} - \sqrt{\dfrac{x}{2}} + \sqrt{\dfrac{x}{32}}$ **112.** $\sqrt[3]{\dfrac{y}{4}} + \sqrt[3]{\dfrac{y}{32}} - \sqrt[3]{\dfrac{y}{500}}$

In Exercises 113–128, simplify each radical.

113. $\sqrt[4]{9}$
114. $\sqrt[6]{27}$
115. $\sqrt[10]{16x^6}$
116. $\sqrt[6]{27x^9}$

DISCOVERY AND WRITING We can often multiply and divide radicals with different indexes. For example, to multiply $\sqrt{3}$ by $\sqrt[3]{5}$, we first write each radical as a sixth root

$$\sqrt{3} = 3^{1/2} = 3^{3/6} = \sqrt[6]{3^3} = \sqrt[6]{27}$$
$$\sqrt[3]{5} = 5^{1/3} = 5^{2/6} = \sqrt[6]{5^2} = \sqrt[6]{25}$$

and then multiply the sixth roots.

$$\sqrt{3}\sqrt[3]{5} = \sqrt[6]{27}\sqrt[6]{25} = \sqrt[6]{(27)(25)} = \sqrt[6]{675}$$

Division is similar. Use this idea to write each of the following expressions with a single radical.

117. $\sqrt{2}\sqrt[3]{2}$
118. $\sqrt{3}\sqrt[3]{5}$
119. $\dfrac{\sqrt[4]{3}}{\sqrt{2}}$
120. $\dfrac{\sqrt[3]{2}}{\sqrt{5}}$

121. For what values of x does $\sqrt[4]{x^4} = x$? Explain.

122. If all of the radicals involved represent real numbers and $y \neq 0$, explain why

$$\sqrt[n]{\dfrac{x}{y}} = \dfrac{\sqrt[n]{x}}{\sqrt[n]{y}}$$

123. If all of the radicals involved represent real numbers and there is no division by 0, explain why

$$\left(\dfrac{x}{y}\right)^{-m/n} = \sqrt[n]{\dfrac{y^m}{x^m}}$$

124. The definition of $x^{m/n}$ requires that $\sqrt[n]{x}$ be a real number. Tell why this is important. (*Hint:* Consider what happens when n is even, m is odd, and x is negative.)

REVIEW

125. Write $-2 < x \leq 5$ using interval notation.

126. Write the expression $|3 - x|$ without using absolute value symbols. Assume that $x > 4$.

127. Write 617,000,000 in scientific notation.

128. Write 0.00235×10^4 in standard notation.

1.4 Polynomials

■ POLYNOMIALS ■ ADDING AND SUBTRACTING POLYNOMIALS ■ MULTIPLYING POLYNOMIALS
■ CONJUGATE BINOMIALS ■ DIVIDING POLYNOMIALS

A **monomial** is a number or the product of a number and one or more variables with whole-number exponents. The number is called the **coefficient** of the variables. Some examples of monomials are

$$3x, \quad 7ab^2, \quad -5ab^2c^4, \quad x^3, \quad \text{and} \quad -12$$

with coefficients of 3, 7, -5, 1, and -12, respectively.

The **degree** of a monomial is the sum of the exponents of its variables. All non-zero constants (except 0) have a degree of 0.

The degree of $3x$ is 1.

The degree of $-5ab^2c^4$ is 7.

The degree of -12 is 0 (since $-12 = -12x^0$).

The degree of $7ab^2$ is 3.

The degree of x^3 is 3.

0 has no defined degree.

■ POLYNOMIALS

A monomial or a sum of monomials is called a **polynomial.** Each monomial in that sum is called a **term** of the polynomial. A polynomial with two terms is called a **binomial,** and a polynomial with three terms is called a **trinomial.**

Monomials	*Binomials*	*Trinomials*
$3x^2$	$2a + 3b$	$x^2 + 7x - 4$
$-25xy$	$4x^3 - 3x^2$	$4y^4 - 2y + 12$
a^2b^3c	$-2x^3 - 4y^2$	$12x^3y^2 - 8xy - 24$

The **degree of a polynomial** is the degree of the term in the polynomial with highest degree. The only polynomial with no defined degree is 0, which is called the **zero polynomial.** Here are some examples.

- $3x^2y^3 + 5xy^2 + 7$ is a trinomial of 5th degree, because its term with highest degree (the first term) is 5.
- $3ab + 5a^2b$ is a binomial of degree 3.
- $5x + 3y^2 + \sqrt[3]{3}z^4 - \sqrt{7}$ is a polynomial, because its variables have whole-number exponents. It is of degree 4.
- $-7y^{1/2} + 3y^2 + \sqrt[5]{3}z$ is not a polynomial, because some of its variables do not have whole-number exponents.

If two terms of a polynomial have the same variables with the same exponents, they are **like** or **similar** terms. To combine the like terms in $3x^2y + 5x^2y$, we use the distributive property:

$$3x^2y + 5x^2y = (3 + 5)x^2y$$
$$= 8x^2y$$

This illustrates that *to combine like terms, we add their coefficients and keep the same variables and the same exponents.*

■ ADDING AND SUBTRACTING POLYNOMIALS

To add and subtract polynomials, we can use the **extended distributive property:**

$$a(b + c + d + e + \cdots) = ab + ac + ad + ae + \cdots$$

With this property, we can remove parentheses enclosing several terms of a polynomial. When the sign preceding the parentheses is $+$, we simply drop the parentheses:

$$+(a + b - c) = +\mathbf{1}(a + b - c)$$
$$= \mathbf{1}a + \mathbf{1}b - \mathbf{1}c$$
$$= a + b - c$$

To add polynomials, we remove parentheses (if necessary) and combine like terms.

EXAMPLE 1   Add $(3x^3y + 5x^2 - 2y) + (2x^3y - 5x^2 + 3x)$.

Solution   To add the polynomials, we remove parentheses and combine like terms.

$$(3x^3y + 5x^2 - 2y) + (2x^3y - 5x^2 + 3x) = 3x^3y + 5x^2 - 2y + 2x^3y - 5x^2 + 3x$$

$$= 3x^3y + 2x^3y + 5x^2 - 5x^2 - 2y + 3x \qquad \text{Use the commutative property to rearrange terms.}$$

$$= 5x^3y - 2y + 3x \qquad \text{Combine like terms.}$$

Self Check   Add $(4x^2 + 3x - 5) + (3x^2 - 5x + 7)$.

Answer $7x^2 - 2x + 2$

To remove parentheses enclosing several terms of a polynomial when the sign preceding the parentheses is $-$, we drop the parentheses and the $-$ sign and change the sign of each term within the parentheses.

$$-(a + b - c) = -\mathbf{1}(a + b - c)$$
$$= -\mathbf{1}a + (-\mathbf{1})b - (-\mathbf{1})c$$
$$= -a - b + c$$

To subtract polynomials, we remove parentheses and combine like terms.

EXAMPLE 2   Subtract $(2x^2 + 3y^2) - (x^2 - 2y^2 + 7)$.

Solution   To subtract the polynomials, we remove parentheses and combine like terms.

$$(2x^2 + 3y^2) - (x^2 - 2y^2 + 7) = 2x^2 + 3y^2 - x^2 + 2y^2 - 7$$

$$= 2x^2 - x^2 + 3y^2 + 2y^2 - 7 \qquad \text{Use the commutative property to rearrange terms.}$$

$$= x^2 + 5y^2 - 7 \qquad \text{Combine like terms.}$$

Self Check Subtract $(4x^2 + 3x - 5) - (3x^2 - 5x + 7)$.

Answer $x^2 + 8x - 12$

To remove parentheses enclosing several terms that are multiplied by a constant, we multiply each term within the parentheses by the constant and drop the parentheses. For example,

$$4(3x^2 - 2x + 6) = 4(3x^2) - 4(2x) + 4(6)$$
$$= 12x^2 - 8x + 24$$

This example suggests that *to add multiples of one polynomial to another, or to subtract multiples of one polynomial from another, we remove parentheses and combine like terms.*

EXAMPLE 3   Simplify $7x(2y^2 + 13x^2) - 5(xy^2 - 13x^3)$.

Solution $7x(2y^2 + 13x^2) - 5(xy^2 - 13x^3)$

$= 14xy^2 + 91x^3 - 5xy^2 + 65x^3$ Use the distributive property to remove parentheses.

$= 14xy^2 - 5xy^2 + 91x^3 + 65x^3$ Use the commutative property to rearrange terms.

$= 9xy^2 + 156x^3$ Combine like terms. ■

Self Check Simplify $3(2b^2 - 3a^2b) + 2b(b + a^2)$.

Answer $8b^2 - 7a^2b$

■ MULTIPLYING POLYNOMIALS

To find the product of $3x^2y^3z$ and $5xyz^2$, we proceed as follows:

$(3x^2y^3z)(5xyz^2) = 3 \cdot x^2 \cdot y^3 \cdot z \cdot 5 \cdot x \cdot y \cdot z^2$

$= 3 \cdot 5 \cdot x^2 \cdot x \cdot y^3 \cdot y \cdot z \cdot z^2$ Use the commutative property to rearrange terms.

$= 15x^3y^4z^3$

This illustrates that *to multiply two monomials, we multiply the coefficients and then multiply the variables.*

To find the product of a monomial and a polynomial, we use the distributive property.

$3xy^2(2xy + x^2 - 7yz) = 3xy^2(2xy) + (3xy^2)(x^2) - (3xy^2)(7yz)$

$= 6x^2y^3 + 3x^3y^2 - 21xy^3z$

This illustrates that *to multiply a polynomial by a monomial, we multiply each term of the polynomial by the monomial.*

To multiply one binomial by another, we use the distributive property twice.

EXAMPLE 4 Multiply **a.** $(x + y)(x + y)$, **b.** $(x - y)(x - y)$, and **c.** $(x + y)(x - y)$.

Solution **a.** $(x + y)(x + y) = (x + y)x + (x + y)y$
$$= x^2 + xy + xy + y^2$$
$$= x^2 + 2xy + y^2$$

b. $(x - y)(x - y) = (x - y)x - (x - y)y$
$$= x^2 - xy - xy + y^2$$
$$= x^2 - 2xy + y^2$$

c. $(x + y)(x - y) = (x + y)x - (x + y)y$
$$= x^2 + xy - xy - y^2$$
$$= x^2 - y^2$$ ∎

Self Check Multiply **a.** $(x + 2)(x + 2)$, **b.** $(x - 3)(x - 3)$, and **c.** $(x + 4)(x - 4)$.
Answers **a.** $x^2 + 4x + 4$, **b.** $x^2 - 6x + 9$, **c.** $x^2 - 4$

The products in Example 4 are **special products**. Because they occur so often, it is worthwhile to learn their forms.

Special Product Formulas
$$(x + y)^2 = (x + y)(x + y) = x^2 + 2xy + y^2$$
$$(x - y)^2 = (x - y)(x - y) = x^2 - 2xy + y^2$$
$$(x + y)(x - y) = x^2 - y^2$$

 WARNING! Remember that $(x + y)^2$ and $(x - y)^2$ have trinomials for their products and that
$$(x + y)^2 \neq x^2 + y^2 \qquad \text{and} \qquad (x - y)^2 \neq x^2 - y^2$$

We can use the **FOIL method** to multiply one binomial by another. The word FOIL is an acronym for **F**irst terms, **O**uter terms, **I**nner terms, and **L**ast terms. To use this method to multiply $(3x - 4)$ by $(2x + 5)$, we write

First terms   Last terms

$$(3x - 4)(2x + 5) = 3x(2x) + 3x(5) - 4(2x) - 4(5)$$

Inner terms

Outer terms

$$= 6x^2 + 15x - 8x - 20$$
$$= 6x^2 + 7x - 20$$

In this example,

- the product of the first terms is $6x^2$,
- the product of the outer terms is $15x$,
- the product of the inner terms is $-8x$, and
- the product of the last terms is -20.

The resulting terms of the product are then combined.

EXAMPLE 5   Use the FOIL method to multiply: $(\sqrt{3} + x)(2 - \sqrt{3}x)$.

Solution
$$(\sqrt{3} + x)(2 - \sqrt{3}x) = 2\sqrt{3} - \sqrt{3}\sqrt{3}x + 2x - x\sqrt{3}x$$
$$= 2\sqrt{3} - 3x + 2x - \sqrt{3}x^2$$
$$= 2\sqrt{3} - x - \sqrt{3}x^2$$

Self Check   Multiply $(2x + \sqrt{3})(x - \sqrt{3})$.
Answer $2x^2 - x\sqrt{3} - 3$

To multiply a polynomial with more than two terms by another polynomial, we multiply each term of one polynomial by each term of the other polynomial and combine like terms whenever possible.

EXAMPLE 6   Multiply   **a.** $(x + y)(x^2 - xy + y^2)$ and   **b.** $(x + y)^3$.

Solution   **a.** $(x + y)(x^2 - xy + y^2) = x^3 - x^2y + xy^2 + yx^2 - xy^2 + y^3$
$$= x^3 + y^3$$

b. $(x + 3)^3 = (x + 3)(x + 3)^2$
$$= (x + 3)(x^2 + 6x + 9)$$
$$= x^3 + 6x^2 + 9x + 3x^2 + 18x + 27$$
$$= x^3 + 9x^2 + 27x + 27$$

Self Check   Multiply $(x + 2)(2x^2 + 3x - 1)$.
Answer $2x^3 + 7x^2 + 5x - 2$

If n is a whole number, the expressions $a^n + 1$ and $2a^n - 3$ are polynomials, and we can use the FOIL method to multiply them together.

$$(a^n + 1)(2a^n - 3) = 2a^{2n} - 3a^n + 2a^n - 3$$
$$= 2a^{2n} - a^n - 3 \qquad \text{Combine like terms.}$$

We can also use the methods previously discussed to multiply expressions, such as $x^{-2} + y$ and $x^2 - y^{-1}$, that are not polynomials.

$$(x^{-2} + y)(x^2 - y^{-1}) = x^{-2+2} - x^{-2}y^{-1} + x^2y - y^{1-1}$$

$$= x^0 - \frac{1}{x^2y} + x^2y - y^0$$

$$= 1 - \frac{1}{x^2y} + x^2y - 1 \qquad x^0 = 1 \text{ and } y^0 = 1.$$

$$= x^2y - \frac{1}{x^2y}$$

■ CONJUGATE BINOMIALS

If the denominator of a fraction is a binomial containing square roots, we can use the product formula $(x + y)(x - y)$ to rationalize the denominator. For example, to rationalize the denominator of

$$\frac{6}{\sqrt{7} + 2}$$

we multiply both the numerator and the denominator of the fraction by $\sqrt{7} - 2$ and simplify.

$$\frac{6}{\sqrt{7} + 2} = \frac{6(\sqrt{7} - 2)}{(\sqrt{7} + 2)(\sqrt{7} - 2)}$$

$$= \frac{6(\sqrt{7} - 2)}{7 - 4}$$

$$= \frac{6(\sqrt{7} - 2)}{3}$$

$$= 2(\sqrt{7} - 2)$$

In this example, we multiplied both the numerator and the denominator of the given fraction by $\sqrt{7} - 2$. This binomial is the same as the denominator of the given fraction $\sqrt{7} + 2$, except for the sign. Such binomials are called *conjugate binomials*.

Conjugate Binomials
Conjugate binomials are binomials that are the same except for the sign between their terms. The conjugate of $a + b$ is $a - b$, and the conjugate of $a - b$ is $a + b$.

EXAMPLE 7   Rationalize the denominator: $\dfrac{\sqrt{3x} - \sqrt{2}}{\sqrt{3x} + \sqrt{2}}$ $(x > 0)$.

Solution   We multiply both the numerator and the denominator by $\sqrt{3x} - \sqrt{2}$ $\Big($the conjugate of $\sqrt{3x} + \sqrt{2}\Big)$ and simplify.

$$\frac{\sqrt{3x} - \sqrt{2}}{\sqrt{3x} + \sqrt{2}} = \frac{\left(\sqrt{3x} - \sqrt{2}\right)\left(\sqrt{3x} - \sqrt{2}\right)}{\left(\sqrt{3x} + \sqrt{2}\right)\left(\sqrt{3x} - \sqrt{2}\right)}$$

$$= \frac{\sqrt{3x}\sqrt{3x} - \sqrt{3x}\sqrt{2} - \sqrt{2}\sqrt{3x} + \sqrt{2}\sqrt{2}}{\left(\sqrt{3x}\right)^2 - \left(\sqrt{2}\right)^2}$$

$$= \frac{3x - \sqrt{6x} - \sqrt{6x} + 2}{3x - 2}$$

$$= \frac{3x - 2\sqrt{6x} + 2}{3x - 2}$$ ∎

Self Check   Rationalize the denominator: $\dfrac{\sqrt{x} + 2}{\sqrt{x} - 2}$.

Answer $\dfrac{x + 4\sqrt{x} + 4}{x - 4}$

In calculus, we often rationalize a numerator.

EXAMPLE 8   Rationalize the numerator: $\dfrac{\sqrt{x + h} - \sqrt{x}}{h}$.

Solution   Multiply the numerator and the denominator by the conjugate of the numerator and simplify.

$$\frac{\sqrt{x + h} - \sqrt{x}}{h} = \frac{\left(\sqrt{x + h} - \sqrt{x}\right)\left(\sqrt{x + h} + \sqrt{x}\right)}{h\left(\sqrt{x + h} + \sqrt{x}\right)}$$

$$= \frac{x + h - x}{h\left(\sqrt{x + h} + \sqrt{x}\right)}$$

$$= \frac{h}{h\left(\sqrt{x + h} + \sqrt{x}\right)}$$

$$= \frac{1}{\sqrt{x + h} + \sqrt{x}}$$ Divide out the common factor of h.   ∎

Self Check   Rationalize the numerator: $\dfrac{\sqrt{4+h}-2}{h}$.

Answer $\dfrac{1}{\sqrt{4+h}+2}$

■ DIVIDING POLYNOMIALS

**Gottfried Wilhelm Leibniz
(1646–1716)**

Leibniz, a German philosopher and logician, is principally known as one of the inventors of calculus, along with Newton. He also developed the binary numeration system, which is basic to modern computers.

To divide monomials, we write the quotient as a fraction and simplify by using the rules of exponents. For example,

$$\frac{6x^2y^3}{-2x^3y} = -3x^{2-3}y^{3-1}$$

$$= -3x^{-1}y^2$$

$$= -\frac{3y^2}{x}$$

To divide a polynomial by a monomial, we write the quotient as a fraction, write the fraction as a sum of separate fractions, and simplify each separate fraction. For example, to divide $8x^5y^4 + 12x^2y^5 - 16x^2y^3$ by $4x^3y^4$, we proceed as follows:

$$\frac{8x^5y^4 + 12x^2y^5 - 16x^2y^3}{4x^3y^4} = \frac{8x^5y^4}{4x^3y^4} + \frac{12x^2y^5}{4x^3y^4} + \frac{16x^2y^3}{4x^3y^4}$$

$$= 2x^2 + \frac{3y}{x} - \frac{4}{xy}$$

To divide two polynomials, we use a process similar to long division of whole numbers. To illustrate, we consider the division

$$\frac{2x^2 + 11x - 30}{x + 7}$$

which can be written in long division form as

$$x + 7\overline{)2x^2 + 11x - 30}$$

The binomial $x + 7$ is called the **divisor,** and the trinomial $2x^2 + 11x - 30$ is called the **dividend.** The final answer, called the **quotient,** will appear above the long division symbol.

We begin the division by asking "What monomial, when multiplied by x, gives $2x^2$?" Because $x \cdot 2x = 2x^2$, the answer is $2x$. We place $2x$ in the quotient, multiply each term of the divisor by $2x$, subtract, and bring down the -30.

$$
\begin{array}{r}
2x \\
x + 7\overline{)2x^2 + 11x - 30} \\
\underline{2x^2 + 14x} \\
-\ 3x - 30
\end{array}
$$

We continue the division by asking "What monomial, when multiplied by x, gives $-3x$?" We place the answer, -3, in the quotient, multiply each term of the divisor by -3, and subtract. This time, there is no number to bring down.

$$
\begin{array}{r}
2x \;-\; 3 \\
x + 7\overline{\smash{\big)}\,2x^2 + 11x - 30} \\
\underline{2x^2 + 14x} \\
-\,3x - 30 \\
\underline{-\,3x - 21} \\
-\,9
\end{array}
$$

Because the degree of the remainder, -9, is less than the degree of the divisor, the division process stops, and we can express the result in $quotient + \frac{remainder}{divisor}$ form. Thus,

1. $\quad \dfrac{2x^2 + 11x - 30}{x + 7} = 2x - 3 + \dfrac{-9}{x + 7}$

If we denote $2x^2 + 11x - 30$ by $P(x)$, $x + 7$ by $Q(x)$, $2x - 3$ by $q(x)$, and -9 by $r(x)$ in Equation 1, we can write the equation in the form

$$
\frac{P(x)}{Q(x)} = q(x) + \frac{r(x)}{Q(x)} \quad (Q(x) \neq 0)
$$

After multiplying both sides by $Q(x)$, we have

2. $\quad P(x) = Q(x)q(x) + r(x) \quad (Q(x) \neq 0)$

Equation 2 illustrates a well-known theorem called the **division algorithm.**

Division Algorithm

If $P(x)$ and $Q(x)$ are polynomials, there are unique polynomials $q(x)$ and $r(x)$ such that

$$
P(x) = Q(x)q(x) + r(x) \quad (Q(x) \neq 0)
$$

where $r(x) = 0$ or is less than the degree of $Q(x)$.

The polynomial $q(x)$ is the quotient, and the polynomial $r(x)$ is the remainder in the division of $P(x)$ by $Q(x)$.

When the remainder $r(x)$ is equal to 0, we say that the polynomial $Q(x)$ divides the polynomial $P(x)$ evenly.

EXAMPLE 9   Divide $6x^3 - 11$ by $2x + 2$.

Solution   We set up the division, leaving spaces for the missing powers of x in the dividend.

$$
2x + 2\overline{\smash{\big)}\,6x^3 -\,11}
$$

The division process continues as usual, with the following results:

$$
\begin{array}{r}
3x^2 - 3x + 3 \\
2x + 2 \overline{)6x^3 \qquad\qquad - 11} \\
\underline{6x^3 + 6x^2} \\
-6x^2 \\
\underline{-6x^2 - 6x} \\
+6x - 11 \\
\underline{+6x + 6} \\
-17
\end{array}
$$

Thus, $\dfrac{6x^3 - 11}{2x + 2} = 3x^2 - 3x + 3 + \dfrac{-17}{2x + 2}$. ■

Self Check Divide $3x + 1 \overline{)9x^2 - 1}$.

Answer $3x - 1$

EXAMPLE 10 Divide $-3x^3 - 3 + x^5 + 4x^2 - x^4$ by $x^2 - 3$.

Solution The division process works best when the terms in the divisor and dividend are written with their exponents in descending order.

$$
\begin{array}{r}
x^3 - x^2 + 1 \\
x^2 \quad 3 \overline{)x^5 \quad x^4 - 3x^3 + 4x^2 - 3} \\
\underline{x^5 \qquad - 3x^3} \\
-x^4 \qquad\qquad + 4x^2 \\
\underline{-x^4 \qquad\qquad + 3x^2} \\
x^2 - 3 \\
\underline{x^2 - 3} \\
0
\end{array}
$$

Thus, $\dfrac{-3x^3 - 3 + x^5 + 4x^2 - x^4}{x^2 - 3} = x^3 - x^2 + 1$. ■

Self Check: Divide $x^2 + 1 \overline{)3x^2 - x + 1 - 2x^3 + 3x^4}$.

Answer $3x^2 - 2x + \dfrac{x + 1}{x^2 + 1}$

EXERCISE 1.4

VOCABULARY AND CONCEPTS *Fill in the blank to make a true statement.*

1. A _____ is a number or the product of a number and one or more _____.

2. The _____ of a monomial is the sum of the exponents of its _____.

3. A _____ is a polynomial with three terms.

4. A _____ is a polynomial with two terms.

5. A monomial is a polynomial with ____ term.

6. The constant 0 is called the ____ polynomial.

7. Terms with the same variables with the same exponents are called ____ terms.

8. The ____ of a polynomial is the same as the degree of its term of highest degree.

9. To combine like terms, we add their ____ and keep the same ____ and the same exponents.

10. The conjugate of $3\sqrt{x} + 2$ is ____.

In Exercises 11–20, tell whether the given expression is polynomial. If so, tell whether it is a monomial, a binomial, or a trinomial, and give its degree.

11. $x^2 + 3x + 4$

12. $x^3 - 5xy$

13. $x^3 + y^{1/2}$

14. $x^{-3} - 5y^{-2}$

15. $\sqrt{5}x^3 + 4x^2$

16. x^2y^3

17. $\sqrt{15}$

18. $\dfrac{5}{x} + \dfrac{x}{5} + 5$

19. 0

20. $3y^3 - 4y^2 + 2y + 2$

PRACTICE   *In Exercises 21–68, do the operations and simplify.*

21. $(x^3 - 3x^2) + (5x^3 - 8x)$

22. $(2x^4 - 5x^3) + (7x^3 - x^4 + 2x)$

23. $(y^5 + 2y^3 + 7) - (y^5 - 2y^3 - 7)$

24. $(3t^7 - 7t^3 + 3) - (7t^7 - 3t^3 + 7)$

25. $2(x^2 + 3x - 1) - 3(x^2 + 2x - 4) + 4$

26. $5(x^3 - 8x + 3) + 2(3x^2 + 5x) - 7$

27. $8(t^2 - 2t + 5) + 4(t^2 - 3t + 2) - 6(2t^2 - 8)$

28. $-3(x^3 - x) + 2(x^2 + x) + 3(x^3 - 2x)$

29. $y(y^2 - 1) - y^2(y + 2) - y(2y - 2)$

30. $-4a^2(a + 1) + 3a(a^2 - 4) - a^2(a + 2)$

31. $xy(x - 4y) - y(x^2 + 3xy) + xy(2x + 3y)$

32. $3mn(m + 2n) - 6m(3mn + 1) - 2n(4mn - 1)$

33. $2x^2y^3(4xy^4)$

34. $-15a^3b(-2a^2b^3)$

35. $-3m^2n(2mn^2)\left(-\dfrac{mn}{12}\right)$

36. $-\dfrac{3r^2s^3}{5}\left(\dfrac{2r^2s}{3}\right)\left(\dfrac{15rs^2}{2}\right)$

37. $-4rs(r^2 + s^2)$

38. $6u^2v(2uv^3 - y)$

39. $6ab^2c(2ac + 3bc^2 - 4ab^2c)$

40. $-\dfrac{mn^2}{2}(4mn - 6m^2 - 8)$

41. $(a + 2)(a + 2)$

42. $(y - 5)(y - 5)$

43. $(a - 6)^2$

44. $(t + 9)^2$

45. $(x + 4)(x - 4)$

46. $(z + 7)(z - 7)$

47. $(x - 3)(x + 5)$

48. $(z + 4)(z - 6)$

49. $(u + 2)(3u - 2)$

50. $(4x + 1)(2x - 3)$

51. $(5x - 1)(2x + 3)$

52. $(4x - 1)(2x - 7)$

53. $(3a - 2b)^2$

54. $(4a + 5b)(4a - 5b)$

55. $(3m + 4n)(3m - 4n)$

56. $(4r + 3s)^2$

57. $(2y - 4x)(3y - 2x)$ **58.** $(-2x + 3y)(3x + y)$ **59.** $(9x - y)(x^2 - 3y)$ **60.** $(8a^2 + b)(a + 2b)$

61. $(5z + 2t)(z^2 - t)$ **62.** $(y - 2x^2)(x^2 + 3y)$ **63.** $(3x - 1)^3$ **64.** $(2x - 3)^3$

65. $(3x + 1)(2x^2 + 4x - 3)$ **66.** $(2x - 5)(x^2 - 3x + 2)$

67. $(3x + 2y)(2x^2 - 3xy + 4y^2)$ **68.** $(4r - 3s)(2r^2 + 4rs - 2s^2)$

In Exercises 69–80, multiply the expressions as you would multiply polynomials.

69. $2y^n(3y^n + y^{-n})$ **70.** $3a^{-n}(2a^n + 3a^{n-1})$

71. $-5x^{2n}y^n(2x^{2n}y^{-n} + 3x^{-2n}y^n)$ **72.** $-2a^{3n}b^{2n}(5a^{-3n}b - ab^{-2n})$

73. $(x^n + 3)(x^n - 4)$ **74.** $(a^n - 5)(a^n - 3)$

75. $(2r^n - 7)(3r^n - 2)$ **76.** $(4z^n + 3)(3z^n + 1)$

77. $x^{1/2}(x^{1/2}y + xy^{1/2})$ **78.** $ab^{1/2}(a^{1/2}b^{1/2} + b^{1/2})$

79. $(a^{1/2} + b^{1/2})(a^{1/2} - b^{1/2})$ **80.** $(x^{3/2} + y^{1/2})^2$

In Exercises 81–92, rationalize each denominator.

81. $\dfrac{2}{\sqrt{3} - 1}$ **82.** $\dfrac{1}{\sqrt{5} + 2}$ **83.** $\dfrac{3x}{\sqrt{7} + 2}$ **84.** $\dfrac{14y}{\sqrt{2} - 3}$

85. $\dfrac{x}{x - \sqrt{3}}$ **86.** $\dfrac{y}{2y + \sqrt{7}}$ **87.** $\dfrac{y + \sqrt{2}}{y - \sqrt{2}}$ **88.** $\dfrac{x - \sqrt{3}}{x + \sqrt{3}}$

89. $\dfrac{\sqrt{2} - \sqrt{3}}{1 - \sqrt{3}}$ **90.** $\dfrac{\sqrt{3} - \sqrt{2}}{1 + \sqrt{2}}$ **91.** $\dfrac{\sqrt{x} - \sqrt{y}}{\sqrt{x} + \sqrt{y}}$ **92.** $\dfrac{\sqrt{2x} + y}{\sqrt{2x} - y}$

In Exercises 93–98, rationalize each numerator.

93. $\dfrac{\sqrt{2} + 1}{2}$ **94.** $\dfrac{\sqrt{x} - 3}{3}$ **95.** $\dfrac{y - \sqrt{3}}{y + \sqrt{3}}$

96. $\dfrac{\sqrt{a} - \sqrt{b}}{\sqrt{a} + \sqrt{b}}$ **97.** $\dfrac{\sqrt{x + 3} - \sqrt{x}}{3}$ **98.** $\dfrac{\sqrt{2 + h} - \sqrt{2}}{h}$

In Exercises 99–106, do each division. Write all answers without using negative exponents.

99. $\dfrac{36a^2b^3}{18ab^6}$ **100.** $\dfrac{-45r^2s^5t^3}{27r^6s^2t^8}$ **101.** $\dfrac{16x^6y^4z^9}{-24x^9y^6z^0}$

102. $\dfrac{32m^6n^4p^2}{26m^6n^7p^2}$ **103.** $\dfrac{5x^3y^2 + 15x^3y^4}{10x^2y^3}$ **104.** $\dfrac{9m^4n^9 - 6m^3n^4}{12m^3n^3}$

105. $\dfrac{24x^5y^7 - 36x^2y^5 + 12xy}{60x^5y^4}$

106. $\dfrac{9a^3b^4 + 27a^2b^4 - 18a^2b^3}{18a^2b^7}$

In Exercises 107–118, do each division. If there is a nonzero remainder, write the answer in quotient $+ \frac{remainder}{divisor}$ form.

107. $x + 3 \overline{)3x^2 + 11x + 6}$

108. $3x + 2 \overline{)3x^2 + 11x + 6}$

109. $2x - 5 \overline{)2x^2 - 19x + 37}$

110. $x - 7 \overline{)2x^2 - 19x + 35}$

111. $x^2 + x - 1 \overline{)x^3 - 2x^2 - 4x + 3}$

112. $x^2 - 3 \overline{)x^3 - 2x^2 - 4x + 5}$

113. $\dfrac{x^5 - 2x^3 - 3x^2 + 9}{x^3 - 2}$

114. $\dfrac{x^5 - 2x^3 - 3x^2 + 9}{x^3 - 3}$

115. $\dfrac{x^5 - 32}{x - 2}$

116. $\dfrac{x^4 - 1}{x + 1}$

117. $11x - 10 + 6x^2 \overline{)36x^4 - 121x^2 + 120 + 72x^3 - 142x}$

118. $x + 6x^2 - 12 \overline{)-121x^2 + 72x^3 - 142x + 120 + 36x^4}$

DISCOVERY AND WRITING

119. Show that a trinomial can be squared by using the formula $(a + b + c)^2 = a^2 + b^2 + c^2 + 2ab + 2bc + 2ca$.

120. Show that $(a + b + c + d)^2 = a^2 + b^2 + c^2 + d^2 + 2ab + 2ac + 2ad + 2bc + 2bd + 2cd$.

121. Explain the FOIL method.

122. Explain how to rationalize the numerator of $\dfrac{\sqrt{x} + 2}{x}$.

123. Explain why $(a + b)^2 \neq a^2 + b^2$.

124. Explain why $\sqrt{a^2 + b^2} \neq \sqrt{a^2} + \sqrt{b^2}$.

REVIEW   *Simplify each expression. Assume that all variables represent positive numbers.*

125. $9^{3/2}$

126. $\left(\dfrac{8}{125}\right)^{-2/3}$

127. $\left(\dfrac{625x^4}{16y^8}\right)^{3/4}$

128. $\sqrt{80x^4}$

129. $\sqrt[3]{16ab^4} - b\sqrt[3]{54ab}$

130. $x\sqrt[4]{1280x} + \sqrt[4]{80x^5}$

1.5 Factoring Polynomials

■ FACTORING OUT A COMMON MONOMIAL ■ FACTORING BY GROUPING ■ FACTORING THE DIFFERENCE OF TWO SQUARES ■ FACTORING TRINOMIALS ■ FACTORING THE SUM AND DIFFERENCE OF TWO CUBES ■ MISCELLANEOUS FACTORING

When two or more polynomials are multiplied together, each polynomial is called a **factor** of the resulting product. For example, the factors of $7(x + 2)(x + 3)$ are

$$7, \quad x + 2, \quad \text{and} \quad x + 3$$

The process of writing a polynomial as the product of several factors is called **factoring.**

In this section, we will discuss factoring where the coefficients of the polynomial and the polynomial factors are integers. If a polynomial cannot be factored by using integers only, we will call it a **prime polynomial.**

■ FACTORING OUT A COMMON MONOMIAL

EXAMPLE 1   Factor $3xy^2 + 6x$.

Solution   We note that each term contains a factor of $3x$:

$$3xy^2 + 6x = \mathbf{3x}(y^2) + \mathbf{3x}(2)$$

We can then use the distributive property to factor out the common factor of $3x$:

$$3xy^2 + 6x = \mathbf{3x}(y^2 + 2) \qquad ■$$

Self Check   Factor $4a^2 - 8ab$.

Answer $4a(a - 2b)$

EXAMPLE 2   Factor $x^2y^2z^2 - xyz$.

Solution   We factor out the common factor of xyz:

$$\begin{aligned} x^2y^2z^2 - xyz &= \mathbf{xyz}(xyz) - \mathbf{xyz}(1) \\ &= \mathbf{xyz}(xyz - 1) \end{aligned}$$

The last term in the expression $x^2y^2z^2 - xyz$ has an understood coefficient of 1. When the xyz is factored out, the 1 must be written.   ■

Self Check   Factor $a^2b^2c^2 + a^3b^3c^3$.

Answer $a^2b^2c^2(1 + abc)$

■ FACTORING BY GROUPING

EXAMPLE 3   Factor $ax + bx + a + b$.

Solution   Although there is no factor common to all four terms, we can factor x out of the first two terms and write the expression as

$$ax + bx + a + b = x(a + b) + (a + b)$$

We can now factor out the common factor of $a + b$.

$$ax + bx + a + b = x(a + b) + (a + b)$$
$$= x(a + b) + 1(a + b)$$
$$= (a + b)(x + 1)$$ ■

Self Check   Factor $x^2 + xy + 2x + 2y$.

Answer $(x + 2)(x + y)$

■ FACTORING THE DIFFERENCE OF TWO SQUARES

EXAMPLE 4   Factor $49x^2 - 4$.

Solution   We observe that each term is a perfect square:

$$49x^2 - 4 = (7x)^2 - 2^2$$

The difference of the squares of two quantities is the product of two factors. One is the sum of the quantities, and the other is the difference of the quantities. Thus, $49x^2 - 4$ factors as

$$49x^2 - 4 = (7x)^2 - 2^2$$
$$= (7x + 2)(7x - 2)$$ ■

Self Check   Factor $9a^2 - 16b^2$.

Answer $(3a + 4b)(3a - 4b)$

Example 4 suggests a formula for factoring the difference of two squares.

Factoring the Difference of Two Squares
$$x^2 - y^2 = (x + y)(x - y)$$

WARNING!   If you are limited to integer coefficients, the sum of two squares cannot be factored. For example, $x^2 + y^2$ is a prime polynomial.

EXAMPLE 5   Factor $16m^4 - n^4$.

Solution   The binomial $16m^4 - n^4$ can be factored as the difference of two squares:

$$16m^4 - n^4 = (4m^2)^2 - (n^2)^2$$
$$= (4m^2 + n^2)(4m^2 - n^2)$$

The second factor is a difference of two squares and can be factored:

$$16m^4 - n^4 = (4m^2 + n^2)[(2m)^2 - n^2]$$
$$= (4m^2 + n^2)(2m + n)(2m - n)$$

With only integers to work with, the factor $4m^2 + n^2$ is prime.   ∎

Self Check   Factor $a^4 - 81b^4$.

Answer $(a^2 + 9b^2)(a + 3b)(a - 3b)$

EXAMPLE 6   Factor $18t^2 - 32$.

Solution   We begin by factoring out the common monomial factor of 2.

$$18t^2 - 32 = 2(9t^2 - 16)$$

Since $9t^2 - 16$ is the difference of two squares, it can be factored.

$$18t^2 - 32 = 2(9t^2 - 16)$$
$$= 2(3t + 4)(3t - 4)$$ ∎

Self Check   Factor $-3x^2 + 12$.

Answer $3(x + 2)(x - 2)$

■ FACTORING TRINOMIALS

Trinomials that are squares of binomials can be factored by using the following formulas.

> **Factoring Trinomial Squares**
> **1.** $x^2 + 2xy + y^2 = (x + y)(x + y) = (x + y)^2$
> **2.** $x^2 - 2xy + y^2 = (x - y)(x - y) = (x - y)^2$

For example, to factor $a^2 - 6a + 9$, we note that $a^2 - 6a + 9$ can be written in the form

$$a^2 - 2(3a) + 3^2$$

which matches the left-hand side of Equation 2 above. Thus,

$$a^2 - 6a + 9 = a^2 - 2(3a) + 3^2$$
$$= (a - 3)(a - 3)$$
$$= (a - 3)^2$$

Factoring trinomials that are not squares of binomials often requires some guesswork. If a trinomial with no common factors is factorable, it must factor into the product of two binomials.

EXAMPLE 7   Factor $x^2 + 3x - 10$.

Solution   To factor $x^2 + 3x - 10$, we must find two binomials $x + a$ and $x + b$ such that

$$x^2 + 3x - 10 = (x + a)(x + b)$$

where the product of a and b is -10 and the sum of a and b is 3.

$$ab = -10 \quad \text{and} \quad a + b = 3$$

To find such numbers, we list the possible factorizations of -10:

$$10(-1) \qquad 5(-2) \qquad -10(1) \qquad -5(2)$$

Only in the factorization $5(-2)$ do the factors have a sum of 3. Thus, $a = 5$ and $b = -2$, and

$$x^2 + 3x - 10 = (x + \boldsymbol{a})(x + \boldsymbol{b})$$
3. $x^2 + 3x - 10 = (x + \boldsymbol{5})(x - \boldsymbol{2})$

Because of the commutative property of multiplication, the order of the factors in Equation 3 is not important. Equation 3 can also be written as

$$x^2 + 3x - 10 = (x - 2)(x + 5)$$ ■

Self Check   Factor $p^2 - 5p - 6$.
Answer $(p - 6)(p + 1)$

EXAMPLE 8   Factor $2x^2 - x - 6$.

Solution   Since the first term is $2x^2$, the first terms of the binomial factors must be $2x$ and x:

$$2x^2 - x - 6 = (2x + ?)(x + ?)$$

The product of the last terms must be -6, and the sum of the products of the outer terms and the inner terms must be $-x$. The only factorization of -6 that will cause this to happen is $3(-2)$.

$$2x^2 - x - 6 = (2x + 3)(x - 2)$$ ■

Self Check   Factor $6x^2 - x - 2$.
Answer $(3x - 2)(2x + 1)$

It is not easy to give specific rules for factoring trinomials, because some guess-work is often necessary. However, the following hints are helpful.

Strategy for Factoring a General Trinomial

1. Write the trinomial in descending powers of one variable.

2. Factor out any greatest common factor, including -1 if that is necessary to make the coefficient of the first term positive.

3. When the sign of the first term of a trinomial is $+$ and the sign of the third term is $+$, the sign between the terms of each binomial factor is the same as the sign of the middle term of the trinomial.

   When the sign of the first term is $+$ and the sign of the third term is $-$, one of the signs between the terms of the binomial factors is $+$ and the other is $-$.

4. Try various combinations of first terms and last terms until you find one that works. If no possibilities work, the trinomial does not factor with integer coefficients.

5. Check the factorization by multiplication.

EXAMPLE 9   Factor $10xy + 24y^2 - 6x^2$.

Solution   We write the trinomial in descending powers of x and then factor out the common factor of -2.

$$10xy + 24y^2 - 6x^2 = -6x^2 + 10xy + 24y^2$$
$$= -2(3x^2 - 5xy - 12y^2)$$

Since the sign of the third term of $3x^2 - 5xy - 12y^2$ is $-$, the signs between the binomial factors will be opposite. Since the first term is $3x^2$, the first terms of the binomial factors must be $3x$ and x:

$$-2(3x^2 - 5xy - 12y^2) = -2(3x \qquad ?)(x \qquad ?)$$

The product of the last terms must be $-12y^2$, and the sum of the outer terms and the inner terms must be $-5xy$. Of the many factorizations of $-12y^2$, only $4y(-3y)$ leads to a middle term of $-5xy$.

$$10xy + 24y^2 - 6x^2 = -6x^2 + 10xy + 24y^2$$
$$= -2(3x^2 - 5xy - 12y^2)$$
$$= -2(3x + 4y)(x - 3y)$$ ∎

Self Check   Factor $-6x^2 - 15xy - 6y^2$.

Answer $-3(x + 2y)(2x + y)$

We can often factor polynomials with variable exponents. For example, if n is a natural number,

$$a^{2n} - 5a^n - 6 = (a^n + 1)(a^n - 6)$$

because

$$(a^n + 1)(a^n - 6) = a^{2n} - 6a^n + a^n - 6$$
$$= a^{2n} - 5a^n - 6 \qquad \text{Combine like terms.}$$

■ FACTORING THE SUM AND DIFFERENCE OF TWO CUBES

Two other types of factoring involve binomials that are either the sum or the difference of two cubes. Like the difference of two squares, they can be factored by using a formula.

Factoring the Sum and Difference of Two Cubes
$$x^3 + y^3 = (x + y)(x^2 - xy + y^2)$$
$$x^3 - y^3 = (x - y)(x^2 + xy + y^2)$$

EXAMPLE 10   Factor $x^3 - 8$.

Solution   This binomial can be written as $x^3 - 2^3$, which is the difference of two cubes. Substituting into the formula for the difference of two cubes gives

$$x^3 - 2^3 = (x - 2)(x^2 + 2x + 2^2)$$
$$= (x - 2)(x^2 + 2x + 4) \qquad ■$$

Self Check   Factor $64 - p^3$.
Answer $(4 - p)(16 + 4p + p^2)$

EXAMPLE 11   Factor $27x^6 + 64y^3$.

Solution   We can write this expression as the sum of two cubes and factor it as follows:

$$27x^6 + 64y^3 = (3x^2)^3 + (4y)^3$$
$$= (3x^2 + 4y)[(3x^2)^2 - (3x^2)(4y) + (4y)^2]$$
$$= (3x^2 + 4y)(9x^4 - 12x^2y + 16y^2) \qquad ■$$

Self Check   Factor $8a^3 + 1000b^6$.
Answer $(2a + 10b^2)(4a^2 - 20ab^2 + 100b^4)$

■ MISCELLANEOUS FACTORING

EXAMPLE 12   Factor $x^2 - y^2 + 6x + 9$.

Solution   Here we will factor a trinomial and a difference of two squares.

$$x^2 - y^2 + 6x + 9 = x^2 + 6x + 9 - y^2 \qquad \text{Use the commutative property to rearrange terms.}$$

$$= (x + 3)^2 - y^2 \qquad \text{Factor } x^2 + 6x + 9.$$
$$= (x + 3 + y)(x + 3 - y) \qquad \text{Factor the difference of two squares.}$$

We could try to factor this expression in another way.

$$x^2 - y^2 + 6x + 9 = (x + y)(x - y) + 3(2x + 3) \qquad \text{Factor } x^2 - y^2 \text{ and } 6x + 9.$$

However, we are unable to finish the factorization. If grouping in one way doesn't work, try various other ways. ■

Self Check   Factor $a^2 + 8a - b^2 + 16$.
Answer $(a + 4 + b)(a + 4 - b)$

EXAMPLE 13   Factor $z^4 - 3z^2 + 1$.

Solution   This trinomial cannot be factored as the product of two binomials, because no combination will give a middle term of $-3z^2$. However, if the middle term were $-2z^2$, the trinomial would be a perfect square, and the factorization would be easy:

$$z^4 - 2z^2 + 1 = (z^2 - 1)(z^2 - 1)$$
$$= (z^2 - 1)^2$$

We can change the middle term in $z^4 - 3z^2 + 1$ to $-2z^2$ by adding z^2 to it. To ensure that adding z^2 does not change the value of the trinomial, however, we must also subtract z^2. We can then proceed as follows.

$$z^4 - 3z^2 + 1 = z^4 - 3z^2 + z^2 + 1 - z^2 \qquad \text{Add and subtract } z^2.$$
$$= z^4 - 2z^2 + 1 - z^2 \qquad \text{Combine } -3z^2 \text{ and } z^2.$$
$$= (z^2 - 1)^2 - z^2 \qquad \text{Factor } z^4 - 2z^2 + 1.$$
$$= (z^2 - 1 + z)(z^2 - 1 - z) \qquad \text{Factor the difference of two squares.}$$

In this type of problem, we will always try to add and subtract a perfect square in hopes of making a perfect square trinomial that will lead to factoring a difference of two squares. ■

Self Check   Factor $x^4 + 3x^2 + 4$.
Answer $(x^2 + 2 + x)(x^2 + 2 - x)$

EXERCISE 1.5

VOCABULARY AND CONCEPTS *Fill in the blank to make a true statement.*

1. When polynomials are multiplied together, each polynomial is a _____ of the product.

2. If a polynomial cannot be factored using _____ coefficients, it is called a _____ polynomial.

Complete each factoring formula.

3. $ax + bx = $ _____ .

4. $x^2 - y^2 = $ _____ .

5. $x^2 + 2xy + y^2 = $ _____ .

6. $x^2 - 2xy + y^2 = $ _____ .

7. $x^3 + y^3 = $ _____ .

8. $x^3 - y^3 = $ _____ .

PRACTICE *In Exercises 9–104, completely factor each expression. If an expression is prime, so indicate.*

9. $3x - 6$

10. $5y - 15$

11. $8x^2 + 4x^3$

12. $9y^3 + 6y^2$

13. $7x^2y^2 + 14x^3y^2$

14. $25y^2z - 15yz^2$

15. $3a^2bc + 6ab^2c + 9abc^2$

16. $5x^3y^3z^3 + 25x^2y^2z^2 - 125xyz$

17. $a(x + y) + b(x + y)$

18. $b(x - y) + a(x - y)$

19. $4a + b - 12a^2 - 3ab$

20. $x^2 + 4x + xy + 4y$

21. $3x^3 + 3x^2 - x - 1$

22. $4x + 6xy - 9y - 6$

23. $2txy + 2ctx - 3ty - 3ct$

24. $2ax + 4ay - bx - 2by$

25. $ax + bx + ay + by + az + bz$

26. $6x^2y^3 + 18xy + 3x^2y^2 + 9x$

27. $4x^2 - 9$

28. $36z^2 - 49$

29. $4 - 9r^2$

30. $16 - 49x^2$

31. $(x + z)^2 - 25$

32. $(x - y)^2 - 9$

33. $25x^4 + 1$

34. $36t^4 + 121$

35. $x^2 - (y - z)^2$

36. $z^2 - (y + 3)^2$

37. $(x - y)^2 - (x + y)^2$

38. $(2a + 3)^2 - (2a - 3)^2$

39. $x^4 - y^4$

40. $z^4 - 81$

41. $3x^2 - 12$

42. $3x^3y - 3xy$

43. $18xy^2 - 8x$

44. $27x^2 - 12$

45. $x^2 + 8x + 16$

46. $a^2 - 12a + 36$

47. $b^2 - 10b + 25$

48. $y^2 + 14y + 49$

49. $m^2 + 4mn + 4n^2$

50. $r^2 - 8rs + 16s^2$

51. $x^2 + 10x + 21$

52. $x^2 + 7x + 10$

53. $x^2 - 4x - 12$

54. $x^2 - 2x - 63$

55. $x^2 - 2x + 15$

56. $x^2 + x + 2$

57. $12x^2 - xy - 6y^2$

58. $8x^2 - 10xy - 3y^2$

59. $-15 + 2a + 24a^2$

60. $-32 - 68x + 9x^2$

61. $6x^2 + 29xy + 35y^2$

62. $10x^2 - 17xy + 6y^2$

63. $12p^2 - 58pq - 70q^2$

64. $3x^2 - 6xy - 9y^2$

65. $-6m^2 + 47mn - 35n^2$

66. $-14r^2 - 11rs + 15s^2$

67. $-6x^3 + 23x^2 + 35x$

68. $-y^3 - y^2 + 90y$

69. $6x^4 - 11x^3 - 35x^2$

70. $12x + 17x^2 - 7x^3$

71. $x^4 + 2x^2 - 15$

72. $x^4 - x^2 - 6$

73. $a^{2n} - 2a^n - 3$

74. $a^{2n} + 6a^n + 8$

75. $6x^{2n} - 7x^n + 2$

76. $9x^{2n} + 9x^n + 2$

77. $4x^{2n} - 9y^{2n}$

78. $8x^{2n} - 2x^n - 3$

79. $10y^{2n} - 11y^n - 6$

80. $16y^{4n} - 25y^{2n}$

81. $8z^3 - 27$

82. $125a^3 - 64$

83. $2x^3 + 2000$

84. $3y^3 + 648$

85. $(x + y)^3 - 64$

86. $(x - y)^3 + 27$

87. $64a^6 - y^6$

88. $a^6 + b^6$

89. $a^3 - b^3 + a - b$

90. $(a^2 - y^2) - 5(a + y)$

91. $64x^6 + y^6$

92. $x^2 + 6x + 9 - 225y^2$

93. $x^2 + 6x + 9 - 144y^2$

94. $x^2 + 2x - 9y^2 + 1$

95. $(a + b)^2 - 3(a + b) - 10$

96. $2(a + b)^2 - 5(a + b) - 3$

97. $x^6 + 7x^3 - 8$

98. $x^6 - 13x^4 + 36x^2$

99. $x^4 + 3x^2 + 4$

100. $x^4 + x^2 + 1$

101. $x^4 + 7x^2 + 16$

102. $y^4 + 2y^2 + 9$

103. $4a^4 + 1 + 3a^2$

104. $x^4 + 25 + 6x^2$

DISCOVERY AND WRITING

105. Explain how to factor the difference of two squares.

106. Explain how to factor the difference of two cubes.

107. Explain how to factor $a^2 - b^2 + a + b$.

108. Explain how to factor $x^2 + 2x + 1$.

In Exercises 109–114, factor the indicated monomial from the given expression.

109. $3x + 2$; 2

110. $a + b$; a

111. $x + x^{1/2}$; $x^{1/2}$

112. $2x + \sqrt{2}y$; $\sqrt{2}$

113. $ab^{3/2} - a^{3/2}b$; ab

114. $ab^2 + b$; b^{-1}

In Exercises 115–118, factor each expression by grouping three terms and two terms.

115. $x^2 + x - 6 + xy - 2y$

116. $2x^2 + 5x + 2 - xy - 2y$

117. $a^4 + 2a^3 + a^2 + a + 1$

118. $a^4 + a^3 - 2a^2 + a - 1$

REVIEW

119. Which natural number is neither prime nor composite?

120. Graph the interval $[-2, 3)$.

121. Simplify: $(x^3x^2)^4$

122. Simplify: $\dfrac{(a^3)^3(a^2)^4}{(a^2a^3)^3}$

123. $\left(\dfrac{3x^4x^3}{6x^{-2}x^4}\right)^0$

124. Simplify: $\sqrt{20x^5}$

1.6 Algebraic Fractions

■ ALGEBRAIC FRACTIONS ■ SIMPLIFYING FRACTIONS ■ MULTIPLYING AND DIVIDING FRACTIONS
■ ADDING AND SUBTRACTING FRACTIONS ■ COMPLEX FRACTIONS

If x and y are real numbers, the quotient $\frac{x}{y}$ ($y \neq 0$) is called a **fraction.** The number x is called the **numerator,** and y is called the **denominator.**

■ ALGEBRAIC FRACTIONS

Algebraic fractions are quotients of algebraic expressions. If the expressions are polynomials, the fraction is a **rational expression.** The first two of the following algebraic fractions are rational expressions.

$$\frac{5y^2 + 2y}{y^2 - 3y - 7} \qquad \frac{8ab^2 - 16c^3}{2x + 3} \qquad \frac{x^{1/2} + 4x}{x^{3/2} - x^{1/2}}$$

 WARNING!   Remember that the denominator of a fraction cannot be zero.

We summarize some of the properties of fractions as follows.

If a, b, and c are real numbers and no denominators are 0, then

Equality of Fractions

$$\frac{a}{b} = \frac{c}{d} \text{ if and only if } ad = bc.$$

Fundamental Property of Fractions

$$\frac{ax}{bx} = \frac{a}{b}$$

Multiplication and Division of Fractions

$$\frac{a}{b} \cdot \frac{c}{d} = \frac{ac}{bd} \qquad \text{and} \qquad \frac{a}{b} \div \frac{c}{d} = \frac{a}{b} \cdot \frac{d}{c} = \frac{ad}{bc}$$

Addition and Subtraction of Fractions

$$\frac{a}{b} + \frac{c}{b} = \frac{a+c}{b} \qquad \text{and} \qquad \frac{a}{b} - \frac{c}{b} = \frac{a-c}{b}$$

The first two examples illustrate each of the previous properties of fractions.

EXAMPLE 1

a. $\dfrac{2a}{3} = \dfrac{4a}{6}$ 　　　　Because $2a(6) = 3(4a)$.

b. $\dfrac{6xy}{10xy} = \dfrac{3(2xy)}{5(2xy)}$ 　　Factor the numerator and denominator and divide out the common factors.

$\qquad\quad = \dfrac{3}{5}$

Self Check 　**a.** Is $\dfrac{3y}{5} = \dfrac{15z}{25}$? 　**b.** Simplify $\dfrac{15a^2b}{25ab^2}$.

Answers 　**a.** no, 　**b.** $\dfrac{3a}{5b}$

EXAMPLE 2

Assume that no denominators are 0.

a. $\dfrac{2r}{7s} \cdot \dfrac{3r}{5s} = \dfrac{2r \cdot 3r}{7s \cdot 5s}$ 　　　　**b.** $\dfrac{3mn}{4pq} \div \dfrac{2pq}{7mn} = \dfrac{3mn}{4pq} \cdot \dfrac{7mn}{2pq}$

$\qquad\quad = \dfrac{6r^2}{35s^2}$ 　　　　　　　　　　　$= \dfrac{21m^2n^2}{8p^2q^2}$

c. $\dfrac{2ab}{5xy} + \dfrac{ab}{5xy} = \dfrac{2ab + ab}{5xy}$ 　　　**d.** $\dfrac{6uv^2}{7w^2} - \dfrac{3uv^2}{7w^2} = \dfrac{6uv^2 - 3uv^2}{7w^2}$

$\qquad\qquad\quad = \dfrac{3ab}{5xy}$ 　　　　　　　　　　　　$= \dfrac{3uv^2}{7w^2}$

Self Check 　Do each operation: 　**a.** $\dfrac{3a}{5b} \cdot \dfrac{2a}{7b}$, 　**b.** $\dfrac{2ab}{3rs} \div \dfrac{2rs}{4ab}$, 　**c.** $\dfrac{5pq}{3t} + \dfrac{3pq}{3t}$, and

d. $\dfrac{5mn^2}{3w} - \dfrac{mn^2}{3w}$.

Answers 　**a.** $\dfrac{6a^2}{35b^2}$, 　**b.** $\dfrac{8a^2b^2}{6r^2s^2}$, 　**c.** $\dfrac{8pq}{3t}$, 　**d.** $\dfrac{4mn^2}{3w}$

To add or subtract fractions with unlike denominators, we write each fraction as an equivalent fraction with a common denominator. We can then add or subtract the fractions. For example,

$$\frac{3x}{5} + \frac{2x}{7} = \frac{3x(7)}{5(7)} + \frac{2x(5)}{7(5)} \qquad \frac{4a^2}{15} - \frac{3a^2}{10} = \frac{4a^2(2)}{15(2)} - \frac{3a^2(3)}{10(3)}$$

$$= \frac{21x}{35} + \frac{10x}{35} \qquad\qquad = \frac{8a^2}{30} - \frac{9a^2}{30}$$

$$= \frac{21x + 10x}{35} \qquad\qquad = \frac{8a^2 - 9a^2}{30}$$

$$= \frac{31x}{35} \qquad\qquad = \frac{-a^2}{30}$$

$$\qquad\qquad\qquad = -\frac{a^2}{30}$$

A fraction is said to be **reduced to lowest terms** if all factors common to the numerator and the denominator have been removed. To **simplify a fraction** means to reduce it to lowest terms.

■ SIMPLIFYING FRACTIONS

To simplify a fraction, we use the fundamental property of fractions. This property enables us to divide out all factors that are common to both the numerator and the denominator.

EXAMPLE 3   Simplify $\frac{x^2 - 9}{x^2 - 3x}$ $(x \neq 0, 3)$.

Solution   We factor the difference of two squares in the numerator, factor out x in the denominator, and divide out the common factor of $x - 3$.

$$\frac{x^2 - 9}{x^2 - 3x} = \frac{(x + 3)\cancel{(x - 3)}}{x\cancel{(x - 3)}}$$

$$= \frac{x + 3}{x}$$ ■

Self Check   Simplify $\frac{a^2 - 4a}{a^2 - a - 12}$.

Answer $\frac{a}{a + 3}$

We will encounter the following properties of fractions in the next examples.

If a and b represent real numbers and there are no divisions by 0, then

$$\frac{a}{1} = a \qquad\qquad \frac{a}{a} = 1$$

$$\frac{a}{b} = \frac{-a}{-b} = -\frac{a}{-b} = -\frac{-a}{b} \qquad -\frac{a}{b} = \frac{a}{-b} = \frac{-a}{b} = -\frac{-a}{-b}$$

EXAMPLE 4 Simplify $\dfrac{x^2 - 2xy + y^2}{y - x}$ $(x \neq y)$.

Solution We factor the trinomial in the numerator, factor -1 from the denominator, and divide out the common factor of $x - y$.

$$\frac{x^2 - 2xy + y^2}{y - x} = \frac{(x - y)\cancel{(x - y)}}{-\cancel{(x - y)}}$$

$$= \frac{x - y}{-1}$$

$$= -\frac{x - y}{1}$$

$$= -(x - y) \qquad\blacksquare$$

Self Check Simplify $\dfrac{a^2 - ab - 2b^2}{2b - a}$.

Answer $-(a + b)$

EXAMPLE 5 Simplify $\dfrac{x^2 - 3x + 2}{x^2 - x - 2}$ $(x \neq 2, -1)$.

Solution We factor the numerator and denominator and divide out the common factor of $x - 2$.

$$\frac{x^2 - 3x + 2}{x^2 - x - 2} = \frac{(x - 1)\cancel{(x - 2)}}{(x + 1)\cancel{(x - 2)}}$$

$$= \frac{x - 1}{x + 1} \qquad\blacksquare$$

Self Check Simplify $\dfrac{a^2 + 3a - 4}{a^2 + 2a - 3}$.

Answer $\dfrac{a + 4}{a + 3}$

■ MULTIPLYING AND DIVIDING FRACTIONS

EXAMPLE 6   Multiply $\dfrac{x^2 - x - 2}{x^2 - 1} \cdot \dfrac{x^2 + 2x - 3}{x - 2}$ $(x \neq 1, -1, 2).$

Solution   To multiply the fractions, we multiply the numerators, multiply the denominators, and divide out the common factors.

$$\frac{x^2 - x - 2}{x^2 - 1} \cdot \frac{x^2 + 2x - 3}{x - 2} = \frac{(x^2 - x - 2)(x^2 + 2x - 3)}{(x^2 - 1)(x - 2)}$$

$$= \frac{(x - 2)(x + 1)(x - 1)(x + 3)}{(x + 1)(x - 1)(x - 2)}$$

$$= x + 3 \qquad ■$$

Self Check   Simplify $\dfrac{x^2 - 9}{x^2 - x} \cdot \dfrac{x - 1}{x^2 - 3x}.$

Answer $\dfrac{x + 3}{x^2}$

EXAMPLE 7   Divide $\dfrac{x^2 - 2x - 3}{x^2 - 4} \div \dfrac{x^2 + 2x - 15}{x^2 + 3x - 10}$ $(x \neq 2, -2, 3, -5).$

Solution   To divide the fractions, we multiply by the reciprocal of the divisor. We then factor the numerator and denominator and divide out common factors.

$$\frac{x^2 - 2x - 3}{x^2 - 4} \div \frac{x^2 + 2x - 15}{x^2 + 3x - 10} = \frac{x^2 - 2x - 3}{x^2 - 4} \cdot \frac{x^2 + 3x - 10}{x^2 + 2x - 15}$$

$$= \frac{(x^2 - 2x - 3)(x^2 + 3x - 10)}{(x^2 - 4)(x^2 + 2x - 15)}$$

$$= \frac{(x - 3)(x + 1)(x - 2)(x + 5)}{(x + 2)(x - 2)(x + 5)(x - 3)}$$

$$= \frac{x + 1}{x + 2} \qquad ■$$

Self Check   Simplify $\dfrac{a^2 - a}{a + 2} \div \dfrac{a^2 - 2a}{a^2 - 4}.$

Answer $a - 1$

EXAMPLE 8   Simplify

$$\frac{2x^2 - 5x - 3}{3x - 1} \cdot \frac{3x^2 + 2x - 1}{x^2 - 2x - 3} \div \frac{2x^2 + x}{3x} \qquad \left(x \neq \frac{1}{3}, -1, 3, 0, -\frac{1}{2}\right).$$

Solution   We can change the division to a multiplication, factor, and simplify.

$$\frac{2x^2 - 5x - 3}{3x - 1} \cdot \frac{3x^2 + 2x - 1}{x^2 - 2x - 3} \div \frac{2x^2 + x}{3x} = \frac{2x^2 - 5x - 3}{3x - 1} \cdot \frac{3x^2 + 2x - 1}{x^2 - 2x - 3} \cdot \frac{3x}{2x^2 + x}$$

$$= \frac{(2x^2 - 5x - 3)(3x^2 + 2x - 1)(3x)}{(3x - 1)(x^2 - 2x - 3)(2x^2 + x)}$$

$$= \frac{(x - 3)(2x + 1)(3x - 1)(x + 1)3x}{(3x - 1)(x + 1)(x - 3)x(2x + 1)}$$

$$= 3 \qquad\qquad \blacksquare$$

Self Check   Simplify $\dfrac{x^2 - 25}{x - 2} \div \dfrac{x^2 - 5x}{x^2 - 2x} \cdot \dfrac{x^2 + 2x}{x^2 + 5x}$.

Answer $x + 2$

■ ADDING AND SUBTRACTING FRACTIONS

To add fractions with like denominators, we add the numerators and keep the common denominator.

EXAMPLE 9   Add $\dfrac{2x + 5}{x + 5} + \dfrac{3x + 20}{x + 5}$ $(x \neq -5)$.

Solution
$$\frac{2x + 5}{x + 5} + \frac{3x + 20}{x + 5} = \frac{5x + 25}{x + 5}$$

$$= \frac{5(x + 5)}{1(x + 5)} \qquad \text{Factor out 5 and divide out the common factor of } x + 5.$$

$$= 5 \qquad\qquad \blacksquare$$

Self Check   Add $\dfrac{3x - 2}{x - 2} + \dfrac{x - 6}{x - 2}$.

Answer   4

To add or subtract fractions with unlike denominators, we must find a common denominator, called the **least** (or lowest) **common denominator (LCD).** Suppose the unlike denominators of three fractions are 12, 20, and 35. To find the LCD, we first find the prime factorization of each number.

$$12 = 4 \cdot 3 \qquad\qquad 20 = 4 \cdot 5 \qquad\qquad 35 = 5 \cdot 7$$
$$= 2^2 \cdot 3 \qquad\qquad = 2^2 \cdot 5$$

Because the LCD is the smallest number that can be divided by 12, 20, and 15, it must contain factors of 2^2, 3, 5, and 7. Thus,

$$\text{LCD} = 2^2 \cdot 3 \cdot 5 \cdot 7 = 420$$

That is, 420 is the smallest number that can be divided without remainder by 12, 20, and 35.

When finding an LCD, we always factor each denominator and then create the LCD by using each factor the greatest number of times that it appears in any one denominator. The product of these factors is the LCD.

EXAMPLE 10 Add $\dfrac{1}{x^2 - 4} + \dfrac{2}{x^2 - 4x + 4}$ $(x \neq 2, -2)$.

Solution We factor each denominator and find the LCD.

$$x^2 - 4 = (x + 2)(x - 2)$$
$$x^2 - 4x + 4 = (x - 2)(x - 2) = (x - 2)^2$$

The LCD is $(x + 2)(x - 2)^2$. We then write each fraction with its denominator in factored form, convert each fraction into an equivalent fraction with a denominator of $(x + 2)(x - 2)^2$, add the fractions, and simplify.

$$\frac{1}{x^2 - 4} + \frac{2}{x^2 - 4x + 4} = \frac{1}{(x + 2)(x - 2)} + \frac{2}{(x - 2)(x - 2)}$$

$$= \frac{1(x - 2)}{(x + 2)(x - 2)(x - 2)} + \frac{2(x + 2)}{(x - 2)(x - 2)(x + 2)}$$

$$= \frac{1(x - 2) + 2(x + 2)}{(x + 2)(x - 2)(x - 2)}$$

$$= \frac{x - 2 + 2x + 4}{(x + 2)(x - 2)(x - 2)}$$

$$= \frac{3x + 2}{(x + 2)(x - 2)^2}$$

WARNING! Always attempt to simplify the final result. In this case, the final fraction is already in lowest terms. ∎

Self Check Add $\dfrac{3}{x^2 - 6x + 9} + \dfrac{1}{x^2 - 9}$.

Answer $\dfrac{4x + 6}{(x + 3)(x - 3)^2}$

EXAMPLE 11 Combine and simplify

$$\frac{x-2}{x^2-1} - \frac{x+3}{x^2+3x+2} + \frac{3}{x^2+x-2} \qquad (x \ne 1, -1, -2)$$

Solution We factor the denominators to find the LCD.

$$x^2 - 1 = (x+1)(x-1)$$
$$x^2 + 3x + 2 = (x+2)(x+1)$$
$$x^2 + x - 2 = (x+2)(x-1)$$

The LCD is $(x+1)(x+2)(x-1)$. We now write each fraction as an equivalent fraction with this LCD, and proceed as follows:

$$\frac{x-2}{x^2-1} - \frac{x+3}{x^2+3x+2} + \frac{3}{x^2+x-2} = \frac{x-2}{(x+1)(x-1)} - \frac{x+3}{(x+1)(x+2)} + \frac{3}{(x-1)(x+2)}$$

$$= \frac{(x-2)(x+2)}{(x+1)(x-1)(x+2)} - \frac{(x+3)(x-1)}{(x+1)(x+2)(x-1)} + \frac{3(x+1)}{(x-1)(x+2)(x+1)}$$

$$= \frac{(x^2-4) - (x^2+2x-3) + (3x+3)}{(x+1)(x+2)(x-1)}$$

$$- \frac{x^2 - 4 - x^2 - 2x + 3 + 3x + 3}{(x+1)(x+2)(x-1)}$$

$$= \frac{x+2}{(x+1)(x+2)(x-1)}$$

$$= \frac{1}{(x+1)(x-1)} \qquad \text{Divide out the common factor of } x+2. \quad \blacksquare$$

Self Check Combine and simplify $\dfrac{4y}{y^2-1} - \dfrac{2}{y+1} + 2$.

Answer $\dfrac{2y}{y-1}$

■ COMPLEX FRACTIONS

A **complex fraction** is a fraction that has a fractional numerator or a fractional denominator.

EXAMPLE 12 Simplify $\dfrac{\dfrac{1}{x} + \dfrac{1}{y}}{\dfrac{x}{y}}$ $\quad (x, y \ne 0)$.

Solution 1   We determine that the LCD of the three fractions in the given complex fraction is xy. We multiply the numerator and denominator of the complex fraction by xy and simplify:

$$\frac{\dfrac{1}{x} + \dfrac{1}{y}}{\dfrac{x}{y}} = \frac{xy\left(\dfrac{1}{x} + \dfrac{1}{y}\right)}{xy\left(\dfrac{x}{y}\right)} = \frac{\dfrac{xy}{x} + \dfrac{xy}{y}}{\dfrac{xxy}{y}} = \frac{y + x}{x^2}$$

Solution 2   We combine the fractions in the numerator of the complex fraction to obtain a single fraction over a single fraction.

$$\frac{\dfrac{1}{x} + \dfrac{1}{y}}{\dfrac{x}{y}} = \frac{\dfrac{1(y)}{x(y)} + \dfrac{1(x)}{y(x)}}{\dfrac{x}{y}} = \frac{\dfrac{y + x}{xy}}{\dfrac{x}{y}}$$

Then we use the fact that any fraction indicates a division:

$$\frac{\dfrac{y + x}{xy}}{\dfrac{x}{y}} = \frac{y + x}{xy} \div \frac{x}{y} = \frac{y + x}{xy} \cdot \frac{y}{x} = \frac{(y + x)y}{xyx} = \frac{y + x}{x^2} \qquad\blacksquare$$

Self Check   Simplify $\dfrac{\dfrac{1}{x} - \dfrac{1}{y}}{\dfrac{1}{x} + \dfrac{1}{y}}$.

Answer $\dfrac{y - x}{y + x}$

EXERCISE 1.6

VOCABULARY AND CONCEPTS   *Fill in the blank to make a true statement.*

1. In the fraction $\frac{a}{b}$, a is called the _____.

2. In the fraction $\frac{a}{b}$, b is called the _____.

3. $\frac{a}{b} = \frac{c}{d}$ if and only if _____.

4. The denominator of a fraction can never be ____.

Complete each formula.

5. $\dfrac{a}{b} \cdot \dfrac{c}{d} = $ ____.

6. $\dfrac{a}{b} \div \dfrac{c}{d} = $ ____.

7. $\dfrac{a}{b} + \dfrac{c}{b} = $ ____.

8. $\dfrac{a}{b} - \dfrac{c}{b} = $ ____.

Tell whether the fractions are equal.

9. $\dfrac{8x}{3y}, \dfrac{16x}{6y}$

10. $\dfrac{3x^2}{4y^2}, \dfrac{12y^2}{16x^2}$

11. $\dfrac{25xyz}{12ab^2c}, \dfrac{50a^2bc}{24xyz}$

12. $\dfrac{15rs^2}{4rs^2}, \dfrac{37.5a^3}{10a^3}$

PRACTICE   *In Exercises 13–14, simplify each fraction.*

13. $\dfrac{7a^2b}{21ab^2}$

14. $\dfrac{35p^3q^2}{49p^4q}$

In Exercises 15–22, do the operations and simplify, whenever possible.

15. $\dfrac{4x}{7} \cdot \dfrac{2}{5a}$

16. $\dfrac{-5y}{2z} \cdot \dfrac{4}{y^2}$

17. $\dfrac{8m}{5n} \div \dfrac{3m}{10n}$

18. $\dfrac{15p}{8q} \div \dfrac{-5p}{16q^2}$

19. $\dfrac{3z}{5c} + \dfrac{2z}{5c}$

20. $\dfrac{7a}{4b} - \dfrac{3a}{4b}$

21. $\dfrac{15x^2y}{7a^2b^3} - \dfrac{x^2y}{7a^2b^3}$

22. $\dfrac{8rst^2}{15m^4t^2} + \dfrac{7rst^2}{15m^4t^2}$

In Exercises 23–30, simplify each fraction.   23-41 odd

23. $\dfrac{2x-4}{x^2-4}$

24. $\dfrac{x^2-16}{x^2-8x+16}$

25. $\dfrac{25-x^2}{x^2+10x+25}$

26. $\dfrac{4-x^2}{x^2-5x+6}$

27. $\dfrac{6x^3+x^2-12x}{4x^3+4x^2-3x}$

28. $\dfrac{6x^4-5x^3-6x^2}{2x^3-7x^2-15x}$

29. $\dfrac{x^3-8}{x^2+ax-2x-2a}$

30. $\dfrac{xy+2x+3y+6}{x^3+27}$

In Exercises 31–72, do the operations and simplify, whenever possible.

31. $\dfrac{x^2-1}{x} \cdot \dfrac{x^2}{x^2+2x+1}$

32. $\dfrac{y^2-2y+1}{y} \cdot \dfrac{y+2}{y^2+y-2}$

33. $\dfrac{3x^2+7x+2}{x^2+2x} \cdot \dfrac{x^2-x}{3x^2+x}$

34. $\dfrac{x^2+x}{2x^2+3x} \cdot \dfrac{2x^2+x-3}{x^2-1}$

35. $\dfrac{x^2+x}{x-1} \cdot \dfrac{x^2-1}{x+2}$

36. $\dfrac{x^2+5x+6}{x^2+6x+9} \cdot \dfrac{x+2}{x^2-4}$

37. $\dfrac{2x^2+32}{8} \div \dfrac{x^2+16}{2}$

38. $\dfrac{x^2+x-6}{x^2-6x+9} \div \dfrac{x^2-4}{x^2-9}$

39. $\dfrac{z^2+z-20}{z^2-4} \div \dfrac{z^2-25}{z-5}$

40. $\dfrac{ax + bx + a + b}{a^2 + 2ab + b^2} \div \dfrac{x^2 - 1}{x^2 - 2x + 1}$

41. $\dfrac{3x^2 + 5x - 2}{x^3 + 2x^2} \div \dfrac{6x^2 + 13x - 5}{2x^3 + 5x^2}$

42. $\dfrac{x^2 + 13x + 12}{8x^2 - 6x - 5} \div \dfrac{2x^2 - x - 3}{8x^2 - 14x + 5}$

43. $\dfrac{x^2 + 7x + 12}{x^3 - x^2 - 6x} \cdot \dfrac{x^2 - 3x - 10}{x^2 + 2x - 3} \cdot \dfrac{x^3 - 4x^2 + 3x}{x^2 - x - 20}$

44. $\dfrac{x^2 - 2x - 3}{21x^2 - 50x - 16} \cdot \dfrac{3x - 8}{x - 3} \div \dfrac{x^2 + 6x + 5}{7x^2 - 33x - 10}$

45. $\dfrac{x^3 + 27}{x^2 - 4} \div \left(\dfrac{x^2 + 4x + 3}{x^2 + 2x} \div \dfrac{x^2 + x - 6}{x^2 - 3x + 9} \right)$

46. $\dfrac{x(x - 2) - 3}{x(x + 7) - 3(x - 1)} \cdot \dfrac{x(x + 1) - 2}{x(x - 7) + 3(x + 1)}$

47. $\dfrac{3}{x + 3} + \dfrac{x + 2}{x + 3}$

48. $\dfrac{3}{x + 1} + \dfrac{x + 2}{x + 1}$

49. $\dfrac{4x}{x - 1} - \dfrac{4}{x - 1}$

50. $\dfrac{6x}{x - 2} - \dfrac{3}{x - 2}$

51. $\dfrac{2}{5 - x} + \dfrac{1}{x - 5}$

52. $\dfrac{3}{x - 6} - \dfrac{2}{6 - x}$

53. $\dfrac{3}{x + 1} + \dfrac{2}{x - 1}$

54. $\dfrac{3}{x + 4} + \dfrac{x}{x - 4}$

55. $\dfrac{a + 3}{a^2 + 7a + 12} + \dfrac{a}{a^2 - 16}$

56. $\dfrac{a}{a^2 + a - 2} + \dfrac{2}{a^2 - 5a + 4}$

57. $\dfrac{x}{x^2 - 4} - \dfrac{1}{x + 2}$

58. $\dfrac{b^2}{b^2 - 4} - \dfrac{4}{b^2 + 2b}$

59. $\dfrac{3x - 2}{x^2 + 2x + 1} - \dfrac{x}{x^2 - 1}$

60. $\dfrac{2t}{t^2 - 25} - \dfrac{t + 1}{t^2 + 5t}$

61. $\dfrac{2}{y^2 - 1} + 3 + \dfrac{1}{y + 1}$

62. $2 + \dfrac{4}{t^2 - 4} - \dfrac{1}{t - 2}$

63. $\dfrac{1}{x - 2} + \dfrac{3}{x + 2} - \dfrac{3x - 2}{x^2 - 4}$

64. $\dfrac{x}{x - 3} - \dfrac{5}{x + 3} + \dfrac{3(3x - 1)}{x^2 - 9}$

65. $\left(\dfrac{1}{x - 2} + \dfrac{1}{x - 3} \right) \cdot \dfrac{x - 3}{2x}$

66. $\left(\dfrac{1}{x + 1} - \dfrac{1}{x - 2} \right) \div \dfrac{1}{x - 2}$

67. $\dfrac{3x}{x - 4} - \dfrac{x}{x + 4} - \dfrac{3x + 1}{16 - x^2}$

68. $\dfrac{7x}{x - 5} + \dfrac{3x}{5 - x} + \dfrac{3x - 1}{x^2 - 25}$

69. $\dfrac{1}{x^2 + 3x + 2} - \dfrac{2}{x^2 + 4x + 3} + \dfrac{1}{x^2 + 5x + 6}$

70. $\dfrac{-2}{x - y} + \dfrac{2}{x - z} - \dfrac{2z - 2y}{(y - x)(z - x)}$

71. $\dfrac{3x - 2}{x^2 + x - 20} - \dfrac{4x^2 + 2}{x^2 - 25} + \dfrac{3x^2 - 25}{x^2 - 16}$

72. $\dfrac{3x + 2}{8x^2 - 10x - 3} + \dfrac{x + 4}{6x^2 - 11x + 3} - \dfrac{1}{4x + 1}$

In Exercises 73–92, simplify each complex fraction.

73. $\dfrac{\dfrac{3a}{b}}{\dfrac{6ac}{b^2}}$

74. $\dfrac{\dfrac{3t^2}{9x}}{\dfrac{t}{18x}}$

75. $\dfrac{3a^2b}{\dfrac{ab}{27}}$

76. $\dfrac{\dfrac{3u^2v}{4t}}{3uv}$

77. $\dfrac{\dfrac{x-y}{ab}}{\dfrac{y-x}{ab}}$

78. $\dfrac{\dfrac{x^2-5x+6}{2x^2y}}{\dfrac{x^2-9}{2x^2y}}$

79. $\dfrac{\dfrac{1}{x}+\dfrac{1}{y}}{xy}$

80. $\dfrac{xy}{\dfrac{11}{x}+\dfrac{11}{y}}$

81. $\dfrac{\dfrac{1}{x}+\dfrac{1}{y}}{\dfrac{1}{x}-\dfrac{1}{y}}$

82. $\dfrac{\dfrac{1}{x}-\dfrac{1}{y}}{\dfrac{1}{x}+\dfrac{1}{y}}$

83. $\dfrac{\dfrac{3a}{b}-\dfrac{4a^2}{x}}{\dfrac{1}{b}+\dfrac{1}{ax}}$

84. $\dfrac{1-\dfrac{x}{y}}{\dfrac{x^2}{y^2}-1}$

85. $\dfrac{x+1-\dfrac{6}{x}}{x+5+\dfrac{6}{x}}$

86. $\dfrac{2z}{1-\dfrac{3}{z}}$

87. $\dfrac{3xy}{1-\dfrac{1}{xy}}$

88. $\dfrac{x-3+\dfrac{1}{x}}{-\dfrac{1}{x}-x+3}$

89. $\dfrac{3x}{x+\dfrac{1}{x}}$

90. $\dfrac{2x^2+4}{2+\dfrac{4x}{5}}$

91. $\dfrac{\dfrac{x}{x+2}-\dfrac{2}{x-1}}{\dfrac{3}{x+2}+\dfrac{x}{x-1}}$

92. $\dfrac{\dfrac{2x}{x-3}+\dfrac{1}{x-2}}{\dfrac{3}{x-3}-\dfrac{x}{x-2}}$

In Exercises 93–96, write each expression without using negative exponents, and simplify the resulting complex fraction.

93. $\dfrac{1}{1+x^{-1}}$

94. $\dfrac{y^{-1}}{x^{-1}+y^{-1}}$

95. $\dfrac{3(x+2)^{-1}+2(x-1)^{-1}}{(x+2)^{-1}}$

96. $\dfrac{2x(x-3)^{-1}-3(x+2)^{-1}}{(x-3)^{-1}(x+2)^{-1}}$

DISCOVERY AND WRITING

97. $\dfrac{x}{1+\dfrac{1}{3x^{-1}}}$

98. $\dfrac{ab}{2+\dfrac{3}{2a^{-1}}}$

99. $\dfrac{1}{1+\dfrac{1}{1+\dfrac{1}{x}}}$

100. $\dfrac{y}{2+\dfrac{2}{2+\dfrac{2}{y}}}$

101. Explain why the formula $\dfrac{a}{b} + \dfrac{c}{d} = \dfrac{ad + bc}{bd}$ is valid.

102. Explain why the formula $\dfrac{a}{b} \div \dfrac{c}{d} = \dfrac{a}{b} \cdot \dfrac{d}{c}$ is valid.

103. Explain the commutative property of addition and tell why it is useful.

104. Explain the distributive property and tell why it is useful.

REVIEW Write each expression without using absolute value symbols.

105. $|-6|$

106. $|5 - x|$, given that $x < 0$

Simplify each expression.

107. $\left(\dfrac{x^3 y^{-2}}{x^{-1} y}\right)^{-3}$

108. $(27x^6)^{2/3}$

109. $\sqrt{20} - \sqrt{45}$

110. $2(x^2 + 4) - 3(2x^2 + 5)$

■ ■ ■ ■ ■ ■ ■ ■ ■ PROBLEMS AND PROJECTS

1. Find the highest power of 2 that can be evaluated with a scientific or graphing calculator.

2. Find the highest power of 7 that can be evaluated with a scientific or graphing calculator.

3. While in Switzerland, you see a Rolex watch that a close friend wants. You send him the following E-mail message:

E-Mail
ROLEX WATCH $3500.   SHOULD I BUY IT FOR YOU?

Your friend responds as follows:

E-Mail
NO PRICE TOO HIGH! REPEAT... NO! PRICE TOO HIGH.

Would you buy the watch? Why or why not? What is wrong with your friend's message? What mathematical principle does this example illustrate?

4. Note that $\dfrac{2}{3} = \dfrac{4}{6} = \dfrac{6}{9} = \dfrac{2 + 4 + 6}{3 + 6 + 9}$, which is $\dfrac{12}{18}$. Is this principle always true? That is, is

$$\frac{a}{b} = \frac{c}{d} = \frac{e}{f} = \frac{a + c + e}{b + d + f}?$$

Can you prove it?

Is $\dfrac{a}{b} = \dfrac{c}{d} = \dfrac{e}{f} = \dfrac{g}{h} = \dfrac{a + c + e + g}{b + d + f + h}?$

For how many fractions is this principle true?

PROJECT 1 Complete the table and work the following problems.

Figure	Name	Perimeter/Circumference
	Square	$P = 4s$
		(π is approximately 3.1416)

1. Find the perimeter of a garden with dimensions of 12 meters by 18 meters.

2. Find the circumference of a circle with a radius of 3.8 feet. Give the answer to the nearest hundredth.

(continued)

PROJECT 2   Complete the table and work the problems that follow.

Figure	Name	Area	Figure	Name	Volume
	Square	$A = s^2$			

*B represents the area of the base.

Give each answer to the nearest hundredth.

1. Find the area of a circle with a diameter of 21 feet. Give the answer to the nearest hundredth.

2. Find the area of a triangle with a base of 21.3 centimeters and a height of 7.5 centimeters.

3. Find the area of a trapezoid with a height of 9.3 inches and bases of 7.2 inches and 10.1 inches.

4. Find the volume of a rectangular solid with dimensions of 8.5 meters, 10.3 meters, and 12.7 meters.

5. Find the volume of a sphere with a radius of 20.5 feet.

6. Find the volume of a 10-foot-long cylinder whose base is a circle with a radius of 1.6 feet.

7. Find the volume of a cone with a circular base 15 centimeters in diameter and a height of 12.5 centimeters.

8. Find the volume of a pyramid with a rectangular base with dimensions of 8.7 meters by 9.3 meters that is 15.8 meters high.

PROJECT 3 A village council is debating whether or not to build an emergency water reservoir for use in drought conditions. The tentative plan calls for a conical reservoir with a diameter of 173 feet and height of 87.2 feet.

During drought conditions, the reservoir will lose water to evaporation as well as supply the village with water. The company that designed the reservoir provided the following information.

If D equals the number of consecutive days the reservoir is used to provide water, the total amount E (in cubic feet) of water lost to evaporation will be

$$E = .1V\left(\frac{D - .7}{D}\right)^2$$

The water left in the reservoir after it has been used for D days will be

| Water left | = | Original volume | − | D | · | (water used per day) | − | E |

Under emergency conditions, the village estimates that it will use about 61,000 cubic feet of water per day. The majority of the council believe that building the reservoir is a good idea if it will supply the city with water for at least 10 days. Otherwise, they will vote against the plan.

1. Find the volume of water that the reservoir will hold. Express the result in scientific notation.

2. Will the reservoir supply the village for 10 days?

3. Would you vote for the plan? Explain.

4. How much water could the village use per day if the supply must last for two weeks?

C H A P T E R   S U M M A R Y

CONCEPT

REVIEW EXERCISES

SECTION 1.1 · Sets of Real Numbers

Real numbers: any number that can be expressed as a decimal.

Natural numbers: $\{1, 2, 3, 4, \ldots\}$

Whole numbers: $\{0, 1, 2, 3, 4, \ldots\}$

Integers: $\{\ldots, -3, -2, -1, 0, 1, 2, 3, \ldots\}$

Rational numbers: $\{x \mid x$ can be written in the form $\frac{a}{b}$ $(b \neq 0)$, where a and b are integers. All decimals that either terminate or repeat.

Irrational numbers: nonrational real numbers. All decimals that neither terminate nor repeat.

Prime numbers: $\{2, 3, 5, 7, 11, 13, \ldots\}$

Composite numbers: $\{4, 6, 8, 9, 10, \ldots\}$

Even integers: $\{\ldots, -4, -2, 0, 2, 4, \ldots\}$

Odd integers: $\{\ldots, -3, -1, 1, 3, 5, \ldots\}$

Associative properties: $(a + b) + c = a + (b + c)$ of addition $(ab)c = a(bc)$ of multiplication

Commutative properties: $a + b = b + a$ of addition $ab = ba$ of multiplication

Distributive property: $a(b + c) = ab + ac$

1. Consider the set $\left\{-6, -3, 0, \frac{1}{2}, 3, \pi, \sqrt{5}, 6, 8\right\}$. List the numbers in this set that are
  a. natural numbers.
  b. whole numbers.
  c. integers.
  d. rational numbers.
  e. irrational numbers.
  f. real numbers.

2. Consider the set $\left\{-6, -3, 0, \frac{1}{2}, 3, \pi, \sqrt{5}, 6, 8\right\}$. List the numbers in this set that are
  a. prime numbers.
  b. composite numbers.
  c. even integers.
  d. odd integers.

3. Tell which property of real numbers justifies each statement.
  a. $(a + b) + 2 = a + (b + 2)$
  b. $a + 7 = 7 + a$
  c. $4(2x) = (4 \cdot 2)x$
  d. $3(a + b) = 3a + 3b$
  e. $(5a)7 = 7(5a)$
  f. $(2x + y) + z = (y + 2x) + z$

Subsets of the real numbers can be graphed on the number line.

4. Graph each subset of the real numbers.

    a. the prime numbers between 10 and 20.

    b. the even integers from 6 to 14.

Open intervals have no endpoints.

Closed intervals have two endpoints.

Half-open intervals have one endpoint.

5. Graph each interval on the number line.

    a. $-3 < x \le 5$

    b. $x \ge 0$ or $x < -1$

    c. $(-2, 4]$

    d. $(-\infty, -4) \cup [6, \infty)$

If $x \ge 0$, then $|x| = x$.
If $x < 0$, then $|x| = -x$.

$|x| \ge 0$

6. Write each expression without absolute value symbols.

    a. $|6|$ **b.** $|-25|$

    c. $|1 - \sqrt{2}|$ **d.** $|\sqrt{3} - 1|$

The **distance between points** a and b on a number line is $d = |a - b|$.

7. On a number line, find the distance between points with coordinates of -5 and 7.

SECTION 1.2   *Integer Exponents and Scientific Notation*

Natural-number exponents:

$$\overbrace{x^n = x \cdot x \cdot x \cdot \,\cdots\, \cdot x}^{n \text{ factors of } x}$$

8. Write each expression without using exponents.

    a. $-5a^3$ **b.** $(-5a)^2$

9. Write each expression using exponents.

    a. $3ttt$ **b.** $(-2b)(3b)$

Rules of exponents: If there are no divisions by 0,

$$x^m x^n = x^{m+n} \qquad (x^m)^n = x^{mn}$$

$$(xy)^n = x^n y^n \qquad \left(\frac{x}{y}\right)^n = \frac{x^n}{y^n}$$

$$x^0 = 1 \qquad x^{-n} = \frac{1}{x^n}$$

$$\frac{x^m}{x^n} = x^{m-n} \qquad \left(\frac{x}{y}\right)^{-n} = \left(\frac{y}{x}\right)^n$$

10. Simplify each expression.

    a. $n^2 n^4$ **b.** $(p^3)^2$

    c. $(x^3 y^2)^4$ **d.** $\left(\dfrac{a^4}{b^2}\right)^3$

    e. $(m^{-3} n^0)^2$ **f.** $\left(\dfrac{p^{-2} q^2}{2}\right)^3$

    g. $\dfrac{a^5}{a^8}$ **h.** $\left(\dfrac{a^2}{b^3}\right)^{-2}$

    i. $\left(\dfrac{3x^2 y^{-2}}{x^2 y^2}\right)^{-2}$ **j.** $\left(\dfrac{a^{-3} b^2}{ab^{-3}}\right)^{-2}$

    k. $\left(\dfrac{-3x^3 y}{xy^3}\right)^{-2}$ **l.** $\left(-\dfrac{2m^{-2} n^0}{4m^2 n^{-1}}\right)^{-3}$

A number is written in **scientific notation** when it is written in the form $N \times 10^n$, where $1 \le N < 10$.

11. If $x = -3$ and $y = 3$, evaluate $-x^2 - xy^2$.

12. Write each number in scientific notation.
  a. 6750
  b. 0.00023

13. Write each number in standard notation.
  a. 4.8×10^2
  b. 0.25×10^{-3}

14. Use scientific notation to simplify $\dfrac{(45,000)(350,000)}{0.000105}$.

SECTION 1.3 — *Fractional Exponents and Radicals*

If $a > 0$, then $a^{1/n}$ is the positive number such that $(a^{1/n})^n = a$.

If $a = 0$, then $a^{1/n} = 0$.

If $a < 0$ and n is odd, then $a^{1/n}$ is the real number such that $(a^{1/n})^n = a$.

If $a < 0$ and n is even, then $a^{1/n}$ is not a real number.

If m and n are positive integers, $\frac{m}{n}$ is in lowest terms, and $a^{1/n}$ is a real number, then
$$a^{m/n} = (a^{1/n})^m = (a^m)^{1/n}$$

$\sqrt[n]{a} = a^{1/n}$

15. Simplify each expression, if possible.
  a. $121^{1/2}$
  b. $\left(\dfrac{27}{125}\right)^{1/3}$
  c. $(32x^5)^{1/5}$
  d. $(81a^4)^{1/4}$
  e. $(-1000x^6)^{1/3}$
  f. $(-25x^2)^{1/2}$
  g. $(x^{12}y^2)^{1/2}$
  h. $\left(\dfrac{x^{12}}{y^4}\right)^{-1/2}$
  i. $\left(\dfrac{-c^{2/3}c^{5/3}}{c^{-2/3}}\right)^{1/3}$
  j. $\left(\dfrac{a^{-1/4}a^{3/4}}{a^{9/2}}\right)^{-1/2}$

16. Simplify each expression.
  a. $64^{2/3}$
  b. $32^{-3/5}$
  c. $\left(\dfrac{16}{81}\right)^{3/4}$
  d. $\left(\dfrac{32}{243}\right)^{2/5}$
  e. $\left(\dfrac{8}{27}\right)^{-2/3}$
  f. $\left(\dfrac{16}{625}\right)^{-3/4}$
  g. $(-216x^3)^{2/3}$
  h. $\dfrac{p^{a/2}p^{a/3}}{p^{a/6}}$

17. Simplify each expression.
  a. $\sqrt{36}$
  b. $-\sqrt{49}$
  c. $\sqrt{\dfrac{9}{25}}$
  d. $\sqrt[3]{\dfrac{27}{125}}$
  e. $\sqrt{x^2y^4}$
  f. $\sqrt[3]{x^3}$
  g. $\sqrt[4]{\dfrac{m^8n^4}{p^{16}}}$
  h. $\sqrt[5]{\dfrac{a^{15}b^{10}}{c^5}}$

If all radicals are real numbers and there are no divisions by 0, then

$$\sqrt[n]{ab} = \sqrt[n]{a}\sqrt[n]{b}$$

$$\sqrt[n]{\frac{a}{b}} = \frac{\sqrt[n]{a}}{\sqrt[n]{b}}$$

$$\sqrt[m]{\sqrt[n]{a}} = \sqrt[n]{\sqrt[m]{a}} = \sqrt[mn]{a}$$

18. Simplify and combine terms.
  a. $\sqrt{50} + \sqrt{8}$
  b. $\sqrt{12} + \sqrt{3} - \sqrt{27}$
  c. $\sqrt[3]{24x^4} - \sqrt[3]{3x^4}$

19. Rationalize each denominator.
  a. $\dfrac{\sqrt{7}}{\sqrt{5}}$ **b.** $\dfrac{8}{\sqrt{8}}$
  c. $\dfrac{1}{\sqrt[3]{2}}$ **d.** $\dfrac{2}{\sqrt[3]{25}}$

20. Rationalize each numerator.
  a. $\dfrac{\sqrt{2}}{5}$ **b.** $\dfrac{\sqrt{5}}{5}$
  c. $\dfrac{\sqrt{2x}}{3}$ **d.** $\dfrac{3\sqrt[3]{7x}}{2}$

SECTION 1.4 — *Polynomials*

Monomial: A polynomial with one term.
Binomial: A polynomial with two terms.
Trinomial: A polynomial with three terms.

The **degree of a monomial** is the sum of the exponents on its variables.

The **degree of a polynomial** is the degree of the term in the polynomial with highest degree.

$$a(b + c + d + \cdots) = ab + ac + ad + \cdots$$

$$(x + y)^2 = x^2 + 2xy + y^2$$

$$(x - y)^2 = x^2 - 2xy + y^2$$

$$(x + y)(x - y) = x^2 - y^2$$

21. Give the degree of each polynomial and tell whether the polynomial is a monomial, binomial, or trinomial.
  a. $x^3 - 8$
  b. $8x^2 - 8x - 8$
  c. $\sqrt{3}x^2$
  d. $4x^4 - 12x^2 + 1$

22. Do the operations and simplify.
  a. $2(x + 3) + 3(x - 4)$
  b. $3x^2(x - 1) - 2x(x + 3) - x^2(x + 2)$
  c. $(3x + 2)(3x + 2)$
  d. $(3x + y)(2x - 3y)$
  e. $(4a + 2b)(2a - 3b)$
  f. $(z + 3)(3z^2 + z - 1)$
  g. $(a^n + 2)(a^n - 1)$

h. $\left(\sqrt{2} + x\right)^2$

i. $\left(\sqrt{2} + 1\right)\left(\sqrt{3} + 1\right)$

j. $\left(\sqrt[3]{3} - 2\right)\left(\sqrt[3]{9} + 2\sqrt[3]{3} + 4\right)$

The **conjugate** of $a + b$ is $a - b$.

23. Rationalize each denominator.

a. $\dfrac{2}{\sqrt{3} - 1}$

b. $\dfrac{-2}{\sqrt{3} - \sqrt{2}}$

c. $\dfrac{2x}{\sqrt{x} - 2}$

d. $\dfrac{\sqrt{x} - \sqrt{y}}{\sqrt{x} + \sqrt{y}}$

24. Rationalize each numerator.

a. $\dfrac{\sqrt{x} + 2}{5}$

b. $\dfrac{1 - \sqrt{a}}{a}$

25. Do each division.

a. $\dfrac{3x^2y^2}{6x^3y}$

b. $\dfrac{4a^2b^3 + 6ab^4}{2b^2}$

c. $2x + 3 \overline{\smash{)}\,2x^3 + 7x^2 + 8x + 3}$

d. $x^2 - 1 \overline{\smash{)}\,x^5 + x^3 - 2x - 3x^2 - 3}$

SECTION 1.5   *Factoring Polynomials*

$x^2 - y^2 = (x + y)(x - y)$
$x^2 + 2xy + y^2 = (x + y)^2$
$x^2 - 2xy + y^2 = (x - y)^2$

$x^3 + y^3 = (x + y)(x^2 - xy + y^2)$
$x^3 - y^3 = (x - y)(x^2 + xy + y^2)$

26. Factor each expression completely.

a. $3t^3 - 3t$

b. $5r^3 - 5$

c. $6x^2 + 7x - 24$

d. $3a^2 + ax - 3a - x$

e. $8x^3 - 125$

f. $6x^2 - 20x - 16$

g. $x^2 + 6x + 9 - t^2$

h. $3x^2 - 1 + 5x$

i. $8z^3 + 343$

j. $1 + 14b + 49b^2$

k. $121z^2 + 4 - 44z$

l. $64y^3 - 1000$

m. $2xy - 4zx - wy + 2zw$

n. $x^8 + x^4 + 1$

SECTION 1.6   *Algebraic Fractions*

If there are no divisions by 0, then

$$\frac{a}{b} = \frac{c}{d} \quad \text{if and only if} \quad ad = bc.$$

$$\frac{a}{b} = \frac{ax}{bx}$$

$$\frac{a}{b} \cdot \frac{c}{d} = \frac{ac}{bd} \qquad \frac{a}{b} \div \frac{c}{d} = \frac{ad}{bc}$$

$$\frac{a}{b} + \frac{c}{b} = \frac{a+c}{b}$$

$$\frac{a}{b} - \frac{c}{b} = \frac{a-c}{b}$$

$$a \cdot 1 = a \qquad \frac{a}{1} = a \qquad \frac{a}{a} = 1$$

$$\frac{a}{b} = \frac{-a}{-b} = -\frac{a}{-b} = -\frac{-a}{b}$$

$$-\frac{a}{b} = \frac{a}{-b} = \frac{-a}{b} = -\frac{-a}{b}$$

27. Do each operation and simplify.

a. $\dfrac{x^2 - 4x + 4}{x + 2} \cdot \dfrac{x^2 + 5x + 6}{x - 2}$

b. $\dfrac{2y^2 - 11y + 15}{y^2 - 6y + 8} \cdot \dfrac{y^2 - 2y - 8}{y^2 - y - 6}$

c. $\dfrac{2t^2 + t - 3}{3t^2 - 7t + 4} \div \dfrac{10t + 15}{3t^2 - t - 4}$

d. $\dfrac{p^2 + 7p + 12}{p^3 + 8p^2 + 4p} \div \dfrac{p^2 - 9}{p^2}$

e. $\dfrac{x^2 + x - 6}{x^2 - x - 6} \cdot \dfrac{x^2 - x - 6}{x^2 + x - 2} \div \dfrac{x^2 - 4}{x^2 - 5x + 6}$

f. $\left(\dfrac{2x + 6}{x + 5} \div \dfrac{2x^2 - 2x - 4}{x^2 - 25} \right) \dfrac{x^2 - x - 2}{x^2 - 2x - 15}$

g. $\dfrac{2}{x - 1} + \dfrac{3x}{x + 5}$

h. $\dfrac{5x}{x - 2} - \dfrac{3x + 7}{x + 2} + \dfrac{2x + 1}{x + 2}$

i. $\dfrac{x}{x - 1} + \dfrac{x}{x - 2} + \dfrac{x}{x - 3}$

j. $\dfrac{x}{x + 1} - \dfrac{3x + 7}{x + 2} + \dfrac{2x + 1}{x + 2}$

k. $\dfrac{3(x + 1)}{x} - \dfrac{5(x^2 + 3)}{x^2} + \dfrac{x}{x + 1}$

l. $\dfrac{3x}{x + 1} + \dfrac{x^2 + 4x + 3}{x^2 + 3x + 2} - \dfrac{x^2 + x - 6}{x^2 - 4}$

28. Simplify each complex fraction.

a. $\dfrac{\dfrac{5x}{2}}{\dfrac{3x^2}{8}}$

b. $\dfrac{\dfrac{3x}{y}}{\dfrac{6x}{y^2}}$

c. $\dfrac{\dfrac{1}{x} + \dfrac{1}{y}}{x - y}$

d. $\dfrac{x^{-1} + y^{-1}}{y^{-1} - x^{-1}}$

■ Chapter 1 Test

Consider the set $\left\{-7, -\frac{2}{3}, 0, 1, 3, \sqrt{10}, 4\right\}$.

1. List the numbers in the set that are odd integers.

2. List the numbers in the set that are prime numbers.

Tell which property justifies each statement.

3. $(a + b) + c = (b + a) + c$

4. $a(b + c) = ab + ac$

Graph each interval on a number line.

5. $-4 < x \le 2$

6. $(-\infty, -3) \cup [6, \infty)$

Write each expression without using absolute value symbols.

7. $|-17|$

8. $|x - 7|$, when $x < 0$.

Find the distance on a number line between points with the following coordinates.

9. -4 and 12

10. -20 and -12

Simplify each expression. Assume that all variables represent positive numbers, and write all answers without using negative exponents.

11. $x^4 x^5 x^2$

12. $\dfrac{r^2 r^3 s}{r^4 s^2}$

13. $\dfrac{(a^{-1} a^2)^{-2}}{a^{-3}}$

14. $\left(\dfrac{x^0 x^2}{x^{-2}}\right)^6$

Write each number in scientific notation.

15. 450,000

16. 0.000345

Write each number in standard notation.

17. 3.7×10^3

18. 1.2×10^{-3}

Simplify each expression. Assume that all variables represent positive numbers, and write all answers without using negative exponents.

19. $(25a^4)^{1/2}$

20. $\left(\dfrac{36}{81}\right)^{3/2}$

21. $\left(\dfrac{8t^6}{27s^9}\right)^{-2/3}$

22. $\sqrt[3]{27a^6}$

23. $\sqrt{12} + \sqrt{27}$

24. $2\sqrt[3]{3x^4} - 3x\sqrt[3]{24x}$

25. Rationalize the denominator: $\dfrac{x}{\sqrt{x} - 2}$.

26. Rationalize the numerator: $\dfrac{\sqrt{x} - \sqrt{y}}{\sqrt{x} + \sqrt{y}}$.

Do each operation.

27. $(a^2 + 3) - (2a^2 - 4)$

28. $(3a^3b^2)(-2a^3b^4)$

29. $(3x - 4)(2x + 7)$

30. $(a^n + 2)(a^n - 3)$

31. $(x^2 + 4)(x^2 - 4)$

32. $(x^2 - x + 2)(2x - 3)$

33. $x - 3\overline{)6x^2 + x - 23}$

34. $2x - 1\overline{)2x^3 + 3x^2 - 1}$

Factor each polynomial.

35. $3x + 6y$

36. $x^2 - 100$

37. $10t^2 - 19tw + 6w^2$

38. $3a^3 - 648$

39. $x^4 - x^2 - 12$

40. $6x^4 + 11x^2 - 10$

Do each operation and simplify if possible.

41. $\dfrac{x}{x + 2} + \dfrac{2}{x + 2}$

42. $\dfrac{x}{x + 1} - \dfrac{x}{x - 1}$

43. $\dfrac{x^2 + x - 20}{x^2 - 16} \cdot \dfrac{x^2 - 25}{x - 5}$

44. $\dfrac{x + 2}{x^2 + 2x + 1} \div \dfrac{x^2 - 4}{x + 1}$

Simplify each complex fraction.

45. $\dfrac{\dfrac{1}{a} + \dfrac{1}{b}}{\dfrac{1}{b}}$

46. $\dfrac{x^{-1}}{x^{-1} + y^{-1}}$

2 Equations and Inequalities

THE TOPIC OF THIS CHAPTER IS EQUATIONS, ONE OF THE MOST IMPORTANT CONCEPTS IN ALGEBRA. EQUATIONS ARE USED IN ALMOST EVERY ACADEMIC DISCIPLINE AND VOCATIONAL AREA, ESPECIALLY IN CHEMISTRY, PHYSICS, MEDICINE, ECONOMICS, AND BUSINESS.

2.1 Linear Equations

■ PROPERTIES OF EQUALITY ■ LINEAR EQUATIONS ■ RATIONAL EQUATIONS ■ FORMULAS

An **equation** is a statement indicating that two quantities are equal. An equation can be either true or false. For example, the equation $2 + 2 = 4$ is true, and the equation $2 + 3 = 6$ is false. An equation such as $3x - 2 = 10$ is true or false depending on the value of x, which is called a **variable.** If $x = 4$, the equation is true, because 4 satisfies the equation.

$$3x - 2 = 10$$
$$3(4) - 2 \overset{?}{=} 10 \qquad \text{Substitute 4 for } x.$$
$$12 - 2 \overset{?}{=} 10$$
$$10 = 10$$

This equation is false for all other replacements of x.

Any number that satisfies an equation is called a **solution** or **root** of the equation. The set of all solutions of an equation is called its **solution set.** We have seen that the solution set of $3x - 2 = 10$ is $\{4\}$. To **solve** an equation means to find its solution set.

There can be restrictions on the values of a variable. For example, in the fraction

$$\frac{x^2 + 4}{x - 2}$$

we cannot replace x with 2, because that would make the denominator equal to 0.

EXAMPLE 1   Find the restrictions on the values of x in the equation $\sqrt{x} = \dfrac{2}{x - 1}$.

Solution   For $\sqrt{x}$ to be a real number, x must be nonnegative, and for $\frac{2}{x-1}$ to be a real number, x cannot be 1. The replacement set for x is restricted to the set of all nonnegative real numbers except 1.   ■

Self Check   Find the restrictions on a:

$$\sqrt{a} = \frac{3}{a - 2}$$

Answer   all nonnegative real numbers but 2

For some equations, called **identities,** every acceptable replacement for the variable is a solution. The equation

$$x^2 - 9 = (x + 3)(x - 3)$$

is an identity, because every number x is a solution. For other equations, called **impossible equations** or **contradictions,** no number is a solution. The equation

$$x = x + 1$$

is an impossible equation. It has no solution, because no real number can be 1 greater than itself.

Equations whose solution sets contain some but not all numbers are called **conditional equations.** The equation

$$3x - 2 = 10$$

is a conditional equation with one solution, the number 4.

If two equations have the same solution set, they are called **equivalent equations.**

■ PROPERTIES OF EQUALITY

There are certain properties of equality that we can use to transform equations into equivalent but less complicated equations. If we use these properties, the resulting equations will be equivalent and have the same solution set.

The Addition and Subtraction Properties
If $a = b$ and c is a real number, then
$$a + c = b + c \qquad \text{and} \qquad a - c = b - c$$

The Multiplication and Division Properties
If $a = b$ and c is a real number, then
$$ac = bc \qquad \text{and} \qquad \frac{a}{c} = \frac{b}{c} \quad (c \neq 0)$$

The Substitution Property
In an equation, a quantity may be substituted for its equal without changing the truth of the equation.

■ LINEAR EQUATIONS

The easiest equations to solve are the **first-degree** or **linear equations.**

> **Linear Equations**
> A **linear equation in one variable** (say, x) is any equation that can be written in the form
>
> $ax + b = 0$ (a and b are real numbers and $a \neq 0$)

To solve the linear equation $2x + 3 = 0$, we subtract 3 from both sides of the equation and then divide both sides by 2 to obtain

$$2x + 3 = 0$$

$$2x + 3 - \mathbf{3} = 0 - \mathbf{3}$$ To undo the addition of 3, subtract 3 from both sides.

$$2x = -3$$

$$\frac{2x}{\mathbf{2}} = -\frac{3}{\mathbf{2}}$$ To undo the multiplication by 2, divide both sides by 2.

$$x = -\frac{3}{2}$$

To show that $-\frac{3}{2}$ satisfies the equation, we substitute $-\frac{3}{2}$ for x and simplify:

$$2x + 3 = 0$$

$$2\left(-\frac{\mathbf{3}}{\mathbf{2}}\right) + 3 \overset{?}{=} 0$$ Substitute $-\frac{3}{2}$ for x.

$$-3 + 3 \overset{?}{=} 0$$ $2\left(-\frac{3}{2}\right) = -3$.

$$0 = 0$$

Because both sides of the equation are equal, the solution checks.

As this example suggests, every linear equation has exactly one solution.

EXAMPLE 2   Find the solution set: $3(x + 2) = 5x + 2$.

Solution   We proceed as follows:

$$3(x + 2) = 5x + 2$$

$$3x + 6 = 5x + 2$$ Use the distributive property and remove parentheses.

$$3x + 6 - \mathbf{3x} = 5x - \mathbf{3x} + 2$$ Subtract $3x$ from both sides.

$$6 = 2x + 2$$ Combine like terms.

$$6 - \mathbf{2} = 2x + 2 - \mathbf{2}$$ Subtract 2 from both sides.

$$4 = 2x$$ Simplify.

$$\frac{4}{\mathbf{2}} = \frac{2x}{\mathbf{2}}$$ Divide both sides by 2.

$$2 = x$$ Simplify.

Because all of the above equations are equivalent, the solution set of the original equation is $\{2\}$. Verify that 2 satisfies the equation.   ■

Self Check   Find the solution set: $4(x - 3) = 7x - 3$.

Answer $\{-3\}$

EXAMPLE 3   Find the solution set: $\dfrac{3}{2}x - \dfrac{2}{3} = \dfrac{1}{5}x$.

Solution   To clear the equation of fractions, we multiply both sides by the LCD of the three fractions and proceed as follows:

$$\frac{3}{2}x - \frac{2}{3} = \frac{1}{5}x$$

$$30\left(\frac{3}{2}x - \frac{2}{3}\right) = 30\left(\frac{1}{5}x\right) \qquad \text{Multiply both sides by 30.}$$

$$45x - 20 = 6x \qquad \text{Remove parentheses and simplify.}$$

$$45x - 20 + 20 = 6x + 20 \qquad \text{Add 20 to both sides.}$$

$$45x = 6x + 20 \qquad \text{Simplify.}$$

$$45x - 6x = 6x - 6x + 20 \qquad \text{Subtract } 6x \text{ from both sides.}$$

$$39x = 20 \qquad \text{Combine like terms.}$$

$$\frac{39x}{39} = \frac{20}{39} \qquad \text{Divide both sides by 39.}$$

$$x = \frac{20}{39} \qquad \text{Simplify.}$$

The solution set is $\left\{\dfrac{20}{39}\right\}$. Verify that $\dfrac{20}{39}$ satisfies the equation.   ■

Self Check   Find the solution set: $\dfrac{2}{3}x - 3 = \dfrac{x}{6}$.

Answer $\{6\}$

EXAMPLE 4   Solve   **a.** $3(x + 5) = 3(1 + x)$ and   **b.** $5 + 5(x + 2) - 2x = 3x + 15$.

Solution   **a.**

$$3(x + 5) = 3(1 + x)$$

$$3x + 15 = 3 + 3x \qquad \text{Remove parentheses.}$$

$$3x - 3x + 15 = 3 + 3x - 3x \qquad \text{Subtract } 3x \text{ from both sides.}$$

$$15 = 3 \qquad \text{Combine like terms.}$$

Since $15 = 3$ is false, the equation has no roots. Its solution set is the empty set, which is denoted as $\emptyset$.

b. $5 + 5(x + 2) - 2x = 3x + 15$

$\qquad 5 + 5x + 10 - 2x = 3x + 15 \qquad\qquad$ Remove parentheses.

$\qquad\qquad\qquad 3x + 15 = 3x + 15 \qquad\qquad$ Simplify.

Because both sides of the final equation are identical, every value of x will make the equation true. The solution set is the set of all real numbers. This equation is an identity. ∎

Self Check   Solve   **a.** $2(x + 1) + 4 = 2(x + 3)$ and   **b.** $-2(x - 4) + 6x = 4(x + 1)$.

Answers   **a.** all real numbers,   **b.** no solution

■ RATIONAL EQUATIONS

Rational equations are equations that contain rational expressions. Some examples of rational equations are

$$\frac{2}{x - 3} = 7, \quad \frac{x + 1}{x - 2} = \frac{3}{x - 2}, \quad \text{and} \quad \frac{x + 2}{x + 3} + \frac{1}{x^2 + 2x - 3} = 1$$

When solving these equations, we will multiply both sides by a quantity containing a variable. When we do this, we could inadvertently multiply both sides of an equation by 0 and obtain a solution that makes the denominator of a fraction 0. In this case, we have found a false solution, called an **extraneous solution.** These solutions do not satisfy the equation and must be discarded.

WARNING!   Be sure to exclude from the solution set of an equation any value that makes the denominator of a fraction equal to 0.

The following equation has an extraneous solution.

$$\frac{x + 1}{x - 2} = \frac{3}{x - 2}$$

$$(x - 2)\left(\frac{x + 1}{x - 2}\right) = (x - 2)\left(\frac{3}{x - 2}\right) \qquad\qquad \text{Multiply both sides by } x - 2.$$

$$x + 1 = 3 \qquad\qquad \frac{x-2}{x-2} = 1$$

$$x = 2 \qquad\qquad \text{Subtract 1 from both sides.}$$

If we check by substituting 2 for x, we obtain 0's in the denominator. Thus, 2 is not a root. The solution set is $\emptyset$.

EXAMPLE 5   Solve $\dfrac{x + 2}{x + 3} + \dfrac{1}{x^2 + 2x - 3} = 1$.

Solution Note that x cannot be -3, because that would cause the denominator of the first fraction to be 0. To find other restrictions, we factor the trinomial in the denominator of the second fraction.

$$x^2 + 2x - 3 = (x + 3)(x - 1)$$

Since the denominator will be 0 when $x = -3$ or $x = 1$, x cannot be -3 or 1.

$$\frac{x + 2}{x + 3} + \frac{1}{x^2 + 2x - 3} = 1$$

$$\frac{x + 2}{x + 3} + \frac{1}{(x + 3)(x - 1)} = 1 \qquad \text{Factor } x^2 + 2x - 3.$$

$$(x + 3)(x - 1)\left[\frac{x + 2}{x + 3} + \frac{1}{(x + 3)(x - 1)}\right] = (x + 3)(x - 1)1 \qquad \begin{array}{l}\text{Multiply both sides}\\ \text{by } (x + 3)(x - 1).\end{array}$$

$$(x + 3)(x - 1)\left(\frac{x + 2}{x + 3}\right) + (x + 3)(x - 1)\frac{1}{(x + 3)(x - 1)} = (x + 3)(x - 1)1 \qquad \text{Remove brackets.}$$

$$(x - 1)(x + 2) + 1 = (x + 3)(x - 1) \qquad \text{Simplify.}$$

$$x^2 + x - 2 + 1 = x^2 + 2x - 3 \qquad \text{Multiply the binomials.}$$

$$x - 1 = 2x - 3 \qquad \begin{array}{l}\text{Subtract } x^2 \text{ from both}\\ \text{sides and combine terms.}\end{array}$$

$$2 = x \qquad \begin{array}{l}\text{Add 3 and subtract}\\ x \text{ from both sides.}\end{array}$$

Because 2 is a meaningful replacement for x, it is a root. However, it is a good idea to check it.

$$\frac{x + 2}{x + 3} + \frac{1}{x^2 + 2x - 3} = 1$$

$$\frac{2 + 2}{2 + 3} + \frac{1}{2^2 + 2(2) - 3} \overset{?}{=} 1 \qquad \text{Substitute 2 for } x.$$

$$\frac{4}{5} + \frac{1}{5} \overset{?}{=} 1$$

$$1 = 1$$

Since 2 satisfies the equation, it is a root. ■

Self Check Solve $\dfrac{3}{5} + \dfrac{7}{x + 2} = 2$.

Answer 3

■ FORMULAS

Many equations, called **formulas,** contain several variables. For example, the formula that converts from degrees Celsius to degrees Fahrenheit is $F = \frac{9}{5}C + 32$. If we need to change a large number of Fahrenheit readings to degrees Celsius, it is tedious to substitute each value of F into the formula and then repeatedly solve for C. It is better to solve the formula for C, substitute the values for F, and evaluate C directly.

EXAMPLE 6   Solve $F = \dfrac{9}{5}C + 32$ for C.

Solution   We use the same methods as for solving linear equations.

$$F = \frac{9}{5}C + 32$$

$$F - 32 = \frac{9}{5}C \qquad \text{Subtract 32 from both sides.}$$

$$\frac{5}{9}(F - 32) = \frac{5}{9}\left(\frac{9}{5}C\right) \qquad \text{Multiply both sides by } \tfrac{5}{9}.$$

$$\frac{5}{9}(F - 32) = C \qquad \text{Simplify.}$$

This result can be written in the alternate form $C \qquad \dfrac{5F - 160}{9}$.   ∎

Self Check   Solve $C = \dfrac{5}{9}(F - 32)$ for F.

Answer $F = \dfrac{9}{5}C + 32$

EXAMPLE 7   The formula $A = p + prt$ is used to find the amount of money in a savings account at the end of a specified time. A represents the amount, p represents the principal (the original deposit), r represents the rate of simple interest per unit of time, and t represents the number of units of time. Solve this formula for p.

Solution   We factor p from both terms on the right-hand side of the equation and proceed as follows:

$$A = p + prt$$
$$A = p(1 + rt) \qquad \text{Factor out } p.$$
$$\frac{A}{(1 + rt)} = p \qquad \text{Divide both sides by } 1 + rt.$$
$$p = \frac{A}{(1 + rt)}$$

∎

Self Check   Solve $pq = fq + fp$ for f.

Answer $f = \dfrac{pq}{q + p}$

EXERCISE 2.1

VOCABULARY AND CONCEPTS   *In Exercises 1–8, fill in the blank to make a true statement.*

1. If a number satisfies an equation, it is called a ____ or ____ of the equation.

2. If an equation is true for all values of its variable, it is called an ____.

3. A contradiction is an equation that is true for ____ values of its variable.

4. A ____ equation is true for some values of its variable and is not true for others.

5. An equation of the form $ax + b = 0$ is called a ____ equation.

6. If an equation contains rational expressions, it is called a ____ equation.

7. A linear equation has ____ root.

8. The ____ of a fraction can never be 0.

PRACTICE   *In Exercises 9–16, each quantity represents a real number. Find all restrictions on x.*

9. $x + 3 = 1$

10. $\dfrac{1}{2}x - 7 = 14$

11. $\dfrac{1}{x} = 12$

12. $\dfrac{3}{x - 2}$

13. $\sqrt{x} = 4$

14. $\sqrt[3]{x} = 64$

15. $\dfrac{1}{x - 3} = \dfrac{5}{x + 2}$

16. $\dfrac{24}{\sqrt{x - 3}}$

In Exercises 17–30, solve each equation, if possible. Classify each one as an identity, a conditional equation, or an equation with no solutions.

17. $2x + 5 = 15$

18. $3x + 2 = x + 8$

19. $2(x + 2) = 2x + 5$

20. $3(x + 2) - x = 2(x + 3)$

21. $\dfrac{x + 7}{2} = 7$

22. $\dfrac{x}{2} - 7 = 14$

23. $2(a + 1) = 3(a - 2) - a$

24. $x^2 = (x + 4)(x - 4) + 16$

25. $3(x - 3) = \dfrac{6x - 18}{2}$

26. $x(x + 2) = (x + 1)^2$

27. $\dfrac{3}{b - 3} = 1$

28. $x^2 - 8x + 15 = (x - 3)(x + 5)$

29. $2x^2 + 5x - 3 = (2x - 1)(x + 3)$

30. $2x^2 + 5x - 3 = 2x\left(x + \dfrac{19}{2}\right)$

In Exercises 31–72, solve each equation. If an equation has no solution, so indicate.

31. $2x + 7 = 10 - x$

32. $9a - 3 = 15 + 3a$

33. $\dfrac{5}{3}z - 8 = 7$

34. $\dfrac{4}{3}y + 12 = -4$

35. $\dfrac{z}{5} + 2 = 4$

36. $\dfrac{3p}{7} - p = -4$

37. $\dfrac{3x - 2}{3} = 2x + \dfrac{7}{3}$

38. $\dfrac{7}{2}x + 5 = x + \dfrac{15}{2}$

39. $5(x - 2) = 2x + 8$

40. $5(r - 4) = -5(r - 4)$

41. $2(2x + 1) - \dfrac{3x}{2} = \dfrac{-3(4 + x)}{2}$

42. $(x - 2)(x - 3) = (x + 3)(x + 4)$

43. $7(2x + 5) - 6(x + 8) = 7$

44. $(t + 1)(t - 1) = (t + 2)(t - 3) + 4$

45. $(x - 2)(x + 5) = (x - 3)(x + 2)$

46. $\dfrac{3x + 1}{20} = \dfrac{1}{2}$

47. $\dfrac{3}{2}(3x - 2) - 10x - 4 = 0$

48. $a(a - 3) + 5 = (a - 1)^2$

49. $x(x + 2) = (x + 1)^2 - 1$

50. $\dfrac{3 + x}{3} + \dfrac{x + 7}{2} = 4x + 1$

51. $\dfrac{(y + 2)^2}{3} = y + 2 + \dfrac{y^2}{3}$

52. $2x - \dfrac{7}{6} + \dfrac{x}{6} = \dfrac{4x + 3}{6}$

53. $2(s + 2) + (s + 3)^2 = s(s + 5) + 2\left(\dfrac{17}{2} + s\right)$

54. $\dfrac{3}{x} + \dfrac{1}{2} = \dfrac{4}{x}$

55. $\dfrac{2}{x + 1} + \dfrac{1}{3} = \dfrac{1}{x + 1}$

56. $\dfrac{3}{x - 2} + \dfrac{1}{x} = \dfrac{3}{x - 2}$

57. $\dfrac{9t + 6}{t(t + 3)} = \dfrac{7}{t + 3}$

58. $x + \dfrac{2(-2x + 1)}{3x + 5} = \dfrac{3x^2}{3x + 5}$

59. $\dfrac{2}{(a - 7)(a + 2)} = \dfrac{4}{(a + 3)(a + 2)}$

60. $\dfrac{2}{n - 2} + \dfrac{1}{n + 1} = \dfrac{1}{n^2 - n - 2}$

61. $\dfrac{2x + 3}{x^2 + 5x + 6} + \dfrac{3x - 2}{x^2 + x - 6} = \dfrac{5x - 2}{x^2 - 4}$

62. $\dfrac{3x}{x^2 + x} - \dfrac{2x}{x^2 + 5x} = \dfrac{x + 2}{x^2 + 6x + 5}$

63. $\dfrac{3x + 5}{x^3 + 8} + \dfrac{3}{x^2 - 4} = \dfrac{2(3x - 2)}{(x - 2)(x^2 - 2x + 4)}$

64. $\dfrac{1}{n + 8} - \dfrac{3n - 4}{5n^2 + 42n + 16} = \dfrac{1}{5n + 2}$

65. $\dfrac{1}{11 - n} - \dfrac{2(3n - 1)}{-7n^2 + 74n + 33} = \dfrac{1}{7n + 3}$

66. $\dfrac{4}{a^2 - 13a - 48} - \dfrac{2}{a^2 - 18a + 32} = \dfrac{1}{a^2 + a - 6}$

67. $\dfrac{5}{y + 4} + \dfrac{2}{y + 2} = \dfrac{6}{y + 2} - \dfrac{1}{y^2 + 6y + 8}$

68. $\dfrac{6}{2a - 6} - \dfrac{3}{3 - 3a} = \dfrac{1}{a^2 - 4a + 3}$

69. $\dfrac{3y}{6 - 3y} + \dfrac{2y}{2y + 4} = \dfrac{8}{4 - y^2}$

70. $\dfrac{3 + 2a}{a^2 + 6 + 5a} - \dfrac{2 - 3a}{a^2 - 6 + a} = \dfrac{5a - 2}{a^2 - 4}$

71. $\dfrac{a}{a + 2} - 1 = -\dfrac{3a + 2}{a^2 + 4a + 4}$

72. $\dfrac{x - 1}{x + 3} + \dfrac{x - 2}{x - 3} = \dfrac{1 - 2x}{3 - x}$

In Exercises 73–92, solve each formula for the specified variable.

73. $k = 2.2p$; p

74. $p = 2l + 2w$; w

75. $A = \dfrac{1}{2}h(b_1 + b_2)$; b_2

76. $V = \dfrac{1}{3}\pi r^2 h$; h

77. $V = \dfrac{1}{3}\pi r^2 h$; r^2

78. $z = \dfrac{x - \mu}{\sigma}$; μ

79. $P_n = L + \dfrac{si}{f}$; s

80. $P_n = L + \dfrac{si}{f}$; f

81. $F = \dfrac{mMg}{r^2}$; m

82. $\dfrac{1}{f} = \dfrac{1}{p} + \dfrac{1}{q}$; f

83. $\dfrac{x}{a} + \dfrac{y}{b} = 1$; y

84. $\dfrac{x}{a} - \dfrac{y}{b} = 1$; a

85. $\dfrac{1}{r} = \dfrac{1}{r_1} + \dfrac{1}{r_2}$; r

86. $\dfrac{1}{r} = \dfrac{1}{r_1} + \dfrac{1}{r_2}$ for r_1

87. $l = a + (n - 1)d$; n

88. $l = a + (n - 1)d$; d

89. $a = (n - 2)\dfrac{180}{n}$; n

90. $S = \dfrac{a - lr}{1 - r}$; a

91. $R = \dfrac{1}{\dfrac{1}{r_1} + \dfrac{1}{r_2} + \dfrac{1}{r_3}}$; r_1

92. $R = \dfrac{1}{\dfrac{1}{r_1} + \dfrac{1}{r_2} + \dfrac{1}{r_3}}$; r_3

DISCOVERY AND WRITING

93. Explain why a linear equation always has exactly one root.

94. Define an extraneous solution and explain how such a solution occurs.

REVIEW   *In Exercises 95–102, simplify each expression. Use absolute value symbols when necessary.*

95. $(25x^2)^{1/2}$

96. $\left(\dfrac{25p^2}{16q^4}\right)^{1/2}$

97. $\left(\dfrac{125x^3}{8y^6}\right)^{-2/3}$

98. $\left(-\dfrac{27y^3}{1000x^6}\right)^{1/3}$

99. $\sqrt{25y^2}$

100. $\sqrt[3]{-125y^9}$

101. $\sqrt[4]{\dfrac{a^4 b^{12}}{z^8}}$

102. $\sqrt[5]{\dfrac{x^{10}y^5}{z^{15}}}$

2.2 Applications of Linear Equations

■ NUMBER PROBLEMS ■ GEOMETRIC PROBLEMS ■ INVESTMENT PROBLEMS ■ BREAK-POINT
ANALYSIS ■ SHARED WORK PROBLEMS ■ MIXTURE PROBLEMS ■ UNIFORM MOTION PROBLEMS

In this section, we will apply equation-solving techniques to solve problems. The following list of steps provides a strategy to follow.

Strategy for Problem Solving

1. Analyze the problem to see what information is given and what you are to find. Often, drawing a diagram or making a table will help you visualize the facts.

2. Pick a variable to represent the quantity that is to be found, and write a sentence telling what that variable represents. Express all other quantities mentioned in the problem as expressions involving this single variable.

3. Find a way to express a quantity in two different ways. This might involve a formula from geometry, finance, or physics.

4. Form an equation indicating that the two quantities found in Step 3 are equal.

5. Solve the equation.

6. Answer the questions asked in the problem.

7. Check the answers in the words of the problem.

This list does not apply to all situations, but it can be applied to a wide range of problems with only slight modifications.

■ NUMBER PROBLEMS

EXAMPLE 1

Averaging grades   A student has scores of 74%, 78%, and 70% on three exams. What score is needed on a fourth exam for the student to earn an average grade of 80%?

Solution   We can let x represent the required grade on the fourth exam. The average grade will be one-fourth of the sum of the four grades, and we know this average is to be 80.

The average of the four grades	=	the required average grade
$\dfrac{74 + 78 + 70 + x}{4}$	=	80

We can solve this equation for x.

$$\frac{222 + x}{4} = 80 \qquad 74 + 78 + 70 = 222.$$

$$222 + x = 320 \qquad \text{Multiply both sides by 4.}$$

$$x = 98 \qquad \text{Subtract 222 from both sides.}$$

To earn an average of 80%, the student must score 98% on the fourth exam.   ■

■ GEOMETRIC PROBLEMS

EXAMPLE 2   **Installing fencing**   A city ordinance requires a man to install a fence around the swimming pool shown in Figure 2-1. If he wants the border around the pool to be of uniform width, and he has 154 feet of fencing, find the width of the border.

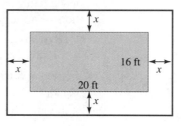

FIGURE 2-1

Solution   We can let x represent the width of the border. The distance around the large rectangle, called its **perimeter,** is given by the formula $P = 2l + 2w$, where l is the length, $20 + 2x$, and w is the width, $16 + 2x$. Since the man has 154 feet of fencing, the perimeter will be 154 feet. We substitute these values into the formula for perimeter.

$P = 2l + 2w$	The formula for the perimeter of a rectangle.
$154 = 2(20 + 2x) + 2(16 + 2x)$	Substitute 154 for P, $20 + 2x$ for l and $16 + 2x$ for w.
$154 = 40 + 4x + 32 + 4x$	Use the distributive property to remove parentheses.
$154 = 72 + 8x$	Combine like terms.
$82 = 8x$	Subtract 72 from both sides.
$10\frac{1}{4} = x$	Divide both sides by 8.

The border will be $10\frac{1}{4}$ feet wide.   ■

■ INVESTMENT PROBLEMS

EXAMPLE 3   **Supplemental income**   A woman invested $10,000, part at 9% and the rest at 14%. The annual income from these two investments was $1275. How much was invested at each rate?

Solution   We can let x represent the amount invested at 9%. Then $10,000 - x$ represents the amount invested at 14%. Since the annual income from any investment is the product of the interest rate and the amount invested, we have the following information.

Type of investment	Rate	Amount invested	Interest earned
9% investment	0.09	x	$0.09x$
14% investment	0.14	$10,000 - x$	$0.14(10,000 - x)$

The total income from these two investments can be expressed in two ways: as $1275 and as the sum of the incomes of the two investments.

The income from the 9% investment		the income from the 14% investment		the total investment
0.09x	+	0.14(10,000 − x)	=	1275

We can solve this equation for x.

$$0.09x + 0.14(10,000 - x) = 1275$$
$$9x + 14(10,000 - x) = 127,500 \qquad \text{Multiply both sides by 100 to eliminate the decimal points.}$$
$$9x + 140,000 - 14x = 127,500 \qquad \text{Use the distributive property to remove parentheses.}$$
$$-5x + 140,000 = 127,500 \qquad \text{Combine like terms.}$$
$$-5x = -12,500 \qquad \text{Subtract 140,000 from both sides.}$$
$$x = 2500 \qquad \text{Divide both sides by } -5.$$

The amount invested at 9% was $2500, and the amount invested at 14% was $7500 ($10,000 − $2500). These amounts are correct, because 9% of $2500 is $225, and 14% of $7500 is $1050, and the sum of these amounts is $1275. ■

■ BREAK-POINT ANALYSIS

Running a machine involves two costs—**setup costs** and **unit costs.** Setup costs include the cost of installing a machine and preparing it to do a job. Unit cost is the cost to manufacture one item, which includes the costs of material and labor.

EXAMPLE 4   **Manufacturing**   Suppose that one machine has a setup cost of $400 and a unit cost of $1.50, while a second machine has a setup cost of $500 and a unit cost of $1.25. Find the **break point** (the number of units manufactured at which the cost on each machine is the same).

George Polya
(1888–1985)

Polya, a Hungarian, became a professor of mathematics at Stanford University. His approach to problem solving made him very popular with faculty and students. His book *How to Solve It* became a best-seller. His problem-solving approach involves four steps:
1. Understand the problem.
2. Devise a plan.
3. Carry out the plan.
4. Check back.

Solution   We can let x represent the number of items to be manufactured. The cost c_1 of using machine 1 is

$$c_1 = 400 + 1.5x$$

and the cost c_2 of using machine 2 is

$$c_2 = 500 + 1.25x$$

The break point occurs when these two costs are equal.

The cost of using machine 1	=	the cost of using machine 2
$400 + 1.5x$	=	$500 + 1.25x$

We can solve this equation for x.

$$400 + 1.5x = 500 + 1.25x$$
$$1.5x = 100 + 1.25x \qquad \text{Subtract 400 from both sides.}$$
$$0.25x = 100 \qquad \text{Subtract } 1.25x \text{ from both sides.}$$
$$x = 400 \qquad \text{Divide both sides by 0.25.}$$

The break point is 400 units. This result is correct, because it will cost the same amount to manufacture 400 units with either machine.

$$c_1 = \$400 + \$1.5(400) = \$1000 \qquad \text{and} \qquad c_2 = \$500 + \$1.25(400) = \$1000 \quad \blacksquare$$

■ SHARED WORK PROBLEMS

EXAMPLE 5   **Road construction**   The Tollway Authority needs to pave 100 miles of interstate highway before freezing temperatures come in about 60 days. Sjostrom and Sons has estimated that it can do the job in 110 days. Scandroli and Sons has estimated that it can do the job in 140 days. If the authority hires both contractors, will the job get done in time?

Solution   We can let n represent the number of days it will take to pave the highway if both contractors are hired. In one day, the contractors working together can do $\frac{1}{n}$ of the job. In one day, Sjostrom can do $\frac{1}{110}$ of the job. In one day, Scandroli can do $\frac{1}{140}$ of the job. The work that they can do together in one day is the sum of what each can do in one day.

The part of the highway Sjostrom can pave in one day	+	the part of the highway Scandroli can pave in one day	=	the part of the highway they can pave together in one day
$\dfrac{1}{110}$	+	$\dfrac{1}{140}$	=	$\dfrac{1}{n}$

We can solve this equation for n.

$$\frac{1}{110} + \frac{1}{140} = \frac{1}{n}$$

$$(110)(140)n\left(\frac{1}{110} + \frac{1}{140}\right) = (110)(140)n\left(\frac{1}{n}\right)$$

Multiply both sides by $(110)(140)n$ to eliminate the fractions.

$$\frac{(110)(140)n}{110} + \frac{(110)(140)n}{140} = \frac{(110)(140)n}{n}$$

Use the distributive property to remove parentheses.

$$140n + 110n = 15{,}400$$

$\frac{110}{110} = 1$, $\frac{140}{140} = 1$, and $\frac{n}{n} = 1$.

$$250n = 15{,}400$$

Combine like terms.

$$n = 61.6$$

Divide both sides by 250.

It will take the contractors about 62 days to pave the highway. If the Tollway Authority is lucky, the job will be done in time. ∎

■ MIXTURE PROBLEMS

EXAMPLE 6

Mixing milk   A container is partially filled with 20 liters of whole milk containing 4% butterfat. How much 1% milk must be added to obtain a mixture that is 2% butterfat?

Solution   See Figure 2-2. We can let l represent the number of liters of 1% milk that must be added to the whole milk. Then $20 + l$ represents the number of liters in the final mixture. The butterfat in the final mixture is the sum of the butterfat in the whole milk and the butterfat in the 1% milk to be added.

The butterfat in the 4% milk	+	the butterfat in the 1% milk	=	the butterfat in the 2% milk
4% of 20 liters	+	1% of l liters	=	2% of $(20 + l)$ liters

We can solve this equation for l.

$$0.04(20) + 0.01(l) = 0.02(20 + l)$$

$$4(20) + l = 2(20 + l)$$ Multiply both sides by 100.

$$80 + l = 40 + 2l$$ Remove parentheses.

$$40 = l$$ Subtract 40 and l from both sides.

To dilute the 20 liters of 4% milk to a 2% mixture, 40 liters of 1% milk must be added. To check, we note that the final mixture contains $0.02(60) = 1.2$ liters of pure butterfat, and that this is equal to the amount of pure butterfat in the 4% milk and the 1% milk: $0.4(20) + 0.01(40) = 1.2$ liters.

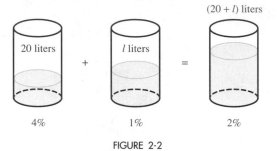

FIGURE 2-2

■ UNIFORM MOTION PROBLEMS

EXAMPLE 7

Catching a car A man leaves home driving at the rate of 50 miles per hour. When his daughter discovers that he has forgotten his wallet, she drives after him at the rate of 65 miles per hour. How long will it take her to catch her dad if he has a 15-minute head start?

Solution Uniform motion problems are based on the formula $d = rt$, where d is the distance, r is the rate, and t is the time. We can organize the information given in the problem in a chart like the one shown in Figure 2-3. In the chart, t represents the number of hours the daughter must drive to overtake her father. Because the father has a fifteen-minute, or $\frac{1}{4}$ hour, head start, he has been on the road for $\left(t + \frac{1}{4}\right)$ hours.

	d	$=$	r	$\cdot$	t
Man	$50\left(t + \frac{1}{4}\right)$		50		$t + \frac{1}{4}$
Daughter	$65t$		65		t

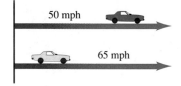

FIGURE 2-3

We can set up the following equation and solve it for t.

The distance the man drives	$=$	the distance the daughter drives
$50\left(t + \dfrac{1}{4}\right)$	$=$	$65t$

We can solve this equation for t.

$$50\left(t + \frac{1}{4}\right) = 65t$$

$$50t + \frac{25}{2} = 65t$$

$$\frac{25}{2} = 15t$$

$$\frac{5}{6} = t$$

It will take the daughter $\frac{5}{6}$ hours, or 50 minutes, to overtake her father. ∎

EXERCISE 2.2

VOCABULARY AND CONCEPTS   *In Exercises 1–6, fill in the blank to make a true statement.*

1. To average n scores, _____ the scores and divide by n.

2. The formula for the _____ of a rectangle is $P = 2l + 2w$.

3. The interest earned on an investment is the product of the interest rate and the _____ invested.

4. The number of units manufactured at which the cost on two machines is equal is called the _____.

5. Distance traveled is the product of the ____ and the ____.

6. 5% of 30 liters is ___ liters.

PRACTICE   *In Exercises 7–52, solve each problem.*

7. **Test scores**   Juan scored 5 points higher on his midterm than his first exam and 13 points higher on his final than his first exam. If his mean (average) score was 90, what did he score on the first exam?

Human development with special needs	82
Assessment	90
Program development and instruction	?
Professional knowledge and legal issues	78
AVERAGE SCORE	86

8. **Test scores**   Sally took four tests in science class. On each successive test, her score improved by 3 points. If her mean score was 69.5%, what did she score on the first test?

9. **Teacher certification**   On the Illinois certification test for teachers specializing in learning disabilities, a teacher earned the following scores. What was the teacher's score in program development?

10. **Golfing**   Par on a golf course is 72. If a professional golfer shot rounds of 76, 68, and 70 in a tournament, what will she need to shoot on the final round to average par?

11. **Replacing locks**   A locksmith charges $40 plus $28 for each lock installed. How many locks can be replaced for $236?

12. **Delivering ads**   A college student earns $20 per day delivering advertising brochures door-to-door, plus 75¢ for each person he interviews. How many people did he interview on a day when he earned $56?

13. **Width of a picture frame**   The picture frame with the dimensions shown in Illustration 1 was built with 14 feet of framing material. Find its width.

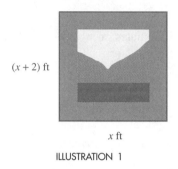

$(x + 2)$ ft

x ft

ILLUSTRATION 1

14. **Fencing a garden**   If a gardener fences in the total rectangular area shown in Illustration 2 instead of just the square area, he will need twice as much fencing to enclose the garden. How much fencing will he need?

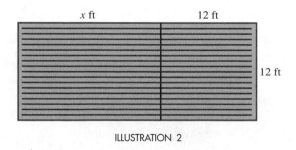

x ft   12 ft

12 ft

ILLUSTRATION 2

15. **Wading pool dimensions**   The area of the triangular swimming pool shown in Illustration 3 is doubled by adding a rectangular wading pool. Find the dimensions of the pool. (*Hint:* The area of a triangle = $\frac{1}{2}bh$, and the area of a rectangle = lw.)

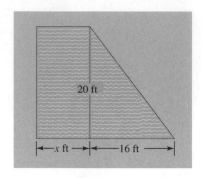

20 ft

x ft   16 ft

ILLUSTRATION 3

16. **House construction**   A builder wants to install a triangular window with the angles shown in Illustration 4. What angles will he have to cut to make the window fit? (*Hint:* The sum of the angles in a triangle equals 180°.)

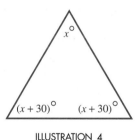

$x°$

$(x + 30)°$ $(x + 30)°$

ILLUSTRATION 4

17. **Length of a living room**   If a carpenter adds a porch with dimensions shown in Illustration 5 to the living room, the living area will be increased by 50%. Find the length of the living room.

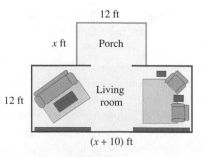

12 ft

x ft   Porch

12 ft   Living room

$(x + 10)$ ft

ILLUSTRATION 5

18. **Depth of water in a trough** The trough in Illustration 6 has a cross-sectional area of 54 square inches. Find the depth, d, of the trough. (*Hint:* Area of a trapezoid = $\frac{1}{2}h(b_1 + b_2)$.)

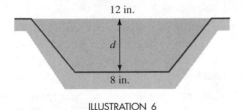

ILLUSTRATION 6

19. **Investment problem** An executive invests $22,000, some at 7% and some at 6% annual interest. If he receives an annual return of $1420, how much is invested at each rate?

20. **Financial planning** After inheriting some money, a woman wants to invest enough to have an annual income of $5000. If she can invest $20,000 at 9% annual interest, how much more will she have to invest at 7% to achieve her goal?

Type	Rate	Amount	Income
9% investment	0.09	20,000	.09(20,000)
7% investment	0.07	x	.07x

21. **Ticket sales** A full-price ticket for a college basketball game costs $2.50, and a student ticket costs $1.75. If 585 tickets were sold, and the total receipts were $1217.25, how many tickets were student tickets?

22. **Ticket sales** Of the 800 tickets sold to a movie, 480 were full-price tickets costing $3 each. If the gate receipts were $2080, what did a student ticket cost?

23. **Investment problem** A woman invests $37,000, part at 8% and the rest at $9\frac{1}{2}$% annual interest. If the $9\frac{1}{2}$% investment provides $452.50 more income than the 8% investment, how much is invested at each rate?

24. **Investment problem** Equal amounts are invested at 6%, 7%, and 8% annual interest. If the three investments yield a total of $2037 annual interest, find the total investment.

25. **Discount** After being discounted 20%, a radio sells for $63.96. Find the original price.

26. **Markup** A merchant increases the wholesale cost of a washing machine by 30% to determine the selling price. If the washer sells for $588.90, find the wholesale cost.

27. **Break-even analysis** A machine to mill a brass plate has setup cost of $600 and a unit cost of $3 for each plate manufactured. A bigger machine has a setup cost of $800 but a unit cost of only $2 for each plate manufactured. Find the break point.

28. **Break-even analysis** A machine to manufacture fasteners has a setup cost of $1200 and a unit cost of $0.005 for each fastener manufactured. A newer machine has a setup cost of $1500 but a unit cost of only $0.0015 for each fastener manufactured. Find the break point.

29. **Computer sales** A computer store has fixed costs of $8925 per month and a unit cost of $850 for every computer it sells. If the store can sell all the computers it can get for $1275 each, how many must be sold for the store to break even? (*Hint:* The break-even point occurs when costs equal income.)

30. **Restaurant management** A restaurant has fixed costs of $137.50 per day and an average unit cost of $4.75 for each meal served. If a typical check is $6, how many customers must eat at the restaurant each day for the owner to make a profit?

31. **Mowing lawns** If a woman can mow a yard with a lawn tractor in 2 hours, and her husband can mow the same lawn with a push mower in 4 hours, how long will it take to mow the lawn if they work together?

32. **Filling a swimming pool** A garden hose can fill a swimming pool in 3 days, and a larger hose can fill the pool in 2 days. How long will it take to fill the pool if both hoses are used?

33. **Filling a swimming pool** An empty swimming pool can be filled in 10 hours. When full, the pool can be drained in 19 hours. How long will it take to fill the empty pool if the drain is left open?

34. Preparing seafood Sam stuffs shrimp in his job as a seafood chef. He can stuff 1000 shrimp in 6 hours. When his sister helps him, they can stuff 1000 shrimp in 4 hours. If Sam gets sick, how long will it take his sister to stuff 500 shrimp?

35. Winterizing a car A car radiator has a 6-liter capacity. If the liquid in the radiator is 40% antifreeze, how much liquid must be replaced with pure antifreeze to bring the mixture up to a 50% solution?

36. Mixing milk If a bottle holding 3 liters of milk contains $3\frac{1}{2}$% butterfat, how much skimmed milk must be added to dilute the milk to 2% butterfat?

37. Preparing solutions A nurse has 1 liter of a solution that is 20% alcohol. How much pure alcohol must she add to bring the solution up to a 25% concentration?

38. Diluting solutions If there are 400 cubic centimeters of a chemical in 1 liter of solution, how many cubic centimeters of water must be added to dilute it to a 25% solution? (*Hint:* 1000 cc = 1 liter.)

39. Cleaning a swimming pool A swimming pool contains 15,000 gallons of water. How many gallons of chlorine must be added to "shock the pool" and bring the water to a $\frac{3}{100}$% solution?

40. Mixing gasolines A new automobile engine can run on a mixture of gasoline and a substitute fuel. If gas costs $1.50 per gallon and the substitute fuel costs 40¢ per gallon, what percent of a mixture must be substitute fuel to bring the cost down to $1 per gallon?

41. Evaporation How many liters of water must evaporate to turn 12 liters of a 24% salt solution into a 36% solution?

42. Preparing medicine A doctor prescribes an ointment that is 2% hydrocortisone. A pharmacist has 1% and 5% concentrations in stock. How much of each should the pharmacist use to make a 1-ounce tube?

43. Driving rates John drove to a distant city in 5 hours. When he returned, there was less traffic, and the trip took only 3 hours. If John averaged 26 mph

faster on the return trip, how fast did he drive each way?

44. Distance problem Suzi drove home at 60 mph, but her brother Jim, who left at the same time, could drive at only 48 mph. When Suzi arrived, Jim still had 60 miles to go. How far did Suzi drive?

45. Distance problem Two cars leave Pima Community College traveling in opposite directions. One car travels at 60 mph and the other at 64 mph. In how many hours will they be 310 miles apart?

46. Bank robbery Some bank robbers leave town, speeding at 70 miles per hour. Ten minutes later, the police give chase, traveling at 78 miles per hour. How long will it take the police to overtake the robbers?

47. Jogging problem Two cross-country runners are 440 yards apart and are running toward each other, one at 8 mph and the other at 10 mph. In how many seconds will they meet?

48. Driving rates One morning, John drove 5 hours before stopping to eat. After lunch, he increased his speed by 10 mph. If he completed a 430-mile trip in 8 hours of driving time, how fast did he drive in the morning?

49. Boating problem A motorboat goes 5 miles upstream in the same time it requires to go 7 miles downstream. If the river flows at 2 miles per hour, find the speed of the boat in still water.

50. Wind velocity A plane can fly 340 mph in still air. If it can fly 200 miles downwind in the same amount of time it can fly 140 miles upwind, find the velocity of the wind.

51. Feeding cattle A farmer wants to mix 2400 pounds of cattle feed that is to be 14% protein. Barley (11.7% protein) will make up 25% of the mixture. The remaining 75% will be made up of oats (11.8% protein) and soybean oil meal (44.5% protein). How many pounds of each will he use?

52. Feeding cattle If the farmer in Exercise 51 wants only 20% of the mixture to be barley, how many pounds of each should he use?

In Exercises 53–54, use a calculator to help solve each problem.

53. Machine tool design 712.51 cubic millimeters of material was removed by drilling the blind hole shown in Illustration 7. Find the depth of the hole. (*Hint:* The volume of a cylinder is given by $V = \pi r^2 h$.)

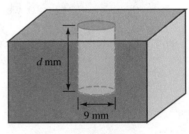

ILLUSTRATION 7

54. Architecture The Norman window with dimensions as shown in Illustration 8 is a rectangle topped by a semicircle. If the area of the window is 68.2 square feet, find its height h.

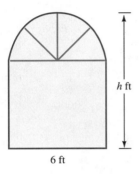

6 ft

ILLUSTRATION 8

DISCOVERY AND WRITING

55. Which type of problem was easiest for you to solve? Why?

56. Which type of problem was hardest for you to solve? Why?

REVIEW *In Exercises 57–62, factor each expression.*

57. $x^2 - 2x - 63$

58. $2x^2 + 11x - 21$

59. $9x^2 - 12x - 5$

60. $9x^2 - 2x - 7$

61. $x^2 + 6x + 9$

62. $x^2 - 10x + 25$

2.3 Quadratic Equations

■ QUADRATIC EQUATIONS ■ COMPLETING THE SQUARE ■ THE QUADRATIC FORMULA
■ THE DISCRIMINANT ■ WRITING EQUATIONS IN QUADRATIC FORM ■ FORMULAS

■ QUADRATIC EQUATIONS

Equations such as $2x^2 - 11x - 21 = 0$ and $3x^2 - x - 2 = 0$ are called *quadratic* or *second-degree* equations.

Quadratic Equations
A **quadratic equation** is an equation that can be written in the form $ax^2 + bx + c = 0$, where a, b, and c are real numbers and $a \neq 0$.

To solve quadratic equations by factoring, we use the following theorem.

> **Zero-Factor Theorem**
> If a and b are real numbers, and if $ab = 0$, then
> $$a = 0 \quad \text{or} \quad b = 0$$

Proof To prove the zero-factor theorem, we suppose that $ab = 0$. If $a = 0$, we are finished, because at least one of a or b is 0.

   If $a \neq 0$, then a has a reciprocal $\frac{1}{a}$, and we can multiply both sides of the equation $ab = 0$ by $\frac{1}{a}$ to obtain

$$ab = 0$$

$$\frac{1}{a}(ab) = \frac{1}{a}(0) \qquad \text{Multiply both sides by } \tfrac{1}{a}.$$

$$\left(\frac{1}{a} \cdot a\right)b = 0 \qquad \text{Use the associative property to group } \tfrac{1}{a} \text{ and } a \text{ together.}$$

$$1b = 0 \qquad \tfrac{1}{a} \cdot a = 1.$$

$$b = 0$$

Thus, if $a \neq 0$, then b must be 0, and the theorem is proved. □

EXAMPLE 1   Solve the quadratic equation $2x^2 - 9x - 35 = 0$.

Solution   The left-hand side can be factored and written as

$$(2x + 5)(x - 7) = 0$$

If either factor is 0, the product will be 0. So we can use the zero-factor theorem and set each factor equal to 0. Then we can solve for x.

$$2x + 5 = 0 \qquad \text{or} \qquad x - 7 = 0$$
$$2x = -5 \qquad\qquad\qquad x = 7$$
$$x = -\frac{5}{2}$$

Because $(2x + 5)(x - 7) = 0$ only if one of its factors is 0, $-\frac{5}{2}$ and 7 are the only solutions of the equation.

   Verify that each one satisfies the equation. ∎

Self Check   Solve $6x^2 + 7x - 3 = 0$.

Answer $\dfrac{1}{3}, -\dfrac{3}{2}$

When the quadratic expression in a quadratic equation factors, the factoring method is easy. However, many quadratic expressions do not factor over the set of integers. For example, the left-hand side of $x^2 - 5x + 3 = 0$ is a prime polynomial and cannot be factored over the set of integers.

To develop a method to solve these equations, we consider the equation $x^2 = c$. If c is positive, its two real roots can be found by adding $-c$ to both sides, factoring the binomial $x^2 - c$, setting each factor equal to 0, and solving for x.

$$x^2 = c$$
$$x^2 - c = 0 \qquad \text{Subtract } c \text{ from both sides.}$$
$$x^2 - \left(\sqrt{c}\right)^2 = 0 \qquad \left(\sqrt{c}\right)^2 = c.$$
$$\left(x - \sqrt{c}\right)\left(x + \sqrt{c}\right) = 0 \qquad \text{Factor the difference of two squares.}$$
$$x - \sqrt{c} = 0 \qquad \text{or} \qquad x + \sqrt{c} = 0 \qquad \text{Set each factor equal to 0.}$$
$$x = \sqrt{c} \qquad\qquad x = -\sqrt{c}$$

The roots of $x^2 = c$ are $x = \sqrt{c}$ and $x = -\sqrt{c}$. This fact is called the **square root property.**

> **Square Root Property**
> If $c > 0$, the equation $x^2 = c$ has two real roots:
> $$x = \sqrt{c} \qquad \text{or} \qquad x = -\sqrt{c}$$

EXAMPLE 2   Solve $x^2 - 8 = 0$.

Solution   We solve for x^2 and apply the square root property.

$$x^2 - 8 = 0$$
$$x^2 = 8$$
$$x = \sqrt{8} \qquad \text{or} \qquad x = -\sqrt{8}$$
$$x = 2\sqrt{2} \qquad\qquad x = -2\sqrt{2} \qquad\qquad \sqrt{8} = \sqrt{4}\sqrt{2} = 2\sqrt{2}.$$

Verify that each root satisfies the equation. ∎

Self Check   Solve $x^2 - 12 = 0$.
Answer $2\sqrt{3}, -2\sqrt{3}$

EXAMPLE 3   Solve $(x + 4)^2 = 1$.

Solution
$$(x + 4)^2 = 1$$
$$x + 4 = \sqrt{1} \qquad \text{or} \qquad x + 4 = -\sqrt{1}$$
$$x + 4 = 1 \qquad\qquad x + 4 = -1$$
$$x = -3 \qquad\qquad x = -5$$

Verify that each root satisfies the equation. ∎

Self Check   Solve $(x + 5)^2 = 4$.

Answer $-3, -7$

■ COMPLETING THE SQUARE

We now discuss another method to solve quadratic equations called **completing the square.** This method is based on the following products:

$$x^2 + 2ax + a^2 = (x + a)^2 \qquad \text{and} \qquad x^2 - 2ax + a^2 = (x - a)^2$$

The trinomials $x^2 + 2ax + a^2$ and $x^2 - 2ax + a^2$ are perfect square trinomials, because each one factors as the square of a binomial. In each case, the coefficient of the first term is 1. If we take one-half of the coefficient of x in the middle term and square it, we obtain the third term.

$$\left[\frac{1}{2}(2a)\right]^2 = a^2 \qquad \text{and} \qquad \left[\frac{1}{2}(-2a)\right]^2 = (-a)^2 = a^2$$

This suggests that to make the binomial $x^2 + bx$ a perfect triomial square, we find one-half of b, square it, and add the result to the binomial. For example, to make $x^2 + 10x$ a perfect square trinomial, we find one-half of 10 to get 5, square 5 to get 25, and add 25 to $x^2 + 10x$.

$$x^2 + 10x + \left[\frac{1}{2}(10)\right]^2 = x^2 + 10x + (5)^2$$
$$= x^2 + 10x + 25$$
$$= (x + 5)^2$$

To make $x^2 - 11x$ a perfect square trinomial, we find one-half of -11 to get $-\frac{11}{2}$, square $-\frac{11}{2}$ to get $\frac{121}{4}$, and add $\frac{121}{4}$ to $x^2 - 11x$.

$$x^2 - 11x + \left[\frac{1}{2}(-11)\right]^2 = x^2 - 11x + \left(-\frac{11}{2}\right)^2$$
$$= x^2 - 11x + \frac{121}{4}$$
$$= \left(x - \frac{11}{2}\right)^2$$

To solve a quadratic equation in x by completing the square, we follow these steps.

Method of Completing the Square

1. If the coefficient of x^2 is not 1, change it to a 1 by dividing both sides of the equation by the coefficient of x^2.

2. If necessary, add a number to both sides of the equation to get the constant on the right-hand side of the equation.

3. Complete the square on x:

   a. Identify the coefficient of x and take one-half of it.

   b. Square the result obtained in part **a.**

   c. Add the number found in part **b.** to both sides of the equation.

4. Factor the perfect square trinomial and combine like terms.

5. Solve the resulting quadratic equation by using the square root property.

To use completing the square to solve the equation $x^2 - 10x + 24 = 0$, we note that the coefficient of x^2 is 1. We move on to Step 2 and subtract 24 from both sides to get the constant term on the right-hand side of the equal sign.

$$x^2 - 10x = -24$$

We then complete the square by adding $\left[\frac{1}{2}(-10)\right]^2 = 25$ to both sides.

$$x^2 - 10x + \mathbf{25} = -24 + \mathbf{25}$$
$$x^2 - 10x + 25 = 1 \qquad \text{Simplify on the right-hand side.}$$

We then factor the perfect square trinomial on the left-hand side.

$$(x - 5)^2 = 1$$

Finally, we use the square root property to solve this equation.

$$x - 5 = 1 \qquad \text{or} \qquad x - 5 = -1$$
$$x = 6 \qquad\qquad\qquad x = 4$$

EXAMPLE 4   Use completing the square to solve $x^2 + 4x - 5 = 0$.

Solution   Here the coefficient of x^2 is already 1. We move on to Step 2 and add 5 to both sides to isolate the binomial $x^2 + 4x$.

$$x^2 + 4x = 5$$

We then find the number to add to both sides to complete the square: one-half of 4 (the coefficient of x) is 2, and $2^2 = 4$. This is the number to add to both sides to complete the square.

$$x^2 + 4x + \mathbf{4} = 5 + \mathbf{4} \qquad\qquad \text{Add 4 to both sides.}$$
$$x^2 + 4x + 4 = 9$$
$$(x + 2)^2 = 9 \qquad\qquad \text{Factor } x^2 + 4x + 4.$$
$$x + 2 = 3 \qquad \text{or} \qquad x + 2 = -3 \qquad \text{Use the square root property.}$$
$$x = 1 \qquad\qquad\qquad x = -5$$

Verify that each root satisfies the original equation.   ∎

Self Check   Solve $x^2 - 2x - 8 = 0$.

Answer $4, -2$

EXAMPLE 5   Solve $x(x + 3) = 2$.

Solution   We remove parentheses to get

$$x^2 + 3x = 2$$

Here the coefficient of x^2 is 1. We move on to Step 2 and find the number to be added to both sides to complete the square: One-half of 3 (the coefficient of x) is $\frac{3}{2}$, and the square of $\frac{3}{2}$ is $\frac{9}{4}$. This is the number to add to both sides.

$$x^2 + 3x + \frac{9}{4} = 2 + \frac{9}{4} \qquad \text{Add } \tfrac{9}{4} \text{ to both sides.}$$

$$\left(x + \frac{3}{2}\right)^2 = \frac{17}{4} \qquad \text{Factor } x^2 + 3x + \tfrac{9}{4}.$$

$$x + \frac{3}{2} = \frac{\sqrt{17}}{2} \qquad \text{or} \qquad x + \frac{3}{2} = -\frac{\sqrt{17}}{2} \qquad \text{Use the square root property.}$$

$$x = \frac{-3 + \sqrt{17}}{2} \qquad\qquad x = \frac{-3 - \sqrt{17}}{2} \qquad \text{Subtract } \tfrac{3}{2} \text{ from both sides.}$$

Verify that each root satisfies the original equation. ■

Self Check   Solve $x(x + 5) = 1$.

Answer $x = \dfrac{-5 \pm \sqrt{29}}{2}$

EXAMPLE 6   Solve $6x^2 + 5x - 6 = 0$.

Solution   We begin by dividing both sides of the equation by 6 to make the coefficient of x^2 equal to 1. Then we proceed as follows:

$$6x^2 + 5x - 6 = 0$$

$$x^2 + \frac{5}{6}x - 1 = 0 \qquad \text{Divide both sides by 6.}$$

$$x^2 + \frac{5}{6}x = 1 \qquad \text{Add 1 to both sides.}$$

$$x^2 + \frac{5}{6}x + \frac{25}{144} = 1 + \frac{25}{144} \qquad \text{Add } \left(\tfrac{1}{2} \cdot \tfrac{5}{6}\right)^2, \text{ or } \tfrac{25}{144}, \text{ to both sides.}$$

$$\left(x + \frac{5}{12}\right)^2 = \frac{169}{144} \qquad \text{Factor } x^2 + \tfrac{5}{6}x + \tfrac{25}{144}.$$

Now we apply the square root property.

$$x + \frac{5}{12} = \sqrt{\frac{169}{144}} \qquad \text{or} \qquad x + \frac{5}{12} = -\sqrt{\frac{169}{144}}$$

$$x + \frac{5}{12} = \frac{13}{12} \qquad\qquad x + \frac{5}{12} = -\frac{13}{12}$$

$$x = \frac{8}{12} \qquad\qquad x = -\frac{18}{12}$$

$$x = \frac{2}{3} \qquad\qquad x = -\frac{3}{2}$$

Verify that each root satisfies the original equation. ∎

Self Check Solve $2x^2 - 5x - 3 = 0$.

Answer $3, -\frac{1}{2}$

■ THE QUADRATIC FORMULA

We can solve the equation $ax^2 + bx + c = 0 \ (a \neq 0)$ by completing the square. The result will be a formula that we can use to solve all quadratic equations.

$$ax^2 + bx + c = 0$$

$$x^2 + \frac{b}{a}x + \frac{c}{a} = \frac{0}{a} \qquad \text{Divide both sides by } a.$$

$$x^2 + \frac{b}{a}x = \frac{c}{a} \qquad \text{Subtract } \tfrac{c}{a} \text{ from both sides.}$$

$$x^2 + \frac{b}{a}x + \frac{b^2}{4a^2} = \frac{b^2}{4a^2} - \frac{4ac}{4aa} \qquad \text{Add } \tfrac{b^2}{4a^2} \text{ to both sides and multiply the numerator and denominator of } \tfrac{c}{a} \text{ by } 4a.$$

$$\left(x + \frac{b}{2a}\right)^2 = \frac{b^2 - 4ac}{4a^2} \qquad \text{Factor the left-hand side and add the fractions on the right-hand side.}$$

We can now apply the square root property.

$$x + \frac{b}{2a} = \sqrt{\frac{b^2 - 4ac}{4a^2}} \qquad \text{or} \qquad x + \frac{b}{2a} = -\sqrt{\frac{b^2 - 4ac}{4a^2}}$$

$$x = -\frac{b}{2a} + \frac{\sqrt{b^2 - 4ac}}{2a} \qquad\qquad x = -\frac{b}{2a} - \frac{\sqrt{b^2 - 4ac}}{2a}$$

$$x = \frac{-b + \sqrt{b^2 - 4ac}}{2a} \qquad\qquad x = \frac{-b - \sqrt{b^2 - 4ac}}{2a}$$

These values of x are the two roots of the equation $ax^2 + bx + c = 0$. They are usually combined into a single expression, called the **quadratic formula.**

Quadratic Formula

The solutions of the general quadratic equation, $ax^2 + bx + c = 0$, are

$$x = \frac{-b \pm \sqrt{b^2 - 4ac}}{2a} \qquad (a \neq 0)$$

The quadratic formula should be read twice, once using the + sign and once using the − sign. The quadratic formula implies that

$$x = \frac{-b + \sqrt{b^2 - 4ac}}{2a} \quad \text{and} \quad x = \frac{-b - \sqrt{b^2 - 4ac}}{2a}$$

WARNING! Be sure to write the quadratic formula correctly. Do not write it as

$$x = -b \pm \frac{\sqrt{b^2 - 4ac}}{2a}$$

EXAMPLE 7 Use the quadratic formula to solve $x^2 - 5x + 3 = 0$.

Solution In this equation $a = 1$, $b = -5$, and $c = 3$.

$$x = \frac{-b \pm \sqrt{b^2 - 4ac}}{2a}$$

$$x = \frac{-(-5) \pm \sqrt{(-5)^2 - 4(1)(3)}}{2(1)} \qquad \text{Substitute 1 for } a, -5 \text{ for } b, \text{ and 3 for } c.$$

$$x = \frac{5 \pm \sqrt{13}}{2} \qquad \qquad (-5)^2 - 4(1)(3) = 25 - 12 = 13.$$

Both values satisfy the original equation. ∎

Self Check Solve $3x^2 - 5x + 1 = 0$.

Answer $x = \dfrac{5 \pm \sqrt{13}}{6}$

EXAMPLE 8 Use the quadratic formula to solve $2x^2 + 8x + 7 = 0$.

Solution Since $a = 2$, $b = 8$, and $c = 7$, we substitute these numbers in the quadratic formula and simplify.

$$x = \frac{-b \pm \sqrt{b^2 - 4ac}}{2a}$$

$$x = \frac{-8 \pm \sqrt{8^2 - 4(2)(7)}}{2(2)} \qquad \text{Substitute 2 for } a, 8 \text{ for } b, \text{ and 7 for } c.$$

$$x = \frac{-8 \pm \sqrt{8}}{4} \qquad \qquad 8^2 - 4(2)(7) = 64 - 56 = 8.$$

$$x = \frac{-8 \pm 2\sqrt{2}}{4}$$

$$x = -2 + \frac{\sqrt{2}}{2} \quad \text{or} \quad x = -2 - \frac{\sqrt{2}}{2}$$

Both values satisfy the original equation. ∎

Self Check   Solve $4x^2 + 16x + 13 = 0$.

Answer $x = -2 \pm \dfrac{\sqrt{3}}{2}$

■ THE DISCRIMINANT

We can predict the type of numbers that will be roots of a quadratic equation before solving it. Suppose that the coefficients a, b, and c in the equation $ax^2 + bx + c = 0$ are real numbers. Then the two roots are given by the quadratic formula

$$x = \frac{-b \pm \sqrt{b^2 - 4ac}}{2a} \quad (a \neq 0)$$

The value of $b^2 - 4ac$, called the **discriminant,** determines the nature of the roots of a quadratic equation. The possibilities are summarized in the following table.

Nature of the Roots of a Quadratic Equation

If a, b, and c are real numbers and $b^2 - 4ac$ is . . .	then the solutions are
positive	unequal real numbers
0	equal real numbers
negative	not real numbers

If a, b, and c are rational numbers and $b^2 - 4ac$ is . . .	then the solutions are
0	equal rational numbers
a nonzero perfect square	unequal rational numbers
a positive nonperfect square	unequal irrational numbers

EXAMPLE 9   Determine the nature of the roots of $3x^2 + 4x + 1 = 0$.

Solution   We calculate the discriminant $b^2 - 4ac$ as follows:

$$b^2 - 4ac = 4^2 - 4(\mathbf{3})(\mathbf{1}) \qquad \text{Substitute 4 for } b, \text{ 3 for } a, \text{ and 1 for } c.$$
$$= 16 - 12$$
$$= 4$$

Since a, b, and c are rational numbers and the discriminant is a perfect square, the two roots will be unequal rational numbers.   ■

Self Check   Determine the nature of the roots of $4x^2 - 3x - 1 = 0$.

Answer   equal rational numbers

EXAMPLE 10   If k is a constant, many quadratic equations are represented by the equation

$$(k - 2)x^2 + (k + 1)x + 4 = 0$$

Find the values of k that will give an equation with roots that are equal real numbers.

Solution   We calculate the discriminant $b^2 - 4ac$ and set it equal to 0.

$$\begin{aligned}
b^2 - 4ac &= (k + 1)^2 - 4(k - 2)(4) \\
0 &= k^2 + 2k + 1 - 16k + 32 \\
0 &= k^2 - 14k + 33 \\
0 &= (k - 3)(k - 11)
\end{aligned}$$

$$\begin{array}{ccc}
k - 3 = 0 & \text{or} & k - 11 = 0 \\
k = 3 & & k = 11
\end{array}$$

When $k = 3$ or $k = 11$, the equation will have equal roots. As a check, we let $k = 3$ and note that the equation $(k - 2)x^2 + (k + 1)x + 4 = 0$ becomes

$$\begin{aligned}
(3 - 2)x^2 + (3 + 1)x + 4 &= 0 \\
x^2 + 4x + 4 &= 0
\end{aligned}$$

The roots of this equation are equal real numbers, as expected:

$$\begin{array}{c}
x^2 + 4x + 4 = 0 \\
(x + 2)(x + 2) = 0
\end{array}$$

$$\begin{array}{ccc}
x + 2 = 0 & \text{or} & x + 2 = 0 \\
x = -2 & & x = -2
\end{array}$$

Similarly, $k = 11$ will give an equation with equal real roots.   ∎

Self Check   Find k such that $(k - 2)x^2 - (k + 3)x + 9 = 0$ will have equal roots.
Answer   3, 27

■ WRITING EQUATIONS IN QUADRATIC FORM

There are many equations that can be written in quadratic form. They can then be solved with techniques used for solving quadratic equations.

EXAMPLE 11   Solve $\dfrac{1}{x - 1} + \dfrac{3}{x + 1} = 2.$

Solution   Since neither denominator can be 0, $x \neq 1$ and $x \neq -1$. If either number appears as a root, it must be discarded.

$$\frac{1}{x-1} + \frac{3}{x+1} = 2$$

$$(x-1)(x+1)\left[\frac{1}{x-1} + \frac{3}{x+1}\right] = (x-1)(x+1)2 \qquad \text{Multiply both sides by } (x-1)(x+1).$$

$$(x+1) + 3(x-1) = 2(x^2-1) \qquad \text{Remove brackets and simplify.}$$

$$4x - 2 = 2x^2 - 2 \qquad \text{Remove parentheses and simplify.}$$

$$0 = 2x^2 - 4x \qquad \text{Add } 2 - 4x \text{ to both sides.}$$

The resulting equation is a quadratic equation that we can solve by factoring.

$$2x^2 - 4x = 0$$
$$2x(x-2) = 0 \qquad \text{Factor } 2x^2 - 4x.$$
$$2x = 0 \quad \text{or} \quad x - 2 = 0$$
$$x = 0 \qquad\qquad x = 2$$

Verify these results by checking each root. ∎

Self Check   Solve $\dfrac{1}{x-1} + \dfrac{2}{x+1} = 1$.

Answer   0, 3

■ FORMULAS

Many formulas involve quadratic equations. For example, if an object is fired straight up into the air with an initial velocity of 88 feet per second, its height is given by the formula $h = 88t - 16t^2$, where h represents height (in feet) and t represents elapsed time (in seconds) since it was fired.

To solve this formula for t, we use the quadratic formula.

$$h = 88t - 16t^2$$
$$16t^2 - 88t + h = 0 \qquad \text{Add } 16t^2 \text{ and } -88t \text{ to both sides.}$$

$$t = \frac{-(-88) \pm \sqrt{(-88)^2 - 4(16)(h)}}{2(16)} \qquad \text{Substitute into the quadratic formula.}$$

$$t = \frac{88 \pm \sqrt{7744 - 64h}}{32} \qquad \text{Simplify.}$$

EXERCISE 2.3

VOCABULARY AND CONCEPTS *In Exercises 1–6, fill in the blank to make a true statement.*

1. A quadratic equation is an equation that can be written in the form _____.

2. If a and b are real numbers and _____, then $a = 0$ or $b = 0$.

3. If $c > 0$, the equation $x^2 = c$ has two roots. They are $x =$ ____ and $x =$ _____.

4. The quadratic formula is _____.

5. If a, b, and c are real numbers and if $b^2 - 4ac = 0$, the roots of the quadratic equation are _____.

6. If a, b, and c are real numbers and $b^2 - 4ac < 0$, the roots of the quadratic equation are _____.

PRACTICE *In Exercises 7–18, solve each equation by factoring. Check all answers.*

7. $x^2 - x - 6 = 0$

8. $x^2 + 8x + 15 = 0$

9. $x^2 - 144 = 0$

10. $x^2 + 4x = 0$

11. $2x^2 + x - 10 = 0$

12. $3x^2 + 4x - 4 = 0$

13. $5x^2 - 13x + 6 = 0$

14. $2x^2 + 5x - 12 = 0$

15. $15x^2 + 16x = 15$

16. $6x^2 - 25x = -25$

17. $12x^2 + 9 = 24x$

18. $24x^2 + 6 = 24x$

In Exercises 19–26, use the square root property to solve each equation. You may need to factor an expression.

19. $x^2 = 9$

20. $x^2 = 20$

21. $y^2 - 50 = 0$

22. $x^2 - 75 = 0$

23. $(x - 1)^2 = 4$

24. $(y + 2)^2 - 49 = 0$

25. $a^2 + 2a + 1 = 9$

26. $x^2 - 6x + 9 = 25$

In Exercises 27–38, complete the square to make each binomial a perfect trinomial square.

27. $x^2 + 6x$

28. $x^2 + 8x$

29. $x^2 - 4x$

30. $x^2 - 12x$

31. $a^2 + 5a$

32. $t^2 + 9t$

33. $r^2 - 11r$

34. $s^2 - 7s$

35. $y^2 + \dfrac{3}{4}y$

36. $p^2 + \dfrac{3}{2}p$

37. $q^2 - \dfrac{1}{5}q$

38. $m^2 - \dfrac{2}{3}m$

In Exercises 39–50, solve each equation by completing the square.

39. $x^2 - 8x + 15 = 0$

40. $x^2 + 10x + 21 = 0$

41. $x^2 + x - 6 = 0$

42. $x^2 - 9x + 20 = 0$

43. $x^2 - 25x = 0$

44. $x^2 + x = 0$

45. $3x^2 + 4x = 4$

46. $2x^2 + 5x = 12$

47. $x^2 + 5 = -5x$

48. $x^2 + 1 = -4x$

49. $3x^2 = 1 - 4x$

50. $2x^2 = 3x + 1$

In Exercises 51–62, use the quadratic formula to solve each equation.

51. $x^2 - 12 = 0$

52. $x^2 - 20 = 0$

53. $2x^2 - x - 15 = 0$

54. $6x^2 + x - 2 = 0$

55. $5x^2 - 9x - 2 = 0$

56. $4x^2 - 4x - 3 = 0$

57. $2x^2 + 2x - 4 = 0$

58. $3x^2 + 18x + 15 = 0$

59. $-3x^2 = 5x + 1$

60. $2x(x + 3) = -1$

61. $5x\left(x + \dfrac{1}{5}\right) = 3$

62. $7x^2 = 2x + 2$

*In Exercises 63–70, use the discriminant to determine the nature of the roots of each equations. **Do not solve the equations.***

63. $x^2 + 6x + 9 = 0$

64. $x^2 - 5x + 2 = 0$

65. $3x^2 - 2x + 5 = 0$

66. $9x^2 + 42x + 49 = 0$

67. $10x^2 + 29x = 21$

68. $10x^2 + x = 21$

69. $-3x^2 + 2x = 21$

70. $-8x^2 - 2x = 13$

71. Does $1492x^2 + 1984x - 1776 = 0$ have any roots that are real numbers?

72. Does $2004x^2 + 10x + 1994 = 0$ have any roots that are real numbers?

73. Find two values of k so that $x^2 + kx + 3k - 5 = 0$ will have two roots that are equal.

74. For what value(s) of b will the solutions of $x^2 - 2bx + b^2 - 0$ be equal?

In Exercises 75–92, change each equation to quadratic form and solve by any method.

75. $x + 1 = \dfrac{12}{x}$

76. $x - 2 = \dfrac{15}{x}$

77. $8x - \dfrac{3}{x} = 10$

78. $15x - \dfrac{4}{x} = 4$

79. $\dfrac{5}{x} = \dfrac{4}{x^2} - 6$

80. $\dfrac{6}{x^2} + \dfrac{1}{x} = 12$

81. $x\left(30 - \dfrac{13}{x}\right) = \dfrac{10}{x}$

82. $x\left(20 - \dfrac{17}{x}\right) = \dfrac{10}{x}$

83. $(a - 2)(a + 4) = 2a(a - 3)$

84. $\dfrac{4 + a}{2a} = \dfrac{a - 2}{3}$

85. $\dfrac{1}{x} + \dfrac{3}{x + 2} = 2$

86. $\dfrac{1}{x - 1} + \dfrac{1}{x - 4} = \dfrac{5}{4}$

87. $\dfrac{1}{x + 1} + \dfrac{5}{2x - 4} = 1$

88. $\dfrac{x(2x + 1)}{x - 2} = \dfrac{10}{x - 2}$

89. $x + 1 + \dfrac{x + 2}{x - 1} = \dfrac{3}{x - 1}$

90. $\dfrac{1}{4 - y} = \dfrac{1}{4} + \dfrac{1}{y + 2}$

91. $\dfrac{24}{a} - 11 = \dfrac{-12}{a + 1}$

92. $\dfrac{36}{b} - 17 = \dfrac{-24}{b + 1}$

In Exercises 93–102, solve each formula for the indicated variable.

93. $h = \dfrac{1}{2} gt^2$; t **94.** $x^2 + y^2 = r^2$; x **95.** $h = 64t - 16t^2$; t **96.** $y = 16x^2 - 4$; x

97. $\dfrac{x^2}{a^2} + \dfrac{y^2}{b^2} = 1$; y **98.** $\dfrac{x^2}{a^2} - \dfrac{y^2}{b^2} = 1$; x **99.** $\dfrac{x^2}{a^2} - \dfrac{y^2}{b^2} = 1$; a **100.** $\dfrac{x^2}{a^2} - \dfrac{y^2}{b^2} = 1$; b

101. $x^2 + xy - y^2 = 0$; x **102.** $x^2 - 3xy + y^2 = 0$; y

DISCOVERY AND WRITING

103. If r_1 and r_2 are the roots of $ax^2 + bx + c = 0$, show that $r_1 + r_2 = -\dfrac{b}{a}$.

104. If r_1 and r_2 are the roots of $ax^2 + bx + c = 0$, show that $r_1 r_2 = \dfrac{c}{a}$.

In Exercises 105–106, a stone is thrown upward, higher than the top of a tree. The stone is even with the top of the tree at times t_1 on the way up and t_2 on the way down. If the height of the tree is h, both t_1 and t_2 are solutions $h = v_0 t - 16t^2$.

105. Show that the tree is $16t_1 t_2$ feet tall.

106. Show that v_0 is $16(t_1 + t_2)$ feet per second.

107. Explain why the zero factor theorem is true.

108. Explain how to complete the square on $x^2 - 17x$.

REVIEW *In Exercises 109–112, simplify each expression.*

109. $5x(x - 2) - x(3x - 2)$

110. $(x + 3)(x - 9) - x(x - 5)$

111. $(m + 3)^2 - (m - 3)^2$

112. $[(y + z)(y - z)]^2$

2.4 Applications of Quadratic Equations

■ GEOMETRIC PROBLEMS ■ UNIFORM MOTION PROBLEMS ■ FLYING OBJECT PROBLEMS
■ BUSINESS PROBLEMS

The solutions of many problems involve quadratic equations.

■ GEOMETRIC PROBLEMS

EXAMPLE 1

The length of a rectangle exceeds its width by 3 feet. If the area of the rectangle is 40 square feet, find its dimensions.

Solution

We can let w represent the width of the rectangle. Then, $w + 3$ will represent its length (see Figure 2-4). Since the formula for the area of a rectangle is $A = lw$ (area = length × width), the area of the rectangle is $(w + 3)w$, which is equal to 40.

The length of the rectangle	·	the width of the rectangle	=	the area of the rectangle
$(w + 3)$	·	w	=	40

We can solve this equation for w.

$$(w + 3)w = 40$$
$$w^2 + 3w = 40$$
$$w^2 + 3w - 40 = 0 \qquad \text{Subtract 40 from both sides.}$$
$$(w - 5)(w + 8) = 0 \qquad \text{Factor.}$$
$$w - 5 = 0 \quad \text{or} \quad w + 8 = 0$$
$$w = 5 \qquad\qquad w = -8$$

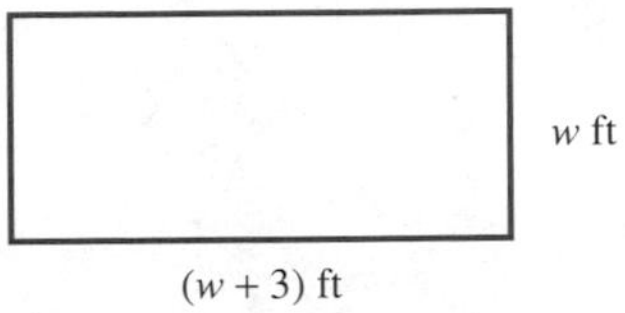

w ft

$(w + 3)$ ft

FIGURE 2-4

When $w = 5$, the length is $w + 3 = 8$. The solution -8 must be discarded, because a rectangle cannot have a negative width.

We can verify that this solution is correct by observing that a rectangle with dimensions of 5 feet by 8 feet does have an area of 40 square feet. ■

■ UNIFORM MOTION PROBLEMS

EXAMPLE 2 A man drives 600 miles to a convention. On the return trip, he is able to increase his speed by 10 miles per hour and save 2 hours of driving time. How fast did he drive in each direction?

Solution We can let s represent the car's speed (in mph) driving to the convention. On the return trip, his speed was $s + 10$ mph. Recall that the distance traveled by an object moving at a constant rate for a certain time is given by the formula $d = rt$. If we divide both sides of this formula by r, we will have a formula for time.

$$t = \frac{d}{r}$$

We can organize the information given in this problem as in Figure 2-5.

	d	=	r	·	t
Outbound trip	600		s		$\dfrac{600}{s}$
Return trip	600		$s + 10$		$\dfrac{600}{s + 10}$

FIGURE 2-5

Although neither the outbound nor the return travel time is given, we know the difference of those times.

The longer time of the outbound trip	−	the shorter time of the return trip	=	the difference in travel times
$\dfrac{600}{s}$	−	$\dfrac{600}{s + 10}$	=	2

We can solve this equation for s.

$$\frac{600}{s} - \frac{600}{s + 10} = 2$$

$$s(s + 10)\left(\frac{600}{s} - \frac{600}{s + 10}\right) = s(s + 10)2 \qquad \text{Multiply both sides by } s(s + 10) \text{ to clear the equation of fractions.}$$

$$600(s + 10) - 600s = 2s(s + 10) \qquad \text{Simplify.}$$

$$600s + 6000 - 600s = 2s^2 + 20s \qquad \text{Remove parentheses.}$$

$$6000 = 2s^2 + 20s \qquad \text{Combine like terms.}$$

$$0 = 2s^2 + 20s - 6000 \qquad \text{Subtract 6000 from both sides.}$$

$$0 = s^2 + 10s - 3000 \qquad \text{Divide both sides by 2.}$$

$$0 = (s - 50)(s + 60) \qquad \text{Factor.}$$

$$s - 50 = 0 \quad \text{or} \quad s + 60 = 0 \qquad \text{Set each factor equal to 0.}$$

$$s = 50 \qquad\qquad s = -60$$

The solution $s = -60$ must be discarded. The man drove 50 mph to the convention and $50 + 10$, or 60 mph on the return trip.

These answers are correct, because a 600-mile trip at 50 mph would take $\frac{600}{50}$, or 12 hours. At 60 mph, the same trip would take only 10 hours, which is 2 hours less time. ∎

■ FLYING OBJECT PROBLEMS

EXAMPLE 3   If an object is thrown straight up into the air with an initial velocity of 144 feet per second, its height is given by the formula $h = 144t - 16t^2$, where h represents its height (in feet) and t represents the time (in seconds) since it was thrown. How long will it take for the object to return to the point from which it was thrown?

Solution   When the object returns to its starting point, its height is again 0. Thus, we can set h equal to 0 and solve for t.

$$h = 144t - 16t^2$$

$$0 = 144t - 16t^2 \qquad \text{Let } h = 0.$$

$$0 = 16t(9 - t) \qquad \text{Factor.}$$

$$16t = 0 \quad \text{or} \quad 9 - t = 0 \qquad \text{Set each factor equal to 0.}$$
$$t = 0 \qquad\qquad\quad t = 9$$

At $t = 0$, the object's height is 0, because it was just released. When $t = 9$, the height is again 0, and the object has returned to its starting point.   ■

■ BUSINESS PROBLEMS

EXAMPLE 4   A bus company shuttles 1120 passengers daily between Rockford, IL and O'Hare airport. The current one-way fare is $10. For each 25¢ increase in the fare, the company predicts that it will lose 48 passengers. What increase in fare will produce daily revenue of $10,208?

Solution   Let q represent the number of quarters the fare will be increased. Then the new fare will be $\$(10 + 0.25q)$. Since the company will lose 48 passengers for each quarter increase, $48q$ passengers will be lost when the rate increases by q quarters. The passenger load will then be $(1120 - 48q)$ passengers.

Since the daily revenue of $10,208 will be the product of the rate and the number of passengers, we have

$$(10 + 0.25q)(1120 - 48q) = 10{,}208$$
$$11{,}200 - 480q + 280q - 12q^2 = 10{,}208 \qquad \text{Remove parentheses.}$$
$$-12q^2 - 200q + 992 = 0 \qquad \begin{array}{l}\text{Combine like terms and subtract} \\ \text{10,208 from both sides.}\end{array}$$
$$3q^2 + 50q - 248 = 0 \qquad \text{Divide both sides by } -4.$$

We can solve this equation with the quadratic formula.

$$q = \frac{-50 \pm \sqrt{50^2 - 4(3)(-248)}}{2(3)} \qquad \begin{array}{l}\text{Substitute 3 for } a, 50 \\ \text{for } b, \text{ and } -248 \text{ for } c.\end{array}$$

$$q = \frac{-50 \pm \sqrt{2500 + 2976}}{6}$$

$$q = \frac{-50 \pm \sqrt{5476}}{6}$$

$$q = \frac{-50 \pm 74}{6}$$

$$q = \frac{-50 + 74}{6} \qquad \text{or} \qquad q = \frac{-50 - 74}{6}$$
$$= \frac{24}{6} \qquad\qquad\qquad\qquad = \frac{-124}{6}$$
$$= 4 \qquad\qquad\qquad\qquad\quad = -\frac{62}{3}$$

Because the number of riders cannot be negative, the result of $-\frac{62}{3}$ must be discarded. To generate $10,208 in daily revenues, the company should raise the fare by 4 quarters, or $1, to $11.   ■

EXERCISE 2.4

VOCABULARY AND CONCEPTS *In Exercises 1–2, fill in the blank to make a true statement.*

1. The formula for the area of a rectangle is _____.

2. The formula that relates distance, rate, and time is _____.

PRACTICE *In Exercises 3–34, solve each problem.*

3. Geometric problem A rectangle is 4 feet longer than it is wide. If its area is 32 square feet, find its dimensions.

4. Geometric problem A rectangle is 5 times as long as it is wide. If the area is 125 square feet, find its perimeter.

5. Geometric problem The side of a square is 4 centimeters shorter than the side of a second square. If the sum of their areas is 106 square centimeters, find the length of one side of the larger square.

6. Geometric problem The base of a triangle is one-third as long as its height. If the area of the triangle is 24 square meters, how long is its base?

7. Metal fabrication A piece of tin, 12 inches on a side, is to have four equal squares cut from its corners, as in Illustration 1. If the edges are then to be folded up to make a box with a floor area of 64 square inches, find the depth of the box.

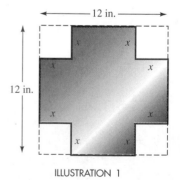

ILLUSTRATION 1

8. Making gutters A piece of sheet metal, 18 inches wide, is bent to form the gutter shown in Illustration 2. If the cross-sectional area is 36 square inches, find the depth of the gutter.

ILLUSTRATION 2

9. Cycling rates A cyclist rides from DeKalb to Rockford, a distance of 40 miles. His return trip takes 2 hours longer because his speed decreases by 10 miles per hour. How fast does he ride each way?

10. Travel time A farmer drives a tractor from one town to another, a distance of 120 kilometers. He drives 10 kilometers per hour faster on the return trip, cutting 1 hour off the time. How fast does he drive each way?

11. Uniform motion problem If the speed were increased by 10 mph, a 420-mile trip would take 1 hour less time. How long will the trip take at the slower speed?

12. Uniform motion problem By increasing her usual speed by 25 kilometers per hour, a bus driver decreases the time on a 25-kilometer trip by 10 minutes. Find the usual speed.

13. Ballistics The height of a projectile fired upward with an initial velocity of 400 feet per second is given by the formula $h = -16t^2 + 400t$, where h is the height in feet and t is the time in seconds. Find the time required for the projectile to return to earth.

14. Ballistics The height of an object tossed upward with an initial velocity of 104 feet per second is given by the formula $h = -16t^2 + 104t$, where h is the height in feet and t is the time in seconds. Find the time required for the object to return to its point of departure.

15. Falling coins   An object will fall s feet in t seconds, where $s = 16t^2$. How long will it take for a penny to hit the ground if it is dropped from the top of the Sears Tower in Chicago? (*Hint:* The tower is 1454 feet tall.)

16. Ballistics   The height of an object thrown upward with an initial velocity of 32 feet per second is given by the formula $h = -16t^2 + 32t$, where t is the time in seconds. How long will it take the object to reach a height of 16 feet?

17. Setting fares   A bus company has 3000 passengers daily, paying a 25¢ fare. For each nickel increase in fare, the company projects that it will lose 80 passengers. What fare increase will produce $994 in daily revenue?

18. Jazz concerts   A jazz group on tour has been drawing average crowds of 500 persons. It is projected that for every $1 increase in the $12 ticket price, the average attendance will decrease by 50. At what ticket price will nightly receipts be $5600?

19. Concert receipts   Tickets for the annual symphony orchestra pops concert cost $15, and the average attendance at the concerts has been 1200 persons. Management projects that for each 50¢ decrease in ticket price, 40 more patrons will attend. How many people attended the concert if the receipts were $17,280?

20. Projecting demand   The *Vilas County News* earns a profit of $20 per year for each of its 3000 subscribers. Management projects that the profit per subscriber would increase by 1¢ for each additional subscriber over the current 3000. How many subscribers are needed to bring a total profit of $120,000?

21. Filling a storage tank   Two pipes are used to fill a water storage tank. The first pipe can fill the tank in 4 hours, and the two pipes together can fill the tank in 2 hours less time than the second pipe alone. How long would it take for the second pipe to fill the tank?

22. Filling a swimming pool   A hose can fill a swimming pool in 6 hours. Another hose needs 3 more hours to fill the pool than the two hoses combined. How long would it take the second hose to fill the pool?

23. Mowing lawns   Kristy can mow a lawn in 1 hour less time than her brother Steven. Together they can finish the job in 5 hours. How long would it take Kristy if she worked alone?

24. Milking cows   Working together, Sarah and Heidi can milk the cows in 2 hours. If they work alone, it takes Heidi 3 hours longer than it takes Sarah. How long would it take Heidi to milk the cows alone?

25. Geometric problem   Is it possible for a rectangle to have a width that is 3 units shorter than its diagonal and a length that is 4 units longer than its diagonal?

26. Geometric problem   If two opposite sides of a square are increased by 10 meters and the other sides are decreased by 8 meters, the area of the rectangle that is formed is 63 square meters. Find the area of the original square.

27. Investment problem   Maude and Matilda each have a bank CD. Maude's is $1000 larger than Matilda's, but the interest rate is 1% less. Last year Maude received interest of $280, and Matilda received $240. Find the rate of interest for each CD.

28. Investment problem   Scott and Laura have both invested some money. Scott invested $3000 more than Laura and at a 2% higher interest rate. If Scott received $800 annual interest and Laura received $400, how much did Scott invest?

29. Buying a microwave oven   Some mathematics professors would like to purchase a $150 microwave oven for the department workroom. If four of the professors don't contribute, everyone's share will increase by $10. How many professors are in the department?

30. Planting a windscreen   A farmer intends to construct a windscreen by planting trees in a quarter-mile row. His daughter points out that 44 fewer trees will be needed if they are planted 1 foot farther apart. If her dad takes her advice, how many trees will be needed? A row start and ends with a tree. (*Hint:* 1 mile = 5280 feet.)

31. Puzzle problem   If a wagon wheel had 10 more spokes, the angle between spokes would decrease by 6°. How many spokes does the wheel have?

32. Puzzle problem A merchant could sell all of his calculators at list price for $180. If he had 3 more calculators, he could sell each one for $10 less and still receive $180. Find the list price of each calculator.

33. Geometric problem If one leg of a right triangle is 14 meters shorter than the other leg, and the hypotenuse is 26 meters, find the length of the two legs.

34. Geometric problem Find the dimensions of a rectangle whose area is 180 cm^2 and whose perimeter is 54 cm.

DISCOVERY AND WRITING

35. Which of the preceding problems did you find the hardest? Why?

36. Which of the preceding problems did you find the easiest? Why?

REVIEW In Exercises 37–40, do the operations and simplify.

37. $\dfrac{2}{x} - \dfrac{1}{x-3}$

38. $\dfrac{1}{x} \cdot \dfrac{x^2 - 5x}{x-3}$

39. $\dfrac{x+3}{x^2 - x - 6} \div \dfrac{x^2 + 3x}{x^2 - 9}$

40. $\dfrac{\dfrac{1}{x} - \dfrac{1}{2}}{x - 2}$

2.5 Complex Numbers

■ IMAGINARY NUMBERS ■ POWERS OF *i* ■ SIMPLIFYING IMAGINARY NUMBERS ■ COMPLEX NUMBERS ■ THE ARITHMETIC OF COMPLEX NUMBERS ■ ABSOLUTE VALUE OF A COMPLEX NUMBER ■ FACTORING THE SUM OF TWO SQUARES

So far, the solution of every quadratic equation has been a real number. This is not always the case. For example, if we use the quadratic formula to solve $x^2 + x + 2 = 0$, we get solutions that are nonreal.

$$x = \frac{-b \pm \sqrt{b^2 - 4ac}}{2a}$$

$$x = \frac{-1 \pm \sqrt{1^2 - 4(1)(2)}}{2(1)} \qquad \text{Substitute 1 for } a, \text{ 1 for } b, \text{ and 2 for } c.$$

$$x = \frac{-1 \pm \sqrt{1 - 8}}{2}$$

$$x = \frac{-1 \pm \sqrt{-7}}{2}$$

Each solution involves $\sqrt{-7}$. This is not a real number, because the square of no real number is -7.

■ IMAGINARY NUMBERS

For years, mathematicians believed that numbers such as

$$\sqrt{-1}, \quad \sqrt{-4}, \quad \sqrt{-5}, \quad \text{and} \quad \sqrt{-7}$$

were nonsense. Even the great English mathematician Sir Isaac Newton (1642–1727) called them "impossible numbers." In the 17th century, these symbols were called **imaginary numbers** by René Descartes. Today, they have important uses, such as describing the behavior of alternating current in electronics.

The imaginary numbers are based on the **imaginary unit i,** where

$$i^2 = -1$$

Because i represents the square root of -1, we also write

$$i = \sqrt{-1}$$

■ POWERS OF i

The powers of i with natural number exponents produce an interesting pattern.

$$i^1 = \sqrt{-1} = i \qquad\qquad i^5 = i^4 i = 1i = i$$
$$i^2 = \left(\sqrt{-1}\right)^2 = -1 \qquad i^6 = i^4 i^2 = 1(-1) = -1$$
$$i^3 = i^2 i = -1i = -i \qquad i^7 = i^4 i^3 = 1(-i) = -i$$
$$i^4 = i^2 i^2 = (-1)(-1) = 1 \qquad i^8 = i^4 i^4 = 1(1) = 1$$

The pattern continues: $i, -1, -i, 1, \ldots$.

EXAMPLE 1   Simplify i^{365}.

Solution   Since $i^4 = 1$, each occurrence of i^4 is a factor of 1. To determine how many factors of i^4 are in i^{365}, we divide 365 by 4. The quotient is 91, and the remainder is 1.

$$i^{365} = (i^4)^{91} \cdot i^1$$
$$\qquad = 1^{91} \cdot i \qquad\qquad i^4 = 1.$$
$$\qquad = i \qquad\qquad 1^{91} = 1 \text{ and } 1 \cdot i = i.$$ ■

Self Check   Simplify i^{1999}.
Answer $-i$

The result of Example 1 illustrates the following theorem.

Powers of i
If n is a natural number that has a remainder of r when divided by 4, then

$$i^n = i^r$$

When n is divisible by 4, the remainder r is 0 and $i^0 = 1$.

We can also simplify powers of i that involve negative integer exponents.

$$i^{-1} = \frac{1}{i} = \frac{1 \cdot i}{i \cdot i} = \frac{i}{-1} = -i$$

$$i^{-2} = \frac{1}{i^2} = \frac{1}{-1} = -1$$

$$i^{-3} = \frac{1}{i^3} = \frac{1 \cdot i}{i^3 \cdot i} = \frac{i}{i^4} = \frac{i}{1} = i$$

$$i^{-4} = \frac{1}{i^4} = \frac{1}{1} = 1$$

The pattern i, -1, $-i$, 1 continues forever.

■ SIMPLIFYING IMAGINARY NUMBERS

Because imaginary numbers follow the rules for exponents, we have

$$(3i)^2 = 3^2 i^2 = 9(-1) = -9$$

Since $(3i)^2 = -9$, $3i$ is a square root of -9, and we can write

$$\sqrt{-9} = 3i$$

This result can also be obtained by using the multiplication property of radicals.

$$\sqrt{-9} = \sqrt{9(-1)}$$
$$= \sqrt{9}\sqrt{-1} \qquad \sqrt{ab} = \sqrt{a}\sqrt{b}.$$
$$= 3i \qquad\qquad \sqrt{9} = 3.$$

We can use the multiplication property of radicals to simplify imaginary numbers.

$$\sqrt{-25} = \sqrt{25(-1)} = \sqrt{25}\sqrt{-1} = 5i$$

$$\sqrt{-7} = \sqrt{7(-1)} = \sqrt{7}\sqrt{-1} = \sqrt{7}i$$

$$\sqrt{\frac{-100}{49}} = \sqrt{\frac{100}{49}(-1)} = \sqrt{\frac{100}{49}}\sqrt{-1} = \frac{10}{7}i$$

WARNING! If a and b are both negative, then $\sqrt{ab} \neq \sqrt{a}\sqrt{b}$. For example, the correct simplification of $\sqrt{-16}\sqrt{-4}$ is

$$\sqrt{-16}\sqrt{-4} = (4i)(2i) = 8i^2 = 8(-1) = -8$$

The following simplification is incorrect, because we get a different result.

$$\sqrt{-16}\sqrt{-4} = \sqrt{(-16)(-4)} = \sqrt{64} = 8 \qquad \text{Here, } a \text{ and } b \text{ are both negative, and the multiplication property of radicals does not apply.}$$

■ COMPLEX NUMBERS

Numbers such as $3 + 4i$, $-5 + 7i$, and $-1 - 9i$ that are the sum or difference of a real number and an imaginary number are called *complex numbers*.

Complex Numbers

A **complex number** is a number that can be written in the form $a + bi$, where a and b are real numbers and $i = \sqrt{-1}$.

The number a is called the **real part,** and b is called the **imaginary part.**

If $b = 0$, the complex number $a + bi$ is the real number a. If $a = 0$ and $b \neq 0$, the complex number $a + bi$ is the imaginary number bi. It follows that the set of real numbers and the set of imaginary numbers are subsets of the set of complex numbers.

Figure 2-6 illustrates how the various sets of numbers are related.

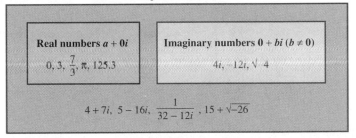

FIGURE 2-6

To decide whether two complex numbers are equal, we will use the following definition.

Equality of Complex Numbers

Two complex numbers are equal if their real parts are equal and their imaginary parts are equal. If $a + bi$ and $c + di$ are two complex numbers, then

$$a + bi = c + di \qquad \text{if and only if} \qquad a = c \quad \text{and} \quad b = d$$

EXAMPLE 2   For what numbers x and y is $3x + 4i = (2y + x) + xi$?

Solution   Since the numbers are equal, their imaginary parts must be equal: $x = 4$. Since their real parts are equal, $3x = 2y + x$. We can solve the system

$$\begin{cases} x = 4 \\ 3x = 2y + x \end{cases}$$

by substituting 4 for x in the second equation and solving for y. We find that $y = 4$. The solution is $x = 4$ and $y = 4$. ∎

Self Check   Find x: $a + (x + 3)i = a - (2x - 1)i$.

Answer $-\frac{2}{3}$

■ THE ARITHMETIC OF COMPLEX NUMBERS

Complex numbers can be added and subtracted as if they were binomials.

Addition and Subtraction of Complex Numbers
Two complex numbers such as $a + bi$ and $c + di$ are added and subtracted as if they were binomials:

$$(a + bi) + (c + di) = (a + c) + (b + d)i$$
$$(a + bi) - (c + di) = (a - c) + (b - d)i$$

Because of the preceding definition, the sum and difference of two complex numbers is another complex number.

EXAMPLE 3   Simplify   **a.** $(3 + 4i) + (2 + 7i)$ and   **b.** $(-5 + 8i) - (2 - 12i)$.

Solution   **a.** $(3 + 4i) + (2 + 7i) = 3 + 4i + 2 + 7i$
$$= 3 + 2 + 4i + 7i$$
$$= 5 + 11i$$

b. $(-5 + 8i) - (2 - 12i) = -5 + 8i - 2 + 12i$
$$= -5 - 2 + 8i + 12i$$
$$= -7 + 20i$$ ∎

Self Check   Simplify   **a.** $(5 - 2i) + (-3 + 9i)$;   and   **b.** $(2 + 5i) - (6 + 7i)$.

Answers   **a.** $2 + 7i$,   **b.** $-4 - 2i$

Complex numbers can be multiplied as if they were binomials.

Multiplication of Complex Numbers
The numbers $a + bi$ and $c + di$ are multiplied as if they were binomials, with $i^2 = -1$:

$$(a + bi)(c + di) = (ac - bd) + (ad + bc)i$$

Because of this definition, the product of two complex numbers is another complex number.

EXAMPLE 4   Multiply   **a.** $(3 + 4i)(2 + 7i)$ and   **b.** $(5 - 7i)(1 + 3i)$.

Solution   **a.** $(3 + 4i)(2 + 7i) = 6 + 21i + 8i + 28i^2$
$$= 6 + 21i + 8i + 28(-1)$$
$$= 6 - 28 + 29i$$
$$= -22 + 29i$$

b. $(5 - 7i)(1 + 3i) = 5 + 15i - 7i - 21i^2$
$$= 5 + 15i - 7i - 21(-1)$$
$$= 5 + 21 + 8i$$
$$= 26 + 8i \qquad ■$$

Self Check   Multiply $(2 - 5i)(3 + 2i)$.
Answer $16 - 11i$

WARNING!   To avoid errors in determining the sign of the result, always express numbers in $a + bi$ form before attempting any algebraic manipulations.

EXAMPLE 5   Multiply $\left(-2 + \sqrt{-16}\right)\left(4 - \sqrt{-9}\right)$.

Solution   We change each number to $a + bi$ form:
$$-2 + \sqrt{-16} = -2 + \sqrt{16}\sqrt{-1} = -2 + 4i$$
$$4 - \sqrt{-9} = 4 - \sqrt{9}\sqrt{-1} = 4 - 3i$$

and then find the product.
$$(-2 + 4i)(4 - 3i) = -8 + 6i + 16i - 12i^2$$
$$= -8 + 6i + 16i - 12(-1)$$
$$= -8 + 12 + 22i$$
$$= 4 + 22i \qquad ■$$

Self Check   Multiply $\left(3 + \sqrt{-25}\right)\left(2 - \sqrt{-9}\right)$.
Answer $21 + i$

Before we discuss the division of complex numbers, we introduce the concept of a complex conjugate.

> **Complex Conjugates**
> The complex numbers $a + bi$ and $a - bi$ are called **complex conjugates** of each other.

For example,

$2 + 5i$ and $2 - 5i$ are complex conjugates.

$-\dfrac{1}{2} + 4i$ and $-\dfrac{1}{2} - 4i$ are complex conjugates.

What makes this concept important is the fact that the product of two complex conjugates is always a real number. For example,

$$
\begin{aligned}
(2 + 5i)(2 - 5i) &= 4 - 10i + 10i - 20i^2 \\
&= 4 - 20(\mathbf{-1}) \\
&= 4 + 20 \\
&= 24
\end{aligned}
$$

In general, we have

$$
\begin{aligned}
(a + bi)(a - bi) &= a^2 - abi + abi - b^2i^2 \\
&= a^2 - b^2(\mathbf{-1}) \\
&= a^2 + b^2
\end{aligned}
$$

Using the concept of a complex conjugate, we can divide complex numbers by rationalizing the denominator.

EXAMPLE 6   Divide $\dfrac{3}{2 + i}$.

Solution   To divide, we multiply the numerator and denominator by the conjugate of the denominator and simplify.

$$
\begin{aligned}
\frac{3}{2 + i} &= \frac{3(\mathbf{2 - i})}{(2 + i)(\mathbf{2 - i})} \\
&= \frac{6 - 3i}{4 - 2i + 2i - i^2} \\
&= \frac{6 - 3i}{4 + 1} \\
&= \frac{6 - 3i}{5} \\
&= \frac{6}{5} - \frac{3}{5}i
\end{aligned}
$$

It is common to accept $\dfrac{6}{5} - \dfrac{3}{5}i$ as a substitute for $\dfrac{6}{5} + \left(-\dfrac{3}{5}\right)i$.   ∎

Self Check   Divide $\dfrac{3}{3-i}$.

Answer $\dfrac{9}{10} + \dfrac{3}{10}i$

EXAMPLE 7   Divide $\dfrac{2 - \sqrt{-16}}{3 + \sqrt{-1}}$.

Solution $\dfrac{2 - \sqrt{-16}}{3 + \sqrt{-1}} = \dfrac{2 - 8i}{3 + i}$ Change each number to $a + bi$ form.

$= \dfrac{(2 - 8i)(3 - i)}{(3 + i)(3 - i)}$ Multiply the numerator and denominator by $3 - i$.

$= \dfrac{6 - 2i - 24i + 8i^2}{9 - 3i + 3i - i^2}$ Remove parentheses.

$= \dfrac{-2 - 26i}{9 + 1}$ Combine like terms; $i^2 = -1$.

$= \dfrac{-2}{10} - \dfrac{26i}{10}$

$= -\dfrac{1}{5} - \dfrac{13}{5}i$ ∎

Self Check   Divide $\dfrac{3 + \sqrt{-25}}{2 - \sqrt{-1}}$.

Answer $\dfrac{1}{5} + \dfrac{13}{5}i$

Examples 6 and 7 illustrate that the quotient of two complex numbers is another complex number.

■ ABSOLUTE VALUE OF A COMPLEX NUMBER

Absolute Value of a Complex Number
If $a + bi$ is a complex number, then
$$|a + bi| = \sqrt{a^2 + b^2}$$

Because of the previous definition, the absolute value of a complex number is a real number. For this reason, i does not appear in the result.

EXAMPLE 8   Find   **a.** $|3 + 4i|$ and   **b.** $|4 - 6i|$.

a. $|3 + 4i| = \sqrt{3^2 + 4^2}$ **b.** $|4 - 6i| = \sqrt{4^2 + (-6)^2}$

$\qquad\qquad\quad = \sqrt{9 + 16}$ $\qquad\qquad\quad = \sqrt{16 + 36}$

$\qquad\qquad\quad = \sqrt{25}$ $\qquad\qquad\quad = \sqrt{52}$

$\qquad\qquad\quad = 5$ $\qquad\qquad\quad = \sqrt{4 \cdot 13}$

$\qquad\qquad\qquad\qquad\qquad\qquad\qquad\quad = \sqrt{4}\sqrt{13}$

$\qquad\qquad\qquad\qquad\qquad\qquad\qquad\quad = 2\sqrt{13}$ ∎

Self Check   Find $|2 - 5i|$.

Answer $\sqrt{29}$

EXAMPLE 9   Find   **a.** $\left|\dfrac{2i}{3 + i}\right|$ and   **b.** $|a + 0i|$.

Solution   **a.** We first write $\dfrac{2i}{3 + i}$ in $a + bi$ form:

$$\frac{2i}{3 + i} = \frac{2i(3 - i)}{(3 + i)(3 - i)} = \frac{6i - 2i^2}{9 - i^2} = \frac{6i + 2}{10} = \frac{1}{5} + \frac{3}{5}i$$

and then find the absolute value of $\frac{1}{5} + \frac{3}{5}i$.

$$\left|\frac{2i}{3 + i}\right| = \left|\frac{1}{5} + \frac{3}{5}i\right| = \sqrt{\left(\frac{1}{5}\right)^2 + \left(\frac{3}{5}\right)^2} = \sqrt{\frac{10}{25}} = \frac{\sqrt{10}}{5}$$

b. $|a + 0i| = \sqrt{a^2 + 0^2} = \sqrt{a^2} = |a|$

From part **b,** we see that $|a| = \sqrt{a^2}$.   ∎

Self Check   Find $\left|\dfrac{3i}{2 - i}\right|$.

Answer $\dfrac{3\sqrt{5}}{5}$

EXAMPLE 10   Solve $x^2 - 4x + 5 = 0$.

Solution   In this equation, $a = 1$, $b = -4$, and $c = 5$.

$$x = \frac{-b \pm \sqrt{b^2 - 4ac}}{2a}$$

$$= \frac{-(-4) \pm \sqrt{(-4)^2 - 4(1)(5)}}{2(1)}$$ Substitute 1 for a, -4 for b, and 5 for c.

$$= \frac{4 \pm \sqrt{16 - 20}}{2}$$

$$= \frac{4 \pm \sqrt{-4}}{2}$$

$$= \frac{4 \pm 2i}{2} \qquad\qquad \sqrt{-4} = \sqrt{4}\sqrt{-1} = 2i.$$

$$= 2 \pm i \qquad\qquad \frac{4 \pm 2i}{2} = \frac{2(2 \pm i)}{2} = 2 \pm i.$$

The roots $x = 2 + i$ and $x = 2 - i$ both satisfy the equation. Note that the roots are complex conjugates. ■

Self Check Solve $x^2 + 3x + 4 = 0$.

Answer $-\dfrac{3}{2} \pm \dfrac{\sqrt{7}}{2}i$

■ FACTORING THE SUM OF TWO SQUARES

Since $i^2 = -1$, it is possible to factor the sum of two squares over the set of complex numbers. For example, to factor $9x^2 + 16y^2$, we proceed as follows:

$$9x^2 + 16y^2 = 9x^2 - (-1)16y^2$$
$$= 9x^2 - i^2(16y^2)$$
$$= 9x^2 - 16y^2 i^2$$
$$= (3x + 4yi)(3x - 4yi) \qquad \text{Factor the difference of two squares.}$$

EXERCISE 2.5

VOCABULARY AND CONCEPTS   *In Exercises 1–8, fill in the blank to make a true statement.*

1. $\sqrt{-3}$, $\sqrt{-9}$, and $\sqrt{-12}$ are examples of _____ numbers.

2. In the complex number $a + bi$, a is the ____ part, and b is the _____ part.

3. If $a = 0$ and $b \neq 0$ in the complex number $a + bi$, the number is an _____ number.

4. If $b = 0$ in the complex number $a + bi$, the number is a ____ number.

5. The complex conjugate of $2 + 5i$ is _____.

6. By definition, $|a + bi| =$ _____.

7. The absolute value of a complex number is a ____ number.

8. The product of two complex conjugates is a ____ number.

PRACTICE   *In Exercises 9–16, simplify each expression.*

9. i^9

10. i^{27}

11. i^{38}

12. i^{99}

13. i^{-6}

14. i^0

15. i^{-10}

16. i^{-31}

In Exercises 17–20, find the values of x and y.

17. $x + (x + y)i = 3 + 8i$

18. $x + 5i = y - yi$

19. $3x - 2yi = 2 + (x + y)i$

20. $\begin{cases} 2 + (x + y)i = 2 - i \\ x + 3i = 2 + 3i \end{cases}$

In Exercises 21–52, do all operations and express all answers in a + bi form.

21. $(2 - 7i) + (3 + i)$

22. $(-7 + 2i) + (2 - 8i)$

23. $(5 - 6i) - (7 + 4i)$

24. $(11 + 2i) - (13 - 5i)$

25. $(14i + 2) + \left(2 - \sqrt{-16}\right)$

26. $\left(5 + \sqrt{-64}\right) - (23i - 32)$

27. $\left(3 + \sqrt{-4}\right) - \left(2 + \sqrt{-9}\right)$

28. $\left(7 - \sqrt{-25}\right) + \left(-8 + \sqrt{-1}\right)$

29. $(2 + 3i)(3 + 5i)$

30. $(5 - 7i)(2 + i)$

31. $(2 + 3i)^2$

32. $(3 - 4i)^2$

33. $\left(11 + \sqrt{-25}\right)\left(2 - \sqrt{-36}\right)$

34. $\left(6 + \sqrt{-49}\right)\left(6 - \sqrt{-49}\right)$

35. $\left(\sqrt{-16} + 3\right)\left(2 + \sqrt{-9}\right)$

36. $\left(12 - \sqrt{-4}\right)\left(-7 + \sqrt{-25}\right)$

37. $\dfrac{1}{i^3}$

38. $\dfrac{3}{i^5}$

39. $\dfrac{-4}{i^{10}}$

40. $\dfrac{-10}{i^{24}}$

41. $\dfrac{1}{2 + i}$

42. $\dfrac{-2}{3 - i}$

43. $\dfrac{2i}{7 + i}$

44. $\dfrac{-3i}{2 + 5i}$

45. $\dfrac{2 + i}{3 - i}$

46. $\dfrac{3 - i}{1 + i}$

47. $\dfrac{4 - 5i}{2 + 3i}$

48. $\dfrac{34 + 2i}{2 - 4i}$

49. $\dfrac{5 - \sqrt{-16}}{-8 + \sqrt{-4}}$

50. $\dfrac{3 - \sqrt{-9}}{2 - \sqrt{-1}}$

51. $\dfrac{2 + i\sqrt{3}}{3 + i}$

52. $\dfrac{3 + i}{4 - i\sqrt{2}}$

In Exercises 53–68, find each absolute value.

53. $|3 + 4i|$

54. $|5 + 12i|$

55. $|2 + 3i|$

56. $|5 - i|$

57. $\left|-7 + \sqrt{-49}\right|$

58. $\left|-2 - \sqrt{-16}\right|$

59. $\left|\dfrac{1}{2} + \dfrac{1}{2}i\right|$

60. $\left|\dfrac{1}{2} - \dfrac{1}{4}i\right|$

61. $|-6i|$

62. $|5i|$

63. $\left|\dfrac{2}{1 + i}\right|$

64. $\left|\dfrac{3}{3 + i}\right|$

65. $\left|\dfrac{-3i}{2 + i}\right|$

66. $\left|\dfrac{5i}{i - 2}\right|$

67. $\left|\dfrac{i + 2}{i - 2}\right|$

68. $\left|\dfrac{2 + i}{2 - i}\right|$

In Exercises 69–76, use the quadratic formula to solve each equation. Simplify all solutions and write them in a + bi form.

69. $x^2 + 2x + 2 = 0$

70. $a^2 + 4a + 8 = 0$

71. $y^2 + 4y + 5 = 0$

72. $x^2 + 2x + 5 = 0$

73. $x^2 - 2x = -5$ **74.** $z^2 - 3z = -8$ **75.** $x^2 - \frac{2}{3}x = -\frac{2}{9}$ **76.** $x^2 + \frac{5}{4} = x$

In Exercises 77–84, factor each expression over the set of complex numbers.

77. $x^2 + 4$ **78.** $16a^2 + 9$ **79.** $25p^2 + 36q^2$ **80.** $100r^2 + 49s^2$

81. $2y^2 + 8z^2$ **82.** $12b^2 + 75c^2$ **83.** $50m^2 + 2n^2$ **84.** $64a^4 + 4b^2$

DISCOVERY AND WRITING

85. Show that the addition of two complex numbers is commutative by adding the complex numbers $a + bi$ and $c + di$ in both orders and observing that the sums are equal.

86. Show that the multiplication of two complex numbers is commutative by multiplying the complex numbers $a + bi$ and $c + di$ in both orders and observing that the products are equal.

87. Show that the addition of complex numbers is associative.

88. Find three examples of complex numbers that are reciprocals of their own conjugates.

89. Tell how to decide whether two complex numbers are equal.

90. Define the complex conjugate of a complex number.

REVIEW *In Exercises 91–94, simplify each expression. Assume that all variables represent nonnegative numbers.*

91. $\sqrt{8x^3}\sqrt{4x}$

92. $\left(\sqrt{x} - 5\right)^2$

93. $\left(\sqrt{x+1} - 2\right)^2$

94. $\left(-3\sqrt{2x+1}\right)^2$

2.6 Polynomial and Radical Equations

■ SOLVING POLYNOMIAL EQUATIONS BY FACTORING ■ OTHER EQUATIONS THAT CAN BE SOLVED BY FACTORING ■ RADICAL EQUATIONS

■ SOLVING POLYNOMIAL EQUATIONS BY FACTORING

The equation $ax^2 + bx + c = 0$ is a polynomial equation of second degree, because its left-hand side contains a second-degree polynomial. Many polynomial equations of higher degree can be solved by factoring.

 b

EXAMPLE 1   Solve **a.** $6x^3 - x^2 - 2x = 0$ and **b.** $x^4 - 5x^2 + 4 = 0$.

Solution **a.** $6x^3 - x^2 - 2x = 0$
$$x(6x^2 - x - 2) = 0 \qquad \text{Factor out } x.$$
$$x(3x - 2)(2x + 1) = 0 \qquad \text{Factor } 6x^2 - x - 2.$$

We set each factor equal to 0.

$$x = 0 \quad \text{or} \quad 3x - 2 = 0 \quad \text{or} \quad 2x + 1 = 0$$

$$x = \frac{2}{3} \qquad\qquad x = -\frac{1}{2}$$

Verify that each solution satisfies the original equation.

b.
$$x^4 - 5x^2 + 4 = 0$$
$$(x^2 - 4)(x^2 - 1) = 0 \qquad \text{Factor } x^4 - 5x^2 + 4 = 0.$$
$$(x + 2)(x - 2)(x + 1)(x - 1) = 0 \qquad \text{Factor each difference of two squares.}$$

We set each factor equal to 0.

$$x + 2 = 0 \quad \text{or} \quad x - 2 = 0 \quad \text{or} \quad x + 1 = 0 \quad \text{or} \quad x - 1 = 0$$

$$x = -2 \qquad\quad x = 2 \qquad\quad x = -1 \qquad\quad x = 1$$

Verify that each solution satisfies the original equation.   ■

Self Check   Solve $2x^3 + 3x^2 - 2x = 0$.

Answer $0, \frac{1}{2}, -2$

■ OTHER EQUATIONS THAT CAN BE SOLVED BY FACTORING

To solve other equations with factoring techniques, we use a property that states that equal powers of equal numbers are equal.

> **Power Property of Real Numbers**
> If a and b are numbers, n is an integer, and $a = b$, then
> $$a^n = b^n$$

When we raise both sides of an equation to the same power, the resulting equation might not be equivalent to the original one. For example, if we raise both sides of

1. $x = 4$ with a solution set of $\{4\}$

to the second power, we obtain

2. $x^2 = 16$ with a solution set of $\{4, -4\}$

Equations 1 and 2 have different solution sets, and the solution -4 of Equation 2 does not satisfy Equation 1. Because raising both sides of an equation to the same power often introduces **extraneous solutions** (false solutions that don't satisfy the original equation), we must check all suspected roots to be certain that they satisfy the original equation.

The following equation has an extraneous solution.

$$x - x^{1/2} - 6 = 0$$
$$(x^{1/2} - 3)(x^{1/2} + 2) = 0 \qquad \text{Factor } x - x^{1/2} - 6.$$
$$x^{1/2} - 3 = 0 \qquad \text{or} \qquad x^{1/2} + 2 = 0 \qquad \text{Set each factor equal to 0.}$$
$$x^{1/2} = 3 \qquad\qquad x^{1/2} = -2$$

Because equal powers of equal numbers are equal, we can square both sides of the previous equations to get

$$(x^{1/2})^2 = (3)^2 \qquad \text{or} \qquad (x^{1/2})^2 = (-2)^2$$
$$x = 9 \qquad\qquad x = 4$$

The number 9 satisfies the equation $x - x^{1/2} - 6 = 0$, but 4 does not, as the following check shows:

If x = 9	*If x = 4*
$x - x^{1/2} - 6 = 0$	$x - x^{1/2} - 6 = 0$
$9 - 9^{1/2} - 6 \stackrel{?}{=} 0$	$4 - 4^{1/2} - 6 \stackrel{?}{=} 0$
$9 - 3 - 6 \stackrel{?}{=} 0$	$4 - 2 - 6 \stackrel{?}{=} 0$
$0 = 0$	$-4 \neq 0$

The number 9 is the only root.

EXAMPLE 2   Solve $2x^{2/5} - 5x^{1/5} - 3 = 0$.

Solution

$$2x^{2/5} - 5x^{1/5} - 3 = 0$$
$$(2x^{1/5} + 1)(x^{1/5} - 3) = 0 \qquad \text{Factor } 2x^{2/5} - 5x^{1/5} - 3.$$
$$2x^{1/5} + 1 = 0 \qquad \text{or} \qquad x^{1/5} - 3 = 0 \qquad \text{Set each factor equal to 0.}$$
$$2x^{1/5} = -1 \qquad\qquad x^{1/5} = 3$$
$$x^{1/5} = -\frac{1}{2}$$

We can raise both sides of each of the previous equations to the fifth power to get

$$(x^{1/5})^5 = \left(-\frac{1}{2}\right)^5 \qquad \text{or} \qquad (x^{1/5})^5 = (3)^5$$
$$x = -\frac{1}{32} \qquad\qquad x = 243$$

Verify that each solution satisfies the original equation.   ∎

Self Check   Solve $x^{2/5} - x^{1/5} - 2 = 0$.

Answer $-1, 32$

■ RADICAL EQUATIONS

Radical equations are equations containing radicals with variables in the radicand. To solve such equations, we will use the power property of real numbers.

 WARNING! Remember to check all roots when solving radical equations, because raising both sides of an equation to a power can introduce extraneous roots.

EXAMPLE 3 Solve $\sqrt{x+3} - 4 = 7$.

Solution

$$\sqrt{x+3} - 4 = 7$$
$$\sqrt{x+3} = 11 \qquad \text{Add 4 to both sides.}$$
$$\left(\sqrt{x+3}\right)^2 = (11)^2 \qquad \text{Square both sides.}$$
$$x + 3 = 121 \qquad \text{Simplify.}$$
$$x = 118 \qquad \text{Subtract 3 from both sides.}$$

Since squaring both sides might introduce extraneous roots, we must check the result of 118.

$$\sqrt{x+3} - 4 = 7$$
$$\sqrt{118 + 3} - 4 \overset{?}{=} 7 \qquad \text{Substitute 118 for } x.$$
$$\sqrt{121} - 4 \overset{?}{=} 7$$
$$11 - 4 \overset{?}{=} 7$$
$$7 = 7$$

Because it checks, 118 is a root of the equation. ■

Self Check Solve $\sqrt{x-3} + 4 = 7$.
Answer 12

EXAMPLE 4 Solve $\sqrt{x+3} = 3x - 1$.

Solution

$$\sqrt{x+3} = 3x - 1$$
$$\left(\sqrt{x+3}\right)^2 = (3x - 1)^2 \qquad \text{Square both sides.}$$
$$x + 3 = 9x^2 - 6x + 1 \qquad \text{Remove parentheses.}$$
$$0 = 9x^2 - 7x - 2 \qquad \text{Add } -x - 3 \text{ to both sides.}$$
$$0 = (9x + 2)(x - 1) \qquad \text{Factor } 9x^2 - 7x - 2.$$

$$9x + 2 = 0 \qquad \text{or} \qquad x - 1 = 0 \qquad \text{Set each factor equal to 0.}$$
$$x = -\frac{2}{9} \qquad\qquad\qquad x = 1$$

Since squaring both sides can introduce extraneous roots, we must check each result.

$$\sqrt{x + 3} = 3x - 1 \qquad \text{or} \qquad \sqrt{x + 3} = 3x - 1$$

$$\sqrt{-\frac{2}{9} + 3} \stackrel{?}{=} 3\left(-\frac{2}{9}\right) - 1 \qquad\qquad \sqrt{1 + 3} \stackrel{?}{=} 3(1) - 1$$

$$\sqrt{\frac{25}{9}} \stackrel{?}{=} -\frac{2}{3} - 1 \qquad\qquad\qquad \sqrt{4} \stackrel{?}{=} 3 - 1$$

$$\frac{5}{3} \neq -\frac{5}{3} \qquad\qquad\qquad\qquad 2 = 2$$

Since $-\frac{2}{9}$ does not satisfy the equation, it is extraneous; 1 is the only solution. ∎

Self Check   Solve $\sqrt{x - 2} = 2x - 10$.

Answer   6

EXAMPLE 5   Solve $\sqrt[3]{x^3 + 56} = x + 2$.

Solution   To eliminate the radical, we cube both sides of the equation.

$$\sqrt[3]{x^3 + 56} = x + 2$$

$$\left(\sqrt[3]{x^3 + 56}\right)^3 = (x + 2)^3 \qquad\qquad \text{Cube both sides.}$$

$$x^3 + 56 = x^3 + 6x^2 + 12x + 8 \qquad \text{Remove parentheses.}$$

$$0 = 6x^2 + 12x - 48 \qquad\qquad\quad \text{Simplify.}$$

$$0 = x^2 + 2x - 8 \qquad\qquad\quad\; \text{Divide both sides by 6.}$$

$$0 = (x + 4)(x - 2) \qquad\qquad \text{Factor } x^2 + 2x - 8.$$

$$x + 4 = 0 \qquad \text{or} \qquad x - 2 = 0 \qquad \text{Set each factor equal to 0.}$$

$$x = -4 \qquad\qquad\qquad x = 2$$

We check each suspected solution to see if either is extraneous.

For $x = -4$	*For* $x = 2$
$\sqrt[3]{x^3 + 56} = x + 2$	$\sqrt[3]{x^3 + 56} = x + 2$
$\sqrt[3]{(-4)^3 + 56} \stackrel{?}{=} -4 + 2$	$\sqrt[3]{2^3 + 56} \stackrel{?}{=} 2 + 2$
$\sqrt[3]{-64 + 56} \stackrel{?}{=} -2$	$\sqrt[3]{8 + 56} \stackrel{?}{=} 4$
$\sqrt[3]{-8} \stackrel{?}{=} -2$	$\sqrt[3]{64} \stackrel{?}{=} 4$
$-2 = -2$	$4 = 4$

Since both values satisfy the equation, both -4 and 2 are roots. ∎

Self Check   Solve $\sqrt[3]{x^3 + 7} = x + 1$.

Answer $1, -2$

EXAMPLE 6   Solve $\sqrt{2x + 3} + \sqrt{x - 2} = 4$.

Solution   We can write the equation in the form

$$\sqrt{2x + 3} = 4 - \sqrt{x - 2} \qquad \text{Subtract } \sqrt{x - 2} \text{ from both sides.}$$

so that the left-hand side contains one radical. We then square both sides to get

$$\left(\sqrt{2x + 3}\right)^2 = \left(4 - \sqrt{x - 2}\right)^2$$

$$2x + 3 = 16 - 8\sqrt{x - 2} + x - 2$$

$$2x + 3 = 14 - 8\sqrt{x - 2} + x \qquad \text{Combine like terms.}$$

$$x - 11 = -8\sqrt{x - 2} \qquad \text{Subtract 14 and } x \text{ from both sides.}$$

We then square both sides again to eliminate the radical.

$$(x - 11)^2 = \left(-8\sqrt{x - 2}\right)^2$$

$$x^2 - 22x + 121 = 64(x - 2)$$

$$x^2 - 22x + 121 = 64x - 128$$

$$x^2 - 86x + 249 = 0$$

$$(x - 3)(x - 83) = 0$$

$$x - 3 = 0 \quad \text{or} \quad x - 83 = 0 \qquad \text{Set each factor equal to 0.}$$

$$x = 3 \qquad\qquad x = 83$$

Substituting these results into the equation will show that 83 doesn't check; it is extraneous. However, 3 does satisfy the equation and is a root. ∎

Self Check   Solve $\sqrt{2x + 1} + \sqrt{x + 5} = 6$.

Answer   4

■ ■ ■ ■ ■ ■ ■ ■ ■   **Highway Engineering**

ACCENT ON
TECHNOLOGY

A highway curve banked at 8° will accommodate traffic traveling s mph if the radius of the curve is r feet, according to the formula $s = 1.45\sqrt{r}$. To find what radius is necessary to accommodate 70-mph traffic, highway engineers substitute 70 for s in the formula and solve for r. See Figure 2-7.

$$s = 1.45\sqrt{r}$$

$$\mathbf{70} = 1.45\sqrt{r} \qquad \text{Substitute 70 for } s.$$

$$\frac{70}{1.45} = \sqrt{r} \qquad \text{Divide both sides by 1.45.}$$

$$\left(\frac{70}{1.45}\right)^2 = r \qquad \text{Square both sides.}$$

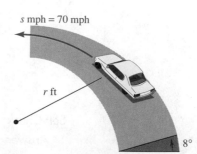

s mph = 70 mph

r ft

8°

FIGURE 2-7

We can use a scientific calculator to find r by entering these numbers and pressing these keys:

$$(\quad 70 \quad \div \quad 1.45 \quad) \quad x^2 \quad =$$

The display will read   2330.558859 . The radius of the curve should be 2331 feet.

■ ■ ■ ■ ■ ■ ■ ■ ■

EXERCISE 2.6

VOCABULARY AND CONCEPTS   *In Exercises 1–4, fill in the blank to make a true statement.*

1. Equal powers of equal real numbers are _____.

2. If a and b are real numbers and $a = b$, then $a^2 = $ ___.

3. False solutions that don't satisfy the equation are called _____ solutions.

4. _____ equations contain radicals with variables in their _____.

PRACTICE   *In Exercises 5–24, solve each equation for real values of the variable by factoring.*

5. $x^3 + 9x^2 + 20x = 0$
   $0, -5, -4$

6. $x^3 + 4x^2 - 21x = 0$

7. $6a^3 - 5a^2 - 4a = 0$
   $0, 4/3, -1/2$

8. $8b^3 - 10b^2 + 3b = 0$

9. $y^4 - 26y^2 + 25 = 0$
   $5, -5, 1$

10. $y^4 - 13y^2 + 36 = 0$

11. $x^4 - 37x^2 + 36 = 0$
   $6, -6, 1$

12. $x^4 - 50x^2 + 49 = 0$

13. $2y^4 - 46y^2 = -180$
   $3\sqrt{2}, -3\sqrt{2}, \sqrt{5}, -\sqrt{5}$

14. $2x^4 - 102x^2 = -196$

15. $z^{3/2} - z^{1/2} = 0$

16. $r^{5/2} - r^{3/2} = 0$

17. $2m^{2/3} + 3m^{1/3} - 2 = 0$

18. $6t^{2/5} + 11t^{1/5} + 3 = 0$

19. $x - 13x^{1/2} + 12 = 0$

20. $p + p^{1/2} - 20 = 0$

21. $2t^{1/3} + 3t^{1/6} - 2 = 0$

22. $z^3 - 7z^{3/2} - 8 = 0$

23. $6p + p^{1/2} - 1 = 0$

24. $3r - r^{1/2} - 2 = 0$

In Exercises 25–56, find all real solutions of each equation.

25. $\sqrt{x - 2} = 5$ 27

26. $3\sqrt{x + 1} = \sqrt{6}$

27. $\sqrt{x + 3} = 2\sqrt{x}$ 1

28. $\sqrt{16x + 4} = \sqrt{x + 4}$

29. $2\sqrt{x^2 + 3} = \sqrt{-16x - 3}$ $-3/2, -5/2$

30. $\sqrt{x^2 + 1} = \dfrac{\sqrt{-7x + 11}}{\sqrt{6}}$

31. $\sqrt[3]{7x + 1} = 4$ 9

32. $\sqrt[3]{11a - 40} = 5$

33. $\sqrt[4]{30t + 25} = 5$ 20

34. $\sqrt[4]{3z + 1} = 2$

35. $\sqrt{x^2 + 21} = x + 3$ 2

36. $\sqrt{5 - x^2} = -(x + 1)$

37. $\sqrt{y + 2} = 4 - y$ 2

38. $\sqrt{3z + 1} = z - 1$

39. $x - \sqrt{7x - 12} = 0$ $3, 4$

40. $x - \sqrt{4x - 4} = 0$

41. $x + 4 = \sqrt{\dfrac{6x + 6}{5}} + 3$ $1/5, -1$

42. $\sqrt{\dfrac{8x + 43}{3}} - 1 = x$

43. $\sqrt{\dfrac{x^2 - 1}{x - 2}} = 2\sqrt{2}$ 3, 5

44. $\dfrac{\sqrt{x^2 - 1}}{\sqrt{3x - 5}} = \sqrt{2}$

45. $\sqrt[3]{x^3 + 7} = x + 1$ -2, 1

46. $\sqrt[3]{x^3 - 7} + 1 = x$

47. $\sqrt[3]{8x^3 + 61} = 2x + 1$ 2 - 5/2

48. $\sqrt[3]{8x^3 - 37} = 2x - 1$

49. $\sqrt{x + 3} = \sqrt{2x + 8} - 1$ -2

50. $\sqrt{x + 2} + 1 = \sqrt{2x + 5}$

51. $\sqrt{y + 8} - \sqrt{y - 4} = -2$ None

52. $\sqrt{r} + \sqrt{r + 2} = 2$

53. $\sqrt{2b + 3} - \sqrt{b + 1} = \sqrt{b - 2}$ 3

54. $\sqrt{a + 1} + \sqrt{3a} = \sqrt{5a + 1}$

55. $\sqrt{\sqrt{b} + \sqrt{b + 8}} = 2$

56. $\sqrt{\sqrt{x + 19} - \sqrt{x - 2}} = \sqrt{3}$

APPLICATIONS

57. Height of a bridge   The distance d (in feet) that an object will fall in t seconds is given by the formula

$$t = \sqrt{\dfrac{d}{16}}$$

To find the height of a bridge above a river, a man drops a stone into the water (see Illustration 1). If it takes the stone 5 seconds to hit the water, how high is the bridge?

ILLUSTRATION 1

58. Horizon distance   The higher a lookout tower, the farther an observer can see. (See Illustration 2.) The distance d called the **horizon distance,** measured in miles) is related to the height h of the observer (measured in feet) by the formula $d = 1.4\sqrt{h}$. How tall must a tower be for the observer to see 30 miles?

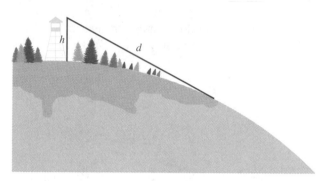

ILLUSTRATION 2

59. Carpentry   During construction, carpenters often brace walls, as shown in Illustration 3. The appropriate length of the brace is given by the formula $l = \sqrt{f^2 + h^2}$. If a carpenter nails a 10-foot brace to the wall 6 feet above the floor, how far from the base of the wall should he nail the brace to the floor?

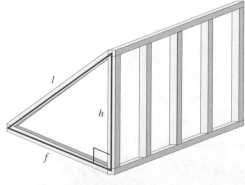

ILLUSTRATION 3

60. Windmills The power generated by a windmill is related to the velocity of the wind by the formula

$$v = \sqrt[3]{\frac{P}{0.02}}$$

where P is the power (in watts) and v is the velocity of the wind (in mph). Find the velocity of the wind when the windmill is generating 600 watts of power.

DISCOVERY AND WRITING

61. Explain why squaring both sides of an equation might introduce extraneous roots.

62. Can cubing both sides of an equation introduce extraneous roots? Explain.

REVIEW *In Exercises 63–64, graph each subset of the real numbers on the number line.*

63. The natural numbers between -4 and 4

64. The integers between -4 and 4

In Exercises 65–70, write each inequality in interval notation and graph the interval.

65. $x \geq 3$

66. $x < -5$

67. $-2 \leq x < 1$

68. $-3 \leq x \leq 3$

69. $x < 1$ or $x \geq 2$

70. $x \leq 1$ or $x > 2$

2.7 Inequalities

■ PROPERTIES OF INEQUALITIES ■ SOLVING LINEAR INEQUALITIES ■ SOLVING COMPOUND INEQUALITIES ■ SOLVING QUADRATIC INEQUALITIES ■ SOLVING RATIONAL INEQUALITIES

We have previously introduced the following symbols.

Symbol	Read as	Examples
$\neq$	"is not equal to"	$8 \neq 10$ and $25 \neq 12$
$<$	"is less than"	$8 < 10$ and $12 < 25$
$>$	"is greater than"	$30 > 10$ and $100 > -5$
$\leq$	"is less than or equal to"	$-6 \leq 12$ and $-8 \leq -8$
$\geq$	"is greater than or equal to"	$12 \geq -5$ and $9 \geq 9$

Since the coordinates of points get larger as we move from left to right on the number line,

$a > b$ if point a lies to the right of point b on a number line.

$a < b$ if point a lies to the left of point b on a number line.

EXAMPLE 1

a. $5 > 3$, because 5 lies to the right of -3 on the number line.

b. $-7 < -2$, because -7 lies to the left of -2 on the number line.

c. $3 \leq 3$, because $3 = 3$.

d. $2 \leq 3$, because $2 < 3$.

e. $x + 1 > x$, because $x + 1$ lies one unit to the right of x on the number line.

■

Self Check Write an inequality symbol to make a true statement: **a.** 25 ▢ 12, **b.** -5 ▢ -5, and **c.** -12 ▢ -20.

Answers **a.** $>$, **b.** $\leq$ or $\geq$, **c.** $>$

■ PROPERTIES OF INEQUALITIES

The Trichotomy Property

For any real numbers a and b, one of the following statements is true.

$$a < b, \qquad a = b, \qquad \text{or} \qquad a > b$$

This property indicates that one of the following statements is true about two real numbers. Either

• the first is less than the second, or
• the first is equal to the second, or
• the first is greater than the second.

The Transitive Property

If a, b, and c are real numbers, then

if $a < b$ and $b < c$, then $a < c$.

if $a > b$ and $b > c$, then $a > c$.

The first part of the transitive property indicates that

• if a first number is less than a second and
• the second number is less than a third, then
• the first number is less than the third.

The second part is similar, with the words "is greater than" substituted for "is less than."

Addition and Subtraction Properties of Inequality

Let a, b, and c represent real numbers.

If $a < b$, then $a + c < b + c$.

If $a < b$, then $a - c < b - c$.

Similar properties exist for $>$, $\leq$, and $\geq$.

This property states that *any real number can be added to (or subtracted from) both sides of an inequality to obtain another inequality with the same order (direction).*

For example, if we add 4 to (or subtract 4 from) both sides of $8 < 12$, we get

$$8 < 12 \qquad\qquad 8 < 12$$
$$8 + 4 < 12 + 4 \qquad 8 - 4 < 12 - 4$$
$$12 < 16 \qquad\qquad 4 < 8$$

and the $<$ symbol is unchanged.

Multiplication and Division Properties of Inequality

Let a, b, and c represent real numbers.

Part 1: If $a < b$ and $c > 0$, then $ca < cb$.

If $a < b$ and $c > 0$, then $\dfrac{a}{c} < \dfrac{b}{c}$.

Part 2: If $a < b$ and $c < 0$, then $ca > cb$.

If $a < b$ and $c < 0$, then $\dfrac{a}{c} > \dfrac{b}{c}$.

Similar properties exist for $>$, $\leq$, and $\geq$.

This property has two parts. Part 1 states that *both sides of an inequality can be multiplied (or divided) by the same positive number to obtain another inequality with the same order.*

For example, if we multiply (or divide) both sides of $8 < 12$ by 4, we get

$$8 < 12 \qquad\qquad 8 < 12$$
$$8(4) < 12(4) \qquad \frac{8}{4} < \frac{12}{4}$$
$$32 < 48 \qquad\qquad$$
$$2 < 3$$

and the $<$ symbol is unchanged.

Hypatia
(370 A.D.–415 A.D.)

Hypatia is the earliest known woman in the history of mathematics. She was a professor at the University of Alexandria.

Part 2 states that *both sides of an inequality can be multiplied (or divided) by the same negative number to obtain another inequality with the opposite order.*

For example, if we multiply (or divide) both sides of $8 < 12$ by -4, we get

$$8 < 12 \qquad\qquad 8 < 12$$
$$8(-4) > 12(-4) \qquad \frac{8}{-4} > \frac{12}{-4}$$
$$-32 > -48 \qquad\qquad -2 > -3$$

and the $<$ symbol is changed to a $>$ symbol.

Because of the previous properties, we can solve inequalities as we do equations. However, we must always remember to change the order of an inequality when multiplying (or dividing) both sides by a negative number.

■ SOLVING LINEAR INEQUALITIES

Linear inequalities are inequalities such as $ax + c < 0$ or $ax - c \geq 0$, where $a \neq 0$. Numbers that make an inequality true when substituted for the variable are solutions of the inequality. Inequalities with the same solution set are called **equivalent inequalities.**

EXAMPLE 2   Solve $3(x + 2) < 8$.

Solution   We proceed as with equations.

$$3(x + 2) < 8$$
$$3x + 6 < 8 \qquad \text{Remove parentheses.}$$
$$3x < 2 \qquad \text{Subtract 6 from both sides.}$$
$$x < \frac{2}{3} \qquad \text{Divide both sides by 3.}$$

FIGURE 2-8

All numbers that are less than $\frac{2}{3}$ are solutions of the inequality. The solution set can be expressed in interval notation as $\left(-\infty, \frac{2}{3}\right)$ and be graphed as in Figure 2-8. ■

Self Check   Solve $5(p - 4) > 25$.

Answer

EXAMPLE 3   Solve $-5(x - 2) \leq 20 + x$.

Solution
$$-5(x - 2) \leq 20 + x$$
$$-5x + 10 \leq 20 + x \qquad \text{Remove parentheses.}$$
$$-6x + 10 \leq 20 \qquad \text{Subtract } x \text{ from both sides.}$$
$$-6x \leq 10 \qquad \text{Subtract 10 from both sides.}$$

We now divide both sides of the inequality by -6, which changes the order of the inequality.

$$x \geq \frac{10}{-6}$$ Divide both sides by -6.

$$x \geq -\frac{5}{3}$$ Simplify the fraction.

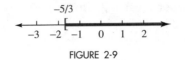

FIGURE 2-9

The graph of the solution set is shown in Figure 2-9. It is the interval $\left[-\frac{5}{3}, \infty\right)$. ■

Self Check   Solve $-4(x + 3) \geq 16$.

Answer

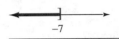

EXAMPLE 4

Hauling corn An empty truck with driver weighs 4350 pounds. It is loaded with feed corn weighing 31 pounds per bushel. Between farm and market is a bridge with a 10,000-pound load limit. How many bushels can the truck legally carry?

Solution   The empty truck with driver weighs 4350 pounds, and corn weighs 31 pounds per bushel. If we let b represent the number of bushels in a legal load, the weight of the corn will be $31b$ pounds. Because the combined weight of the truck, driver, and cargo cannot exceed 10,000 pounds, we can form the following inequality.

The weight of the empty truck	plus	the weight of the corn	must be less than or equal to	10,000 pounds.
4350	+	31b	≤	10,000

We can solve the inequality as follows.

$$4350 + 31b \leq 10,000$$
$$31b \leq 5650 \qquad \text{Subtract 4350 from each side.}$$
$$b \leq 182.2580645 \qquad \text{Divide both sides by 31.}$$

The truck cannot legally carry more than $182\frac{1}{4}$ bushels. ■

■ SOLVING COMPOUND INEQUALITIES

The statement that x is between 2 and 5 implies two inequalities

$$x > 2 \qquad \text{and} \qquad x < 5$$

It is customary to write both inequalities as one **compound inequality:**

$$2 < x < 5 \qquad \text{Read as "2 is less than } x \text{ and } x \text{ is less than 5."}$$

WARNING!   Remember that $2 < x < 5$ means that $x > 2$ and $x < 5$. The word *and* indicates that both inequalities must be true at the same time.

To express that x is not between 2 and 5, we must convey the idea that either x is greater than or equal to 5, or that x is less than or equal to 2. This is equivalent to the compound inequality

$$x \geq 5 \qquad \text{or} \qquad x \leq 2$$

This inequality is satisfied by all numbers x that satisfy one or both of its parts.

 WARNING!   It is incorrect to write $x \geq 5$ or $x \leq 2$ as $2 \geq x \geq 5$, because this would mean that $2 \geq 5$, which is false.

EXAMPLE 5   Solve $5 < 3x - 7 \leq 8$.

Solution   We can isolate x between the inequality symbols by adding 7 to each part of the inequality to get

$$5 + 7 < 3x - 7 + 7 \leq 8 + 7 \qquad \text{Add 7 to each part.}$$
$$12 < 3x \leq 15 \qquad \text{Do the additions.}$$

and dividing all parts by 3 to get

$$4 < x \leq 5$$

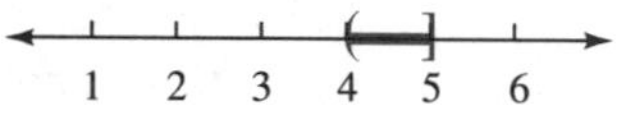

The solution set is the interval $(4, 5]$, whose graph appears in Figure 2-10.   ∎

FIGURE 2-10

Self Check   Solve $-5 \leq 2x + 1 < 9$.

Answer

 EXAMPLE 6   Solve $3 + x \leq 3x + 1 < 7x - 2$.

Solution   Because it is impossible to isolate x between the inequality symbols, we must solve each part of the inequality separately.

$$3 + x \leq 3x + 1 \qquad \text{and} \qquad 3x + 1 < 7x - 2$$
$$3 \leq 2x + 1 \qquad\qquad\qquad 1 < 4x - 2$$
$$2 \leq 2x \qquad\qquad\qquad\quad 3 < 4x$$
$$1 \leq x \qquad\qquad\qquad\quad \frac{3}{4} < x$$
$$x \geq 1 \qquad\qquad\qquad\quad x > \frac{3}{4}$$

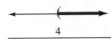

FIGURE 2-11

Since the connective in this inequality is "and," the solution set is the intersection of the intervals $[1, \infty)$ and $\left(\frac{3}{4}, \infty\right)$, which is $[1, \infty)$. The graph is shown in Figure 2-11. ∎

Self Check   Solve $x + 1 < 2x - 3 \leq 3x - 5$.

Answer

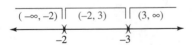

It is possible for an inequality to be true for all values of its variable. It is also possible for an inequality to have no solutions.

$x < x + 1$ is true for all numbers x.

$x > x + 1$ is true for no numbers x.

■ SOLVING QUADRATIC INEQUALITIES

If $a \neq 0$, inequalities like $ax^2 + bx + c < 0$ and $ax^2 + bx + c > 0$ are called **quadratic inequalities.** We will begin by discussing two methods for solving the quadratic inequality $x^2 - x - 6 > 0$.

EXAMPLE 7   Solve $x^2 - x - 6 > 0$.

Solution 1   First we solve the equation $x^2 - x - 6 = 0$.

$$x^2 - x - 6 = 0$$
$$(x + 2)(x - 3) = 0$$
$$x + 2 = 0 \quad \text{or} \quad x - 3 = 0$$
$$x = -2 \qquad\qquad x = 3$$

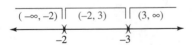

FIGURE 2-12

The graphs of these solutions establish the three intervals shown in Figure 2-12. To decide which intervals are solutions, we test a number in each one and see if it satisfies the inequality.

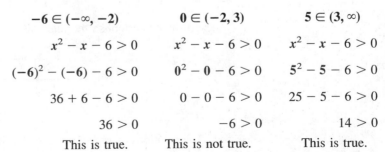

FIGURE 2-13

The solutions are in the intervals $(-\infty, 2)$ and $(3, \infty)$, as shown in Figure 2-13.

Solution 2   Another method relies on the number line and a notation that keeps track of the signs of the factors of $x^2 - x - 6$, which are $(x - 3)(x + 2)$. First, we consider the factor $x - 3$.

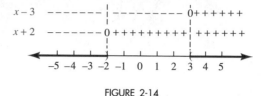

If $x = 3$, then $x - 3 = 0$.

If $x < 3$, then $x - 3$ is negative.

If $x > 3$, then $x - 3$ is positive.

FIGURE 2-14

Then we consider the factor $x + 2$.

If $x = -2$, then $x + 2 = 0$.

If $x < -2$, then $x + 2$ is negative.

If $x > -2$, then $x + 2$ is positive.

We place this information on the **sign graph** shown in Figure 2-14 by using $+$ and $-$ signs. Only to the left of -2 and to the right of 3 do the signs of both factors agree. Only there is the product positive. ∎

Self Check   Solve $x(x + 1) - 2 \geq 4$.

Answer

EXAMPLE 8   Solve $x(x + 3) < -2$.

We remove parentheses and add 2 to both sides to make the right-hand side equal to 0. We then solve $x^2 + 3x + 2 < 0$.

Solution 1   First we solve the equation $x^2 + 3x + 2 = 0$.

$$x^2 + 3x + 2 = 0$$
$$(x + 2)(x + 1) = 0$$
$$x + 2 = 0 \qquad \text{or} \qquad x + 1 = 0$$
$$x = -2 \qquad\qquad\qquad x = -1$$

These solutions establish the intervals shown in Figure 2-15.

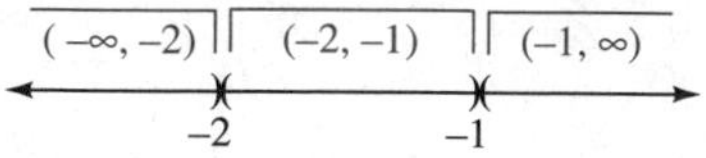

$$(-\infty, -2) \quad\| \quad (-2, -1) \quad \| \quad (-1, \infty)$$

$$\begin{array}{ccc} & \times & \times \\ & -2 & -1 \end{array}$$

FIGURE 2-15

The solutions of $x^2 + 3x + 2 < 0$ will be the numbers in one or more of these intervals. To decide which ones are solutions, we test a number in each interval to see if it satisfies the inequality.

$$-7 \in (-\infty, -2) \qquad\qquad -\tfrac{3}{2} \in (-2, -1) \qquad\qquad 0 \in (-1, \infty)$$

$$x^2 + 3x + 2 < 0 \qquad\qquad x^2 + 3x + 2 < 0 \qquad\quad x^2 + 3x + 2 < 0$$

$$(-7)^2 + 3(-7) + 2 < 0 \qquad \left(-\frac{3}{2}\right)^2 + 3\left(-\frac{3}{2}\right) + 2 < 0 \qquad 0^2 + 3(0) + 2 < 0$$

$$49 - 21 + 2 < 0 \qquad\qquad \frac{9}{4} + \left(-\frac{9}{2}\right) + 2 < 0 \qquad 0 + 0 + 2 < 0$$

$$30 < 0 \qquad\qquad\qquad -\frac{1}{4} < 0 \qquad\qquad\qquad 2 < 0$$

This is not true.   This is true.   This is not true.

The solution is the interval $(-2, -1)$, whose graph appears in Figure 2-16.

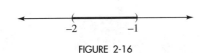

FIGURE 2-16

Solution 2   We construct the sign graph shown in Figure 2-17. Only between -2 and -1 do the factors have opposite signs. Here, the product is negative.

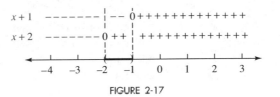

FIGURE 2-17

   ■

Self Check   Solve $x^2 - 5x - 6 < 0$.

Answer

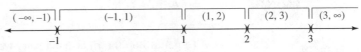

■ SOLVING RATIONAL INEQUALITIES

An inequality that contains a fraction with a polynomial denominator is called a **rational inequality.** To solve it, we can use the same techniques used to solve quadratic inequalities.

EXAMPLE 9   Solve the rational inequality $\dfrac{x^2 - x - 2}{x^2 - 4x + 3} \leq 0$.

Solution 1   The intervals are found by solving $x^2 - x - 2 = 0$ and $x^2 - 4x + 3 = 0$. The solutions of the first equation are -1 and 2, and the solutions of the second equation are 1 and 3. These solutions establish the five intervals shown in Figure 2-18.

$$(-\infty, -1) \mid\mid \quad (-1, 1) \quad \mid\mid \quad (1, 2) \quad \mid\mid \quad (2, 3) \quad \mid\mid \quad (3, \infty)$$

FIGURE 2-18

We test some numbers to see that numbers in the intervals $(-1, 1)$ and $(2, 3)$ satisfy the inequality, but numbers in the intervals $(-\infty, -1)$, $(1, 2)$, and $(3, \infty)$ do not. The graph of the solution set is shown in Figure 2-19.

Because $x = -1$ and $x = 2$ make the numerator 0, they satisfy the inequality. Thus, their graphs are drawn with brackets to show that -1 and 2 are included. Because 1 and 3 give 0's in the denominator, the parentheses at $x = 1$ and $x = 3$ show that 1 and 3 are not in the solution set.

FIGURE 2-19

Solution 2   We factor each trinomial and write the inequality in the form

$$\frac{(x - 2)(x + 1)}{(x - 3)(x - 1)} \le 0$$

We then construct the sign graph shown in Figure 2-20. The value of the fraction will be 0 when $x = 2$ and $x = -1$. The value will be negative when there are an odd number of negative factors. This happens between -1 and 1 and between 2 and 3.

The graph of the solution set appears in Figure 2-20. The brackets at -1 and 2 show that these numbers are in the solution set. The parentheses at 1 and 3 show that these numbers are not in the solution set.

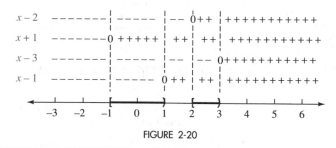

FIGURE 2-20 ■

Self Check   Solve $\dfrac{x^2 + 2x - 3}{x^2 + 4x + 3} > 0$.

Answer

EXAMPLE 10   Solve $\dfrac{6}{x} > 2$.

Solution   To get a 0 on the right-hand side, we add -2 to both sides. We then combine like terms on the left-hand side.

$$\frac{6}{x} > 2$$

$$\frac{6}{x} - 2 > 0$$

$$\frac{6}{x} - \frac{2x}{x} > 0$$

$$\frac{6 - 2x}{x} > 0$$

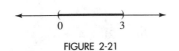

FIGURE 2-21

The inequality now has the form of a rational inequality. The intervals are found by solving $6 - 2x = 0$ and $x = 0$. The solution of the first equation is 3, and the solution of the second equation is 0. This determines the intervals $(-\infty, 0)$, $(0, 3)$, and $(3, \infty)$. Because only the numbers in the interval $(0, 3)$ satisfy the original inequality, the solution set is $(0, 3)$. The graph is shown in Figure 2-21.

We could construct a sign graph as in Figure 2-22 and obtain the same solution set.

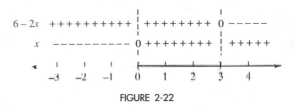

FIGURE 2-22

Self Check   Solve $\frac{2}{x} < 4$.

Answer

WARNING!   It is tempting to solve Example 10 by multiplying both sides by x and solving the inequality $6 > 2x$. However, multiplying both sides by x gives $6 > 2x$ only when x is positive. If x is negative, multiplying both sides by x will reverse the direction of the $>$ symbol, and the inequality $\frac{6}{x} > 2$ will be equivalent to $6 < 2x$. If you fail to consider both cases, you will get a wrong answer.

EXERCISE 2.7

VOCABULARY AND CONCEPTS   *In Exercises 1–12, fill in the blank to get a true statement. Assume that all variables represent real numbers.*

1. If $x > y$, then x lies to the _____ of y on a number line.

2. $a < b$, _____, or $a > b$.

3. If $a < b$ and $b < c$, then _____.

4. If $a < b$, then $a + c <$ _____.

5. If $a < b$, then $a - c <$ _____.

6. If $a < b$ and $c > 0$, then ac ___ bc.

7. If $a < b$ and $c < 0$, then ac ___ bc.

8. If $a < b$ and $c < 0$, then $\frac{a}{c}$ ___ $\frac{b}{c}$.

9. $3x - 5 < 12$ and $ax + c > 0$ are examples of _____ inequalities.

10. $ax^2 + bx - c \geq 0$ and $3x^2 - 6x < 0$ are examples of _____ inequalities.

11. If two inequalities have the same solution set, they are called _____ inequalities.

12. An inequality that contains a fraction with a polynomial denominator is called a _____ inequality.

PRACTICE *In Exercises 13–32, solve each inequality, graph the solution set, and write the answer in interval notation.*

13. $3x + 2 < 5$

$3x < 3$

$x < 3$

14. $-2x + 4 < 6$

15. $3x + 2 \geq 5$

$3x \geq 3$

$x \geq 1$

16. $-2x + 4 \geq 6$

17. $-5x + 3 > -2$

$-5x > 5$

$x > -1$

18. $4x - 3 > -4$

19. $-5x + 3 \leq -2$

$-5x \leq -5$

$x \leq -1$

20. $4x - 3 \leq -4$

21. $2(x - 3) \leq -2(x - 3)$

22. $3(x + 2) \leq 2(x + 5)$

23. $\frac{3}{5}x + 4 > 2$

$3x + 20 > 10$

$3x > -10$

$x > \frac{-10}{3}$

24. $\frac{1}{4}x - 3 > 5$

25. $\frac{x + 3}{4} < \frac{2x - 4}{3}$

$3 \cdot 9x+9 < 4 \cdot 2 \, 4x+14$

26. $\frac{x + 2}{5} > \frac{x - 1}{2}$

27. $\frac{6(x - 4)}{5} \geq \frac{3(x + 2)}{4}$

$24(x-4) \geq 15(x+2)$

$24x-96 \geq 15x+30$

$9x \geq 126$

$x \geq 14$

28. $\frac{3(x + 3)}{2} < \frac{2(x + 7)}{3}$

29. $\frac{5}{9}(a + 3) - a \geq \frac{4}{3}(a - 3) - 1$

30. $\frac{2}{3}y - y \leq -\frac{3}{2}(y - 5)$

31. $\frac{2}{3}a - \frac{3}{4}a < \frac{3}{5}\left(a + \frac{2}{3}\right) + \frac{1}{3}$

32. $\frac{1}{4}b + \frac{2}{3}b - \frac{1}{2} > \frac{1}{2}(b + 1) + b$

In Exercises 33–52, solve each inequality, write the answer in interval notation, and graph the solution set.

33. $4 < 2x - 8 \le 10$

$12 < 2x \le 18$

$6 < x \le 9$

34. $3 \le 2x + 2 < 6$

$1 \le 2x < 4$

$\frac{1}{2} \le x < 2$

35. $9 \ge \dfrac{x-4}{2} > 2$

36. $5 < \dfrac{x-2}{6} < 6$

37. $0 \le \dfrac{4-x}{3} \le 5$

38. $0 \ge \dfrac{5-x}{2} \ge -10$

39. $-2 \ge \dfrac{1-x}{2} \ge -10$

40. $-2 \le \dfrac{1-x}{2} < 10$

41. $-3x > -2x > -x$

42. $-3x < -2x < -x$

43. $x < 2x < 3x$

44. $x > 2x > 3x$

45. $2x + 1 < 3x - 2 < 12$

46. $2 - x < 3x + 5 < 18$

47. $2 + x < 3x - 2 < 5x + 2$

48. $x > 2x + 3 > 4x - 7$

49. $3 + x > 7x - 2 > 5x - 10$

50. $2 - x < 3x + 1 < 10x$

51. $x \le x + 1 \le 2x + 3$

52. $-x \ge -2x + 1 \ge -3x + 1$

In Exercises 53–80, solve each inequality, write the answer in interval notation, and graph the solution set.

53. $x^2 + 7x + 12 < 0$

54. $x^2 - 13x + 12 \le 0$

55. $x^2 - 5x + 6 \ge 0$

56. $6x^2 + 5x - 6 > 0$

57. $x^2 + 5x + 6 < 0$

58. $x^2 + 9x + 20 \ge 0$

59. $6x^2 + 5x + 1 \ge 0$

60. $x^2 + 9x + 20 < 0$

61. $6x^2 - 5x < -1$ **62.** $9x^2 + 24x > -16$ **63.** $2x^2 \geq 3 - x$ **64.** $9x^2 \leq 24x - 16$

65. $\dfrac{x + 3}{x - 2} < 0$ **66.** $\dfrac{x + 3}{x - 2} > 0$ **67.** $\dfrac{x^2 + x}{x^2 - 1} > 0$

68. $\dfrac{x^2 - 4}{x^2 - 9} < 0$ **69.** $\dfrac{x^2 + 5x + 6}{x^2 + x - 6} \geq 0$ **70.** $\dfrac{x^2 + 10x + 25}{x^2 - x - 12} \leq 0$

71. $\dfrac{6x^2 - x - 1}{x^2 + 4x + 4} > 0$ **72.** $\dfrac{6x^2 - 3x - 3}{x^2 - 2x - 8} < 0$ **73.** $\dfrac{3}{x} > 2$

74. $\dfrac{3}{x} < 2$ **75.** $\dfrac{6}{x} < 4$ **76.** $\dfrac{6}{x} > 4$ **77.** $\dfrac{3}{x - 2} \leq 5$

78. $\dfrac{3}{x + 2} \leq 4$ **79.** $\dfrac{6}{x^2 - 1} < 1$ **80.** $\dfrac{6}{x^2 - 1} > 1$

APPLICATIONS *In Exercises 81–88, solve each problem.*

81. Long distance   A long-distance telephone call costs 36¢ for the first three minutes and 11¢ for each additional minute. How long can a person talk for less than $2?

82. Buying computers   A student who can afford to spend up to $2000 sees the ad shown in Illustration 1. If she buys a computer, how many CD-ROMs can she buy?

83. Buying CDs   Andy can spend up to $275 on a CD player and some CDs. If he can buy a disk player for $150 and disks for $9.75, what is the greatest number of disks that he can buy?

ILLUSTRATION 1

84. Buying videotapes Mary wants to spend less than $600 for a VCR and some videotapes. If the VCR of her choice costs $425 and videotapes cost $7.50, how many tapes can she buy?

85. Renting a rototiller The cost of renting a rototiller is $17.50 for the first hour and $8.95 for each additional hour. How long can a person have the rototiller if the cost must be less than $75?

86. Geometry The perimeter of a rectangle is to be between 180 inches and 200 inches. Find the range of values for its length when its width is 40 inches.

87. Geometry The perimeter of an equilateral triangle is to be between 50 centimeters and 60 centimeters. Find the range of lengths of one side.

88. Geometry The perimeter of a square is to be from 25 meters to 60 meters. Find the range of values for its area.

DISCOVERY AND WRITING

89. Mathematics Express the relationship $20 < l < 30$ in terms of P, where $P = 2l + 2w$.

90. Changing temperature scales Express the relationship $10 < C < 20$ in terms of F, where $F = \frac{9}{5}C + 32$.

91. The techniques used for solving linear equations and linear inequalities are similar, yet different. Explain.

92. Explain why the relation $\geq$ is transitive.

REVIEW *In Exercises 93–98,* $A = \left\{-9, -\pi, -2, -\frac{1}{2}, 0, 1, 2, \sqrt{7}, \frac{21}{2}\right\}.$

93. Which numbers are even integers?

94. Which numbers are natural numbers?

95. Which numbers are prime numbers?

96. Which numbers are irrational numbers?

97. Which numbers are real numbers?

98. Which numbers are rational numbers?

2.8 Absolute Value

■ ABSOLUTE VALUE ■ EQUATIONS OF THE FORM $|x| = k$ ■ EQUATIONS WITH TWO ABSOLUTE VALUES ■ INEQUALITIES OF THE FORM $|x| < k$ ■ INEQUALITIES OF THE FORM $|x| > k$ ■ INEQUALITIES WITH TWO ABSOLUTE VALUES

In this section, we review absolute value and examine its consequences in greater detail.

■ ABSOLUTE VALUE

Absolute Value
The **absolute value** of the real number x, denoted as $|x|$, is defined as

If $|x| \geq 0$, then $|x| = x$.
If $|x| < 0$, then $|x| = -x$.

This definition provides a way to associate a nonnegative real number with any real number.

- If $x \geq 0$, then x (which is positive or 0) is its own absolute value.
- If $x < 0$, then $-x$ (which is positive) is the absolute value.

Either way, $|x|$ is positive or 0:

$$|x| \geq 0 \quad \text{for all real numbers } x$$

EXAMPLE 1   Find   **a.** $|7|$,   **b.** $|-3|$,   **c.** $-|-7|$,   and   **d.** $|x - 2|$.

Solution   **a.** Because 7 is positive, $|7| = 7$.

b. Because -3 is negative, $|-3| = -(-3) = 3$.

c. The expression $-|-7|$ means "the negative of the absolute value of -7." Thus, $-|-7| = -(7) = -7$.

d. To denote the absolute value of a variable quantity, we must give a conditional answer.

$$\text{If } x - 2 \geq 0, \text{ then } |x - 2| = x - 2.$$
$$\text{If } x - 2 < 0, \text{ then } |x - 2| = -(x - 2) = 2 - x. \qquad ■$$

Self Check   Find   **a.** $|0|$,   **b.** $|-17|$,   and   **c.** $|x + 5|$.

Answers   **a.** 0;   **b.** $|17|$;   **c.** if $x + 5 \geq 0$, $|x + 5| = x + 5$; if $x + 5 < 0$, $|x + 5| = -x - 5$

■ EQUATIONS OF THE FORM $|x| = k$

In the equation $|x| = 8$, x can be either 8 or -8, because $|8| = 8$ and $|-8| = 8$. In general, the following is true.

Absolute Value Equations
If $k \geq 0$, then
$$|x| = k \quad \text{is equivalent to} \quad x = k \quad \text{or} \quad x = -k$$

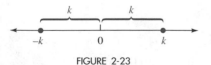

FIGURE 2-23

The absolute value of a number represents the distance on the number line that a point is from the origin. The solutions of $|x| = k$ are the coordinates of the two points that lie exactly k units from the origin. (See Figure 2-23.)

The equation $|x - 3| = 7$ indicates that a point on the number line with a coordinate of $x - 3$ is 7 units from the origin. Thus, $|x - 3|$ can be 7 or -7.

$$x - 3 = 7 \qquad \text{or} \qquad x - 3 = -7$$
$$x = 10 \qquad\qquad\qquad x = -4$$

The solutions of 10 and -4 are shown in Figure 2-24. Both of these numbers satisfy the equation.

$$|x - 3| = 7 \quad \text{and} \quad |x - 3| = 7$$
$$|\mathbf{10} - 3| = 7 \qquad\qquad |\mathbf{-4} - 3| = 7$$
$$|7| = 7 \qquad\qquad\qquad |-7| = 7$$
$$7 = 7 \qquad\qquad\qquad\quad 7 = 7$$

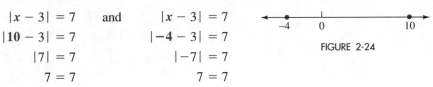

FIGURE 2-24

EXAMPLE 2

Solve $|3x - 5| = 7$.

Solution

The equation $|3x - 5| = 7$ is equivalent to two equations

$$3x - 5 = 7 \quad \text{or} \quad 3x - 5 = -7$$

which can be solved separately:

$$3x - 5 = 7 \quad \text{or} \quad 3x - 5 = -7$$
$$3x = 12 \qquad\qquad\qquad 3x = -2$$
$$x = 4 \qquad\qquad\qquad x = -\frac{2}{3}$$

The solution set consists of the points shown in Figure 2-25.

FIGURE 2-25 ■

Self Check

Solve $|2x + 3| = 7$.

Answer

■ EQUATIONS WITH TWO ABSOLUTE VALUES

The equation $|a| = |b|$ is true when $a = b$ or when $a = -b$. For example,

$$|3| = |3| \quad \text{or} \quad |3| = |-3|$$
$$3 = 3 \qquad\qquad 3 = 3$$

In general, the following is true.

Equations with Two Absolute Values
If a and b represent algebraic expressions, the equation $|a| = |b|$ is equivalent to

$$a = b \quad \text{or} \quad a = -b$$

EXAMPLE 3 $|2x| = |x - 3|$.

Solution   The equation $|2x| = |x - 3|$ will be true when $2x$ and $x - 3$ are equal or when they are negatives. This gives two equations, which can be solved separately:

$$2x = x - 3 \qquad \text{or} \qquad 2x = -(x - 3)$$
$$x = -3 \qquad\qquad\qquad 2x = -x + 3$$
$$3x = 3$$
$$x = 1$$

Verify that -3 and 1 satisfy the equation.   ∎

Self Check   Solve $|3x + 1| = |5x - 3|$.

Answer $2, \frac{1}{4}$

FIGURE 2-26

■ INEQUALITIES OF THE FORM $|x| < k$

The inequality $|x| < 5$ indicates that a point with coordinate x is less than 5 units from the origin. (See Figure 2-26.) Thus, x is between -5 and 5, and

$$|x| < 5 \quad \text{is equivalent to} \quad -5 < x < 5$$

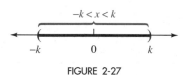

FIGURE 2-27

This suggests that if $k > 0$, the solution set of $|x| < k$ includes the coordinates of the points on the number line that are less than k units from the origin. (See Figure 2-27.)

In general, the following is true.

> **Inequalities of the Form $|x| < k$**
> If $k > 0$, then
> $$|x| < k \quad \text{is equivalent to} \quad -k < x < k$$
> If $k \geq 0$,
> $$|x| \leq k \quad \text{is equivalent to} \quad -k \leq x \leq k$$

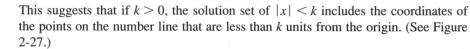

EXAMPLE 4   Solve $|x - 2| < 7$.

Solution   The inequality $|x - 2| < 7$ is equivalent to

$$-7 < x - 2 < 7$$

We can add 2 to each part of this inequality to get

$$-5 < x < 9$$

FIGURE 2-28

The solution set is the interval $(-5, 9)$, shown in Figure 2-28.   ∎

Self Check	Solve $|x + 3| < 9$.
Answer	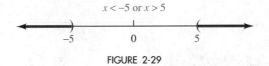

■ INEQUALITIES OF THE FORM $|x| > k$

The inequality $|x| > 5$ indicates that a point with coordinate x is more than 5 units from the origin. (See Figure 2-29.) Thus, $x < -5$ or $x > 5$.

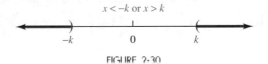

FIGURE 2-29

In general, the inequality $|x| > k$ indicates that a point with coordinate x is more than k units from the origin. (See Figure 2-30.)

FIGURE 2-30

Inequalities of the Form $|x| > k$
If $k \geq 0$, then
$$|x| > k \quad \text{is equivalent to} \quad x < -k \text{ or } x > k$$
$$|x| \geq k \quad \text{is equivalent to} \quad x \leq -k \text{ or } x \geq k$$

EXAMPLE 5   Solve $\left| \dfrac{2x + 3}{2} \right| + 7 \geq 12$.

Solution   We begin by subtracting 7 from both sides of the inequality to isolate the absolute value on the left-hand side.
$$\left| \frac{2x + 3}{2} \right| \geq 5$$

This result is equivalent to two inequalities that can be solved separately.

$$\frac{2x + 3}{2} \leq -5 \qquad \text{or} \qquad \frac{2x + 3}{2} \geq 5$$
$$2x + 3 \leq -10 \qquad\qquad 2x + 3 \geq 10$$
$$2x \leq -13 \qquad\qquad\quad 2x \geq 7$$
$$x \leq -\frac{13}{2} \qquad\qquad\quad x \geq \frac{7}{2}$$

The solution set is the union of the intervals $\left(-\infty, -\frac{13}{2}\right]$ and $\left[\frac{7}{2}, \infty\right)$. Its graph appears in Figure 2-31.

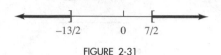

$$-13/2 \qquad 0 \qquad 7/2$$

FIGURE 2-31 ■

Self Check Solve $\left| \dfrac{3x - 6}{3} \right| + 2 \geq 12$.

Answer

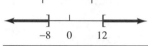

$$-8 \quad 0 \quad 12$$

EXAMPLE 6 Solve $0 < |x - 5| \leq 3$.

Solution The inequality $0 < |x - 5| \leq 3$ consists of two inequalities that can be solved separately. The inequality $0 < |x - 5|$ is true for all x except 5. The inequality $|x - 5| \leq 3$ is equivalent to the inequality

$$-3 \leq x - 5 \leq 3$$
$$2 \leq x \leq 8 \qquad \text{Add 5 to each part.}$$

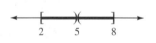

$$2 \quad 5 \quad 8$$

FIGURE 2-32

The solution set is the interval $[2, 8]$, except 5. This is the union of the intervals $[2, 5)$ and $(5, 8]$, as shown in Figure 2-32. ■

Self Check Solve $0 < |x + 2| \leq 5$.

Answer

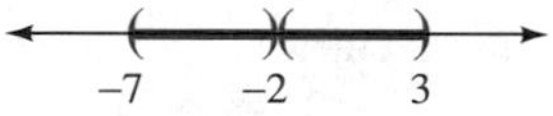

$$-7 \quad -2 \quad 3$$

■ INEQUALITIES WITH TWO ABSOLUTE VALUES

In Example 9 of Section 2.5, we saw that $|a|$ could be defined as

$$|a| = \sqrt{a^2}$$

We will use this fact in the next example.

EXAMPLE 7 Solve $|x + 2| > |x + 1|$.

Solution
$$|x + 2| > |x + 1|$$
$$\sqrt{(x + 2)^2} > \sqrt{(x + 1)^2} \qquad \text{Use } |a| = \sqrt{a^2}.$$
$$(x + 2)^2 > (x + 1)^2 \qquad \text{Square both sides.}$$
$$x^2 + 4x + 4 > x^2 + 2x + 1 \qquad \text{Expand each binomial.}$$

$$4x > 2x - 3 \qquad \text{Subtract } x^2 \text{ and 4 from both sides.}$$
$$2x > -3 \qquad \text{Subtract } 2x \text{ from both sides.}$$
$$x > -\frac{3}{2} \qquad \text{Divide both sides by 2.}$$

The solution set is the interval $\left(-\frac{3}{2}, \infty\right)$. Check several solutions to verify that this interval does not contain any extraneous solutions. ■

Self Check   Solve $|x - 3| \le |x + 2|$.
Answer $\left[\frac{1}{2}, \infty\right)$

Three other properties of absolute value are sometimes useful.

Properties of Absolute Value

If a and b are real numbers, then

1. $|ab| = |a||b|$

2. $\left|\dfrac{a}{b}\right| = \dfrac{|a|}{|b|}$ $(b \ne 0)$

3. $|a + b| \le |a| + |b|$

Properties 1 and 2 state that the absolute value of a product (or a quotient) is the product (or the quotient) of the absolute values.

Property 3 states that the absolute value of a sum is either equal to or less than the sum of the absolute values.

EXERCISE 2.8

VOCABULARY AND CONCEPTS   *In Exercises 1–8, fill in the blank to make a true statement.*

1. If $x \ge 0$, then $|x| = $ ___.

2. If $x < 0$, then $|x| = $ ____.

3. $|x| = k$ is equivalent to _____.

4. $|a| = |b|$ is equivalent to $a = b$ or _____.

5. $|x| < k$ is equivalent to _____.

6. $|x| > k$ is equivalent to _____.

7. $|x| \ge k$ is equivalent to _____.

8. $\sqrt{a^2} = $ ____.

PRACTICE   *In Exercises 9–20, write each expression without using absolute value symbols.*

9. $|7|$

10. $|-9|$

11. $|0|$

12. $|3 - 5|$

13. $|5| - |-3|$

14. $|-3| + |5|$

15. $|\pi - 2|$

16. $|\pi - 4|$

17. $|x - 5|$ and $x > 5$

18. $|x - 5|$ and $x < 5$

19. $|x^3|$

20. $|2x|$

In Exercises 21–44, solve for x.

21. $|x + 2| = 2$

22. $|2x + 5| = 3$

23. $|3x - 1| = 5$

24. $|7x - 5| = 3$

25. $\left|\dfrac{3x - 4}{2}\right| = 5$

26. $\left|\dfrac{10x + 1}{2}\right| = \dfrac{9}{2}$

27. $\left|\dfrac{2x - 4}{5}\right| = 2$

28. $\left|\dfrac{3x + 11}{7}\right| = 1$

29. $\left|\dfrac{x - 3}{4}\right| = -2$

30. $\left|\dfrac{x - 5}{3}\right| = 0$

31. $\left|\dfrac{4x - 2}{x}\right| = 3$

32. $\left|\dfrac{2(x - 3)}{3x}\right| = 6$

33. $|x| = x$

34. $|x| + x = 2$

35. $|x + 3| = |x|$

36. $|x + 5| = |5 - x|$

37. $|x - 3| = |2x + 3|$

38. $|x - 2| = |3x + 8|$

39. $|x + 2| = |x - 2|$

40. $|2x - 3| = |3x - 5|$

41. $\left|\dfrac{x + 3}{2}\right| = |2x - 3|$

42. $\left|\dfrac{x - 2}{3}\right| = |6 - x|$

43. $\left|\dfrac{3x - 1}{2}\right| = \left|\dfrac{2x + 3}{3}\right|$

44. $\left|\dfrac{5x + 2}{3}\right| = \left|\dfrac{x - 1}{4}\right|$

In Exercises 45–60, solve each inequality, express the solution set in interval notation, and graph it.

45. $|x - 3| < 6$

46. $|x - 2| \geq 4$

47. $|x + 3| > 6$

48. $|x + 2| \leq 4$

49. $|2x + 4| \geq 10$

50. $|5x - 2| < 7$

51. $|3x + 5| + 1 \leq 9$

52. $|2x - 7| - 3 > 2$

53. $|x + 3| > 0$

54. $|x - 3| \leq 0$

55. $\left|\dfrac{5x + 2}{3}\right| < 1$

56. $\left|\dfrac{3x + 2}{4}\right| > 2$

57. $3\left|\dfrac{3x - 1}{2}\right| > 5$

58. $2\left|\dfrac{8x + 2}{5}\right| \leq 1$

59. $\dfrac{|x - 1|}{-2} > -3$

60. $\dfrac{|2x - 3|}{-3} < -1$

In Exercises 61–70, solve each inequality, express the solution set in interval notation, and graph it.

61. $0 < |2x + 1| < 3$

62. $0 < |2x - 3| < 1$

63. $8 > |3x - 1| > 3$

64. $8 > |4x - 1| > 5$

65. $2 < \left| \dfrac{x-5}{3} \right| < 4$

66. $3 < \left| \dfrac{x-3}{2} \right| < 5$

67. $10 > \left| \dfrac{x-2}{2} \right| > 4$

68. $5 \geq \left| \dfrac{x+2}{3} \right| > 1$

69. $2 \leq \left| \dfrac{x+1}{3} \right| < 3$

70. $8 > \left| \dfrac{3x+1}{2} \right| > 2$

In Exercises 71–78, solve each inequality and express the solution using interval notation.

71. $|x + 1| \geq |x|$

72. $|x + 1| < |x + 2|$

73. $|2x + 1| < |2x - 1|$

74. $|3x - 2| \geq |3x + 1|$

75. $|x + 1| < |x|$

76. $|x + 2| \leq |x + 1|$

77. $|2x + 1| \geq |2x - 1|$

78. $|3x - 2| < |3x + 1|$

APPLICATIONS

79. Finding temperature ranges The temperatures on a sunny summer day satisfy the inequality $|t - 78°| \leq 8°$, where t is the temperature in degrees Fahrenheit. Express this range as an inequality.

80. Finding operating temperatures A car CD player has an operating temperature of $|t - 40°| < 80°$, where t is the temperature in degrees Fahrenheit. Express this range as an inequality.

81. Range of camber angles The specifications for a certain car state that the camber angle c of its wheels should be $0.6° \pm 0.5°$. Express this range with an inequality containing an absolute value.

82. Tolerance of a sheet of steel A sheet of steel is to be 0.25 inch thick, with a tolerance of 0.015 inch. Express this specification with an inequality containing an absolute value.

DISCOVERY AND WRITING

83. Explain how to find the absolute value of a number.

84. Explain why the equation $|x| + 9 = 0$ has no solution.

85. Explain the use of parentheses and brackets when graphing inequalities.

86. If $k > 0$, explain the differences between the solution sets of $|x| < k$ and $|x| > k$.

REVIEW *In Exercises 87–88, write each number in scientific notation.*

87. 37,250

88. 0.0003725

In Exercises 89–90, write each number in standard notation.

89. 5.23×10^5

90. 7.9×10^{-4}

■ ■ ■ ■ ■ ■ ■ ■ ■ ■ PROBLEMS AND PROJECTS

1. A total of 736 people paid $7772 to attend a concert. Senior citizen tickets cost $7, and all others cost $12. For the rights to perform the music, a 5% royalty on the receipts from senior citizen tickets and a 9% royalty on the remaining receipts must be paid. How much is owed in royalties?

2. Find the distance x required to balance the lever shown in Illustration 1. Research beginning and intermediate algebra books to find the principles from physics that are necessary to solve the problem.

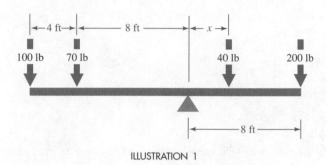

ILLUSTRATION 1

3. Present to a friend the proof of the quadratic formula given in Section 2.3. Then present the following proof.

$$ax^2 + bx + c = 0$$
$$ax^2 + bx = -c$$
$$4a(ax^2) + 4a(bx) = 4a(-c)$$
$$4a^2x^2 + 4abx = -4ac$$
$$4a^2x^2 + 4abx + b^2 = b^2 - 4ac$$
$$(2ax + b)^2 = b^2 - 4ac$$
$$2ax + b = \pm\sqrt{b^2 - 4ac}$$
$$2ax = -b \pm \sqrt{b^2 - 4ac}$$
$$x = \frac{-b \pm \sqrt{b^2 - 4ac}}{2a}$$

Which proof did your friend understand better? Why? Which proof did you prefer?

4. A researcher wants to estimate the mean real estate tax paid by homeowners living in Rockford, IL. To do so, he decides to select a random sample of homeowners and compute the mean tax paid by the homeowners in that sample. How large must the sample be for the researcher to be 95% certain that his computed sample mean will be within $35 of the true population mean—that is, within $35 of the mean tax paid by all homeowners in the city? Assume that the standard deviation, σ, of all tax bills in the city is $120.

From statistics, the researcher has the formula

$$\frac{3.84\sigma^2}{N} < E^2$$

where E is the maximum acceptable error and N is sample size.

PROJECT 1   Use a geometry textbook to find the necessary facts to find x in each problem.

1.

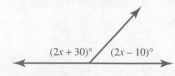

$(2x + 30)°$ $(2x - 10)°$

2.

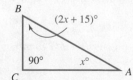

B
$(2x + 15)°$
$90°$ $x°$
C A

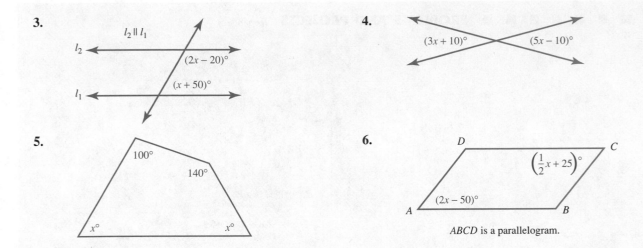

3.

$l_2 \parallel l_1$

l_2

$(2x - 20)°$

$(x + 50)°$

l_1

4.

$(3x + 10)°$ $(5x - 10)°$

5.

100°

140°

$x°$ $x°$

6.

D C

$\left(\frac{1}{2}x + 25\right)°$

$(2x - 50)°$

A B

ABCD is a parallelogram.

PROJECT 2   Find the lengths of sides *a*, *b*, *c*, and *d* shown in Illustration 2. What pattern do you notice? How many triangles will be necessary to produce a side of length 3?

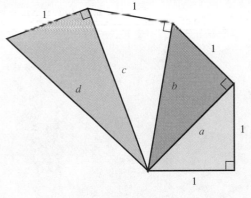

1   1

1

c

d b 1

a

1

1

1

ILLUSTRATION 2

PROJECT 3   Carlos and Mike agree to have a bicycle race. Each will leave his own house at the same time and ride to the other's house, whereupon the winner will call his own house and leave a message for the loser. A map of the race is shown in Illustration 3. Carlos stays on the highway and averages 21 mph. Mike knows that the two are evenly matched when biking on the highway, so he cuts across country for the first part of his trip, averaging 15 mph. When he reaches the highway at point *A*, he turns right and follows the highway, averaging 21 mph. Carlos and Mike never meet during the race, and amazingly, the race is a tie.

(continued)

■ ■ ■ ■ ■ ■ ■ ■ ■ ■   **PROBLEMS AND PROJECTS**   *(continued)*

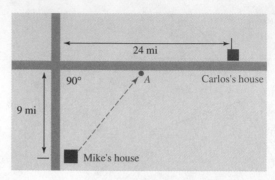

ILLUSTRATION 3

a. To the nearest second, how long did it take each person to complete the race?

b. How far from the intersection of the two highways is point A?

c. Show that if Mike had started straight for Carlos's house, he would have lost the race.

d. Show that if Mike had biked to a point 9 miles from the intersection of the two highways, he would have won.

C H A P T E R   S U M M A R Y

CONCEPT   **REVIEW EXERCISES**

| SECTION 2.1 | *Linear Equations* |

If $a = b$ and c is a number, then
$$a + c = b + c$$
$$a - c = b - c$$
$$ac = bc \quad \text{and} \quad \frac{a}{c} = \frac{b}{c} \ (c \neq 0)$$

A **linear equation** is an equation that can be written in the form
$$ax + b = 0$$

1. Find the restrictions on x, if any.

   a. $3x + 7 = 4$ **b.** $x + \dfrac{1}{x} = 2$

   c. $\sqrt{x} = 4$ **d.** $\dfrac{1}{x - 2} = \dfrac{2}{x - 3}$

2. Solve each equation and classify it as an identity, a conditional equation, or an equation with no solution.

   a. $3(9x + 4) = 28$ **b.** $\dfrac{3}{2}a = 7(a + 11)$

   c. $8(3x - 5) - 4(x + 3) = 12$ **d.** $\dfrac{x + 3}{x + 4} + \dfrac{x + 3}{x + 2} = 2$

   e. $\dfrac{3}{x - 1} = \dfrac{1}{2}$ **f.** $\dfrac{8x^2 + 72x}{9 + x} = 8x$

   g. $\dfrac{3x}{x - 1} - \dfrac{5}{x + 3} = 3$ **h.** $x + \dfrac{1}{2x - 3} = \dfrac{2x^2}{2x - 3}$

3. Solve each formula for the indicated variable.

   a. $C = \dfrac{5}{9}(F - 32); F$ **b.** $P_n = l + \dfrac{si}{f}; f$

   c. $\dfrac{1}{f} = \dfrac{1}{f_1} + \dfrac{1}{f_2}; f_1$ **d.** $S = \dfrac{a - lr}{1 - r}; l$

| SECTION 2.2 | *Applications of Linear Equations* |

Use the following steps to solve problems:

1. Read the problem.
2. Pick a variable to represent the quantity to be found.

4. Preparing a solution A liter of fluid is 50% alcohol. How much water must be added to dilute it to a 20% solution?

5. Washing windows Scott can wash 37 windows in 3 hours, while Bill can wash 27 windows in 2 hours. How long will it take the two of them to wash 100 windows?

3. Form an equation.
4. Solve the equation.
5. Check the solution in the words of the problem.

6. Filling a tank   A tank can be filled in 9 hours by one pipe and in 12 hours by another. How long will it take both pipes to fill the empty tank?

7. Producing brass   How many ounces of pure zinc must be alloyed with 20 ounces of brass that is 30% zinc and 70% copper to produce brass that is 40% zinc?

8. Lending money   A bank lends $10,000, part of it at 11% annual interest and the rest at 14%. If the annual income is $1265, how much was lent at each rate?

9. Producing oriental rugs   An oriental rug manufacturer can use one loom with a setup cost of $750 that can weave a rug for $115. Another loom, with a setup cost of $950, can produce a rug for $95. How many rugs are produced if the costs are the same on each loom?

SECTION 2.3   *Quadratic Equations*

Zero-factor theorem:
If $ab = 0$, then $a = 0$ or $b = 0$.

Square root property:
If $c > 0$, $x^2 = c$ has two real roots:

$$x = \sqrt{c} \quad \text{and} \quad x = -\sqrt{c}$$

Quadratic formula:

$$x = \frac{-b \pm \sqrt{b^2 - 4ac}}{2a} \quad (a \neq 0)$$

Discriminant:
If $b^2 - 4ac > 0$, the roots of $ax^2 + bx + c = 0$ are unequal real numbers.

If $b^2 - 4ac = 0$, the roots of $ax^2 + bx + c = 0$ are equal real numbers.

If $b^2 - 4ac < 0$, the roots of $ax^2 + bx + c = 0$ are nonreal numbers.

10. Solve each equation by factoring.
  a. $2x^2 - x - 6 = 0$
  b. $12x^2 + 13x = 4$
  c. $5x^2 - 8x = 0$
  d. $27x^2 = 30x - 8$

11. Solve each equation by completing the square.
  a. $x^2 - 8x + 15 = 0$
  b. $3x^2 + 18x = -24$
  c. $5x^2 - x - 1 = 0$
  d. $5x^2 - x = 0$

12. Use the quadratic formula to solve each equation.
  a. $x^2 + 5x - 14 = 0$
  b. $3x^2 - 25x = 18$
  c. $5x^2 = 1 - x$
  d. $-5 = a^2 + 2a$

13. Find the value of k that will make the roots of $kx^2 + 4x + 12 = 0$ equal.

14. Find the values of k that will make the roots of $4y^2 + (k + 2)y = 1 - k$ equal.

15. Solve: $\dfrac{4}{a - 4} + \dfrac{4}{a - 1} = 5$.

| **SECTION 2.4** | *Applications of Quadratic Equations* |

16. Fencing a field   A farmer wishes to enclose a rectangular garden with 300 yards of fencing. A river runs along one side of the garden, so no fencing is needed there. What will be the dimensions of the rectangle if the area is 10,450 square yards?

17. Flying rates   A jet plane, flying 120 mph faster than a propeller-driven plane, travels 3,520 miles in 3 hours less time than the propeller plane requires to fly the same distance. How fast does each plane fly?

18. Flight of a ball   A ball thrown into the air reaches a height h (in feet) according to the formula $h = -16t^2 + 64t$, where t is the time elapsed since the ball was thrown. Find the shortest time it will take the ball to reach a height of 48 feet.

19. Width of a walk   A man built a walk of uniform width around a rectangular pool. If the area of the walk is 117 square feet and the dimensions of the pool are 16 feet by 20 feet, how wide is the walk?

| **SECTION 2.5** | *Complex Numbers* |

$a + bi = c + di$
if and only if
$a = c$ and $b = d$

$(a + bi) + (c + di) =$
$\quad (a + c) + (b + d)i$

$(a + bi) - (c + di) =$
$\quad (a - c) + (b - d)i$

$(a + bi)(c + di) =$
$\quad (ac - bd) + (ad + bc)i$

$|a + bi| = \sqrt{a^2 + b^2}$

The **complex conjugate** of $a + bi$ is $a - bi$.

20. Do all operations and express all answers in $a + bi$ form.

   a. Simplify i^{53}

   b. Simplify i^{103}

   c. $(2 - 3i) + (-4 + 2i)$

   d. $(2 - 3i) - (4 + 2i)$

   e. $\left(3 - \sqrt{-36}\right) + \left(\sqrt{-16} + 2\right)$

   f. $\left(3 + \sqrt{-9}\right)\left(2 - \sqrt{-25}\right)$

   g. $\dfrac{3}{i}$

   h. $-\dfrac{2}{i^3}$

   i. $\dfrac{3}{1 + i}$

   j. $\dfrac{2i}{2 - i}$

   k. $\dfrac{3 + i}{3 - i}$

   l. $\dfrac{3 - 2i}{1 + i}$

   m. $|3 - i|$

   n. $\left|\dfrac{1 + i}{1 - i}\right|$

SECTION 2.6	*Polynomial and Radical Equations*

If $a = b$, then $a^2 = b^2$.

21. Solve each equation.

a. $\dfrac{3x}{2} - \dfrac{2x}{x-1} = x - 3$ **b.** $\dfrac{12}{x} - \dfrac{x}{2} = x - 3$

c. $x^4 - 2x^2 + 1 = 0$ **d.** $x^4 + 36 = 37x^2$

e. $a - a^{1/2} - 6 = 0$ **f.** $x^{2/3} + x^{1/3} - 6 = 0$

g. $\sqrt{x-1} + x = 7$ **h.** $\sqrt{a+9} - \sqrt{a} = 3$

i. $\sqrt{5-x} + \sqrt{5+x} = 4$ **j.** $\sqrt{y+5} + \sqrt{y} = 1$

SECTION 2.7	*Inequalities*

If a, b, and c are real numbers:

If $a < b$, $a + c < b + c$ and $a - c < b - c$.

If $a < b$ and $c > 0$, $ac < bc$ and $\dfrac{a}{c} < \dfrac{b}{c}$.

If $a < b$ and $c < 0$, $ac > bc$ and $\dfrac{a}{c} > \dfrac{b}{c}$.

If $a < b$ and $b < c$, then $a < c$.

22. Solve each inequality.

a. $2x - 9 < 5$ **b.** $5x + 3 \geq 2$

c. $\dfrac{5(x-1)}{2} < x$ **d.** $0 \leq \dfrac{3+x}{2} < 4$

e. $(x+2)(x-4) > 0$ **f.** $(x-1)(x+4) < 0$

g. $x^2 - 2x - 3 < 0$ **h.** $2x^2 + x - 3 > 0$

i. $\dfrac{x+2}{x-3} \geq 0$ **j.** $\dfrac{x-1}{x+4} \leq 0$

k. $\dfrac{x^2 + x - 2}{x - 3} \geq 0$ **l.** $\dfrac{5}{x} < 2$

SECTION 2.8 *Absolute Value*

$|x| = \begin{cases} x \text{ when } x \geq 0 \\ -x \text{ when } x < 0 \end{cases}$

If $a \geq 0$, then $|x| = a$ is equivalent to $x = a$ or $x = -a$.

If $a > 0$, then $|x| < a$ is equivalent to $-a < x < a$.

If $a > 0$, then $|x| > a$ is equivalent to $x > a$ or $x < -a$.

$|a| = \sqrt{a^2}$

Properties of absolute value:
1. $|ab| = |a||b|$
2. $\left| \dfrac{a}{b} \right| = \dfrac{|a|}{|b|}$
3. $|a + b| \leq |a| + |b|$

23. Solve each equation or inequality.

a. $|x + 1| = 6$

b. $|2x - 1| = |2x + 1|$

c. $|x + 3| < 3$

d. $|3x - 7| \geq 1$

e. $\left| \dfrac{x + 2}{3} \right| < 1$

f. $\left| \dfrac{x - 3}{4} \right| > 8$

g. $1 < |2x + 3| < 4$

h. $0 < |3x - 4| < 7$

■ Chapter 2 Test

Find all restrictions on x.

1. $\dfrac{x}{x(x-1)}$

2. $\sqrt{x}$

Solve each equation.

3. $7(2a+5)-7=6(a+8)$

4. $\dfrac{3}{x^2-5x-14}=\dfrac{4}{x^2+5x+6}$

5. Solve for x: $z=\dfrac{x-\mu}{\sigma}$.

6. Solve for a: $\dfrac{1}{a}=\dfrac{1}{b}+\dfrac{1}{c}$.

7. A student's average on three tests is 75. If the final is to count as two one-hour tests, what grade must the student make to bring the average up to 80?

8. A woman invests part of $20,000 at 6% interest and the rest at 7%. If her annual interest is $1260, how much did she invest at 6%?

Solve each equation.

9. $4x^2-8x+3=0$

10. $2b^2-12=-5b$

11. Write the quadratic formula.

12. Use the quadratic formula to solve $3x^2-5x-9=0$.

13. Find k such that $x^2+(k+1)x+k+4=0$ will have two equal roots.

14. The height of a projectile shot up into the air is given by the formula $h=-16t^2+128t$. Find the time t required for the projectile to return to its starting point.

Simplify each expression.

15. i^{13}

16. i^0

Do each operation and write all answers in $a+bi$ form.

17. $(4-5i)-(-3+7i)$

18. $(4-5i)(3-7i)$

19. $\dfrac{2}{2-i}$

20. $\dfrac{1+i}{1-i}$

Find each absolute value.

21. $|5-12i|$

22. $\left|\dfrac{1}{3+i}\right|$

Solve each equation.

23. $z^4-13z^2+36=0$

24. $2p^{2/5}-p^{1/5}-1=0$

25. $\sqrt{x+5}=12$

26. $\sqrt{2z+3}=1-\sqrt{z+1}$

Solve each inequality.

27. $5x - 3 \leq 7$

28. $\dfrac{x+3}{4} > \dfrac{2x-4}{3}$

29. $1 + x < 3x - 3 < 4x - 2$

30. $\dfrac{x+2}{x-1} \leq 0$

Solve each equation.

31. $\left| \dfrac{3x+2}{2} \right| = 4$

32. $|x+3| = |x-3|$

Solve each inequality and graph the solution set.

33. $|2x - 5| > 2$

34. $\left| \dfrac{2x+3}{3} \right| \leq 5$

■ Cumulative Review Exercises

In Exercises 1–2, $A = \left\{ -5, -3, -2, 0, 1, \sqrt{2}, 2, \frac{5}{2}, 6, 11 \right\}$.

1. Which numbers are even integers?

2. Which numbers are prime numbers?

In Exercises 3–4, write each inequality as an interval and graph it.

3. $-4 \leq x < 7$

4. $x \geq 2$ or $x < 0$

In Exercises 5–6, tell which property of the real numbers justifies each expression.

5. $(a + b) + c = c + (a + b)$

6. If $x < 3$ and $3 < y$, then $x < y$.

In Exercises 7–14, simplify each expression. Assume that all variables represent positive numbers. Give all answers with positive exponents.

7. $(81a^4)^{1/2}$

8. $81(a^4)^{1/2}$

9. $(a^{-3}b^{-2})^{-2}$

10. $\left(\dfrac{4x^4}{12x^2y} \right)^{-2}$

11. $\left(\dfrac{4x^0y^2}{x^2y}\right)^{-2}$ **12.** $\left(\dfrac{4x^{-5}y^2}{6x^{-2}y^{-3}}\right)^{2}$ **13.** $(a^{1/2}b)^2(ab^{1/2})^2$ **14.** $(a^{1/2}b^{1/2}c)^2$

In Exercises 15–18, rationalize each denominator and simplify.

15. $\dfrac{3}{\sqrt{3}}$ **16.** $\dfrac{2}{\sqrt[3]{4x}}$ **17.** $\dfrac{3}{y-\sqrt{3}}$ **18.** $\dfrac{3x}{\sqrt{x}-1}$

In Exercises 19–22, simplify each expression and combine like terms.

19. $\sqrt{75}-3\sqrt{5}$ **20.** $\sqrt{18}+\sqrt{8}-2\sqrt{2}$ **21.** $\left(\sqrt{2}-\sqrt{3}\right)^2$ **22.** $\left(3-\sqrt{5}\right)\left(3+\sqrt{5}\right)$

In Exercises 23–28, do the operations and simplify when necessary.

23. $(3x^2-2x+5)-3(x^2+2x-1)$ **24.** $5x^2(2x^2-x)+x(x^2-x^3)$

25. $(3x-5)(2x+7)$ **26.** $(z+2)(z^2-z+2)$

27. $3x+2\overline{)6x^3+x^2+x+2}$ **28.** $x^2+2\overline{)3x^4+7x^2-x+2}$

In Exercises 29–32, factor each polynomial.

29. $3t^2-6t$ **30.** $3x^2-10x-8$ **31.** x^8-2x^4+1 **32.** x^6-1

In Exercises 33–38, do the operations and simplify.

33. $\dfrac{x^2-4}{x^2+5x+6}\cdot\dfrac{x^2-2x-15}{x^2+3x-10}$ **34.** $\dfrac{6x^3+x^2-x}{x+2}\div\dfrac{3x^2-x}{x^2+4x+4}$

35. $\dfrac{2}{x+3}+\dfrac{5x}{x-3}$ **36.** $\dfrac{x-2}{x+3}\left(\dfrac{x+3}{x^2-4}-1\right)$

37. $\dfrac{\dfrac{1}{a}+\dfrac{1}{b}}{\dfrac{1}{ab}}$ **38.** $\dfrac{x^{-1}-y^{-1}}{x-y}$

In Exercises 39–40, solve each equation.

39. $\dfrac{3x}{x+5}=\dfrac{x}{x-5}$ **40.** $8(2x-3)-3(5x+2)=4$

In Exercises 41–42, solve each formula for the indicated variable.

41. $\dfrac{1}{R} = \dfrac{1}{R_1} + \dfrac{1}{R_2}; R$

42. $S = \dfrac{a - lr}{1 - r}; r$

43. Gardening A gardener wishes to enclose her rectangular raspberry patch with 40 feet of fencing. The raspberry bushes are planted along the garage, so no fencing is needed on that side. What will be the dimensions if the total area is 192 square feet?

44. Financial planning A college student invests part of a $25,000 inheritance at 7% interest and the rest at 6%. If his annual interest is $1670, how much did he invest at 6%?

In Exercises 45–48, do the operations and express all answers in a + bi form.

45. $\dfrac{2 + i}{2 - i}$

46. $\dfrac{i(3 - i)}{(1 + i)(1 + i)}$

47. $|3 + 4i|$

48. $\dfrac{5}{i^7} + 5i$

In Exercises 49–52, solve each equation.

49. $\dfrac{x + 3}{x - 1} - \dfrac{6}{x} = 1$

50. $x^4 + 36 = 13x^2$

51. $\sqrt{y + 2} + \sqrt{11 - y} = 5$

52. $z^{2/3} - 13z^{1/3} + 36 = 0$

In Exercises 53–58, graph the solution set of each inequality.

53. $5x - 7 \leq 4$

54. $x^2 - 8x + 15 > 0$

55. $\dfrac{x^2 + 4x + 3}{x - 2} \geq 0$

56. $\dfrac{9}{x} > x$

57. $|2x - 3| \geq 5$

58. $\left| \dfrac{3x - 5}{2} \right| < 2$

3 The Rectangular Coordinate System and Graphs of Equations

MATHEMATICAL EXPRESSIONS OFTEN INDICATE RELATIONSHIPS BETWEEN TWO VARIABLES. TO VISUALIZE THESE RELATIONSHIPS, WE DRAW GRAPHS OF THEIR EQUATIONS.

3.1 The Rectangular Coordinate System

■ THE RECTANGULAR COORDINATE SYSTEM ■ GRAPHING LINEAR EQUATIONS ■ GRAPHING HORIZONTAL AND VERTICAL LINES ■ APPLICATIONS ■ THE DISTANCE FORMULA ■ THE MIDPOINT FORMULA

René Descartes
(1596–1650)

Descartes is famous for his work in philosophy as well as for his work in mathematics. His philosophy is expressed in the words, "I think, therefore I am." He is best known in mathematics for his invention of a coordinate system and his work with conic sections.

In algebra, we often present information in tables and graphs. For example, several solutions of the equation $y = -\frac{1}{2}x + 4$ are listed in the table shown in Figure 3-1.

$y = -\frac{1}{2}x + 4$

x	y
-4	6
-2	5
0	4
2	3
4	2

The solutions to equations with two variables are pairs of real numbers.

FIGURE 3-1

Each pair of x- and y-values in the table is a solution of the equation, because each pair satisfies it. To show that $x = 2$ and $y = 3$ satisfies the equation, we substitute 2 for x and 3 for y and simplify.

$$y = -\frac{1}{2}x + 4$$

$$3 = -\frac{1}{2}(2) + 4 \qquad \text{Substitute 2 for } x \text{ and 3 for } y.$$

$$3 = -1 + 4$$

$$3 = 3$$

Since the left-hand side equals the right-hand side, the equation is satisfied.

Before we can present the information shown in Figure 3-1 in graphical form, we need to discuss the rectangular coordinate system.

■ THE RECTANGULAR COORDINATE SYSTEM

The **rectangular coordinate system** consists of two perpendicular number lines that divide the plane into four **quadrants,** numbered as in Figure 3-2. The horizontal number line is called the **x-axis,** and the vertical number line is called the **y-axis.** These axes intersect at the **origin,** which is the 0 point on each axis. The positive direction on the x-axis is to the right, the positive direction on the y-axis is upward, and the same unit distance is used on both axes.

To plot (or graph) the point associated with the pair $x = 2$ and $y = 3$, denoted as $(2, 3)$, we start at the origin, count 2 units to the right, and then count 3 units up. (See Figure 3-3.) Point P (which lies in the first quadrant) is the graph of the pair $(2, 3)$. The pair $(2, 3)$ gives the **coordinates** of point P.

To plot point Q with coordinates $(-4, 6)$, we start at the origin, count 4 units to the left, and then count 6 units up. Point Q lies in the second quadrant. Point R with coordinates $(6, -4)$ lies in the fourth quadrant.

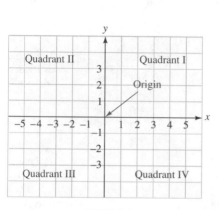

FIGURE 3-2

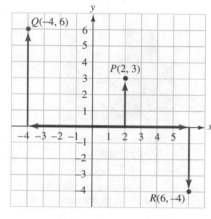

FIGURE 3-3

WARNING!   The pairs $(-4, 6)$ and $(6, -4)$ represent different points. One is in the second quadrant, and one is in the fourth quadrant. Since order is important when graphing pairs of real numbers, such pairs are called **ordered pairs.**

The first coordinate in the ordered pair $(-4, 6)$ is the **x-coordinate** or **abscissa.** The second coordinate is the **y-coordinate** or **ordinate.**

■ GRAPHING LINEAR EQUATIONS

The **graph of the equation** $y = -\frac{1}{2}x + 4$ is the graph of all points (x, y) on the rectangular coordinate system whose coordinates satisfy the equation. To graph $y = -\frac{1}{2}x + 4$, we plot the pairs listed in the table shown in Figure 3-4. These points lie on the line shown in the figure.

When we say that the graph of an equation is a line, we imply two things:

1. Every point with coordinates that satisfy the equation will lie on the line.
2. Any point on the line will have coordinates that satisfy the equation.

When the graph of an equation is a line, we call the equation a **linear equation.** These equations are often written in the form $Ax + By = C$, where A, B, and C stand for specific numbers (called **constants**) and x and y are variables.

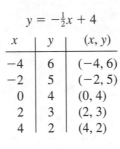

$$y = -\tfrac{1}{2}x + 4$$

x	y	(x, y)
-4	6	$(-4, 6)$
-2	5	$(-2, 5)$
0	4	$(0, 4)$
2	3	$(2, 3)$
4	2	$(4, 2)$

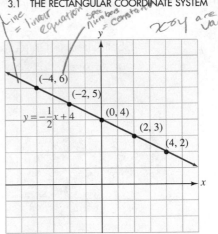

FIGURE 3-4

EXAMPLE 1   Graph $x + 2y = 5$.

Solution   We pick values for x or y, substitute them into the equation, and solve for the other variable. If $y = -1$, we find x as follows.

$$x + 2y = 5$$
$$x + 2(-1) = 5 \qquad \text{Substitute } -1 \text{ for } y.$$
$$x - 2 = 5 \qquad \text{Simplify}$$
$$x = 7 \qquad \text{Add 2 to both sides.}$$

The pair $(7, -1)$ satisfies the equation. To find another, we let $x = 0$ and find y.

$$x + 2y = 5$$
$$0 + 2y = 5 \qquad \text{Substitute 0 for } x.$$
$$2y = 5 \qquad \text{Simplify.}$$
$$y = \frac{5}{2} \qquad \text{Divide both sides by 2.}$$

The pair $\left(0, \tfrac{5}{2}\right)$ also satisfies the equation.

These pairs and others that satisfy the equation are shown in Figure 3-5. We plot the points and join them with a line to get the graph of the equation.

$$x + 2y = 5$$

x	y	(x, y)
7	-1	$(7, -1)$
0	$\frac{5}{2}$	$\left(0, \frac{5}{2}\right)$
5	0	$(5, 0)$
3	1	$(3, 1)$
1	2	$(1, 2)$

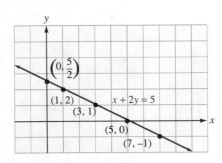

FIGURE 3-5

Self Check   Graph $3x - 2y = 6$.

Answer

In Example 1, the graph crossed the y-axis at the point with coordinates $\left(0, \frac{5}{2}\right)$, called the **$y$-intercept.** It crossed the x-axis at the point with coordinates $(5, 0)$, called the **x-intercept.**

Intercepts of a Line

The **y-intercept** of a line is the point $(0, b)$, where the line intersects the y-axis. To find b, substitute 0 for x in the equation of the line and solve for y.

The **x-intercept** of a line is the point $(a, 0)$, where the line intersects the x-axis. To find a, substitute 0 for y in the equation of the line and solve for x.

EXAMPLE 2   Use the x- and y-intercepts to graph the equation $3x + 2y = 12$.

Solution   To find the y-intercept, we substitute 0 for x and solve for y.

$$3x + 2y = 12$$
$$3(\mathbf{0}) + 2y = 12 \qquad \text{Substitute 0 for } x.$$
$$2y = 12 \qquad \text{Simplify.}$$
$$y = 6 \qquad \text{Divide both sides by 2.}$$

The y-intercept is the point $(0, 6)$. To find the x-intercept, we substitute 0 for y and solve for x.

$$3x + 2y = 12$$
$$3x + 2(\mathbf{0}) = 12 \qquad \text{Substitute 0 for } y.$$
$$3x = 12 \qquad \text{Simplify.}$$
$$x = 4 \qquad \text{Divide both sides by 3.}$$

The x-intercept is the point $(4, 0)$. Although two points are enough to draw the line, it is a good idea to find and plot a third point as a check. If we let $x = 2$, we will find that $y = 3$.

We plot each pair (as in Figure 3-6) and join them with a line to get the graph of the equation.

$3x + 2y = 12$

x	y	(x, y)
0	6	$(0, 6)$
4	0	$(4, 0)$
2	3	$(2, 3)$

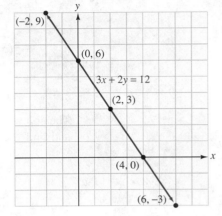

FIGURE 3-6

■

Self Check   Graph $2x - 3y = 12$.

Answer

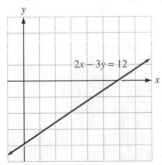

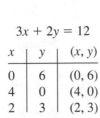

EXAMPLE 3   Graph $3(y + 2) = 2x - 3$.

Solution   To make it easier to find pairs (x, y) that satisfy the equation, we solve the equation for y.

$$3(y + 2) = 2x - 3$$
$$3y + 6 = 2x - 3 \qquad \text{Remove parentheses.}$$
$$3y = 2x - 9 \qquad \text{Subtract 6 from both sides.}$$
$$y = \frac{2}{3}x - 3 \qquad \text{Divide both sides by 3.}$$

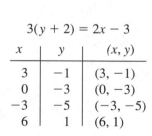

Solve for y, but not x
Plug in random(?) numbers
as values for x to
solve for y again
you get when x =
y = pairs

We now substitute numbers for x to find the corresponding values of y. If we let $x = 3$, we get

$$y = \frac{2}{3}x - 3$$

$$y = \frac{2}{3}(3) - 3 \qquad \text{Substitute 3 for } x.$$

$$y = 2 - 3 \qquad \text{Simplify.}$$

$$y = -1$$

The point $(3, -1)$ lies on the graph. To find the y-intercept of the graph, we let $x = 0$ and find y:

$$y = \frac{2}{3}x - 3$$

$$y = \frac{2}{3}(0) - 3 \qquad \text{Substitute 0 for } x.$$

$$y = -3 \qquad \text{Simplify.}$$

The point $(0, -3)$ lies on the graph. We plot these points and others, as in Figure 3-7, and draw the line.

$$3(y + 2) = 2x - 3$$

x	y	(x, y)
3	−1	$(3, -1)$
0	−3	$(0, -3)$
−3	−5	$(-3, -5)$
6	1	$(6, 1)$

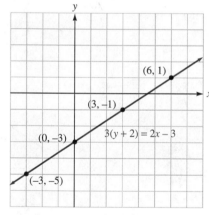

FIGURE 3-7

Self Check
Answer

Graph $2(x - 1) = 6 - 8y$.

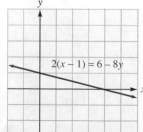

$2x - 2 = 6 - 8y$
$2x + 8y = 6 + 2$
$2x + 8y = 8$

$x = 4$
$y = 1$

■ GRAPHING HORIZONTAL AND VERTICAL LINES

EXAMPLE 4   Graph   **a.** $y = 2$ and   **b.** $x = -3$.

Solution   **a.** In the equation $y = 2$, the value of y is always 2. After plotting the pairs shown in Figure 3-8, we see that the graph is a horizontal line, parallel to the x-axis and having a y-intercept of $(0, 2)$. The line has no x-intercept.

b. In the equation $x = -3$, the value of x is always -3. After plotting the pairs shown in Figure 3-8, we see that the graph is a vertical line, parallel to the y-axis and having an x-intercept of $(-3, 0)$. The line has no y-intercept.

$y = 2$

x	y	(x, y)
-3	2	$(-3, 2)$
0	2	$(0, 2)$
2	2	$(2, 2)$
4	2	$(4, 2)$

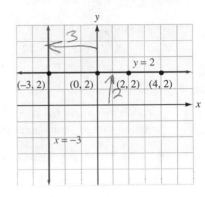

$x = -3$

x	y	(x, y)
-3	-2	$(-3, -2)$
-3	0	$(-3, 0)$
-3	2	$(-3, 2)$
-3	3	$(-3, 3)$

FIGURE 3-8   ■

Self Check   Graph   **a.** $x = 2$ and   **b.** $y = -3$.

Answer

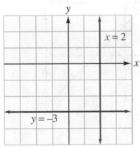

Example 4 suggests the following facts.

Equations of Vertical and Horizontal Lines
If a and b are real numbers, then

• The graph of the equation $x = a$ is a vertical line with x-intercept of $(a, 0)$. If $a = 0$, the line is the y-axis.

• The graph of the equation $y = b$ is a horizontal line with y-intercept of $(0, b)$. If $b = 0$, the line is the x-axis.

■ ■ ■ ■ ■ ■ ■ ■ ■ ■   **Graphing Devices**

ACCENT ON
TECHNOLOGY

We can graph equations with a graphing calculator or computer software. These devices all have a window to display graphs. (See Figure 3-9.) To see a graph, we must choose the minimum and maximum values for the x- and y-coordinates. A window with standard settings of

$$\text{Xmin} = -10 \qquad \text{Xmax} = 10 \qquad \text{Ymin} = -10 \qquad \text{Ymax} = 10$$

will produce a graph where the value of x is in the interval $[-10, 10]$, and y is in the interval $[-10, 10]$.

To use a graphing calculator to graph $3x + 2y = 12$, we must first solve the equation for y.

$$3x + 2y = 12$$
$$2y = -3x + 12 \qquad \text{Subtract } 3x \text{ from both sides.}$$
$$y = -\frac{3}{2}x + 6 \qquad \text{Divide both sides by 2.}$$

We then enter the right-hand side after a symbol like "$Y_1 =$" or "$f(x) =$." The window should look something like this:

$$Y_1 = -(3/2)X + 6 \qquad \text{or} \qquad f(x) = -(3/2)X + 6$$

We then use a GRAPH command to see the graph shown in Figure 3-10(a). To show more detail, we can draw the graph in a different window. A window with settings of $[-1, 5]$ for x and $[-2, 7]$ for y will give the graph shown in Figure 3-10(b).

FIGURE 3-9
(Courtesy of Texas Instruments)

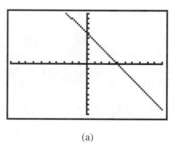 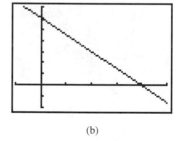

(a)   (b)

FIGURE 3-10

We can use a TRACE command to find the coordinates of any point on a graph. After pressing the ⬛TRACE⬛ key, a flashing cursor will appear on the screen. The coordinates of the cursor will also appear at the bottom of the screen.

To find the x-intercept of the graph of $2y = -5x - 7$, we graph the equation using $[-10, 10]$ for x and $[-10, 10]$ for y and use TRACE to get Figure 3-11(a). We can then move the cursor along the line toward the x-intercept until we arrive at a point with the coordinates shown in Figure 3-11(b).

To get better results, we can zoom in to get a magnified picture, trace again, and move the cursor to the point with coordinates shown in Figure 3-11(c). Since the y-coordinate is almost 0, this point is nearly the x-intercept.

We can achieve better results with repeated zooms.

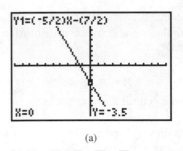
(a)

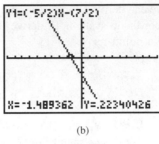
(b)

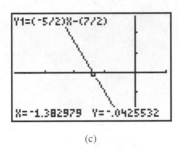
(c)

FIGURE 3-11

■ APPLICATIONS

EXAMPLE 5

Depreciation   A computer purchased for $4750 is expected to depreciate according to the formula $y = -950x + 4750$, where y is the value of the computer after x years. When will the computer have no value?

Solution   The computer will have no value when its value (y) is 0. To find x when $y = 0$, we substitute 0 for y and solve for x.

$$y = -950x + 4750$$
$$0 = -950x + 4750$$
$$-4750 = -950x \qquad \text{Subtract 4750 from both sides.}$$
$$5 = x \qquad \text{Divide both sides by } -950.$$

The computer will have no value in 5 years.   ■

■ ■ ■ ■ ■ ■ ■ ■ ■ **Depreciation**

ACCENT ON
TECHNOLOGY

To solve Example 5 with a graphing calculator, we graph $y = -950x + 4750$ in the window $X = [-10, 10]$ and $Y = [-10, 5000]$, as shown in Figure 3-12(a). We chose the maximum y-value of 5000 because y is almost 5000 when $x = 0$. We then trace to get Figure 3-12(b). We then zoom and trace again to get Figure 3-12(c), which shows that $y = 0$ when $x = 5$.

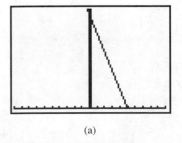
(a)

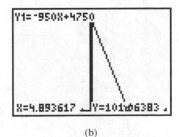
(b)

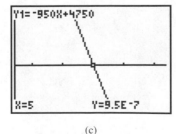
(c)

FIGURE 3-12

■ THE DISTANCE FORMULA

We can derive a formula to find the distance between any two points in a rectangular coordinate system. To derive this formula, we use **subscript notation** and denote two points as

$P(x_1, y_1)$ Read as "point P with coordinates of x sub 1 and y sub 1."

$Q(x_2, y_2)$ Read as "point Q with coordinates of x sub 2 and y sub 2."

If $P(x_1, y_1)$ and $Q(x_2, y_2)$ are two points in Figure 3-13 and point R has coordinates (x_2, y_1), triangle PQR is a right triangle. By the Pythagorean theorem, the square of the hypotenuse of right triangle PQR is equal to the sum of the squares of the two legs. Because leg RQ is vertical, the square of its length is $(y_2 - y_1)^2$. Since leg PR is horizontal, the square of its length is $(x_2 - x_1)^2$. Thus, we have

1. $d^2 = (x_2 - x_1)^2 + (y_2 - y_1)^2$

Because equal positive numbers have equal positive square roots, we can take the positive square root of both sides of Equation 1 to obtain the **distance formula.**

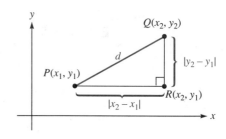

FIGURE 3-13

The Distance Formula
The distance d between points (x_1, y_1) and (x_2, y_2) is given by

$$d = \sqrt{(x_2 - x_1)^2 + (y_2 - y_1)^2} \quad \left(\text{diff of } x\text{'s}\right)^2 + \left(\text{diff of } y\text{'s}\right)^2$$

EXAMPLE 6   Find the distance between $P(-1, -2)$ and $Q(-7, 8)$.

Solution   If we let $P(-1, -2) = P(x_1, y_1)$ and $Q(-7, 8) = Q(x_2, y_2)$, we can substitute -1 for x_1, -2 for y_1, -7 for x_2, and 8 for y_2 into the formula and simplify.

$$d(PQ) = \sqrt{(x_2 - x_1)^2 + (y_2 - y_1)^2}$$
$$d(PQ) = \sqrt{[-7 - (-1)]^2 + [8 - (-2)]^2}$$
$$= \sqrt{(-6)^2 + (10)^2}$$
$$= \sqrt{36 + 100}$$
$$= \sqrt{136}$$
$$= \sqrt{4 \cdot 34}$$
$$= 2\sqrt{34} \qquad\qquad \sqrt{4 \cdot 34} = \sqrt{4}\sqrt{34} = 2\sqrt{34}.$$

■

Self Check   Find the distance between $P(-2, -5)$ and $Q(3, 7)$.

Answer   13

■ THE MIDPOINT FORMULA

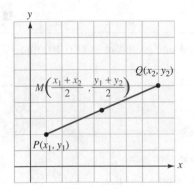

FIGURE 3-14

If point M in Figure 3-14 lies midway between points $P(x_1, y_1)$ and $Q(x_2, y_2)$, point M is called the **midpoint** of segment PQ. To find the coordinates of M, we find the average of the x-coordinates and the average of the y-coordinates of P and Q.

The Midpoint Formula
The midpoint of the line with endpoints at $P(x_1, y_1)$ and $Q(x_2, y_2)$ is the point M with coordinates of

$$M\left(\frac{x_1 + x_2}{2}, \frac{y_1 + y_2}{2}\right)$$

(average of x's, average of y's)

EXAMPLE 7   Find the midpoint of the segment joining $P(-7, 2)$ and $Q(1, -4)$.

Solution   We substitute $P(-7, 2)$ for $P(x_1, y_1)$ and $Q(1, -4)$ for $Q(x_2, y_2)$ into the midpoint formula to get

$$x_M = \frac{x_1 + x_2}{2} \qquad \text{and} \qquad y_M = \frac{y_1 + y_2}{2}$$

$$= \frac{-7 + 1}{2} \qquad\qquad\qquad = \frac{2 + (-4)}{2}$$

$$= \frac{-6}{2} \qquad\qquad\qquad\qquad = \frac{-2}{2}$$

$$= -3 \qquad\qquad\qquad\qquad\quad = -1$$

The midpoint is $M(-3, -1)$.   ■

Self Check   Find the midpoint of the segment joining $P(-7, -8)$ and $Q(-2, 10)$.

Answer $M\left(-\frac{9}{2}, 1\right)$

EXAMPLE 8   The midpoint of the segment joining $P(-3, 2)$ and $Q(x_2, y_2)$ is $M(1, 4)$. Find the coordinates of Q.

Solution   We can let $P(x_1, y_1) = P(-3, 2)$ and $M(x_M, y_M) = M(1, 4)$, and then find the coordinates x_2 and y_2 of point $Q(x_2, y_2)$.

$$x_M = \frac{x_1 + x_2}{2} \quad \text{and} \quad y_M = \frac{y_1 + y_2}{2}$$

$$1 = \frac{-3 + x_2}{2} \qquad\qquad 4 = \frac{2 + y_2}{2}$$

$$2 = -3 + x_2 \qquad\qquad 8 = 2 + y_2 \qquad \text{Multiply both sides by 2.}$$

$$5 = x_2 \qquad\qquad 6 = y_2$$

The coordinates of point Q are $(5, 6)$. ∎

Self Check If the midpoint of a segment PQ is $M(2, -5)$ and one endpoint is $Q(6, 9)$, find P.
Answer $(-2, -19)$

EXERCISE 3.1

VOCABULARY AND CONCEPTS   *In Exercises 1–14, fill in the blank to make a true statement.*

1. The coordinate axes divide the plane into four <u>quadrants</u>

2. The coordinate axes intersect at the <u>origin</u>.

3. The positive direction on the x-axis is <u>right</u>.

4. The positive direction on the y-axis is <u>up</u>.

5. The x-coordinate is also called the _____.

6. The y-coordinate is also called the _____.

7. A <u>linear</u> equation is an equation whose graph is a line.

8. The point where a line intersects the <u>y-axis</u> is called the y-intercept.

9. The point where a line intersects the x-axis is called the <u>x intersect</u>

10. The graph of the equation $x = a$ will be a <u>vertical</u> line.

11. The graph of the equation $y = b$ will be a <u>horizontal</u> line.

12. Complete the distance formula: _____
$$d = \underline{\sqrt{(x_1 - x_2)^2 + (y_1 - y_2)^2}}$$

13. If a point divides a segment into two equal segments, the point is called the <u>midpoint</u> of the segment.

14. The midpoint of the segment joining $P(x_1, y_1)$ and $Q(x_2, y_2)$ is

_____.

PRACTICE   *In Exercises 15–22, refer to Illustration 1 and determine the coordinates of each point.*

15. *A*
16. *B*
17. *C*
18. *D*

19. *E*
20. *F*
21. *G*
22. *H*

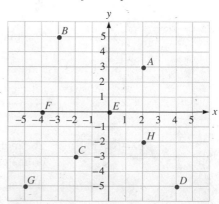

ILLUSTRATION 1

In Exercises 23–30, graph each point. Indicate the quadrant in which the point lies, or the axis on which it lies.

23. $(2, 5)$ **24.** $(-3, 4)$ **25.** $(-4, -5)$ **26.** $(6, 2)$

27. $(5, 2)$ **28.** $(3, -4)$ **29.** $(4, 0)$ **30.** $(0, 2)$

In Exercises 31–38, find the x- and y-intercepts and use them to graph each equation.

31. $x + y = 5$ **32.** $x - y = 3$ **33.** $2x - y = 4$ **34.** $3x + y = 9$

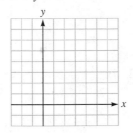 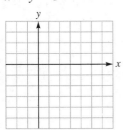 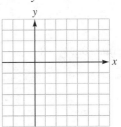 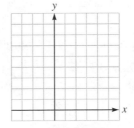

35. $3x + 2y = 6$ **36.** $2x - 3y = 6$ **37.** $4x - 5y = 20$ **38.** $3x - 5y = 15$

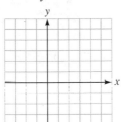 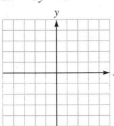 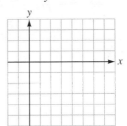 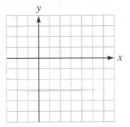

In Exercises 39–48, solve each equation for y and graph the equation. Then check your graph with a graphing calculator.

39. $y - 2x = 7$ **40.** $y + 3 = -4x$ **41.** $y + 5x = 5$ **42.** $y - 3x = 6$

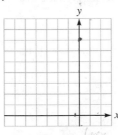 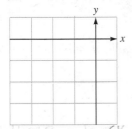 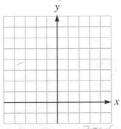 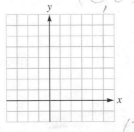

43. $6x - 3y = 10$ **44.** $4x + 8y - 1 = 0$ **45.** $3x = 6y - 1$ **46.** $2x + 1 = 4y$

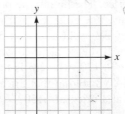 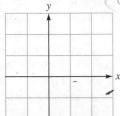  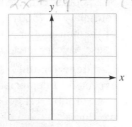

47. $2(x - y) = 3x + 2$

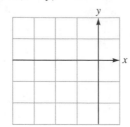

48. $5(x + 2) = 3y - x$

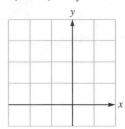

In Exercises 49–56, graph each equation.

49. $y = 3$

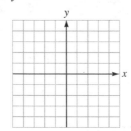

50. $x = -4$

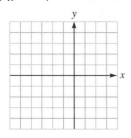

51. $3x + 5 = -1$

· $3x = y^{-}6 \quad x = -2$

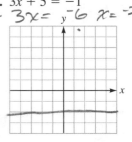

52. $7y - 1 = 6$

$7y = 7 \quad y = 1$

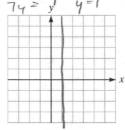

53. $3(y + 2) = y$

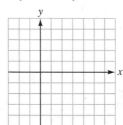

54. $4 + 3y = 3(x + y)$

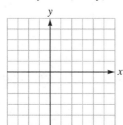

55. $3(y + 2x) = 6x + y$

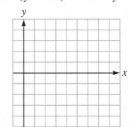

56. $5(y - x) = x + 5y$

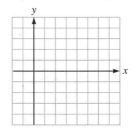

In Exercises 57–60, use a graphing calculator to graph each equation and then find the x-coordinate of the x-intercept to the nearest hundredth.

57. $y = 3.7x - 4.5$

58. $y = \dfrac{3}{5}x + \dfrac{5}{4}$

59. $1.5x - 3y = 7$

60. $.3x + y = 7.5$

61. House appreciation   A house purchased for $125,000 is expected to appreciate according to the formula $y = 7500x + 125{,}000$, where y is the value of the house after x years. Find the value of the house 5 years later.

62. Car depreciation   A car purchased for $17,000 is expected to depreciate according to the formula $y = -1360x + 17{,}000$. When will the car be worthless?

63. Demand equation   The number of television sets that consumers buy depends on price. The higher the price, the fewer TVs people will buy. The equation that relates price to the number of TVs sold at that price is called a **demand equation.** If the demand equation for a 13-inch TV is $p = -\frac{1}{10}q + 170$, where p is the price and q is the number of TVs sold at that price, how many TVs will be sold at a price of $150?

64. Supply equation The number of television sets that manufacturers produce depends on price. The higher the price, the more TVs manufacturers will produce. The equation that relates price to the number of TVs produced at that price is called a **supply equation.** If the supply equation for a 13-inch TV is $p = \frac{1}{10}q + 130$, where p is the price and q is the number of TVs produced for sale at that price, how many TVs will be produced if the price is $150?

ILLUSTRATION 2

65. Meshing gears The rotational speed, V, of a large gear (with N teeth) is related to the speed, v, of the smaller gear (with n teeth) by the equation $V = \frac{nv}{N}$. If the larger gear in Illustration 2 is making 60 revolutions per minute, how fast is the smaller gear spinning?

66. Crime prevention The number, n, of incidents of family violence requiring police response appears to be related to d, the money spent on crisis intervention, by the equation $n = 430 - 0.005d$. What expenditure would reduce the number of incidents to 350?

In Exercises 67–74, find the distance between P and O(0, 0).

67. $P(4, -3)$ **68.** $P(-5, 12)$ **69.** $P(-3, 2)$ **70.** $P(5, 0)$

71. $P(1, 1)$ **72.** $P(6, -8)$ **73.** $P\left(\sqrt{3}, 1\right)$ **74.** $P\left(\sqrt{7}, \sqrt{2}\right)$

In Exercises 75–84, find the distance between P and Q.

75. $P(3, 7); Q(6, 3)$ **76.** $P(4, 9); Q(9, 21)$ **77.** $P(4, -6); Q(-1, 6)$ **78.** $P(0, 5); Q(6, -3)$

79. $P(-2, -15); Q(-9, -39)$ **80.** $P(-7, 11); Q(3, -13)$ **81.** $P(3, -3); Q(-5, 5)$

82. $P(6, -3); Q(-3, 2)$ **83.** $P(\pi, -2); Q(\pi, 5)$ **84.** $P\left(\sqrt{5}, 0\right); Q(0, 2)$

In Exercises 85–92, find the midpoint of the line segment PQ.

85. $P(2, 4); Q(6, 8)$ **86.** $P(3, -6); Q(-1, -6)$ **87.** $P(2, -5); Q(-2, 7)$ **88.** $P(0, 3); Q(-10, -13)$

89. $P(-8, 5); Q(8, -5)$ **90.** $P(3, -2); Q(2, -3)$ **91.** $P(0, 0); Q\left(\sqrt{5}, \sqrt{5}\right)$ **92.** $P\left(\sqrt{3}, 0\right); Q\left(0, -\sqrt{5}\right)$

In Exercises 93–96, one endpoint P and the midpoint M of line segment PQ are given. Find the coordinates of the other endpoint, Q.

93. $P(1, 4); M(3, 5)$ **94.** $P(2, -7); M(-5, 6)$ **95.** $P(5, -5); M(5, 5)$ **96.** $P(-7, 3); M(0, 0)$

97. Show that a triangle with vertices at $(13, -2)$, $(9, -8)$, and $(5, -2)$ is isosceles.

98. Show that a triangle with vertices at $(-1, 2)$, $(3, 1)$, and $(4, 5)$ is isosceles.

99. In Illustration 3, points M and N are the midpoints of AC and BC, respectively. Find the length of MN.

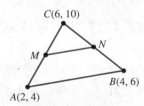

ILLUSTRATION 3

100. In Illustration 4, points M and N are the midpoints of AC and BC, respectively. Show that $d(MN) = \frac{1}{2}[d(AB)]$.

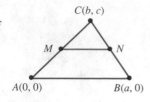

ILLUSTRATION 4

101. In Illustration 5, point M is the midpoint of the hypotenuse of right triangle AOB. Show that the area of rectangle $OLMN$ is one-half of the area of triangle AOB.

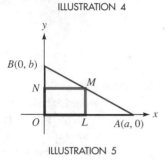

ILLUSTRATION 5

102. Rectangle $ABCD$ in Illustration 6 is twice as long as it is wide, and its sides are parallel to the coordinate axes. If the perimeter is 42, find the coordinates of point C.

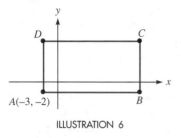

ILLUSTRATION 6

APPLICATIONS

103. See Illustration 7. An ocean liner is located 23 miles east and 72 miles north of Pigeon Cove Lighthouse, and its home port is 47 miles west and 84 miles south of the lighthouse. How far is the ship from port?

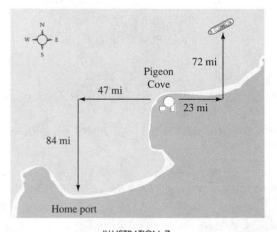

ILLUSTRATION 7

104. Two holes are to be drilled at locations specified by the engineering drawings shown in Illustration 8. Find the distance between the centers of the holes.

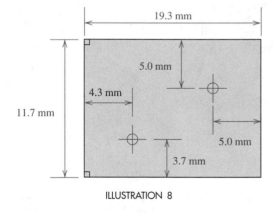

ILLUSTRATION 8

DISCOVERY AND WRITING *Write a paragraph using your own words.*

105. Explain how to graph a line using the intercept method.

106. Explain how to determine the quadrant in which the point $P(a, b)$ lies.

REVIEW *Graph each interval on the number line.*

107. $[-3, 2) \cup (-2, 3]$ **108.** $(-1, 4) \cap [-2, 2]$ **109.** $[-3, -2) \cap (2, 3]$ **110.** $[-4, -3) \cup (2, 3]$

3.2 The Slope of a Nonvertical Line

■ SLOPE OF A LINE ■ APPLICATIONS OF SLOPE ■ HORIZONTAL AND VERTICAL LINES ■ SLOPES OF PARALLEL LINES ■ SLOPES OF PERPENDICULAR LINES

■ SLOPE OF A LINE

A service offered by a computer online company costs \$2 per month plus \$3 for each hour of connect time. The table shown in Figure 3-15(a) gives the cost (y) for certain numbers of hours (x) of connect time. If we construct a graph from this data, we get the line shown in Figure 3-15(b).

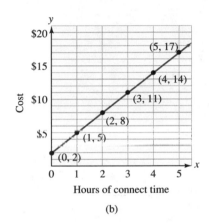

Hours of connect time						
x	0	1	2	3	4	5
y	2	5	8	11	14	17
Cost						

(a)

(b)

FIGURE 3-15

From the graph, we can see that if x changes from 0 to 1, y changes from 2 to 5. As x changes from 1 to 2, y changes from 5 to 8, and so on. The ratio of the change in y divided by the change in x is the constant 3.

$$\frac{\text{Change in } y}{\text{Change in } x} = \frac{5-2}{1-0} = \frac{8-5}{2-1} = \frac{11-8}{3-2} = \frac{14-11}{4-3} = \frac{17-14}{5-4} = \frac{3}{1} = 3$$

The ratio of the change in y divided by the change in x between any two points on any line is always a constant. This constant rate of change is called the **slope** of the line.

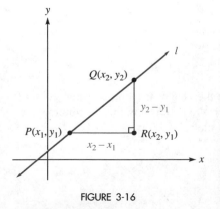

FIGURE 3-16

The Slope of a Nonvertical Line
The **slope of the nonvertical line** (see Figure 3-16) passing through points $P(x_1, y_1)$ and $Q(x_2, y_2)$ is

$$m = \frac{\text{change in } y}{\text{change in } x} = \frac{y_2 - y_1}{x_2 - x_1} \quad (x_2 \neq x_1)$$

$m = (\text{mountain}) \, \text{slope}$

EXAMPLE 1   Find the slope of the line passing through $P(-1, -2)$ and $Q(7, 8)$. (See Figure 3-17.)

Solution   We can let $P(x_1, y_1) = P(-1, -2)$ and $Q(x_2, y_2) = Q(7, 8)$. Then we substitute -1 for x_1, -2 for y_1, 7 for x_2, and 8 for y_2 to get

$$m = \frac{\text{change in } y}{\text{change in } x}$$

$$m = \frac{y_2 - y_1}{x_2 - x_1}$$

$$= \frac{8 - (-2)}{7 - (-1)}$$

$$= \frac{10}{8}$$

$$= \frac{5}{4}$$

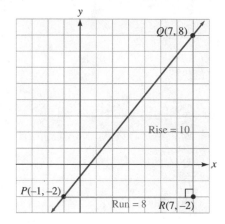

FIGURE 3-17

The slope of the line is $\frac{5}{4}$. We would have obtained the same result if we had let $P(x_1, y_1) = P(7, 8)$ and $Q(x_2, y_2) = Q(-1, -2)$.   ■

Self Check   Find the slope of the line passing through $P(-3, -4)$ and $Q(5, 9)$.

Answer $\frac{13}{8}$

WARNING!   When calculating slope, always subtract the y values and the x values in the same order.

$$m = \frac{y_2 - y_1}{x_2 - x_1} \qquad \text{or} \qquad m = \frac{y_1 - y_2}{x_1 - x_2}$$

However,

$$m \neq \frac{y_2 - y_1}{x_1 - x_2} \qquad \text{and} \qquad m \neq \frac{y_1 - y_2}{x_2 - x_1}$$

The change in y (often denoted as Δy) is the **rise** of the line between points P and Q. The change in x (often denoted as Δx) is the **run.** Using this terminology, we can define slope to be the ratio of the rise to the run:

$$m = \frac{y_2 - y_1}{x_2 - x_1} = \frac{\Delta y}{\Delta x} = \frac{\text{rise}}{\text{run}} \quad (\Delta x \neq 0)$$

EXAMPLE 2   Find the slope of the line determined by $5x + 2y = 10$. (See Figure 3-18.)

Solution   We first find the coordinates of two points on the line. Two convenient points are the
y- and x-intercepts.

- If $y = 0$, then $x = 2$, and the point $(2, 0)$ lies on the line.
- If $x = 0$, then $y = 5$, and the point $(0, 5)$ lies on the line.

We then find the slope of the line between $P(2, 0)$ and $Q(0, 5)$.

$$m = \frac{\text{change in } y}{\text{change in } x}$$

$$m = \frac{y_2 - y_1}{x_2 - x_1}$$

$$= \frac{5 - 0}{0 - 2}$$

$$= -\frac{5}{2}$$

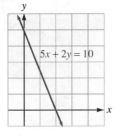

FIGURE 3-18

The slope is $-\frac{5}{2}$.   ■

Self Check   Find the slope of the line determined by $3x - 2y = 9$.

Answer $\frac{3}{2}$

■ APPLICATIONS OF SLOPE

EXAMPLE 3   **Cost of carpet**   If carpet costs \$25 per square yard, the total cost c of n
square yards is the price per square yard times the number of square yards purchased:

$c =$	cost per square yard	·	the number of square yards purchased
$c =$	25	·	n

Graph the equation $c = 25n$ and interpret the slope of the line.

Solution   We can graph the equation on a coordinate system with a vertical c-axis and a horizontal n-axis. Figure 3-19 shows a table of ordered pairs and the graph.

$$c = 25n$$

x	y	(x, y)
10	250	(10, 250)
20	500	(20, 500)
30	750	(30, 750)
40	1000	(40, 1000)
50	1250	(50, 1250)

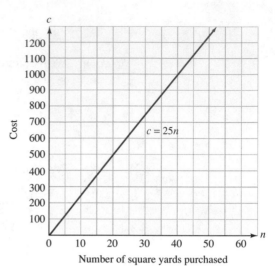

FIGURE 3-19

If we pick the points (30, 750) and (50, 1250) to find the slope, we have

$$m = \frac{\Delta c}{\Delta n}$$

$$= \frac{c_2 - c_1}{n_2 - n_1}$$

$$= \frac{1250 - 750}{50 - 30} \qquad \text{Substitute 1250 for } c_2, \text{ 750 for } c_1, \text{ 50 for } n_2, \text{ and 30 for } n_1.$$

$$= \frac{500}{20}$$

$$= 25$$

The slope of 25 (in dollars/square yard) is the ratio of the change in the cost to the change in the number of square yards purchased. This is the cost per square yard. ∎

EXAMPLE 4

Rate of descent   It takes a skier 25 minutes to complete the course shown in Figure 3-20. Find his average rate of descent in feet per minute.

Solution   To find the average rate of descent, we must find the ratio of the change in altitude to the change in time. To find this ratio, we calculate the slope of the line passing through the points (0, 12,000) and (25, 8500).

$$\begin{aligned} \text{Average rate} \\ \text{of descent} \end{aligned} = \frac{12,000 - 8500}{0 - 25}$$

$$= \frac{3500}{-25}$$

$$= -140$$

The average rate of descent is -140 ft/min.

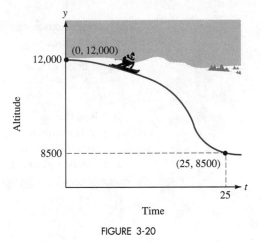

FIGURE 3-20

■ HORIZONTAL AND VERTICAL LINES

If $P(x_1, y_1)$ and $Q(x_2, y_2)$ are points on the horizontal line shown in Figure 3-21(a), then $y_1 = y_2$, and the numerator of the fraction

$$\frac{y_2 - y_1}{x_2 - x_1} \qquad \text{On a horizontal line, } x_2 \neq x_1.$$

is 0. Thus, the value of the fraction is 0, and the slope of the horizontal line is 0.

If $P(x_1, y_1)$ and $Q(x_2, y_2)$ are points on the vertical line shown in Figure 3-21(b), then $x_1 = x_2$, and the denominator of the fraction

$$\frac{y_2 - y_1}{x_2 - x_1} \qquad \text{On a vertical line, } y_2 \neq y_1.$$

is 0. Since the denominator of a fraction cannot be 0, a vertical line has no defined slope.

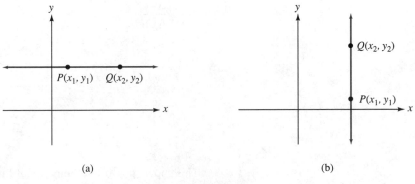

(a)   (b)

FIGURE 3-21

Horizontal and Vertical Lines

Horizontal lines (lines with equations of the form $y = b$) have a slope of 0.

Vertical lines (lines with equations of the form $x = a$) have no defined slope.

If a line rises as we follow it from left to right, as in Figure 3-22(a), its slope is positive. If a line drops as we follow it from left to right, as in Figure 3-22(b), its slope is negative.

If a line is horizontal, as in Figure 3-22(c), its slope is 0. If a line is vertical, as in Figure 3-22(d), it has no defined slope.

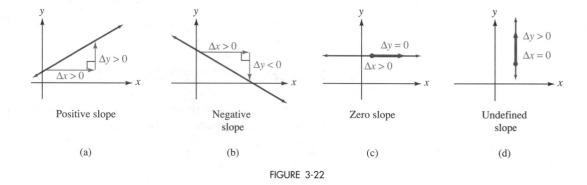

FIGURE 3-22

■ SLOPES OF PARALLEL LINES

To see a relationship between parallel lines and their slopes, we refer to the parallel lines l_1 and l_2 shown in Figure 3-23, with slopes of m_1 and m_2, respectively. Because right triangles ABC and DEF are similar, it follows that

$$
\begin{aligned}
m_1 &= \frac{\Delta y \text{ of } l_1}{\Delta x \text{ of } l_1} \\
&= \frac{\Delta y \text{ of } l_2}{\Delta x \text{ of } l_2} \\
&= m_2
\end{aligned}
$$

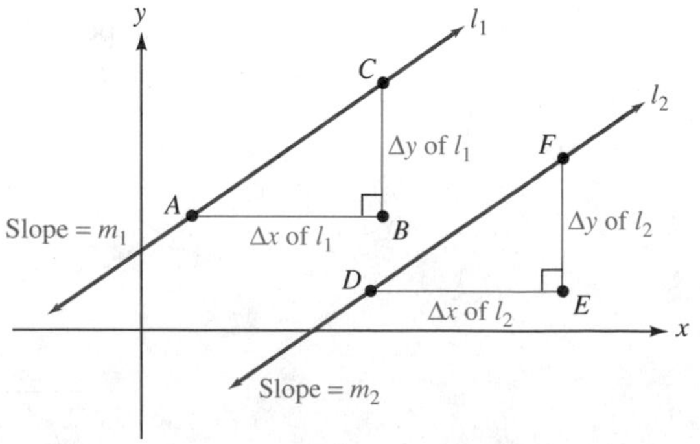

FIGURE 3-23

This shows that if two nonvertical lines are parallel, they have the same slope. It is also true that when two lines have the same slope, they are parallel.

Slopes of Parallel Lines

Nonvertical parallel lines have the same slope, and lines having the same slope are parallel.

Since vertical lines are parallel, lines with no defined slope are parallel.

EXAMPLE 5   The lines in Figure 3-24 are parallel. Find y.

Solution   Since the lines are parallel, their slopes are equal. To find y, we find the slope of each line, set them equal, and solve the resulting equation.

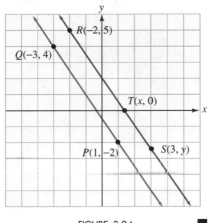

Slope of PQ = slope of RS

$$\frac{-2 - 4}{1 - (-3)} = \frac{y - 5}{3 - (-2)}$$

$$\frac{-6}{4} = \frac{y - 5}{5} \qquad \text{Simplify.}$$

$$-30 = 4(y - 5) \qquad \text{Multiply both sides by 20.}$$

$$-30 = 4y - 20 \qquad \text{Remove parentheses and simplify.}$$

$$-10 = 4y \qquad \text{Add 20 to both sides.}$$

$$-\frac{5}{2} = y \qquad \text{Divide both sides by 4 and simplify.}$$

Thus, $y = -\frac{5}{2}$.

FIGURE 3-24   ■

Self Check   Find x in Figure 3-24.

Answer $\frac{4}{3}$

■ SLOPES OF PERPENDICULAR LINES

Two real numbers a and b are called **negative reciprocals** if $ab = -1$. For example,

$$\frac{7}{3} \qquad \text{and} \qquad -\frac{3}{7}$$

are negative reciprocals, because $\frac{7}{3}\left(-\frac{3}{7}\right) = -1$. The following theorem relates perpendicular lines and their slopes.

Slopes of Perpendicular Lines

If two nonvertical lines are perpendicular, their slopes are negative reciprocals.

If the slopes of two lines are negative reciprocals, the lines are perpendicular.

Proof   Suppose that l_1 and l_2 are lines with slopes of m_1 and m_2 that intersect at the origin. Let $P(a, b)$ be a point on l_1, and let $Q(c, d)$ be a point on l_2. Neither point P nor point Q can be the origin. See Figure 3-25.

First, we suppose that l_1 and l_2 are perpendicular. Then triangle POQ is a right triangle with its right angle at O, and by the Pythagorean theorem,

$$d(OP)^2 + d(OQ)^2 = d(PQ)^2$$
$$(a - 0)^2 + (b - 0)^2 + (c - 0)^2 + (d - 0)^2 = (a - c)^2 + (b - d)^2$$
$$a^2 + b^2 + c^2 + d^2 = a^2 - 2ac + c^2 + b^2 - 2bd + d^2$$
$$0 = -2ac - 2bd$$
$$bd = -ac$$

1. $$\frac{b}{a} \cdot \frac{d}{c} = -1 \qquad \text{Divide both sides by } ac.$$

The coordinates of P are (a, b), and the coordinates of O are $(0, 0)$. Using the definition of slope, we have

$$m_1 = \frac{b - 0}{a - 0} = \frac{b}{a}$$

Similarly, we have

$$m_2 = \frac{d}{c}$$

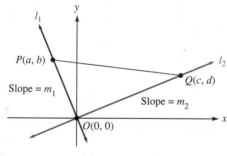

FIGURE 3-25

We substitute m_1 for $\frac{b}{a}$ and m_2 for $\frac{d}{c}$ in Equation 1 to obtain

$$m_1 m_2 = -1$$

Hence, if lines l_1 and l_2 are perpendicular, they have slopes that are negative reciprocals of each other.

Conversely, we suppose that the slopes of lines l_1 and l_2 are negative reciprocals of each other. Because the steps of the previous proof are reversible, we have $d(OP)^2 + d(OQ)^2 = d(PQ)^2$. By the Pythagorean theorem, triangle POQ is a right triangle. Thus, l_1 and l_2 are perpendicular.   □

It is also true that a line with a slope of 0 is horizontal and thus is perpendicular to a vertical line that has no defined slope.

EXAMPLE 6   Are the lines shown in Figure 3-26 perpendicular?

Solution   We find the slopes of lines and see whether or not they are negative reciprocals.

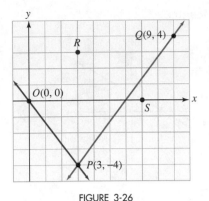

FIGURE 3-26

$$\text{Slope of } OP = \frac{\Delta y}{\Delta x}$$

$$= \frac{y_2 - y_1}{x_2 - x_1}$$

$$= \frac{-4 - 0}{3 - 0}$$

$$= -\frac{4}{3}$$

$$\text{Slope of } PQ = \frac{\Delta y}{\Delta x}$$

$$= \frac{y_2 - y_1}{x_2 - x_1}$$

$$= \frac{4 - (-4)}{9 - 3}$$

$$= \frac{8}{6}$$

$$= \frac{4}{3}$$

Since the slopes are not negative reciprocals, the lines are not perpendicular. ■

Self Check   Is either line in Figure 3-26 perpendicular to the line passing through R and S?

Answer   yes

EXERCISE 3.2

VOCABULARY AND CONCEPTS   *In Exercises 1–10, fill in the blank to make a true statement.*

1. The slope of a nonvertical line is defined to be the change in y ~~divided~~ by the change in x.

2. The change in ~~y~~ is often called the rise.

3. The change in x is often called the ~~run~~.

4. When computing the slope from the coordinates of two points, always subtract the y values and the x values in the ~~same order~~.

5. The symbol Δy means ~~change in~~ y.

6. The slope of a ~~horizontal~~ line is 0.

7. The slope of a ~~vertical~~ line is undefined.

8. If the slopes of two lines are equal, the lines are ~~parallel~~.

9. Numbers whose product is -1 are called ~~negative~~ reciprocals.

10. If two lines are ~~perpendicular~~, their slopes are negative ~~reciprocals~~.

PRACTICE   *In Exercises 11–22, find the slope of the line passing through each pair of points, if possible.*

11. $P(2, 5)$; $Q(3, 10)$ $\frac{10-5}{3-2}$ $\frac{5}{5}$

12. $P(3, -1)$; $Q(5, 3)$ $m = \frac{2}{4}\left(\frac{1}{2}\right)$

13. $P(3, -2)$; $Q(-1, 5)$ $\frac{7}{-4}$

14. $P(3, 7)$; $Q(6, 16)$

15. $P(8, -7)$; $Q(4, 1)$ $m = 5$

16. $P(5, 17)$; $Q(17, 17)$

17. $P(-4, 3)$; $Q(-4, -3)$ $\frac{1}{2}$

18. $P(2, \sqrt{7})$; $Q(\sqrt{7}, 2)$

19. $P\left(\frac{3}{2}, \frac{2}{3}\right)$; $Q\left(\frac{5}{2}, \frac{7}{3}\right)$ $\frac{\frac{5}{3}}{\frac{3}{2}1}$ $= \frac{5}{3}$ $m = \frac{5}{3}$

20. $P\left(-\frac{2}{5}, \frac{1}{3}\right)$; $Q\left(\frac{3}{5}, -\frac{5}{3}\right)$

21. $P(a + b, c)$; $Q(b + c, a)$ assume $a \neq c$

22. $P(b, 0)$; $Q(a + b, a)$ assume $a \neq 0$

$\left(\frac{5}{2}, 0\right)\left(0, \frac{5}{8}\right)$

In Exercises 23–30, find two points on the line and find the slope of the line.

23. $y = 3x + 2$

24. $y = 5x - 8$

25. $5x - 10y = 3$

26. $8y + 2x = 5$

27. $3(y + 2) = 2x - 3$

28. $4(x - 2) = 3y + 2$

29. $3(y + x) = 3(x - 1)$

30. $2x + 5 = 2(y + x)$

In Exercises 31–36, tell whether the slope of the line is positive, negative, 0, or undefined.

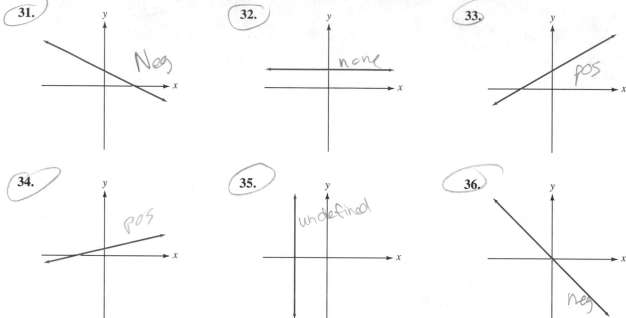

31.   *Neg*

32.   *none*

33.   *pos*

34.   *pos*

35.   *undefined*

36.   *neg*

APPLICATIONS

37. Rate of growth   When a college started an aviation program, the administration agreed to predict enrollments using a straight-line method. If the enrollment during the first year was 12, and the enrollment during the fifth year was 26, find the rate of growth per year (the slope of the line). See Illustration 1.

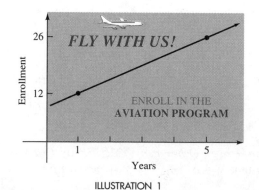

FLY WITH US!

ENROLL IN THE
AVIATION PROGRAM

Enrollment — 26, 12
Years — 1, 5

ILLUSTRATION 1

38. Rate of growth   A small business predicts sales according to a straight-line method. If sales were $50,000 in the first year and $110,000 in the third

year, find the rate of growth in dollars per year (the slope of the line).

39. Rate of decrease   The price of computer technology has been dropping steadily for the past ten years. If a desktop PC cost $6700 ten years ago, and the same computing power cost $2200 three years ago, find the rate of decrease per year. (Assume a straight-line model).

40. Hospital costs   Illustration 2 shows the changing mean daily cost for a hospital room. Find the rate of change per year of the portion of the room cost that is absorbed by the hospital.

	Cost passed on to patient	Total cost to the hospital
1980	$130	$245
1985	214	459
1990	295	670

ILLUSTRATION 2

41. Charting temperature changes   The following Fahrenheit temperature readings were recorded over a four-hour period.

Time	12:00	1:00	2:00	3:00	4:00
Temperature	47°	53°	59°	65°	71°

Let t represent the time (in hours), with 12:00 corresponding to $t = 0$. Let T represent the temperature. Plot the points (t, T), and draw the line through those points. Explain the meaning of $\frac{\Delta T}{\Delta t}$.

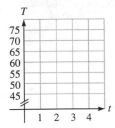

42. Tracking the Dow   The Dow Jones Industrial Averages at the close of trade on three consecutive days were as follows:

Day	Monday	Tuesday	Wednesday
Close	7981	7964	7947

Let d represent the day, with $d = 0$ corresponding to Monday, and let D represent the Dow Jones average. Plot the points (d, D), and draw the graph. Explain the meaning of $\frac{\Delta D}{\Delta d}$.

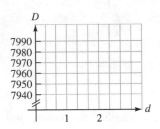

43. Speed of an airplane   A pilot files a flight plan indicating her intention to fly at a constant speed of 590 mph. Write an equation that expresses the distance traveled in terms of the flying time. Then graph the equation, and interpret the slope of the line. (*Hint: d = rt.*)

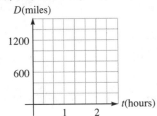

44. Growth of savings   A student deposits $25 each month into a Holiday Club account at her bank. The account pays no interest. Write an equation that expresses the amount in her account in terms of the number of deposits. Then graph the line, and interpret the slope of the line.

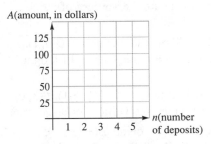

PRACTICE   *In Exercises 45–54, determine whether the lines with the given slopes are parallel, perpendicular, or neither.*   ? ?

45. $m_1 = 3$; $m_2 = -\dfrac{1}{3}$

46. $m_1 = \dfrac{2}{3}$; $m_2 = \dfrac{3}{2}$ neither

47. $m_1 = \sqrt{8}$; $m_2 = 2\sqrt{2}$

48. $m_1 = 1$; $m_2 = -1$ perp

49. $m_1 = -\sqrt{2}$; $m_2 = \dfrac{\sqrt{2}}{2}$

50. $m_1 = 2\sqrt{7}$; $m_2 = \sqrt{28}$

51. $m_1 = -0.125$; $m_2 = 8$

52. $m_1 = 0.125$; $m_2 = \dfrac{1}{8}$

53. $m_1 = ab^{-1}$; $m_2 = -a^{-1}b$ $(a, b \neq 0)$

54. $m_1 = \left(\dfrac{a}{b}\right)^{-1}$; $m_2 = -\dfrac{b}{a}$ $(a, b \neq 0)$

In Exercises 55–60, determine whether the line through the given points and the line through $R(-3, 5)$ and $S(2, 7)$ are parallel, perpendicular, or neither.

55. $P(2, 4)$; $Q(7, 6)$

56. $P(-3, 8)$; $Q(-13, 4)$

57. $P(-4, 6)$; $Q(-2, 1)$

58. $P(0, -9)$; $Q(4, 1)$

59. $P(a, a)$; $Q(3a, 6a)$ $(a \neq 0)$

60. $P(b, b)$; $Q(-b, 6b)$ $(b \neq 0)$

In Exercises 61–64, find the slopes of lines PQ and PR, and determine whether points P, Q, and R lie on the same line.

61. $P(-2, 8)$; $Q(-6, 9)$; $R(2, 5)$

62. $P(1, -1)$; $Q(3, -2)$; $R(-3, 0)$

63. $P(-a, a)$; $Q(0, 0)$; $R(a, -a)$

64. $P(a, a + b)$; $Q(a + b, b)$; $R(a - b, a)$

In Exercises 65–70, determine which, if any, of the three lines PQ, PR, and QR are perpendicular.

65. $P(5, 4)$; $Q(2, -5)$; $R(8, -3)$

66. $P(8, -2)$; $Q(4, 6)$; $R(6, 7)$

67. $P(1, 3)$; $Q(1, 9)$; $R(7, 3)$

68. $P(2, -3)$; $Q(-3, 2)$; $R(3, 8)$

69. $P(0, 0)$; $Q(a, b)$; $R(-b, a)$

70. $P(a, b)$; $Q(-b, a)$; $R(a - b, a + b)$

71. Right triangles Show that the points $A(-1, -1)$, $B(-3, 4)$, and $C(4, 1)$ are the vertices of a right triangle.

72. Right triangles Show that the points $D(0, 1)$, $E(-1, 3)$, and $F(3, 5)$ are the vertices of a right triangle.

73. Squares Show that the points $A(1, -1)$, $B(3, 0)$, $C(2, 2)$, and $D(0, 1)$ are the vertices of a square.

74. Squares Show that the points $E(-1, -1)$, $F(3, 0)$, $G(2, 4)$, and $H(-2, 3)$ are the vertices of a square.

75. Parallelograms Show that the points $A(-2, -2)$, $B(3, 3)$, $C(2, 6)$, and $D(-3, 1)$ are the vertices of a parallelogram. (Show that both pairs of opposite sides are parallel.)

76. Trapezoids Show that points $E(1, -2)$, $F(5, 1)$, $G(3, 4)$, and $H(-3, 4)$ are the vertices of a trapezoid. (Show that only one pair of opposite sides are parallel.)

77. Geometry In Illustration 3, points M and N are midpoints of CB and BA, respectively. Show that MN is parallel to AC.

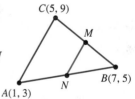

ILLUSTRATION 3

78. Geometry In Illustration 4, $d(AB) = d(AC)$. Show that AD is perpendicular to BC.

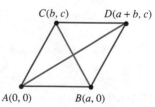

ILLUSTRATION 4

DISCOVERY AND WRITING

79. Explain why a vertical line has no defined slope.

80. Explain how to determine whether two lines are parallel, perpendicular, or neither.

REVIEW *In Exercises 81–84, solve each equation for y and simplify.*

81. $3x + 7y = 21$

82. $y - 3 = 5(x + 2)$

83. $\dfrac{x}{5} + \dfrac{y}{2} = 1$

84. $x - 5y = 15$

3.3 Writing Equations of Lines

■ POINT–SLOPE FORM OF THE EQUATION OF A LINE ■ SLOPE–INTERCEPT FORM OF THE EQUATION OF A LINE ■ GRAPHING EQUATIONS WRITTEN IN SLOPE–INTERCEPT FORM ■ GENERAL FORM OF THE EQUATION OF A LINE ■ STRAIGHT-LINE DEPRECIATION

We have seen that if we have an equation in two variables, we can find its graph. In this section, we will begin with a graph and find its equation.

■ POINT–SLOPE FORM OF THE EQUATION OF A LINE

Suppose that line l in Figure 3-27 has a slope of m and passes through the point $P(x_1, y_1)$. If $Q(x, y)$ is a second point on line l, we have

$$m = \frac{y - y_1}{x - x_1}$$

If we multiply both sides by $x - x_1$, we have

1. $y - y_1 = m(x - x_1)$

Because Equation 1 displays the coordinates of the point (x_1, y_1) on the line and the slope m of the line, it is called the **point–slope form** of the equation of a line.

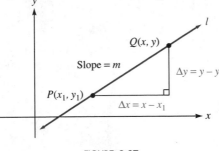

FIGURE 3-27

Point–Slope Form

The equation of the line passing through $P(x_1, y_1)$ and with slope m is

$$y - y_1 = m(x - x_1)$$

EXAMPLE 1   Write the equation of the line with slope $-\frac{5}{3}$ and passing through $P(3, -1)$.

Solution   We substitute $-\frac{5}{3}$ for m, 3 for x_1, and -1 for y in the point–slope form and simplify.

$$y - y_1 = m(x - x_1)$$

$$y - (-1) = -\frac{5}{3}(x - 3) \qquad \text{Substitute } -\frac{5}{3} \text{ for } m, 3 \text{ for } x_1, \text{ and } -1 \text{ for } y_1.$$

$$y + 1 = -\frac{5}{3}x + 5 \qquad \text{Remove parentheses.}$$

$$y = -\frac{5}{3}x + 4 \qquad \text{Subtract 1 from both sides.}$$

The equation of the line is $y = -\frac{5}{3}x + 4$.   ∎

Self Check   Write the equation of the line with slope $-\frac{2}{3}$ and passing through $P(-4, 5)$.

Answer $y = -\frac{2}{3}x + \frac{7}{3}$

EXAMPLE 2   Find the equation of the line passing through $P(3, 7)$ and $Q(-5, 3)$.

Solution   First we find the slope of the line.

$$m = \frac{y_2 - y_1}{x_2 - x_1}$$

$$= \frac{3 - 7}{-5 - 3} \qquad \text{Substitute 3 for } y_2, 7 \text{ for } y_1, -5 \text{ for } x_2, \text{ and 3 for } x_1.$$

$$= \frac{-4}{-8}$$

$$= \frac{1}{2}$$

We can choose either point P or point Q and substitute its coordinates into the point–slope form. If we choose $P(3, 7)$, we substitute $\frac{1}{2}$ for m, 3 for x_1, and 7 for y_1.

$$y - y_1 = m(x - x_1)$$

$$y - 7 = \frac{1}{2}(x - 3) \qquad \text{Substitute } \frac{1}{2} \text{ for } m, 3 \text{ for } x_1, \text{ and 7 for } y_1.$$

$$y = \frac{1}{2}x - \frac{3}{2} + 7 \qquad \text{Remove parentheses and add 7 to both sides.}$$

$$y = \frac{1}{2}x + \frac{11}{2} \qquad -\frac{3}{2} + 7 = -\frac{3}{2} + \frac{14}{2} = \frac{11}{2}.$$

The equation of the line is $y = \frac{1}{2}x + \frac{11}{2}$.   ∎

Self Check   Find the equation of the line passing through $P(-5, 4)$ and $Q(8, -6)$.

Answer $y = -\frac{10}{13}x + \frac{2}{13}$

■ SLOPE–INTERCEPT FORM OF THE EQUATION OF A LINE

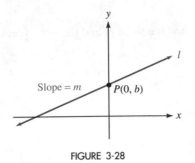

FIGURE 3-28

Since the y-intercept of the line shown in Figure 3-28 is the point $(0, b)$, we can write the equation of the line by substituting 0 for x_1 and b for y_1 in the point–slope form and simplifying.

$$y - y_1 = m(x - x_1)$$
$$y - b = m(x - 0) \qquad \text{Substitute 0 for } x_1 \text{ and } b \text{ for } y_1.$$
$$y - b = mx \qquad x - 0 = x.$$
2. $$y = mx + b \qquad \text{Add } b \text{ to both sides.}$$

Because Equation 2 displays the slope m and the y-coordinate b of the y-intercept, it is called the **slope–intercept form** of the equation of a line.

> **Slope–Intercept Form**
> The equation of the line with slope m and y-intercept $(0, b)$ is
> $$y = mx + b$$

EXAMPLE 3 Use slope–intercept form to write the equation of the line with slope 4 that passes through $P(5, 9)$.

Solution Since we know that $m = 4$ and that the pair $(5, 9)$ satisfies the equation, we substitute 4 for m, 5 for x, and 9 for y, in the equation $y = mx + b$ and solve for b.

$$y = mx + b$$
$$9 = 4(5) + b \qquad \text{Substitute 4 for } m, 5 \text{ for } x, \text{ and 9 for } y.$$
$$9 = 20 + b \qquad \text{Simplify.}$$
$$-11 = b \qquad \text{Subtract 20 from both sides.}$$

Because $m = 4$ and $b = -11$, the equation is $y = 4x - 11$. ■

Self Check Use slope–intercept form to write the equation of the line with slope $\frac{7}{3}$ and passing through $(3, 1)$.

Answer $y = \frac{7}{3}x - 6$

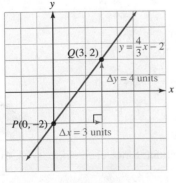

FIGURE 3-29

■ GRAPHING EQUATIONS WRITTEN IN SLOPE–INTERCEPT FORM

It is easy to graph a linear equation when it is written in slope–intercept form. For example, to graph $y = \frac{4}{3}x - 2$, we note that $b = -2$ and that the y-intercept is $(0, b) = (0, -2)$. (See Figure 3-29.)

Because the slope is $\frac{\Delta y}{\Delta x} = \frac{4}{3}$, we can locate another point Q on the line by starting at point P and counting 3 units to the right and 4 units up. The change in x from point P to point Q is $\Delta x = 3$, and the corresponding change in y is $\Delta y = 4$. The line joining points P and Q is the graph of the equation.

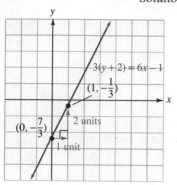

FIGURE 3-30

EXAMPLE 4   Find the slope and the y-intercept of the line with equation $3(y + 2) = 6x - 1$. Then graph it.

Solution   We write the equation in the form $y = mx + b$ to find the slope m and the y-intercept $(0, b)$.

$$3(y + 2) = 6x - 1$$
$$3y + 6 = 6x - 1 \qquad \text{Remove parentheses.}$$
$$3y = 6x - 7 \qquad \text{Subtract 6 from both sides.}$$
$$y = 2x - \frac{7}{3} \qquad \text{Divide both sides by 3.}$$

The slope is 2, and the y-intercept is $\left(0, -\frac{7}{3}\right)$. We plot the y-intercept. Then we find a second point on the line by moving 1 unit to the right and 2 units up to the point $\left(1, -\frac{1}{3}\right)$. To get the graph, we draw a line through the two points, as shown in Figure 3-30. ∎

Self Check   Find the slope and the y-intercept of the line with equation $2(x - 3) = -3(y + 5)$. Then graph it.

Answer $-\frac{2}{3}$, $(0, -3)$

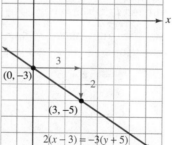

EXAMPLE 5   Show that the lines represented by $4x + 8y = 10$ and $2x = 12 - 4y$ are parallel.

Solution   We solve each equation for y to see that the lines are distinct and that their slopes are equal.

$$4x + 8y = 10 \qquad\qquad 2x = 12 - 4y$$
$$8y = -4x + 10 \qquad\qquad 4y = -2x + 12$$
$$y = \frac{-4x}{8} + \frac{10}{8} \qquad\qquad y = \frac{-2x}{4} + \frac{12}{4}$$
$$y = -\frac{1}{2}x + \frac{5}{4} \qquad\qquad y = -\frac{1}{2}x + 3$$

Since the values of b are different, the lines are distinct. Since each slope is $-\frac{1}{2}$, the lines are parallel. ∎

Self Check   Are the lines represented by $y = 3x + 2$ and $6x + 2y = 5$ parallel?

Answer   no

EXAMPLE 6   Show that the lines represented by $4x + 8y = 10$ and $4x - 2y = 21$ are perpendicular.

Solution   We solve each equation for y to see that the slopes of their straight-line graphs are negative reciprocals.

$$4x + 8y = 10 \qquad\qquad 4x - 2y = 21$$
$$8y = -4x + 10 \qquad\qquad -2y = -4x + 21$$
$$y = \frac{-4x}{8} + \frac{10}{8} \qquad\qquad y = \frac{-4x}{-2} + \frac{21}{-2}$$
$$y = -\frac{1}{2}x + \frac{5}{4} \qquad\qquad y = 2x - \frac{21}{2}$$

Since the slopes, $-\frac{1}{2}$ and 2, are negative reciprocals, the lines are perpendicular.

Self Check   Are the lines represented by $3x + 2y - 7$ and $y = \frac{2}{3}x + 3$ perpendicular?

Answer   yes

EXAMPLE 7   Write the equation of the line passing through $P(-2, 5)$ and parallel to the line $y = 8x - 3$.

Solution   The slope of the line given by $y = 8x - 3$ is 8, the coefficient of x. Since the graph of the desired equation is to be parallel to the graph of $y = 8x - 3$, its slope must also be 8.

We substitute -2 for x_1, 5 for y_1, and 8 for m in the point–slope form and simplify.

$$y - y_1 = m(x - x_1)$$
$$y - 5 = 8[x - (-2)] \qquad \text{Substitute 5 for } y_1, \text{ 8 for } m, \text{ and } -2 \text{ for } x_1.$$
$$y - 5 = 8(x + 2) \qquad -(-2) = 2.$$
$$y - 5 = 8x + 16 \qquad \text{Use the distributive property to remove parentheses.}$$
$$y = 8x + 21 \qquad \text{Add 5 to both sides.}$$

The equation of the desired line is $y = 8x + 21$.

Self Check   Write the equation of the line passing through $Q(1, 2)$ and parallel to the line $y = 8x - 3$.

Answer $y = 8x - 6$

EXAMPLE 8

Write the equation of the line passing through $P(-2, 5)$ and perpendicular to the line $y = 8x - 3$.

Solution

Because the slope of the given line is 8, the slope of the desired perpendicular line must be $-\frac{1}{8}$, which is the negative reciprocal of 8.

We substitute -2 for x_1, 5 for y_1, and $-\frac{1}{8}$ for m into the point–slope form and simplify.

$$y - y_1 = m(x - x_1)$$

$$y - 5 = -\frac{1}{8}[x - (-2)] \qquad \text{Substitute 5 for } y_1, -\frac{1}{8} \text{ for } m, \text{ and } -2 \text{ for } x_1.$$

$$y - 5 = -\frac{1}{8}(x + 2) \qquad -(-2) = 2.$$

$$8y - 40 = -(x + 2) \qquad \text{Multiply both sides by 8.}$$

$$8y - 40 = -x - 2 \qquad \text{Use the distributive property to remove parentheses.}$$

$$8y = -x + 38 \qquad \text{Add 40 to both sides.}$$

$$y = -\frac{1}{8}x + \frac{19}{4} \qquad \text{Divide both sides by 8 and simplify.}$$

The equation of the line is $y = -\frac{1}{8}x + \frac{19}{4}$. ■

Self Check

Write the equation of the line passing through $Q(1, 2)$ and perpendicular to $y = 8x - 3$.

Answer $y = -\frac{1}{8}x + \frac{17}{8}$

■ GENERAL FORM OF THE EQUATION OF A LINE

We have shown that the graph of any equation of the form $y = mx + b$ is a line with slope m and y-intercept $(0, b)$. The graph of any equation of the form $Ax + By = C$ (where A and B are not *both* zero) is also a line. To see why, we look at three possibilities.

- If $A \neq 0$ and $B \neq 0$, the equation $Ax + By = C$ can be written in slope–intercept form.

$$Ax + By = C$$

$$By = -Ax + C \qquad \text{Subtract } Ax \text{ from both sides.}$$

$$y = -\frac{A}{B}x + \frac{C}{B} \qquad \text{Divide both sides by } B.$$

This is the equation of a line with slope $-\frac{A}{B}$ and y-intercept $\left(0, \frac{C}{B}\right)$.

- If $A = 0$ and $B \neq 0$, the equation $Ax + By = C$ can be written in the form $y = \frac{C}{B}$. This is the equation of a horizontal line with y-intercept $\left(0, \frac{C}{B}\right)$.

- If $A \neq 0$ and $B = 0$, the equation $Ax + By = C$ can be written in the form $x = \frac{C}{A}$. This is the equation of a vertical line.

Recall that $Ax + By = C$ is called the **general form of the equation of a line.**

WARNING! When writing equations in general form, we usually clear the equation of fractions and make A positive. For example, the equation $-x + \frac{5}{2}y = 2$ can be changed to $2x - 5y = -4$ by multiplying both sides by -2. We will also divide out any common integer factors of A, B, and C. For example, we would write $4x + 8y = 12$ as $x + 2y = 3$.

> ### General Form of the Equation of a Line
> If A, B, and C are real numbers and $B \neq 0$, the graph of
>
> $$Ax + By = C$$
>
> is a nonvertical line with slope of $-\frac{A}{B}$ and a y-intercept of $\left(0, \frac{C}{B}\right)$.
>
> If $B = 0$, the graph is a vertical line with x-intercept of $\left(\frac{C}{A}, 0\right)$.

EXAMPLE 9 Find the slope and the y-intercept of the graph of $3x - 2y = 5$.

Solution The equation $3x - 2y = 5$ is in general form, with $A = 3$, $B = -2$, and $C = 5$. By the previous theorem, the slope of the graph is

$$m = -\frac{A}{B} = -\frac{3}{-2} = \frac{3}{2}$$

and the y-intercept is

$$\left(0, \frac{C}{B}\right) = \left(0, \frac{5}{-2}\right)$$

The slope is $\frac{3}{2}$, and the y-intercept is $\left(0, -\frac{5}{2}\right)$. ∎

Self Check Find the slope and the y-intercept of the graph of $3x - 4y = 12$.

Answer $\frac{3}{4}$, $(0, -3)$

We summarize the various forms of the equation of a line as follows.

General form	$Ax + By = C$ A and B cannot both be 0.
Slope–intercept form	$y = mx + b$ The slope is m, and the y-intercept is $(0, b)$.
Point–slope form	$y - y_1 = m(x - x_1)$ The slope is m, and the line passes through (x_1, y_1).
A horizontal line	$y = b$ The slope is 0, and the y-intercept is $(0, b)$.
A vertical line	$x = a$ There is no defined slope, and the x-intercept is $(a, 0)$.

■ STRAIGHT-LINE DEPRECIATION

For tax purposes, many businesses use *straight-line depreciation* to find the declining value of aging equipment.

EXAMPLE 10

Value of a lathe   A machine shop buys a lathe for $1970 and expects it to last for ten years. It can then be sold as scrap for a *salvage value* of $270. If y is the value of the lathe after x years of use, and y and x are related by the equation of a line,
a. Find the equation of the line.   **b.** Find the value of the lathe after $2\frac{1}{2}$ years.
c. Find the economic meaning of the y-intercept of the line.   **d.** Find the economic meaning of the slope of the line.

Solution   **a.** We find the slope and use the point–slope form to find the equation of the line. (See Figure 3-31.)

When the lathe is new, its age x is 0, and its value y is $1970. When the lathe is 10 years old, $x = 10$ and $y = \$270$. Since the line passes through the points $(0, 1970)$ and $(10, 270)$, the slope of the line is

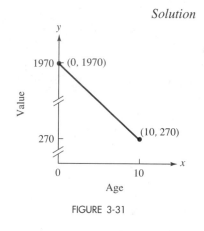

FIGURE 3-31

$$m = \frac{y_2 - y_1}{x_2 - x_1}$$

$$= \frac{270 - 1970}{10 - 0} \qquad \text{Substitute 270 for } y_2, \text{ 1970 for } y_1,$$
$$\text{10 for } x_2, \text{ and 0 for } x_1.$$

$$= \frac{-1700}{10}$$

$$= -170$$

To find the equation of the line, we substitute -170 for m, 0 for x_1, and 1970 for y_1 in the point–slope form and simplify.

$$y - y_1 = m(x - x_1)$$
$$y - 1970 = -170(x - 0)$$
3. $$\qquad y = -170x + 1970$$

The value y of the lathe is related to its age x by the equation $y = -170x + 1970$.

b. To find the value after $2\frac{1}{2}$ years, we substitute 2.5 for x in Equation 3 and solve for y.

$$y = -170x + 1970$$
$$= -170(2.5) + 1970 \qquad \text{Substitute 2.5 for } x.$$
$$= -425 + 1970$$
$$= 1545$$

In $2\frac{1}{2}$ years, the lathe will be worth $1545.

c. The y-intercept of the graph is $(0, b)$, where b is the value of y when $x = 0$.

$$y = -170x + 1970$$
$$y = -170(0) + 1970 \qquad \text{Substitute 0 for } x.$$
$$y = 1970$$

The y-coordinate b of the y-intercept is the value of a 0-year-old lathe, which is the lathe's original cost.

d. Each year, the value decreases by \$170, because the slope of the line is -170. The slope of the depreciation line is the *annual depreciation rate*. ∎

EXERCISE 3.3

VOCABULARY AND CONCEPTS *In Exercises 1–6, fill in the blank to make a true statement.*

1. The formula for the point–slope form of a line is
_____.

2. In the equation $y = mx + b$, ___ is the slope of the graph of the line, and $(0, b)$ is the _____.

3. The equation $y = mx + b$ is called the
_____ form of the equation of a line.

4. The general form of the equation of a line is
_____.

5. The slope of the graph of the equation $Ax + By = C$ is ___.

6. The y-intercept of the graph of the equation $Ax + By = C$ is _____.

PRACTICE *In Exercises 7–12, use point–slope form to write the equation of the line with the given properties. Write each equation in general form.*

7. $m = 2$, passing through $P(2, 4)$

8. $m = -3$, passing through $P(3, 5)$

9. $m = 2$, passing through $P\left(-\frac{3}{2}, \frac{1}{2}\right)$

10. $m = -6$, passing through $P\left(\frac{1}{4}, -2\right)$

11. $m = \pi$, passing through $P(\pi, 0)$

12. $m = \pi$, passing through $P(0, \pi)$

In Exercises 13–14, use the point–slope form to write the equation of each line. Write the equation in general form.

13.

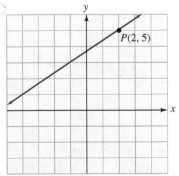

$P(2, 5)$

14.

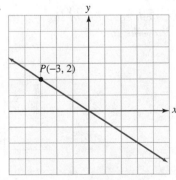

$P(-3, 2)$

In Exercises 15–18, use point–slope form to write the equation of the line passing through the two given points. Write each equation in slope–intercept form.

15. $P(0, 0), Q(4, 4)$ **16.** $P(-5, -5), Q(0, 0)$

17. $P(3, 4), Q(0, -3)$ **18.** $P(4, 0), Q(6, -8)$

In Exercises 19–20, use point–slope form to write the equation of each line. Write each answer in slope–intercept form.

19.

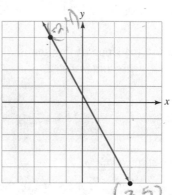

20.

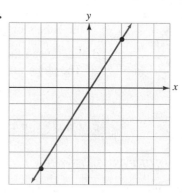

In Exercises 21–28, use the slope–intercept form to write the equation of the line with the given properties. Write each equation in slope–intercept form.

21. $m = 3; b = -2$ **22.** $m = -\dfrac{1}{3}; b = \dfrac{2}{3}$ **23.** $m = 5; b = -\dfrac{1}{5}$ **24.** $m = \sqrt{2}; b = \sqrt{2}$

25. $m = a; b = \dfrac{1}{a}$ **26.** $m = a; b = 2a$ **27.** $m = a; b = a$ **28.** $m = \dfrac{1}{a}; b = a$

In Exercises 29–34, use the slope–intercept form to write the equation of a line passing through the given point and having the given slope. Express the answer in general form.

29. $P(0, 0); m = \dfrac{3}{2}$ **30.** $P(-3, -7); m = -\dfrac{2}{3}$ **31.** $P(-3, 5); m = -3$ **32.** $P(-5, 1); m = 1$

33. $P(0, \sqrt{2}); m = \sqrt{2}$ **34.** $P(-\sqrt{3}, 0); m = 2\sqrt{3}$

In Exercises 35–40, write each equation in slope–intercept form to find the slope and the y-intercept. Then use the slope and y-intercept to draw the line.

35. $y + 1 = x$ **36.** $x + y = 2$ **37.** $x = \dfrac{3}{2}y - 3$

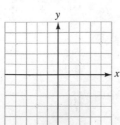
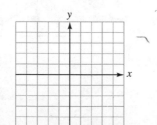
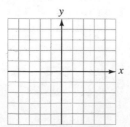

38. $x = -\dfrac{4}{5}y + 2$

39. $3(y - 4) = -2(x - 3)$

40. $-4(2x + 3) = 3(3y + 8)$

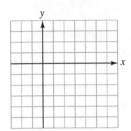

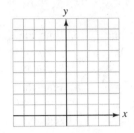

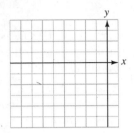

In Exercises 41–46, find the slope and the y-intercept of the line determined by the given equation.

41. $3x - 2y = 8$

42. $-2x + 4y = 12$

43. $-2(x + 3y) = 5$

44. $5(2x - 3y) = 4$

45. $x = \dfrac{2y - 4}{7}$

46. $3x + 4 = -\dfrac{2(y - 3)}{5}$

In Exercises 47–58, tell whether the graphs of each pair of equations are parallel, perpendicular, or neither.

47. $y = 3x + 4,\ y = 3x - 7$

48. $y = 4x - 13,\ y = \dfrac{1}{4}x + 13$

49. $x + y = 2,\ y = x + 5$

50. $x = y + 2,\ y = x + 3$

51. $y = 3x + 7,\ 2y = 6x - 9$

52. $2x + 3y = 9,\ 3x - 2y = 5$

53. $x = 3y + 4,\ y = -3x + 7$

54. $3x + 6y = 1,\ y = \dfrac{1}{2}x$

55. $y = 3,\ x = 4$

56. $y = -3,\ y = -7$

57. $x = \dfrac{y - 2}{3},\ 3(y - 3) + x = 0$

58. $2y = 8,\ 3(2 + x) = 3(y + 2)$

In Exercises 59–64, write the equation of the line that passes through the given point and is parallel to the given line. Write the answer in slope–intercept form.

59. $P(0, 0),\ y = 4x - 7$

60. $P(0, 0),\ x = -3y - 12$

61. $P(2, 5),\ 4x - y = 7$

62. $P(-6, 3),\ y + 3x = -12$

63. $P(4, -2),\ x = \dfrac{5}{4}y - 2$

64. $P(1, -5),\ x = -\dfrac{3}{4}y + 5$

In Exercises 65–70, write the equation of the line that passes through the given point and is perpendicular to the given line. Write the answer in slope–intercept form.

65. $P(0, 0),\ y = 4x - 7$

66. $P(0, 0),\ x = -3y - 12$

67. $P(2, 5),\ 4x - y = 7$

68. $P(-6, 3),\ y + 3x = -12$

69. $P(4, -2),\ x = \dfrac{5}{4}y - 2$

70. $P(1, -5),\ x = -\dfrac{3}{4}y + 5$

In Exercises 71–74, use the method of Example 9 to find the slope and the y-intercept of the graph of each equation.

71. $4x + 5y = 20$

72. $9x - 12y = 17$

73. $2x + 3y = 12$

74. $5x + 6y = 30$

75. Find the equation of the line perpendicular to the line $y = 3$ and passing through the midpoint of the segment joining $(2, 4)$ and $(-6, 10)$.

76. Find the equation of the line parallel to the line $y = -8$ and passing through the midpoint of the segment joining $(-4, 2)$ and $(-2, 8)$.

77. Find the equation of the line parallel to the line $x = 3$ and passing through the midpoint of the segment joining $(2, -4)$ and $(8, 12)$.

78. Find the equation of the line perpendicular to the line $x = 3$ and passing through the midpoint of the segment joining $(-2, 2)$ and $(4, -8)$.

APPLICATIONS   *In Exercises 79–89, assume straight-line depreciation or straight-line appreciation.*

79. Depreciation   A taxicab was purchased for $24,300. Its salvage value at the end of its 7-year useful life is expected to be $1900. Find the depreciation equation.

80. Depreciation   A small business purchases the computer system shown in Illustration 1. It will be depreciated over a 4-year period, when its salvage value will be $300. Find the depreciation equation.

expect the property to double in value in 10 years. Find the appreciation equation.

82. Appreciation   A house purchased for $112,000 is expected to double in value in 12 years. Find its appreciation equation.

83. Depreciation   Find the depreciation equation for the TV in the want ad in Illustration 2.

For Sale: 3-year-old 54-inch TV, $1,900 new. Asking $1,190. Call 875-5555. Ask for Mike.

ILLUSTRATION 2

84. Depreciation   A word processor cost $555 when new and is expected to be worth $80 after 5 years. What will it be worth after 3 years?

85. Salvage value   A copier cost $1050 when new and will be depreciated at the rate of $120 per year. If the useful life of the copier is 8 years, find its salvage value.

86. Rate of depreciation   A truck that cost $27,600 when new will have no salvage value after 12 years. Find its annual rate of depreciation.

87. Value of antiques   An antique table is expected to appreciate $40 each year. If the table will be worth $450 in 2 years, what will it be worth in 13 years?

88. Value of antiques   An antique clock is expected to be worth $350 after 2 years and $530 after 5 years. What will the clock be worth after 7 years?

$3900

ILLUSTRATION 1

81. Appreciation   An apartment building was purchased for $475,000, excluding the cost of land. The owners

89. Purchase price of real estate A cottage that was purchased 3 years ago is now appraised at $47,700. If the property has been appreciating $3500 per year, find its original purchase price.

90. Computer repair A computer repair company charges a fixed amount, plus an hourly rate, for a service call. Use the information in Illustration 3 to find the hourly rate.

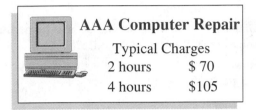

AAA Computer Repair

Typical Charges

2 hours	$ 70
4 hours	$105

ILLUSTRATION 3

91. Automobile repair An auto repair shop charges an hourly rate, plus the cost of parts. If the cost of labor for a $1\frac{1}{2}$-hour radiator repair is $69, find the cost of labor for a 5-hour transmission overhaul.

92. Printer charges A printer charges a fixed setup cost, plus $1 for every 100 copies. If 700 copies cost $52, how much will it cost to print 1000 copies?

93. Predicting fires A local fire department recognizes that city growth and the number of reported fires are related by a linear equation. City records show that 300 fires were reported in a year when the local population was 57,000 persons, and 325 fires were reported in a year when the population was 59,000 persons. How many fires can be expected in the year when the population reaches 100,000 persons?

94. Estimating the cost of rain gutter A neighbor tells you that an installer of rain gutter charges $60, plus a dollar amount per foot. If the neighbor paid $435 for the installation of 250 feet of gutter, how much will it cost you to have 300 feet installed?

95. Converting temperatures Water freezes at 32° Fahrenheit, or 0° Celsius. Water boils at 212° F, or 100° C. Find a formula for converting a temperature from degrees Fahrenheit to degrees Celsius.

96. Converting units A speed of 1 mile per hour is equal to 88 feet per minute, and of course, 0 miles per hour is 0 feet per minute. Find an equation for converting a speed x, in miles per hour, to the corresponding speed y, in feet per minute.

97. Predicting stock prices The value of the stock of ABC Corporation has been increasing by the same fixed dollar amount each year. The pattern is expected to continue. Let 1996 be the base year corresponding to $x = 0$, with $x = 1, 2, 3, \ldots$ corresponding to later years. ABC stock was selling at $37\frac{1}{2}$ in 1996 and at $45 in 1998. If $-y$ represents the price of ABC stock, find the equation $y = mx + b$ that relates x and y, and predict the price in the year 2000.

98. Estimating inventory Inventory of unsold goods showed a surplus of 375 units in January and 264 in April. Assume that the relationship between inventory and time is given by the equation of a line, and estimate the expected inventory in March. Because March lies between January and April, this estimation is called **interpolation.**

99. Oil depletion When a Petroland oil well was first brought on line, it produced 1900 barrels of crude oil per day. In each later year, owners expect its daily production to drop by 70 barrels. Find the daily production after $3\frac{1}{2}$ years.

100. Waste management The corrosive waste in industrial sewage limits the useful life of the piping in a waste processing plant to 12 years. The piping system was originally worth $137,000, and it will cost the company $33,000 to remove it at the end of its 12-year useful life. Find the depreciation equation.

DISCOVERY AND WRITING *Write a paragraph using your own words.*

101. Explain how to find the equation of a line passing through two given points.

102. In straight-line depreciation, explain why the slope of the line is called the *rate of depreciation.*

103. Prove that the equation of a line with x-intercept of $(a, 0)$ and y-intercept of $(0, b)$ can be written in the form

$$\frac{x}{a} + \frac{y}{b} = 1$$

104. Find the x- and y-intercepts of the line $bx + ay = ab$.

In Exercises 105–110, investigate the properties of slope and the y-intercept by experimenting with the following problems.

105. Graph $y = mx + 2$ for several positive values of m. What do you notice?

106. Graph $y = mx + 2$ for several negative values of m. What do you notice?

107. Graph $y = 2x + b$ for several increasing positive values of b. What do you notice?

108. Graph $y = 2x + b$ for several decreasing negative values of b. What do you notice?

109. How will the graph of $y = \frac{1}{2}x + 5$ compare to the graph of $y = \frac{1}{2}x - 5$?

110. How will the the graph of $y = \frac{1}{2}x - 5$ compare to the graph of $y = \frac{1}{2}x$?

REVIEW *Simplify each expression.*

111. $x^7 x^3 x^{-5}$

112. $\dfrac{y^3 y^{-4}}{y^{-5}}$

113. $\left(\dfrac{81}{25}\right)^{-3/2}$

114. $\sqrt[3]{27x^7}$

115. $\sqrt{27} - 2\sqrt{12}$

116. $\dfrac{5}{\sqrt{5}}$

3.4 Graphs of Equations

■ INTERCEPTS OF GRAPHS ■ SYMMETRIES OF GRAPHS ■ MISCELLANEOUS GRAPHS ■ CIRCLES
■ GRAPHING EQUATIONS OF CIRCLES ■ SOLVING EQUATIONS BY GRAPHING

The graphs of many equations are curves. If we plot several points (x, y) that satisfy such an equation, the shape of the graph will usually become evident. We can then sketch the graph by joining these points with a smooth curve.

■ INTERCEPTS OF GRAPHS

In Figure 3-32(a), the x-intercepts of the graph are $(a, 0)$ and $(b, 0)$, the points where the graph intersects the x-axis. In Figure 3-32(b), the **y-intercept** is $(0, c)$, the point where the graph intersects the y-axis.

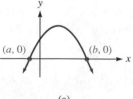

(a)   (b)

FIGURE 3-32

To graph the equation $y = x^2 - 4$, we first find the x- and y-intercepts. To find the x-intercepts, we let $y = 0$ and solve for x.

$$y = x^2 - 4$$
$$0 = x^2 - 4 \qquad \text{Substitute 0 for } y.$$
$$0 = (x + 2)(x - 2) \qquad \text{Factor } x^2 - 4.$$
$$x + 2 = 0 \quad \text{or} \quad x - 2 = 0 \qquad \text{Set each factor equal to 0.}$$
$$x = -2 \qquad\qquad x = 2$$

Since $y = 0$ when $x = -2$ and $x = 2$, the x-intercepts are $(-2, 0)$ and $(2, 0)$. (See Figure 3-33.)

To find the y-intercept, we let $x = 0$ and solve for y.

$$y = x^2 - 4$$
$$y = 0^2 - 4 \qquad \text{Substitute 0 for } x.$$
$$y = -4$$

Since $y = -4$ when $x = 0$, the y-intercept is $(0, -4)$.

We can find other pairs (x, y) that satisfy the equation by substituting numbers for x and finding the corresponding values of y. For example, if $x = -3$, then $y = (-3)^2 - 4$, or 5, and the point $(-3, 5)$ lies on the graph.

The coordinates of the intercepts and other points appear in Figure 3-33. If we plot the points and draw a curve through them, we will obtain the graph of the equation.

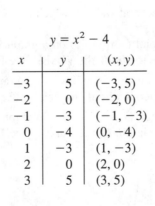

$$y = x^2 - 4$$

x	y	(x, y)
-3	5	$(-3, 5)$
-2	0	$(-2, 0)$
-1	-3	$(-1, -3)$
0	-4	$(0, -4)$
1	-3	$(1, -3)$
2	0	$(2, 0)$
3	5	$(3, 5)$

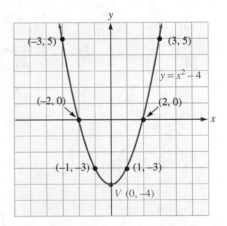

FIGURE 3-33

This graph is called a **parabola**. Its lowest point, $V(0, -4)$, is called the **vertex**. Because the y-axis divides the parabola into two congruent halves, the y-axis is called an **axis of symmetry**. We say that the parabola is **symmetric about the y-axis.**

■ ■ ■ ■ ■ ■ ■ ■ ■ **Graphing $y = x^2 - 4$**

ACCENT ON
TECHNOLOGY

To use a graphing calculator to graph the equation $y = x^2 - 4$, we enter the right-hand side of the equation after the symbol $Y_1 =$. The display will show the equation:

$$Y_1 = X \wedge 2 - 4$$

If we set use window settings of $[-10, 10]$ for x and $[-10, 10]$ for y and press the GRAPH key, we will obtain the graph shown in Figure 3-34(a). If we use settings of $[-4, 4]$ for x and $[-4, 4]$ for y, we will obtain the graph shown in Figure 3-34(b).

From the graph, we can see that the x-intercepts are $(-2, 0)$ and $(2, 0)$ and that the y-intercept is $(0, -4)$.

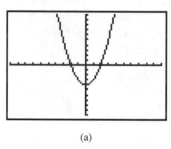 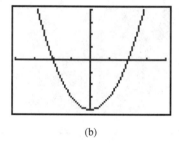

(a)   (b)

■ ■ ■ ■ ■ ■ ■ ■ ■

FIGURE 3-34

■ SYMMETRIES OF GRAPHS

There are several ways in which a graph can have symmetry.

1. If the point $(-x, y)$ lies on a graph whenever the point (x, y) does, the graph is **symmetric about the y-axis.** (See Figure 3-35(a).)

2. If the point $(-x, -y)$ lies on the graph whenever the point (x, y) does, the graph is **symmetric about the origin.** (See Figure 3-35(b).)

3. If the point $(x, -y)$ lies on the graph whenever the point (x, y) does, the graph is **symmetric about the x-axis.** (See Figure 3-35(c).)

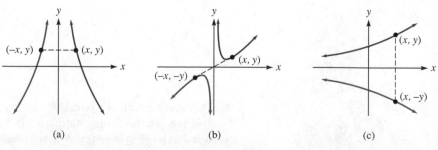

(a)   (b)   (c)

FIGURE 3-35

■ MISCELLANEOUS GRAPHS

We will now graph many equations. To graph each one, we will find the x- and y-intercepts, test for symmetries, plot points, and join the points with a smooth curve.

EXAMPLE 1 Graph $y = |x|$.

Solution **x-intercepts.** To find the x-intercepts, we let $y = 0$ and solve for x.

$$y = |x|$$
$$0 = |x| \qquad \text{Substitute 0 for } y.$$
$$|x| = 0$$
$$x = 0$$

The x-intercept is $(0, 0)$. (See Figure 3-36.)

y-intercepts. To find the y-intercepts, we let $x = 0$ and solve for y.

$$y = |x|$$
$$y = |0| \qquad \text{Substitute 0 for } x.$$
$$y = 0$$

The y-intercept is $(0, 0)$.

Symmetry. To test for y-axis symmetry, we replace x with $-x$.

1. $y = |x|$ \qquad The original equation.
$\quad\ y = |-x|$ \qquad Replace x with $-x$.
2. $y = |x|$ \qquad\quad $|-x| = |x|$.

Since Equations 1 and 2 are the same, the graph is symmetric about the y-axis.
To test for symmetry about the origin, we replace x with $-x$ and y with $-y$.

1. $\quad y = |x|$ \qquad The original equation.
$\ -y = |-x|$ \qquad Replace x with $-x$ and y with $-y$.
3. $-y = |x|$ \qquad $|-x| = |x|$.

Since Equations 1 and 3 are different, the graph is not symmetric about the origin. To test for x-axis symmetry, we replace y with $-y$.

1. $y = |x|$ The original equation.

4. $-y = |x|$ Replace y with $-y$.

Since Equations 1 and 4 are different, the graph is not symmetric about the x-axis.

To graph the equation, we plot the x- and y-intercepts and several other pairs (x, y) with positive values of x. We can use the property of y-axis symmetry to draw the graph for negative values of x. See Figure 3-36.

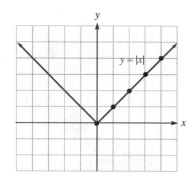

$y = |x|$

x	y	(x, y)
0	0	$(0, 0)$
1	1	$(1, 1)$
2	2	$(2, 2)$
3	3	$(3, 3)$
4	4	$(4, 4)$

FIGURE 3-36   ■

Self Check   Graph $y = -|x|$.

Answer

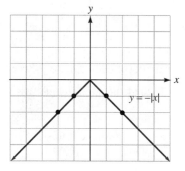

■ ■ ■ ■ ■ ■ ■ ■ ■   **Graphing $y = |x|$**

**ACCENT ON
TECHNOLOGY**

To use a graphing calculator to graph $y = |x|$, we enter the right-hand side of the equation after the symbol $Y_1 =$. The display will look like

$Y_1 = \text{abs} (X)$

If we use window settings of $[-10, 10]$ for x and $[-10, 10]$ for y and press the GRAPH key, we will obtain the graph shown in Figure 3-37.

From the graph, we see that there is one x- and y-intercept, the point $(0, 0)$.

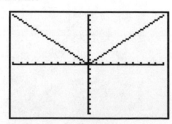

FIGURE 3-37

■ ■ ■ ■ ■ ■ ■ ■ ■

EXAMPLE 2   Graph $y = x^3 - x$.

Solution   **x-intercepts.**   To find the x-intercepts, we let $y = 0$ and solve for x.

$$y = x^3 - x$$
$$0 = x^3 - x \qquad \text{Substitute 0 for } y.$$
$$0 = x(x^2 - 1) \qquad \text{Factor out } x.$$
$$0 = x(x + 1)(x - 1) \qquad \text{Factor } x^2 - 1.$$
$$x = 0 \quad \text{or} \quad x + 1 = 0 \quad \text{or} \quad x - 1 = 0 \qquad \text{Set each factor equal to 0.}$$
$$x = -1 \qquad\qquad x = 1$$

The x-intercepts are $(0, 0)$, $(-1, 0)$, and $(1, 0)$.

y-intercepts.   To find the y-intercepts, we let $x = 0$ and solve for y.

$$y = x^3 - x$$
$$y = 0^3 - 0 \qquad \text{Substitute 0 for } x.$$
$$y = 0$$

The y-intercept is $(0, 0)$.

Symmetry.   To test for y-axis symmetry, we replace x with $-x$.

1. $\quad y = x^3 - x \qquad\qquad\qquad$ The original equation.
$\qquad y = (-x)^3 - (-x) \qquad\qquad$ Replace x with $-x$.
2. $\quad y = -x^3 + x \qquad\qquad\qquad$ Simplify.

Since Equations 1 and 2 are different, the graph is not symmetric about the y-axis.
To test for symmetry about the origin, we replace x with $-x$ and y with $-y$.

1. $\quad y = x^3 - x \qquad\qquad\qquad$ The original equation.
$\qquad -y = (-x)^3 - (-x) \qquad\qquad$ Replace x with $-x$ and y with $-y$.
$\qquad -y = -x^3 + x \qquad\qquad\qquad$ Simplify.
3. $\quad y = x^3 - x \qquad\qquad\qquad$ Multiply both sides by -1.

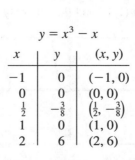

$$y = x^3 - x$$

x	y	(x, y)
-1	0	$(-1, 0)$
0	0	$(0, 0)$
$\frac{1}{2}$	$-\frac{3}{8}$	$\left(\frac{1}{2}, -\frac{3}{8}\right)$
1	0	$(1, 0)$
2	6	$(2, 6)$

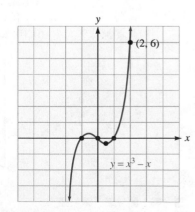

FIGURE 3-38

Since Equations 1 and 3 are the same, the graph is symmetric about the origin.

We test for symmetry about the x-axis by replacing y with $-y$.

1. $y = x^3 - x$ The original equation.

      $-y = x^3 - x$ Replace y with $-y$.

4. $y = -x^3 + x$ Multiply both sides by -1.

Since Equations 1 and 4 are different, the graph is not symmetric about the x-axis.

To graph the equation, we plot the x- and y-intercepts and several other pairs (x, y) with positive values of x. We can use the property of symmetry about the origin to draw the graph for negative values of x. See Figure 3-38. ∎

Self Check   Graph $y = x^3 - 9x$.

Answer

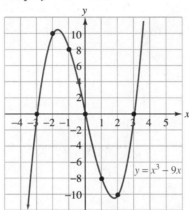

Graphing $y = x^3 - x$

ACCENT ON TECHNOLOGY

To use a graphing calculator to graph $y = x^3 - x$, we enter the right-hand side of the equation after the symbol $Y_1 =$. The display will show the equation:

$$Y_1 = X \wedge 3 - X$$

If we set use window settings of $[-10, 10]$ for x and $[-10, 10]$ for y and press the GRAPH key, we will obtain the graph shown in Figure 3-39(a).

If we zoom to get Figure 3-39(b), we can see that the x-intercepts are $(-1, 0)$, $(0, 0)$, and $(1, 0)$ and that the only y-intercept is $(0, 0)$.

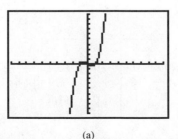

(a)

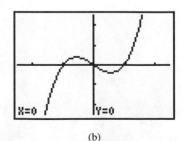

(b)

FIGURE 3-39

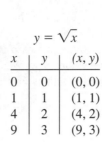

EXAMPLE 3   Graph $y = \sqrt{x}$.

Solution   Because the x- and y-intercepts are both $(0, 0)$, the graph passes through the origin.
We now test for symmetries. Since $x \geq 0$ in the radical $\sqrt{x}$, there is no point in replacing x with $-x$. Since $y = \sqrt{x}$ and $\sqrt{x} \geq 0$, there is also no point in replacing y with $-y$. The graph of the equation has no symmetries. We plot several points to obtain the graph in Figure 3-40.

$y = \sqrt{x}$

x	y	(x, y)
0	0	$(0, 0)$
1	1	$(1, 1)$
4	2	$(4, 2)$
9	3	$(9, 3)$

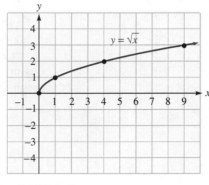

FIGURE 3-40

Self Check   Graph $y = -\sqrt{x}$.

Answer

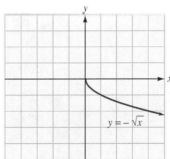

■ ■ ■ ■ ■ ■ ■ ■ ■   **Graphing $y = \sqrt{x}$**

ACCENT ON
TECHNOLOGY

To use a graphing calculator to graph $y = \sqrt{x}$, we enter the right-hand side of the equation after the symbol $Y_1 =$. The display will show the equation

$$Y_1 = \sqrt{}(x)$$

If we set use window settings of $[-10, 10]$ for x and $[-10, 10]$ for y and press the GRAPH key, we will obtain the graph shown in Figure 3-41.

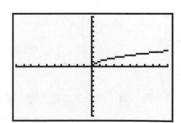

FIGURE 3-41

EXAMPLE 4   Graph $y^2 = x$.

Solution   Since the x- and y-intercepts are both $(0, 0)$, the graph passes through the origin.
To test for x-axis symmetry, we replace y with $-y$.

1. $y^2 = x$ The original equation.

   $(-y)^2 = x$ Replace y with $-y$.

2. $y^2 = x$ Simplify.

Since Equations 1 and 2 are the same, the graph is symmetric about the x-axis.
There are no other symmetries.

To graph the equation, we plot the x- and y-intercept and several other pairs
(x, y) with positive values of y. We can use the property of x-axis symmetry to draw
the graph for negative values of y. (See Figure 3-42.)

$y^2 = x$

x	y	(x, y)
0	0	$(0, 0)$
1	1	$(1, 1)$
4	2	$(4, 2)$
9	3	$(9, 3)$

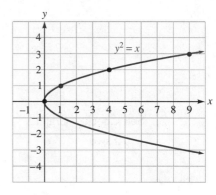

FIGURE 3-42   ∎

Self Check   Graph $y^2 = -x$.

Answer

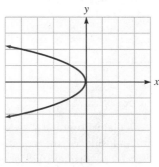

■ ■ ■ ■ ■ ■ ■ ■ ■   **Graphing $y^2 = x$**

ACCENT ON   To use a graphing calculator to graph $y^2 = x$, we must first solve the equation for y.
TECHNOLOGY $$y^2 = x$$
                     $$y = \pm\sqrt{x}$$
              $$y = \sqrt{x} \quad \text{or} \quad y = -\sqrt{x}$$

We graph both equations on the same coordinate axes by entering the first equation as Y_1 and the second equation as Y_2.

$$Y_1 = \sqrt{x} \quad \text{and} \quad Y_2 = -\sqrt{x}$$

If we use window settings of $[-10, 10]$ for x and $[-10, 10]$ for y and press the GRAPH key, we will obtain the graph shown in Figure 3-43.

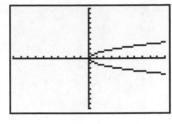

FIGURE 3-43

The graphs in Examples 3 and 4 are related. We have solved the equation $y^2 = x$ for y, and two equations resulted:

$$y = \sqrt{x} \quad \text{and} \quad y = -\sqrt{x}$$

The first equation, $y = \sqrt{x}$, was graphed in Example 3. It is the top half of the parabola shown in Example 4. The second equation, $y = -\sqrt{x}$, is the bottom half.

■ CIRCLES

Circles
A **circle** is the set of all points in a plane that are a fixed distance from a point called its **center**. The fixed distance is the **radius of the circle.**

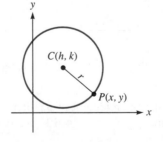

FIGURE 3-44

To find the equation of a circle with radius r and center at $C(h, k)$, we must find all points $P(x, y)$ in the xy-plane such that the length of line segment PC is r. (See Figure 3-44.) We can use the distance formula to find the length of CP, which is r:

$$r = \sqrt{(x - h)^2 + (y - k)^2}$$

After squaring both sides, we get

$$r^2 = (x - h)^2 + (y - k)^2$$

This equation is called the **standard equation of a circle.**

The Standard Equation of a Circle with Center at (h, k)
The graph of any equation that can be written in the form
$$(x - h)^2 + (y - k)^2 = r^2$$
is a circle with radius r and center at point (h, k).

If $r = 0$, the circle is a single point called a **point circle.** If the center of a circle is the origin, then $(h, k) = (0, 0)$, and we have the following result.

> **The Standard Equation of a Circle with Center at (0, 0)**
> The graph of any equation that can be written in the form
> $$x^2 + y^2 = r^2$$
> is a circle with radius r and center at the origin.

If we square the binomials in $(x - h)^2 + (y - k)^2 = r^2$, we obtain an equation of the form

$$x^2 + y^2 + cx + dy + e = 0$$

where c, d, and e are real numbers. This form is called the **general form of the equation of a circle.**

EXAMPLE 5   Find the general form of the equation of the circle with radius 5 and center at $(3, 2)$.

Solution   We substitute 5 for r, 3 for h, and 2 for k in the standard equation of a circle and simplify:

$$(x - \boldsymbol{h})^2 + (y - \boldsymbol{k})^2 = \boldsymbol{r}^2$$
$$(x - \boldsymbol{3})^2 + (y - \boldsymbol{2})^2 = \boldsymbol{5}^2$$

$$x^2 - 6x + 9 + y^2 - 4y + 4 = 25 \qquad \text{Remove parentheses.}$$
$$x^2 + y^2 - 6x - 4y - 12 = 0 \qquad \text{Subtract 25 from both sides and simplify.}$$

The general form is $x^2 + y^2 - 6x - 4y - 12 = 0$. ∎

Self Check   Find the general form of the equation of a circle with radius 6 and center at $(-2, 5)$.
Answer $x^2 + y^2 + 4x - 10y - 7 = 0$

EXAMPLE 6   Find the general form of the equation of the circle with endpoints of its diameter at $(8, -3)$ and $(-4, 13)$.

Solution   We can find the center $O(h, k)$ of the circle by finding the midpoint of its diameter. By the midpoint formulas and $(x_1, y_1) = (8, -3)$ and $(x_2, y_2) = (-4, 13)$, we have

$$h = \frac{\boldsymbol{x_1 + x_2}}{2} \qquad\qquad k = \frac{\boldsymbol{y_1 + y_2}}{2}$$

$$h = \frac{\boldsymbol{8 + (-4)}}{2} \qquad\qquad k = \frac{\boldsymbol{-3 + 13}}{2}$$

$$= \frac{4}{2} \qquad\qquad\qquad = \frac{10}{2}$$

$$= 2 \qquad\qquad\qquad\quad = 5$$

The center is $O(h, k) = O(2, 5)$.

To find the radius, we find the distance between the center and one endpoint of the diameter. The center is $O(2, 5)$, and one endpoint is $(8, -3)$.

$$r = \sqrt{(x_2 - x_1)^2 + (y_2 - y_1)^2}$$ The distance formula.

$$r = \sqrt{(2 - 8)^2 + [5 - (-3)]^2}$$ Substitute 8 for x_1, -3 for y_1, 2 for x_2, and 5 for y_2.

$$= \sqrt{(-6)^2 + (8)^2}$$

$$= \sqrt{36 + 64}$$

$$= 10$$ $\sqrt{36 + 64} = \sqrt{100} = 10.$

To find the equation of a circle with center at $(2, 5)$ and radius 10, we substitute 2 for h, 5 for k, and 10 for r in the standard equation of the circle and simplify:

$$(x - h)^2 + (y - k)^2 = r^2$$
$$(x - 2)^2 + (y - 5)^2 = 10^2$$
$$x^2 - 4x + 4 + y^2 - 10y + 25 = 100$$ Remove parentheses.
$$x^2 + y^2 - 4x - 10y - 71 = 0$$ Subtract 100 from both sides and simplify. ■

Self Check Find the equation of a circle with endpoints of its diameter at $(-2, 2)$ and $(6, 8)$.

Answer $x^2 + y^2 - 4x - 10y + 4 = 0$

■ GRAPHING EQUATIONS OF CIRCLES

EXAMPLE 7 Graph the circle whose equation is $x^2 + y^2 - 4x + 2y = 20$.

Solution To find the coordinates of the center and the radius, we write the equation in standard form by completing the square on both x and y:

$$x^2 + y^2 - 4x + 2y = 20$$
$$x^2 - 4x + y^2 + 2y = 20$$
$$x^2 - 4x + 4 + y^2 + 2y + 1 = 20 + 4 + 1$$ Add 4 and 1 to both sides to complete the square.
$$(x - 2)^2 + (y + 1)^2 = 25$$ Factor $x^2 - 4x + 4$ and $y^2 + 2y + 1$.
$$(x - 2)^2 + [y - (-1)]^2 = 5^2$$

From the equation of the circle, we see that its radius is 5 and that the coordinates of its center are $h = 2$ and $k = -1$. The graph is shown in Figure 3-45.

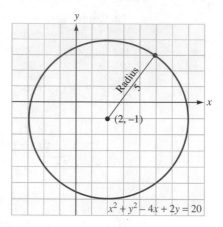

FIGURE 3-45 ■

Self Check Graph $x^2 + y^2 + 2x - 4y = -1$.

Answer

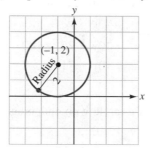

■ ■ ■ ■ ■ ■ ■ ■ ■ **Graphing Circles**

ACCENT ON
TECHNOLOGY

To use a graphing calculator to graph the equation $(x - 2)^2 + (y + 1)^2 = 25$, we must solve the equation for y:

$$(x - 2)^2 + (y + 1)^2 = 25$$
$$(y + 1)^2 = 25 - (x - 2)^2$$
$$y + 1 = \pm\sqrt{25 - (x - 2)^2}$$
$$y = -1 \pm \sqrt{25 - (x - 2)^2}$$

This last expression represents two equations: $y = -1 + \sqrt{25 - (x - 2)^2}$ and $y = -1 - \sqrt{25 - (x - 2)^2}$. We graph both of these equations separately on the same coordinate axes by entering the first equation as Y_1 and the second as Y_2:

$$Y_1 = -1 + \sqrt{}(25 - (x - 2) \wedge 2)$$
$$Y_2 = -1 - \sqrt{}(25 - (x - 2) \wedge 2)$$

Depending on the setting of the maximum and minimum values of x and y, the graph may not appear to be a circle, and it may look like Figure 3-46(a). Most graphing calculators can be set to display equal-sized divisions on the x- and y-axes, producing a better graph—like that shown in Figure 3-46(b).

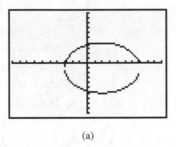

(a)

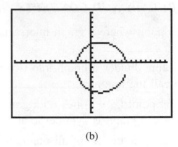

(b)

FIGURE 3-46

■ SOLVING EQUATIONS BY GRAPHING

It is easy to verify that the solutions of $x^2 - 3 = 0$ are $\sqrt{3}$ and $-\sqrt{3}$, because either of these numbers makes the left-hand side of the equation equal to 0. In the equation $y = x^2 - 3$, those same values of x will cause y to equal 0. The solutions of the equation $x^2 - 3 = 0$ are the x values of the x-intercepts of the graph of the equation $y = x^2 - 3$.

This idea provides a method for using a graphing calculator to solve equations.

EXAMPLE 8   Find the positive solution of $x^2 - 3 = 0$.

Solution   We graph $y = x^2 - 3$ and trace to move the cursor near the positive x-intercept, as in Figure 3-47(a). We then zoom and trace three times to get Figure 3-47(b). A value of x similar to that shown in Figure 3-47(b) provides an approximation to the solution. The solution provided in Figure 3-47(b) agrees to three decimal places with the actual solution, $\sqrt{3} = \mathbf{1.732}051$. Repeated zooms will provide more accurate results.

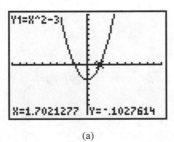

(a)

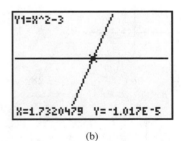

(b)

FIGURE 3-47   ■

Self Check   Find the negative solution of $x^2 - 5 = 0$. Give the answer to the nearest hundredth.
Answer -2.24

EXERCISE 3.4

VOCABULARY AND CONCEPTS   *In Exercises 1–10, fill in the blank to make a true statement.*

1. The point where a graph intersects the x-axis is called the _____.

2. The y-intercept is the point where a graph intersects the _____.

3. If a line divides a graph into two congruent halves, we call the line an _____.

4. If the point $(-x, y)$ lies on a graph whenever (x, y) does, the graph is symmetric about the _____.

5. If the point $(x, -y)$ lies on a graph whenever (x, y) does, the graph is symmetric about the _____.

6. If the point $(-x, -y)$ lies on a graph whenever (x, y) does, the graph is symmetric about the _____.

7. A _____ is the set of all points in a plane that are a fixed distance from a point called its _____.

8. A _____ is the distance from the center of a circle to a point on the circle.

9. The standard equation of a circle with center at the origin and radius r is _____.

10. The standard equation of a circle with center at (h, k) and radius r is _____.

PRACTICE   *In Exercises 11–22, find the x- and y-intercepts of each graph.* ***Do not graph the equation.***

11. $y = x^2 - 4$

12. $y = x^2 - 9$

13. $y = 4x^2 - 2x$

14. $y = 2x - 4x^2$

15. $y = x^2 - 4x - 5$

16. $y = x^2 - 10x + 21$

17. $y = x^2 + x - 2$

18. $y = x^2 + 2x - 3$

19. $y = x^3 - 9x$

20. $y = x^3 + x$

21. $y = x^4 - 1$

22. $y = x^4 - 25x^2$

In Exercises 23–30, graph each parabola. Check your work with a graphing calculator.

23. $y = x^2$

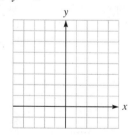

24. $y = -x^2$

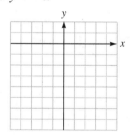

25. $y = -x^2 + 2$

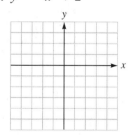

26. $y = x^2 - 1$

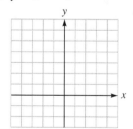

27. $y = x^2 - 4x$

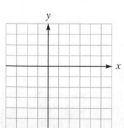

28. $y = x^2 + 2x$

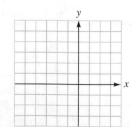

29. $y = \dfrac{1}{2}x^2 - 2x$

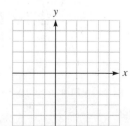

30. $y = \dfrac{1}{2}x^2 + 3$

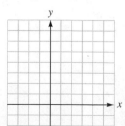

In Exercises 31–46, find the symmetries, if any, of the graph of each equation. **Do not graph the equation.**

31. $y = x^2 + 2$

32. $y = 3x + 2$

33. $y^2 + 1 = x$

34. $y^2 + y = x$

35. $y^2 = x^2$

36. $y = 3x + 7$

37. $y = 3x^2 + 7$

38. $x^2 + y^2 = 1$

39. $y = 3x^3 + 7$

40. $y = 3x^3 + 7x$

41. $y^2 = 3x$

42. $y = 3x^4 + 7$

43. $y = |x|$

44. $y = |x + 1|$

45. $|y| = x$

46. $|y| = |x|$

In Exercises 47–62, graph each equation. Be sure to find any intercepts and symmetries. Check your work with a graphing calculator.

47. $y = x^2 + 4x$

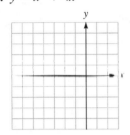

48. $y = x^2 - 6x$

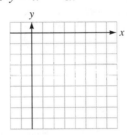

49. $y = x^3$

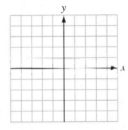

50. $y = x^3 + x$

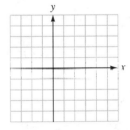

51. $y = |x - 2|$

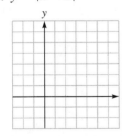

52. $y = |x| - 2$

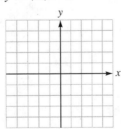

53. $y = 3 - |x|$

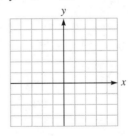

54. $y = 3|x|$

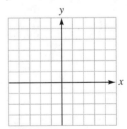

55. $y^2 = -x$

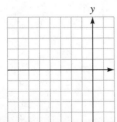

56. $y^2 = 4x$

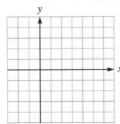

57. $y^2 = 9x$

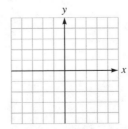

58. $y^2 = -4x$

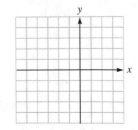

59. $y = \sqrt{x} - 1$

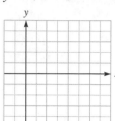

60. $y = 1 - \sqrt{x}$

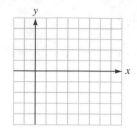

61. $xy = 4$

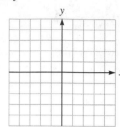

62. $xy = -9$

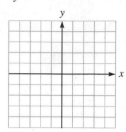

In Exercises 63–72, find the equation in general form of each circle with the given properties.

63. Center at the origin; $r = 1$

64. Center at the origin; $r = 4$

65. Center at $(6, 8)$; $r = 4$

66. Center at $(5, 3)$; $r = 2$

67. Center at $(3, -4)$; $r = \sqrt{2}$

68. Center at $(-9, 8)$; $r = 2\sqrt{3}$

69. Ends of diameter at $(3, -2)$ and $(3, 8)$

70. Ends of diameter at $(5, 9)$ and $(-5, -9)$

71. Center at $(-3, 4)$ and passing through the origin

72. Center at $(-2, 6)$ and passing through the origin

In Exercises 73–82, graph each equation.

73. $x^2 + y^2 - 25 = 0$

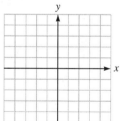

74. $x^2 + y^2 - 8 = 0$

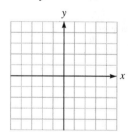

75. $(x - 1)^2 + (y + 2)^2 = 4$

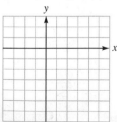

76. $(x + 1)^2 + (y - 2)^2 = 9$

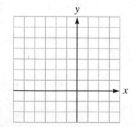

77. $x^2 + y^2 + 2x - 24 = 0$

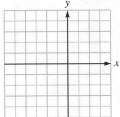

78. $x^2 + y^2 - 4y = 12$

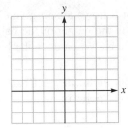

79. $9x^2 + 9y^2 - 12y = 5$

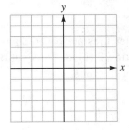

80. $4x^2 + 4y^2 + 4y = 15$

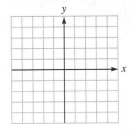

81. $4x^2 + 4y^2 - 4x + 8y + 1 = 0$

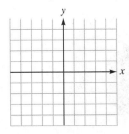

82. $9x^2 + 9y^2 - 6x + 18y + 1 = 0$

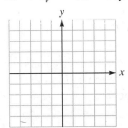

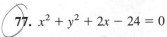 *In Exercises 83–86, use a graphing calculator to graph each equation and find the coordinates of the vertex. Round to the nearest hundredth.*

83. $y = 2x^2 - x + 1$ **84.** $y = x^2 + 5x - 6$ **85.** $y = 7 + x - x^2$ **86.** $y = 2x^2 - 3x + 2$

In Exercises 87–90, use a graphing calculator to solve each equation. Round to the nearest hundredth.

87. $x^2 - 7 = 0$ **88.** $x^2 - 3x + 2 = 0$ **89.** $x^3 - 3 = 0$ **90.** $3x^3 - x^2 - x = 0$

APPLICATIONS

91. Golfing   A golfer's tee shot follows a path given by $y = 64t - 16t^2$, where y is the height of the ball (in feet) after t seconds of flight. When does the ball strike the ground?

92. Golfing   Halfway through its flight, the golf ball of Exercise 91 reaches the highest point of its trajectory. How high is that?

93. 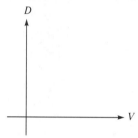 **Stopping distance**   The stopping distance D (in feet) for a car moving V miles per hour is given by $D = 0.08V^2 + 0.9V$. Graph the equation for velocities between 0 and 60 mph.

94. 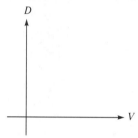 **Stopping distance**   How much farther does it take to stop at 60 mph than at 30 mph?

95. CB radio   The CB radio of a trucker covers the circular area shown in Illustration 1. Find the equation of that circle, in general form.

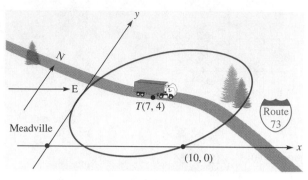

ILLUSTRATION 1

96. Tires   Two 24-inch-diameter tires are rolled against a wall, as shown in Illustration 2. Find the equations of the circular boundaries of the tires.

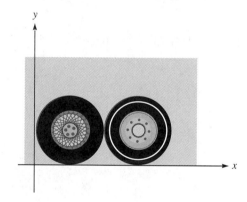

ILLUSTRATION 2

DISCOVERY AND WRITING   *The solution of the inequality $P(x) < 0$ consists of those numbers x for which the graph of $y = P(x)$ lies below the x-axis. To solve $P(x) < 0$, we graph $y = P(x)$ and trace to find numbers x that produce negative values of y. In Exercises 97–98, solve each inequality.*

97. $x^2 + x - 6 < 0$

98. $x^2 - 3x - 10 > 0$

REVIEW   *Solve each equation.*

99. $3(x + 2) + x = 5x$

100. $12b + 6(3 - b) = b + 3$

101. $\dfrac{5(2 - x)}{3} - 1 = x + 5$

102. $\dfrac{r - 1}{3} = \dfrac{r + 2}{6} + 2$

103. Mixing alloys In 60 ounces of alloy for watch cases, there are 20 ounces of gold. How much copper must be added to the alloy so that a watch case weighing 4 ounces, made from the new alloy, will contain exactly 1 ounce of gold?

104. Mixing coffee To make a mixture of 80 pounds of coffee worth $272, a grocer mixes coffee worth $3.25 a pound with coffee worth $3.85 a pound. How many pounds of cheaper coffee should the grocer use?

3.5 Proportion and Variation

■ PROPORTIONS ■ SOLVING PROPORTIONS ■ DIRECT VARIATION ■ INVERSE VARIATION ■ JOINT VARIATION ■ COMBINED VARIATION

The quotient of two numbers is often called a **ratio.** For example, the fraction $\frac{3}{2}$ can be read as "the ratio of 3 to 2." Some more examples of ratios are

$$\frac{3}{5}, \quad \frac{x+1}{9}, \quad \frac{a}{b}, \quad \text{and} \quad \frac{x^2-4}{x+5}$$

■ PROPORTIONS

An equation indicating that two ratios are equal is called a **proportion.** Some examples of proportions are

$$\frac{2}{3} = \frac{4}{6}, \quad \frac{x}{y} = \frac{3}{5}, \quad \text{and} \quad \frac{x^2+8}{2(x+3)} = \frac{17(x+3)}{2}$$

In the proportion $\frac{a}{b} = \frac{c}{d}$, the numbers a and d are called the **extremes,** and the numbers b and c are called the **means.**

To develop an important property of proportions, we suppose that

$$\frac{a}{b} = \frac{c}{d}$$

and multiply both sides by bd to get

$$bd\left(\frac{a}{b}\right) = bd\left(\frac{c}{d}\right)$$
$$\frac{bda}{b} = \frac{bdc}{d}$$
$$da = bc$$

Thus, if $\frac{a}{b} = \frac{c}{d}$, then $ad = bc$. This proves the following statement.

Property of Proportions
In any proportion, the product of the extremes is equal to the product of the means.

We can use this property to solve proportions.

■ SOLVING PROPORTIONS

EXAMPLE 1 Solve the proportion $\dfrac{x}{5} = \dfrac{2}{x + 3}$.

Solution

$$\frac{x}{5} = \frac{2}{x + 3}$$

$$x(x + 3) = 5 \cdot 2 \qquad \text{The product of the extremes equals the product of the means.}$$

$$x^2 + 3x = 10 \qquad \text{Remove parentheses and simplify.}$$

$$x^2 + 3x - 10 = 0 \qquad \text{Subtract 10 from both sides.}$$

$$(x - 2)(x + 5) = 0 \qquad \text{Factor the trinomial.}$$

$$x - 2 = 0 \quad \text{or} \quad x + 5 = 0 \qquad \text{Set each factor equal to 0.}$$

$$x = 2 \qquad\qquad x = -5$$

Thus, $x = 2$ or $x = -5$. Verify each solution. ■

Self Check Solve $\dfrac{2}{5} = \dfrac{3}{x - 4}$.

Answer $\frac{23}{2}$

EXAMPLE 2 Gasoline and oil for a lawn mower are to be mixed in a 50-to-1 ratio. How many ounces of oil should be mixed with 6 gallons of gasoline?

Solution We first express 6 gallons as $6 \cdot 128$ ounces $= 768$ ounces. We then let x represent the number of ounces of oil needed, set up the proportion, and solve it.

$$\frac{50}{1} = \frac{768}{x}$$

$$50x = 768 \qquad \text{The product of the extremes equals the product of the means.}$$

$$x = \frac{768}{50} \qquad \text{Divide both sides by 50.}$$

$$x \approx 15.36 \qquad \text{Read} \approx \text{as "is approximately equal to."}$$

Approximately 15 ounces of oil should be added to 6 gallons of gasoline. ■

Self Check   How many ounces of oil should be mixed with 6 gallons of gas if the ratio is to be 40 to 1?

Answer   19.2 oz

■ DIRECT VARIATION

Two variables are said to **vary directly** or be **directly proportional** if their ratio is a constant. The variables x and y vary directly when

$$\frac{y}{x} = k \qquad \text{or equivalently} \qquad y = kx \quad (k \text{ is a constant})$$

> **Direct Variation**
> The words **"y varies directly with x,"** or **"y is directly proportional to x,"** mean that $y = kx$ for some real-number constant k.
>
> The number k is called the **constant of proportionality.**

EXAMPLE 3   If a car travels 70 miles at 30 miles per hour, how far will it travel in the same time at 45 miles per hour?

Solution   Distance traveled in a given time varies directly with the speed. The phrase "distance varies directly with speed" translates into the formula $d = ks$, where d represents the distance traveled and s represents the speed. The constant of proportionality k can be found by substituting 70 for d and 30 for s in the equation $d = ks$:

$$d = ks$$
$$70 = k(30)$$
$$k = \frac{7}{3}$$

To evaluate the distance d traveled at 45 miles per hour, we substitute $\frac{7}{3}$ for k and 45 for s into the formula $d = ks$.

$$d = \frac{7}{3}s$$
$$= \frac{7}{3}(45)$$
$$= 105$$

In the time it takes to go 70 miles at 30 miles per hour, the car could travel 105 miles at 45 miles per hour.   ■

Self Check   How far will the car travel in the same time if its speed is 60 mph?

Answer   140 mi

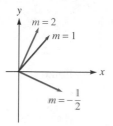

FIGURE 3-48

The statement "y varies directly with x" is equivalent to the equation $y = mx$, where m is the constant of proportionality. Because the equation is in the form $y = mx + b$ with $b = 0$, its graph is a line with slope m and y-intercept 0. The graphs of $y = mx$ for several values of m are shown in Figure 3-48.

The graph of the relationship of direct variation is always a line that passes through the origin.

■ INVERSE VARIATION

Two variables are said to **vary inversely** or be **inversely proportional** if their product is a constant.

$$xy = k \qquad \text{or equivalently} \qquad y = \frac{k}{x} \quad (k \text{ is a constant})$$

> **Inverse Variation**
> The words **"y varies inversely with x,"** or **"y is inversely proportional to x,"** mean that $y = \frac{k}{x}$ for some real-number constant k.

EXAMPLE 4   If the intensity of a light source is 100 lumens at a distance of 20 feet, find the intensity at 30 feet.

Solution   Intensity of illumination from a light source varies inversely with the square of the distance from the source. If I is the intensity and d is the distance from the light source, the phrase "intensity varies inversely with the square of the distance" translates into the formula

$$I = \frac{k}{d^2}$$

We can evaluate k by substituting 100 for I and 20 for d in the formula and solving for k.

$$I = \frac{k}{d^2}$$

$$100 = \frac{k}{20^2}$$

$$k = 40,000$$

To find the intensity at a distance of 30 feet, we substitute 40,000 for k and 30 for d in the formula

$$I = \frac{k}{d^2}$$

$$I = \frac{40,000}{30^2}$$

$$= \frac{400}{9}$$

At 30 feet, the intensity of light would be $\frac{400}{9}$ lumens per square centimeter.   ■

Self Check   Find the intensity at 50 feet.

Answer   16 lumens

The statement "*y* is inversely proportional to *x*" is equivalent to the equation $y = \frac{k}{x}$, where *k* is a constant. Figure 3-49 shows the graphs of $y = \frac{k}{x}$ ($x > 0$) for three values of *k*. In each case, the equation determines one branch of a curve called a **hyperbola.** Verify these graphs with a graphing calculator.

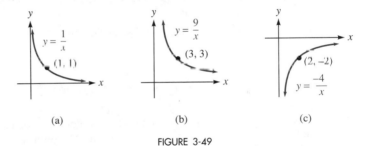

(a)   (b)   (c)

FIGURE 3-49

■ JOINT VARIATION

Joint Variation
The words **"*y* varies jointly with *w* and *x*"** mean that $y = kwx$ for some real-number constant *k*.

EXAMPLE 5   A 25-gram mass moving at the rate of 30 centimeters per second has a kinetic energy of 11,250 dyne-centimeters. Find the kinetic energy of a 10-gram mass that is moving at 40 centimeters per second.

Solution   Kinetic energy of an object varies jointly with its mass and the square of its velocity. If we let *E*, *m*, and *v* represent the kinetic energy, mass, and velocity, respectively, the phrase "energy varies jointly with the mass and the square of its velocity" translates into the formula

$$E = kmv^2$$

The constant k can be evaluated by substituting 11,250 for E, 25 for m, and 30 for v in the formula.

$$E = kmv^2$$
$$11{,}250 = k(25)(30)^2$$
$$11{,}250 = 22{,}500k$$
$$k = \frac{1}{2}$$

We can now substitute $\frac{1}{2}$ for k, 10 for m, and 40 for v in the formula and evaluate E.

$$E = kmv^2$$
$$= \frac{1}{2}(10)(40)^2$$
$$= 8000$$

A 10-gram mass that is moving at 40 centimeters per second has a kinetic energy of 8000 dyne-centimeters. ■

Self Check   Find the kinetic energy of a 25-gram mass that is moving at 100 centimeters per second.

Answer   125,000 dyne-centimeters

■ COMBINED VARIATION

The preceding terminology can be used in various combinations. In each of the following statements, the formula on the left translates into the words on the right.

$$y = \frac{kx}{z}$$ y varies directly with x and inversely with z.

$$y = kx^2\sqrt[3]{z}$$ y varies jointly with the square of x and the cube root of z.

$$y = \frac{kx\sqrt{z}}{\sqrt[3]{t}}$$ y varies jointly with x and the square root of z and inversely with the cube root of t.

$$y = \frac{k}{xz}$$ y varies inversely with the product of x and z.

EXAMPLE 6   If it takes 100 workers 4 weeks to build 2 miles of highway, how long will it take 80 workers to build 10 miles of highway?

Solution   The time it takes to build a highway varies directly with the length of the road, but inversely with the number of workers. We can let t represent the time in weeks, l represent the length in miles, and w represent the number of workers. The relationship between these variables can be expressed by the equation

$$t = \frac{kl}{w}$$

We substitute 4 for t, 100 for w, and 2 for l to find k:

$$4 = \frac{k(2)}{100}$$

$400 = 2k$ Multiply both sides by 100.

$200 = k$ Divide both sides by 2.

We now substitute 80 for w, 10 for l, and 200 for k in the equation $t = \frac{kl}{w}$ and simplify:

$$t = \frac{kl}{w}$$

$$t = \frac{200(10)}{80}$$

$$= 25$$

It will take 25 weeks for 80 workers to build 10 miles of highway. ■

Self Check   How long will it take 100 workers to build 20 miles of highway?

Answer   40 weeks

EXERCISE 3.5

VOCABULARY AND CONCEPTS   *In Exercises 1–10, fill in the blank to make a true statement.*

1. A ratio is the _____ of two numbers.

2. A proportion is a statement that two _____ are equal.

3. In the proportion $\frac{a}{b} = \frac{c}{d}$, b and c are called the _____.

4. In the proportion $\frac{a}{b} = \frac{c}{d}$, a and d are called the _____.

5. In a proportion, the product of the _____ is equal to the product of the _____.

6. Direct variation translates into the equation _____.

7. The equation $y = \frac{k}{x}$ indicates _____ variation.

8. In the equation $y = kx$, k is called the _____ of proportionality.

9. The equation $y = kxz$ represents _____ variation.

10. If the equation $y = \frac{kx^2}{z^3}$, y varies directly with ___ and inversely with ___.

PRACTICE   *In Exercises 11–14, solve each proportion.*

11. $\dfrac{4}{x} = \dfrac{2}{7}$

12. $\dfrac{5}{2} = \dfrac{x}{6}$

13. $\dfrac{x}{2} = \dfrac{3}{x+1}$

14. $\dfrac{x+5}{6} = \dfrac{7}{8-x}$

In Exercises 15–16, set up and solve a proportion to answer each question.

15. The ratio of women to men in a mathematics class is 3 to 5. How many women are in the class if there are 30 men?

16. The ratio of lime to sand in mortar is 3 to 7. How much lime must be mixed with 21 bags of sand to make mortar?

In Exercises 17–22, find the constant of proportionality.

17. y is directly proportional to x. If $x = 30$, then $y = 15$.

18. z is directly proportional to t. If $t = 7$, then $z = 21$.

19. I is inversely proportional to R. If $R = 20$, then $I = 50$.

20. R is inversely proportional to the square of I. If $I = 25$, then $R = 100$.

21. E varies jointly with I and R. If $R = 25$ and $I = 5$, then $E = 125$.

22. z is directly proportional to the sum of x and y. If $x = 2$ and $y = 5$, then $z = 28$.

In Exercises 23–26, solve each problem.

23. y is directly proportional to x. If $y = 15$ when $x = 4$, find y when $x = \frac{7}{5}$.

24. w is directly proportional to z. If $w = -6$ when $z = 2$, find w when $z = -3$.

25. P varies jointly with r and s. If $P = 16$ when $r = 5$ and $s = -8$, find P when $r = 2$ and $s = 10$.

26. m varies jointly with the square of n and the square root of q. If $m = 24$ when $n = 2$ and $q = 4$, find m when $n = 5$ and $q = 9$.

In Exercises 27–30, decide if the graph could represent direct variation, inverse variation, or neither.

27.

28.

29.

30.

APPLICATIONS

31. Gas laws The volume of a gas varies directly with the temperature and inversely with the pressure. When the temperature of a certain gas is 330°, the pressure is 40 pounds per square inch and the volume is 20 cubic feet. Find the volume when the pressure increases 10 pounds per square inch and the temperature decreases to 300°.

32. Hooke's law The force f required to stretch a spring a distance d is directly proportional to d. A force of 5 newtons stretches a spring 0.2 meter. What force will stretch the spring 0.35 meter?

33. Free-falling objects The distance that an object will fall in t seconds varies directly with the square of t. An object falls 16 feet in 1 second. How long will it take to fall 144 feet?

34. Heat dissipation The power, in watts, dissipated as heat in a resistor varies jointly with the resistance, in ohms, and the square of the current, in amperes. A 10-ohm resistor carrying a current of 1 ampere dissipates 10 watts. How much power is dissipated in a 5-ohm resistor carrying a current of 3 amperes?

35. Heat dissipation The power, in watts, dissipated as heat in a resistor varies directly with the square of the voltage and inversely with the resistance. If 20 volts are placed across a 20-ohm resistor, it will dissipate 20 watts. What voltage across a 10-ohm resistor will dissipate 40 watts?

36. Period of a pendulum The time required for one complete swing of a pendulum is called the **period** of the pendulum. The period varies directly with the square of its length. If a 1-meter pendulum has a period of 1 second, find the length of a pendulum with a period of 2 seconds.

37. Frequency of vibration The **pitch,** or **frequency,** of a vibrating string varies directly with the square root of the tension. If a string vibrates at a frequency of 144 hertz due to a tension of 2 pounds, find the frequency when the tension is 18 pounds.

38. Gravitational attraction   The gravitational attraction between two massive objects varies jointly with their masses and inversely with the square of the distance between them. What happens to this force if each mass is tripled and the distance between them is doubled?

39. Plane geometry   The area of an equilateral triangle varies directly with the square of the length of a side. Find the constant of proportionality.

40. Solid geometry   The diagonal of a cube varies directly with the length of a side. Find the constant of proportionality.

DISCOVERY AND WRITING

41. Explain the terms *extremes* and *means*.

43. Explain the term *joint variation*.

45. Explain why $xy = k$ indicates that y varies inversely with x.

42. Distinguish between a *ratio* and a *proportion*.

44. Explain why $\frac{y}{x} = k$ indicates that y varies directly with x.

46. As temperature increases on the Fahrenheit scale, it also increases on the Celsius scale. Is this direct variation? Explain.

REVIEW   *Do each operation and simplify.*

47. $\dfrac{1}{x + 2} + \dfrac{2}{x + 1}$

48. $\dfrac{x^2 - 1}{x + 1} \cdot \dfrac{x - 1}{x^2 - 2x + 1}$

49. $\dfrac{x^2 + 3x - 4}{x^2 - 5x + 4} \div \dfrac{x - 1}{x^2 - 3x - 4}$

50. $\dfrac{x + 2}{3x - 3} \div (2x + 4)$

51. $\dfrac{x^2 + 4 - (x + 2)^2}{4x^2}$

52. $\dfrac{\dfrac{1}{x} - \dfrac{1}{3}}{\dfrac{1}{x} - 1}$

■ ■ ■ ■ ■ ■ ■ ■ ■ PROBLEMS AND PROJECTS

1. Seven years ago, the ABC Corporation bought a network server for $10,350. Now, its salvage value is $250. The current price of a replacement computer is $4500, but it will be worthless in four years. Seven years ago, XYZ Incorporated also bought a server, but paid $17,500. They expect five more years of useful life, but then it will be worthless. Find the depreciation equation of each computer. Which company made the best purchasing choice, and why?

2. Two lines pass through the origin. One passes through (a, b) and the other through (c, d). The lines are also perpendicular. Show that $ac + bd = 0$.

(continued)

■ ■ ■ ■ ■ ■ ■ ■ ■ **PROBLEMS AND PROJECTS** *(continued)*

3. The highway department plans to extend Diagonal Road to 15th Avenue, as in Illustration 1. Where will it cross 11th Street?

4. A photographer uses the graph in Illustration 2 to determine how much concentrate to use to make a batch of developer. Find the equation of the line and graph it using a graphing calculator. Use both the equation and the calculator graph to find the amount of concentrate needed to make 40 ounces of solution. Do the graph and equation give the same answer?

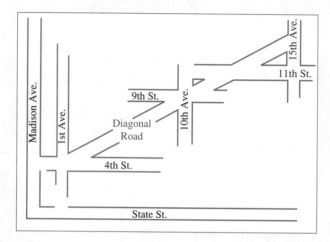

ILLUSTRATION 1

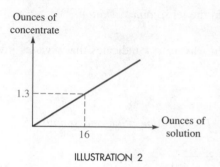

ILLUSTRATION 2

PROJECT 1   The time required to develop film depends on temperature. A photographer would like to use graphs like those in Illustration 3 to determine proper development times for temperatures between 60° and 80° F. The film manufacturer supplied the information in the table. Find the equations of the three lines and graph them. For which contrast is the correct processing temperature most critical? Why?

Temperature	Processing time (minutes)		
	High contrast	**Average contrast**	**Low contrast**
60°	21	15	10
80°	8	$5\frac{1}{2}$	4

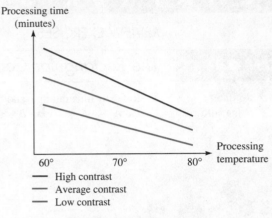

Processing time
(minutes)

Processing temperature

60° 70° 80°

— High contrast
— Average contrast
— Low contrast

ILLUSTRATION 3

PROJECT 2 It is the weight of the water that keeps objects from sinking. For example, the weights of the two buoys in Illustration 4 are supported by forces exactly equal to the weight of the volume of water displaced by the submerged portions.

For each buoy, find an equation relating d, the depth it is submerged, to w, the weight. Graph each equation. Are there restrictions on the variable d? How can you determine, from the equations and from the graphs, the greatest floatable weights? Each cubic foot of fresh water weighs 62.4 pounds.

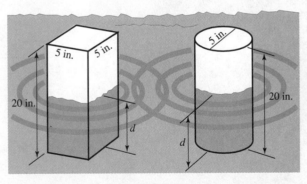

ILLUSTRATION 4

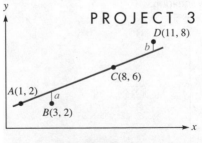

ILLUSTRATION 5

PROJECT 3 Any two points lie on a line, but more than two points, picked at random, will probably not lie on the same line. There is, however, a line that comes as close as possible to passing through several points. The four points in Illustration 5 do not lie on the same line. Find equations of the six lines determined by pairs of points: lines AB, AC, AD, BC, BD, and CD. One of these lines is drawn. Which of the six do you think is the "best fit" of all four points, and why? (You might consider the deviations of the y values, marked a and b on the graph.) Is there another line that you think is a better fit than any of these six? Why?

| **SECTION 3.1** | *The Rectangular Coordinate System* |

The rectangular coordinate system divides the plane into four quadrants.

1. Refer to Illustration 1 and find the coordinates of each point.

    a. *A*   **b.** *B*   **c.** *C*   **d.** *D*

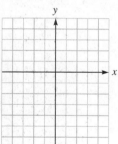

ILLUSTRATION 1

2. Graph each point. Indicate the quadrant in which the point lies, or the axis on which it lies.

    a. $(-3, 5)$ **b.** $(5, -3)$ **c.** $(0, -7)$ **d.** $\left(-\frac{1}{2}, 0\right)$

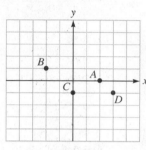

The graph of an equation in x and y is the set of all points (x, y) that satisfy the equation.

3. Use the x- and the y-intercepts to graph each equation.

    a. $3x - 5y = 15$ **b.** $x + y = 7$

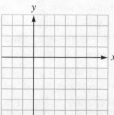 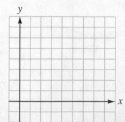

c. $x + y = -7$

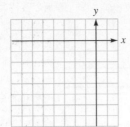

d. $x - 5y = 5$

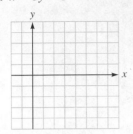

Equation of a vertical line through (a, b):

$x = a$

Equation of a horizontal line through (a, b):

$y = b$

4. Graph each equation.

a. $y = 4$

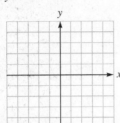

b. $x = -2$

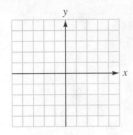

5. Depreciation   A car purchased for $18,750 is expected to depreciate according to the formula $y = -2200x + 18,750$. Find its value after 3 years.

The distance formula:

$d = \sqrt{(x_2 - x_1)^2 + (y_2 - y_1)^2}$

6. Find the length of the segment PQ.
   a. $P(-3, 7)$; $Q(3, -1)$
   c. $P(\sqrt{3}, 9)$; $Q(\sqrt{3}, 7)$
   b. $P(0, 5)$; $Q(-12, 10)$
   d. $P(a, -a)$; $Q(-a, a)$

The midpoint formula:
The midpoint of the line segment joining (x_1, y_1) and (x_2, y_2) is the point M with coordinates

$\left(\dfrac{x_1 + x_2}{2}, \dfrac{y_1 + y_2}{2} \right)$

7. Find the midpoint of the segment PQ.
   a. $P(-3, 7)$; $Q(3, -1)$
   c. $P(\sqrt{3}, 9)$; $Q(\sqrt{3}, 7)$
   b. $P(0, 5)$; $Q(-12, 10)$
   d. $P(a, -a)$; $Q(-a, a)$

SECTION 3.2   *The Slope of a Nonvertical Line*

The slope, m, of a line passing through (x_1, y_1) and (x_2, y_2) is given by

$m = \dfrac{y_2 - y_1}{x_2 - x_1} \quad (x_2 \neq x_1)$

A vertical line has no defined slope.

8. Find the slope of the line PQ, if possible.
   a. $P(3, -5)$; $Q(1, 7)$
   c. $P(b, a)$; $Q(a, b)$
   b. $P(2, 7)$; $Q(-5, -7)$
   d. $P(a + b, b)$; $Q(b, b - a)$

9. Rate of descent   If an airlane descends 3000 feet in 15 minutes, what is the average rate of descent in feet per minute?

10. Tell whether the slope of each line is 0 or undefined.

a.

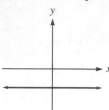

b.

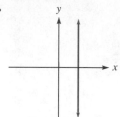

11. Tell whether the slope of each line is positive or negative.

a.

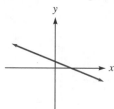

b.

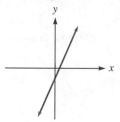

Nonvertical parallel lines have the same slope.

Perpendicular lines have slopes that are negative reciprocals of each other, provided neither line is vertical.

12. A line passes through $(-2, 5)$ and $(6, 10)$. A line parallel to it passes through $(2, 2)$ and $(10, y)$. Find y.

13. A line passes through $(-2, 5)$ and $(6, 10)$. A line perpendicular to it passes through $(-2, 5)$ and $(x, -3)$. Find x.

SECTION 3.3   *Writing Equations of Lines*

Point–slope form:

$$y - y_1 = m(x - x_1)$$

14. Use the point–slope form to write the equation of each line.

a. The line passes through the origin and the point $(-5, 7)$.

b. The line passes through $(-2, 1)$ and has a slope of -4.

c. The line passes through $(7, -5)$ and $(4, 1)$.

Slope–intercept form:

$$y = mx + b$$

15. Use the slope–intercept form to write the equation of each line.

a. The line has a slope of $\frac{2}{3}$ and a y-intercept of 3.

b. The slope is $-\frac{3}{2}$ and the line passes through $(0, -5)$.

16. Use the slope–intercept form to graph each equation.

    a. $y = \dfrac{3}{5}x - 2$ **b.** $y = -\dfrac{4}{3}x + 3$

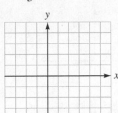 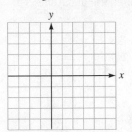

17. Write the equation of each line.
    a. The line passes through $(7, -2)$ and is parallel to the line segment joining $(2, 4)$ and $(4, -10)$.
    b. The line passes through $(7, -2)$ and is perpendicular to the line segment joining $(2, 4)$ and $(4, -10)$.
    c. The line is parallel to $3x - 4y = 7$ and passes through $(2, 0)$.
    d. The line passes through $(0, 5)$ and is perpendicular to the line $3y + x - 4 = 0$.

General form of a line:
$$Ax + By = C$$

18. Find the slope and the y-intercept of the graph of each line.
    a. $5x + 2y = 7$ **b.** $3x - 4y = 14$

Equation of a horizontal line through (a, b):
$$y = b$$

19. Write the equation of each line.
    a. The line has a slope of 0 and passes through $(-5, 17)$.
    b. The line has no defined slope and passes through $(-5, 17)$.

Equation of a vertical line through (a, b):
$$x = a$$

SECTION 3.4   *Graphs of Equations*

Intercepts of a graph:

To find the **x-intercepts,** let $y = 0$ and solve for x.

To find the **y-intercepts,** let $x = 0$ and solve for y.

Tests for symmetry:
If $(-x, y)$ lies on the graph whenever (x, y) does, the graph is symmetric about the y-axis.

20. Graph each equation. Find all intercepts and symmetries.
    a. $y = x^2 + 2$ **b.** $y = x^3 - 2$

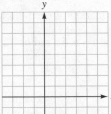 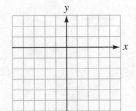

If $(x, -y)$ lies on the graph whenever (x, y) does, the graph is symmetric about the x-axis.

If $(-x, -y)$ lies on the graph whenever (x, y) does, the graph is symmetric about the origin.

c. $y = \dfrac{1}{2}|x|$

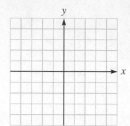

d. $y = -\sqrt{x - 4}$

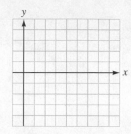

e. $y = \sqrt{x} + 2$

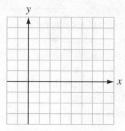

f. $y = |x + 1| + 2$

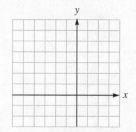

21. Use a graphing calculator to graph each equation.

   a. $y = |x - 4| + 2$ **b.** $y = -\sqrt{x + 2} + 3$

22. Use a graphing calculator to graph each equation.

   a. $y = x + 2|x|$ **b.** $y^2 = x - 3$

Equation of a circle:
Center at (h, k):
   $(x - h)^2 + (y - k)^2 = r^2$
Center at $(0, 0)$:
   $x^2 + y^2 = r^2$

23. Write the equation of each circle.

   a. Center at $(-3, 4)$; radius 12   **b.** Ends of diameter at $(-6, -3)$ and $(5, 8)$

24. Graph each equation.

   a. $x^2 + y^2 - 2y = 15$

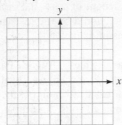

   b. $x^2 + y^2 - 4x + 2y = 4$

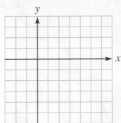

25. 🖩 Use a graphing calculator to solve each equation. If an answer is not exact, round to the nearest hundredth.

   a. $x^2 - 11 = 0$

   b. $x^3 - x = 0$

   c. $|x^2 - 2| - 1 = 0$

   d. $x^2 - 3x = 5$

| **SECTION 3.5** | *Proportion and Variation* |

In any proportion, the product of the extremes is equal to the product of the means.

26. Solve each proportion.

   a. $\dfrac{x + 3}{10} = \dfrac{x - 1}{x}$

   b. $\dfrac{x - 1}{2} = \dfrac{12}{x + 1}$

y varies directly with *x*:

$$y = kx$$

27. Hooke's law   The force required to stretch a spring is directly proportional to the amount of stretch. If a 3-pound force stretches a spring 5 inches, what force would stretch the spring 3 inches?

28. Kinetic energy   A moving body has a kinetic energy directly proportional to the square of its velocity. By what factor does the kinetic energy of an automobile increase if its speed increases from 30 miles per hour to 50 miles per hour?

y varies inversely with *x*:

$$y = \dfrac{k}{x}$$

29. Gas laws   The volume of gas in a balloon varies directly as the temperature and inversely as the pressure. If the volume is 400 cubic centimeters when the temperature is 300°K and the pressure is 25 dynes per square centimeter, find the volume when the temperature is 200°K and the pressure is 20 dynes per square centimeter.

y varies jointly with *w* and *x*:

$$y = kwx$$

30. The area of a rectangle varies jointly with its length and width. Find the constant of proportionality.

31. 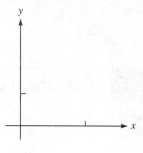 **Electrical resistance**   The resistance of a wire varies directly as the length of the wire and inversely as the square of its diameter. A 1000-foot length of wire, 0.05 inches in diameter, has a resistance of 200 ohms. What would be the resistance of a 1500-foot length of wire that is 0.08 inches in diameter?

32. Billing for services   Angie's Painting and Decorating Service charges a fixed amount for accepting a wallpapering job, and adds a fixed dollar amount for each roll hung. If they bill a customer $177 to hang 11 rolls, and $294 to hang 20 rolls, find the cost to hang 27 rolls.

33. Paying for college   Rolf must earn $5040 for next semester's tuition. Assume he works x hours tutoring algebra at $14 per hour and y hours tutoring Spanish at $18 per hour, and makes his goal. Write an equation expressing the relationship between x and y, and graph the equation. If Rolf tutors algebra for 180 hours, how long must he tutor Spanish?

■ Chapter 3 Test

Indicate the quadrant in which the point lies, or the axis on which it lies.

1. $(-3, \pi)$

2. $(0, -8)$

In Questions 3–4, find the x- and the y-intercepts and use them to graph the equation.

3. $x + 3y = 6$

4. $2x - 5y = 10$

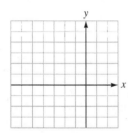

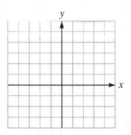

In Questions 5–8, graph each equation.

5. $2(x + y) = 3x + 5$

6. $3x - 5y = 3(x - 5)$

7. $\dfrac{1}{2}(x - 2y) = y - 1$

8. $\dfrac{x + y - 5}{7} = 3x$

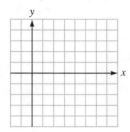

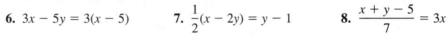

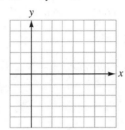

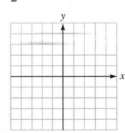

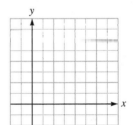

In Questions 9–10, find the distance between points P and Q.

9. $P(1, -1)$; $Q(-3, 4)$

10. $P(0, \pi)$; $Q(-\pi, 0)$

In Questions 11–12, find the midpoint of the line segment PQ.

11. $P(3, -7)$; $Q(-3, 7)$

12. $P\left(0, \sqrt{2}\right)$; $Q\left(\sqrt{8}, \sqrt{18}\right)$

In Questions 13–14, find the slope of the line PQ.

13. $P(3, -9)$; $Q(-5, 1)$

14. $P\left(\sqrt{3}, 3\right)$; $Q\left(-\sqrt{12}, 0\right)$

In Questions 15–16, determine if the two lines are parallel, perpendicular, or neither.

15. $y = 3x - 2$; $y = 2x - 3$

16. $2x - 3y = 5$; $3x + 2y = 7$

In Questions 17–22, write the equation of the line with the given properties.

17. passing through $(3, -5)$; $m = 2$

18. $m = 3$; $b = \dfrac{1}{2}$

19. parallel to $2x - y = 3$; $b = 5$

20. perpendicular to $2x - y = 3$; $b = 5$

21. passing through $\left(2, -\frac{3}{2}\right)$ and $\left(3, \frac{1}{2}\right)$

22. parallel to the y-axis and passing through $(3, -4)$

In Questions 23–24, find the x- and y-intercepts of each graph.

23. $y = x^3 - 16x$

24. $y = |x - 4|$

In Questions 25–26, find the symmetries of each graph.

25. $y^2 = x - 1$

26. $y = x^4 + 1$

In Questions 27–30, graph each equation.

27. $y = x^2 - 9$

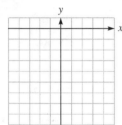

28. $x = |y|$

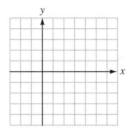

29. $y = 2\sqrt{x}$

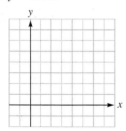

30. $x = y^3$

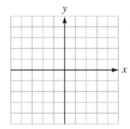

In Questions 31–32, write the equation of each circle.

31. Center at $(5, 7)$; radius of 8

32. Center at $(2, 4)$; passing through $(6, 8)$

In Questions 33–34, graph each equation.

33. $x^2 + y^2 = 9$

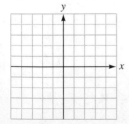

34. $x^2 - 4x + y^2 + 3 = 0$

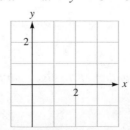

In Questions 35–36, write each statement as an equation.

35. y varies directly as the square of z.

36. w varies jointly with r and the square of s.

37. P varies directly with Q. $P = 7$ when $Q = 2$. Find P when $Q = 5$.

38. y is directly proportional to x and inversely proportional to the square of z. $y = 16$ when $x = 3$ and $z = 2$. Find x when $y = 2$ and $z = 3$.

In Questions 39–40, use a graphing calculator to find the positive root of each equation.

39. $x^2 - 7 = 0$

40. $x^2 - 5x - 5 = 0$

4 *Functions*

In this chapter, we will discuss one of the most important concepts in mathematics, the concept of function.

4.1 Functions and Function Notation

■ FUNCTIONS ■ FUNCTION NOTATION ■ GRAPHS OF FUNCTIONS ■ THE VERTICAL LINE TEST
■ LINEAR FUNCTIONS

In Chapter 3, we saw that an equation in x and y sets up a correspondence between numbers x and values y. Correspondences between the elements of two sets is a common occurrence in everyday life. For example,

- To every house, there corresponds an address.
- To every car, there corresponds a license plate.
- To every state, there corresponds a governor.

In this chapter, we will discuss situations where one quantity corresponds to (or depends on) another quantity according to some specific rule. For example, the equation $y = x^2 - 1$ sets up a correspondence where each number x determines one value y, according to the rule *square x and subtract 1*. In this case, the value y depends on x.

Since this equation is a rule specifying how one value of y corresponds to each number x, it is an example of a *function*. We can think of the numbers x as inputs into the function and the corresponding values of y as outputs. The rule of the function determines what output will result from each input. This idea of inputs and outputs is shown in Figure 4-1(a). In the equation $y = x^2 - 1$, if the input x is 2, the output y is

$$y = 2^2 - 1 = 3$$

This is illustrated in Figure 4-1(b).

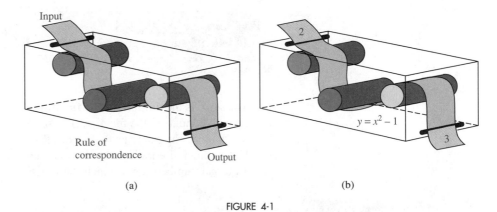

FIGURE 4-1

The inputs and outputs of functions discussed in this text will usually be real numbers, and the rule of the function will usually be given as an equation. However, functions can also be indicated by a table of values or by a graph.

The equation $y = x^2 - 1$ determines the table of ordered pairs and the graph shown in Figure 4-2. To see how the table determines the correspondence, we simply find an input in the x-column and then read across to find the corresponding output in the y-column. If we select $x = 2$ as an input, we get $y = 3$ for the output.

To see how the graph determines the correspondence, we draw a vertical and horizontal line through any point (say, point P) on the graph shown in Figure 4-2. Because these lines intersect the x-axis at 2 and the y-axis at 3, the point $P(2, 3)$ associates 3 on the y-axis with 2 on the x-axis.

$$y = x^2 - 1$$

x	y	(x, y)
-2	3	$(-2, 3)$
-1	0	$(-1, 0)$
0	-1	$(0, -1)$
1	0	$(1, 0)$
2	3	$(2, 3)$

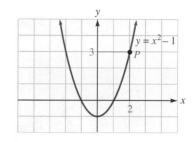

FIGURE 4-2

■ FUNCTIONS

Any rule that assigns one value of y to each number x is called a **function.** The value of y that corresponds to x is called the **image of x,** and we say that the function is **defined at x.** For the function determined by the equation $y = x^2 - 1$, we have seen that the image of 2 is 3.

Since the value of y depends on the number x, we call y the **dependent variable** and x the **independent variable.**

Functions

A **function** from one set **X** to another set **Y** is a rule that assigns to each element in **X** one element in **Y.**

The set **X** is the **domain of the function.** We can think of the domain as the set of all possible inputs to the function. The set of all images of x in set **Y** is the **range.** We can think of the range as the set of all possible outputs of the function.

Unless otherwise indicated, we will assume that the domain of a function is the set of all real numbers x for which the function is defined.

EXAMPLE 1   Decide whether the following equations define y to be a function of x. If so, find the domain and range.   **a.** $7x + y = 5$ and   **b.** $y = \dfrac{3}{x - 2}$.

Solution   **a.** First, we solve $7x + y = 5$ for y to get

$$y = 5 - 7x$$

Since there is one value of y for each number we substitute for x, the equation determines y to be a function of x.

Because x can be any real number, the domain is the interval $(-\infty, \infty)$.
Because y can be any real number, the range is the interval $(-\infty, \infty)$.

b. Since each input x (except 2) gives one output y, the equation $y = \frac{3}{x-2}$ defines y to be a function of x.

The domain is the interval $(-\infty, 2) \cup (2, \infty)$, which is the set of all real numbers except 2. We cannot substitute 2 for x, because division by 0 is undefined.

The range is the interval $(-\infty, 0) \cup (0, \infty)$, which is the set of all real numbers except 0. Here y cannot be 0, because the numerator of the fraction is 3. ∎

Self Check   Decide whether each equation defines y to be a function of x. If so, find the domain and range.   **a.** $y = |x|$ and   **b.** $y = \sqrt{x}$.

Answers   **a.** a function, $(-\infty, \infty)$, $[0, \infty)$;   **b.** a function, $[0, \infty)$, $[0, \infty)$

EXAMPLE 2   Find the domain and range of the function defined by each equation.

a. $y = \sqrt{3x - 2}$ and   **b.** $y = \dfrac{1}{x^2 - 5x - 6}$.

Solution   **a.** The radicand must be nonnegative, so we have

$$3x - 2 \geq 0$$
$$3x \geq 2 \qquad \text{Add 2 to both sides.}$$
$$x \geq \frac{2}{3} \qquad \text{Divide both sides by 3.}$$

The domain is the interval $\left[\frac{2}{3}, \infty\right)$.

Because the radical sign calls for the positive square root, the range is the interval $[0, \infty)$.

b. We can factor the denominator to see what values of x will give 0's in the denominator. These values are not in the domain.

$$x^2 - 5x - 6 = 0$$
$$(x - 6)(x + 1) = 0$$
$$x - 6 = 0 \qquad \text{or} \qquad x + 1 = 0$$
$$x = 6 \qquad\qquad\qquad x = -1$$

The domain is the interval $(-\infty, -1) \cup (-1, 6) \cup (6, \infty)$. Since the numerator of the fraction is 1, y cannot be 0. The range is the interval $(-\infty, 0) \cup (0, \infty)$. ∎

Self Check   Find the domain and range of the function defined by $y = \sqrt[3]{x + 2}$.
Answer   Both are $(-\infty, \infty)$.

EXAMPLE 3   Find the domain and range of the function defined by the equation $y = \dfrac{x + 1}{x - 2}$.

Solution   Here, x cannot be 2, because that would give a 0 in the denominator. So the domain is the interval $(-\infty, 2) \cup (2, \infty)$.

One way to find the restrictions on y is to solve the equation for x.

$$y = \frac{x + 1}{x - 2}$$

$y(x - 2) = x + 1$ Multiply both sides by $x - 2$.

$yx - 2y = x + 1$ Remove parentheses.

$yx - x = 2y + 1$ Add $2y$ and subtract x from both sides.

$x(y - 1) = 2y + 1$ Factor out x.

$x = \dfrac{2y + 1}{y - 1}$ Divide both sides by $y - 1$.

From the last equation, we can see that y cannot be 1. So the range is $(-\infty, 1) \cup (1, \infty)$. ∎

Self Check   Find the domain and range of the function defined by $y = \dfrac{x - 2}{x + 1}$.

Answers $(-\infty, -1) \cup (-1, \infty)$; $(-\infty, 1) \cup (1, \infty)$

■ FUNCTION NOTATION

To indicate that y is a function of x, we often use **function notation** and write

$$y = f(x)$$

Technically, y is the image of x under the function f, and the rule of correspondence is the function. However, it is common to read $y = f(x)$ as "y is a function of x." Functions are often denoted by letters other than f.

Function notation provides a way of indicating the image of a certain number x. The symbol $f(2)$, read as "f of 2" or "f at 2," indicates the image of 2 under the function f. To evaluate it, we substitute 2 for x.

$f(x) = 5 - 7x$

$f(2) = 5 - 7(2)$ Substitute 2 for x.

$\quad\;\; = -9$

If $x = 2$, then $y = f(2) = -9$.

The image of -5 is denoted by $f(-5)$. To evaluate it, we substitute -5 for x.

$f(x) = 5 - 7x$

$f(-5) = 5 - 7(-5)$ Substitute -5 for x.

$\quad\;\;\; = 40$

If $x = -5$, then $y = f(-5) = 40$.

EXAMPLE 4   Let $g(x) = 3x^2 + x - 4$. Find   **a.** $g(-3)$,   **b.** $g(k)$,   **c.** $g(-t^3)$, and
d. $g(k + 1)$.

Solution   **a.** $g(x) = 3x^2 + x - 4$ **b.** $g(x) = 3x^2 + x - 4$
$g(-3) = 3(-3)^2 + (-3) - 4$ $g(k) = 3k^2 + k - 4$
$= 3(9) - 3 - 4$
$= 20$

c. $g(x) = 3x^2 + x - 4$ **d.** $g(x) = 3x^2 + x - 4$
$g(-t^3) = 3(-t^3)^2 + (-t^3) - 4$ $g(k + 1) = 3(k + 1)^2 + (k + 1) - 4$
$= 3t^6 - t^3 - 4$ $= 3(k^2 + 2k + 1) + k + 1 - 4$
$= 3k^2 + 6k + 3 + k + 1 - 4$
$= 3k^2 + 7k$ ■

Self Check   Find   **a.** $g(0)$,   **b.** $g(2)$, and   **c.** $g(k - 1)$.
Answers   **a.** -4,   **b.** 10,   **c.** $3k^2 - 5k - 2$

The fraction $\dfrac{f(x + h) - f(x)}{h}$ is called the **difference quotient** and is important
in calculus.

EXAMPLE 5   If $f(x) = 2x + 1$, evaluate $\dfrac{f(x + h) - f(x)}{h}$.

Solution   First, we find $f(x + h)$.

$f(x + h) = 2(x + h) + 1$
$= 2x + 2h + 1$

Then we substitute $2x + 2h + 1$ for $f(x + h)$ and $2x + 1$ for $f(x)$ in the difference
quotient.

$$\frac{f(x + h) - f(x)}{h} = \frac{2x + 2h + 1 - (2x + 1)}{h}$$

$$= \frac{2x + 2h + 1 - 2x - 1}{h}$$

$$= \frac{2h}{h}$$

$$= 2$$

The value is 2.   ■

Self Check   If $f(x) = x^2 + 2$, evaluate the difference quotient.
Answer $2x + h$

■ GRAPHS OF FUNCTIONS

If f is a function whose domain and range are sets of real numbers, then its graph is the set of all points $(x, f(x))$ in the xy-plane. In other words, the graph of f is the graph of the equation $y = f(x)$. For example, the graph of the function $y = f(x) = -7x + 5$ is a line with slope -7 and y-intercept $(0, 5)$. (See Figure 4-3.)

More formally, we have this definition.

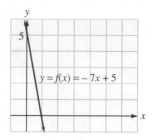

FIGURE 4-3

> **Graph of a Function**
> The **graph** of a function in the xy-plane is the set of all points (x, y) where x is in the domain of f and y is in the range of f.

In Figure 4-4, we see the graph of the absolute value function, the square root function, and a general function.

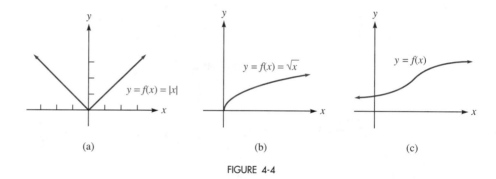

(a)   (b)   (c)

FIGURE 4-4

It is easy to read the domain and range from the graph of a function, as shown in Figure 4-5.

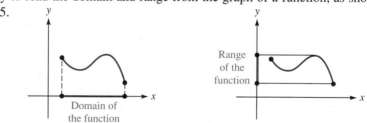

FIGURE 4-5

■ THE VERTICAL LINE TEST

A test, called the **vertical line test,** can be applied to a graph to determine whether it represents a function. If each vertical line that intersects the graph does so exactly once, then each number x determines exactly one value of y, and the graph represents a function. (See Figure 4-6(a).)

If any vertical line intersects the graph more than once, then to some numbers x there correspond more than one value of y, and the graph does not represent a function. (See Figure 4-6(b).)

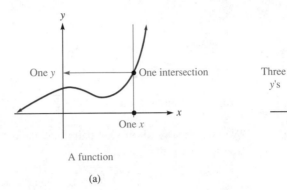

A function

(a)

Not a function

(b)

FIGURE 4-6

EXAMPLE 6   Decide which of the following graphs represent functions.

a.   **b.** 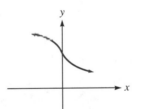   **c.**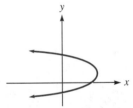

Solution   **a.** This graph fails the vertical line test, so it does not represent a function.

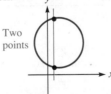

b. This graph passes the vertical line test, so it does represent a function.

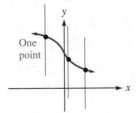

c. This graph fails the vertical line test, so it does not represent a function.

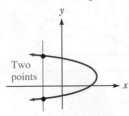

■

Self Check   Does the following graph represent a function?

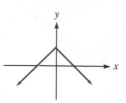

Answer   yes

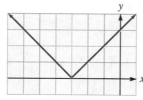

EXAMPLE 7   Graph the function $f(x) = \sqrt{x-2}$.

Solution   We choose several ordered pairs that satisfy the equation, plot them, and draw the graph. (See Figure 4-7.)

$$f(x) = \sqrt{x-2}$$

x	y	(x, y)
2	0	(2, 0)
6	2	(6, 2)
11	3	(11, 3)
18	4	(18, 4)

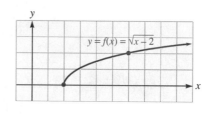

FIGURE 4-7

Self Check   Graph $f(x) = |x + 3|$.
Answer

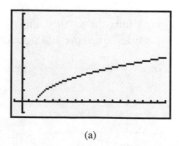

ACCENT ON TECHNOLOGY   The calculator graphs of $f(x) = \sqrt{x-2}$ and $f(x) = |x + 3|$ are shown in Figure 4-8.

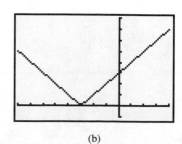

(a)   (b)

FIGURE 4-8

■ LINEAR FUNCTIONS

The equation of a nonvertical line defines a linear function—an important function in mathematics and its applications.

> **Linear Functions**
> A **linear function** is a function determined by an equation of the form
>
> $$f(x) = mx + b$$

EXAMPLE 8

The cost of electricity in Eagle River is a linear function of x, the number of kilowatt-hours used. If the cost of 100 kwh is $17 and the cost of 500 kwh is $57, find an equation that expresses the function.

Solution

Since c (the cost of electricity) is given to be a linear function of x, there are constants m and b such that

1. $c = mx + b$

Because $c = 17$ when $x = 100$, the point $(x_1, c_1) = (100, 17)$ lies on the straight-line graph of this function. Since $c = 57$ when $x = 500$, the point $(x_2, c_2) = (500, 57)$ also lies on that line. The slope of the line is

$$
\begin{aligned}
m &= \frac{c_2 - c_1}{x_2 - x_1} \\
&= \frac{57 - 17}{500 - 100} \\
&= \frac{40}{400} \\
&= 0.10
\end{aligned}
$$

Thus, $m = 0.10$. To determine b, we can substitute 0.10 for m and the coordinates of point $P(100, 17)$ in the equation $c = mx + b$ and solve for b.

$$
\begin{aligned}
c &= mx + b \\
17 &= 0.10(100) + b \\
17 &= 10 + b \\
7 &= b
\end{aligned}
$$

Thus, $c = 0.10x + 7$. The electric company charges $7 plus 10¢ per kilowatt-hour used. ■

Self Check

Find the cost of using 400 kwh of electricity.

Answer $47

In the previous examples, most functions have been represented with equations. However, it important to note that not all equations define functions. For example, the equation $x = |y|$ does not define a function, because two values of y can correspond to one number x. For example, if $x = 2$, then y can be either 2 or -2. The graph of the equation is shown in Figure 4-9. Note that the graph does not pass the vertical line test and thus cannot represent a function.

**Srinivasa Ramanujan
(1899–1920)**

Ramanujan was one of the most prominent Indian mathematicians. He taught at Cambridge University in England. He was a self-taught genius in pure mathematics.

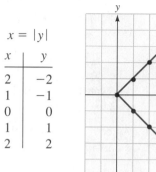

$x = |y|$

x	y
2	-2
1	-1
0	0
1	1
2	2

FIGURE 4-9

Another way to visualize the definitions in this section is to consider the diagram shown in Figure 4-10(a). The function f that assigns the element y to the element x is represented by an arrow leaving x and pointing to y. The set of all those elements in **X** from which arrows originate is the domain of the function. The set of all elements of **Y** to which arrows point is the range.

To constitute a function, each element of the domain must have exactly one image in the range. However, the same value of y could be the image of several numbers x. In the function shown in Figure 4-10(b), the single image y corresponds to the three numbers x_1, x_2, and x_3 in the domain.

The correspondence shown in Figure 4-10(c) is not a function, because two values of y correspond to the same number x.

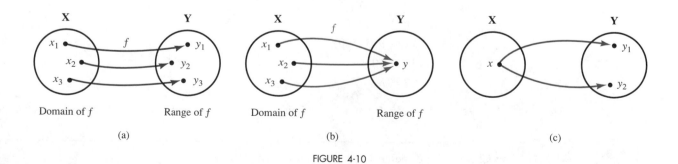

FIGURE 4-10

EXERCISES 4.1

VOCABULARY AND CONCEPTS *In Exercises 1–10, the variable x comes from a set **X**, and the variable y comes from a set **Y**. Fill in the blank to make a true statement.*

1. A rule that assigns one value of y to any number x is called a _____.

2. The value of y that corresponds to a certain number x is called the _____ of x.

3. The set of replacements for x is called the _____ of the function.

4. The range is the set of all _____ of x.

5. The statement "y is a function of x" can be written as the equation _____.

6. The graph of a function in the xy-plane is the set of all points _____, where x is in the _____ of f and y is in the _____ of f.

7. In the function of Exercise 5, __ is called the independent variable.

8. In the function of Exercise 5, y is called the _____ variable.

9. If every _____ line that intersects a graph does so _____, the graph represents a function.

10. A function that can be written in the form $y = mx + b$ is called a _____ function.

PRACTICE *In Exercises 11–22, assume that all variables represent real numbers. Indicate whether each equation determines y to be a function of x.*

11. $y = x$

12. $y - 2x = 0$

13. $y^2 = x$

14. $|y| = x$

15. $y = x^2$

16. $y - 7 = 7$

17. $y^2 - 4x = 1$

18. $|x - 2| = y$

19. $|x| = |y|$

20. $x = 7$

21. $y = 7$

22. $|x + y| = 7$

In Exercises 23–34, let the function f be defined by the equation $y = f(x)$, where x and f(x) are real numbers. Find the domain and range of each function.

23. $f(x) = 3x + 5$

24. $f(x) = -5x + 2$

25. $f(x) = x^2$

26. $f(x) = x^3$

27. $f(x) = \dfrac{3}{x + 1}$

28. $f(x) = \dfrac{-7}{x + 3}$

29. $f(x) = \sqrt{x}$

30. $f(x) = \sqrt{x^2 - 1}$

31. $f(x) = \dfrac{4}{x^2 + 3x + 2}$

32. $f(x) = \dfrac{7}{x^2 - 3x - 4}$

33. $f(x) = \dfrac{x - 2}{x + 3}$

34. $f(x) = \dfrac{x + 2}{x - 1}$

In Exercises 35–46, let the function f be defined by y = f(x), where x and f(x) are real numbers. Find f(2), f(−3), f(k),
and f(k² − 1).

35. $f(x) = 3x - 2$

36. $f(x) = 5x + 7$

37. $f(x) = \dfrac{1}{2}x + 3$

38. $f(x) = \dfrac{2}{3}x + 5$

39. $f(x) = x^2$

40. $f(x) = 3 - x^2$

41. $f(x) = \dfrac{2}{x + 4}$

42. $f(x) = \dfrac{3}{x - 5}$

43. $f(x) = \dfrac{1}{x^2 - 1}$

44. $f(x) = \dfrac{3}{x^2 + 3}$

45. $f(x) = \sqrt{x^2 + 1}$

46. $f(x) = \sqrt{x^2 - 1}$

In Exercises 47–50, evaluate the difference quotient for each function f(x).

47. $f(x) = 3x + 1$ **48.** $f(x) = 5x - 1$ **49.** $f(x) = x^2 + 1$ **50.** $f(x) = x^2 - 3$

In Exercises 51–52, indicate the domain and range of each function as intervals on the x- and y-axes.

51.

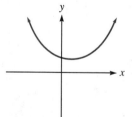

52.

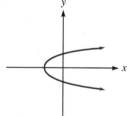

In Exercises 53–58, which graphs represent functions?

53.

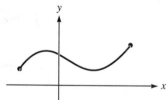

54.

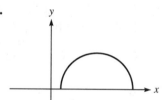

55.

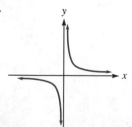

56.

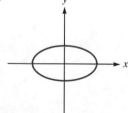

57.

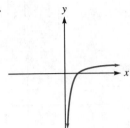

58.

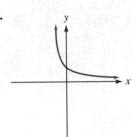

In Exercises 59–74, graph each function.

59. $f(x) = 2x + 3$

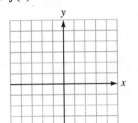

60. $f(x) = 3x + 2$

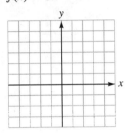

61. $2x = 3y - 3$

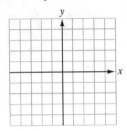

62. $3x = 2(y + 1)$

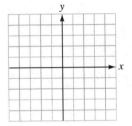

63. $f(x) = -\sqrt{x}$

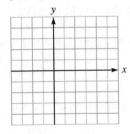

64. $f(x) = \sqrt{x + 1}$

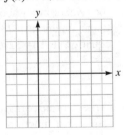

65. $f(x) = -\sqrt{x + 1}$

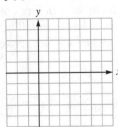

66. $f(x) = \sqrt{x} + 2$

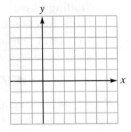

67. $f(x) = -|x|$

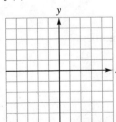

68. $f(x) = -|x| - 3$

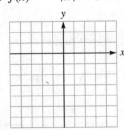

69. $f(x) = |x - 2|$

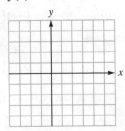

70. $f(x) = -|x - 2|$

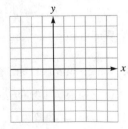

71. $f(x) = \sqrt{2x - 4}$ **72.** $f(x) = -\sqrt{2x - 4}$ **73.** $f(x) = \left|\dfrac{1}{2}x + 3\right|$ **74.** $f(x) = -\left|\dfrac{1}{2}x + 3\right|$

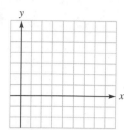

In Exercises 75–78, use a graphing calculator to graph each function. Then give the domain and range of the function.

75. $f(x) = \sqrt{2x - 5}$ **76.** $f(x) = |3x + 2|$ **77.** $f(x) = \sqrt[3]{5x - 1}$ **78.** $f(x) = -\sqrt[3]{3x + 2}$

APPLICATIONS

79. Temperature conversion The Fahrenheit temperature reading F is a linear function of the Celsius reading C. If $C = 0$ when $F = 32$ and the readings are the same at $-40°$, express F as a function of C.

80. Free-falling objects The velocity of a falling object is a linear function of the time t it has been falling. If $v = 15$ when $t = 0$ and $v = 79$ when $t = 2$, express v as a function of t.

81. Water billing The cost c of water is a linear function of n, the number of gallons used. If 1000 gallons cost \$4.70 and 9000 gallons cost \$14.30, express c as a function of n.

82. Simple interest The amount A of money on deposit for t years in an account earning simple interest is a linear function of t. Express that function as an equation if $A = \$272$ when $t = 3$ and $A = \$320$ when $t = 5$.

DISCOVERY AND WRITING

83. Write a brief paragraph explaining how to find the domain of a function.

84. Write a brief paragraph explaining how to find the range of a function.

85. Find a function whose graph has an x-intercept of $\sqrt{5}$.

86. Use a graphing calculator to graph the function of Exercise 85, and use TRACE and ZOOM to find $\sqrt{5}$ to 3 decimal places.

REVIEW *In Exercises 87–90, consider this set:* $\left\{-3, -1, 0, 0.5, \frac{3}{4}, 1, \pi, 7, 8\right\}$.

87. Which numbers are natural numbers?

88. Which numbers are rational numbers?

89. Which numbers are prime numbers?

90. Which numbers are even numbers?

In Exercises 91–92, write each interval in interval notation.

91.

$-4 \qquad 7$

92.

$-3 \qquad 5$

In Exercises 93–94, graph each interval.

93. $(-3, 5) \cup [6, \infty)$

94. $(-\infty, 0) \cup (0, \infty)$

4.2 Quadratic Functions

■ QUADRATIC FUNCTIONS ■ FINDING THE VERTEX OF A PARABOLA ■ MAXIMIZING AREA
■ MAXIMIZING REVENUE

The linear function defined by the equation $y = f(x) = mx + b$ is a first-degree polynomial function, because its left-hand side is a first-degree polynomial in the variable x. In this section, we will discuss functions involving polynomials of second degree.

■ QUADRATIC FUNCTIONS

A function defined by a polynomial of second degree is called a **quadratic function.**

> **Quadratic Functions**
> A **quadratic function** is a second-degree polynomial function in one variable. It is defined by an equation of the form $y = f(x) = ax^2 + bx + c$, where a, b, and c are constants and $a \neq 0$.

EXAMPLE 1   Graph the function $y = f(x) = x^2 - 2x - 3$.

Solution   We can plot several points with coordinates that satisfy the equation and join them with a smooth curve to obtain the graph shown in Figure 4-11.

$$y = x^2 - 2x - 3$$

x	y	(x, y)
-2	5	$(-2, 5)$
-1	0	$(-1, 0)$
0	-3	$(0, -3)$
1	-4	$(1, -4)$
2	-3	$(2, -3)$
3	0	$(3, 0)$
4	5	$(4, 5)$

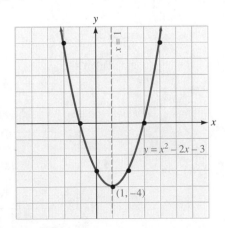

FIGURE 4-11

Self Check Graph $y = f(x) = x^2 + 2x$.

Answer

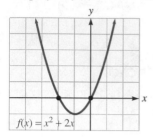

$f(x) = x^2 + 2x$

EXAMPLE 2 Graph the function $y = f(x) = -\dfrac{1}{3}x^2 + 3$.

Solution We can plot several points with coordinates that satisfy the equation and join them with a smooth curve to obtain the graph shown in Figure 4-12.

$$y = -\frac{1}{3}x^2 + 3$$

x	y	(x, y)
-4	$-\frac{7}{3}$	$\left(-4, -\frac{7}{3}\right)$
-3	0	$(-3, 0)$
-1	$\frac{8}{3}$	$\left(-1, \frac{8}{3}\right)$
0	3	$(0, 3)$
1	$\frac{8}{3}$	$\left(1, \frac{8}{3}\right)$
3	0	$(3, 0)$
4	$-\frac{7}{3}$	$\left(4, -\frac{7}{3}\right)$

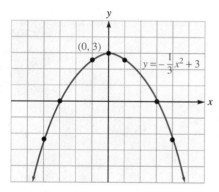

FIGURE 4-12

Self Check Graph $y = f(x) = -3x^2 + 2$.

Answer

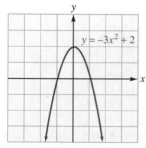

ACCENT ON
TECHNOLOGY
We could also draw the graphs in Examples 1 and 2 with a graphing calculator. The graph of $y = f(x) = x^2 - 2x - 3$ appears in Figure 4-13(a), and the graph of $y = f(x) = -\frac{1}{3}x^2 + 3$ appears in Figure 4-13(b).

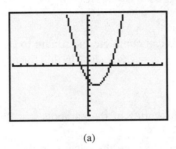

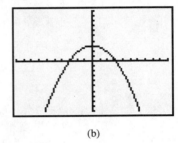

(a)   (b)

FIGURE 4-13

The vertex of the parabola shown in Figure 4-11 is the point $(1, -4)$. The axis of symmetry is the line $x = 1$, the vertical line that passes through the vertex.

The vertex of the parabola in Figure 4-12 is the point $(0, 3)$. Its axis is the y-axis, the line $x = 0$.

■ FINDING THE VERTEX OF A PARABOLA

If a, h, and k are constants and $a \neq 0$, the graph of the equation

$$y - k = a(x - h)^2$$

is a parabola, because if the parentheses were removed and the equation solved for y, the right-hand side would be a second-degree polynomial. If $a > 0$, the parabola opens upward, as in Figure 4-11. If $a < 0$, the parabola opens downward, as in Figure 4-12.

The values (h, k) shown in the equation $y - k = a(x - h)^2$ are the coordinates of the vertex of its parabolic graph, as the following discussion will show.

If $a > 0$, the graph of $y - k = a(x - h)^2$ is a parabola opening upward. The vertex of this parabola is the point on the graph that has the least possible y-coordinate. Since a is positive and $(x - h)^2$ is never negative, the minimum value of $a(x - h)^2$ is 0, and this occurs when $x = h$.

When the right-hand side of the equation is 0, the left-hand side is also 0, and the corresponding value of y is k. Therefore, the parabola's vertex is the point (h, k).

A similar argument will show that when $a < 0$, the vertex is the highest point on the graph, which is the point (h, k).

> **Vertex of a Parabola**
> The graph of the equation
> $$y - k = a(x - h)^2 \quad (a \neq 0)$$
> is a parabola with its vertex at the point (h, k). The parabola opens upward if $a > 0$ and downward if $a < 0$.

We can complete the square to change the equation $y = ax^2 + bx + c$ into the form $y - k = a(x - h)^2$. In this form, we can read the coordinates (h, k) of the vertex.

EXAMPLE 3   Find the vertex of the parabola $y = 2x^2 - 5x - 3$.

Solution   We can complete the square on the right-hand side:

$$y = 2x^2 - 5x - 3$$

$$y + 3 = 2x^2 - 5x \qquad\qquad \text{Add 3 to both sides.}$$

$$y + 3 = 2\left(x^2 - \frac{5}{2}x \qquad \right) \qquad \begin{array}{l}\text{To make the coefficient of the } x^2 \text{ term}\\ \text{equal to 1, factor a 2 from the terms on}\\ \text{the right-hand side.}\end{array}$$

To complete the square within the parentheses, we add the square of one-half of the coefficient of x.

- First, we find half the coefficient of x, which is $\dfrac{1}{2}\left(-\dfrac{5}{2}\right) = -\dfrac{5}{4}$.
- Then we square $-\dfrac{5}{4}$: $\left(-\dfrac{5}{4}\right)^2 = \dfrac{25}{16}$.

We add $\frac{25}{16}$ within the parentheses. Because of the factor of 2, however, we are really adding twice $\frac{25}{16}$, or $\frac{25}{8}$, to the right-hand side. We must also add $\frac{25}{8}$ to the left-hand side.

$$y + 3 = 2\left(x^2 - \frac{5}{2}x \qquad \right)$$

$$y + 3 + \frac{25}{8} = 2\left(x^2 - \frac{5}{2}x + \frac{25}{16} \right) \qquad \text{Add } \tfrac{25}{8} \text{ to both sides.}$$

$$y + \frac{49}{8} = 2\left(x - \frac{5}{4} \right)^2 \qquad \text{Combine like terms and factor.}$$

$$y - \left(-\frac{49}{8} \right) = 2\left(x - \frac{5}{4} \right)^2$$

The equation is now in the form

$$y - k = a(x - h)^2$$

with $h = \frac{5}{4}$ and $k = -\frac{49}{8}$. The vertex is at $(h, k) = \left(\frac{5}{4}, -\frac{49}{8}\right)$. ∎

Self Check   Find the vertex of the graph of $y = 4x^2 - 16x + 19$.

Answer $(2, 3)$

To find the vertex of any parabola defined by $y = ax^2 + bx + c$, we write the equation in the form $y - k = a(x - h)^2$ by completing the square:

$$y = ax^2 + bx + c$$

$$y = a\left(x^2 + \frac{b}{a}x \qquad\right) + c \qquad \text{Factor out } a.$$

$$y + \frac{b^2}{4a} = a\left(x^2 + \frac{b}{a}x + \frac{b^2}{4a^2}\right) + c \qquad \text{Add } \frac{b^2}{4a} \text{ to both sides.}$$

$$y + \frac{b^2}{4a} - c = a\left(x^2 + \frac{b}{a}x + \frac{b^2}{4a^2}\right) \qquad \text{Subtract } c \text{ from both sides.}$$

$$y - \left(c - \frac{b^2}{4a}\right) = a\left(x - \frac{-b}{2a}\right)^2 \qquad \text{Factor.}$$

We compare this equation to the standard form $y - h = a(x - k)^2$ to determine that $h = -\frac{b}{2a}$ and $k = c - \frac{b^2}{4a}$. This proves the following theorem.

Vertex of a Parabola

The graph of the equation

$$y = ax^2 + bx + c \quad (a \neq 0)$$

is a parabola with vertex $\left(-\dfrac{b}{2a}, c - \dfrac{b^2}{4a}\right)$.

■ MAXIMIZING AREA

EXAMPLE 4   A farmer has 400 feet of fencing to enclose a rectangular corral. To save money and fencing, he intends to use the bank of a river as one boundary of the corral, as in Figure 4-14. Find the dimensions that would enclose the largest area.

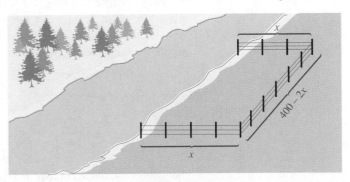

FIGURE 4-14

Solution We let x represent the width of the fenced area. Then $400 - 2x$ represents the length. Because the area A of a rectangle is the product of the length and the width, we have

$$A = (400 - 2x)x \qquad \text{or} \qquad A = -2x^2 + 400x$$

The graph of this equation is a parabola, and since the coefficient of x^2 is negative, the parabola opens downward. Since its vertex is its highest point, the A-coordinate of the vertex represents the maximum area, and the x-coordinate represents the width of the corral with that maximum area. We compare the equations

$$A = -2x^2 + 400x \qquad \text{and} \qquad y = ax^2 + bx + c$$

to determine that $a = -2$, $b = 400$, and $c = 0$. The vertex of the parabola is the point

$$\left(-\frac{b}{2a}, c - \frac{b^2}{4a}\right) = \left(-\frac{400}{2(-2)}, 0 - \frac{400^2}{4(-2)}\right) = (100, 20{,}000)$$

If the farmer's fence runs 100 feet out from the river, 200 feet parallel to the river, and 100 feet back to the river, it will enclose the largest possible area, which is 20,000 square feet. ∎

■ MAXIMIZING REVENUE

To determine a selling price that will maximize revenue, manufacturers must consider the economic principle of supply and demand: Increasing the number of units manufactured decreases the price that can be charged for each unit.

EXAMPLE 5 A manufacturer of automobile airbags has determined that x units can be manufactured and sold each week at $\$(384 - 0.1x)$ each. Find the weekly production level that will maximize the revenue from sales, and find that revenue.

Solution We let y represent the revenue from sales. Because the revenue is the product of the number of units sold and the price charged for each unit, we have

$$y = x(384 - 0.1x) \qquad \text{or} \qquad y = -0.1x^2 + 384x$$

The graph of this equation is a parabola. Since the coefficient of x^2 is negative, it opens downward, and its vertex represents its highest point. The x-coordinate of the vertex is the production level that will maximize revenue, and the y-coordinate is that maximum revenue. We compare the equations

$$y = -0.1x^2 + 384x \qquad \text{and} \qquad y = ax^2 + bx + c$$

to see that $a = -0.1$, $b = 384$, and $c = 0$. Thus, the vertex of the parabola is the point

$$\left(-\frac{b}{2a}, c - \frac{b^2}{4a}\right) = \left(-\frac{384}{2(-0.1)}, 0 - \frac{384^2}{4(-0.1)}\right) = (1920, 368{,}640)$$

The greatest possible revenue from sales is $\$368{,}640$, which is attained at a weekly production level of 1920 units. ∎

EXERCISE 4.2

VOCABULARY AND CONCEPTS   *In Exercises 1–4, fill in the blank to make a true statement.*

1. A quadratic function is defined by the equation
_____ ($a \neq 0$).

2. The vertex of the parabolic graph of the equation
$y - 5 = 2(x - 3)^2$ will be at _____.

3. The x-coordinate of the vertex of the parabolic graph
of $y = ax^2 + bx + c$ is ____.

4. The y-coordinate of the vertex of the parabolic graph
of $y = ax^2 + bx + c$ is _____.

PRACTICE   *In Exercises 5–12, graph each quadratic function.*

5. $y = x^2 - x$

6. $y = x^2 + 2x$

7. $y = -3x^2 + 2$

8. $y = -3x^2 + 4$

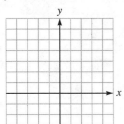
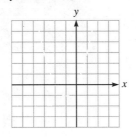
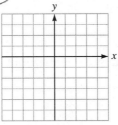
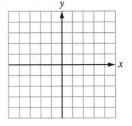

9. $y = \dfrac{1}{2}x^2 + 3$

10. $y = \dfrac{1}{2}x^2 - 2$

11. $y = x^2 - 4x + 1$

12. $y = -x^2 - 4x + 1$

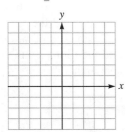
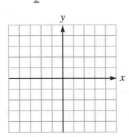
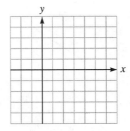
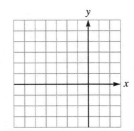

In Exercises 13–20, find the vertex of each parabola.

13. $y = x^2 - 1$

14. $y = -x^2 + 2$

15. $y = x^2 - 4x + 4$

16. $y = x^2 - 10x + 25$

17. $y = x^2 + 6x - 3$

18. $y = -x^2 + 9x - 2$

19. $y = -2x^2 + 12x - 17$

20. $y = 2x^2 + 16x + 33$

APPLICATIONS

21. Architecture   A parabolic arch has an equation
of $x^2 + 20y - 400 = 0$, where x is measured in
feet. Find the maximum height of the arch.

22. Ballistics   An object is thrown from the origin of a
coordinate system with the x-axis along the ground
and the y-axis vertical. Its path, or **trajectory,** is

given by the equation $y = 400x - 16x^2$. Find the
object's maximum height.

23. Ballistics   A child throws a ball up a hill that makes
an angle of 45° with the horizontal. The ball lands
100 feet up the hill. Its trajectory is a parabola with
equation $y = -x^2 + ax$ for some number a. Find a.
(See Illustration 1.)

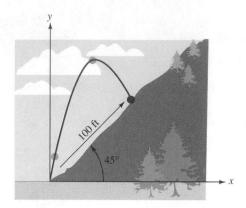

ILLUSTRATION 1

24. Maximizing area   The rectangular garden in Illustration 2 has a width of x and a perimeter of 100 feet. Find x such that the area of the rectangle is maximum.

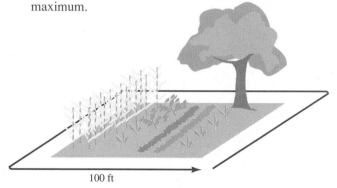

ILLUSTRATION 2

25. Maximizing storage area   A farmer wants to partition a rectangular feed storage area in a corner of his barn. The barn walls form two sides of the stall, and the farmer has 50 feet of partition for the remaining two sides. What dimensions will maximize the area? (See Illustration 3.)

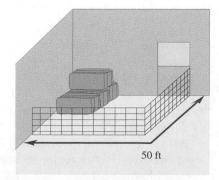

50 ft

ILLUSTRATION 3

26. Maximizing grazing area   A rancher wishes to enclose a rectangular partitioned corral with 1800 feet of fencing. (See Illustration 4.) What dimensions of the corral would enclose the largest possible area? Find the maximum area.

ILLUSTRATION 4

27. Sheet metal fabrication   A 24-inch-wide sheet of metal is to be bent into a rectangular trough with the cross section shown in Illustration 5. Find the dimensions that will maximize the amount of water the trough can hold. That is, find the dimensions that will maximize the cross-sectional area.

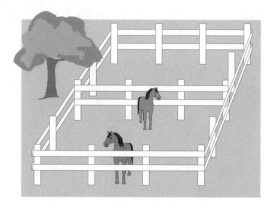

Depth   24 in.   Width

ILLUSTRATION 5

28. Landscape design   A gardener will use D feet of edging to border a rectangular plot of ground. Show that the maximum area will be enclosed if the rectangle is a square.

29. Selling television sets   A wholesaler of appliances finds that she can sell $(1200 - p)$ television sets each week when the price is p dollars. What price will maximize revenue?

30. Finding mass transit fares The Municipal Transit Authority serves 150,000 commuters daily when the fare is $1.80. Market research has determined that every penny decrease in the fare will result in 1000 new riders. What fare will maximize revenue?

31. Finding hotel rates A 300-room hotel is two-thirds filled when the nightly room rate is $90. Experience has shown that each $5 increase in cost results in 10 fewer occupied rooms. Find the nightly rate that will maximize income.

32. Selling concert tickets Tickets for a concert are cheaper when purchased in quantity. The first 100 tickets are priced at $10 each, but each additional block of 100 tickets purchased decreases the cost of each ticket by 50¢. How many blocks of tickets should be sold to maximize the revenue?

In Exercises 33–36, use this information: At a time t seconds after an object is tossed vertically upward, it reaches a height s in feet given by the equation $s = 80t - 16t^2$.

33. In how many seconds does the object reach its maximum height?

34. In how many seconds does the object return to the point from which it was thrown?

35. What is the maximum height reached by the object?

36. Show that it takes the same amount of time for the object to reach its maximum height as it does to return from that height to the point from which it was thrown.

In Exercises 37–40, use a graphing calculator to determine the coordinates of the vertex of each parabola. You will have to select appropriate viewing windows.

37. $y = 2x^2 + 9x - 56$

38. $y = 14x - \dfrac{x^2}{5}$

39. $y = (x - 7)(5x + 2)$

40. $y = -x(0.2 + 0.1x)$

DISCOVERY AND WRITING

41. Find the dimensions of the largest rectangle that can be inscribed in the right triangle *ABC* shown in Illustration 6.

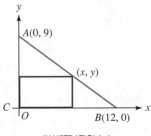

ILLUSTRATION 6

42. Point *P* lies in the first quadrant and on the line $x + y = 1$ in such position that the area of triangle *OPA* is maximum. Find the coordinate of *P*. (See Illustration 7.)

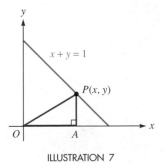

ILLUSTRATION 7

43. The sum of two numbers is 6, and the sum of the squares of those two numbers is as small as possible. What are the numbers?

44. What number most exceeds its square?

REVIEW *In Exercises 45–50, find f(a) and f(−a).*

45. $f(x) = x^2 - 3x$

46. $f(x) = x^3 - 3x$

47. $f(x) = (5 - x)^2$

48. $f(x) = \dfrac{1}{x^2 - 4}$

49. $f(x) = 7$

50. $f(x) = -|x|$

4.3 Polynomial and Other Functions

■ GRAPHING POLYNOMIAL FUNCTIONS ■ EVEN AND ODD FUNCTIONS ■ INCREASING AND DECREASING FUNCTIONS ■ PIECEWISE-DEFINED FUNCTIONS ■ THE GREATEST INTEGER FUNCTION

So far, we have discussed two types of polynomial functions—first-degree (or linear functions) and second-degree (or quadratic functions). In this section, we will discuss polynomial functions of higher degree.

> **Polynomial Functions**
> A **polynomial function in one variable (say, x)** is defined by an equation of the form $y = P(x)$, where $P(x)$ is a polynomial in the variable x.
>
> The **degree of the polynomial function** $y = P(x)$ is the degree of $P(x)$.

Some examples of polynomial functions are

$$y = f(x) = 3x^3 + 2x^2 - x + 4 \quad \text{(with a degree of 3)}$$

$$y = f(x) = \frac{1}{2}x^4 - 3x^2 + 2 \quad \text{(with a degree of 4)}$$

■ GRAPHING POLYNOMIAL FUNCTIONS

EXAMPLE 1 Graph the function $f: y = f(x) = x^3 - 4x$.

Solution To check for symmetry about the y-axis, we replace x with $-x$.

1. $\quad y = x^3 - 4x \qquad\qquad$ The original equation.

$\qquad y = (-x)^3 - 4(-x) \qquad$ Replace x with $-x$.

2. $\quad y = -x^3 + 4x \qquad\qquad$ Simplify.

Since Equations 1 and 2 are not equivalent, there is no symmetry about the y-axis. To test for symmetry about the origin, we replace x with $-x$ and y with $-y$.

1. $y = x^3 - 4x$ The original equation.

$-y = (-x)^3 - 4(-x)$ Replace x with $-x$ and y with $-y$.

$-y = -x^3 + 4x$ Simplify.

3. $y = x^3 - 4x$ Multiply both sides by -1.

Since Equations 3 and 1 are identical, the graph is symmetric about the origin. There can be no symmetry about the x-axis, because the equation is a function.

To find the x-intercepts, we let $y = 0$ and solve for x:

$$x^3 - 4x = 0$$
$$x(x^2 - 4) = 0$$
$$x(x + 2)(x - 2) = 0$$
$$x = 0 \quad \text{or} \quad x + 2 = 0 \quad \text{or} \quad x - 2 = 0$$
$$\qquad\qquad x = -2 \qquad\qquad x = 2$$

The x-intercepts are $(0, 0)$, $(-2, 0)$, and $(2, 0)$. If we let $x = 0$, we can see that the y-intercept is also $(0, 0)$.

Finally, we plot the intercepts and a few points for positive x and use the symmetry to draw the rest of the graph, as shown in Figure 4-15.

$y = x^3 - 4x$

x	y	(x, y)
-2	0	$(-2, 0)$
0	0	$(0, 0)$
1	-3	$(1, -3)$
2	0	$(2, 0)$
3	15	$(3, 15)$

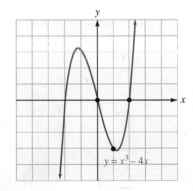

FIGURE 4-15

Self Check

Answer

Graph $y = f(x) = x^3 - 9x$.

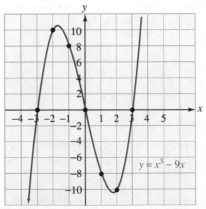

EXAMPLE 2   Graph the function f: $y = f(x) = x^4 - 5x^2 + 4$.

Solution   Because x appears with only even exponents, $y = f(x)$ is equivalent to $y = f(-x)$, and the graph is symmetric about the y-axis. The graph is not symmetric about the origin or the x-axis, however.

To find the x-intercepts, we let $y = 0$ and solve for x.

$$0 = x^4 - 5x^2 + 4$$
$$0 = (x^2 - 4)(x^2 - 1)$$
$$0 = (x + 2)(x - 2)(x + 1)(x - 1)$$

$$x + 2 = 0 \quad \text{or} \quad x - 2 = 0 \quad \text{or} \quad x + 1 = 0 \quad \text{or} \quad x - 1 = 0$$
$$x = -2 \quad | \quad x = 2 \quad | \quad x = -1 \quad | \quad x = 1$$

The x-intercepts are $(-2, 0)$, $(2, 0)$, $(-1, 0)$, and $(1, 0)$. To find the y-intercept, we let $x = 0$ and see that the y-intercept is $(0, 4)$.

Finally, we plot the intercepts and a few points for positive x and use the symmetry to draw the rest of the graph, as shown in Figure 4-16.

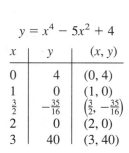

$$y = x^4 - 5x^2 + 4$$

x	y	(x, y)
0	4	$(0, 4)$
1	0	$(1, 0)$
$\frac{3}{2}$	$-\frac{35}{16}$	$\left(\frac{3}{2}, -\frac{35}{16}\right)$
2	0	$(2, 0)$
3	40	$(3, 40)$

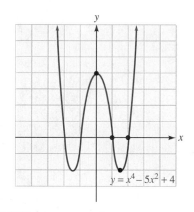

FIGURE 4-16

Self Check   Graph $y = f(x) = x^4 - 10x^2 + 9$.

Answer

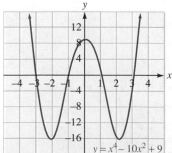

■ ■ ■ ■ ■ ■ ■ ■ ■

ACCENT ON
TECHNOLOGY

If we use a graphing calculator to draw the graphs of Examples 1 and 2, we get the graphs shown in Figure 4-17.

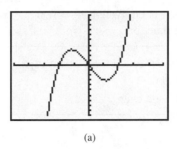

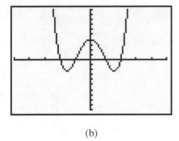

(a)   (b)

■ ■ ■ ■ ■ ■ ■ ■ ■

FIGURE 4-17

■ EVEN AND ODD FUNCTIONS

A function with a graph that is symmetric about the y-axis is called an **even function.** A function with a graph that is symmetric about the origin is called an **odd function.**

> **Even and Odd Functions**
> A function f with the property that $f(-x) = f(x)$ for all x in the domain of f is called an **even function.**
>
> A function f with the property that $f(-x) = -f(x)$ for all x in the domain of f is called an **odd function.**

Since the graph in Example 1 is symmetric about the origin, it represents an odd function. Since the graph in Example 2 is symmetric about the y-axis, it represents an even function.

EXAMPLE 3

Decide whether the functions are even or odd: **a.** $y = f(x) = x^2$ and **b.** $y = f(x) = x^3$.

Solution   **a.** We find $f(-x)$ and compare it to $f(x)$:

$$y = f(-x) = (-x)^2 = x^2$$

which is equivalent to $y = f(x) = x^2$. Since $f(-x) = f(x)$, the function is an even function.

b. We find $f(-x)$ and compare it to $f(x)$:

$$y = f(-x) = (-x)^3 = -x^3$$

which is not equivalent to $y = f(x) = x^3$. Since $f(-x) \neq f(x)$, the function is not an even function.

To determine whether it is an odd function, we find $f(-x)$ and compare it to $-f(x)$:

$$y = f(-x) = (-x)^3 = -x^3$$

which is equivalent to $-f(x) = -x^3$. Since $f(-x) = -f(x)$, the function is an odd function. ■

Self Check   Classify the functions as even or odd: **a.** $y = f(x) = x^3 + x$ and
b. $y = f(x) = x^2 + 4$.

Answer   **a.** odd,   **b.** even

■ INCREASING AND DECREASING FUNCTIONS

If the values $f(x)$ increase as x increases on an interval, we say that the function is **increasing on the interval.** (See Figure 4-18(a).) If the values $f(x)$ decrease as x increases on an interval, we say that the function is **decreasing on the interval.** (See Figure 4-18(b).) If the values $f(x)$ remain unchanged as x increases on an interval, we say that the function is **constant on the interval.** (See Figure 4-18(c).)

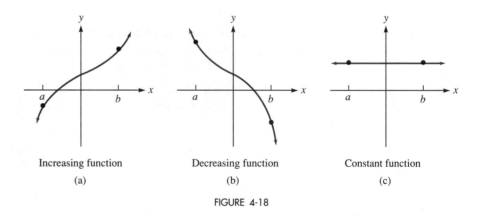

Increasing function   Decreasing function   Constant function
(a)   (b)   (c)

FIGURE 4-18

■ PIECEWISE-DEFINED FUNCTIONS

Piecewise-defined functions are defined by using different equations for different intervals in their domains. To illustrate, we will graph the piecewise-defined function f given by

$$y = f(x) = \begin{cases} -2 & \text{if } x \le 0 \\ x + 1 & \text{if } x > 0 \end{cases}$$

For each number x, we decide which part of the definition to use. If $x \le 0$, the corresponding value of y is -2. In the interval $(-\infty, 0]$, the function is constant.

If $x > 0$, the corresponding value of y is $x + 1$. In the interval $(0, \infty)$, the function is increasing. The graph appears in Figure 4-19.

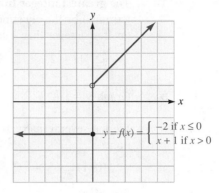

$$y = f(x) = \begin{cases} -2 & \text{if } x \leq 0 \\ x+1 & \text{if } x > 0 \end{cases}$$

FIGURE 4-19

EXAMPLE 4   Graph the function defined by

$$y = f(x) = \begin{cases} -x & \text{if } x < 0 \\ x^2 & \text{if } 0 \leq x \leq 1 \\ 1 & \text{if } x > 1 \end{cases}$$

Solution   For each number x, we decide which part of the definition to use. If $x < 0$, the value of y is determined by the equation $y = -x$. In the interval $(-\infty, 0)$, the function is decreasing.

If $0 \leq x \leq 1$, the value of y is x^2. In the interval $(0, 1)$, the function is increasing.

If $x > 1$, the value of y is 1. In the interval $(1, \infty)$, the function is constant. The graph appears in Figure 4-20.

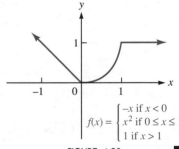

$$f(x) = \begin{cases} -x & \text{if } x < 0 \\ x^2 & \text{if } 0 \leq x \leq 1 \\ 1 & \text{if } x > 1 \end{cases}$$

FIGURE 4-20   ■

Self Check   Graph $y = f(x) = \begin{cases} 2x & \text{if } x \leq 0 \\ x - 1 & \text{if } x > 0 \end{cases}$.

Answer

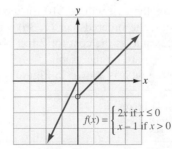

$$f(x) = \begin{cases} 2x & \text{if } x \leq 0 \\ x - 1 & \text{if } x > 0 \end{cases}$$

■ THE GREATEST INTEGER FUNCTION

The **greatest integer function** is important in computer science. This function is determined by the equation $y = [\![x]\!]$, where the value of y that corresponds to x is the greatest integer that is less than or equal to x. For example,

$$[\![2.71]\!] = 2, \qquad \left[\!\left[23\tfrac{1}{2}\right]\!\right] = 23, \qquad [\![10]\!] = 10, \qquad [\![\pi]\!] = 3 \qquad [\![-2.5]\!] = -3$$

EXAMPLE 5   Graph $y = [\![x]\!]$.

Solution   We list several intervals and the corresponding values of the greatest integer function:

$[0, 1)$	$y = [\![x]\!] = 0$	For numbers from 0 to 1 (not including 1), the greatest integer in the interval is 0.
$[1, 2)$	$y = [\![x]\!] = 1$	For numbers from 1 to 2 (not including 2), the greatest integer in the interval is 1.
$[2, 3)$	$y = [\![x]\!] = 2$	For numbers from 2 to 3 (not including 3), the greatest integer in the interval is 2.

Within each interval, the values of y are constant, but they jump by 1 at integer values of x. The graph is shown in Figure 4-21. From the graph, we can see that the domain of the greatest integer function is the interval $(-\infty, \infty)$. The range is the set of integers. ■

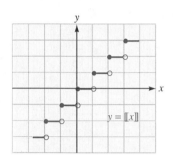

FIGURE 4-21

Self Check   Find   **a.** $[\![7.61]\!]$ and   **b.** $[\![-3.75]\!]$.

Answers   **a.** 7,   **b.** -4

Since the greatest integer function is made up of a series of horizontal line segments, it is an example of a group of functions that are called **step functions.**

EXAMPLE 6   To print a business form, a printer charges $10 for the order, plus $20 for each box containing 500 forms. The printer counts any portion of a box as a full box. Graph this step function.

Solution   If we order the forms and then change our minds before the forms are printed, the cost will be $10. Thus, the ordered pair $(0, 10)$ will be on the graph. If we purchase up to one full box, the cost will be $10 for the order and $20 for the printing, for a total of $30. Thus, the ordered pair $(1, 30)$ will be on the graph. The cost for $1\tfrac{1}{2}$ boxes will be the same as the cost for 2 full boxes, or $50. Thus, the ordered pairs $(1.5, 50)$ and $(2, 50)$ are on the graph.

The complete graph is shown in Figure 4-22. ■

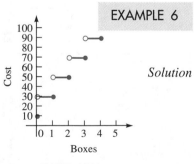

FIGURE 4-22

Self Check   Find the cost of $4\tfrac{1}{2}$ boxes.

Answer $110

EXERCISE 4.3

VOCABULARY AND CONCEPTS *In Exercises 1–8, fill in the blank to make a true statement.*

1. The degree of the function $y = f(x) = x^4 - 3$ is ___.

2. If the graph of a function is symmetric about the _____, it is called an *even* function.

3. If the graph of a function is symmetric about the origin, it is called an ____ function.

4. If the values of $f(x)$ get larger as x increases on an interval, we say that the function is _____ on the interval.

5. _____ functions are defined by different equations for different intervals in their domains.

6. If the values of $f(x)$ get smaller as x increases on an interval, we say that the function is _____ on the interval.

7. $[\![3.69]\!] = $ ___.

8. If the values of $f(x)$ do not change as x increases on an interval, we say that the function is _____ on the interval.

PRACTICE *In Exercises 9–16, graph each polynomial function.*

9. $f(x) = x^3$

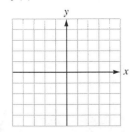

10. $f(x) = x^4$

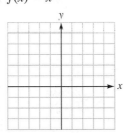

11. $f(x) = x^3 - x$

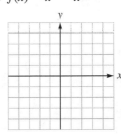

12. $f(x) = x^3 + x^2$

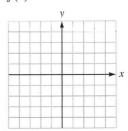

13. $f(x) = -x^3$

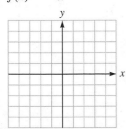

14. $f(x) = -x^3 + 1$

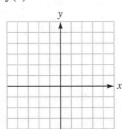

15. $f(x) = x^4 - 2x^2 + 1$

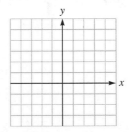

16. $f(x) = x^4 - 5x^2 + 4$

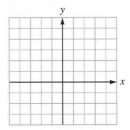

In Exercises 17–24, tell whether each function is even or odd. If it is neither, so indicate.

17. $y = x^4 + x^2$

18. $y = x^3 - 2x$

19. $y = x^3 + x^2$

20. $y = x^6 - x^2$

21. $y = x^5 + x^3$

22. $y = x^3 - x^2$

23. $y = 2x^3 - 3x$

24. $y = 4x^2 - 5$

In Exercises 25–30, tell where each function is increasing, decreasing, or constant.

25.

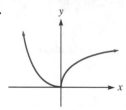

26.

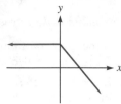

27.

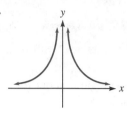

28.

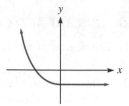

29. $y = f(x) = x^2 - 4x + 4$

30. $y = f(x) = 4 - x^2$

In Exercises 31–36, graph each piecewise-defined function.

31. $y = f(x) = \begin{cases} x + 2 & \text{if } x < 0 \\ 2 & \text{if } x \geq 0 \end{cases}$

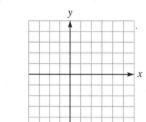

32. $y = f(x) = \begin{cases} 2x & \text{if } x < 0 \\ -2x & \text{if } x \geq 0 \end{cases}$

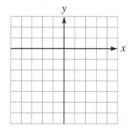

33. $y = f(x) = \begin{cases} -x & \text{if } x < 0 \\ x^2 & \text{if } x \geq 0 \end{cases}$

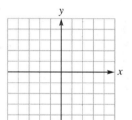

34. $y = f(x) = \begin{cases} |x| & \text{if } x < 0 \\ \sqrt{x} & \text{if } x \geq 0 \end{cases}$

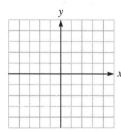

35. $y = f(x) =$

$\begin{cases} 0 & \text{if } x < 0 \\ x^2 & \text{if } 0 \leq x \leq 2 \\ 4 - 2x & \text{if } x > 2 \end{cases}$

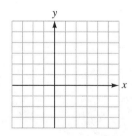

36. $y = f(x) = \begin{cases} 2 & \text{if } x < 0 \\ 2 - x & \text{if } 0 \leq x < 2 \\ x & \text{if } x \geq 2 \end{cases}$

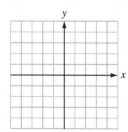

In Exercises 37–40, graph each function.

37. $y = [\![2x]\!]$

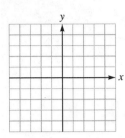

38. $y = \left[\!\!\left[\dfrac{1}{3}x + 3\right]\!\!\right]$

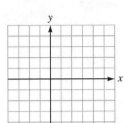

39. $y = [\![x]\!] - 1$

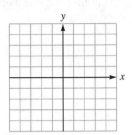

40. $y = [\![x + 2]\!]$

APPLICATIONS

41. Renting a car   A rental company charges $20 to rent a car for one day, plus $2 for every 100 miles (or portion of 100 miles) that it is driven. Graph the ordered pairs (m, c), where m represents the miles driven and c represents the cost. Find the cost if the car is driven 275 miles.

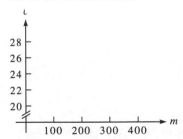

42. Riding in a taxi   A taxicab company charges $3 for a trip up to 1 mile, and $2 for every extra mile (or portion of a mile). Graph the ordered pairs (m, c), where m represents the miles traveled and c represents the cost. Find the cost to ride $10\frac{1}{4}$ miles.

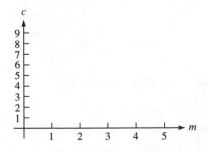

43. Computer communications   A computer information network charges for connect time at a rate of $12 per hour, computed for every minute or fraction of a minute. Graph the points (t, c), where c is the cost of t minutes of connect time. Find the cost of $7\frac{1}{2}$ minutes.

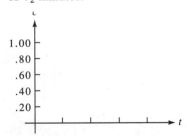

44. Plumbing repair   A plumber charges $30, plus $40 per hour (or fraction of an hour), to install a new bathtub. Graph the points (t, c), where t is the time it takes to do the job and c is the cost. If the job took 4 hours, how much did it cost?

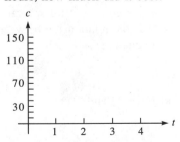

45. Rounding numbers Measurements are rarely exact; they are often *rounded* to an appropriate precision. Graph the points (x, y), where y is the result of rounding the number x to the nearest ten.

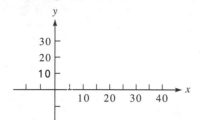

46. Signum function Computer programmers often use the following function, denoted by $y = \text{sgn } x$. Graph this function and find its domain and range.

$$y = \begin{cases} -1 & \text{if } x < 0 \\ 0 & \text{if } x = 0 \\ 1 & \text{if } x > 0 \end{cases}$$

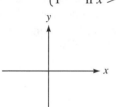

47. Graph the function defined by $y = \frac{|x|}{x}$ and compare it to the graph in Exercise 46. Are the graphs the same?

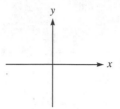

48. Graph the function $y = x + |x|$.

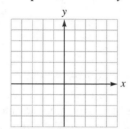

DISCOVERY AND WRITING *In Exercises 49–52, use a graphing calculator to explore the properties of graphs of polynomial functions. Write a paragraph summarizing your observations.*

49. Graph the function $y = x^2 + ax$ for several values of a. How does the graph change?

50. Graph the function $y = x^3 + ax$ for several values of a. How does the graph change?

51. Graph the function $y = (x - a)(x - b)$ for several values of a and b. What is the relationship between the x-intercepts and the equation?

52. Use the insight you gained in Exercise 51 to factor $x^3 - 3x^2 - 4x + 12$.

REVIEW

53. If $f(x) = 3x + 2$, find $f(x + 1)$ and $f(x) + 1$.

54. If $f(x) = x^2$, find $f(x - 2)$ and $f(x) - 2$.

55. If $f(x) = \dfrac{3x + 1}{5}$, find $f(x - 3)$ and $f(x) - 3$.

56. If $f(x) = 8$, find $f(x + 8)$ and $f(x) + 8$.

4.4 Translating and Stretching Graphs

■ VERTICAL TRANSLATIONS ■ HORIZONTAL TRANSLATIONS ■ SHIFTS INVOLVING TWO TRANSLATIONS ■ REFLECTIONS ABOUT THE X- AND Y-AXES ■ VERTICAL AND HORIZONTAL STRETCHINGS

■ VERTICAL TRANSLATIONS

The graphs of different equations may be identical except for their position in the xy-plane. For example, Figure 4-23 shows the graph of $y = x^2 + k$ for three values of k. If $k = 0$, we have the graph of $y = x^2$. The graph of $y = x^2 + 2$ is identical to the graph of $y = x^2$ except that it is shifted 2 units upward. The graph of $y = x^2 - 3$ is identical to the graph of $y = x^2$ except that it is shifted 3 units downward. These shifts are called **vertical translations.**

In general, we can make the following observations.

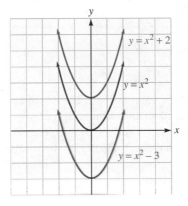

FIGURE 4-23

Vertical Translations

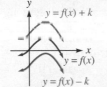

If f is a function and k is a positive number, then

- The graph of $y = f(x) + k$ is identical to the graph of $y = f(x)$ except that it is translated k units upward.
- The graph of $y = f(x) - k$ is identical to the graph of $y = f(x)$ except that it is translated k units downward.

EXAMPLE 1   Graph the function $y = |x| - 2$.

Solution   The graph is identical to the graph of $y = |x|$ except that it is translated 2 units downward. (See Figure 4-24.)

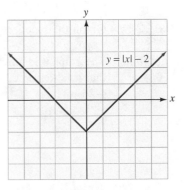

FIGURE 4-24

Self Check Graph $y = |x| + 3$.

Answer

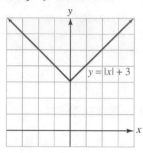

■ HORIZONTAL TRANSLATIONS

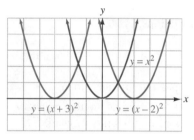

FIGURE 4-25

Figure 4-25 shows the graph of $y = (x + h)^2$ for three values of h. If $h = 0$, we have the graph of $y = x^2$. The graph of $y = (x - 2)^2$ is identical to the graph of $y = x^2$ except that it is shifted 2 units to the right. The graph of $y = (x + 3)^2$ is identical to the graph of $y = x^2$ except that it is shifted 3 units to the left. These shifts are called **horizontal translations.**

In general, we can make the following observations.

Horizontal Translations

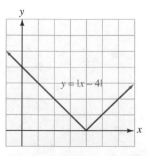

If f is a function and k is a positive number, then

- The graph of $y = f(x - k)$ is identical to the graph of $y = f(x)$, except that it is translated k units to the right.
- The graph of $y = f(x + k)$ is identical to the graph of $y = f(x)$, except that it is translated k units to the left.

EXAMPLE 2   Graph the function $y = |x - 4|$.

Solution   The graph is identical to the graph of $y = |x|$ except that it is translated 4 units to the right. (See Figure 4-26.)

FIGURE 4-26 ■

Self Check Graph $y = |x + 2|$.

Answer

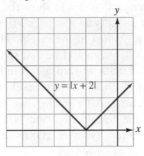

■ SHIFTS INVOLVING TWO TRANSLATIONS

EXAMPLE 3 Graph $y = (x - 5)^3 + 4$.

Solution The graph of $y = (x - 5)^3 + 4$ is identical to the graph of $y = x^3$ except that it has been translated 4 units upward and 5 units to the right. (See Figure 4-27.)

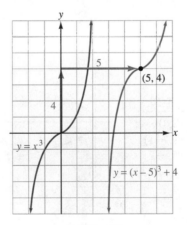

FIGURE 4-27 ■

Self Check Graph $y = (x + 2)^2 - 2$.

Answer

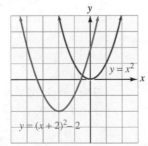

■ ■ ■ ■ ■ ■ ■ ■

ACCENT ON TECHNOLOGY

We can use a graphing calculator to show the effect of vertical and horizontal translations by graphing $y = x^2$ and $y = (x - 3)^2 + 2$. The graph of $y = x^2$ is a parabola opening upward with vertex at the origin. The graph of $y = (x - 3)^2 + 2$ should be that same parabola translated 2 units upward and 3 units to the right. Figure 4-28 shows the result of graphing these functions. After tracing, we see that the vertex of the translated graph is the point $(3, 2)$, as expected. (Further zooms and traces will give more exact results.)

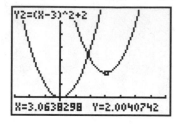

■ ■ ■ ■ ■ ■ ■ ■

FIGURE 4-28

■ **REFLECTIONS ABOUT THE X- AND Y-AXES**

Figure 4-29(a) shows that the graph of $y = -\sqrt{x}$ is identical to the graph of $y = \sqrt{x}$ except that it is reflected about the x-axis. Figure 4-29(b) shows that the graph of $y = \sqrt{-x}$ is identical to the graph of $y = \sqrt{x}$ except that it is reflected about the y-axis.

$y = -\sqrt{x}$

x	y	(x, y)
0	0	$(0, 0)$
1	-1	$(1, -1)$
4	-2	$(4, -2)$

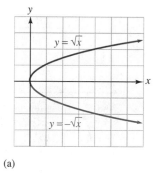

(a)

$y = \sqrt{-x}$

x	y	(x, y)
0	0	$(0, 0)$
-1	1	$(-1, 1)$
-4	2	$(-4, 2)$

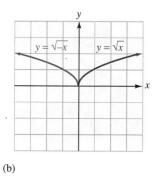

(b)

FIGURE 4-29

In general, we can make the following observations.

Reflections

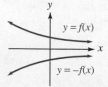

• The graph of $y = -f(x)$ is identical to the graph of $y = f(x)$ except that it is reflected about the x-axis.

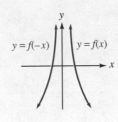

- The graph of $y = f(-x)$ is identical to the graph of $y = f(x)$ except that it is reflected about the y-axis.

EXAMPLE 4   Graph the function $y = -|x + 1|$.

Solution   The graph is identical to the graph of $y = |x + 1|$ except that it is reflected about the x-axis. (See Figure 4-30.)

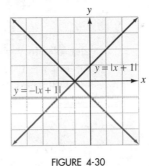

FIGURE 4-30   ■

Self Check   Graph $y = |-x + 1|$.

Answer

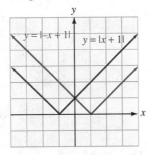

■ **VERTICAL AND HORIZONTAL STRETCHINGS**

Figure 4-31 shows the graphs of $y = x^2 - 1$ and $y = 3(x^2 - 1)$. Because each value of $y = 3(x^2 - 1)$ is 3 times greater than the value of $y = x^2 - 1$, its graph is stretched vertically by a factor of 3. The x-intercepts of both graphs are $(1, 0)$ and $(-1, 0)$.

$$y = x^2 - 1$$

x	y	(x, y)
-2	3	$(-2, 3)$
-1	0	$(-1, 0)$
0	-1	$(0, -1)$
1	0	$(1, 0)$
2	3	$(2, 3)$

$$y = 3(x^2 - 1)$$

x	y	(x, y)
-2	9	$(-2, 9)$
-1	0	$(-1, 0)$
0	-3	$(0, 3)$
1	0	$(1, 0)$
2	9	$(2, 9)$

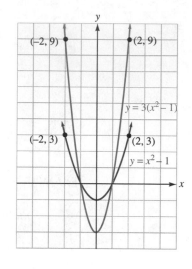

FIGURE 4-31

Figure 4-32 shows the graphs of $y = x^2 - 1$ and $y = (3x)^2 - 1$. Because any value of x in $y = (3x)^2 - 1$ is $\frac{1}{3}$ of the corresponding value of x in $y = x^2 - 1$, the graph of $y = (3x)^2 - 1$ is stretched horizontally by a factor of $\frac{1}{3}$. The y-intercepts of both graphs are $(0, 1)$.

$$y = x^2 - 1$$

x	y	(x, y)
-2	3	$(-2, 3)$
-1	0	$(-1, 0)$
0	-1	$(0, -1)$
1	0	$(1, 0)$
2	3	$(2, 3)$

$$y = (3x)^2 - 1$$

x	y	(x, y)
$-\frac{2}{3}$	3	$(-\frac{2}{3}, 3)$
$-\frac{1}{3}$	0	$(-\frac{1}{3}, 0)$
0	-1	$(0, -1)$
$\frac{1}{3}$	0	$(\frac{1}{3}, 0)$
$\frac{2}{3}$	3	$(\frac{2}{3}, 3)$

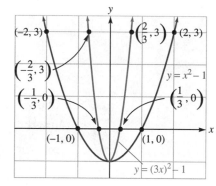

FIGURE 4-32

In general, we can make the following observations.

Vertical Stretching

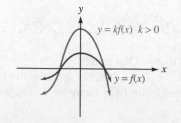

If k is a positive number, then

• The graph of $y = kf(x)$ can be obtained by stretching the graph of $y = f(x)$ vertically by a factor of k.

Horizontal Stretching

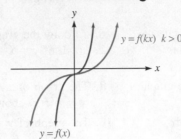

- The graph of $y = f(kx)$ can be obtained by stretching the graph of $y = f(x)$ horizontally by a factor of $\frac{1}{k}$.

■ ■ ■ ■ ■ ■ ■ ■ ■

ACCENT ON TECHNOLOGY

We can use a graphing calculator to show the effect of a vertical stretching and reflection by graphing $y = x^2 - 4$ and $y = -2(x^2 - 4)$. The graph of $y = x^2 - 4$ is a parabola opening upward with vertex at $(0, -4)$. If that graph is stretched vertically by a factor of 2 and then reflected about the x-axis, the graph of $y = -2(x^2 - 4)$ should be the result. That is what happens, as Figure 4-33 shows. Notice that the x-intercepts of the graphs are the same.

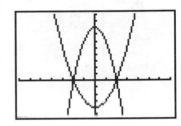

FIGURE 4-33

■ ■ ■ ■ ■ ■ ■ ■ ■

We can summarize the ideas in the section as follows.

Translations and Reflections

If $y = f(x)$ is a function and k represents a positive number, then

The graph of	*can be obtained by graphing $y = f(x)$ and*
$y = f(x) + k$	translating the graph k units up.
$y = f(x) - k$	translating the graph k units down.
$y = f(x + k)$	translating the graph k units to the left.
$y = f(x - k)$	translating the graph k units to the right.
$y = -f(x)$	reflecting the graph about the x-axis.
$y = f(-x)$	reflecting the graph about the y-axis.
$y = kf(x)$	stretching the graph vertically by a factor of k.
$y = f(kx)$	stretching the graph horizontally by a factor of $\frac{1}{k}$.

EXERCISE 4.4

VOCABULARY AND CONCEPTS *In Exercises 1–10, fill in the blank to make a true statement.*

1. The graph of $y = f(x) + 5$ is identical to the graph of $y = f(x)$ except that it is translated 5 units _____.

2. The graph of $y = $ _____ is identical to the graph of $y = f(x)$ except that it is translated 7 units downward.

3. The graph of $y = f(x - 3)$ is identical to the graph of $y = f(x)$ except that it is translated 3 units _____.

4. The graph of $y = f(x + 2)$ is identical to the graph of $y = f(x)$ except that it is translated 2 units _____.

5. To draw the graph of $y = (x + 2)^2 - 3$, translate the graph of $y = x^2$ ___ units to the left and 3 units _____.

6. To draw the graph of $y = (x - 3)^2 + 1$, translate the graph of $y = x^2$ 3 units to the _____ and 1 unit ___.

7. The graph of $y = f(-x)$ is a reflection of the graph of $y = f(x)$ about the _____.

8. The graph of _____ is a reflection of the graph of $y = f(x)$ about the x-axis.

9. The graph of $y = f(4x)$ stretches the graph of $y = f(x)$ _____ by a factor of $\frac{1}{4}$.

10. The graph of $y = 8f(x)$ stretches the graph of $y = f(x)$ _____ by a factor of 8.

In Exercises 11–18, the graph of each equation is a translation of the graph of $y = x^2$. Graph each equation.

11. $y = x^2 - 2$
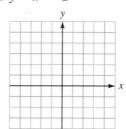

12. $y = (x - 2)^2$
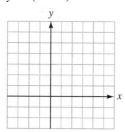

13. $y = (x + 3)^2$
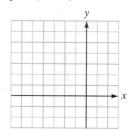

14. $y = x^2 + 3$
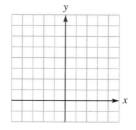

15. $y = (x + 1)^2 + 2$
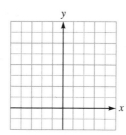

16. $y = (x - 3)^2 - 1$
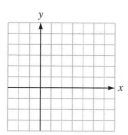

17. $y = \left(x + \dfrac{1}{2}\right)^2 - \dfrac{1}{2}$
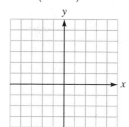

18. $y = \left(x - \dfrac{3}{2}\right)^2 + \dfrac{5}{2}$
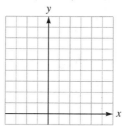

In Exercises 19–26, the graph of each equation is a translation of the graph of $y = x^3$. Graph each equation.

19. $y = x^3 + 1$
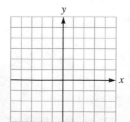

20. $y = x^3 - 3$
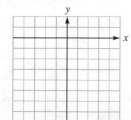

21. $y = (x - 2)^3$
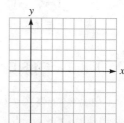

22. $y = (x + 3)^3$
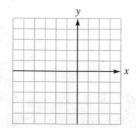

23. $y = (x - 2)^3 - 3$

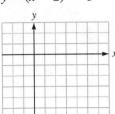

24. $y = (x + 1)^3 + 4$

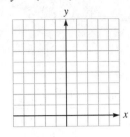

25. $y + 2 = x^3$

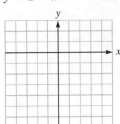

26. $y - 7 = (x - 5)^3$

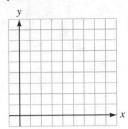

In Exercises 27–30, the graph of each equation is a reflection of the graph of $y = x^2$, $y = x^3$, or $y = (x - 1)^2$. Graph each equation.

27. $y = -x^2$

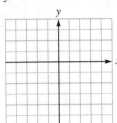

28. $y = (-x)^3$

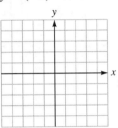

29. $y = -x^3$

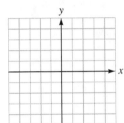

30. $y = (-x - 1)^2$

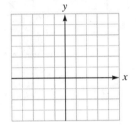

In Exercises 31–34, the graph of each equation is a stretching of the graph of $y = x^2$. Graph each equation.

31. $y = 2x^2$

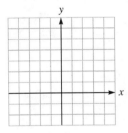

32. $y = \frac{1}{2}x^2$

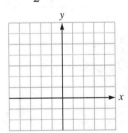

33. $y = -3x^2$

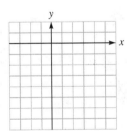

34. $y = -\frac{1}{3}x^2$

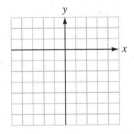

In Exercises 35–38, the graph of each equation is a stretching of the graph of $y = x^3$. Graph each equation.

35. $y = \left(\frac{1}{2}x\right)^3$

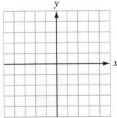

36. $y = \frac{1}{8}x^3$

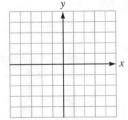

37. $y = -8x^3$

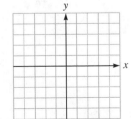

38. $y = (-2x)^3$

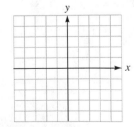

In Exercises 39–48, graph each equation.

39. $y = |x - 2| + 1$

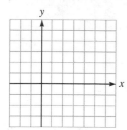

40. $y = |x + 5| - 2$

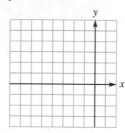

41. $y = \sqrt{x - 2} + 1$ $(x \geq 2)$

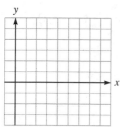

42. $y = \sqrt{x + 5} - 2$ $(x \geq -5)$

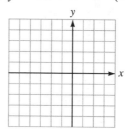

43. $y = |3x|$

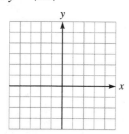

44. $y = 3|x|$

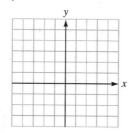

45. $y = 2\sqrt{x} + 3$ $(x \geq 0)$

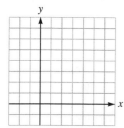

46. $y = 2\sqrt{x + 3}$ $(x \geq -3)$

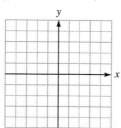

47. $y = -2|x| + 3$

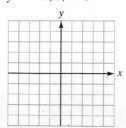

48. $y = -2|x + 3|$

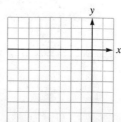

DISCOVERY AND WRITING   *In Exercises 49–52, use a graphing calculator to perform each experiment. Write a brief paragraph describing your findings.*

49. Investigate the translations of the graph of a function by graphing the parabola $y = (x - k)^2 + k$ for several values of k. What do you observe about successive positions of the vertex?

50. Investigate the translations of the graph of a function by graphing the parabola $y = (x - k)^2 + k^2$ for several values of k. What do you observe about successive positions of the vertex?

51. Investigate the horizontal stretching of the graph of a function by graphing $y = \sqrt{ax}$ for several values of a. What do you observe?

52. Investigate the vertical stretching of the graph of a function by graphing $y = b\sqrt{x}$ for several values of b. What do you observe? Are these graphs different from the graphs in Exercise 51?

Write a paragraph using your own words.

53. Explain why the effect of vertically stretching a graph by a factor of -1 is to reflect the graph about the x-axis.

54. Explain why the effect of horizontally stretching a graph by a factor of -1 is to reflect the graph about the y-axis.

REVIEW   *In Exercises 55–56, simplify each fraction.*

55. $\dfrac{x^2 + x - 6}{x^2 + 5x + 6}$

56. $\dfrac{2x^2 + 3x}{2x^2 + x - 3}$

In Exercises 57–58, find the domain of each function.

57. $f(x) = \dfrac{x + 7}{x - 3}$

58. $f(x) = \dfrac{x^2 + 1}{x^2 + 3x + 2}$

In Exercises 59–60, do each division and write the answer in quotient $+ \frac{remainder}{divisor}$ form.

59. $\dfrac{x^2 + 3x}{x + 1}$

60. $\dfrac{x^2 + 3}{x + 1}$

4.5 Rational Functions

■ AVERAGE HOURLY COST ■ RATIONAL FUNCTIONS ■ VERTICAL AND HORIZONTAL ASYMPTOTES
■ FINDING ASYMPTOTES ■ GRAPHING RATIONAL FUNCTIONS ■ GRAPHS WITH DISCONTINUITIES

■ AVERAGE HOURLY COST

Rational expressions often define functions. For example, if the cost of subscribing to an on-line computer service is $6 per month plus $1.50 per hour of access time, the average (mean) hourly cost of the service is the total monthly cost, divided by the number of hours of access time:

$$\bar{c} = \frac{C}{n} = \frac{1.50n + 6}{n}$$ $\bar{c}$ is the mean hourly cost, C is the total monthly cost, and n is the number of hours the service is used.

The function

1. $\bar{c} = f(n) = \dfrac{1.50n + 6}{n}$ $(n > 0)$

gives the mean hourly cost of using the service for n hours per month. Since $n > 0$, the domain of this function is the interval $(0, \infty)$.

■ ■ ■ ■ ■ ■ ■ ■ ■

ACCENT ON TECHNOLOGY

If we use a graphing calculator with window settings of $[0, 10]$ for x and $[0, 10]$ for y to graph the function $f(n) = \frac{1.50n + 6}{n}$, we will obtain the graph shown in Figure 4-34. Note that the graph of the function passes the vertical line test, as expected.

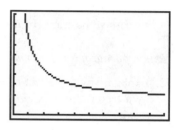

FIGURE 4-34

■ ■ ■ ■ ■ ■ ■ ■ ■

From the graph, we can see that the mean hourly cost decreases as the number of hours of access time increases. Since the cost of each extra hour of access time is $1.50, the mean hourly cost can approach $1.50 but never drop below it. The graph of the function approaches the line $y = 1.5$ as n increases without bound. When a graph approaches a line but never touches it, we call the line an **asymptote**. The line $y = 1.5$ is a **horizontal asymptote** of the graph.

As n gets smaller and approaches 0, the graph approaches the y-axis but never touches it. The y-axis is called a **vertical asymptote** of the graph.

EXAMPLE 1

Find the mean hourly cost when the service described above is used for   **a.** 3 hours and   **b.** 10.4 hours.

Solution   **a.** To find the mean hourly cost for 3 hours of access time, we substitute 3 for n in Equation 1 and simplify:

$$\bar{c} = f(\mathbf{3}) = \frac{1.50(\mathbf{3}) + 6}{\mathbf{3}} = 3.5$$

The mean hourly cost for 3 hours of access time is $3.50.

b. To find the mean hourly cost for 10.4 hours of access time, we substitute 10.4 for n in Equation 1 and simplify:

$$\bar{c} = f(\mathbf{10.4}) = \frac{1.50(\mathbf{10.4}) + 6}{\mathbf{10.4}} = 2.076923$$

The mean hourly cost for 10.4 hours of access time is approximately $2.08.   ■

Self Check   Find the mean hourly cost when the service is used for $\frac{1}{2}$ hour.

Answer $13.50

■ RATIONAL FUNCTIONS

The function $f(n) = \frac{1.50n + 6}{n}$ $(n > 0)$ is an example of a **rational function.** Such functions are defined by equations of the form

$$y = \frac{P(x)}{Q(x)}$$

where $P(x)$ and $Q(x)$ are polynomials. Because $Q(x)$ appears in the denominator of a fraction, it cannot equal 0. Thus, the domain of a rational function must exclude all values of x for which $Q(x) = 0$.

Here are some examples of rational functions and their domains:

$$y = f(x) = \frac{3}{x - 2} \qquad \text{Domain: } (-\infty, 2) \cup (2, \infty)$$

$$y = f(x) = \frac{5x + 2}{x^2 - 4} \qquad \text{Domain: } (-\infty, -2) \cup (-2, 2) \cup (2, \infty)$$

$$y = f(x) = \frac{3x^2 - 2}{(2x + 1)(x - 3)} \qquad \text{Domain: } \left(-\infty, -\frac{1}{2}\right) \cup \left(-\frac{1}{2}, 3\right) \cup (3, \infty)$$

EXAMPLE 2   Find the domain of $f(x) = \dfrac{3x + 2}{x^2 - 7x + 12}$.

Solution   To find the numbers x that make the denominator 0, we set $x^2 - 7x + 12$ equal to 0 and solve for x.

$$x^2 - 7x + 12 = 0$$
$$(x - 4)(x - 3) = 0 \qquad \text{Factor } x^2 - 7x + 12 = 0.$$
$$x - 4 = 0 \quad \text{or} \quad x - 3 = 0 \qquad \text{Set each factor equal to 0.}$$
$$x = 4 \qquad\qquad x = 3 \qquad \text{Solve each linear equation.}$$

The domain is the interval $(-\infty, 3) \cup (3, 4) \cup (4, \infty)$. ■

Self Check   Find the domain of $f(x) = \dfrac{2x - 3}{x^2 - x - 2}$.

Answer $(-\infty, -1) \cup (-1, 2) \cup (2, \infty)$

■ ■ ■ ■ ■ ■ ■ ■ ■

ACCENT ON TECHNOLOGY   We can use graphing calculators to find domains and ranges of rational functions. If we use settings of $[-10, 10]$ for x and $[-10, 10]$ for y and graph $f(x) = \frac{2x + 1}{x - 1}$, we will obtain Figure 4-35.

From the graph, we can see that every real number x except 1 gives a value of y. Thus, the domain of the function is $(-\infty, 1) \cup (1, \infty)$. We can also see that y can be any value except 2. The range of the function is $(-\infty, 2) \cup (2, \infty)$.

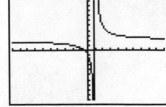

FIGURE 4-35

■ ■ ■ ■ ■ ■ ■ ■

■ VERTICAL AND HORIZONTAL ASYMPTOTES

We have seen that if a graph approaches a line but never touches it, we call the line an *asymptote*. There are two asymptotes that are apparent in Figure 4-35.

■ ■ ■ ■ ■ ■ ■ ■

ACCENT ON TECHNOLOGY

From Figure 4-35, we can see that

- As x approaches 1 from the left, the values of y decrease, and the graph approaches the vertical line $x = 1$.
- As x approaches 1 from the right, the values of y increase, and the graph approaches the vertical line $x = 1$.

For this reason, the line $x = 1$ is a *vertical asymptote*.

Although the vertical line in Figure 4-35 appears to be the asymptote, it is not. Graphing calculators draw graphs by connecting dots whose x-coordinates are close together. Often when two such points straddle a vertical asymptote and their y-coordinates are far apart, the calculator draws a line between them anyway, producing what appears to be the vertical asymptote shown in the figure. If you set your calculator to dot mode instead of connected mode, the vertical line will not appear.

From the figure, we can also see that

- As x increases to the right of 1, the values of y decrease and approach the value $y = 2$.
- As x decreases to the left of 1, the values of y increase and approach the value $y = 2$.

■ ■ ■ ■ ■ ■ ■ ■

The line $y = 2$ is a *horizontal asymptote*. Graphing calculators do not draw lines that appear to be horizontal asymptotes.

In the following definition, we read $f(x) \to \infty$ as "$f(x)$ approaches ∞."

Vertical and Horizontal Asymptotes
The line $x = a$ is a **vertical asymptote** for the graph of a function f when

$$f(x) \to \infty \quad \text{or} \quad f(x) \to -\infty \quad \text{as} \quad x \to a$$

The line $y = b$ is a **horizontal asymptote** for the graph of a function f when

$$f(x) \to b \quad \text{as} \quad x \to \infty \quad \text{or as} \quad x \to -\infty$$

Figure 4-36 shows several graphs and their asymptotes.

Vertical asymptote at x = a   *Horizontal asymptote at y = b*

$f(x) \to \infty$ $f(x) \to -\infty$ $f(x) \to b$

as $x \to a$ from the left   as $x \to a$ from the right   as $x \to \infty$ or $x \to -\infty$

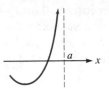
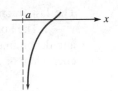
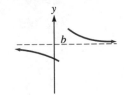

FIGURE 4-36

■ FINDING ASYMPTOTES

EXAMPLE 3   Find the asymptotes of each function: **a.** $y = \dfrac{x+2}{x^2-1}$ and
b. $y = \dfrac{2x^2+x+2}{x^2-1}$.

Solution   Each function has a vertical asymptote at $x = 1$ and $x = -1$, because at these values, their denominators are 0. We must also find the horizontal asymptotes, if any.

a. Because the degree of the numerator is less than the degree of the denominator, we can find the horizontal asymptote by dividing the numerator and the denominator by x^2, which is the largest power of x in the denominator.

$$y = \frac{x+2}{x^2-1} = \frac{\dfrac{x}{x^2}+\dfrac{2}{x^2}}{\dfrac{x^2}{x^2}-\dfrac{1}{x^2}} = \frac{\dfrac{1}{x}+\dfrac{2}{x^2}}{1-\dfrac{1}{x^2}}$$

Since $\dfrac{1}{x}$, $\dfrac{2}{x^2}$, and $\dfrac{1}{x^2}$ all approach 0 as $x \to \infty$, y approaches

$$\frac{0+0}{1-0} = 0$$

The horizontal asymptote is the line $y = 0$. If the degree of the numerator is less than the degree of the denominator, the horizontal asymptote is always $y = 0$.

b. To find the horizontal asymptote, we do a long division and write the result in *quotient* $+ \frac{remainder}{divisor}$ form.

$$
\begin{array}{r}
2 \\
x^2-1 \overline{)\, 2x^2 + x + 2} \\
\underline{2x^2 - 2} \\
x + 4
\end{array}
$$

Thus,

$$y = \frac{2x^2 + x + 2}{x^2 - 1} = 2 + \frac{x + 4}{x^2 - 1}$$

The last fraction approaches 0 as $x \to \infty$, for reasons discussed in part **a.** So, $y \to 2$. The horizontal asymptote is the line $y = 2$. If the degrees of the numerator and denominator are the same, the horizontal asymptote is always $y =$ the lead coefficient of the numerator divided by the lead coefficient of the denominator.

∎

Self Check   Find the asymptotes of $y = \dfrac{x^2}{x^2 - 4}$.

Answer   Vertical asymptote at $x = 2$ and $x = -2$; horizontal asymptote at $y = 1$

EXAMPLE 4   Find the asymptotes of each function.   **a.** $y = \dfrac{3x^3 + 2x^2 + 2}{x^2 - 1}$ and
b. $y = \dfrac{x^4 + x + 2}{x^2 - 1}$

Solution   Again, the vertical asymptotes are at $x = 1$ and $x = -1$. We must also find the horizontal asymptotes.

a. To find the horizontal asymptote, we do a long division and write the result in $quotient + \frac{remainder}{divisor}$ form.

$$
\begin{array}{r}
3x + 2 \\
x^2 - 1\,\overline{)\,3x^3 + 2x^2 + 2} \\
\underline{3x^3 - 3x} \\
2x^2 + 3x \\
\underline{2x^2 - 2} \\
+ 3x + 4
\end{array}
$$

Thus,

$$y = \frac{3x^3 + 2x^2 + 2}{x^2 - 1} = 3x + 2 + \frac{3x + 4}{x^2 - 1}$$

The last fraction approaches 0 as $x \to \infty$, and the graph approaches the line $y = 3x + 2$. Since the line is not vertical, we call the asymptote a **slant asymptote.** If the degree of the numerator is 1 more than the degree of the denominator, there will be a slant asymptote.

b. To find the horizontal asymptote, we do a long division and write the result in $quotient + \frac{remainder}{divisor}$ form.

$$
\begin{array}{r}
x^2 + 1 \\
x^2 - 1\,\overline{)\,x^4 + x + 2} \\
\underline{x^4 - x^2 } \\
x^2 + x \\
\underline{x^2 - 1} \\
x + 3
\end{array}
$$

Thus,

$$y = \frac{x^4 + x + 2}{x^2 - 1} = x^2 + 1 + \frac{x + 3}{x^2 - 1}$$

The last fraction approaches 0 as $x \to \infty$, so the curve approaches the parabola $y = x^2 + 1$. Since a parabola is not a line, the graph has no horizontal or slant asymptotes. ∎

Self Check Find the asymptotes of $y = \dfrac{2x^3 - 3x + 1}{x^2 - 4}$.

Answer Vertical asymptote at $x = 2$ and $x = -2$; slant asymptote at $y = 2x$

■ GRAPHING RATIONAL FUNCTIONS

We follow these steps to graph the rational function $y = \frac{P(x)}{Q(x)}$, where $P(x)$ and $Q(x)$ are polynomials written in descending powers of x and $\frac{P(x)}{Q(x)}$ is in simplest form.

> **Strategy for Graphing Rational Functions**
>
> **Check for symmetry.** If $P(x)$ and $Q(x)$ involve only even powers of x, then $f(x) = f(-x)$, and the graph is symmetric about the y-axis. Otherwise, y-axis symmetry does not exist. Check for symmetry about the origin.
>
> **Look for vertical asymptotes.** The real roots of $Q(x) = 0$, if any, determine the vertical asymptotes of the graph.
>
> **Look for y-intercepts.** Let $x = 0$. The resulting value of y, if any, is the y-intercept of the graph.
>
> **Look for x-intercepts.** The real roots of $P(x) = 0$, if any, are the x-intercepts of the graph.
>
> **Look for horizontal asymptotes.**
>
> • If the degree of $P(x)$ is less than the degree of $Q(x)$, the line $y = 0$ is a horizontal asymptote.
>
> • If the degrees of $P(x)$ and $Q(x)$ are equal, the line $y = \frac{p}{q}$, where p and q are the lead coefficients of $P(x)$ and $Q(x)$, is a horizontal asymptote.
>
> • If the degree of $P(x)$ is greater than the degree of $Q(x)$, there is no horizontal asymptote.
>
> **Look for slant asymptotes.** If the degree of $P(x)$ is 1 greater than the degree of $Q(x)$, there is a slant asymptote. To find it, divide $P(x)$ by $Q(x)$ and ignore the remainder.

WARNING! A graph might intersect a horizontal asymptote when $|x|$ is small. However, it will approach but never touch a horizontal asymptote as $x \to \infty$.

EXAMPLE 5 Graph the function $y = \dfrac{x^2 - 4}{x^2 - 1}$.

Solution *Symmetry:* Because x appears to even powers only, there is symmetry about the y-axis. There is no symmetry about the origin.

Vertical asymptotes: To find the vertical asymptotes, we set the denominator equal to 0 and solve for x.

$$x^2 - 1 = 0$$

$$x + 1 = 0 \qquad \text{or} \qquad x - 1 = 0$$

$$x = -1 \qquad\qquad\qquad x = 1$$

There will be vertical asymptotes at $x = -1$ and $x = 1$.

y-intercept: We can find the y-intercept by setting x equal to 0 and solving for y:

$$y = \frac{x^2 - 4}{x^2 - 1} = \frac{0^2 - 4}{0^2 - 1} = \frac{-4}{-1} = 4$$

The y-intercept is $(0, 4)$.

x-intercepts: We can find the x-intercepts by setting the numerator equal to 0 and solving for x:

$$x^2 - 4 = 0$$

$$(x + 2)(x - 2) = 0$$

$$x + 2 = 0 \qquad \text{or} \qquad x - 2 = 0$$

$$x = -2 \qquad\qquad\qquad x = 2$$

The x-intercepts are $(2, 0)$ and $(-2, 0)$.

Horizontal asymptotes: Since the degrees of the numerator and denominator polynomials are the same, the line

$$y = \frac{1}{1} \qquad \begin{array}{l}\text{1 is the lead coefficient of the numerator.}\\ \text{1 is the lead coefficient of the denominator.}\end{array}$$

is a horizontal asymptote. The horizontal asymptote is the line $y = 1$. The graph is shown in Figure 4-37. ∎

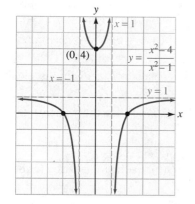

FIGURE 4-37

Self Check Use a graphing calculator to graph the function in Example 5.

Answer

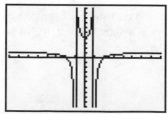

EXAMPLE 6   Graph the function $y = \dfrac{3x}{x-2}$.

Solution   *Symmetry:* Because x appears to an odd power, there is no symmetry about the y-axis. There is no symmetry about the origin either.

Vertical asymptotes: To find the vertical asymptotes, we set the denominator equal to 0 and solve for x. Since the solution is 2, there will be a vertical asymptote at $x = 2$.

y-intercept: We can find the y-intercept by setting x equal to 0 and solving for y. Since the solution is 0, the y-intercept is $(0, 0)$. The graph passes through the origin.

x-intercepts: We can find the x-intercepts by setting the numerator equal to 0 and solving for x:

$$3x = 0$$
$$x = 0$$

The only x-intercept is $(0, 0)$.

Horizontal asymptotes: Since the degrees of the numerator and denominator polynomials are the same, the line

$$y = \frac{3}{1} \qquad \begin{array}{l}\text{3 is the lead coefficient of the numerator.}\\ \text{1 is the lead coefficient of the denominator.}\end{array}$$

is a horizontal asymptote. The horizontal asymptote is the line $y = 3$.

To find what happens when x is greater than 2, we pick a value of x greater than 2 and find the corresponding value of y. If $x = 3$, then $y = 9$. After plotting the point $(3, 9)$, we use the intercepts and asymptotes to sketch the graph shown in Figure 4-38. ∎

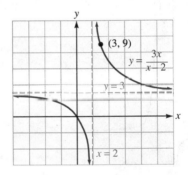

FIGURE 4-38

Self Check   Use a graphing calculator to graph the function in Example 6.
Answer

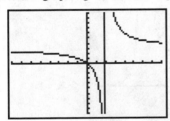

EXAMPLE 7   Graph the function $y = \dfrac{1}{x(x-1)^2}$.

Solution   *Symmetry:* The expansion of $x(x-1)^2$ gives a third-degree polynomial in which x appears to an odd power. So there is no symmetry about the y-axis. Because $y = f(x)$ and $-y = f(-x)$ are not equivalent, there is no symmetry about the origin.

Vertical asymptotes: Since 0 and 1 make the denominator 0, the vertical asymptotes are the lines $x = 0$ and $x = 1$.

y-intercepts: Since x cannot be 0, the graph has no y-intercept.

x-intercepts: Since y cannot be 0, the graph has no x-intercept.

Horizontal asymptotes: As $x \to \infty$, the denominator of the fraction becomes large, and the corresponding values of y approach 0. Thus, the line $y = 0$ is a horizontal asymptote.

    The table in Figure 4-39 gives three points lying between the asymptotes. The values of y change sign at $x = 0$: to the left of the y-axis, $y < 0$, and to the right of the y-axis, $y > 0$. This happens because x appears to an odd power as a factor in the denominator. The value of y does not change sign at $x = 1$: To the left and to the right of the asymptote $x = 1$, y is positive. The function behaves this way because $(x - 1)$ appears to an even power in the denominator. The graph of the function appears in Figure 4-39.

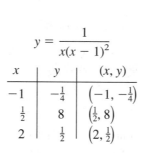

$$y = \dfrac{1}{x(x-1)^2}$$

x	y	(x, y)
-1	$-\frac{1}{4}$	$\left(-1, -\frac{1}{4}\right)$
$\frac{1}{2}$	8	$\left(\frac{1}{2}, 8\right)$
2	$\frac{1}{2}$	$\left(2, \frac{1}{2}\right)$

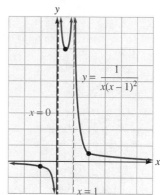

FIGURE 4-39

    ■

Self Check
Answer

Use a graphing calculator to graph the function in Example 7.

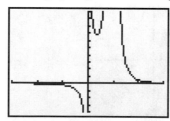

EXAMPLE 8

Graph the function $y = \dfrac{1}{x^2 + 1}$.

Solution

Symmetry: Because x appears only to an even power, the graph will be symmetric about the y-axis.

Vertical asymptotes: Since no numbers x make the denominator 0, the graph has no vertical asymptotes.

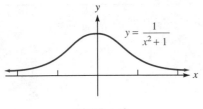

FIGURE 4-40

y-intercepts: Since $f(0) = 1$, the y-intercept of the graph is $(0, 1)$.

x-intercepts: Because the denominator is always positive, the graph lies entirely above the x-axis. There are no x-intercepts.

Horizontal asymptotes: Since $f(x) \to 0$ as $x \to \infty$, the graph has $y = 0$ as a horizontal asymptote.

  The graph appears in Figure 4-40. ∎

Self Check Answer

Use a graphing calculator to graph the function in Example 8.

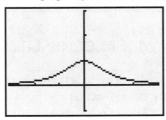

EXAMPLE 9

Graph the function $y = \dfrac{x^2 + x - 2}{x - 3}$.

Solution

We first factor the numerator of the expression

$$y = \frac{(x - 1)(x + 2)}{x - 3}$$

Symmetry: The function is not symmetric about the y-axis or the origin.

Vertical asymptotes: The vertical asymptote is the line $x = 3$.

y-intercepts: The y-intercept is $\left(0, \frac{2}{3}\right)$.

x-intercepts: The x-intercepts are $(1, 0)$ and $(-2, 0)$.

Horizontal asymptotes: We do a long division and write the expression as

$$y = \frac{x^2 + x - 2}{x - 3} = x + 4 + \frac{10}{x - 3}$$

The fraction $\frac{10}{x-3} \to 0$ as $x \to \infty$. This time the graph does not approach a constant. It approaches the line $y = x + 4$. Because this line is not horizontal, the graph has no horizontal asymptote. However, it does have a slant asymptote: the line $y = x + 4$. The graph appears in Figure 4-41.

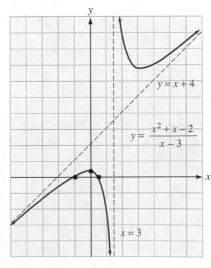

FIGURE 4-41 ∎

Self Check   Use a graphing calculator to graph the function in Example 9.

Answer

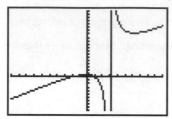

■ GRAPHS WITH DISCONTINUITIES

We have discussed rational functions where the fraction is in simplified form. We now consider a rational function $y = \frac{P(x)}{Q(x)}$ where $P(x)$ and $Q(x)$ have a common factor. Graphs of such functions contain gaps or **discontinuities** that are not the result of vertical asymptotes.

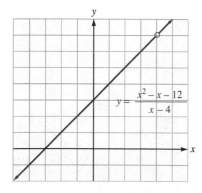

EXAMPLE 10   Find the domain of the function $y = \dfrac{x^2 - x - 12}{x - 4}$ and graph it.

Solution   Since a denominator cannot be 0, $x \neq 4$. The domain is the set of all real numbers except 4.

When we factor the numerator of the expression, we see that the numerator and denominator have a common factor of $x - 4$.

$$y = \frac{x^2 - x - 12}{x - 4}$$

$$= \frac{(x + 3)(x - 4)}{x - 4}$$

If $x \neq 4$, the common factor of $x - 4$ can be divided out. The resulting function is equivalent to the original function only when we keep the restriction that $x \neq 4$. Thus,

$$y = \frac{(x + 3)(x - 4)}{x - 4} = x + 3 \quad \text{(provided that } x \neq 4\text{)}$$

When $x = 4$, the function is not defined. The graph of the function appears in Figure 4-42. It is a line with the point with x-coordinate of 4 missing.   ■

FIGURE 4-42

Self Check   Use a graphing calculator to graph the function in Example 10.

Answer

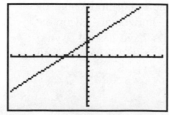

Note that the calculator graph does not indicate the excluded point.

EXERCISE 4.5

VOCABULARY AND CONCEPTS   *In Exercises 1–10, fill in the blank to make a true statement.*

1. When a graph approaches a line but never touches it, we call the line an _____.

2. A rational function is a function with a polynomial numerator and a _____ polynomial denominator.

3. To find a _____ asymptote, set the denominator polynomial equal to 0 and solve the equation.

4. To find the _____ of a rational function, let $x = 0$ and solve for y.

5. To find the _____ of a rational function, set the numerator equal to 0 and solve the equation.

6. In the function $\frac{P(x)}{Q(x)}$, if the degree of $P(x)$ is less than the degree of $Q(x)$, the horizontal asymptote is _____.

7. In the function $y = \frac{P(x)}{Q(x)}$, if the degree of $P(x)$ and $Q(x)$ are _____, the horizontal asymptote is

$$y = \frac{\text{the lead coefficient of the numerator}}{\text{the lead coefficient of the denominator}}$$

8. In a rational function, if the degree of the numerator is 1 greater than the degree of the denominator, the graph will have a _____.

9. A graph can cross a _____ asymptote only when x is small.

10. A graph with discontinuities will have _____ points.

PRACTICE   *In Exercises 11–14, the time, t, it takes to travel 600 miles is a function of the mean rate of speed, r:*

$$t = f(r) = \frac{600}{r}$$

Find t for the given values of r.

11. 30 mph   **12.** 40 mph   **13.** 50 mph   **14.** 60 mph

In Exercises 15–18, suppose the cost (in dollars) of removing p% of the pollution in a river is given by the function

$$c = f(p) = \frac{50{,}000p}{100 - p} \quad (0 \le p \le 100)$$

Find the cost of removing each percent of pollution.

15. 10%   **16.** 30%   **17.** 50%   **18.** 80%

In Exercises 19–24, a service club wants to publish a directory of its members. Some investigation shows that the cost of typesetting and photography will be $700, and the cost of printing each directory will be $1.25.

19. Find a function that gives the total cost c of printing x directories.

20. Find a function that gives the mean cost per directory $\bar{c}$ of printing x directories.

21. Find the total cost of printing 500 directories.

22. Find the mean cost per directory if 500 directories are printed.

23. Find the mean cost per directory if 1000 directories are printed.

24. Find the mean cost per directory if 2000 directories are printed.

In Exercises 25–30, an electric company charges $7.50 per month plus 9¢ for each kilowatt hour (kwh) of electricity used.

25. Find a function that gives the total cost c of n kwh of electricity.

26. Find a function that gives the mean cost per kwh, $\bar{c}$, when using n kwh.

27. Find the total cost for using 775 kwh.

28. Find the mean cost per kwh when 775 kwh are used.

29. Find the mean cost per kwh when 1000 kwh are used.

30. Find the mean cost per kwh when 1200 kwh are used.

In Exercises 31–38, find the domain of each rational function. **Do not graph the function.**

31. $y = \dfrac{x^2}{x - 2}$

32. $y = \dfrac{x^3 - 3x^2 + 1}{x + 3}$

33. $y = \dfrac{2x^2 + 7x - 2}{x^2 - 25}$

34. $y = \dfrac{5x^2 + 1}{x^2 + 5}$

35. $y = \dfrac{x - 1}{x^3 - x}$

36. $y = \dfrac{x + 2}{2x^2 - 9x + 9}$

37. $y = \dfrac{3x^2 + 5}{x^2 + 1}$

38. $y = \dfrac{7x^2 - x + 2}{x^4 + 4}$

In Exercises 39–66, find all vertical, horizontal, and slant asymptotes, x- and y-intercepts, and symmetries, and then graph each function. Check your work with a graphing calculator.

39. $y = \dfrac{1}{x - 2}$

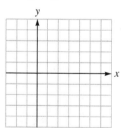

40. $y = \dfrac{3}{x + 3}$

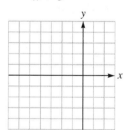

41. $y = \dfrac{x}{x - 1}$

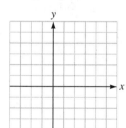

42. $y = \dfrac{x}{x + 2}$

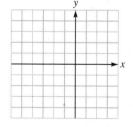

43. $y = \dfrac{x + 1}{x + 2}$

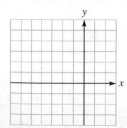

44. $y = \dfrac{x - 1}{x - 2}$

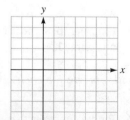

45. $y = \dfrac{2x - 1}{x - 1}$

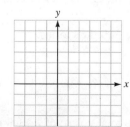

46. $y = \dfrac{3x + 2}{x^2 - 4}$

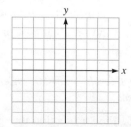

47. $y = \dfrac{x^2 - 9}{x^2 - 4}$

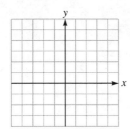

48. $y = \dfrac{x^2 - 4}{x^2 - 9}$

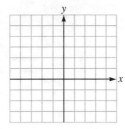

49. $y = \dfrac{x^2 - x - 2}{x^2 - 4x + 3}$

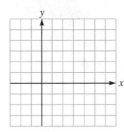

50. $y = \dfrac{x^2 + 7x + 12}{x^2 - 7x + 12}$

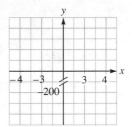

51. $y = \dfrac{x^2 + 2x - 3}{x^3 - 4x}$

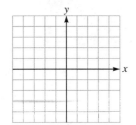

52. $y = \dfrac{3x^2 - 4x + 1}{2x^3 + 3x^2 + x}$

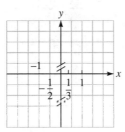

53. $y = \dfrac{x^2 - 9}{x^2}$

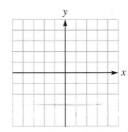

54. $y = \dfrac{3x^2 - 12}{x^2}$

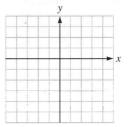

55. $y = \dfrac{x}{(x + 3)^2}$

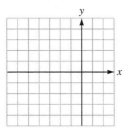

56. $y = \dfrac{x}{(x - 1)^2}$

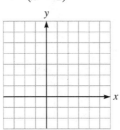

57. $y = \dfrac{x + 1}{x^2(x - 2)}$

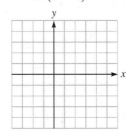

58. $y = \dfrac{x - 1}{x^2(x + 2)^2}$

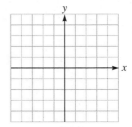

59. $y = \dfrac{x}{x^2 + 1}$

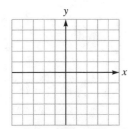

60. $y = \dfrac{x - 1}{x^2 + 2}$

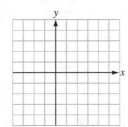

61. $y = \dfrac{3x^2}{x^2 + 1}$

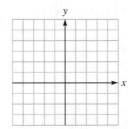

62. $y = \dfrac{x^2 - 9}{2x^2 + 1}$

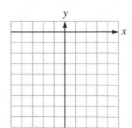

63. $y = \dfrac{x^2 - 2x - 8}{x - 1}$

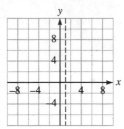

64. $y = \dfrac{x^2 + x - 6}{x + 2}$

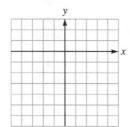

65. $y = \dfrac{x^3 + x^2 + 6x}{x^2 - 1}$

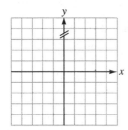

66. $y = \dfrac{x^3 - 2x^2 + x}{x^2 - 4}$

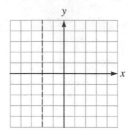

In Exercises 67–74, the numerator and denominator of the fraction share a common factor. Graph each rational function.

67. $y = \dfrac{x^2}{x}$

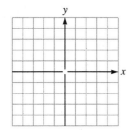

68. $y = \dfrac{x^2 - 1}{x - 1}$

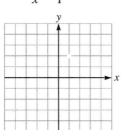

69. $y = \dfrac{x^3 + x}{x}$

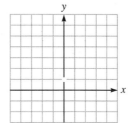

70. $y = \dfrac{x^3 - x^2}{x - 1}$

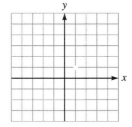

71. $y = \dfrac{x^2 - 2x + 1}{x - 1}$

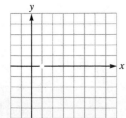

72. $y = \dfrac{2x^2 + 3x - 2}{x + 2}$

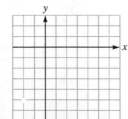

73. $y = \dfrac{x^3 - 1}{x - 1}$

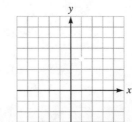

74. $y = \dfrac{x^2 - x}{x^2}$

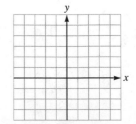

DISCOVERY AND WRITING

75. Can a rational function have two horizontal asymptotes? Explain.

76. Can a rational function have two slant asymptotes? Explain.

In Exercises 77–80, a, b, c, and d are nonzero constants.

77. Show that the graph of

$$y = \frac{ax + b}{cx^2 + d}$$

has the horizontal asymptote $y = 0$.

78. Show that the graph of

$$y = \frac{ax^2 + b}{cx^2 + d}$$

has the horizontal asymptote $y = \frac{a}{c}$.

79. Show that the graph of

$$y = \frac{ax^3 + b}{cx^2 + d}$$

has the slant asymptote $y = \frac{a}{c}x$.

80. Graph the rational function

$$y = \frac{x^3 + 1}{x}$$

and explain why the curve is said to have a *parabolic asymptote*.

In Exercises 81–84, use a graphing calculator to perform each experiment. Write a brief paragraph describing your findings.

81. Investigate the positioning of the vertical asymptotes of a rational function by graphing $y = \frac{x}{x-k}$ for several values of k. What do you observe?

82. Investigate the positioning of the vertical asymptotes of a rational function by graphing $y = \frac{x}{x^2 - k}$ for $k = 4, 1, -1,$ and 0. What do you observe?

83. Find the range of the rational function $y = \frac{kx^2}{x^2 + 1}$ for several values of k. What do you observe?

84. Investigate the positioning of the x-intercepts of a rational function by graphing $y = \frac{x^2 - k}{x}$ for $k = 1, -1,$ and 0. What do you observe?

REVIEW *In Exercises 85–88, do each operation.*

85. $(2x^2 + 3x) + (x^2 - 2x)$

86. $(3x + 2) - (x^2 + 2)$

87. $(5x + 2)(2x + 5)$

88. $\dfrac{2x^2 + 3x + 1}{x + 1}$

89. If $f(x) = 3x + 2$, find $f(x + 1)$.

90. If $f(x) = x^2 + x$, find $f(2x + 1)$.

4.6 Operations on Functions

■ ALGEBRA OF FUNCTIONS ■ COMPOSITION OF FUNCTIONS ■ THE IDENTITY FUNCTION
■ PROBLEM SOLVING

■ ALGEBRA OF FUNCTIONS

With the following definitions, it is possible to perform arithmetic operations on algebraic functions.

Adding, Subtracting, Multiplying, and Dividing Functions

If the ranges of the functions f and g are subsets of the real numbers, then

1. The **sum** of f and g, denoted as $f + g$, is defined by

$$(f + g)(x) = f(x) + g(x)$$

2. The **difference** of f and g, denoted as $f - g$, is defined by

$$(f - g)(x) = f(x) - g(x)$$

3. The **product** of f and g, denoted as $f \cdot g$, is defined by

$$(f \cdot g)(x) = f(x)g(x)$$

4. The **quotient** of f and g, denoted as f/g, is defined by

$$(f/g)(x) = \frac{f(x)}{g(x)} \quad (g(x) \neq 0)$$

The domain of each function, unless otherwise restricted, is the set of real numbers x that are in the domains of both f and g. In the case of the quotient f/g, there is the further restriction that $g(x) \neq 0$.

EXAMPLE 1   Let $f(x) = 3x + 1$ and $g(x) = 2x - 3$. Find each of the following functions and its domain: **a.** $f + g$ and   **b.** $f - g$.

Solution   **a.** $(f + g)(x) = f(x) + g(x)$

$$= (3x + 1) + (2x - 3)$$
$$= 5x - 2$$

Since the domain of both f and g is the set of real numbers, the domain of $f + g$ is the interval $(-\infty, \infty)$.

b. $(f - g)(x) = f(x) - g(x)$

$$= (3x + 1) - (2x - 3)$$
$$= x + 4$$

Since the domain of both f and g is the set of real numbers, the domain of $f - g$ is the interval $(-\infty, \infty)$.   ∎

Self Check   Find $g - f$.
Answer $(g - f)(x) = -x - 4$

EXAMPLE 2   Let $f(x) = 3x + 1$ and $g(x) = 2x - 3$. Find each of the following functions and its domain: **a.** $f \cdot g$ and   **b.** f/g.

Solution   **a.** $(f \cdot g)(x) = f(x) \cdot g(x)$

$$= (3x + 1)(2x - 3)$$
$$= 6x^2 - 7x - 3$$

Since the domain of both f and g is the set of real numbers, the domain of $f \cdot g$ is the interval $(-\infty, \infty)$.

b. $(f/g)(x) = \dfrac{f(x)}{g(x)}$

$= \dfrac{3x + 1}{2x - 3} \quad (2x - 3 \neq 0)$

Since $2x - 3 \neq 0$, the domain of f/g is the set of all real numbers but $\frac{3}{2}$. This is the interval $\left(-\infty, \frac{3}{2}\right) \cup \left(\frac{3}{2}, \infty\right)$. ∎

Self check   Find g/f and its domain.

Answer $(g/f)(x) = \frac{2x-3}{3x+1}, \left(-\infty, \frac{1}{3}\right) \cup \left(\frac{1}{3}, \infty\right)$

EXAMPLE 3   Let $f(x) = x^2 - 4$ and $g(x) = \sqrt{x}$. Find each function and its domain: **a.** $f + g$, **b.** $f \cdot g$,   **c.** f/g,   and   **d.** g/f.

Solution   Because all real numbers can be squared, the domain of f is the interval $(-\infty, \infty)$. Because $\sqrt{x}$ is to be a real number, the domain of g is the interval $[0, \infty)$.

a. $(f + g)(x) = f(x) + g(x)$

$= x^2 - 4 + \sqrt{x}$

The domain consists of the numbers x that are in the domain of both f and g. This is the interval $(-\infty, \infty) \cap [0, \infty)$, which is $[0, \infty)$. The domain of $f + g$ is $[0, \infty)$.

b. $(f \cdot g)(x) = f(x)g(x)$

$= (x^2 - 4)\sqrt{x}$

$= x^2\sqrt{x} - 4\sqrt{x}$

The domain consists of the numbers x that are in the domain of both f and g. The domain of $f \cdot g$ is $[0, \infty)$.

c. $(f/g)(x) = \dfrac{f(x)}{g(x)}$

$= \dfrac{x^2 - 4}{\sqrt{x}}$

The domain consists of the numbers x that are in the domain of both f and g, except 0 (because division by 0 is undefined). The domain of f/g is $(0, \infty)$.

d. $(g/f)(x) = \dfrac{g(x)}{f(x)}$

$= \dfrac{\sqrt{x}}{x^2 - 4}$

The domain consists of the numbers x that are in the domain of both f and g, except 2 (because division by 0 is undefined). The domain of f/g is $[0, 2) \cup (2, \infty)$. ∎

Self Check Find $g - f$ and its domain.

Answer $\sqrt{x} - x^2 + 4,\ [0, \infty)$

EXAMPLE 4 Find $(f + g)(3)$ when $f(x) = x^2 + 1$ and $g(x) = 2x + 1$.

Solution We first find $(f + g)(x)$.

$$(f + g)(x) = f(x) + g(x)$$
$$= x^2 + 1 + 2x + 1$$
$$= x^2 + 2x + 2$$

We then find $(f + g)(3)$.

$$(f + g)(x) = x^2 + 2(x) + 2$$
$$(f + g)(3) = 3^2 + 2(3) + 2 \qquad \text{Substitute 3 for } x.$$
$$= 9 + 6 + 2$$
$$= 17 \qquad\qquad\qquad\qquad \blacksquare$$

Self Check Find $(f \cdot g)(-2)$.

Answer -15

EXAMPLE 5 Let $h(x) = x^2 + 3x + 2$. Find two functions f and g such that **a.** $f + g = h$ and **b.** $f \cdot g = h$.

Solution **a.** There are many possibilities. One is $f(x) = x^2$ and $g(x) = 3x + 2$, for then

$$(f + g)(x) = f(x) + g(x)$$
$$= (x^2) + (3x + 2)$$
$$= x^2 + 3x + 2$$
$$= h(x)$$

Another possibility is $f(x) = x^2 + 2x$ and $g(x) = x + 2$.

b. Again, there are many possibilities. One is suggested by factoring $x^2 + 3x + 2$:

$$x^2 + 3x + 2 = (x + 1)(x + 2)$$

If we let $f(x) = x + 1$ and $g(x) = x + 2$, then

$$(f \cdot g)(x) = f(x) \cdot g(x)$$
$$= (x + 1)(x + 2)$$
$$= x^2 + 3x + 2$$
$$= h(x)$$

Another possibility is $f(x) = 3$ and $g(x) = \dfrac{x^2}{3} + x + \dfrac{2}{3}$. $\blacksquare$

Self Check   Find two functions f and g such that $f - g = h$.

Answer   One possibility is $f(x) = 2x^2$ and $g(x) = x^2 - 3x - 2$.

■ COMPOSITION OF FUNCTIONS

Often one quantity is a function of a second quantity that depends, in turn, on a third quantity. For example, the cost of a car trip is a function of the gasoline consumed. The amount of gasoline consumed, in turn, is a function of the number of miles driven. Such chains of dependence are analyzed mathematically as *composition of functions*.

Suppose that $y = f(x)$ and $y = g(x)$ define two functions. Any number x in the domain of g will produce a corresponding value $g(x)$ in the range of g. If $g(x)$ is in the domain of function f, then $g(x)$ can be substituted into f, and a corresponding value $f(g(x))$ will be determined. This two-step process defines a new function, called a **composite function,** denoted by $f \circ g$. (See Figure 4-43.)

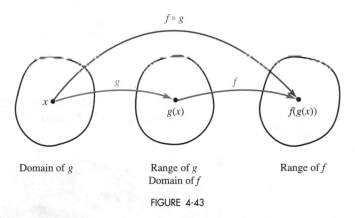

FIGURE 4-43

Composite Functions

The **composite function $f \circ g$** is defined by

$$(f \circ g)(x) = f(g(x))$$

The **domain** of $f \circ g$ consists of all those numbers in the domain of g for which $g(x)$ is in the domain of f.

For example, if $f(x) = 5x + 1$ and $g(x) = 4x - 3$, then

$(f \circ g)(x) = f(g(x))$	$(g \circ f)(x) = g(f(x))$
$= f(4x - 3)$	$= g(5x + 1)$
$= 5(4x - 3) + 1$	$= 4(5x + 1) - 3$
$= 20x - 14$	$= 20x + 1$

WARNING!   Note that in the previous example, $(f \circ g)(x) \neq (g \circ f)(x)$. This shows that the composition of functions is not commutative.

We have seen that a function can be represented by a machine. If we put a number from the domain into the machine (the input), a number from the range comes out (the output). For example, if we put 2 into the machine shown in Figure 4-44(a), the number $f(2) = 5(2) - 2 = 8$ comes out. In general, if we put x into the machine shown in Figure 4-44(b), the value $f(x)$ comes out.

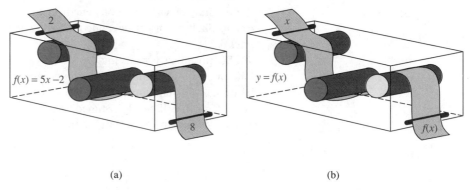

(a)   (b)

FIGURE 4-44

The function machines shown in Figure 4-45 illustrate the composition $f \circ g$. When we put a number x into the function g, $g(x)$ comes out. The value $g(x)$ goes into function f, which transforms $g(x)$ into $f(g(x))$.

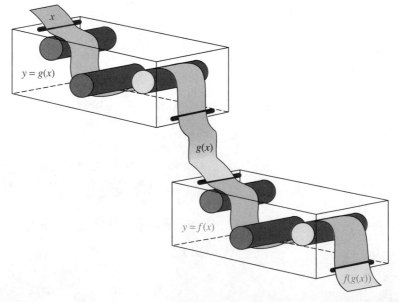

FIGURE 4-45

To be in the domain of the composite function $f \circ g$, a number x has to be in the domain of g, and the output of g must be in the domain of f. Thus, the domain of $f \circ g$ consists of those inputs x that are in the domain of g, and for which $g(x)$ is in the domain of f.

To further illustrate these ideas, suppose we let $f(x) = 2x + 1$ and $g(x) = x - 4$.

- $(f \circ g)(9)$ means $f(g(9))$. In Figure 4-46(a), function g receives the number 9, subtracts 4, and releases the number $g(x) = 5$. The 5 goes into the f function, which doubles it and adds 1. The final result, 11, is the output of the composite function $f \circ g$:

$$(f \circ g)(9) = f(g(9)) = f(5) = 2(5) + 1 = 11$$

- $(f \circ g)(x)$ means $f(g(x))$. In Figure 4-46(a), function g receives the number x, subtracts 4, and releases the number $x - 4$. The $x - 4$ goes into the f function, which doubles it and adds 1. The final result, $2x - 7$, is the output of the composite function $f \circ g$.

$$(f \circ g)(x) = f(g(x)) = f(x - 4) = 2(x - 4) + 1 = 2x - 7$$

- $(g \circ f)(-2)$ means $g(f(-2))$. In Figure 4-46(b), function f receives the number -2, doubles it and adds 1, and releases -3 into the g function. Function g subtracts 4 from -3 and releases a final result of -7. Thus,

$$(g \circ f)(-2) = g(f(-2)) = g(-3) = -3 - 4 = -7$$

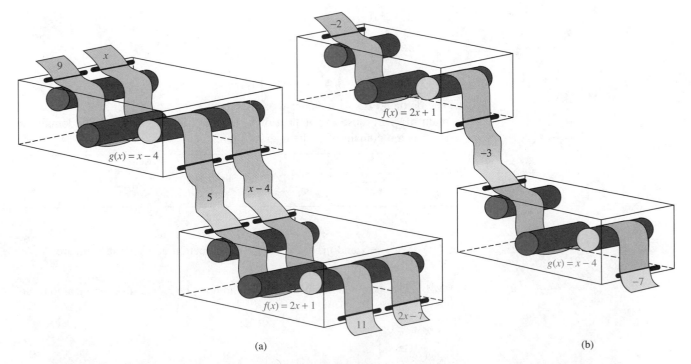

(a)   (b)

FIGURE 4-46

EXAMPLE 6 If $f(x) = 2x + 7$ and $g(x) = 4x + 1$, find **a.** $(f \circ g)(x)$ and **b.** $(g \circ f)(x)$.

Solution **a.** $(f \circ g) = f(g(x)) = f(4x + 1)$
$$= 2(4x + 1) + 7$$
$$= 8x + 9$$

b. $(g \circ f)(x) = g(f(x)) = g(2x + 7)$
$$= 4(2x + 7) + 1$$
$$= 8x + 29 \qquad \blacksquare$$

Self Check   If $h(x) = x + 1$, find $(f \circ h)(x)$.
Answer $2x + 9$

EXAMPLE 7 Let $f(x) = \sqrt{x}$ and $g(x) = x - 3$. Find the domain of each function: **a.** f, **b.** g, **c.** $f \circ g$, and **d.** $g \circ f$.

Solution **a.** For $\sqrt{x}$ to be a real number, the domain of f must be the interval $[0, \infty)$.

b. The domain of g is the interval $(-\infty, \infty)$.

c. Because $g(x)$ must be in the domain of f, the value $g(x)$ must be nonnegative. So we have

$$g(x) \geq 0$$
$$x - 3 \geq 0 \qquad \text{Because } g(x) = x - 3.$$
$$x \geq 3$$

The domain of $f \circ g$ is the interval $[3, \infty)$.

d. The domain of $g \circ f$ contains those numbers x for which $f(x)$ is also in the domain of g. Because the domain of g includes all real numbers, the domain of $g \circ f$ is the domain of f, the interval $[0, \infty)$. $\qquad \blacksquare$

Self Check   Find $f \circ g$.
Answer $(f \circ g)(x) = \sqrt{x - 3}$

EXAMPLE 8 Let $f(x) = \sqrt{1 - x}$ and $g(x) = x^2$. Find each function and give its domain: **a.** $(f \circ g)(x)$ and **b.** $(g \circ f)(x)$.

Solution For $\sqrt{1 - x}$ to be a real number, the radicand must be greater than or equal to 0. So we have

$$1 - x \geq 0$$
$$-x \geq -1$$
$$x \leq 1$$

The domain of f is the interval $(-\infty, 1]$.

Because the square of any real number is a real number, the domain of g is the interval $(-\infty, \infty)$.

a. $(f \circ g)(x) = f(g(x))$

$\qquad\qquad = f(x^2)$

$\qquad\qquad = \sqrt{1 - x^2}$

The domain of $f \circ g$ consists of those numbers x for which $g(x)$ is in the domain of f. Thus, $g(x)$ must be less than or equal to 1.

$\qquad g(x) \leq 1$

$\qquad\quad x^2 \leq 1 \qquad$ Because $g(x) = x^2$.

If we solve the inequality $x^2 - 1 \leq 0$, we will see that the domain of $f \circ g$ is $[-1, 1]$.

b. $(g \circ f)(x) = g(f(x))$

$\qquad\qquad = g\left(\sqrt{1 - x}\right) \quad (x \leq 1)$

$\qquad\qquad = \left(\sqrt{1 - x}\right)^2 \quad (x \leq 1)$

$\qquad\qquad = 1 - x \quad (x \leq 1)$

The domain of $g \circ f$ consists of those numbers x for which $f(x)$ is in the domain of g. Because the domain of g is $(-\infty, \infty)$, the domain of $g \circ f$ is just the domain of f: $(-\infty, 1]$. ∎

Self Check   If $h(x) = x + 2$, find $(f \circ h)(x)$ and its domain.

Answer $(f \circ h)(x) = \sqrt{-x - 1}, (-\infty, -1]$

■ THE IDENTITY FUNCTION

The **identity function** is defined by the equation $I(x) = x$. Under this function, the value that corresponds to any real number x is x itself. If f is any function, the composition of f with the identity function is the function f:

$$(f \circ I)(x) = (I \circ f)(x) = f(x)$$

EXAMPLE 9   Let f be a function and I be the identity function. Show that   **a.** $(f \circ I)(x) = f(x)$ and   **b.** $(I \circ f)(x) = f(x)$.

Solution   **a.** $(f \circ I)(x)$ means $f(I(x))$. Because $I(x) = x$, we have

$$(f \circ I)(x) = f(I(x)) = f(x)$$

b. $(I \circ f)(x)$ means $I(f(x))$. Because I passes any number through unchanged, we have $I(f(x)) = f(x)$ and

$$(I \circ f)(x) = I(f(x)) = f(x)$$ ∎

Self Check   Find   **a.** $(f \circ I)(5)$ and   **b.** $(I \circ f)(5)$.
Answer $f(5), f(5)$

■ PROBLEM SOLVING

EXAMPLE 10   **Temperature change**   A laboratory sample is removed from a cooler at a temperature of 15° F. Technicians warm the sample at the rate of 3° F per hour. Express the sample's temperature in degrees Celsius as a function of the time, t, since it was removed from the cooler.

Solution   The temperature of the sample is 15° F when $t = 0$. Because it warms at 3° F per hour after that, it warms $3t°$ per hour, and the Fahrenheit temperature is given by the function

$$F(t) = 3t + 15$$

The Celsius temperature is a function of the Fahrenheit temperature t, given by the formula

$$C(t) = \frac{5}{9}(t - 32)$$

To express the sample's Celsius temperature as a function of time, we find the composition function

$$(C \circ F)(t) = C(F(t))$$

$$= \frac{5}{9}(F(t) - 32) \qquad \text{Substitute } F(t) \text{ for } F \text{ in } \tfrac{5}{9}(F - 32).$$

$$= \frac{5}{9}[(3t + 15) - 32] \qquad \text{Substitute } 3t + 15 \text{ for } F(t).$$

$$= \frac{5}{9}(3t - 17) \qquad \text{Simplify.}$$

■

EXERCISE 4.6

VOCABULARY AND CONCEPTS   *In Exercises 1–10, fill in the blank to make a true statement.*

1. $(f + g)(x) =$ _____.

2. $(f - g)(x) =$ _____.

3. $(f \cdot g)(x) =$ _____.

4. $(f/g)(x) =$ _____, where $g(x) \neq 0$.

5. The set of real numbers is the interval _____.

6. $(f \circ g)(x) =$ _____.

7. $(g \circ f)(x) =$ _____.

8. $(I \circ g)(x) =$ _____.

9. The function $I(x) = x$ is called the _____ function.

10. $(f \circ I)(x) = (I \circ f)(x) =$ _____.

PRACTICE   *In Exercises 11–14, let $f(x) = 2x + 1$ and $g(x) = 3x - 2$. Find each function and its domain.*

11. $f + g$ **12.** $f - g$ **13.** $f \cdot g$ **14.** f/g

In Exercises 15–18, let $f(x) = x^2 + x$ and $g(x) = x^2 - 1$. Find each function and its domain.

15. $f - g$ **16.** $f + g$ **17.** f/g **18.** $f \cdot g$

In Exercises 19–26, let $f(x) = x^2 - 1$ and $g(x) = 3x - 2$. Find each value, if possible.

19. $(f + g)(2)$ **20.** $(f + g)(-3)$ **21.** $(f - g)(0)$ **22.** $(f - g)(-5)$

23. $(f \cdot g)(2)$ **24.** $(f \cdot g)(-1)$ **25.** $(f/g)\left(\dfrac{2}{3}\right)$ **26.** $(f/g)(t)$

In Exercises 27–34, find two functions f and g such that $h(x)$ can be expressed as the function indicated. Several answers are possible.

27. $h(x) = 3x^2 + 2x;\ f + g$ **28.** $h(x) = 3x^2;\ f \cdot g$

29. $h(x) = \dfrac{3x^2}{x^2 - 1};\ f/g$ **30.** $h(x) = 5x + x^2;\ f - g$

31. $h(x) = x(3x^2 + 1);\ f - g$ **32.** $h(x) = (3x - 2)(3x + 2);\ f + g$

33. $h(x) = x^2 + 7x - 18;\ f \cdot g$ **34.** $h(x) = 5x^5;\ f/g$

In Exercises 35–38, let $f(x) = 2x - 5$ and $g(x) = 5x - 2$. Find each value.

35. $(f \circ g)(2)$ **36.** $(g \circ f)(-3)$ **37.** $(f \circ f)\left(-\dfrac{1}{2}\right)$ **38.** $(g \circ g)\left(\dfrac{3}{5}\right)$

In Exercises 39–42, let $f(x) = 3x^2 - 2$ and $g(x) = 4x + 4$. Find each value.

39. $(f \circ g)(-3)$ **40.** $(g \circ f)(3)$ **41.** $(f \circ f)(\sqrt{3})$ **42.** $(g \circ g)(-4)$

In Exercises 43–46, let $f(x) = 3x$ and $g(x) = x + 1$. Find each composite function and determine its domain.

43. $f \circ g$ **44.** $g \circ f$ **45.** $f \circ f$ **46.** $g \circ g$

In Exercises 47–50, let $f(x) = x^2$ and $g(x) = 2x$. Find each composite function and determine its domain.

47. $g \circ f$ **48.** $f \circ g$ **49.** $g \circ g$ **50.** $f \circ f$

In Exercises 51–54, let $f(x) = \sqrt{x}$ and $g(x) = x^2 + 1$. Find each composite function and determine its domain.

51. $f \circ g$ **52.** $g \circ f$ **53.** $f \circ f$ **54.** $g \circ g$

In Exercises 55–58, let $f(x) = \sqrt{x + 1}$ and $g(x) = x^2 - 1$. Find each composite function and determine its domain.

55. $g \circ f$ **56.** $f \circ g$ **57.** $g \circ g$ **58.** $f \circ f$

In Exercises 59–70, find two functions f and g such that the composition $f \circ g$ expresses the given correspondence. Several answers are possible.

59. $y = 3x - 2$ **60.** $y = 7x - 5$ **61.** $y = x^2 - 2$ **62.** $y = x^3 - 3$

63. $y = (x - 2)^2$ **64.** $y = (x - 3)^3$ **65.** $y = \sqrt{x + 2}$ **66.** $y = \dfrac{1}{x - 5}$

67. $y = \sqrt{x} + 2$ **68.** $y = \dfrac{1}{x} - 5$ **69.** $y = x$ **70.** $y = 3$

DISCOVERY AND WRITING

71. Let $f(x) = 3x$. Show that $(f + f)(x) = f(x + x)$.

72. Let $g(x) = x^2$. Show that $(g + g)(x) \neq g(x + x)$.

73. Let $f(x) = \dfrac{x - 1}{x + 1}$. Find $(f \circ f)(x)$.

74. Let $g(x) = \dfrac{x}{x - 1}$. Find $(g \circ g)(x)$.

In Exercises 75–78, let $y = f(x) = x^2 - x$, $g(x) = x - 3$, and $h(x) = 3x$. Use a graphing calculator to graph both functions on the same axis. Write a brief paragraph summarizing your observations.

75. f and $f \circ g$ **76.** f and $g \circ f$ **77.** f and $f \circ h$ **78.** f and $h \circ f$

REVIEW *In Exercises 79–82, solve each equation for y.*

79. $x = 3y - 7$ **80.** $x = \dfrac{7}{y}$ **81.** $x = \dfrac{y}{y + 3}$ **82.** $x = \dfrac{y - 1}{y}$

4.7 Inverse Functions

■ ONE-TO-ONE FUNCTIONS ■ THE HORIZONTAL LINE TEST ■ FINDING THE INVERSE OF A FUNCTION ■ THE RELATIONSHIP BETWEEN THE GRAPHS OF f AND f^{-1}

The linear function defined by $C = \frac{5}{9}(F - 32)$ gives a formula to convert degrees Fahrenheit to degrees Celsius. If we substitute a Fahrenheit reading into the formula, a Celsius reading comes out. For example, if we substitute $41°$ for F, we obtain a Celsius reading of $5°$:

$$C = \frac{5}{9}(F - 32)$$

$$= \frac{5}{9}(41 - 32)$$

$$= \frac{5}{9}(9)$$

$$= 5$$

If we want to find a Fahrenheit reading from a Celsius reading, we need a formula into which we can substitute a Celsius reading and have a Fahrenheit reading come out. Such a formula is $F = \frac{9}{5}C + 32$, which takes the Celsius reading of 5° and turns it back into a Fahrenheit reading of 41°.

$$F = \frac{9}{5}C + 32$$

$$= \frac{9}{5}(5) + 32$$

$$= 41$$

The functions defined by these two formulas do opposite things. The first turns 41° F into 5° Celsius, and the second turns 5° Celsius back into 41° F. Such functions are called *Inverse functions.*

Before we discuss inverse functions, we need to make some definitions.

■ ONE-TO-ONE FUNCTIONS

Recall that each element x in the domain of a function has a single image y. For some functions, different numbers x in the domain can have the same image. (See Figure 4-47(a).) For other functions, called **one-to-one functions,** different numbers x have different images. (See Figure 4-47(b).)

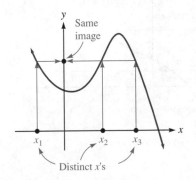

Not a one-to-one function

(a)

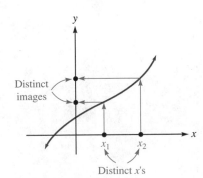

A one-to-one function

(b)

FIGURE 4-47

One-to-One Functions

A function f from a set **X** to a set **Y** is called **one-to-one** if and only if different numbers in the domain of f have different images in the range of f.

The previous definition implies that if x_1 and x_2 are two numbers in the domain of f and $x_1 \neq x_2$, then $f(x_1) \neq f(x_2)$.

EXAMPLE 1 Determine whether the functions   **a.** $f(x) = x^2$ and   **b.** $f(x) = x^3$ are one-to-one.

Solution **a.** The function $f(x) = x^2$ is not one-to-one, because different numbers in the domain have the same image. For example, 2 and -2 have the same image: $f(2) = f(-2) = 4$.

b. The function $f(x) = x^3$ is one-to-one, because different numbers x produce different images $f(x)$. This is because different numbers have different cubes. ∎

Self Check Determine whether $f(x) = \sqrt{x}$ is one-to-one.

Answer yes

■ THE HORIZONTAL LINE TEST

A **horizontal line test** can be used to determine whether the graph of a function represents a one-to-one function. If every horizontal line that intersects the graph of a function does so only once, the function is one-to-one. (See Figure 4-48(a).) If any horizontal line intersects the graph of a function more than once, the function is not one-to-one. (See Figure 4-48(b).)

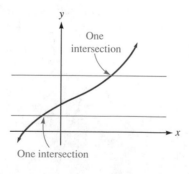

A one-to-one function

(a)

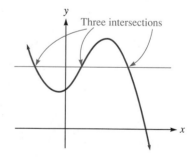

Not a one-to-one function

(b)

FIGURE 4-48

Figure 4-49(a) illustrates a function f from set **X** to set **Y.** Since three arrows point to a single y, the function f is not one-to-one. If the arrows in Figure 4-49(a) were reversed, the diagram would not represent a function, because three values of x in set **X** would correspond to a value y in set **Y.**

If the arrows of the one-to-one function f in Figure 4-49(b) were reversed as in Figure 4-49(c), the diagram would represent a function. This function is called the **inverse of function f** and is denoted by the symbol f^{-1}.

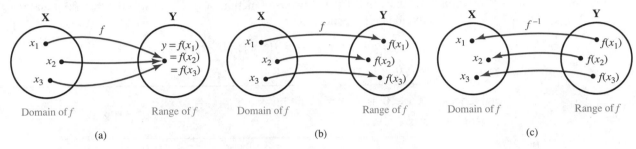

(a) (b) (c)

FIGURE 4-49

Inverse Functions

If f is a one-to-one function with domain **X** and range **Y**, then f^{-1}, called **inverse function of f,** has domain **Y** and range **X.** The function f^{-1} defined by

$$f^{-1}(y) = x \qquad \text{is equivalent to} \qquad f(x) = y$$

for all y in **Y.**

 WARNING! The -1 in the notation for inverse function is not an exponent. Remember that

$$f^{-1}(x) \neq \frac{1}{f(x)}$$

Figure 4-50 shows a one-to-one function f and its inverse f^{-1}. To the number x in the domain of f, there corresponds its image $f(x)$ in the range of f. Since $f(x)$ is also in the domain of f^{-1}, the image of $f(x)$ under the function f^{-1} is $f^{-1}(f(x)) = x$. Thus,

$$(f^{-1} \circ f)(y) = f^{-1}(f(y)) = x$$

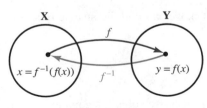

FIGURE 4-50

Furthermore, if y is any number in the domain of f^{-1}, then

$$(f \circ f^{-1})(y) = f(f^{-1}(y)) = y$$

Properties of a One-to-One Function

If f is a one-to-one function with domain **X** and range **Y,** there is a one-to-one function f^{-1} with domain **Y** and range **X** such that

$$(f^{-1} \circ f)(x) = x \qquad \text{and} \qquad (f \circ f^{-1})(y) = y$$

To show that one function is the inverse of another, we must show that their composition is the *identity function*.

EXAMPLE 2   Show that $f(x) = x^3$ is the inverse function of $g(x) = \sqrt[3]{x}$.

Solution   To show that f is the inverse function of g, we must show that $f \circ g$ and $g \circ f$ are the identity function.

$$(f \circ g)(x) = f(g(x)) = f(\sqrt[3]{x}) = (\sqrt[3]{x})^3 = x$$
$$(g \circ f)(x) = g(f(x)) = g(x^3) = \sqrt[3]{x^3} = x$$ ∎

Self Check   If $x \geq 0$, is $f(x) = x^2$ the inverse of $g(x) = \sqrt{x}$?

Answer   yes

■ FINDING THE INVERSE OF A FUNCTION

If f is the one-to-one function $y = f(x)$, then f^{-1} reverses the correspondence of f. That is, if $f(a) = b$, then $f^{-1}(b) = a$. To determine f^{-1}, we follow these steps.

Strategy for Finding f^{-1}

1. Write the function in the form $y = f(x)$.
2. Interchange the positions of variables x and y in the equation $y = f(x)$. The resulting equation, $x = f(y)$, defines the inverse function f^{-1}.
3. Solve the equation $x = f(y)$ for y, if possible. The result is $f^{-1}(x)$.

EXAMPLE 3   Find the inverse of $y = f(x) = \frac{3}{2}x + 2$ and verify the result.

Solution   Since the function is written in the form $y = f(x)$, we can find the inverse function by interchanging the positions of x and y.

$$x = \frac{3}{2}y + 2$$

To write the inverse function in the form $y = f^{-1}(x)$, we solve the equation for y.

$$x = \frac{3}{2}y + 2$$
$$2x = 3y + 4 \qquad \text{Multiply both sides by 2.}$$
$$2x - 4 = 3y \qquad \text{Subtract 4 from both sides.}$$
$$y = \frac{2x - 4}{3} \qquad \text{Divide both sides by 3.}$$

The inverse of $y = f(x) = \dfrac{3}{2}x + 2$ is $y = f^{-1}(x) = \dfrac{2x - 4}{3}$.
To verify that

$$f(x) = \frac{3}{2}x + 2 \qquad \text{and} \qquad f^{-1}(x) = \frac{2x - 4}{3}$$

are inverses, we must show that $f \circ g$ and $g \circ f$ are the identity function.

$$\begin{aligned}
(f \circ f^{-1})(x) &= f(f^{-1}(x)) \\
&= f\left(\frac{2x - 4}{3}\right) \\
&= \frac{3}{2}\left(\frac{2x - 4}{3}\right) + 2 \\
&= x - 2 + 2 \\
&= x
\end{aligned}$$

$$\begin{aligned}
(f^{-1} \circ f)(x) &= f^{-1}(f(x)) \\
&= f^{-1}\left(\frac{3}{2}x + 2\right) \\
&= \frac{2(\frac{3}{2}x + 2) - 4}{3} \\
&= \frac{3x + 4 - 4}{3} \\
&= x
\end{aligned}$$

Self Check Find $f(2)$. Then find $f^{-1}(5)$. Explain the significance of the results.

Answer 5, 2

■ THE RELATIONSHIP BETWEEN THE GRAPHS OF f AND f^{-1}

Because we interchange the positions of x and y to find the inverse of a function, the point (b, a) lies on the graph of $y = f^{-1}(x)$ whenever the point (a, b) lies on the graph of $y = f(x)$. Thus, the graph of a function and its inverse are reflections of each other about the line $y = x$.

EXAMPLE 4 Find the inverse of $y = f(x) = x^3 + 3$. Graph the function and its inverse on the same set of coordinate axes.

Solution We find the inverse of the function $y = f(x) = x^3 + 3$ by interchanging x and y.

$$x = y^3 + 3 \qquad \text{The inverse function.}$$

To write the inverse in the form $f^{-1}(x)$, we solve the equation for y to get

$$x - 3 = y^3$$
$$y = \sqrt[3]{x - 3} \text{ ·}$$

Thus, $f^{-1}(x) = \sqrt[3]{x - 3}$. In the graph that appears in Figure 4-51, the line $y = x$ is the axis of symmetry.

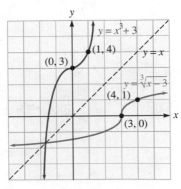

FIGURE 4-51

Self Check Find $f(2)$. Then find $f^{-1}(11)$. Explain the significance of the result.

Answer 11, 2

EXAMPLE 5 The function $y = f(x) = x^2 + 3$ is not one-to-one. However, it becomes one-to-one when we restrict its domain to the interval $(-\infty, 0]$. Under this restriction,
a. Find the range of f. b. Find the inverse of f along with its domain and range.
c. Graph each function.

Solution a. The function f is defined by $y = f(x) = x^2 + 3$, with domain $(-\infty, 0]$. If x is replaced with numbers in this interval, y ranges over the values 3 and above. (See Figure 4-52.) Thus, the range of f is the interval $[3, \infty)$.

b. To find the inverse of f, we interchange x and y in the equation that defines f and solve for y.

$$y = x^2 + 3 \quad (x \le 0)$$
$$x = y^2 + 3 \quad (y \le 0) \qquad \text{Interchange } x \text{ and } y.$$
$$x - 3 = y^2 \quad (y \le 0) \qquad \text{Subtract 3 from both sides.}$$

To solve this equation for y, we take the square root of both sides. Because $y \le 0$, we have

$$-\sqrt{x - 3} = y \quad (y \le 0)$$

The inverse of f is defined by

$$y = f^{-1}(x) = -\sqrt{x - 3}$$

It has domain $[3, \infty)$ and range $(-\infty, 0]$. (See Figure 4-52.)

c. The graphs of the functions appear in Figure 4-52. Note that the line of symmetry is $y = x$. ∎

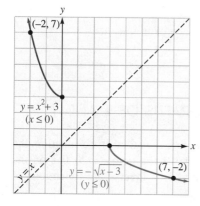

FIGURE 4-52

Self Check Find the inverse of f when its domain is restricted to the interval $[0, \infty)$.

Answer $y = f^{-1}(x) = \sqrt{x - 3}$

If a function is defined by the equation $y = f(x)$, we can often find the domain of f by inspection. Finding the range can be more difficult. One way to find the range of f is to find the domain of f^{-1}.

EXAMPLE 6 Find the domain and range of $y = f(x) = \dfrac{2}{x} + 3$.

Solution Because x cannot be 0, the domain of f is the interval $(-\infty, 0) \cup (0, \infty)$.
To find the range of f, we find the domain of f^{-1} as follows:

$$y = \frac{2}{x} + 3$$

$$x = \frac{2}{y} + 3 \qquad \text{Interchange } x \text{ and } y.$$

$$xy = 2 + 3y \qquad \text{Multiply both sides by } y.$$

$$xy - 3y = 2 \qquad \text{Subtract } 3y \text{ from both sides.}$$

$$y(x - 3) = 2 \qquad \text{Factor out } y.$$

$$y = \frac{2}{x - 3} \qquad \text{Divide both sides by } x - 3.$$

This final equation defines f^{-1} whose domain is the interval $(-\infty, 3) \cup (3, \infty)$. Because the range of f is the domain of f^{-1}, the range of f is the interval $(-\infty, 3) \cup (3, \infty)$. ∎

Self Check   Find the range of $y = f(x) = \frac{3}{x} - 1$.

Answer $(-\infty, -1) \cup (-1, \infty)$

EXERCISE 4.7

VOCABULARY AND CONCEPTS   *In Exercises 1–4, fill in the blank to make a true statement.*

1. If different numbers in the domain of a function have different images, the function is called a _____ function.

2. If every _____ line intersects the graph of a function only once, the function is one-to-one.

3. To find the inverse of $y = f(x)$, we _____ the variables x and y.

4. The graph of a function and its inverse are reflections of each other about the line _____.

PRACTICE   *In Exercises 5–16, determine whether each function is one-to-one.*

5. $y = 3x$

6. $y = \frac{1}{2}x$

7. $y = x^2 + 3$

8. $y = x^4 - x^2$

9. $y = x^3 - x$

10. $y = x^2 - x$

11. $y = |x|$

12. $y = |x - 3|$

13. $y = 5$

14. $y = \sqrt{x - 5}; x \geq 5$

15. $y = (x - 2)^2; x \geq 2$

16. $y = \frac{1}{x}$

In Exercises 17–20, use the horizontal line test to determine whether each graph represents a one-to-one function.

17.

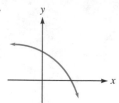

18.

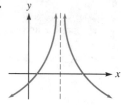

19.

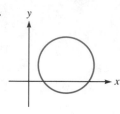

20.

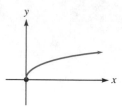

In Exercises 21–24, verify that the functions are inverses by showing that $f \circ g$ and $g \circ f$ are identity functions.

21. $f(x) = 5x$ and $g(x) = \dfrac{1}{5}x$

22. $f(x) = 3x + 2$ and $g(x) = \dfrac{x-2}{3}$

23. $f(x) = \dfrac{x+1}{x}$ and $g(x) = \dfrac{1}{x-1}$

24. $f(x) = \dfrac{x+1}{x-1}$ and $g(x) = \dfrac{x+1}{x-1}$

In Exercises 25–32, each equation defines a one-to-one function f. Determine f^{-1} and verify that $f \circ f^{-1}$ and $f^{-1} \circ f$ are the identity function.

25. $y = 3x$

26. $y = \dfrac{1}{3}x$

27. $y = 3x + 2$

28. $y = 2x - 5$

29. $y = \dfrac{1}{x+3}$

30. $y = \dfrac{1}{x-2}$

31. $y = \dfrac{1}{2x}$

32. $y = \dfrac{1}{x^3}$

In Exercises 33–44, find the inverse of each one-to-one function and graph both the function and its inverse on the same set of coordinate axes.

33. $y = 5x$

34. $y = \dfrac{3}{2}x$

35. $y = 2x - 4$

36. $y = \dfrac{3}{2}x - 2$

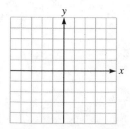

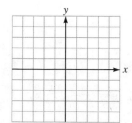

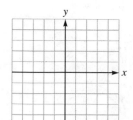

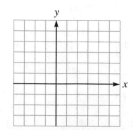

37. $x - y = 2$

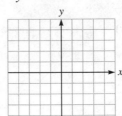

38. $x + y = 0$

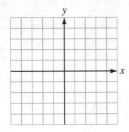

39. $2x + y = 4$

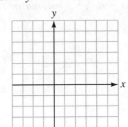

40. $3x + 2y = 6$

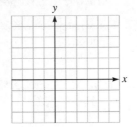

41. $y = \dfrac{1}{2x}$

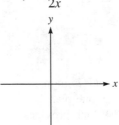

42. $y = \dfrac{1}{x - 3}$

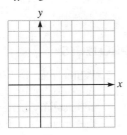

43. $y = \dfrac{x + 1}{x - 1}$

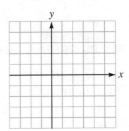

44. $y = \dfrac{x - 1}{x}$

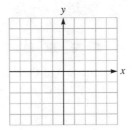

In Exercises 45–50, the function f defined by the given equation is one-to-one on the given domain. Find $f^{-1}(x)$.

45. $f(x) = x^2 - 3 \quad \{x \mid x \le 0\}$

46. $f(x) = \dfrac{1}{x^2} \quad \{x \mid x > 0\}$

47. $f(x) = x^4 - 8 \quad \{x \mid x \ge 0\}$

48. $f(x) = \dfrac{-1}{x^4} \quad \{x \mid x < 0\}$

49. $f(x) = \sqrt{4 - x^2} \quad \{x \mid 0 \le x \le 2\}$

50. $f(x) = \sqrt{x^2 - 1} \quad \{x \mid x \le -1\}$

In Exercises 51–54, find the domain and the range of f. Find the range by finding the domain of f^{-1}.

51. $f(x) = \dfrac{x}{x - 1}$

52. $f(x) = \dfrac{x - 2}{x + 3}$

53. $f(x) = \dfrac{1}{x} - 2$

54. $f(x) = \dfrac{3}{x} - \dfrac{1}{2}$

DISCOVERY AND WRITING

55. Write a brief paragraph to explain why the range of f is the domain of f^{-1}.

56. Write a brief paragraph to explain why the graphs of a function and its inverse are reflections in the line $y = x$.

 *In Exercises 57–58, use a graphing calculator to graph each function for various values of a.*

57. For what values of a is $f(x) = x^3 + ax$ a one-to-one function?

58. For what values of a is $f(x) = x^3 + ax^2$ a one-to-one function?

REVIEW *In Exercises 59–66, simplify each expression.*

59. $16^{3/4}$

60. $25^{-1/2}$

61. $(-8)^{2/3}$

62. $-8^{2/3}$

63. $\left(\dfrac{64}{125}\right)^{-1/3}$

64. $49^{3/2}$

65. $49^{-1/2}$

66. $\left(\dfrac{9}{25}\right)^{-3/2}$

■ ■ ■ ■ ■ ■ ■ ■ ■ **PROBLEMS AND PROJECTS**

1. To encourage their sales staffs, managers of two jewelry stores offer bonuses that are functions of weekly sales, as shown in Illustration 1. For each plan, draw a graph, using the same set of axes. Place the earned bonus on the vertical axis and sales on the horizontal axis.

2. Which plan in Problem 1 do you think is more fair? For any given level of sales, is one plan consistently better than the other? Which provides the greater incentive to sell more? Write brief paragraphs to explain your choices.

Sales	Bonus	
	Golden Gifts	Glitter and Co.
Sales < $3000	1% of sales	zero
$3000 ≤ sales < $10,000	3% of sales	5% of sales over $3000
$10,000 ≤ sales	5% of sales	$350, plus 7% of sales over $10,000

ILLUSTRATION 1

3. For the **modular function** $y = \text{mod}(x)$, define y to be the remainder obtained when the integer x is divided by 5. For example, $\text{mod}(9) = 4$, because when 9 is divided by 5, the remainder is 4. Similarly, $\text{mod}(15) = 0$, and $\text{mod}(3) = 3$.

Graph this function. (*Hint:* Your graph will consist of unconnected points. What is the domain and range of the function?)

4. Look at your graph of the modular function in Problem 3. How would you extend the domain to negative values of x? How does your choice fit with the definition ("remainder when x is divided by 5")?

PROJECT 1 Understanding the behavior of a liquid flowing in a pipe has applications in many systems, including home plumbing, oil pipelines, veins and arteries, and hydraulic control lines in aircraft. The *velocity profile* in Illustration 2 shows that fluid next to the pipe wall is moving slowly, but nearer the center of the pipe, flow becomes more rapid.

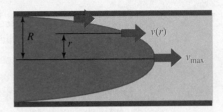

ILLUSTRATION 2

The velocity at a distance r from the pipe's center is given by

$$v(r) = v_{max}\left(1 - \frac{r}{R}\right)^{1/k}$$

where R is the radius of the pipe, v_{max} is the greatest velocity, and k depends on properties of the fluid. For one particular fluid in a 2-inch-diameter pipe, $v_{max} = 8$ and $k = 10$.

a. Graph the function. In the context of this application, what is the domain of the function? (*Hint:* Restrictions on r come from the pipe, not the equation.)

b. Where is the velocity greatest? Does your graph show that this greatest velocity is v_{max}?

c. Graph the function again for $k = 4$. Which graph describes the flow of a thick, syrupy fluid, and which describes a fluid more like water? Why?

PROJECT 2 Illustration 3 shows the geometric basis of the algebraic process of completing the square. Study the figure, where each algebraic expression is related to the area of the corresponding geometric figure.
What is the area of the missing piece that will "complete the square"?

Draw a similar figure that shows geometrically how to complete the square on the expression $x^2 + 8x$.

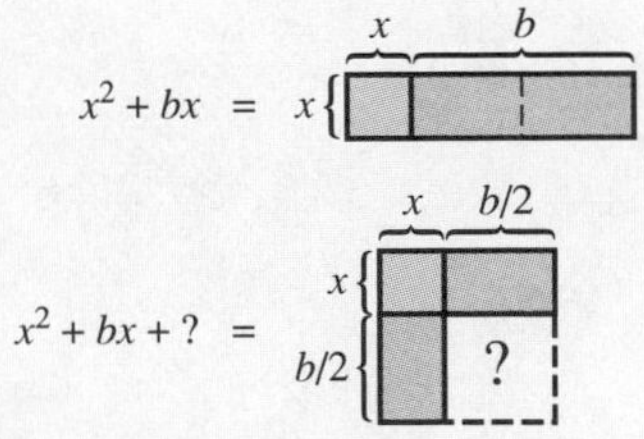

$$x^2 + bx = x\left\{\vphantom{\Big|}\right.$$

$$x^2 + bx + ? =$$

ILLUSTRATION 3

(continued)

PROJECT 3   The surface area of a sphere is a function of the radius, given by

1. $A = 4\pi r^2$

Its volume is also a function of the radius, given by

2. $V = \dfrac{4}{3}\pi r^2$

a. Express the surface area of a sphere as a function of its volume. (*Hint:* Solve Equation 2 for r and substitute into Equation 1.)

b. Use a graphing calculator to graph the function you found in part **a.**

c. Insects don't have lungs; they breathe through their skin. Oxygen requirements increase with an insect's volume, but oxygen intake depends on the insect's surface area. Refer to your graph and write a paragraph explaining why insects must be limited in size.

C H A P T E R   S U M M A R Y

CONCEPT

REVIEW EXERCISES

SECTION 4.1

Functions and Function Notation

A **function** is a correspondence that assigns to each number x in some set **X** a single value y in some set **Y**.

1. Indicate whether the equation defines y as a function of x. Assume that all variables represent real numbers.
 a. $y = 3$ **b.** $y + 5x^2 = 2$ **c.** $y^2 - x = 5$ **d.** $y = |x| + x$

2. Find the domain and the range of each function.

 a. $f(x) = 3x^2 - 5$ **b.** $f(x) = \dfrac{3x}{x - 5}$

 c. $f(x) = \sqrt{x - 1}$ **d.** $f(x) = \sqrt{x^2 + 1}$

3. Find $f(2)$, $f(-3)$, and $f(0)$.

 a. $f(x) = 5x - 2$ **b.** $f(x) = \dfrac{6}{x \quad 5}$

 c. $f(x) = |x - 2|$ **d.** $f(x) = \dfrac{x^2 - 3}{x^2 + 3}$

SECTION 4.2

Quadratic Functions

The graph of $y - k = a(x - h)^2$ $(a \neq 0)$ is a parabola with its vertex at (h, k). It opens upward if $a > 0$ and downward if $a < 0$.

4. Graph each parabola and find its vertex.
 a. $y = x^2 - x$ **b.** $y = x - x^2$

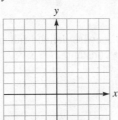

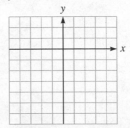

 c. $y = x^2 - 3x - 4$ **d.** $y = 3x^2 - 8x - 3$

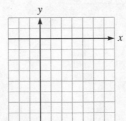

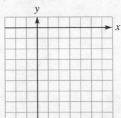

The graph of $y = ax^2 + bx + c$ $(a \neq 0)$ is a parabola with vertex
$$\left(-\frac{b}{2a}, c - \frac{b^2}{4a}\right)$$

5. Architecture A parabolic arch has an equation of $3x^2 + y - 300 = 0$. Find the maximum height of the arch.

6. Puzzle problem The sum of two numbers is 1, and their product is as small as possible. Find the numbers.

| **SECTION 4.3** | *Polynomial and Other Functions* |

If $f(-x) = f(x)$ for all x in the domain, then f is an **even function.**

If $f(-x) = -f(x)$ for all x in the domain, then f is an **odd function.**

7. Graph each polynomial function and tell whether it is even, odd, or neither.

a. $y = x^3 - x$

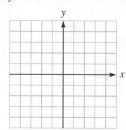

b. $y = x^2 - 4x$

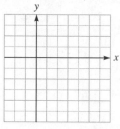

c. $y = x^3 - x^2$

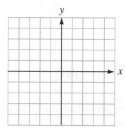

d. $y = 1 - x^4$

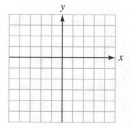

8. Graph each piecewise-defined function and tell when it is increasing, decreasing, or constant.

a. $y = f(x) = \begin{cases} x + 5 & \text{if } x \leq 0 \\ 5 - x & \text{if } x > 0 \end{cases}$

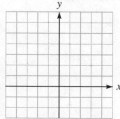

b. $y = f(x) = \begin{cases} x + 3 & \text{if } x \leq 0 \\ 3 & \text{if } x > 0 \end{cases}$

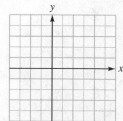

| **SECTION 4.4** | *Translating and Stretching Graphs* |

Vertical translations:

If $k > 0$, the graph of
$\begin{cases} y = f(x) + k \\ y = f(x) - k \end{cases}$

is identical to the graph of
$y = f(x)$, except

it is translated k units
$\begin{cases} \text{upward} \\ \text{downward} \end{cases}$

Horizontal translations:

If $k > 0$, the graph of
$\begin{cases} y = f(x - k) \\ y = f(x + k) \end{cases}$

is identical to the graph of
$y = f(x)$, except

it is translated k units to the
$\begin{cases} \text{right} \\ \text{left} \end{cases}$

Vertical stretchings:

If $k > 0$, the graph of
$y = kf(x)$ can be obtained by
stretching the graph of
$y = f(x)$ vertically by a factor
of k.

If $k < 0$, the graph of
$y = kf(x)$ can be obtained by
stretching the reflection of the
graph of $y = f(x)$ in the x-axis
vertically by a factor of $|k|$.

Horizontal stretchings:

If $k > 0$, the graph of
$y = f(kx)$ can be obtained by
stretching the graph of
$y = f(x)$ horizontally by a
factor of $\frac{1}{k}$.

If $k < 0$, the graph of
$y = f(kx)$ can be obtained by
stretching the reflection of the
graph of $y = f(x)$ in the y-axis
horizontally by a factor of $\left|\frac{1}{k}\right|$.

9. Each function is a translation of a simpler function. Graph both on one set of coordinate axes.

a. $y = \sqrt{x + 2} + 3$

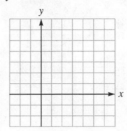

b. $y = |x - 4| + 2$

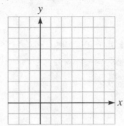

10. Each function is a stretching of a simpler function. Graph both on one set of coordinate axes.

a. $y = \dfrac{1}{3}x^3$

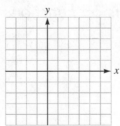

b. $y = (-5x)^3$

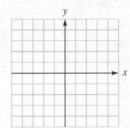

SECTION 4.5 | Rational Functions

To graph the rational function $y = \frac{P(x)}{Q(x)}$:

1. Check symmetries.

2. Look for vertical asymptotes.

3. Look for the y-intercept.

4. Look for x-intercepts.

5. Look for horizontal asymptotes.

6. Look for slant asymptotes.

11. Graph each rational function.

a. $y = \dfrac{x}{(x-1)^2}$

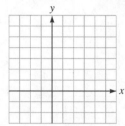

b. $y = \dfrac{(x-1)^2}{x}$

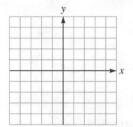

c. $y = \dfrac{x^2 - x - 2}{x^2 + x - 2}$

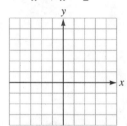

d. $y = \dfrac{x^3 + x}{x^2 - 4}$

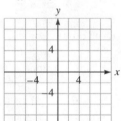

SECTION 4.6 | Operations on Functions

If the ranges of the functions f and g are subsets of the real numbers, then

1. $(f + g)(x) = f(x) + g(x)$

2. $(f - g)(x) = f(x) - g(x)$

3. $(f \cdot g)(x) = f(x) \cdot g(x)$

4. $(f/g)(x) = \dfrac{f(x)}{g(x)}$

The domain of each function, unless otherwise restricted, is the set of real numbers x that are in the domain of both f and g. In the case of the quotient f/g, there is the further restriction that $g(x) \neq 0$.

12. Let $f(x) = x^2 - 1$ and $g(x) = 2x + 1$. Find each function.

a. $f + g$

b. $f \cdot g$

c. $f - g$

d. f/g

Composition of functions:

$(f \circ g)(x) = f(g(x))$

The domain of $f \circ g$ is the set of all x in the domain of g for which $g(x)$ is in the domain of f.

e. $f \circ g$

f. $g \circ f$

Inverse Functions

If f is a one-to-one function with domain **X** and range **Y**, there is a one-to-one function f^{-1} with domain **Y** and range **X** such that

$(f^{-1} \circ f)(x) = x$ and

$(f \circ f^{-1})(y) = y$

The graph of a function is symmetric to the graph of its inverse. The axis of symmetry is the line $y = x$.

13. Each equation defines a one-to-one function. Find f^{-1}.

a. $y = 7x - 1$

b. $y = \dfrac{1}{2 - x}$

c. $y = \dfrac{x}{1 - x}$

d. $y = \dfrac{3}{x^3}$

■ Chapter 4 Test

In Questions 1–2, find the domain and range of each function.

1. $f(x) = \dfrac{3}{x - 5}$

2. $f(x) = \sqrt{x + 3}$

In Questions 3–4, find $f(-1)$ and $f(2)$.

3. $f(x) = \dfrac{x}{x - 1}$

4. $f(x) = \sqrt{x + 7}$

In Questions 5–8, find the vertex of each parabola.

5. $y = 3(x - 7)^2 - 3$

6. $y = x^2 - 2x - 3$

7. $y = 3x^2 - 24x + 38$

8. $y = 5 - 4x - x^2$

In Questions 9–10, graph each polynomial function.

9. $y = x^4 - x^2$

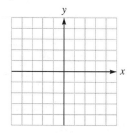

10. $y = x^5 - x^3$

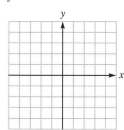

In Questions 11–12, assume that an object tossed vertically upward reaches a height of h feet after t seconds, where $h = 100t - 16t^2$.

11. In how many seconds does the object reach its maximum height?

12. What is that maximum height?

13. Suspension bridge   The cable of a suspension bridge is in the shape of the parabola $x^2 - 2500y + 25,000 = 0$ in the coordinate system shown in Illustration 1. (Distances are in feet.) How far above the roadway is the cable's lowest point?

14. Refer to Question 13. How far above the roadway does the cable attach to the vertical pillars?

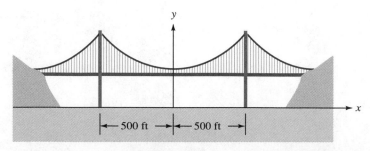

ILLUSTRATION 1

In Questions 15–16, find all asymptotes of the graph of each rational function. Do not graph the function.

15. $y = \dfrac{x - 1}{x^2 - 9}$

16. $y = \dfrac{x^2 - 5x - 14}{x - 3}$

In Questions 17–18, graph each rational function. Check for asymptotes, intercepts, and symmetry.

17. $y = \dfrac{x^2}{x^2 - 9}$

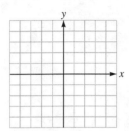

18. $y = \dfrac{x}{x^2 + 1}$

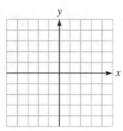

In Questions 19–20, graph each rational function. The numerator and denominator share a common factor.

19. $y = \dfrac{2x^2 - 3x - 2}{x - 2}$

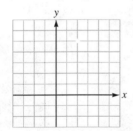

20. $y = \dfrac{x}{x^2 - x}$

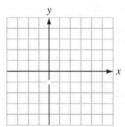

In Questions 21–24, let $f(x) = 3x$ and $g(x) = x^2 + 2$. Find each function.

21. $f + g$

22. $g \circ f$

23. f/g

24. $f \circ g$

In Questions 25–26, $f(x)$ is one-to-one. Find f^{-1}.

25. $f(x) = \dfrac{x + 1}{x - 1}$

26. $f(x) = x^3 - 3$

In Questions 27–28, graph each function.

27. $y = (x - 3)^2 + 1$

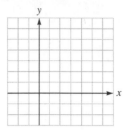

28. $y = \sqrt{x - 1} + 5$

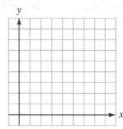

In Questions 29–30, find the range of f by finding the domain of f^{-1}.

29. $y = \dfrac{3}{x} - 2$

30. $y = \dfrac{3x - 1}{x - 3}$

■ Cumulative Review Exercises

In Exercises 1–2, use the x- and y-intercepts to graph each equation.

1. $5x - 3y = 15$

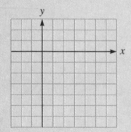

2. $3x + 2y = 12$

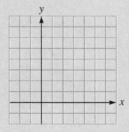

In Exercises 3–4, find the length, the midpoint, and the slope of the line segment PQ.

3. $P\left(-2, \dfrac{7}{2}\right); Q\left(3, -\dfrac{1}{2}\right)$

4. $P(3, 7); Q(-7, 3)$

In Exercises 5–8, write the equation of the line with the given properties. Give the answer in slope-intercept form.

5. The line passes through $(-3, 5)$ and $(3, -7)$.

6. The line passes through $\left(\dfrac{3}{2}, \dfrac{5}{2}\right)$ and has a slope of $\dfrac{7}{2}$.

7. The line is parallel to $3x - 5y = 7$ and passes through $(-5, 3)$.

8. The line is perpendicular to $x - 4y = 12$ and passes through the origin.

In Exercises 9–12, graph each equation. Make use of intercepts and symmetries.

9. $x^2 = y - 2$

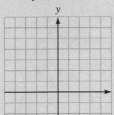

10. $y^2 = x - 2$

11. $x^2 + y^2 = 100$

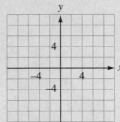

12. $x^2 - 2x + y^2 = 8$

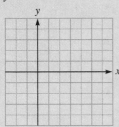

In Exercises 13–14, solve each proportion.

13. $\dfrac{x - 2}{x} = \dfrac{x - 6}{5}$

14. $\dfrac{x + 2}{x - 6} = \dfrac{3x + 1}{2x - 11}$

15. Dental billing The billing schedule for dental X rays specifies a fixed amount for the office visit plus a fixed amount for each X ray exposure. If 2 X rays cost \$37 and 4 cost \$54, find the cost of 5 exposures.

16. Automobile collisions The energy dissipated in an automobile collision varies directly with the square of the speed. By what factor does the energy increase in a 50-mile-per-hour collision compared with a 20-mph collision?

In Exercises 17–20, decide whether each equation defines a function.

17. $y = 3x - 1$

18. $y = x^2 + 3$

19. $y = \dfrac{1}{x - 2}$

20. $y^2 = 4x$

In Exercises 21–24, find the domain and range of each function.

21. $y = f(x) = x^2 + 5$

22. $y = f(x) = \dfrac{7}{x + 2}$

23. $y = f(x) = -\sqrt{x - 2}$

24. $y = f(x) = \sqrt{x + 4}$

In Exercises 25–26, find the vertex of the parabolic graph of each equation.

25. $y = x^2 + 5x - 6$

26. $y = -x^2 + 5x + 6$

In Exercises 27–30, graph the function defined by each equation.

27. $y = x^2 - 4$

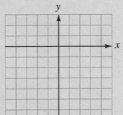

28. $y = -x^2 + 4$

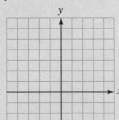

29. $y = x^3 + x$

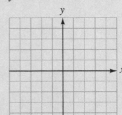

30. $y = -x^4 + 2x^2 + 1$

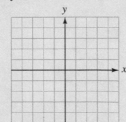

In Exercises 31–32, graph the function defined by each equation. Show all asymptotes.

31. $y = \dfrac{x}{x-3}$

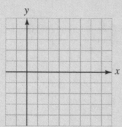

32. $y = \dfrac{x^2-1}{x^2-9}$

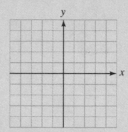

In Exercises 33–36, $f(x) = 3x - 4$ and $g(x) = x^2 + 1$. Find each function and its domain.

33. $(f+g)(x)$ **34.** $(f-g)(x)$ **35.** $(f \cdot g)(x)$ **36.** $(f/g)(x)$

In Exercises 37–40, $f(x) = 3x - 4$ and $g(x) = x^2 + 1$. Find each value.

37. $(f \circ g)(2)$ **38.** $(g \circ f)(2)$ **39.** $(f \circ g)(x)$ **40.** $(g \circ f)(x)$

In Exercises 41–44, find the inverse of the function defined by each equation.

41. $y = 3x + 2$ **42.** $y = \dfrac{1}{x-3}$ **43.** $y = x^2 + 5 \ (x \geq 0)$ **44.** $3x - y = 1$

In Exercises 45–46, write each sentence as an equation.

45. y varies directly with the product of w and z

46. y varies directly with x and inversely with the square of t.

5 Exponential and Logarithmic Functions

IN THIS CHAPTER, WE WILL DISCUSS EXPONENTIAL FUNCTIONS, WHICH ARE OFTEN USED IN BANKING, ECOLOGY, AND SCIENCE. WE WILL ALSO DISCUSS LOGARITHMIC FUNCTIONS, WHICH ARE OFTEN APPLICABLE IN CHEMISTRY, GEOLOGY, AND ENVIRONMENTAL SCIENCE.

5.1   Exponential Functions

■ IRRATIONAL EXPONENTS ■ EXPONENTIAL FUNCTIONS ■ VERTICAL AND HORIZONTAL
TRANSLATIONS ■ RADIOACTIVE DECAY ■ INVESTMENT INTEREST ■ OCEANOGRAPHY

**Sir Isaac Newton
(1642–1727)**

Newton made contributions in
many of the sciences. He is
best known in mathematics for
developing calculus and in
physics for discovering the
laws of motion.

The graph in Figure 5-1 shows the balance in a bank account in which $5000 was
invested in 1990 at 8% compounded monthly. The graph shows that in the year
2005, the value of the account will be approximately $17,000, and in the year 2030,
the value will be approximately $121,000.

The curve in Figure 5-1
is the graph of a function
called an *exponential function*.

Before we can discuss
exponential functions, we
must define irrational exponents.

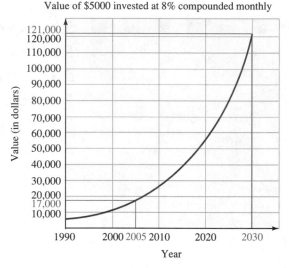

FIGURE 5-1

■   IRRATIONAL EXPONENTS

We have discussed expressions of the form b^x, where x is a rational number.

5^2	means	"the square of 5."
$4^{1/3}$	means	"the cube root of 4."
$6^{-2/5} = \dfrac{1}{6^{2/5}}$	means	"the reciprocal of the fifth root of 6^2."

To give meaning to b^x when x is an irrational exponent, we consider the expression

$3^{\sqrt{2}}$ where $\sqrt{2}$ is the irrational number 1.414213562. . .

Because each exponent is a rational number, each number in the following list is defined.

$$3^{1.4}, \quad 3^{1.41}, \quad 3^{1.414}, \quad 3^{1.4142}, \quad 3^{1.41421}, \ldots$$

Since the exponents are getting closer to $\sqrt{2}$, the numbers in this list are successively better approximations of $3^{\sqrt{2}}$. We can use a calculator to obtain a very good
approximation.

■ ■ ■ ■ ■ ■ ■ ■ ■   **Evaluating Exponential Expressions**

ACCENT ON
TECHNOLOGY

To find the value of $3^{\sqrt{2}}$, we can use a calculator. With a scientific calculator, we enter these numbers and press these keys:

3 $\boxed{y^x}$ 2 $\boxed{\sqrt{}}$ $\boxed{=}$

The display will read $\boxed{4.728804388}$.

With a graphing calculator, we enter these numbers and press these keys:

3 $\boxed{\wedge}$ $\boxed{\sqrt{}}$ 2 $\boxed{\text{Enter}}$

■ ■ ■ ■ ■ ■ ■ ■ ■

The display will read $3\wedge\sqrt{}2$
                                          4.728804388

If b is a positive number and x is a real number, the exponential expression b^x always represents a positive number. It is also true that the familiar properties of exponents hold for irrational exponents.

EXAMPLE 1   Use properties of exponents to simplify   **a.** $\left(3^{\sqrt{2}}\right)^{\sqrt{2}}$ and   **b.** $a^{\sqrt{8}} \cdot a^{\sqrt{2}}$.

Solution   **a.** $\left(3^{\sqrt{2}}\right)^{\sqrt{2}} = 3^{\sqrt{2}\sqrt{2}}$ Keep the base and multiply the exponents.
$= 3^2$ $\sqrt{2}\sqrt{2} = \sqrt{4} = 2$.
$= 9$

b. $a^{\sqrt{8}} \cdot a^{\sqrt{2}} = a^{\sqrt{8}+\sqrt{2}}$ Keep the base and add the exponents.
$= a^{2\sqrt{2}+\sqrt{2}}$ $\sqrt{8} = \sqrt{4}\sqrt{2} = 2\sqrt{2}$.
$= a^{3\sqrt{2}}$ $2\sqrt{2} + \sqrt{2} = 3\sqrt{2}$.   ■

Self Check   Simplify   **a.** $\left(2^{\sqrt{3}}\right)^{\sqrt{12}}$ and   **b.** $x^{\sqrt{20}} \cdot x^{\sqrt{5}}$.
Answers   **a.** 64,   **b.** $x^{3\sqrt{5}}$

■ EXPONENTIAL FUNCTIONS

If $b > 0$ and $b \ne 1$, the function $f(x) = b^x$ is an **exponential function.** Since x can be any real number, its domain is the set of real numbers. Since the base b is always positive, $f(x)$ is always positive, and the range is the set of positive numbers.

Since $b \ne 1$, an exponential function cannot be the constant function $f(x) = 1^x$, in which $f(x) = 1$ for every real number x.

> **Exponential Functions**
> An **exponential function with base b** is defined by the equation
>
> $y = f(x) = b^x$ ($b > 0$, $b \ne 1$, and x is a real number)
>
> The **domain of any exponential function** is the interval $(-\infty, \infty)$. The **range of any exponential function** is the interval $(0, \infty)$.

Since the domain and range of $y = f(x) = b^x$ are sets of real numbers, we can graph exponential functions on a coordinate system. To construct the graph of

$$y = f(x) = 2^x$$

we find several points (x, y) whose coordinates satisfy the equation, plot the points, and join them with a smooth curve, as in Figure 5-2.

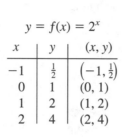

$$y = f(x) = 2^x$$

x	y	(x, y)
-1	$\frac{1}{2}$	$\left(-1, \frac{1}{2}\right)$
0	1	$(0, 1)$
1	2	$(1, 2)$
2	4	$(2, 4)$

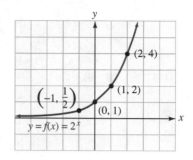

FIGURE 5-2

By looking at the graph, we can see that the domain is the interval $(-\infty, \infty)$ and that the range is the interval $(0, \infty)$.

EXAMPLE 2   Graph $y = f(x) = 4^x$.

Solution   We find several points (x, y) that satisfy the equation, plot the points, and join them with a smooth curve. (See Figure 5-3.)

$$y = f(x) = 4^x$$

x	y	(x, y)
-1	$\frac{1}{4}$	$\left(-1, \frac{1}{4}\right)$
0	1	$(0, 1)$
1	4	$(1, 4)$

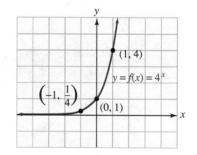

FIGURE 5-3   ∎

Self Check   Graph $y = f(x) = 3^x$.

Answer

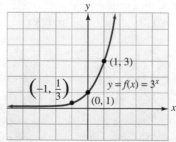

The graph in Example 2 has the following properties:

1. It passes through the point $(0, 1)$.
2. It (the graph of $y = 4^x$) passes through the point $(1, 4)$.
3. It approaches (but never touches) the x-axis. The x-axis is an asymptote.
4. The domain is the interval $(-\infty, \infty)$, and the range is the interval $(0, \infty)$.

EXAMPLE 3   Graph $y = f(x) = \left(\dfrac{1}{2}\right)^x$.

Solution   We find several points (x, y), plot the points, and join them with a smooth curve. (See Figure 5-4.)

$$y = f(x) = \left(\tfrac{1}{2}\right)^x$$

x	y	(x, y)
-2	4	$(-2, 4)$
-1	2	$(-1, 2)$
0	1	$(0, 1)$
1	$\tfrac{1}{2}$	$\left(1, \tfrac{1}{2}\right)$

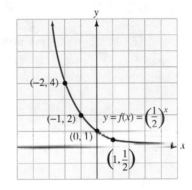

FIGURE 5-4   ■

Self Check   Graph $y = f(x) = \left(\dfrac{1}{3}\right)^x$.

Answer

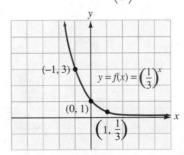

The graph in Example 3 has the following properties:

1. It passes through the point $(0, 1)$.
2. It (the graph of $y = \left(\tfrac{1}{2}\right)^x$) passes through the point $\left(1, \tfrac{1}{2}\right)$.
3. The x-axis is an asymptote of the graph.
4. The domain is the interval $(-\infty, \infty)$, and the range is the interval $(0, \infty)$.

These examples illustrate the following general properties of exponential functions.

> **Properties of Exponential Functions**
> The **domain of the exponential function** $y = f(x) = b^x$ is $(-\infty, \infty)$, the set of real numbers.
> The **range** is $(0, \infty)$, the set of positive real numbers.
> The graph has a y-intercept at $(0, 1)$.
> The x-axis is an asymptote of the graph.
> The graph of $y = \boldsymbol{b}^x$ passes through the point $(1, \boldsymbol{b})$.

EXAMPLE 4   From the graph of $y = f(x) = b^x$, find the value of b.

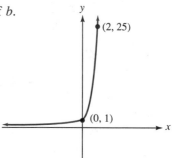

Solution   We substitute 25 for y and 2 for x in the equation $y = b^x$ to get

$$y = b^x$$
$$\mathbf{25} = b^2$$
$$5 = b \qquad b \text{ must be positive. } 5^2 = 25.$$

The base b is 5.   ∎

Self Check   Is this the graph of an exponential function?

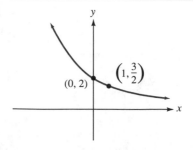

Answer   no; it doesn't pass through $(0, 1)$

In Example 2 (where $b = 4$), the values of y increase as the values of x increase. Since the graph rises as we move to the right, the function is an increasing function.

In Example 3 $\left(\text{where } b = \frac{1}{2}\right)$, the values of y decrease as the values of x increase. Since the graph drops as we move to the right, the function is a decreasing function.

In general, the following is true.

Increasing and Decreasing Functions

If $b > 1$, then $f(x) = b^x$ is an **increasing function**.

If $0 < b < 1$, then $f(x) = b^x$ is a **decreasing function**.

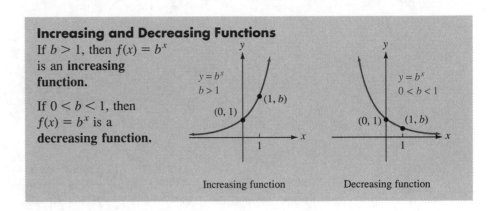

Increasing function   Decreasing function

An exponential function with base b is either increasing (for $b > 1$) or decreasing (for $0 < b < 1$). Because different real numbers x always determine different values b^x, the exponential function is one-to-one.

The exponential function defined by
$$y = f(x) = b^x, \quad \text{where } b > 0 \text{ and } b \neq 1$$
is one-to-one. This implies that

1. If $b^r = b^s$, then $r = s$.
2. If $r \neq s$, then $b^r \neq b^s$.

■ VERTICAL AND HORIZONTAL TRANSLATIONS

We have seen that when $k > 0$, the graph of

1. $y = f(x) + k$ 2. $y = f(x) - k$
3. $y = f(x - k)$ 4. $y = f(x + k)$

is identical to the graph of $y = f(x)$, except that it is translated k units

1. upward   2. downward
3. to the right   4. to the left

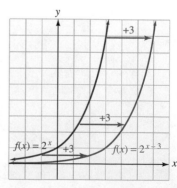

EXAMPLE 5 On one set of axes, graph $y = f(x) = 2^x$ and $y = f(x) = 2^x + 3$.

Solution The graph of $y = f(x) = 2^x + 3$ is identical to the graph of $y = f(x) = 2^x$, except that it is translated 3 units upward. (See Figure 5-5.)

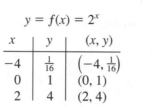

$y = f(x) = 2^x$		
x	y	(x, y)
-4	$\frac{1}{16}$	$\left(-4, \frac{1}{16}\right)$
0	1	$(0, 1)$
2	4	$(2, 4)$

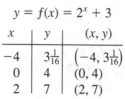

$y = f(x) = 2^x + 3$		
x	y	(x, y)
-4	$3\frac{1}{16}$	$\left(-4, 3\frac{1}{16}\right)$
0	4	$(0, 4)$
2	7	$(2, 7)$

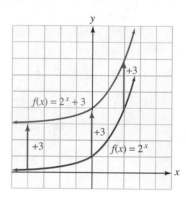

FIGURE 5-5

Self Check Graph $y = f(x) = 3^x$ and $y = f(x) = 3^x - 2$.

Answer

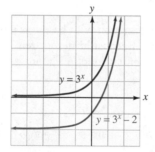

EXAMPLE 6 On one set of axes, graph $y = f(x) = 2^x$ and $y = f(x) = 2^{x-3}$.

Solution The graph of $y = 2^{x-3}$ is identical to the graph of $y = 2^x$, except that it is translated 3 units to the right. (See Figure 5-6.)

$y = 2^x$		
x	y	(x, y)
0	1	$(0, 1)$
2	4	$(2, 4)$
4	16	$(4, 16)$

$y = 2^{x-3}$		
x	y	(x, y)
0	$\frac{1}{8}$	$\left(0, \frac{1}{8}\right)$
2	$\frac{1}{2}$	$\left(2, \frac{1}{2}\right)$
4	2	$(4, 2)$

FIGURE 5-6

Self Check
Answer

On one set of axes, graph $y = f(x) = 3^x$ and $y = f(x) = 3^{x+1}$.

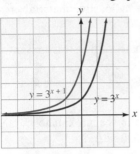

The graphs of $y = f(x) = kb^x$ and $y = f(x) = b^{kx}$ are vertical and horizontal stretchings of the graph of $y = f(x) = b^x$. To graph these functions, we can plot several points and join them with a smooth curve, or use a graphing calculator.

■ ■ ■ ■ ■ ■ ■ ■ ■   **Graphing Exponential Functions**

ACCENT ON
TECHNOLOGY

To use a graphing calculator to graph the exponential function $y = f(x) = 2(3^{x/2})$, we enter the right-hand side of the equation after the symbol $Y_1 =$. The display will show the equation

$$Y_1 = 2(3 \wedge (X/2))$$

If we use window settings of $[-10, 10]$ for x and $[-2, 18]$ for y and press the graph key, we will obtain the graph shown in Figure 5-7.

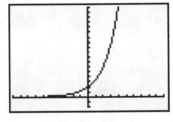

FIGURE 5-7

■ ■ ■ ■ ■ ■ ■ ■ ■

A mathematical description of an observed event is called a **model** of that event. Some events that change with time can be modeled by exponential functions of the form

$$y = f(t) = ab^{kt} \qquad \text{Remember that } ab^{kt} \text{ means } a(b^{kt}).$$

where a, b, and k are constants and t represents time. If f is an increasing function, we say that y *grows exponentially.* If f is a decreasing function, we say that y *decays exponentially.*

■ RADIOACTIVE DECAY

The atomic structure of a radioactive material changes as the material emits radiation. Uranium, for example, changes (decays) into thorium, then into radium, and eventually into lead.

Experiments have determined the time it takes for one-half of a sample of a given radioactive element to decompose. That time is a constant, called the element's **half-life.** The amount present decays exponentially according to the following formula.

Radioactive Decay Formula

The amount A of radioactive material present at time t is given by

$$A = A_0 2^{-t/h}$$

where A_0 is the amount that was present initially (at $t = 0$) and h is the material's half-life.

EXAMPLE 7 The half-life of radium is approximately 1600 years. How much of a 1-gram sample will remain after 1000 years?

Solution In this example, $A_0 = 1$, $h = 1600$, and $t = 1000$. We substitute these values into the formula for radioactive decay and simplify.

$$A = A_0 2^{-t/h}$$
$$A = 1 \cdot 2^{-1000/1600}$$
$$\approx 0.648419777 \qquad \text{Use a calculator.}$$

After 1000 years, approximately 0.65 gram of radium will remain. ∎

Self Check After 800 years, how much radium will remain?
Answer about 0.71 g

■ INVESTMENT INTEREST

Banks pay **interest** for using their customers' money. Interest is calculated as a percent of the amount in the account and can be paid annually, quarterly (four times per year), or daily. Interest left on deposit in a bank account will also earn interest. Such accounts earn **compound interest.**

Compound Interest Formula

If P dollars are deposited in an account earning interest at an annual rate r, compounded k times each year, the amount A in the account after t years is given by

$$A = P\left(1 + \frac{r}{k}\right)^{kt}$$

The value $i = \frac{r}{k}$ is called the **periodic interest rate.**

EXAMPLE 8 The parents of a newborn child deposit $8000 in an investment plan that earns 9% interest, compounded quarterly. If the money is left untouched, how much will the child have in 55 years?

Solution We substitute 8000 for P, 0.09 for r, and 55 for t into the formula for compound interest. Because quarterly means four times per year, $k = 4$.

$$A = P\left(1 + \frac{r}{k}\right)^{kt}$$

$$A = 8000\left(1 + \frac{0.09}{4}\right)^{4 \cdot 55}$$

$$= 8000(1.0225)^{220}$$

$$= 1{,}069{,}103.27 \qquad \text{Use a calculator.}$$

In 55 years, the account will be worth \$1,069,103.27. ∎

Self Check Would \$20,000 invested at 7% interest, compounded monthly, have provided more income at age 55?

Answer no

In financial calculations, the initial amount deposited is often called the **present value,** denoted by PV. The amount to which the account will grow is called the **future value,** denoted by FV. The interest rate for each compounding period is called the **periodic interest rate,** i, and the number of times interest is compounded is the **number of compounding periods,** n. Using these definitions, an alternate formula for compound interest is as follows.

Compound Interest Formula
$$FV = PV(1 + i)^n$$

This alternate formula appears on many business calculators. To use this formula to solve Example 8, we proceed as follows:

$$FV = PV(1 + i)^n$$

$$FV = 8000(1 + 0.0225)^{220} \qquad i = \tfrac{0.09}{4} = 0.0225, \text{ and } n = 4(55) = 220.$$

$$= 8000(1.0225)^{220}$$

$$= 1{,}069{,}103.27 \qquad \text{Use a calculator.}$$

■ OCEANOGRAPHY

EXAMPLE 9 The intensity I of light at a distance x meters below the surface of a body of water decreases exponentially according to the formula

$$I = I_0 k^x$$

where I_0 is the intensity of light above the water and k is a constant that depends on the clarity of the water. For a certain area of the Atlantic Ocean, $I_0 = 12$ and $k = 0.6$. Find the intensity of light at a depth of 5 meters.

Solution   After substituting 12 for I_0 and 0.6 for k, we have the formula

$$I = 12(0.6)^x$$

We substitute 5 for x and calculate I.

$$I = 12(0.6)^5$$
$$I \approx 0.93312$$

At a depth of 5 meters, the intensity of the light is slightly less than 1. ∎

Self Check   Find the intensity at a depth of 10 meters.
Answer   0.073

EXERCISE 5.1

VOCABULARY AND CONCEPTS   *In Exercises 1–10, fill in the blank to make a true statement.*

1. $y = b^x$ ($b > 0$, $b \neq 1$) represents an _____ function.

2. If $y = b^x$ represents an increasing function, then $b > $ __.

3. The *domain* of the exponential function is _____.

4. The number b is called the _____ of the exponential function $y = b^x$.

5. The *range* of the exponential function is _____.

6. The graphs of all exponential functions have the same __-intercept, the point _____.

7. The graph of $y = b^x$ approaches the x-axis, which is called an _____ of the curve.

8. If $y = b^x$ represents a decreasing function, then __ $< b <$ __.

9. If $b > 1$, then $y = b^x$ defines a(n) _____ function.

10. The graph of an exponential function $y = b^x$ always passes through the point $(1, $__$)$.

PRACTICE   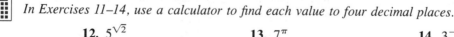   *In Exercises 11–14, use a calculator to find each value to four decimal places.*

11. $4^{\sqrt{3}}$

12. $5^{\sqrt{2}}$

13. 7^{π}

14. $3^{-\pi}$

In Exercises 15–18, simplify each expression.

15. $5^{\sqrt{2}}5^{\sqrt{2}}$

16. $\left(5^{\sqrt{2}}\right)^{\sqrt{2}}$

17. $\left(a^{\sqrt{8}}\right)^{\sqrt{2}}$

18. $a^{\sqrt{12}}a^{\sqrt{3}}$

In Exercises 19–22, tell whether the graph could represent an exponential function.

19.

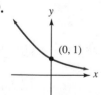

20.

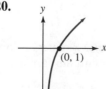

21.

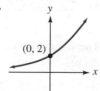

22.

In Exercises 23–30, find the value of b, if any, that would cause the graph of $y = b^x$ to look like the graph indicated.

23.

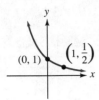

24.

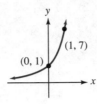

25.

26.

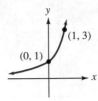

27.

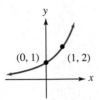

28.

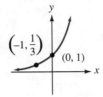

29.

30.

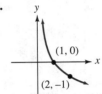

In Exercises 31–38, graph each exponential function.

31. $y = 3^x$

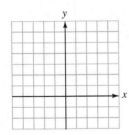

32. $y = 5^x$

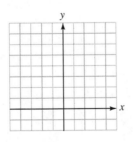

33. $y = \left(\dfrac{1}{5}\right)^x$

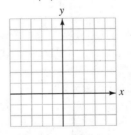

34. $y = \left(\dfrac{1}{3}\right)^x$

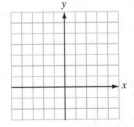

35. $y = -2^x$

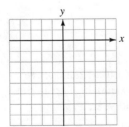

36. $y = -3^x$

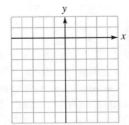

37. $y = \left(\dfrac{3}{4}\right)^x$

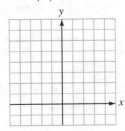

38. $y = \left(\dfrac{4}{3}\right)^x$

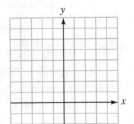

In Exercises 39–54, graph each function. Do not use a graphing calculator.

39. $y = 3^x - 1$

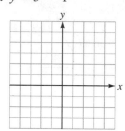

40. $y = 2^x + 3$

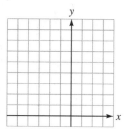

41. $y = 2^x + 1$

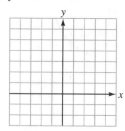

42. $y = 4^x - 4$

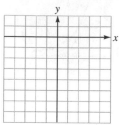

43. $y = 3^{x-1}$

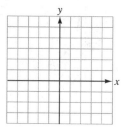

44. $y = 2^{x+3}$

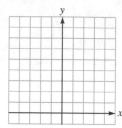

45. $y = 3^{x+1}$

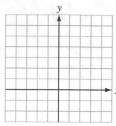

46. $y = 2^{x-3}$

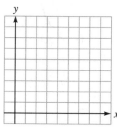

47. $y = 2^{x+1} - 2$

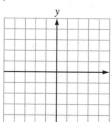

48. $y = 3^{x-1} + 2$

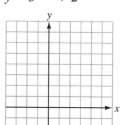

49. $y = 3^{x-2} + 1$

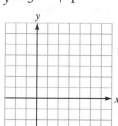

50. $y = 3^{x+2} - 1$

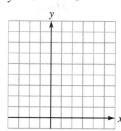

51. $y = 5(2^x)$

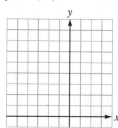

52. $y = 2(5^x)$

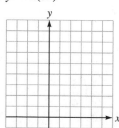

53. $y = 3^{-x}$

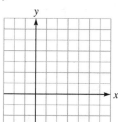

54. $y = 2^{-x}$

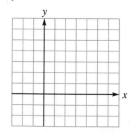

APPLICATIONS 🖩 *Use a calculator to help solve each problem.*

55. Tritium decay Tritium, a radioactive isotope of hydrogen, has a half-life of 12.4 years. Of an initial sample of 50 grams, how much will remain after 100 years?

56. Carbon-14 decay The half-life of radioactive carbon-14 is 5700 years. How much of an initial sample will remain after 3000 years?

57. Plutonium decay   One of the isotopes of plutonium, ^{237}Pu, decays with a half-life of 40 days. How much of an initial sample will remain after 60 days?

58. Comparing radioactive decay   One isotope of holmium, ^{162}Ho, has a half-life of 22 minutes. The half-life of a second isotope, ^{164}Ho, is 37 minutes. Starting with a sample containing equal amounts, find the ratio of the amounts of ^{162}Ho to ^{164}Ho after one hour.

59. Drug absorption in smokers   The biological half-life of the asthma medication theophylline is 4.5 hours for smokers. Find the amount of the drug retained in a smoker's system 12 hours after a dose of 1 unit is taken.

60. Drug absorption in nonsmokers   For a nonsmoker, the biological half-life of theophylline is 8 hours. Find the amount of the drug retained in a nonsmoker's system 12 hours after taking a one-unit dose.

In Exercises 61–62, assume there are no deposits or withdrawals.

61. Compound interest   An initial deposit of $500 earns 8% interest, compounded quarterly. How much will be in the account in 10 years?

62. Compound interest   An initial deposit of $1000 earns 9% interest, compounded monthly. How much will be in the account in $4\frac{1}{2}$ years?

63. Comparing interest rates   How much more interest could $500 earn in 5 years, compounded semiannually (two times a year), if the annual interest rate were $5\frac{1}{2}$% instead of 5%?

64. Comparing savings plans   One bank guarantees to pay interest at 7.25%, compounded monthly. Another bank offers 7.35%, compounded annually. Which bank provides the better investment?

65. 360/365 method   Some financial institutions pay daily interest, compounded by the 360/365 method, by using the formula

$$A = A_0\left(1 + \frac{r}{360}\right)^{365t}$$

where t is in years. Using this method, what will an initial investment of $1000 be worth in 5 years, assuming a 7% annual interest rate?

66. Carrying charge   A college student takes advantage of the ad shown in Illustration 1 and buys a bedroom set for $1100. He plans to pay the $1100 plus interest when his income tax refund comes in 8 months. At that time, what will he need to pay?

ILLUSTRATION 1

67. Credit card interest   A bank credit card charges interest at the rate of 21% per year, compounded monthly. If a college senior charges her last tuition bill of $1500 and intends to pay it in one year, what will she have to pay?

68. Bluegill population   A northern Wisconsin lake is stocked with 10,000 bluegill. The population is expected to grow exponentially according to the model $P = P_0 2^{t/2}$. How many bluegill will be in the lake in 5 years?

69. Community growth   The population of Eagle River is growing exponentially according to the model $P = 375(1.3)^t$, where t is measured in years from the present date. Find the population in 3 years.

70. Oceanography   The intensity I of light (in lumens) at a distance x meters below the surface is given by

$$I = I_0 k^x$$

where I_0 is the intensity at the surface and k depends on the clarity of the water.

At one location in the Arctic Ocean, $I_0 = 8$ and $k = 0.5$. Find the intensity at a depth of 2 meters.

71. Oceanography At one location in the Atlantic Ocean, $I_0 = 12$ and $k = 0.6$. Find the intensity at a depth of 5 meters. (See Exercise 70.)

72. Newton's law of cooling A bucket of water, initially at 100°C, is placed in a room at temperature 40°C. The temperature, T, of the water after t hours is given by $T = 40 + 60(0.75)^t$. Find the temperature in $3\frac{1}{2}$ hours.

DISCOVERY AND WRITING

73. Financial planning To have P available in n years, A can be invested now in an account paying interest at an annual rate r, compounded annually. Show that

$$A = P(1 + r)^{-n}$$

74. a. If $2^{t+4} = k2^t$, find k.
  b. If $5^{3t} = k^t$, find k.

REVIEW   *Factor each expression completely.*

75. $x^2 + 9x^4$ **76.** $x^2 - 9x^4$ **77.** $x^2 + x - 12$ **78.** $x^3 + 27$

5.2   Base-e Exponential Functions

■ THE VALUE OF e ■ GRAPHING THE EXPONENTIAL FUNCTION ■ MALTHUSIAN POPULATION GROWTH ■ BIOLOGY ■ EPIDEMIOLOGY ■ THE MALTHUSIAN THEORY

■ THE VALUE OF e

In mathematical models of natural events, the number $e = 2.71828182845904$. . . appears often as the base of an exponential function. We introduce this important number by recalling the formula for compound interest

$$A = P\left(1 + \frac{r}{k}\right)^{kt}$$

and allowing k, representing the number of compounding periods per year, to become very large. To see what happens, we let $k = rp$, where p is a new variable.

$$A = P\left(1 + \frac{r}{k}\right)^{kt}$$

$$A = P\left(1 + \frac{r}{rp}\right)^{rpt} \qquad \text{Substitute } rp \text{ for } k.$$

$$A = P\left(1 + \frac{1}{p}\right)^{rpt} \qquad \text{Simplify } \frac{r}{rp}.$$

$$A = P\left[\left(1 + \frac{1}{p}\right)^p\right]^{rt} \qquad \text{Remember that } (x^m)^n = x^{mn}.$$

Leonhard Euler
(1707–1783)

Euler first used the letter *i* to represent $\sqrt{-1}$, the letter *e* for the base of natural logarithms, and the symbol Σ for summation. Euler was one of the most prolific mathematicians of all time, contributing to almost all areas of mathematics. Much of his work was accomplished after he became blind.

Because the annual rate *r* is a positive constant and $k = rp$, it follows that as *k* becomes very large, so does *p*. The question of what happens to the value of *A* becomes tied to the question: "What happens to the value of

$$\left(1 + \frac{1}{p}\right)^p$$

as *p* becomes very large?"

Some results calculated for increasing values of *p* appear in Table 5-1.

p	$\left(1 + \dfrac{1}{p}\right)^p$
1	2
10	2.5937425
100	2.7048138
1,000	2.7169239
1,000,000	2.7182805
1,000,000,000	2.7182818

TABLE 5-1

As the results in this table suggest, as *p* increases, the value of

$$\left(1 + \frac{1}{p}\right)^p$$

approaches the number *e*.

If interest earned on an amount P_0 is compounded more and more often, the number *p* grows large without bound, and the formula

$$A = P\left[\left(1 + \frac{1}{p}\right)^p\right]^{rt}$$

becomes

$$A = Pe^{rt} \qquad \text{Substitute } e \text{ for } \left(1 + \frac{1}{p}\right)^p.$$

When the amount invested grows exponentially according to the formula $A = Pe^{rt}$, interest is said to be **compounded continuously.**

Continuous Compound Interest Formula
$$A = Pe^{rt}$$

EXAMPLE 1   If $25,000 accumulates interest at an annual rate of 8% compounded continuously, find the balance in the account in 50 years.

Solution

$$A = Pe^{rt}$$
$$A = \mathbf{25{,}000}e^{(0.08)(50)}$$
$$= 25{,}000e^4$$
$$\approx 1{,}364{,}953.75 \qquad \text{Use a calculator.}$$

In 50 years, the balance will be $1,364,953.75—over a million dollars. ■

Self Check   Find the balance in 60 years.
Answer $3,037,760.44

The exponential function $y = f(x) = e^x$ is so important that it is often called **the exponential function.** The exponential function is often denoted as exp. Thus,

$$\mathbf{exp}(x) = e^x$$

■ GRAPHING THE EXPONENTIAL FUNCTION

To graph the exponential function, we plot several points and join them with a smooth curve (as in Figure 5-8(a)) or use a graphing calculator (as in Figure 5-8(b)).

$y = f(x) = e^x$

x	y	(x, y)
-1	0.37	$(-1, 0.37)$
0	1	$(0, 1)$
1	2.7	$(1, 2.7)$
2	7.4	$(2, 7.4)$

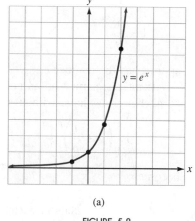

(a)

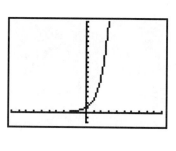

(b)

FIGURE 5-8

■ ■ ■ ■ ■ ■ ■ ■ ■   **Graphing Exponential Functions**

ACCENT ON TECHNOLOGY   To use a graphing calculator to graph the exponential function $y = f(x) = 3e^{-x/2}$, we enter the right-hand side of the equation after the symbol $Y_1 =$. The display will show the equation

$$Y_1 = 3(e \wedge (-x/2))$$

If we use window settings of $[-10, 10]$ for x and $[-10, 10]$ for y and press the GRAPH key, we will obtain the graph shown in Figure 5-9.

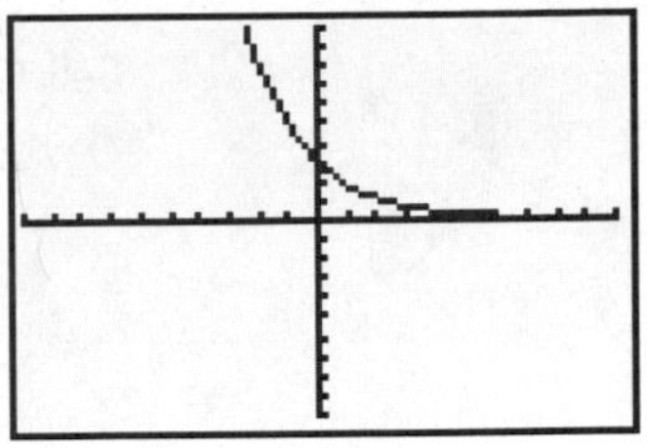

FIGURE 5-9

■ ■ ■ ■ ■ ■ ■ ■ ■

■ MALTHUSIAN POPULATION GROWTH

An equation based on the exponential function provides a model for **population growth.** One such model, called the **Malthusian model of population growth,** assumes a constant birth rate and a constant death rate. In this model, the population P grows exponentially according to the following formula.

Malthusian Model of Population Growth

If b is the annual birth rate, d is the annual death rate, t is the time (in years), P_0 is the initial population at $t = 0$, and P is the current population, then

$$P = P_0 e^{kt}$$

where $k = b - d$ is the **annual growth rate,** the difference between the annual birth rate and death rate.

EXAMPLE 2

The population of the United States is approximately 275 million people. Assuming that the annual birth rate is 19 per 1000 and the annual death rate is 7 per 1000, what does the Malthusian model predict the U.S. population will be in 50 years?

Solution   Since k is the difference between the birth and death rates, we have

$$k = b - d$$

$$k = \frac{19}{1000} - \frac{7}{1000} \qquad \text{Substitute } \tfrac{19}{1000} \text{ for } b \text{ and } \tfrac{7}{1000} \text{ for } d.$$

$$k = 0.019 - 0.007$$

$$= 0.012$$

We can now substitute 275,000,000 for P_0, 50 for t, and 0.012 for k in the formula for the Malthusian model of population growth and simplify.

$$P = P_0 e^{kt}$$

$$P = (275,000,000)e^{(0.012)(50)}$$

$$= (275,000,000)e^{0.6}$$

$$\approx 501082670.1 \qquad\qquad \text{Use a calculator.}$$

After 50 years, the U.S. population will exceed 501 million people.   ■

Self Check In Example 2, find the population in 100 years.

Answer more than 913 million

■ BIOLOGY

EXAMPLE 3 A population of 1000 bacteria doubles every 8 hours. Applying the Malthusian model, find the population in 12 hours.

Solution We can substitute 1000 for P_0, 2000 for P, and 8 for t into the formula for the Malthusian model of population growth to get

$$P = P_0 e^{kt}$$
$$2000 = 1000 e^{k(8)}$$

$$2 = e^{k(8)} \qquad \text{Divide both sides by 1000.}$$

$$2^{1/8} = [e^{k(8)}]^{1/8} \qquad \text{Raise both sides to the } \tfrac{1}{8} \text{ power.}$$

1. $\quad 2^{1/8} = e^{k} \qquad \text{Simplify.}$

We know that the population grows according to the formula

$$P = 1000 e^{kt}$$
$$= 1000(2^{1/8})^t \qquad \text{See Equation 1 and substitute } 2^{1/8} \text{ for } e^k.$$
$$= 1000(2^{t/8})$$

To find the population after 12 hours, we substitute 12 for t and simplify.

$$P = 1000(2^{t/8})$$
$$= 1000(2^{12/8})$$
$$= 1000(2^{3/2})$$
$$\approx 2828.427125 \qquad \text{Use a calculator.}$$

After 12 hours, there will be approximately 2800 bacteria. ■

Self Check How many will there be in 24 hours?

Answer 8000

■ EPIDEMIOLOGY

Many infectious diseases, including some caused by viruses, spread most rapidly when they first infect a population, but then more slowly as the number of uninfected individuals decreases. These situations are often modeled by a function, called a **logistic function,** of the form

$$P = \dfrac{M}{1 + \left(\dfrac{M}{P_0} - 1 \right) e^{-kt}}$$

where P is the size of the infected population at any time t, P_0 is the infected population size at $t = 0$, and k is a constant determined by how contagious the virus is in a given environment. M is the theoretical maximum size of the population P.

EXAMPLE 4   In a city with a population of 1,200,000, there are currently 1000 cases of infection with the HIV virus. If the spread of the disease is projected by the formula

$$P = \frac{1,200,000}{1 + (1200 - 1)e^{-0.4t}}$$

how many people will be infected in 3 years?

Solution   We can substitute 3 for t in the given formula and calculate P.

$$P = \frac{1,200,000}{1 + (1200 - 1)e^{-0.4t}}$$

$$P = \frac{1,200,000}{1 + (1199)e^{-0.4(3)}}$$

$$\approx 3314$$

In 3 years, approximately 3300 people are expected to have the disease.   ■

Self Check   How many will be infected in 10 years?

Answer   about 52,000

■ THE MALTHUSIAN THEORY

The English economist Thomas Robert Malthus (1766–1834) pioneered in population study. He believed that poverty and starvation were unavoidable, because the human population tends to grow exponentially, whereas the food supply tends to grow linearly.

EXAMPLE 5   Suppose that a country with a population of 1000 people is growing exponentially according to the formula

$$P = 1000e^{0.02t}$$

where t is in years. Furthermore, assume that the food supply, measured in adequate food per day per person, is growing linearly according to the formula

$$y = 30.625x + 2000$$

In how many years will the population outstrip the food supply?

Solution   We can use a graphing calculator with window settings of $[0, 100]$ for x and $[0, 10,000]$ for y. After graphing the function, we trace and zoom to find the point where the two graphs intersect. (See Figure 5-10.) From the graph, we can see that the food supply will be adequate for about 71 years. At that time, the population of approximately 4200 people will have problems.

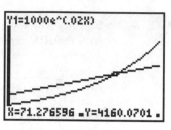

FIGURE 5-10

EXERCISE 5.2

VOCABULARY AND CONCEPTS *In Exercises 1–6, fill in the blank to make a true statement.*

1. To two decimal places, the value of e is _____.

2. The continuous compound interest formula is $A =$ _____.

3. Since $e > 1$, the base-e exponential function is a(n) _____ function.

4. The graph of the exponential function $y = e^x$ passes through the points $(0, 1)$ and _____.

5. The Malthusian model assumes a constant _____ rate and a constant _____ rate.

6. The Malthusian prediction is pessimistic, because a _____ grows exponentially, but food supplies grow _____.

PRACTICE *In Exercises 7–14, graph each function. Check your work with a graphing calculator.*

7. $y = -e^x$

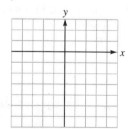

8. $y = e^{-x}$

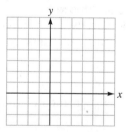

9. $y = e^{-0.5x}$

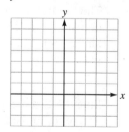

10. $y = -e^{2x}$

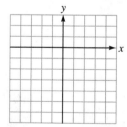

11. $y = 2e^{-x}$

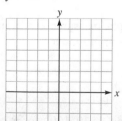

12. $y = -3e^x$

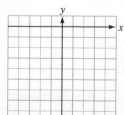

13. $y = e^x + 1$

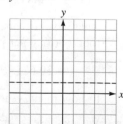

14. $y = e^x - 2$

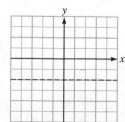

In Exercises 15–22, tell whether the graph of $y = e^x$ could look like the graph indicated.

15.

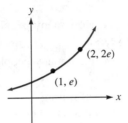

16.

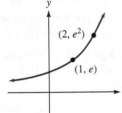

17.

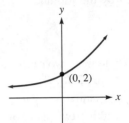

18.

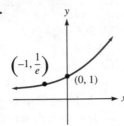

19.

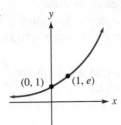

20.

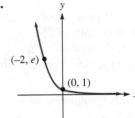

21.

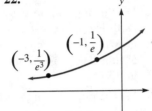

22.

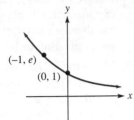

APPLICATIONS *In Exercises 23–28, assume that there are no deposits or withdrawals.*

23. Continuous compound interest An initial investment of $5000 earns 8.2% interest, compounded continuously. What will the investment be worth in 12 years?

24. Continuous compound interest An initial investment of $2000 earns 8% interest, compounded continuously. What will the investment be worth in 15 years?

25. Determining the initial deposit An account now contains $11,180 and has been accumulating interest at a 7% annual rate, compounded continuously, for 7 years. Find the initial deposit.

26. Determining a previous balance An account now contains $3610 and has been accumulating interest at

8% annual interest, compounded continuously. How much was in the account 1 year ago?

27. Comparison of compounding methods An initial deposit of $5000 grows at an annual rate of 8.5%, for 5 years. Compare the final balances resulting from continuous compounding and annual compounding.

28. Comparison of compounding methods An initial deposit of $30,000 grows at an annual rate of 8%, for 20 years. Compare the final balances resulting from continuous compounding and annual compounding.

In Exercises 29–46, solve each problem.

29. Population growth The growth of a town's population is modeled by

$$P = 173e^{0.03t}$$

How large will the population be when $t = 20$?

30. Population decline The decline of a city's population is modeled by

$$P = 1.2 \times 10^6 e^{-0.008t}$$

How large will the population be when $t = 30$?

31. **Epidemics**   The spread of ungulate fever through a herd of cattle can be modeled by the formula

$$P = P_0 e^{0.27t}$$

If a rancher does not act quickly to treat two cases, how many cattle will have the disease in 12 days?

32. **Alcohol absorption**   In one individual, the percent alcohol level t minutes after drinking two shots of whiskey is given by $P = 0.3(1 - e^{-0.05t})$. Find the blood alcohol level after 20 minutes.

33. **World population growth**   The population of the earth is approximately 6 billion people and is growing at an annual rate of 1.9%. Assuming a Malthusian growth model, find the world population in 30 years.

34. **World population growth**   The population of the earth is approximately 6 billion people and is growing at an annual rate of 1.9%. Assuming a Malthusian growth model, find the world population in 40 years.

35. **World population growth**   Assuming a Malthusian growth model and an annual growth rate of 1.9%, by what factor will the current population of the earth increase in 50 years? (See Exercise 33.)

36. **World population growth**   Assuming a Malthusian growth model and an annual growth rate of 1.9%, by what factor will the current population of the earth increase in 100 years? (See Exercise 33.)

37. **Growth of a nation**   The population of a country with 2×10^5 people is expected to double every 20 years. Assuming a Malthusian model, find the population in 35 years.

38. **Town planning**   The population of a small town is now 140 persons and is expected to grow exponentially, tripling every 15 years. Assuming a Malthusian model, what do the city planners project the population to be in 5 years?

39. **Medicine**   The concentration, x, of a certain drug in an organ after t minutes is given by $x = 0.08(1 - e^{-0.1t})$. Find the concentration of the drug after 30 minutes.

40. **Medicine**   Refer to Exercise 39. Find the initial concentration of the drug.

41. **Epidemics**   Refer to Example 4. How many people will have the HIV virus in 5 years?

42. **Epidemics**   Refer to Example 4. How many people will have the HIV virus in 8 years?

43. **Drug absorption**   The amount A of a drug remaining in a person's bloodstream after t hours is given by the formula

$$A = A_0 e^{kt}$$

where A_0 is the initial dose. After 2.3 hours, one-half of an initial dose of triazolam (a drug for treating insomnia) will remain. What percent will remain after 24 hours?

44. **Skydiving**   Before the parachute opens, the velocity v (in meters per second) of a skydiver is given by $v = 50(1 - e^{-0.2t})$. Find the initial velocity.

45. **Skydiving**   Refer to Exercise 44 and find the velocity after 20 seconds.

46. **Free-falling objects**   After t seconds, a certain falling object has a velocity v given by $v = 50(1 - e^{-0.3t})$. Which is falling faster after 2 seconds, this object or the skydiver in Exercise 44?

In Exercises 47–48, use a graphing calculator to solve each problem.

47. **Improved farming**   In Example 5, suppose that better farming methods change the formula for food growth to $y = 31x + 2000$. How long will the food supply be adequate?

48. **Population control**   In Example 5, suppose that a birth control program changed the formula for population growth to $P = 1000e^{0.01t}$. How long will the food supply be adequate?

DISCOVERY AND WRITING

49. The value of e can be calculated to any degree of accuracy by adding the first several terms of the following list:

$$1, 1, \frac{1}{2}, \frac{1}{2 \cdot 3}, \frac{1}{2 \cdot 3 \cdot 4}, \frac{1}{2 \cdot 3 \cdot 4 \cdot 5}, \cdots$$

The more terms that are added, the closer the sum will be to e. Add the first six numbers in the preceding list. To how many decimal places is the sum accurate?

51. Graph the logistic function in Example 4:

$$P = \frac{1{,}200{,}000}{1 + (1199)e^{-0.4t}}$$

Use window settings of $[0, 20]$ for x and $[0, 1{,}500{,}000]$ for y.

53. If $e^{t+3} = ke^t$, find k.

50. Graph the function defined by the equation

$$y = f(x) = \frac{e^x + e^{-x}}{2}$$

from $x = -2$ to $x = 2$. The graph will look like a parabola, but it is not. The graph, called a **catenary,** is important in the design of power distribution networks, because it represents the shape of a uniform flexible cable whose ends are suspended from the same height. The function is called the **hyperbolic cosine function.**

52. Use the trace capabilities of your graphing calculator to explore the logistic function of Example 4 and Exercise 51. As time passes, what value does P approach? How many years does it take for 20% of the population to become infected? for 80%?

54. If $e^{3t} = k^t$, find k.

REVIEW *Find the value of x that makes each statement true.*

55. $2^3 = x$

56. $3^x = 9$

57. $x^3 = 27$

58. $3^{-2} = x$

59. $x^{-3} = \frac{1}{8}$

60. $3^x = \frac{1}{3}$

61. $9^{1/2} = x$

62. $x^{1/3} = 3$

5.3 Logarithmic Functions

■ LOGARITHMS ■ GRAPHS OF LOGARITHMIC FUNCTIONS ■ VERTICAL AND HORIZONTAL TRANSLATIONS ■ BASE-10 LOGARITHMS ■ ELECTRICAL ENGINEERING ■ GEOLOGY

■ LOGARITHMS

Because an exponential function defined by $y = b^x$ is one-to-one, it has an inverse function that is defined by the equation $x = b^y$. To express this inverse function in the form $y = f^{-1}(x)$, we must solve the equation $x = b^y$ for y. For this, we need the following definition.

> **Logarithmic Functions**
> If $b > 0$ and $b \neq 1$, the **logarithmic function with base b** is defined by
>
> $$y = \log_b x \qquad \text{if and only if} \qquad x = b^y$$
>
> The **domain of the logarithmic function** is the interval $(0, \infty)$. The **range of the logarithmic function** is the interval $(-\infty, \infty)$.

Since the function $y = \log_b x$ is the inverse of the one-to-one exponential function $y = b^x$, the logarithmic function is also one-to-one.

WARNING!   Since the domain of the logarithmic function is the set of positive numbers, it is impossible to find the logarithm of 0 or the logarithm of a negative number.

The previous definition guarantees that any pair (x, y) that satisfies the equation $y = \log_b x$ also satisfies the equation $x = b^y$.

$\log_b x = \mathbf{y}$	because	$x = b^y$
$\log_5 25 = \mathbf{2}$	because	$25 = 5^{\mathbf{2}}$
$\log_7 1 = \mathbf{0}$	because	$1 = 7^{\mathbf{0}}$
$\log_{16} 4 = \dfrac{\mathbf{1}}{\mathbf{2}}$	because	$4 = 16^{1/2}$
$\log_2 \dfrac{1}{8} = \mathbf{-3}$	because	$\dfrac{1}{8} = 2^{\mathbf{-3}}$

In each of these examples, the logarithm of a number is an exponent. In fact,

$\log_b x$ **is the exponent to which b is raised to get x.**

To express this as an equation, we write

$$b^{\log_b x} = x$$

EXAMPLE 1   Find y in each equation: **a.** $\log_2 8 = y$,   **b.** $\log_5 1 = y$,   and   **c.** $\log_7 \dfrac{1}{49} = y$.

Solution   **a.** The equation $\log_2 8 = y$ is equivalent to $8 = 2^y$. Since $8 = 2^3$, we have $2^3 = 2^y$ and $y = 3$.

b. The equation $\log_5 1 = y$ is equivalent to $1 = 5^y$. Since $1 = 5^0$, we have $5^0 = 5^y$ and $y = 0$.

c. $\log_7 \frac{1}{49} = y$ is equivalent to $\frac{1}{49} = 7^y$. Since $\frac{1}{49} = 7^{-2}$, we have $7^{-2} = 7^y$ and $y = -2$. ∎

Self Check   Find y in each equation: **a.** $\log_3 9 = y$,   **b.** $\log_2 16 = y$,   and   **c.** $\log_5 \dfrac{1}{25} = y$.

Answers   **a.** 2,   **b.** 4,   **c.** -2

EXAMPLE 2   Find a in each equation: **a.** $\log_a 32 = 5$,   **b.** $\log_9 a = -\dfrac{1}{2}$,   and   **c.** $\log_9 3 = a$.

Solution   **a.** $\log_a 32 = 5$ is equivalent to $a^5 = 32$. Since $2^5 = 32$, we have $a^5 = 2^5$ and $a = 2$.

b. $\log_9 a = -\frac{1}{2}$ is equivalent to $9^{-1/2} = a$. Since $9^{-1/2} = \frac{1}{3}$, it follows that $a = \frac{1}{3}$.

c. $\log_9 3 = a$ is equivalent to $3 = 9^a$. Since $3 = \sqrt{9} = 9^{1/2}$, we have $9^{1/2} = 9^a$ and $a = \frac{1}{2}$. ∎

Self Check   Find d in each equation: **a.** $\log_4 \dfrac{1}{16} = d$,   **b.** $\log_d 36 = 2$,   and   **c.** $\log_8 d = -\dfrac{1}{3}$.

Answers   **a.** -2,   **b.** 6,   **c.** $\frac{1}{2}$

■ GRAPHS OF LOGARITHMIC FUNCTIONS

To graph the logarithmic function $y = f(x) = \log_2 x$, we calculate and plot several points with coordinates (x, y) that satisfy the equation $x = 2^y$. After joining these points with a smooth curve, we have the graph shown in Figure 5-11(a).

To graph $y = f(x) = \log_{1/2} x$, we calculate and plot several points with coordinates (x, y) that satisfy the equation $x = \left(\frac{1}{2}\right)^y$. After joining these points with a smooth curve, we have the graph shown in Figure 5-11(b).

$y = \log_2 x$

x	y	(x, y)
$\frac{1}{4}$	-2	$\left(\frac{1}{4}, -2\right)$
$\frac{1}{2}$	-1	$\left(\frac{1}{2}, -1\right)$
1	0	$(1, 0)$
2	1	$(2, 1)$
4	2	$(4, 2)$
8	3	$(8, 3)$

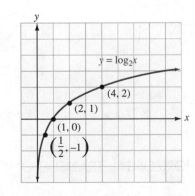

$y = \log_{1/2} x$

x	y	(x, y)
$\frac{1}{4}$	2	$\left(\frac{1}{4}, 2\right)$
$\frac{1}{2}$	1	$\left(\frac{1}{2}, 1\right)$
1	0	$(1, 0)$
2	-1	$(2, -1)$
4	-2	$(4, -2)$
8	-3	$(8, -3)$

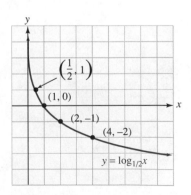

FIGURE 5-11

The graphs of all logarithmic functions are similar to those in Figure 5-11. If $b > 1$, the logarithmic function is increasing, as in Figure 5-12(a). If $0 < b < 1$, the logarithmic function is decreasing, as in Figure 5-12(b).

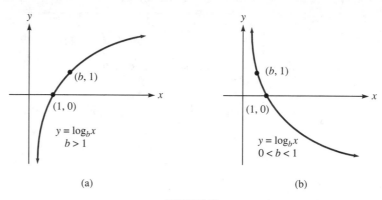

(a)   (b)

FIGURE 5-12

The graph of $y = f(x) = \log_b x$ has the following properties.

1. It passes through the point $(1, 0)$.
2. It passes through the point $(b, 1)$.
3. The y-axis is an asymptote.
4. The domain is $(0, \infty)$, and the range is $(-\infty, \infty)$.

The exponential and logarithmic functions are inverses of each other and therefore have symmetry about the line $y = x$. The graphs of $y = \log_b x$ and $y = b^x$ are shown in Figure 5-13(a) when $b > 1$, and in Figure 5-13(b) when $0 < b < 1$.

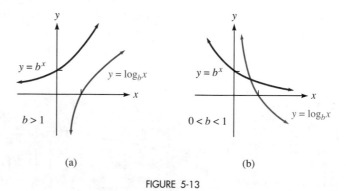

(a)   (b)

FIGURE 5-13

■ VERTICAL AND HORIZONTAL TRANSLATIONS

The graphs of many functions involving logarithms are translations of the basic logarithmic graphs.

| EXAMPLE 3 | Graph the function defined by $y = 3 + \log_2 x$. |

Solution The graph of $y = 3 + \log_2 x$ is identical to the graph of $y = \log_2 x$, except that it is translated 3 units upward. (See Figure 5-14.)

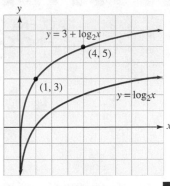

FIGURE 5-14 ■

Self Check Graph $y = \log_3 x - 2$.

Answer

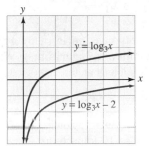

| EXAMPLE 4 | Graph $y = \log_{1/2} (x - 1)$. |

Solution The graph of $y = \log_{1/2} (x - 1)$ is identical to the graph of $y = \log_{1/2} x$, except that it is translated 1 unit to the right. (See Figure 5-15.)

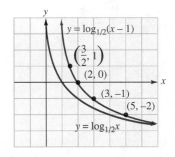

FIGURE 5-15 ■

Self Check Graph $y = \log_{1/3} (x + 2)$.

Answer

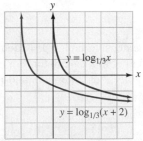

Graphing Logarithmic Functions

Graphing calculators can draw graphs of logarithmic functions if the base of the logarithmic function is 10 or e. To use a calculator to graph the logarithmic function $y = f(x) = -2 + \log_{10}\left(\frac{1}{2}x\right)$, we enter the right-hand side of the equation after the symbol $Y_1 =$. The display will show the equation

$$Y_1 = -2 + \log(1/2 \ast x)$$

If we use window settings of $[-1, 5]$ for x and $[-4, 1]$ for y and press the graph key, we will obtain the graph shown in Figure 5-16.

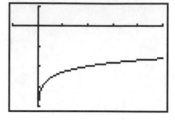

FIGURE 5-16

■ BASE-10 LOGARITHMS

For computational purposes and in many applications, we will use base-10 logarithms (also called **common logarithms**). When the base b is not indicated in the notation $\log x$, we assume that $b = 10$:

log x means   **$\log_{10}$** x

Because base-10 logarithms appear so often, it is a good idea to become familiar with the following base-10 logarithms:

$$\log_{10} \frac{1}{100} = -2 \qquad \text{because} \qquad 10^{-2} = \frac{1}{100}$$

$$\log_{10} \frac{1}{10} = -1 \qquad \text{because} \qquad 10^{-1} = \frac{1}{10}$$

$$\log_{10} 1 = 0 \qquad \text{because} \qquad 10^0 = 1$$

$$\log_{10} 10 = 1 \qquad \text{because} \qquad 10^1 = 10$$

$$\log_{10} 100 = 2 \qquad \text{because} \qquad 10^2 = 100$$

$$\log_{10} 1000 = 3 \qquad \text{because} \qquad 10^3 = 1000$$

In general, we have

$$\log_{10} 10^x = x$$

Before calculators, extensive tables provided logarithms of numbers. Today, logarithms are easy to find with a calculator. For example, to find $\log 2.34$ with a scientific calculator, we enter these numbers and press these keys:

2.34 LOG

The display will read .369215857. To four decimal places, $\log 2.34 = 0.3692$.
    To use a graphing calculator, we enter these numbers and press these keys:

LOG 2.34 ENTER

The display will read log 2.34
.3692158574

EXAMPLE 5 Find x in the equation $\log x = 0.7482$. Round to four decimal places.

Solution The equation $\log x = 0.7482$ is equivalent to $10^{0.7482} = x$. To find x with a scientific calculator, we enter these numbers and press these keys:

10 y^x .7482 =

The display will read 5.600154388. To four decimal places,

$x = 5.6002$

If your calculator has a 10^x key, enter .7482 and press it to get the same result.

■

Self Check Solve $\log x = 1.87737$.
Answer 75.3998

■ ELECTRICAL ENGINEERING

Electronic engineers use common logarithms to measure the voltage gain of devices such as amplifiers or lengths of transmission line. The unit of gain, called the **decibel,** is defined by a logarithmic function:

Decibel Voltage Gain
If E_O is the output voltage of a device and E_I is the input voltage, the **decibel voltage gain** is given by

$$\text{db gain} = 20 \log \frac{E_O}{E_I}$$

EXAMPLE 6 Find the db gain of an amplifier if its input is 0.5 volt and its output is 40 volts.

Solution The decibel voltage gain is found by substituting these values into the formula for db gain:

$$\text{db voltage gain} = 20 \log \frac{E_O}{E_I}$$

$$\text{db voltage gain} = 20 \log \frac{40}{0.5}$$

$$= 20 \log 80$$

$$\approx 38 \qquad \text{Use a calculator.}$$

The db gain is 38 decibels.

■

Self Check   Find the db gain if the input is 0.7 volt.

Answer   about 35 decibels

■ GEOLOGY

Seismologists measure the intensity of earthquakes on the **Richter scale,** which is based on a logarithmic function.

> ### Richter Scale
> If R is the intensity of an earthquake, A is the amplitude (measured in micrometers), and P is the period (the time of one oscillation of the earth's surface, measured in seconds), then
>
> $$R = \log \frac{A}{P}$$

EXAMPLE 7   Find the intensity of an earthquake with amplitude of 5000 micrometers $\left(\frac{1}{2}\text{ centimeter}\right)$ and a period of 0.07 second.

Solution   We substitute 5000 for A and 0.07 for P into the Richter scale formula and simplify:

$$R = \log \frac{A}{P}$$

$$R = \log \frac{5000}{0.07}$$

$$\approx \log 71,428.57143 \qquad \text{Use a calculator.}$$

$$\approx 4.9$$

The earthquake measures 4.9.   ■

Self Check   Find the intensity of an aftershock with the same period but one-half of the amplitude.

Answer   about 4.6

EXERCISE 5.3

VOCABULARY AND CONCEPTS   *In Exercises 1–12, fill in the blank to make a true statement.*

1. The equation $y = \log_b x$ is equivalent to _____.

2. The domain of the logarithmic function is the interval _____.

3. The _____ of the logarithmic function is the interval $(-\infty, \infty)$.

4. $b^{\log_b x} = $ ___.

5. Because the exponential function is one-to-one, it has an _____ function.

6. The inverse of the exponential function is called a _____ function.

7. $\log_b x$ is the _____ to which b is raised to get x.

8. The y-axis is an _____ to the graph of $y = f(x) = \log_b x$.

9. The graph of $y = f(x) = \log_b x$ passes through the points _____ and _____.

10. $\log_{10} 10^x = $ ___.

11. db gain = _____.

12. The intensity of an earthquake is measured by the formula $R = $ _____.

PRACTICE *In Exercises 13–20, write each equation in exponential form.*

13. $\log_3 81 = 4$

14. $\log_7 7 = 1$

15. $\log_{1/2} \dfrac{1}{8} = 3$

16. $\log_{1/5} 1 = 0$

17. $\log_4 \dfrac{1}{64} = -3$

18. $\log_6 \dfrac{1}{36} = -2$

19. $\log_\pi \pi = 1$

20. $\log_7 \dfrac{1}{49} = -2$

In Exercises 21–28, write each equation in logarithmic form.

21. $8^2 = 64$

22. $10^3 = 1000$

23. $4^{-2} = \dfrac{1}{16}$

24. $3^{-4} = \dfrac{1}{81}$

25. $\left(\dfrac{1}{2}\right)^{-5} = 32$

26. $\left(\dfrac{1}{3}\right)^{-3} = 27$

27. $x^y = z$

28. $m^n = p$

In Exercises 29–60, find each value of x.

29. $\log_2 8 = x$

30. $\log_3 9 = x$

31. $\log_4 64 = x$

32. $\log_6 216 = x$

33. $\log_{1/2} \dfrac{1}{8} = x$

34. $\log_{1/3} \dfrac{1}{81} = x$

35. $\log_9 3 = x$

36. $\log_{125} 5 = x$

37. $\log_{1/2} 8 = x$

38. $\log_{1/2} 16 = x$

39. $\log_8 x = 2$

40. $\log_7 x = 0$

41. $\log_7 x = 1$

42. $\log_2 x = 8$

43. $\log_{25} x = \dfrac{1}{2}$

44. $\log_4 x = \dfrac{1}{2}$

45. $\log_5 x = -2$

46. $\log_3 x = -4$

47. $\log_{36} x = -\dfrac{1}{2}$

48. $\log_{27} x = -\dfrac{1}{3}$

49. $\log_x 5^3 = 3$

50. $\log_x 5 = 1$

51. $\log_x \dfrac{9}{4} = 2$

52. $\log_x \dfrac{\sqrt{3}}{3} = \dfrac{1}{2}$

53. $\log_x \dfrac{1}{64} = -3$

54. $\log_x \dfrac{1}{100} = -2$

55. $\log_x \dfrac{9}{4} = -2$

56. $\log_x \dfrac{\sqrt{3}}{3} = \dfrac{-1}{2}$

57. $2^{\log_2 5} = x$

58. $3^{\log_3 4} = x$

59. $x^{\log_4 6} = 6$

60. $x^{\log_3 8} = 8$

In Exercises 61–64, find the value of b, if any, that would cause the graph of $y = \log_b x$ to look like the graph indicated.

61.

62.

63.

64.
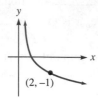

In Exercises 65–76, graph each function.

65. $y = \log_3 x$
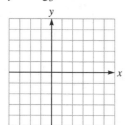

66. $y = \log_4 x$
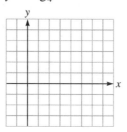

67. $y = \log_{1/3} x$
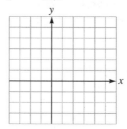

68. $y = \log_{1/4} x$
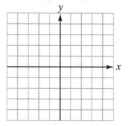

69. $y = 2 + \log_2 x$
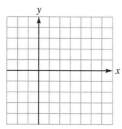

70. $y = \log_2 (x - 1)$
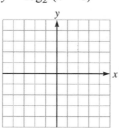

71. $y = \log_3 (x + 2)$
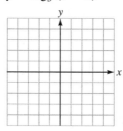

72. $y = -3 + \log_3 x$
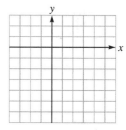

73. $y = \log_3 (3x)$
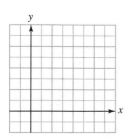

74. $y = \log_3 \left(\dfrac{x}{3} \right)$
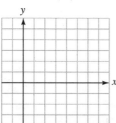

75. $y = \log_2 (-x)$
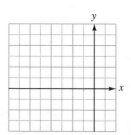

76. $y = -\log_2 x$
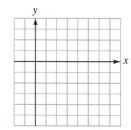

In Exercises 77–80, use a calculator to find each value to four decimal places.

77. $\log 3.25$

78. $\log 0.57$

79. $\log 0.00467$

80. $\log 375.876$

In Exercises 81–82, use a calculator to find y to four decimal places.

81. $\log y = 1.4023$

82. $\log y = 0.926$

83. $\log y = -3.71$

84. $\log y = \log \pi$

APPLICATIONS 📟 *Use a calculator to help solve each problem.*

85. **Gain of an amplifier**   An amplifier produces an output of 17 volts when the input signal is 0.03 volt. Find the decibel voltage gain.

86. **Transmission lines**   A 4.9 volt input to a long transmission line decreases to 4.7 volts at the other end. Find the decibel voltage loss.

87. **Gain of an amplifier**   Find the db gain of an amplifier whose input voltage is 0.71 volt and whose output voltage is 20 volts.

88. **Gain of a amplifier**   Find the db gain of an amplifier whose output voltage is 2.8 volts and whose input voltage is 0.05 volt.

89. **Earthquakes**   An earthquake has amplitude of 5000 micrometers and a period of 0.2 second. Find its measure on the Richter scale.

90. **Earthquakes**   An earthquake has an amplitude of 8000 micrometers and a period of 0.008 second. Find its measure on the Richter scale.

91. **Earthquakes**   An earthquake with a period of $\frac{1}{4}$ second has an amplitude of 2500 μm. Find its measure on the Richter scale.

92. **Earthquakes**   An earthquake has a period of $\frac{1}{2}$ second and an amplitude of 50,000 μm. Find its measure on the Richter scale.

93. **Depreciation**   In business, equipment is often depreciated using the double declining-balance method. In this method, a piece of equipment with a life expectancy of N years, costing $\$C$, will depreciate to a value of $\$V$ in n years, where n is given by the formula

$$n = \frac{\log V - \log C}{\log \left(1 - \dfrac{2}{N}\right)}$$

A computer that cost $\$37,000$ has a life expectancy of 5 years. If it has depreciated to a value of $\$8000$, how old is it?

94. **Depreciation**   A typewriter worth $\$470$ when new had a life expectancy of 12 years. If it is now worth $\$189$, how old is it? (See Exercise 93.)

95. **Time for money to grow twentyfold**   If $\$P$ are invested at the end of each year in an annuity earning interest at an annual rate r, then the amount in the account will be $\$A$ after n years, where

$$n = \frac{\log \left[\dfrac{Ar}{P} + 1\right]}{\log (1 + r)}$$

If $\$1000$ is invested each year in an annuity earning 12% annual interest, when will the account be worth $\$20,000$?

96. **Time for money to grow tenfold**   If $\$5000$ is invested each year in an annuity earning 8% annual interest, when will the account be worth $\$50,000$? (See Exercise 95.)

DISCOVERY AND WRITING

97. Graphs of $y = \log_a x$ and $y = \log_b x$ are shown in Illustration 1. Which is larger, a or b, and why?

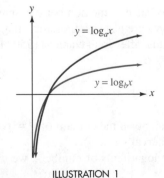

ILLUSTRATION 1

98. Graphs of $y = \log_a x$ and $y = \log_b x$ are shown in Illustration 2. Which is larger, a or b, and why?

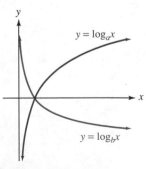

ILLUSTRATION 2

99. Pick two numbers, and add their logarithms. Then find the logarithm of the product of those two numbers. What do you observe? Does it work for three numbers?

100. If $\log_a b = 7$, find $\log_b a$.

REVIEW *In Exercises 101–102, find the vertex of each parabola.*

101. $y = x^2 + 7x + 3$

102. $y = 3x^2 - 8x - 1$

103. Fencing a pasture A farmer will use 3400 feet of fencing to enclose and divide the pasture in Illustration 3. What dimensions will enclose the greatest area?

104. Selling appliances When the price is p dollars, an appliance dealer can sell $(2200 - p)$ refrigerators. What price will maximize revenue?

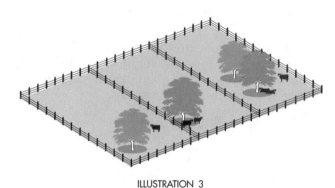

ILLUSTRATION 3

5.4 Base-e Logarithms

■ BASE-*e* LOGARITHMS ■ GRAPHING BASE-*e* LOGARITHMS ■ CHARGING BATTERIES ■ POPULATION GROWTH ■ ISOTHERMAL EXPANSION

■ BASE-*e* LOGARITHMS

We have seen the importance of the number e in mathematical models of events in nature. Base-*e* logarithms are just as important. They are called **natural logarithms** or **Napierian logarithms,** after John Napier (1550–1617). They are usually written as ln x, rather than $\log_e x$:

ln x   means $\log_e x$

Like all logarithmic functions, the domain of $y = f(x) = \ln x$ is the interval $(0, \infty)$, and the range is the interval $(-\infty, \infty)$.

To find the base-*e* logarithms of numbers, we can use a calculator.

■ ■ ■ ■ ■ ■ ■ ■ ■

ACCENT ON
TECHNOLOGY

To use a scientific calculator to find the value of ln 2.34, we enter these numbers and press these keys:

2.34 `LN`

The display will read `.850150929`. To four decimal places, ln 2.34 = 0.8502.
To use a graphing calculator, we enter these numbers and press these keys:

`LN` 2.34 `ENTER`

The display will read `ln 2.34`
                                  `.8501509294`

■ ■ ■ ■ ■ ■ ■ ■ ■

EXAMPLE 1

Use a calculator to find each value: **a.** ln 17.32 and **b.** ln (log 0.05).

Solution   **a.** Enter these numbers and press these keys:

Scientific calculator   *Graphing calculator*
17.32 `LN`   `LN` 17.32 `ENTER`

Either way, the result is 2.851861903.

b. Enter these numbers and press these keys:

Scientific calculator   *Graphing calculator*
0.05 `LOG` `LN`   `LN` `(` `LOG` 0.05 `)` `ENTER`

Either way, we obtain an error, because log 0.05 is a negative number, and we cannot take the logarithm of a negative number. ■

Self Check   Find each value to four decimal places: **a.** ln π and **b.** ln $\left(\log \frac{1}{3}\right)$.
Answers   **a.** 1.1447, **b.** no value

EXAMPLE 2

Solve each equation: **a.** ln x = 1.335 and **b.** ln x = log 5.5. Give each result to four decimal places.

Solution   **a.** The equation ln x = 1.335 is equivalent to $e^{1.335} = x$. To use a scientific calculator to find x, enter these numbers and press these keys:

1.335 e^x

The display will read 3.799995946. To four decimal places,

$x = 3.8000$

b. The equation ln x = log 5.5 is equivalent to $e^{\log 5.5} = x$. To use a scientific calculator to find x, enter these numbers and press these keys:

5.5 `LOG` e^x

The display will read 2.096695826. To four decimal places,

$x = 2.0967$ ■

Self Check   Solve   **a.** $\ln x = 1.9344$ and   **b.** $\log x = \ln 3.2$. Give each result to four decimal places.

Answers   **a.** 6.9199,   **b.** 14.5596

■ GRAPHING BASE-e LOGARITHMS

The equation $y = \ln x$ is equivalent to the equation $x = e^y$. To get the graph of $\ln x$, we can plot points that satisfy the equation $x = e^y$ and join them with a smooth curve, as shown in Figure 5-17(a). In Figure 5-17(b), we show a calculator graph.

$y = \ln x$

x	y	(x, y)
$\frac{1}{e} \approx 0.4$	-1	$(0.4, -1)$
1	0	$(1, 0)$
$e \approx 2.7$	1	$(2.7, 1)$
$e^2 \approx 7.4$	2	$(7.4, 2)$

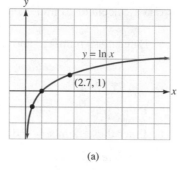

(a)

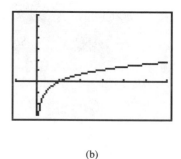

(b)

FIGURE 5-17

Many graphs of logarithmic functions involve translations of the graph of $y = f(x) = \ln x$. For example, Figure 5-18 shows a calculator graph of the functions $y = \ln x$, $y = \ln x + 2$, and $y = \ln x - 3$.

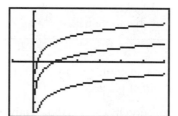

The graph of $y = \ln x + 2$ is 2 units above the graph of $y = \ln x$.

The graph of $y = \ln x - 3$ is 3 units below the graph of $y = \ln x$.

FIGURE 5-18

Figure 5-19 shows the calculator graph of the functions $y = \ln x$, $y = \ln (x - 2)$, and $y = \ln (x + 3)$.

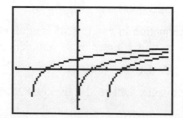

The graph of $y = \ln (x - 2)$ is 2 units to the right of the graph of $y = \ln x$.

The graph of $y = \ln (x + 3)$ is 3 units to the left of the graph of $y = \ln x$.

FIGURE 5-19

John Napier
(1550–1617)

Napier is famous for his work with natural logarithms. In fact, natural logarithms are often called Napierian logarithms. He also invented a device called Napier's rods, which did multiplications mechanically. This was a forerunner of modern-day computers.

Base-*e* logarithms have many applications; some are explored in the subsections that follow.

■ CHARGING BATTERIES

A battery charges at a rate that depends on how close it is to being fully charged—it charges fastest when it is most discharged. The charge C at any instant is modeled by the following formula.

> **Charging Batteries**
> If M is the theoretical maximum charge that a battery can hold and k is a positive constant that depends on the battery and the charger, then
> $$C = M(1 - e^{-kt})$$

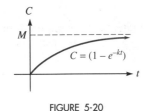

FIGURE 5-20

Plotting the variable C against the variable t gives a curve like that shown in Figure 5-20. The graph shows that a full charge, M, is never attained. However, the actual charge can come very close to M, provided the battery is charged long enough.

To find how long it will take a battery to reach a given charge C, we solve the equation $C = M(1 - e^{-kt})$ for t.

$$C = M(1 - e^{-kt})$$

$$\frac{C}{M} = 1 - e^{-kt} \qquad \text{Divide both sides by } M.$$

$$\frac{C}{M} - 1 = -e^{-kt} \qquad \text{Subtract 1 from both sides.}$$

$$1 - \frac{C}{M} = e^{-kt} \qquad \text{Multiply both sides by } -1.$$

$$\ln\left(1 - \frac{C}{M}\right) = -kt \qquad \begin{array}{l}\text{Change the exponential equation to}\\ \text{logarithmic form.}\end{array}$$

$$-\frac{1}{k}\ln\left(1 - \frac{C}{M}\right) = t \qquad \text{Multiply both sides by } -\tfrac{1}{k}.$$

The formula that determines the time t required to charge a battery to a given level C is

1. $\quad t = -\dfrac{1}{k}\ln\left(1 - \dfrac{C}{M}\right)$

EXAMPLE 3 How long will it take to bring a battery to 90% of full charge? Assume that $k = 0.025$ and that time is measured in minutes.

Solution 90% of full charge means 90% of M. From the previous discussion, we know that we can substitute $0.90M$ for C and 0.025 for k into Equation 1.

$$t = -\frac{1}{k} \ln\left(1 - \frac{C}{M}\right)$$

$$t = -\frac{1}{0.025} \ln\left(1 - \frac{0.9\,M}{M}\right)$$

$$= -40 \ln(1 - 0.9)$$

$$= -40 \ln(0.1)$$

$$\approx 92 \qquad\qquad\qquad \text{Use a calculator.}$$

The battery will reach 90% charge in about 92 minutes.   ■

Self Check   How long will it take the battery in Example 3 to reach 80% of full charge?

Answer   about 64 min

■ POPULATION GROWTH

If a population grows exponentially at a certain annual rate, the time required for the population to double is called the **doubling time** and is given by the following formula.

Population Doubling Time
If r is the annual rate and t is the time required for a population to double, then

$$t = \frac{\ln 2}{r}$$

EXAMPLE 4   The population of the earth is growing at approximately 2% per year. If this rate continues, when will the population double?

Solution   Because the population is growing at the rate of 2% per year, we substitute 0.02 for r into the formula for doubling time and simplify.

$$t = \frac{\ln 2}{r}$$

$$t = \frac{\ln 2}{0.02}$$

$$\approx 34.65735903$$

The population will double in about 35 years.   ■

Self Check   If the world population's annual growth rate were reduced to 1.5% per year, what would be the doubling time?

Answer   46 years

■ ISOTHERMAL EXPANSION

When energy is added to a gas, its temperature and volume could increase. In **iso-thermal expansion,** the temperature remains constant—only the volume changes. The energy required is calculated as follows.

> **Isothermal Expansion**
> If the temperature T is constant, the energy E required to increase the volume of one mole of gas from an initial volume V_i to a final volume V_f is given by
>
> $$E = RT \ln \left(\frac{V_f}{V_i} \right)$$
>
> E is measured in joules and T in degrees Kelvin, and R is the universal gas constant 8.314 joules/mole/K.

EXAMPLE 5   Find the amount of energy that must be supplied to triple the volume of 1 mole of gas at a constant temperature of 300K.

Solution   We substitute 8.314 for R and 300 for T into the formula. Because the final volume is to be three times the initial volume, we also substitute $3V_i$ for V_f.

$$E = RT \ln \left(\frac{V_f}{V_i} \right)$$

$$E = (8.314)(300) \ln \left(\frac{3V_i}{V_i} \right)$$

$$\approx 2494.2 \ln 3$$

$$\approx 2740.15877$$

Approximately 2740 joules of energy must be added.   ■

Self Check   What energy is required to double the volume?

Answer   1729 joules

EXERCISE 5.4

VOCABULARY AND CONCEPTS   *In Exercises 1–10, fill in the blank to make a true statement.*

1. $y = \ln x$ means $y =$ _____.

2. The domain of the function $y = f(x) = \ln x$ is the interval _____.

3. The range of the function $y = f(x) = \ln x$ is the interval _____.

4. The graph of $y = f(x) = \ln x$ has the _____ as an asymptote.

5. In the expression $\log x$, the base is understood to be ____.

6. In the expression $\ln x$, the base is understood to be ____.

7. The formula for charging batteries is
_____ .

8. If a population grows exponentially at a rate r, the time it will take for the population to double is given by the formula _____ .

9. The formula for isothermal expansion is

_____

10. The logarithm of a negative number is _____ .

PRACTICE *In Exercises 11–18, use a calculator to find each value, if possible. Express all answers to four decimal places.*

11. ln 35.15

12. ln 0.675

13. ln 7.896

14. ln 0.00465

15. log (ln 1.7)

16. ln (log 9.8)

17. ln (log 0.1)

18. log (ln 0.01)

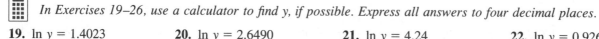 *In Exercises 19–26, use a calculator to find y, if possible. Express all answers to four decimal places.*

19. ln $y = 1.4023$

20. ln $y = 2.6490$

21. ln $y = 4.24$

22. ln $y = 0.926$

23. ln $y = -3.71$

24. ln $y = -0.28$

25. log $y = $ ln 8

26. ln $y = $ log 7

In Exercises 27–30, tell whether the graph could represent the graph of $y = \ln x$.

27.

28.

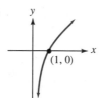

29.

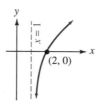

30.

In Exercises 31–34, use a graphing calculator to graph each function.

31. $y = \ln \left(\dfrac{1}{2} x \right)$

32. $y = \ln x^2$

33. $y = \ln (-x)$

34. $y = \ln (3x)$

APPLICATIONS   *Use a calculator to solve each problem.*

35. Battery charge   If a battery can reach half of its full charge in 6 hours, how long will it take the battery to reach a 90% charge? Assume that the battery was fully discharged when it began charging.

36. Battery charge   A battery reaches 80% of a full charge in 8 hours. If it started charging when it was fully discharged, how long did it take to reach a 40% charge?

37. **Population growth**   A town's population grows at the rate of 12% per year. If this growth rate remains constant, how long will it take the population to double?

38. **Population growth**   A population growing at an annual rate r will triple in a time t given by the formula

$$t = \frac{\ln 3}{r}$$

How long will it take the population of the town in Exercise 37 to triple?

39. **Isothermal expansion**   One mole of gas expands isothermally to triple its volume. If the gas temperature is 400K, what energy is absorbed?

40. **Isothermal expansion**   One mole of gas expands isothermally to double its volume. If the gas temperature is 300K, what energy is absorbed?

41. **Breakdown voltage**   The coaxial power cable shown in Illustration 1 has a central wire with radius

$R_1 = 0.25$ centimeter. It is insulated by a surrounding shield with inside radius $R_2 = 2$ centimeters. The maximum voltage the cable can withstand is called the **breakdown voltage** V of the insulation. V is given by the formula

$$V = ER_1 \ln \frac{R_2}{R_1}$$

where E is the **dielectric strength** of the insulation. If $E = 400{,}000$ volts/centimeter, find V.

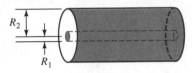

ILLUSTRATION 1

42. **Breakdown voltage**   In Exercise 41, if the inside diameter of the shield were doubled, what voltage could the cable withstand?

DISCOVERY AND WRITING

43. One form of the logistic function is given by the equation

$$y = \frac{1}{1 + e^{-2x}}$$

Explain how you would find the y-intercept of its graph.

44. Graph the function $y = \ln |x|$. Explain why the graph looks the way it does.

REVIEW   *Write the equation of the required line.*

45. parallel to $y = 5x - 8$ and passing through the origin

46. having a slope of 7 and a y-intercept of 3

47. passing through the point $(3, 2)$ and perpendicular to the line $y = \frac{2}{3}x - 12$

48. parallel to the line $3x + 2y = 9$ and passing through the point $(-3, 5)$

49. vertical line through the point $(2, 3)$

50. horizontal line through the point $(2, 3)$

Simplify each expression.

51. $\dfrac{2(x+2)-1}{4x^2-9}$

52. $\dfrac{x+1}{x} + \dfrac{x-1}{x+1}$

53. $\dfrac{x^2+3x+2}{3x+9} \cdot \dfrac{x+3}{x^2-4}$

54. $\dfrac{1+\dfrac{y}{x}}{\dfrac{y}{x}-1}$

5.5 Properties of Logarithms

■ PROPERTIES OF LOGARITHMS ■ THE CHANGE-OF-BASE FORMULA ■ pH SCALE ■ ELECTRONICS
■ PHYSIOLOGY

■ PROPERTIES OF LOGARITHMS

Since logarithms are exponents, the properties of exponents have counterparts in the theory of logarithms. We begin with four basic properties.

> **Properties of Logarithms**
> If b is a positive number and $b \neq 1$, then
> **1.** $\log_b 1 = 0$ **2.** $\log_b b = 1$
> **3.** $\log_b b^x = x$ **4.** $b^{\log_b x} = x$ $(x > 0)$

Properties 1 through 4 follow directly from the definition of logarithm.

1. $\log_b 1 = 0$, because $b^0 = 1$.

2. $\log_b b = 1$, because $b^1 = b$.

3. $\log_b b^x = x$, because $b^x = b^x$.

4. $b^{\log_b x} = x$, because $\log_b x$ is the exponent to which b is raised to get x.

Properties 3 and 4 also indicate that the composition of the exponential and logarithmic functions (in both directions) is the identity function. This is expected, because the exponential and logarithmic functions are inverse functions.

EXAMPLE 1

Simplify each expression: **a.** $\log_3 1$,   **b.** $\log_4 4$,   **c.** $\log_7 7^3$, and   **d.** $b^{\log_b 3}$.

Solution   **a.** By Property 1, $\log_3 1 = \mathbf{0}$, because $3^0 = 1$.

b. By Property 2, $\log_4 4 = \mathbf{1}$, because $4^1 = 4$.

c. By Property 3, $\log_7 7^3 = \mathbf{3}$, because $7^3 = 7^3$.

d. By Property 4, $b^{\log_b 3} = \mathbf{3}$, because $\log_b 3$ is the power to which b is raised to get 3.   ∎

Self Check   Simplify   **a.** $\log_4 1$,   **b.** $\log_3 3$,   **c.** $\log_2 2^4$,   and   **d.** $5^{\log_5 2}$.

Answers   **a.** 0,   **b.** 1,   **c.** 4,   **d.** 2

The next two properties state that

The logarithm of a product is the sum of the logarithms.

The logarithm of a quotient is the difference of the logarithms.

Properties of Logarithms

If M, N, and b are positive numbers and $b \neq 1$, then

5. $\log_b MN = \log_b M + \log_b N$ **6.** $\log_b \dfrac{M}{N} = \log_b M - \log_b N$

Proof   To prove Property 5, we let $x = \log_b M$ and $y = \log_b N$. We use the definition of logarithm to write each equation in exponential form.

$$M = b^x \qquad \text{and} \qquad N = b^y$$

Then $MN = b^x b^y$ and a property of exponents gives

$$MN = b^{x+y} \qquad b^x b^y = b^{x+y}: \text{Keep the base and add the exponents.}$$

We write this exponential equation in logarithmic form as

$$\log_b MN = x + y$$

Substituting the values of x and y completes the proof.

$$\log_b MN = \log_b M + \log_b N$$ □

The proof of Property 6 is similar. You will be asked to do it in an exercise.

 WARNING!   By Property 5 of logarithms, the logarithm of a *product* is equal to the *sum* of the logarithms. The logarithm of a sum or a difference usually does not simplify. In general,

$$\log_b (M + N) \neq \log_b M + \log_b N \qquad \text{and} \qquad \log_b (M - N) \neq \log_b M - \log_b N$$

By Property 6, the logarithm of a *quotient* is equal to the *difference* of the logarithms. The logarithm of a quotient is not the quotient of the logarithms:

$$\log_b \frac{M}{N} \neq \frac{\log_b M}{\log_b N}$$

■ ■ ■ ■ ■ ■ ■ ■

ACCENT ON TECHNOLOGY

We can use a calculator to illustrate Property 5 of logarithms by showing that

$$\ln\,[(3.7)(15.9)] = \ln 3.7 + \ln 15.9$$

We calculate the left- and right-hand sides of the equation separately and compare the results. To use a scientific calculator to find $\ln\,[(3.7)(15.9)]$, we enter these numbers and press these keys:

3.7 ☒ 15.9 = LN

The display will read 4.074651929.
 To find $\ln 3.7 + \ln 15.9$, we enter these numbers and press these keys:

3.7 LN + 15.9 LN =

The display will read 4.074651929. Since the left- and right-hand sides are equal, the equation is true.

■ ■ ■ ■ ■ ■ ■ ■ ■

Two more properties state that

The logarithm of a power is the power times the logarithm.

If the logarithms of two numbers are equal, the numbers are equal.

Properties of Logarithms
If M, p, and b are positive numbers and $b \neq 1$, then

7. $\log_b M^p = p \log_b M$ **8.** If $\log_b x = \log_b y$, then $x = y$.

Proof To prove Property 7, we let $x = \log_b M$, write the expression in exponential form, and raise both sides to the pth power:

$$M = b^x$$
$$(M)^p = (b^x)^p \qquad \text{Raise both sides to the } p\text{th power.}$$
$$M^p = b^{px} \qquad \text{Keep the base and multiply the exponents.}$$

Using the definition of logarithms gives

$$\log_b M^p = px$$

Substituting the value for x completes the proof.

$$\log_b M^p = p \log_b M \qquad\qquad\qquad \square$$

Property 8 follows from the fact that the logarithmic function is a one-to-one function. Property 8 will be important in the next section when we solve logarithmic equations.

We can use the properties of logarithms to write a logarithm as the sum or difference of several logarithms.

EXAMPLE 2 Assume that x, y, and z are positive numbers. Write each expression in terms of the logarithms of x, y, and z: **a.** $\log_b xyz$ and **b.** $\log_b \dfrac{x}{yz}$.

Solution **a.** $\log_b xyz = \log_b (xy)z$

$\qquad\qquad = \log_b (xy) + \log_b z$ The log of a product is the sum of the logs.

$\qquad\qquad = \log_b x + \log_b y + \log_b z$ The log of a product is the sum of the logs.

b. $\log_b \dfrac{x}{yz} = \log_b x - \log_b (yz)$ The log of a quotient is the difference of the logs.

$\qquad\qquad = \log_b x - (\log_b y + \log_b z)$ The log of a product is the sum of the logs.

$\qquad\qquad = \log_b x - \log_b y - \log_b z$ Remove parentheses. ■

Self Check Find $\log_b \dfrac{xy}{z}$.

Answer $\log_b x + \log_b y - \log_b z$

EXAMPLE 3 Assume that x, y, and z are positive numbers. Write each expression in terms of the logarithms of x, y, and z: **a.** $\log_b (x^3 y^2 z)$ and **b.** $\log_b \dfrac{y^2 \sqrt{z}}{x}$.

Solution **a.** $\log_b (x^3 y^2 z) = \log_b x^3 + \log_b y^2 + \log_b z$ The log of a product is the sum of the logs.

$\qquad\qquad = 3 \log_b x + 2 \log_b y + \log_b z$ The log of a power is the power times the log.

b. $\log_b \dfrac{y^2 \sqrt{z}}{x} = \log_b \left(y^2 \sqrt{z}\right) - \log_b x$ The log of a quotient is the difference of the logs.

$\qquad\qquad = \log_b y^2 + \log_b z^{1/2} - \log_b x$ The log of a product is the sum of the logs; $\sqrt{z} = z^{1/2}$.

$\qquad\qquad = 2 \log_b y + \dfrac{1}{2} \log_b z - \log_b x$ The log of a power is the power times the log. ■

Self Check Find $\log_b \sqrt[3]{\dfrac{x^2 y}{z}}$.

Answer $\frac{1}{3}(2 \log_b x + \log_b y - \log_b z)$

We can use the properties of logarithms to combine several logarithms into one logarithm.

EXAMPLE 4 Assume that x, y, z, and b are positive numbers and $b \neq 0$. Write each expression as one logarithm: **a.** $2 \log_b x + \dfrac{1}{3} \log_b y$ **b.** $\dfrac{1}{2} \log_b (x - 2) - \log_b y + 3 \log_b z$.

Solution   **a.** $2 \log_b x + \dfrac{1}{3} \log_b y = \log_b x^2 + \log_b y^{1/3}$ A power times a log is the log of the power.

$$= \log_b (x^2 y^{1/3})$$ The sum of two logs is the log of the product.

b. $\dfrac{1}{2} \log_b (x - 2) - \log_b y + 3 \log_b z$

$$= \log_b (x - 2)^{1/2} - \log_b y + \log_b z^3$$ A power times a log is the log of the power.

$$= \log_b \frac{(x - 2)^{1/2}}{y} + \log_b z^3$$ The difference of two logs is the log of the quotient.

$$= \log_b \frac{z^3 \sqrt{x - 2}}{y}$$ The sum of two logs is the log of the product. ∎

Self Check   Write the expression as one logarithm: $2 \log_b x + \frac{1}{2} \log_b y - 3 \log_b (x - y)$.

Answer $\log_b \dfrac{x^2 \sqrt{y}}{(x - y)^3}$

We summarize the eight properties of logarithms as follows.

Properties of Logarithms

If b, M, and N are positive numbers and $b \neq 1$, then

1. $\log_b 1 = 0$ **2.** $\log_b b = 1$

3. $\log_b b^x = x$ **4.** $b^{\log_b x} = x$

5. $\log_b MN = \log_b M + \log_b N$ **6.** $\log_b \dfrac{M}{N} = \log_b M - \log_b N$

7. $\log_b M^p = p \log_b M$ **8.** If $\log_b x = \log_b y$, then $x = y$.

EXAMPLE 5 Given that $\log_{10} 2 \approx 0.3010$ and $\log_{10} 3 \approx 0.4771$, find approximations for **a.** $\log_{10} 18$ and   **b.** $\log_{10} 2.5$.

Solution   **a.** $\log_{10} 18 = \log_{10} (2 \cdot 3^2)$

$$= \log_{10} 2 + \log_{10} 3^2$$ The log of a product is the sum of the logs.

$$= \log_{10} 2 + 2 \log_{10} 3$$ The log of a power is the power times the log.

$$\approx 0.3010 + 2(0.4771)$$

$$\approx 1.2552$$

b. $\log_{10} 2.5 = \log_{10} \left(\dfrac{5}{2} \right)$

$\qquad\qquad = \log_{10} 5 - \log_{10} 2$ The log of a quotient is the difference of the logs.

$\qquad\qquad = \log_{10} \dfrac{10}{2} - \log_{10} 2$ Write 5 as $\frac{10}{2}$.

$\qquad\qquad = \log_{10} 10 - \log_{10} 2 - \log_{10} 2$ The log of a quotient is the difference of the logs.

$\qquad\qquad = 1 - 2 \log_{10} 2$ $\log_{10} 10 = 1.$

$\qquad\qquad \approx 1 - 2(0.3010)$

$\qquad\qquad \approx 0.3980$ ■

Self Check   Use the information given in Example 5 to find $\log_{10} 0.75$.

Answer -0.1249

■ THE CHANGE-OF-BASE FORMULA

If we know the base-a logarithm of a number, we can find its logarithm to some other base b with a formula called the **change-of-base formula**.

> **Change-of-Base Formula**
> If a, b, and x are real numbers, then
> $$\log_b x = \frac{\log_a x}{\log_a b}$$

To prove this formula, we begin with the equation $\log_b x = y$.

$\qquad\qquad y = \log_b x$

$\qquad\qquad x = b^y$ Change the equation from logarithmic to exponential form.

$\qquad\qquad \mathbf{\log_a x = \log_a b^y}$ Take the base-a logarithm of both sides.

$\qquad\qquad \log_a x = y \log_a b$ The log of a power is the power times the log.

$\qquad\qquad y = \dfrac{\log_a x}{\log_a b}$ Divide both sides by $\log_a b$.

$\qquad\qquad \log_b x = \dfrac{\log_a x}{\log_a b}$ Refer to the first equation and substitute $\log_b x$ for y.

If we know logarithms to base a (for example, $a = 10$), we can find the logarithm of x to a new base b. We simply divide the base-a logarithm of x by the base-a logarithm of b.

WARNING! $\dfrac{\log_a x}{\log_a b}$ means that one logarithm is to be divided by the other. They are not to be subtracted.

EXAMPLE 6   Find $\log_3 5$.

Solution   We can substitute 3 for b, 10 for a, and 5 for x into the change-of-base formula:

$$\log_b x = \frac{\log_a x}{\log_a b}$$

$$\log_3 5 = \frac{\log_{10} 5}{\log_{10} 3}$$

$$\approx 1.464973521$$

To four decimal places, $\log_3 5 = 1.4650$.

Self Check   Find $\log_5 3$.
Answer   0.6826

■   **pH SCALE**

The more acidic a chemical solution, the greater the concentration of hydrogen ions. Chemists measure this concentration indirectly by the **pH scale,** or the **hydrogen ion index.**

> **pH of a Solution**
> If $[H^+]$ is the hydrogen ion concentration in gram-ions per liter, then
> $$pH = -\log\,[H^+]$$

Since pure water has approximately 10^{-7} gram-ions per liter, its pH is

$$pH = -\log\,[H^+]$$
$$pH = -\log\,\mathbf{10^{-7}}$$
$$= -(-7)\log 10 \qquad \text{The log of a power is the power times the log.}$$
$$= -(-7) \cdot 1 \qquad \text{Use Property 2 of logarithms: } \log_b b = 1.$$
$$= 7$$

EXAMPLE 7   Seawater has a pH of approximately 8.5. Find its hydrogen ion concentration.

Solution   We can substitute 8.5 for pH and solve the equation $pH = -\log[H^+]$ for $[H^+]$.

$$8.5 = -\log\,[H^+]$$
$$-8.5 = \log\,[H^+]$$
$$[H^+] = 10^{-8.5} \qquad \text{Change the equation from logarithmic form to} $$
$$\qquad\qquad\qquad \text{exponential form.}$$

We use a calculator to find that $[H^+] \approx 3.2 \times 10^{-9}$ gram-ions per liter.   ■

Self Check The pH of a solution is 5.7. Find the hydrogen ion concentration.

Answer 2×10^{-6}

■ ELECTRONICS

Recall that if E_O is the output voltage of a device and E_I is the input voltage, the decibel voltage gain is given by

$$\text{db gain} = 20 \log \frac{E_O}{E_I}$$

If input and output are measured in watts instead of volts, a different formula is needed.

EXAMPLE 8 Show that the formula for db voltage gain is

$$\text{db gain} = 10 \log \frac{P_O}{P_I}$$

where P_I is the power input and P_O is the power output.

Solution Power is directly proportional to the square of the voltage. So for some constant k,

$$P_I = k(E_I)^2 \quad \text{and} \quad P_O = k(E_O)^2$$

and

$$\frac{P_O}{P_I} = \frac{k(E_O)^2}{k(E_I)^2} = \left(\frac{E_O}{E_I}\right)^2$$

We raise both sides to the $\frac{1}{2}$ power to get

$$\frac{E_O}{E_I} = \left(\frac{P_O}{P_I}\right)^{1/2}$$

which we substitute into the formula for db gain.

$$\text{db gain} = 20 \log \frac{E_O}{E_I}$$

$$= 20 \log \left(\frac{P_O}{P_I}\right)^{1/2}$$

$$= 20 \cdot \frac{1}{2} \log \frac{P_O}{P_I} \qquad \text{The log of a power is the power times the log.}$$

$$\text{db gain} = 10 \log \frac{P_O}{P_I} \qquad \text{Simplify.}$$

■

Self Check Find the db gain of a device if $P_O = 30$ watts and $P_I = 2$ watts.

Answer 11.76

■ PHYSIOLOGY

In physiology, experiments suggest that the relationship between the loudness and the intensity of sound is a logarithmic one known as the Weber–Fechner law.

> **Weber–Fechner Law**
> If L is the apparent loudness of a sound and I is the intensity, then
> $$L = k \ln I$$

EXAMPLE 9   What increase in the intensity of a sound is necessary to cause a doubling of the apparent loudness?

Solution   We use the formula $L = k \ln I$. To double the apparent loudness, we multiply both sides of the equation by 2 and use Property 7 of logarithms.

$$L = k \ln t$$
$$2L = 2k \ln I$$
$$= k \ln I^2$$

To double the apparent loudness, we must square the intensity.   ■

Self Check   What increase is necessary to triple the apparent loudness?
Answer   Cube the intensity.

EXERCISE 5.5

VOCABULARY AND CONCEPTS   *In Exercises 1–10, fill in the blank to make a true statement.*

1. $\log_b 1 =$ ___

2. $\log_b b =$ ___

3. $\log_b MN = \log_b$ ___ $+ \log_b$ ___

4. $b^{\log_b x} =$ ___

5. If $\log_b x = \log_b y$, then ___ = ___.

6. $\log_b \dfrac{M}{N} = \log_b M$ ___ $\log_b N$

7. $\log_b x^p = p \cdot \log_b$ ___

8. $\log_b b^x =$ ___

9. $\log_b (A + B)$ ___ $\log_b A + \log_b B$

10. $\log_b A + \log_b B$ ___ $\log_b AB$

In Exercises 11–18, simplify each expression.

11. $\log_4 1 =$ ___

12. $\log_4 4 =$ ___

13. $\log_4 4^7 =$ ___

14. $4^{\log_4 8} =$ ___

15. $5^{\log_5 10} =$ ___

16. $\log_5 5^2 =$ ___

17. $\log_5 5 =$ ___

18. $\log_5 1 =$ ___

PRACTICE      *In Exercises 19–24, use a calculator to verify each equation.*

19. $\log [(3.7)(2.9)] = \log 3.7 + \log 2.9$

20. $\ln \dfrac{9.3}{2.1} = \ln 9.3 - \ln 2.1$

21. $\ln (3.7)^3 = 3 \ln 3.7$

22. $\log\sqrt{14.1} = \dfrac{1}{2} \log 14.1$

23. $\log 3.2 = \dfrac{\ln 3.2}{\ln 10}$

24. $\ln 9.7 = \dfrac{\log 9.7}{\log e}$

In Exercises 25–36, assume that x, y, z, and b are positive numbers. Use the properties of logarithms to write each expression in terms of the logarithms of x, y, and z.

25. $\log_b 2xy$

26. $\log_b 3xz$

27. $\log_b \dfrac{2x}{y}$

28. $\log_b \dfrac{x}{yz}$

29. $\log_b x^2 y^3$

30. $\log_b x^3 y^2 z$

31. $\log_b (xy)^{1/3}$

32. $\log_b x^{1/2} y^3$

33. $\log_b x\sqrt{z}$

34. $\log_b \sqrt{xy}$

35. $\log_b \dfrac{\sqrt[3]{x}}{\sqrt[3]{yz}}$

36. $\log_b \sqrt[4]{\dfrac{x^3 y^2}{z^4}}$

In Exercises 37–44, assume that x, y, and z are positive numbers. Use the properties of logarithms to write each expression as the logarithm of one quantity.

37. $\log_b (x + 1) - \log_b x$

38. $\log_b x + \log_b (x + 2) - \log_b 8$

39. $2 \log_b x + \dfrac{1}{3} \log_b y$

40. $-2 \log_b x - 3 \log_b y + \log_b z$

41. $-3 \log_b x - 2 \log_b y + \dfrac{1}{2} \log_b z$

42. $3 \log_b (x + 1) - 2 \log_b (x + 2) + \log_b x$

43. $\log_b \left(\dfrac{x}{z} + x \right) - \log_b \left(\dfrac{y}{z} + y \right)$

44. $\log_b (xy + y^2) - \log_b (xz + yz) + \log_b z$

In Exercises 45–64, tell whether each statement is true or false.

45. $\log_b ab = \log_b a + 1$

46. $\log_b \dfrac{1}{a} = -\log_b a$

47. $\log_b 0 = 1$

48. $\log_b 2 = \log_2 b$

49. $\log_b (x + y) \neq \log_b x + \log_b y$

50. $\log_b xy = (\log_b x)(\log_b y)$

51. If $\log_a b = c$, then $\log_b a = c$.

52. If $\log_a b = c$, then $\log_b a = \dfrac{1}{c}$.

53. $\log_7 7^7 = 7$

54. $7^{\log_7 7} = 7$

55. $\log_b (-x) = -\log_b x$

56. If $\log_b a = c$, then $\log_b a^p = pc$.

57. $\dfrac{\log_b A}{\log_b B} = \log_b A - \log_b B$

58. $\log_b (A - B) = \dfrac{\log_b A}{\log_b B}$

59. $\log_b \dfrac{1}{5} = -\log_b 5$

60. $3 \log_b \sqrt[3]{a} = \log_b a$

61. $\dfrac{1}{3} \log_b a^3 = \log_b a$

62. $\log_{4/3} y = -\log_{3/4} y$

63. $\log_b y + \log_{1/b} y = 0$

64. $\log_{10} 10^3 = 3(10^{\log_{10} 3})$

In Exercises 65–76, assume that $\log_{10} 4 = 0.6021$, $\log_{10} 7 = 0.8451$, and $\log_{10} 9 = 0.9542$. Use these values and the properties of logarithms to find each value. **Do not use a calculator.**

65. $\log_{10} 28$

66. $\log_{10} \dfrac{7}{4}$

67. $\log_{10} 2.25$

68. $\log_{10} 36$

69. $\log_{10} \dfrac{63}{4}$

70. $\log_{10} \dfrac{4}{63}$

71. $\log_{10} 252$

72. $\log_{10} 49$

73. $\log_{10} 112$

74. $\log_{10} 324$

75. $\log_{10} \dfrac{144}{49}$

76. $\log_{10} \dfrac{324}{63}$

 In Exercises 77–84, use a calculator and the change-of-base formula to find each logarithm.

77. $\log_3 7$

78. $\log_7 3$

79. $\log_\pi 3$

80. $\log_3 \pi$

81. $\log_3 8$

82. $\log_5 10$

83. $\log_{\sqrt{2}} \sqrt{5}$

84. $\log_\pi e$

APPLICATIONS

85. pH of a solution Find the pH of a solution with a hydrogen ion concentration of 1.7×10^{-5} gram-ions per liter.

86. pH of calcium hydroxide Find the hydrogen ion concentration of a saturated solution of calcium hydroxide whose pH is 13.2.

87. pH of apples The pH of apples can range from 2.9 to 3.3. Find the range in the hydrogen ion concentration.

88. pH of sour pickles The hydrogen ion concentration of sour pickles is 6.31×10^{-4}. Find the pH.

89. db gain An amplifier produces a 40-watt output with a $\frac{1}{2}$-watt input. Find the db gain.

90. db loss Losses in a long telephone line reduce a 12-watt input signal to an output of 3 watts. Find the db gain. (Because it is a loss, the "gain" will be negative.)

91. Weber–Fechner law What increase in intensity is necessary to quadruple the loudness?

92. Weber–Fechner law What decrease in intensity is necessary to make a sound half as loud?

93. Isothermal expansion If a certain amount E of energy is added to one mole of a gas, it expands from an initial volume of 1 liter to a final volume V without changing its temperature according to the formula

$$E = 8300 \ln (V)$$

Find the volume if twice that energy is added to the gas.

94. Richter scale By what factor must the amplitude of an earthquake change to increase its severity by 1 point on the Richter scale? Assume that the period remains constant. The Richter scale is given by

$$R = \log \dfrac{A}{P}$$

where A is the amplitude and P the period of the tremor.

DISCOVERY AND WRITING

95. Prove Property 6 of logarithms:

$$\log_b \frac{M}{N} = \log_b M - \log_b N$$

96. Show that $-\log_b x = \log_{1/b} x$.

97. Show that $e^{x \ln a} = a^x$.

98. Show that $e^{\ln x} = x$.

99. Show that $\ln(e^x) = x$.

100. If $\log_b 3x = 1 + \log_b x$, find b.

101. Explain why $\ln(\log 0.9)$ is undefined.

102. Explain why $\log_b(\ln 1)$ is undefined.

If A and B are both negative, then AB and $\frac{A}{B}$ are positive, and log AB and log $\frac{A}{B}$ are defined.

103. Is it still true that $\log AB = \log A + \log B$? Explain.

104. Is it still true that $\log \frac{A}{B} = \log A - \log B$? Explain.

REVIEW   *Decide whether each equation defines a function.*

105. $y = 3x - 1$

106. $y = \dfrac{x + 3}{x - 1}$

107. $y^2 = 4x$

108. $y = 4x^2$

Find the domain of each function.

109. $f(x) = x^2 - 4$

110. $f(x) = \dfrac{1}{x^2 - 4}$

111. $f(x) = \sqrt{x^2 + 4}$

112. $f(x) = \sqrt{x^2 - 4}$

5.6 Exponential and Logarithmic Equations

■ SOLVING EXPONENTIAL EQUATIONS ■ SOLVING LOGARITHMIC EQUATIONS ■ CARBON-14 DATING ■ POPULATION GROWTH

An **exponential equation** is an equation with a variable in one of its exponents. Some examples of exponential equations are

$$3^x = 5, \quad 6^{x-3} = 2^x, \quad \text{and} \quad 3^{2x+1} - 10(3^x) + 3 = 0$$

A **logarithmic equation** is an equation with logarithmic expressions that contain a variable. Some examples of logarithmic equations are

$$\log 2x = 25, \quad \ln x - \ln(x - 12) = 24, \quad \text{and} \quad \log x = \log \frac{1}{x} + 4$$

In this section, we will learn how to solve many of these equations.

■ SOLVING EXPONENTIAL EQUATIONS

EXAMPLE 1   Solve the exponential equation $3^x = 5$.

Solution   Since logarithms of equal numbers are equal, we can take the common logarithm of each side of the equation. We can then move the variable x from its position as an exponent to a position as a coefficient.

$$3^x = 5$$

$$\log 3^x = \log 5 \qquad \text{Take the common logarithm of each side.}$$

$$x \log 3 = \log 5 \qquad \text{The log of a power is the power times the log.}$$

1. $$x = \frac{\log 5}{\log 3} \qquad \text{Divide both sides by log 3.}$$

$$\approx 1.464973521 \qquad \text{Use a calculator.}$$

To four decimal places, $x = 1.4650$.   ■

Self Check   Solve $5^x = 3$.

Answer $x = 0.6826$

WARNING!   A careless reading of Equation 1 leads to a common error. The right-hand side of Equation 1 calls for a division, not a subtraction.

$$\frac{\log 5}{\log 3} \qquad \text{means} \qquad (\log 5) \div (\log 3)$$

It is the expression $\log \frac{5}{3}$ that means $\log 5 - \log 3$.

EXAMPLE 2   Solve the exponential equation $6^{x-3} = 2^x$.

Solution
$$6^{x-3} = 2^x$$

$$\log 6^{x-3} = \log 2^x \qquad \text{Take the common logarithm of each side.}$$

$$(x - 3) \log 6 = x \log 2 \qquad \text{The log of a power is the power times the log.}$$

$$x \log 6 - 3 \log 6 = x \log 2 \qquad \text{Use the distributive property.}$$

$$x \log 6 - x \log 2 = 3 \log 6 \qquad \text{Add 3 log 6 and subtract } x \log 2 \text{ from both sides.}$$

$$x(\log 6 - \log 2) = 3 \log 6 \qquad \text{Factor out } x \text{ on the left-hand side.}$$

$$x = \frac{3 \log 6}{\log 6 - \log 2} \qquad \text{Divide both sides by log 6 − log 2.}$$

$$x \approx 4.892789261 \qquad \text{Use a calculator.}$$ ■

Self Check Solve $5^{x+3} = 3^x$.

Answer -9.451980309

EXAMPLE 3 Solve the exponential equation $2^{x^2+2x} = \frac{1}{2}$.

Solution Since $\frac{1}{2} = 2^{-1}$, we can write the equation in the form

$$2^{x^2+2x} = 2^{-1}$$

Since equal quantities with equal bases have equal exponents, we have

$$
\begin{aligned}
x^2 + 2x &= -1 \\
x^2 + 2x + 1 &= 0 \qquad &\text{Add 1 to both sides.} \\
(x + 1)(x + 1) &= 0 \qquad &\text{Factor the trinomial.} \\
x + 1 = 0 \quad \text{or} \quad x + 1 &= 0 \qquad &\text{Set each factor equal to 0.} \\
x = -1 \qquad\qquad x &= -1
\end{aligned}
$$

Verify that -1 satisfies the equation. ∎

Self Check Solve $3^{x^2+2x} = 27$.

Answer $1, \quad 3$

■ SOLVING LOGARITHMIC EQUATIONS

In each of the following examples, we use the properties of logarithms to change a logarithmic equation into an algebraic equation.

EXAMPLE 4 Solve $\log_b (3x + 2) - \log_b (2x - 3) = 0$.

Solution

$$
\begin{aligned}
\log_b (3x + 2) - \log_b (2x - 3) &= 0 \\
\log_b (3x + 2) &= \log_b (2x - 3) \qquad &\text{Add } \log_b (2x - 3) \text{ to both sides.} \\
3x + 2 &= 2x - 3 \qquad &\text{If the logs of two numbers are equal, the numbers are equal.} \\
x &= -5 \qquad &\text{Subtract } 2x \text{ and 2 from both sides.}
\end{aligned}
$$

Check:
$$
\begin{aligned}
\log_b (3x + 2) - \log_b (2x - 3) &= 0 \\
\log_b [3(-5) + 2] - \log_b [2(-5) - 3] &\overset{?}{=} 0 \\
\log_b (-13) - \log_b (-13) &\overset{?}{=} 0
\end{aligned}
$$

Since the logarithm of a negative number does not exist, -5 is extraneous and must be discarded. The equation has no roots. ∎

Self Check Solve $\log_b (5x + 2) - \log_b (6x + 1) = 0$.
Answer 1

WARNING! Example 4 illustrates that you *must* check the solutions of a logarithmic equation.

EXAMPLE 5 Solve $\log x + \log (x - 3) = 1$.

Solution

$$\log x + \log (x - 3) = 1$$

$$\log x(x - 3) = 1 \qquad \text{The sum of two logs is the log of a product.}$$

$$x(x - 3) = 10^1 \qquad \text{Use the definition of logarithms to change the equation to exponential form.}$$

$$x^2 - 3x - 10 = 0 \qquad \text{Remove parentheses and subtract 10 from both sides.}$$

$$(x + 2)(x - 5) = 0 \qquad \text{Factor the trinomial.}$$

$$x + 2 = 0 \qquad \text{or} \qquad x - 5 = 0$$

$$x = -2 \qquad\qquad x = 5$$

Check: The number -2 is not a solution, because it does not satisfy the equation (a negative number does not have a logarithm). We check the remaining number, 5.

$$\log x + \log (x - 3) = 1$$

$$\log 5 + \log (5 - 3) \overset{?}{=} 1 \qquad \text{Substitute 5 for } x.$$

$$\log 5 + \log 2 \overset{?}{=} 1$$

$$\log 10 \overset{?}{=} 1 \qquad \text{The sum of two logs is the log of a product.}$$

$$1 = 1 \qquad \log_b b = 1.$$

Since 5 does check, it is a root. ■

Self Check Solve $\log x + \log (x - 15) = 2$.
Answer 20; -5 is extraneous

EXAMPLE 6 Solve $\dfrac{\log (5x - 6)}{\log x} = 2$.

Solution We can multiply both sides by $\log x$ to get

$$\log (5x - 6) = 2 \log x$$

and apply Property 7 of logarithms to get

$$\log (5x - 6) = \log x^2$$

By Property 8 of logarithms, $5x - 6 = x^2$, because they have equal logarithms. So

$$x^2 = 5x - 6$$
$$x^2 - 5x + 6 = 0$$
$$(x - 3)(x - 2) = 0$$
$$x - 3 = 0 \quad \text{or} \quad x - 2 = 0$$
$$x = 3 \qquad\qquad x = 2$$

Verify that both 2 and 3 satisfy the equation.   ■

Self Check Solve $\dfrac{\log (8x - 15)}{\log x} = 2$.

Answer   3, 5

■ ■ ■ ■ ■ ■ ■ ■ ■

ACCENT ON TECHNOLOGY

We can use a graphing calculator to solve logarithmic equations. For example, to solve $\log x + \log (x - 3) = 1$, we subtract 1 from both sides to get

$$\log x + \log (x - 3) - 1 = 0$$

and then graph the corresponding function

$$y = \log x + \log (x - 3) - 1$$

with settings of $[0, 10]$ for x and $[-2, 2]$ for y, to obtain the graph in Figure 5-21.

Since the root of the equation is the x-intercept, we can find the root by tracing to find the value of the x-intercept. The root is $x = 5$. (See Figure 5-22.)

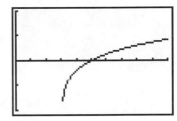

FIGURE 5-21

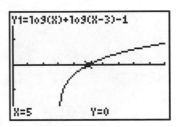

FIGURE 5-22

■ ■ ■ ■ ■ ■ ■ ■ ■

■ CARBON-14 DATING

When a living organism dies, the oxygen/carbon dioxide cycle common to all living things ceases; then carbon-14, a radioactive isotope with a half-life of 5700 years, is no longer absorbed. By measuring the amount of carbon-14 present in ancient objects, archaeologists can estimate the object's age.

The amount A of radioactive material present at time t is given by the model

$$A = A_0 2^{-t/h}$$

where A_0 is the amount present initially and h is the half-life of the material.

EXAMPLE 7   How old is a wooden statue that now has only one-third of its original carbon-14 content?

Solution   To find the time t when $A = \frac{1}{3}A_0$, we substitute $\frac{A_0}{3}$ for A and 5700 for h into the radioactive decay formula and solve for t:

$$A = A_0 2^{-t/h}$$

$$\frac{A_0}{3} = A_0 2^{-t/5700}$$

$$1 = 3(2^{-t/5700}) \qquad \text{Divide both sides by } A_0.$$

$$\log 1 = \log 3(2^{-t/5700}) \qquad \text{Take the common logarithm of each side.}$$

$$0 = \log 3 + \log 2^{-t/5700} \qquad \text{The log of a product is the sum of the logs.}$$

$$-\log 3 = -\frac{t}{5700}\log 2 \qquad \text{Subtract log 3 from both sides and use Property 7 of logarithms.}$$

$$5700\left(\frac{\log 3}{\log 2}\right) = t \qquad \text{Multiply both sides by } -\frac{5700}{\log 2}.$$

$$t \approx 9034.286254 \qquad \text{Use a calculator.}$$

The wooden statue is approximately 9000 years old.   ∎

Self Check   How old is an artifact that now has 60% of its original carbon-14 content?
Answer   about 4200 years old

■ **POPULATION GROWTH**

When there is sufficient food and space, populations of living organisms tend to increase exponentially according to the Malthusian growth model

$$P = P_0 e^{kt}$$

where P_0 is the initial population at $t = 0$ and k depends on the rate of growth.

EXAMPLE 8   The bacteria in a laboratory culture increased from an initial population of 500 to 1500 in 3 hours. Find the time it will take for the population to reach 10,000.

Solution
$$P = P_0 e^{kt}$$
$$1500 = 500(e^{k3}) \qquad \text{Substitute 1500 for } P, \text{ 500 for } P_0, \text{ and 3 for } t.$$
$$3 = e^{3k} \qquad \text{Divide both sides by 500.}$$
$$3k = \ln 3 \qquad \text{Change the equation from exponential to logarithmic form.}$$
$$k = \frac{\ln 3}{3} \qquad \text{Divide both sides by 3.}$$

To find out when the population will reach 10,000, we substitute 10,000 for P, 500 for P_0, and $\frac{\ln 3}{3}$ for k in the equation $P = P_0 e^{kt}$ and solve for t:

$$P = P_0 e^{kt}$$
$$10{,}000 = 500 e^{[\ln 3/3]t}$$

$$20 = e^{[\ln 3/3]t} \qquad \text{Divide both sides by 500.}$$

$$\left(\frac{\ln 3}{3}\right)t = \ln 20 \qquad \text{Change the equation to logarithmic form.}$$

$$t = \frac{3 \ln 20}{\ln 3} \qquad \text{Multiply both sides by } \frac{3}{\ln 3}.$$

$$\approx 8.180499084 \qquad \text{Use a calculator.}$$

The culture will reach 10,000 bacteria in a little more than 8 hours. ∎

Self Check If the population increases from 1000 to 3000 in 3 hours, how long will it take to reach 20,000?

Answer about 8 hours

EXERCISE 5.6

VOCABULARY AND CONCEPTS *In Exercises 1–4, fill in the blank to make a true statement.*

1. An equation with a variable in its exponent is called a(n) _____ equation.

2. An equation with a logarithmic expression that contains a variable is a(n) _____ equation.

3. The formula for carbon dating is $A =$ _____.

4. The formula for population growth is $P =$ _____.

PRACTICE   *In Exercises 5–28, solve each equation. If an answer is not exact, give the answer to four decimal places.*

5. $4^x = 5$

6. $7^x = 12$

7. $13^{x-1} = 2$

8. $5^{x+1} = 3$

9. $2^{x+1} = 3^x$

10. $5^{x-3} = 3^{2x}$

11. $2^x = 3^x$

12. $3^{2x} = 4^x$

13. $7^{x^2} = 10$

14. $8^{x^2} = 11$

15. $8^{x^2} = 9^x$

16. $5^{x^2} = 2^{5x}$

17. $2^{x^2-2x} = 8$

18. $5^{x^2-3x} = 625$

19. $3^{x^2+4x} = \dfrac{1}{81}$

20. $7^{x^2+3x} = \dfrac{1}{49}$

21. $4^{x+2} - 4^x = 15$ (*Hint:* $4^{x+2} = 4^x 4^2$.)

22. $3^{x+3} + 3^x = 84$ (*Hint:* $3^{x+3} = 3^x 3^3$.)

23. $2(3^x) = 6^{2x}$

24. $2(3^{x+1}) = 3(2^{x-1})$

25. $2^{2x} - 10(2^x) + 16 = 0$ (*Hint:* Let $y = 2^x$.)

26. $3^{2x} - 10(3^x) + 9 = 0$ (*Hint:* Let $y = 3^x$.)

27. $2^{2x+1} - 2^x = 1$ (*Hint:* $2^{a+b} = 2^a 2^b$.)

28. $3^{2x+1} - 10(3^x) + 3 = 0$ (*Hint:* $3^{a+b} = 3^a 3^b$.)

In Exercises 29–54, solve each equation.

29. $\log(2x - 3) = \log(x + 4)$

30. $\log(3x + 5) - \log(2x + 6) = 0$

31. $\log \dfrac{4x + 1}{2x + 9} = 0$

32. $\log \dfrac{5x + 2}{2(x + 7)} = 0$

33. $\log x^2 = 2$

34. $\log x^3 = 3$

35. $\log x + \log(x - 48) = 2$

36. $\log x + \log(x + 9) = 1$

37. $\log x + \log(x - 15) = 2$

38. $\log x + \log(x + 21) = 2$

39. $\log(x + 90) = 3 - \log x$

40. $\log(x - 6) - \log(x - 2) = \log \dfrac{5}{x}$

41. $\log(x - 1) - \log 6 = \log(x - 2) - \log x$

42. $\log(2x - 3) - \log(x - 1) = 0$

43. $\log x^2 = (\log x)^2$

44. $\log(\log x) = 1$

45. $\dfrac{\log(3x - 4)}{\log x} = 2$

46. $\dfrac{\log(8x - 7)}{\log x} = 2$

47. $\dfrac{\log(5x + 6)}{2} = \log x$

48. $\dfrac{1}{2}\log(4x + 5) = \log x$

49. $\log_3 x = \log_3\left(\dfrac{1}{x}\right) + 4$

50. $\log_5(7 + x) + \log_5(8 - x) - \log_5 2 = 2$

51. $2\log_2 x = 3 + \log_2(x - 2)$

52. $2\log_3 x - \log_3(x - 4) = 2 + \log_3 2$

53. $\log(7y + 1) = 2\log(y + 3) - \log 2$

54. $2\log(y + 2) = \log(y + 2) - \log 12$

In Exercises 55–58, use a graphing calculator to solve each equation.

55. $\log x + \log(x - 15) = 2$

56. $\log x + \log(x + 3) = 1$

57. $2^{x+1} = 7$

58. $\ln(2x + 5) - \ln 3 = \ln(x - 1)$

APPLICATIONS   *Use a calculator to help solve each problem.*

59. Tritium decay   The half-life of tritium is 12.4 years. How long will it take for 25% of a sample of tritium to decompose?

60. Radioactive decay   In two years, 20% of a radioactive element decays. Find its half-life.

61. Thorium decay   An isotope of thorium, ^{227}Th, has a half-life of 18.4 days. How long will it take 80% of the sample to decompose?

62. Lead decay   An isotope of lead, ^{201}Pb, has a half-life of 8.4 hours. How many hours ago was there 30% more of the substance?

63. Carbon-14 dating   A cloth fragment is found in an ancient tomb. It contains 60% of the carbon-14 that it is assumed to have had initially. How old is the cloth?

64. Carbon-14 dating   Only 10% of the carbon-14 in a small wooden bowl remains. How old is the bowl?

65. Compound interest   If $500 is deposited in an account paying 8.5% annual interest, compounded semiannually, how long will it take for the account to increase to $800?

66. Continuous compound interest   In Exercise 65, how long will it take if the interest is compounded continuously?

67. Compound interest   If $1300 is deposited in a savings account paying 9% interest, compounded quarterly, how long will it take the account to increase to $2100?

68. Compound interest A sum of $5000 deposited in an account grows to $7000 in 5 years. Assuming annual compounding, what interest rate is being paid?

69. Rule of seventy A rule of thumb for finding how long it takes an investment to double is called the **rule of seventy.** To apply the rule, divide 70 by the interest rate (expressed as a percent). At 5%, it takes $\frac{70}{5} = 14$ years to double the investment. At 7%, it takes $\frac{70}{7} = 10$ years. Explain why this formula works.

70. Bacterial growth A bacterial culture grows according to the formula

$$P = P_0 a^t$$

If it takes 5 days for the culture to triple in size, how long will it take to double in size?

71. Oceanography The intensity I of a light a distance x meters beneath the surface of a lake decreases exponentially. If the light intensity at 6 meters is 70% of the intensity at the surface, at what depth will the intensity be 20%?

72. Rodent control The rodent population in a city is currently estimated at 30,000. If it is expected to double every 5 years, when will the population reach 1 million?

73. Newton's law of cooling Water whose temperature is at 100°C is left to cool in a room where the temperature is 60°C. After 3 minutes, the water temperature is 90°. If the water temperature T is a function of time t given by $T = 60 + 40e^{kt}$, find k.

74. Newton's law of cooling Refer to Exercise 73 and find the time for the water temperature to reach 70°C.

75. Newton's law of cooling A block of steel, initially at 0°C, is placed in an oven heated to 300°C. After 5 minutes, the temperature of the steel is 100°C. If the steel temperature T is a function of time t given by $T = 300 - 300e^{kt}$, find the value of k.

76. Newton's law of cooling Refer to Exercise 75 and find the time for the steel temperature to reach 200°C.

DISCOVERY AND WRITING

77. Explain why it is necessary to check the solutions of a logarithmic equation.

78. Invent a logarithmic equation with one solution: $x = 2$.

79. Invent an exponential equation with one solution: $x = 2$.

80. Can you solve $x = \log x$ algebraically? Can you find an approximate solution?

REVIEW *Find the inverse of the function defined by each equation.*

81. $y = 3x + 2$

82. $y = \dfrac{1}{x - 3}$

Let $f(x) = 5x - 1$ and $g(x) = x^2$. Find each value.

83. $(f \circ g)(2)$

84. $(g \circ f)(2)$

85. $(f \circ g)(x)$

86. $(g \circ f)(x)$

■ ■ ■ ■ ■ ■ ■ ■ ■ **PROBLEMS AND PROJECTS**

1. Match each equation with its graph.

____ **a.** $y = 3^x$

____ **b.** $y = 3^x - 1$

____ **c.** $y = 3^{x-1}$

____ **d.** $y = \log_3 x$ ▪

____ **e.** $y = \log_3 (x - 1)$

____ **f.** $y = \log_{1/3} x$

(continued)

■ ■ ■ ■ ■ ■ ■ ■ ■ ■ **PROBLEMS AND PROJECTS** *(continued)*

s.
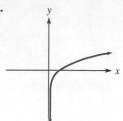

t.
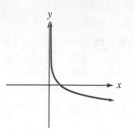

u.
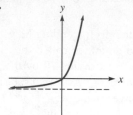

v.
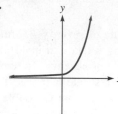

w.
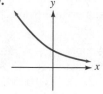

x.
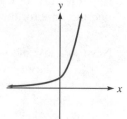

y.
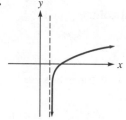

z.
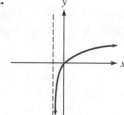

2. On a graphing calculator, graph the function $y = \ln(e^x)$. Explain why the graph is a line. What is a simpler form of the equation of that line?

3. Graph the function $y = e^{\ln x}$ on a graphing calculator. Is its graph a line, or just part of a line? Explain.

4. The tank in Illustration 1 initially contains 20 gallons of pure water. A brine solution containing 0.5 pounds of salt per gallon is pumped into the tank, and the well-stirred mixture leaves at the same rate. The amount A of salt in the tank after t minutes is given by

$$A = 10(1 - e^{-0.03t})$$

a. Graph this function.

b. What is A when $t = 0$? Explain why that value is expected.

c. What is A after 2 minutes? After 10 minutes?

d. What value does A approach after a long time (as t becomes large)? Explain why this is the value you would expect.

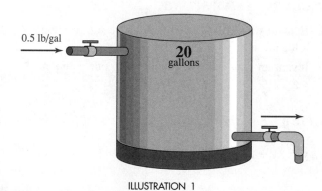

ILLUSTRATION 1

PROJECT 1   Graphing calculators graph base-e and base-10 logarithmic functions easily, because logarithms to these bases are built-in. Find a way of graphing the function $y = \log_3 x$, even though there is no $\boxed{\log_3}$ key.

PROJECT 2   For positive values of x, the graphs of the polynomial function $y = x^3$ and the exponential function $y = e^x$ are both increasing, as shown in Illustration 2. Which is increasing faster? At $x = 2$, 3, and 4, the graph of the polynomial is winning, but when $x = 5$, the graph of the exponential has caught up to and passed the polynomial graph.

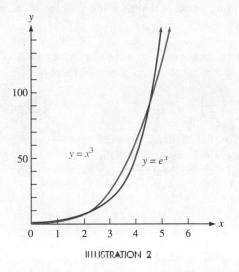

ILLUSTRATION 2

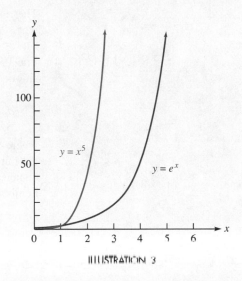

ILLUSTRATION 3

The higher a polynomial's degree, the faster its graph rises. Illustration 3 shows the graphs of $y = x^5$ and $y = e^x$. It looks as if the graph of the polynomial is rising more rapidly. Will the exponential graph ever catch up? In a race between a polynomial and an exponential function, which function will eventually win?

Experiment with a graphing calculator, and graph $y = x^5$ and $y = e^x$. Then try a race between $y = x^{20}$ and $y = e^x$. (*Hint:* Useful viewing windows are difficult to find. For $y = x^{20}$, try $80 \leq x \leq 100$ and $0 \leq Y \leq 5 \times 10^{39}$.) Write a brief report of your conclusions.

PROJECT 3   When graphing a function $y = f(x)$, various numbers x are used to produce corresponding values of y, and many pairs (x, y) are plotted to produce the graph. In **parametric equations,** various values of a third variable, called a **parameter,** are used to generate both x- and y- values. Set your graphing calculator for parametric equations (try the MODE key, or consult the manual). Set the range of the parameter, t:

Tmin = 0

Tmax = 10

Tstep = 0.1

(continued)

■ ■ ■ ■ ■ ■ ■ ■ ■ ■ **PROBLEMS AND PROJECTS** *(continued)*

Then graph the parametric equations

$$X_{1_T} = \ln t$$
$$Y_{1_T} = t$$

The resulting graph is that of $y = e^x$, which is the inverse of $y = \ln x$. Explain why.

PROJECT 4   The 19 frets on the neck of a classical guitar such as that in Illustration 4 are positioned according to the exponential function

$$d = L\left(\frac{1}{2}\right)^{n/12}$$

where L is the distance between the nut and the bridge and d is the distance from the bridge to fret number n.

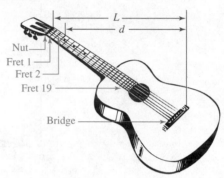

ILLUSTRATION 4

- Find a classical guitar and measure its nut-to-bridge distance, L. Use this value in the equation and calculate the distance to frets 1 and 19. Do the calculated values agree with the measured distances?

- Calculate and measure the distance to fret 12, which produces a tone one octave above that of the open string. What fractional part of L is this distance?

- Calculate the distance to fret zero. From the value calculated, where is fret zero?

- Does this formula work for an electric guitar?

C H A P T E R   S U M M A R Y

CONCEPT

REVIEW EXERCISES

| SECTION 5.1 | *Exponential Functions* |

An exponential function with base b is defined by the equation

$$y = f(x) = b^x$$
$$(b > 0, b \neq 1)$$

1. Use properties of exponents to simplify.

    a. $5^{\sqrt{2}} \cdot 5^{\sqrt{2}}$ **b.** $\left(2^{\sqrt{5}}\right)^{\sqrt{2}}$

2. Graph the function defined by each equation.

    a. $y = 3^x$ **b.** $y = \left(\dfrac{1}{3}\right)^x$

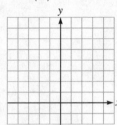

3. The graph of $f(x) = 6^x$ will pass through the points $(0, p)$ and $(1, q)$. Find p and q.

4. Give the domain and range of the function $f(x) = b^x$.

5. Graph each function by using a translation.

    a. $y = f(x) = \left(\dfrac{1}{2}\right)^x - 2$ **b.** $y = f(x) = \left(\dfrac{1}{2}\right)^{x+2}$

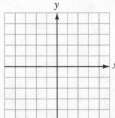

Radioactive decay:
$$A = A_0 2^{-t/h}$$

Compound interest:
$$A = P\left(1 + \dfrac{r}{k}\right)^{kt}$$

Intensity of light:
$$I = I_0 k^x$$

6. 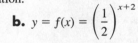 One-third of a radioactive material decays in 20 years. Find its half-life.

7. 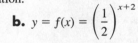 How much will \$10,500 become if it earns 9% per year for 60 years, compounded quarterly?

8. 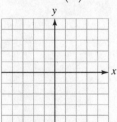 Find the intensity of light at a depth of 12 meters if $I_0 = 14$ and $k = 0.7$

| **SECTION 5.2** | *Base-e Exponential Functions* |

$e = 2.71828182845904...$

Continuous compound interest:

$A = Pe^{rt}$

9. 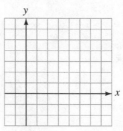 If \$10,500 accumulates interest at an annual rate of 9% compounded continuously, how much will be in the account in 60 years?

10. Graph each function.
   a. $y = f(x) = e^x + 1$ **b.** $y = f(x) = e^{x-3}$

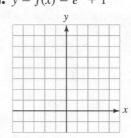

Malthusian population growth:

$P = P_0 e^{kt}$

11. 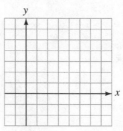 The population of the United States is approximately 275,000,000 people. Find the population in 50 years if $k = 0.015$.

| **SECTION 5.3** | *Logarithmic Functions* |

If $b > 0$ and $b \ne 1$, then

$y = \log_b x$ means $x = b^y$

12. Give the domain and range of the logarithmic function.

13. Find each value.
   a. $\log_3 9$ **b.** $\log_9 \dfrac{1}{3}$

   c. $\log_x 1$ **d.** $\log_5 0.04$

   e. $\log_a \sqrt{a}$ **f.** $\log_a \sqrt[3]{a}$

14. Find x.
   a. $\log_2 x = 5$ **b.** $\log_{\sqrt{3}} x = 4$

   c. $\log_{\sqrt{2}} x = 6$ **d.** $\log_{0.1} 10 = x$

   e. $\log_x 2 = -\dfrac{1}{3}$ **f.** $\log_x 32 = 5$

   g. $\log_{0.25} x = -1$ **h.** $\log_{0.125} x = -\dfrac{1}{3}$

   i. $\log_{\sqrt{2}} 32 = x$ **j.** $\log_{\sqrt{5}} x = -4$

   k. $\log_{\sqrt{3}} 9\sqrt{3} = x$ **l.** $\log_{\sqrt{5}} 5\sqrt{5} = x$

15. Graph each function.

a. $y = f(x) = \log(x - 2)$

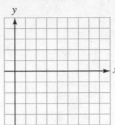

b. $y = f(x) = 3 + \log x$

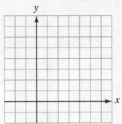

16. Graph each pair of equations on one set of coordinate axes.

a. $y = 4^x$ and $y = \log_4 x$

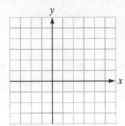

b. $y = \left(\dfrac{1}{3}\right)^x$ and $y = \log_{1/3} x$

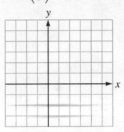

Decibel voltage gain:

$$\text{db gain} = 20 \log \frac{E_O}{E_I}$$

17. An amplifier has an output of 18 volts when the input is 0.04 volt. Find the db gain.

Richter scale:

$$R = \log \frac{A}{P}$$

18. An earthquake had a period of 0.3 second and an amplitude of 7500 micrometers. Find its measure on the Richter scale.

SECTION 5.4	*Base-e Logarithms*

$\ln x$ means $\log_e x$.

19. 🖩 Use a calculator to find each value to four decimal places.

a. $\ln 452$

b. $\ln(\log 7.85)$

20. Solve each equation. Round each answer to four decimal places.

a. $\ln x = 2.336$

b. $\ln x = \log 8.8$

21. Graph each function.

a. $y = f(x) = 1 + \ln x$

b. $y = f(x) = \ln (x + 1)$

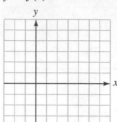

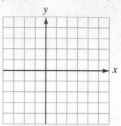

Charging batteries:

$C = M(1 - e^{-kt})$

22. A dead battery is charged for $9\frac{1}{2}$ hours. What percent of maximum charge is attained? (Assume $k = 0.17$.)

Population doubling time:

$t = \dfrac{\ln 2}{r}$

23. How long will it take the population of the United States to double if the growth rate is 3% per year?

Isothermal expansion:

$E = RT \ln \left(\dfrac{V_f}{V_i} \right)$

24. Find the amount of energy that must be supplied to double the volume of 1 mole of gas at a constant temperature of 350K. (*Hint: R = 8.314.*)

| SECTION 5.5 | *Properties of Logarithms* |

Properties of logarithms:
If b is a positive number and $b \neq 1$,

1. $\log_b 1 = 0$
2. $\log_b b = 1$
3. $\log_b b^x = x$
4. $b^{\log_b x} = x$
5. $\log_b MN = \log_b M + \log_b N$
6. $\log_b \dfrac{M}{N} = \log_b M - \log_b N$
7. $\log_b M^p = p \log_b M$
8. If $\log_b x = \log_b y$, then $x = y$

25. Simplify each expression.
  a. $\log_7 1$
  b. $\log_7 7$
  c. $\log_7 7^3$
  d. $7^{\log_7 4}$

26. Simplify each expression.
  a. $\ln e^4$
  b. $\ln 1$
  c. $10^{\log_{10} 7}$
  d. $e^{\log_e 3}$
  e. $\log_b b^4$
  f. $\ln e^9$

27. Write each expression in terms of the logarithms of x, y, and z.

  a. $\log_b \dfrac{x^2 y^3}{z^4}$

  b. $\log_b \sqrt{\dfrac{x}{yz^2}}$

28. Write each expression as the logarithm of one quantity.

  a. $3 \log_b x - 5 \log_b y + 7 \log_b z$

  b. $\dfrac{1}{2}(\log_b x + 3 \log_b y) - 7 \log_b z$

29. Assume that $\log a = 0.6$, $\log b = 0.36$, and $\log c = 2.4$. Find the value of each expression.

a. $\log abc$

b. $\log a^2 b$

c. $\log \dfrac{ac}{b}$

d. $\log \dfrac{a^2}{c^3 b^2}$

Change-of-base formula:

$$\log_b y = \frac{\log_a y}{\log_a b}$$

30. ⊞ To four decimal places, find $\log_5 17$.

pH scale:

$$pH = -\log [H^+]$$

31. pH of grapefruit   The pH of grapefruit juice is about 3.1. Find its hydrogen ion concentration.

Weber–Fechner law:

$$L = k \ln I$$

32. Find the decrease in loudness if the intensity is cut in half.

SECTION 5.6	*Exponential and Logarithmic Equations*

33. Solve each equation for x.

a. $3^x = 7$

b. $5^{x+2} = 625$

c. $2^x = 3^{x-1}$

d. $2^{x^2+4x} = \dfrac{1}{8}$

34. Solve each equation for x.

a. $\log x + \log (29 - x) = 2$

b. $\log_2 x + \log_2 (x - 2) = 3$

c. $\log_2 (x + 2) + \log_2 (x - 1) = 2$

d. $\dfrac{\log (7x - 12)}{\log x} = 2$

e. $\log x + \log (x - 5) = \log 6$

f. $\log 3 - \log (x - 1) = -1$

g. $e^{x \ln 2} = 9$

h. $\ln x = \ln (x - 1)$

i. $\ln x = \ln (x - 1) + 1$

j. $\ln x = \log_{10} x$ (*Hint:* Use the change-of-base formula.)

Carbon dating:

$$A = A_0 2^{-t/h}$$

35. Carbon-14 dating   A wooden statue found in Egypt has a carbon-14 content that is two-thirds of that found in living wood. If the half-life of carbon-14 is 5700 years, how old is the statue?

■ Chapter 5 Test

In Questions 1–2, graph each function.

1. $y = 2^x + 1$

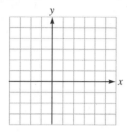

2. $y = 2^{-x}$

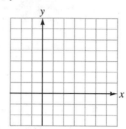

In Questions 3–4, solve each equation.

3. A radioactive material decays according to the formula $A = A_0(2)^{-t}$. How much of a 3-gram sample will be left in 6 years?

4. An initial deposit of $1000 earns 6% interest, compounded twice a year. How much will be in the account in one year?

In Questions 5–6, graph each function.

5. $y = e^x$

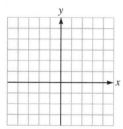

6. $y = 2e^x$

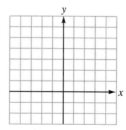

7. An account contains $2000 and has been earning 8% interest, compounded continuously. How much will be in the account in 10 years?

In Questions 8–12, find each value.

8. $\log_7 343$

9. $\log_3 \dfrac{1}{27}$

10. $\log_{10} 10^{12} + 10^{\log_{10} 5}$

11. $\log_{3/2} \dfrac{9}{4}$

12. $\log_{2/3} \dfrac{27}{8}$

In Questions 13–14, graph each function.

13. $y = \log (x - 1)$

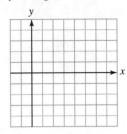

14. $y = 2 + \ln x$

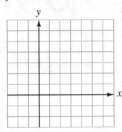

In Questions 15–16, write each expression in terms of the logarithms of a, b, and c.

15. $\log a^2 b c^3$

16. $\ln \sqrt{\dfrac{a}{b^2 c}}$

In Questions 17–18, write each expression as a logarithm of a single quantity.

17. $\dfrac{1}{2} \log (a + 2) + \log b - 2 \log c$

18. $\dfrac{1}{3}(\log a - 2 \log b) - \log c$

In Questions 19–20, assume that $\log 2 = 0.3010$ and $\log 3 = 0.4771$. Find each value. **Do not use a calculator.**

19. $\log 24$

20. $\log \dfrac{8}{3}$

In Questions 21–22, use the change-of-base formula to find each logarithm. **Do not attempt to simplify the answer.**

21. $\log_7 3$

22. $\log_\pi e$

In Questions 23–26, tell whether each statement is true or false.

23. $\log_a ab = 1 + \log_a b$

24. $\dfrac{\log a}{\log b} = \log a - \log b$

25. $\log a^{-3} = \dfrac{1}{3 \log a}$

26. $\ln (-x) = -\ln x$

27. Find the pH of a solution with a hydrogen ion concentration of 3.7×10^{-7}. (*Hint:* $pH = -\log [H^+]$.)

28. Find the db gain of an amplifier when $E_O = 60$ volts and $E_I = 0.3$ volt. (*Hint:* db gain $= 20 \log (E_O/E_I)$.)

In Questions 29–32, solve each equation.

29. $3^{x-1} = 100^x$

30. $3^{x^2 - 2x} = 27$

31. $\log (5x + 2) = \log (2x + 5)$

32. $\log x + \log (x - 9) = 1$

6 Solving Polynomial Equations

A GOAL THROUGHOUT ALL OF ALGEBRA HAS BEEN TO SOLVE EQUATIONS. IN THIS CHAPTER, WE WILL DEVELOP METHODS TO SOLVE POLYNOMIAL EQUATIONS OF ANY DEGREE.

6.1   The Remainder and Factor Theorems; Synthetic Division

■ ZEROS OF A POLYNOMIAL ■ THE REMAINDER THEOREM ■ THE FACTOR THEOREM ■ SYNTHETIC DIVISION ■ USING SYNTHETIC DIVISION TO EVALUATE POLYNOMIALS

We have seen that many polynomial equations can be solved by factoring. For example, to solve the equation $x^3 - 3x^2 + 2x = 0$, we proceed as follows:

$$x^3 - 3x^2 + 2x = 0$$
$$x(x^2 - 3x + 2) = 0 \qquad \text{Factor out } x.$$
$$x(x - 1)(x - 2) = 0 \qquad \text{Factor } x^2 - 3x + 2x.$$
$$x = 0 \quad \text{or} \quad x - 1 = 0 \quad \text{or} \quad x - 2 = 0 \qquad \text{Set each factor equal to 0.}$$
$$x = 1 \qquad\qquad x = 2$$

The solution set of this equation is $\{0, 1, 2\}$.

■ ZEROS OF A POLYNOMIAL

In general, a **polynomial equation** is an equation that can be written in the form $P(x) = 0$, where

$$P(x) = a_n x^n + a_{n-1} x^{n-1} + a_{n-2} x^{n-2} + \cdots + a_1 x + a_0$$

where n is a natural number and the polynomial is of degree n. A **zero of the polynomial** $P(x)$ is any number r for which $P(r) = 0$. It follows that a zero of $P(x)$ is a root of the polynomial equation $P(x) = 0$. For example, 0, 1, and 2 are zeros of the polynomial

$$P(x) = x^3 - 3x^2 + 2x$$

because

$$P(0) = 0^3 - 3(0)^2 + 2(0) \qquad P(1) = 1^3 - 3(1)^2 + 2(1) \qquad P(2) = 2^3 - 3(2)^2 + 2(2)$$
$$= 0 - 0 + 0 \qquad\qquad\quad = 1 - 3 + 2 \qquad\qquad\quad = 8 - 12 + 4$$
$$= 0 \qquad\qquad\qquad\qquad = 0 \qquad\qquad\qquad\qquad = 0$$

We began the section by showing that the zeros of $x^3 - 3x^2 + 2x$ are the roots of the polynomial equation

$$x^3 - 3x^2 + 2x = 0$$

■ THE REMAINDER THEOREM

There is a relationship between a zero r of a polynomial $P(x)$ and the results of a long division of $P(x)$ by $x - r$. This relationship is best shown with an example.

EXAMPLE 1   Let $P(x) = 3x^3 - 5x^2 + 3x - 10$. Find   **a.** $P(1)$ and   **b.** divide $P(x)$ by $x - 1$.

Solution   **a.** $P(1) = 3(1)^3 - 5(1)^2 + 3(1) - 10$

$$= 3 - 5 + 3 - 10$$

$$= -9$$

b.
$$
\begin{array}{r}
3x^2 - 2x + 1 \\
x - 1{\overline{\smash{\big)}\,3x^3 - 5x^2 + 3x - 10}} \\
\underline{3x^3 - 3x^2} \\
-2x^2 + 3x \\
\underline{-2x^2 + 2x} \\
+ x - 10 \\
\underline{x - 1} \\
- 9
\end{array}
$$

Note that the remainder is equal to $P(1)$. ∎

Self Check   Let $P(x) = 2x^2 - 3x + 5$. Find $P(2)$ and divide $P(x)$ by $x - 2$. What do you notice about the results?

Answer $P(2)$ is the remainder.

EXAMPLE 2   Let $P(x) = 3x^3 - 5x^2 + 3x - 10$.   **a.** Find $P(-2)$ and   **b.** divide $P(x)$ by $x + 2$.

Solution   **a.** $P(-2) = 3(-2)^3 - 5(-2)^2 + 3(-2) - 10$

$$= 3(-8) - 5(4) + 3(-2) - 10$$

$$= -24 - 20 - 6 - 10$$

$$= -60$$

b.
$$
\begin{array}{r}
3x^2 - 11x + 25 \\
x + 2{\overline{\smash{\big)}\,3x^3 - 5x^2 + 3x - 10}} \\
\underline{3x^3 + 6x^2} \\
-11x^2 + 3x \\
\underline{-11x^2 - 22x} \\
25x - 10 \\
\underline{25x + 50} \\
- 60
\end{array}
$$

Note that the remainder is equal to $P(-2)$. ∎

Self Check   Let $P(x) = 2x^2 - 3x + 5$. Find $P(-3)$ and divide $P(x)$ by $x + 3$. What do you notice about the results?

Answer $P(-3)$ is the remainder.

When $P(x)$ was divided by $x - 1$ in Example 1, the remainder was $P(1) = -9$. When $P(x)$ was divided by $x + 2$, or $x - (-2)$, in Example 2, the remainder was $P(-2) = -60$. These results are not coincidental. The **remainder theorem** states that a division of any polynomial $P(x)$ by $x - r$ gives $P(r)$ as the remainder.

> **Remainder Theorem**
>
> If $P(x)$ is a polynomial, r is a real or nonreal complex number, and $P(x)$ is divided by $x - r$, then the remainder is $P(r)$.

Proof   To divide $P(x)$ by $x - r$, we must find a quotient $Q(x)$ and a remainder $R(x)$ such that

$$\text{dividend} = \text{divisor} \cdot \text{quotient} + \text{remainder}$$
$$\downarrow \qquad \downarrow \qquad \downarrow \qquad \downarrow$$
$$P(x) \quad = (x - r) \cdot \quad Q(x) \quad + \quad R(x)$$

Since the degree of the remainder $R(x)$ must be less than the degree of the divisor $x - r$, and the degree of $x - r$ is 1, $R(x)$ must be a constant R.

In the equation

$$P(x) = (x - r)Q(x) + R$$

the polynomial on the left-hand side is the same as the polynomial on the right-hand side, and the values that they assume for any number x are equal. If we replace x with r, we have

$$P(r) = (r - r)Q(r) + R$$
$$= (0)Q(r) + R$$
$$= R$$

Thus, $P(r) = R$.   □

EXAMPLE 3   Find the remainder that will occur when $P(x) = 2x^4 - 10x^3 + 17x^2 - 14x - 3$ is divided by $x - 3$.

Solution   By the remainder theorem, the remainder will be $P(3)$.

$$P(x) = 2x^4 - 10x^3 + 17x^2 - 14x - 3$$
$$P(3) = 2(3)^4 - 10(3)^3 + 17(3)^2 - 14(3) - 3 \qquad \text{Substitute 3 for } x.$$
$$= 0$$

The remainder will be 0. Although this calculation is tedious, it is easy to do with a calculator.   ■

Self Check   Find the remainder when $P(x)$ is divided by $x - 2$.
Answer -11

■ THE FACTOR THEOREM

If $R = P(r) = 0$ in the equation $P(x) = (x - r)Q(x) + R$, then $P(x)$ factors as $(x - r)Q(x)$. This fact can help us factor polynomials.

Factor Theorem
Let $P(x)$ be any polynomial, and let r be a real or nonreal complex number. Then

If $P(r) = 0$, then $x - r$ is a factor of $P(x)$.

If $x - r$ is a factor of $P(x)$, then $P(r) = 0$.

Proof *Part 1:* First, we assume that $P(r) = 0$ and prove that $x - r$ is a factor of $P(x)$. If $P(r) = 0$, then $R = 0$, and the equation $P(x) = (x - r)Q(x) + R$ becomes

$$P(x) = (x - r)Q(x) + 0$$
$$P(x) = (x - r)Q(x)$$

Therefore, $x - r$ divides $P(x)$ evenly, and $x - r$ is a factor of $P(x)$.

Part 2: Conversely, we assume that $x - r$ is a factor of $P(x)$ and prove that $P(r) = 0$. Because, by assumption, $x - r$ is a factor of $P(x)$, $x - r$ divides $P(x)$ evenly, and the division has a remainder of 0. By the remainder theorem, this remainder is $P(r)$. Hence, $P(r) = 0$. ☐

We can state the factor theorem in a slightly different way:

If r is a zero of the polynomial $P(x)$, then $x - r$ is a factor of $P(x)$.

If $x - r$ is a factor of $P(x)$, then r is a zero of the polynomial.

EXAMPLE 4   Determine whether $x + 2$ is a factor of $P(x) = x^4 - 7x^2 - 6x$. If so, find the other factor by long division.

Solution By the factor theorem, $x + 2$, or $x - (-2)$, is a factor of $P(x)$ if $P(-2) = 0$. So we find the value of $P(-2)$.

$$P(x) = x^4 - 7x^2 - 6x$$
$$P(-2) = (-2)^4 - 7(-2)^2 - 6(-2) \qquad \text{Substitute } -2 \text{ for } x.$$
$$= 16 - 28 + 12$$
$$= 0$$

Since $P(-2) = 0$, we know that $x - (-2)$, or $x + 2$, is a factor of $P(x)$.
   To find the other factor, we divide $x^4 - 7x^2 - 6x$ by $x + 2$.

$$
\begin{array}{r}
x^3 - 2x^2 - 3x \\
x + 2 \overline{)x^4 \qquad\quad - 7x^2 - 6x} \\
\underline{x^4 + 2x^3} \\
-2x^3 - 7x^2 \\
\underline{-2x^3 - 4x^2} \\
-3x^2 - 6x \\
\underline{-3x^2 - 6x} \\
0
\end{array}
$$

$x^4 - 7x^2 - 6x$ factors as $(x + 2)(x^3 - 2x^2 - 3x)$. ∎

Self Check   Is $x + 3$ a factor of $P(x)$?

Answer   no

EXAMPLE 5   Let $P(x) = 3x^3 - 5x^2 + 3x - 10$. Show that $P(2) = 0$ and use this fact to help factor $P(x)$.

Solution   We find $P(2)$.

$$P(x) = 3x^3 - 5x^2 + 3x - 10$$
$$P(2) = 3(2)^3 - 5(2)^2 + 3(2) - 10 \qquad \text{Substitute 2 for } x.$$
$$= 3(8) - 5(4) + 6 - 10$$
$$= 0$$

Since $P(2) = 0$, it follows that $x - 2$ is a factor of $P(x)$. To find the other factor, we divide $P(x)$ by $x - 2$.

$$
\begin{array}{r}
3x^2 + x + 5 \\
x - 2 \overline{)\, 3x^3 - 5x^2 + 3x - 10} \\
\underline{3x^3 - 6x^2} \\
x^2 + 3x \\
\underline{x^2 - 2x} \\
5x - 10 \\
\underline{5x - 10} \\
0
\end{array}
$$

$3x^3 - 5x^2 + 3x - 10$ factors as $(x - 2)(3x^2 + x + 5)$. ∎

Self Check   Show that $P(1) = 0$ and use this fact to help factor $x^3 - 1$.

Answer $(x - 1)(x^2 + x + 1)$

EXAMPLE 6   Solve $3x^3 - 5x^2 + 3x - 10 = 0$.

Solution   From Example 5, we have

$$3x^3 - 5x^2 + 3x - 10 = 0$$
$$(x - 2)(3x^2 + x + 5) = 0$$

To solve for x, we set each factor equal to 0 and apply the quadratic formula to the equation $3x^2 + x + 5 = 0$.

$$x - 2 = 0 \qquad \text{or} \qquad 3x^2 + x + 5 = 0$$

$$x = 2 \qquad\qquad x = \frac{-1 \pm \sqrt{1^2 - 4(3)(5)}}{2(3)}$$

$$x = \frac{-1 \pm i\sqrt{59}}{6}$$

The solution set is $\left\{ 2, -\dfrac{1}{6} + \dfrac{\sqrt{59}}{6}i, -\dfrac{1}{6} - \dfrac{\sqrt{59}}{6}i \right\}$. ∎

Self Check   Solve $x^3 - 1$.

Answer $1, \dfrac{-1 \pm i\sqrt{3}}{2}$

EXAMPLE 7   Find a polynomial $P(x)$ with zeros of 3, 3, and -5.

Solution   We have proved that if r is a zero of a polynomial, then $x - r$ is a factor of the polynomial. In particular, if 3, 3, and -5 are zeros of $P(x)$, then $x - 3$, $x - 3$, and $x - (-5)$ are factors of $P(x)$.

$$P(x) = (x - 3)(x - 3)(x + 5)$$
$$P(x) = (x^2 - 6x + 9)(x + 5) \qquad \text{Multiply } x - 3 \text{ and } x - 3.$$
$$P(x) = x^3 - x^2 - 21x + 45$$

The polynomial $P(x) = x^3 - x^2 - 21x + 45$ has zeros of 3, 3, and -5.   ■

Self Check   Find a polynomial $P(x)$ with zeros of -2, 2, and 3.

Answer $x^3 - 3x^2 - 4x + 12$

Because 3 occurs twice as a zero in Example 7, we say that 3 is a **zero of multiplicity 2.**

■ SYNTHETIC DIVISION

Synthetic division is an easy way to divide higher-degree polynomials by binomials of the form $x - r$. To see how it works, we consider the following long division. On the left is a complete division. On the right is a modified version in which the variables have been removed.

$$
\begin{array}{r}
2x^2 + 10x\ + 27 \\
x - 3\overline{)2x^3 + \ 4x^2 - \ \ 3x + 10} \\
\underline{2x^3 - \ 6x^2} \\
10x^2 - \ \ 3x \\
\underline{10x^2 - 30x} \\
27x + 10 \\
\underline{27x - 81} \\
\text{(remainder)} \quad 91
\end{array}
\qquad
\begin{array}{r}
2 + 10 + 27 \\
1 - 3\overline{)2 + \ 4 - \ \ 3 + 10} \\
\underline{2 - \ 6} \\
10 - \ 3 \\
\underline{10 - 30} \\
27 + 10 \\
\underline{27 - 81} \\
\text{(remainder)} \quad 91
\end{array}
$$

We can shorten the work even more by omitting the numbers printed in color.

$$
\begin{array}{r}
2 + 10 + 27 \\
-3\overline{)2 + 4 - 3 + 10} \\
\underline{- 6} \\
10 \\
\underline{- 30} \\
27 \\
\underline{- 81} \\
\text{(remainder)}\quad 91
\end{array}
$$

We can then compress the work vertically to get

$$
\begin{array}{r}
2 + 10 + 27 \\
-3\overline{)2 + 4 - 3 + 10} \\
\underline{- 6 - 30 - 81} \\
10 \quad 27 \quad 91
\end{array}
$$

If we write the 2 in the quotient on the bottom line, the bottom line gives both the coefficients of the quotient and the remainder. The top line can now be eliminated and the division appears as

$$
\begin{array}{r}
-3 2 + 4 - 3 + 10 \\
\underline{- 6 - 30 - 81} \\
2 \quad 10 \quad 27 \quad 91
\end{array}
$$

The bottom line was obtained by subtracting the middle line from the top line. If we replace the -3 in the divisor with $+3$, the signs of each number in the middle line will be reversed in the division process. Then the bottom line can be obtained by addition, and we have the final form of the synthetic division.

$$
\begin{array}{r}
+3 2 + 4 - 3 + 10 \\
\underline{+ 6 + 30 + 81} \\
\mathbf{2 \quad 10 \quad 27 \quad 91}
\end{array}
$$
     The coefficients of the dividend.

     The coefficients of the quotient and the remainder.

Thus,

$$
\frac{2x^3 + 4x^2 - 3x + 10}{x - 3} = 2x^2 + 10x + 27 + \frac{91}{x - 3}
$$

EXAMPLE 8   Use synthetic division to divide $10x + 3x^4 - 8x^3 + 3$ by $x - 2$.

Solution   We first write the terms of $P(x)$ in descending powers of x:

$$
3x^4 - 8x^3 + 10x + 3
$$

We then write the coefficients of the dividend, with its terms in descending powers of x, and the 2 from the divisor in the following form:

$$
2 \quad 3 \quad -8 \quad \mathbf{0} \quad 10 \quad 3
$$
     Write 0 for the coefficient of the missing x^2 term.

Then we follow these steps:

$$\begin{array}{r|rrrrr} 2 & 3 & -8 & 0 & 10 & 3 \\ & \downarrow & & & & \\ \hline & 3 & & & & \end{array}$$

Bring down the 3.

$$\begin{array}{r|rrrrr} 2 & 3 & -8 & 0 & 10 & 3 \\ & & 6 & & & \\ \hline & 3 & -2 & & & \end{array}$$

Multiply 2 and 3 together to get 6, and add 6 and -8 to get -2.

$$\begin{array}{r|rrrrr} 2 & 3 & -8 & 0 & 10 & 3 \\ & & 6 & -4 & & \\ \hline & 3 & -2 & -4 & & \end{array}$$

Multiply 2 and -2 together to get -4, and add -4 and 0 to get -4.

$$\begin{array}{r|rrrrr} 2 & 3 & -8 & 0 & 10 & 3 \\ & & 6 & -4 & -8 & \\ \hline & 3 & -2 & -4 & 2 & \end{array}$$

Multiply 2 and -4 together to get -8, and add -8 and 10 to get 2.

$$\begin{array}{r|rrrrr} 2 & 3 & -8 & 0 & 10 & 3 \\ & & 6 & -4 & -8 & 4 \\ \hline & 3 & -2 & -4 & 2 & 7 \end{array}$$

Multiply 2 and 2 together to get 4, and add 4 and 3 to get 7.

The coefficients of the quotient and the remainder.

Thus,

$$\frac{10x + 3x^4 - 8x^3 + 3}{x - 2} = 3x^3 - 2x^2 - 4x + 2 + \frac{7}{x - 2}$$ ∎

Self Check   Divide $2x^3 - 4x^2 + 5x - 7$ by $x - 3$.

Answer $2x^2 + 2x + 11 + \dfrac{26}{x - 3}$

■ USING SYNTHETIC DIVISION TO EVALUATE POLYNOMIALS

EXAMPLE 9   Use synthetic division to find $P(-2)$ when $P(x) = 5x^3 + 3x^2 - 21x - 1$.

Solution   Because of the remainder theorem, $P(-2)$ is the remainder when $P(x)$ is divided by $x - (-2)$. We use synthetic division to find the remainder.

$$\begin{array}{r|rrrr} -2 & 5 & 3 & -21 & -1 \\ & & -10 & & \\ \hline & 5 & -7 & & \end{array}$$

$-2(5) = -10; \; 3 + (-10) = -7.$

$$\begin{array}{r|rrrr} -2 & 5 & 3 & -21 & -1 \\ & & -10 & 14 & \\ \hline & 5 & -7 & -7 & \end{array}$$

$-2(-7) = 14; \; -21 + 14 = -7.$

$$\begin{array}{r|rrrr} -2 & 5 & 3 & -21 & -1 \\ & & -10 & 14 & 14 \\ \hline & 5 & -7 & -7 & 13 \end{array} \qquad -2(-7) = 14; \ -1 + 14 = 13.$$

Because the remainder is 13, $P(-2) = 13$. ∎

Self Check   Find $P(3)$.

Answer   98

EXAMPLE 10   If $P(x) = x^3 - x^2 + x - 1$, find $P(i)$, where $i = \sqrt{-1}$.

Solution   We can use synthetic division.

$$\begin{array}{r|rrrr} i & 1 & -1 & +1 & -1 \\ & & i & -1-i & +1 \\ \hline & 1 & i-1 & -i & 0 \end{array}$$

Since the remainder is 0, $P(i) = 0$, and i is a zero of $P(x)$. ∎

Self Check   Find $P(-i)$.

Answer   0

EXERCISE 6.1

VOCABULARY AND CONCEPTS   *In Exercises 1–6, fill in the blank to make a true statement.*

1. The variables in a polynomial have _____-number exponents.

2. A zero of $P(x)$ is any number r for which _____.

3. The remainder theorem holds when r is a real or _____ number.

4. If $P(x)$ is a polynomial and $P(x)$ is divided by _____, the remainder will be $P(r)$.

5. If $P(x)$ is a polynomial, then $P(r) = 0$ if and only if $x - r$ is a _____ of $P(x)$.

6. A shortcut method for dividing a polynomial by a binomial of the form $x - r$ is called _____ division.

PRACTICE   *In Exercises 7–10, let $P(x) = 2x^4 - 2x^3 + 5x^2 - 1$. Find each value by substituting the given value of x into the polynomial and simplifying. Then find the value by doing a long division and finding the remainder.*

7. $P(1)$ **8.** $P(2)$ **9.** $P(-2)$ **10.** $P(-1)$

In Exercises 11–14, find the remainder that occurs when $P(x) = 3x^4 + 5x^3 - 4x^2 - 2x + 1$ is divided by each binomial. ***Do not use long division or synthetic division.***

11. $x - 12$ **12.** $x + 15$ **13.** $x + 3.25$ **14.** $x - 7.12$

In Exercises 15–22, use the factor theorem to decide whether each statement is true. If the statement is not true, so indicate.

15. $x - 1$ is a factor of $x^7 - 1$.

16. $x - 2$ is a factor of $x^3 - x^2 + 2x - 8$.

17. $x - 1$ is a factor of $3x^5 + 4x^2 - 7$.

18. $x + 1$ is a factor of $3x^5 + 4x^2 - 7$.

19. $x + 3$ is a factor of $2x^3 - 2x^2 + 1$.

20. $x - 3$ is a factor of $3x^5 - 3x^4 + 5x^2 - 13x - 6$.

21. $x - 1$ is a factor of $x^{1984} - x^{1776} + x^{1492} - x^{1066}$.

22. $x + 1$ is a factor of $x^{1984} + x^{1776} - x^{1492} - x^{1066}$.

In Exercises 23–28, a partial solution set is given for each equation. Find the complete solution set.

23. $x^3 + 3x^2 - 13x - 15 = 0$; $\{-1\}$

24. $x^3 + 6x^2 + 5x - 12 = 0$; $\{1\}$

25. $x^4 - 2x^3 - 2x^2 + 6x - 3 = 0$; $\{1, 1\}$

26. $x^5 + 4x^4 + 4x^3 - x^2 - 4x - 4 = 0$; $\{1, -2, -2\}$

27. $x^4 - 5x^3 + 7x^2 - 5x + 6 = 0$; $\{2, 3\}$

28. $x^4 + 2x^3 - 3x^2 - 4x + 4 = 0$; $\{1, -2\}$

In Exercises 29–40, find a polynomial with the given zeros.

29. 4, 5

30. $-3, 5$

31. 1, 1, 1

32. 1, 0, -1

33. 2, 4, 5

34. 7, 6, 3

35. $1, -1, \sqrt{2}, -\sqrt{2}$

36. $0, 0, 0, \sqrt{3}, -\sqrt{3}$

37. $\sqrt{2}, i, -i$

38. $i, i, 2$

39. $0, 1 + i, 1 - i$

40. $i, 2 + i, 2 - i$

In Exercises 41–48, use synthetic division to express $P(x) = 3x^3 - 2x^2 - 6x - 4$ in the form (divisor)(quotient) + remainder for each divisor.

41. $x - 1$

42. $x - 2$

43. $x - 3$

44. $x - 4$

45. $x + 1$

46. $x + 2$

47. $x + 3$

48. $x + 4$

In Exercises 49–56, use synthetic division to do each division.

49. $\dfrac{x^3 + x^2 + x - 3}{x - 1}$

50. $\dfrac{x^3 - x^2 - 5x + 6}{x - 2}$

51. $\dfrac{7x^3 - 3x^2 - 5x + 1}{x + 1}$

52. $\dfrac{2x^3 + 4x^2 - 3x + 8}{x - 3}$

53. $\dfrac{4x^4 - 3x^3 - x + 5}{x - 3}$

54. $\dfrac{x^4 + 5x^3 - 2x^2 + x - 1}{x + 1}$

55. $\dfrac{3x^5 - 768x}{x - 4}$

56. $\dfrac{x^5 - 4x^2 + 4x + 4}{x + 3}$

In Exercises 57–60, $P(x) = 5x^3 + 2x^2 - x + 1$. Use synthetic division to find each value of $P(x)$.

57. $P(2)$

58. $P(-2)$

59. $P(-5)$

60. $P(3)$

In Exercises 61–64, $P(x) = 2x^4 - x^2 + 2$. Use synthetic division to find each value of $P(x)$.

61. $P\left(\dfrac{1}{2}\right)$ **62.** $P\left(\dfrac{1}{3}\right)$ **63.** $P(i)$ **64.** $P(-i)$

In Exercises 65–68, $P(x) = x^4 - 8x^3 + 8x + 14x^2 - 15$. Write the terms of $P(x)$ in descending powers of x and use synthetic division to find each value of $P(x)$.

65. $P(1)$ **66.** $P(0)$ **67.** $P(-3)$ **68.** $P(-1)$

In Exercises 69–72, $P(x) = 8 - 8x^2 + x^5 - x^3$. Write the terms of $P(x)$ in descending powers of x and use synthetic division to find each value of $P(x)$.

69. $P(i)$ **70.** $P(-i)$ **71.** $P(-2i)$ **72.** $P(2i)$

DISCOVERY AND WRITING

73. If 0 is a zero of $P(x) = a_n x^n + a_{n-1} x^{n-1} + \cdots a_1 x + a_0$, find a_0.

74. If 0 occurs twice as a zero of $P(x) = a_n x^n + a_{n-1} x^{n-1} + \cdots + a_1 x + a_0$, find a_1.

75. If $P(2) = 0$ and $P(-2) = 0$, explain why $x^2 - 4$ is a factor of $P(x)$.

76. If $P(x) = x^4 - 3x^3 + kx^2 + 4x - 1$ and $P(2) = 11$, find k.

REVIEW *Find the quadrant in which each point lies.*

77. $P(3, -2)$ **78.** $Q(-2, -5)$

79. $R(8, \pi)$ **80.** $S(-9, 9)$

Find the distance between each pair of points.

81. $A(3, -3)$ and $B(-5, 3)$ **82.** $C(-8, 2)$ and $D(2, -22)$

Find the slope of the line passing through each pair of points.

83. $E(3, 5)$ and $F(-5, -3)$ **84.** $G\left(3, \dfrac{3}{5}\right)$ and $H\left(-\dfrac{3}{5}, 1\right)$

6.2 Descartes' Rule of Signs and Bounds on Roots

■ THE FUNDAMENTAL THEOREM OF ALGEBRA ■ CONJUGATE PAIRS THEOREM ■ DESCARTES' RULE OF SIGNS ■ BOUNDS ON ROOTS

The remainder theorem and synthetic division provide a way of verifying that a particular number is a root of a polynomial equation, but they do not provide the roots. We need some guidelines to indicate how many roots to expect, the kind of roots to expect, and where they are located. This section develops several theorems that provide such guidelines.

■ THE FUNDAMENTAL THEOREM OF ALGEBRA

Before attempting to find the roots of a polynomial equation, it would be useful to know if any roots exist. This question was answered by Carl Friedrich Gauss (1777–1855) when he proved the **fundamental theorem of algebra.**

Fundamental Theorem of Algebra

If $P(x)$ is a polynomial with positive degree, then $P(x)$ has at least one zero.

The fundamental theorem of algebra guarantees that polynomials such as

$$2x^3 + 3 \qquad \text{and} \qquad 32.75x^{1984} + ix^3 - (2 + i)x - 5$$

all have zeros. Since all polynomials with positive degree have zeros, their corresponding polynomial equations all have roots. They are the zeros of the polynomial.

The next theorem will help us show that every nth-degree polynomial equation has exactly n roots.

Polynomial Factorization Theorem

If $n > 0$ and $P(x)$ is an nth-degree polynomial, then $P(x)$ has exactly n linear factors:

$$P(x) = a_n(x - r_1)(x - r_2)(x - r_3) \cdots (x - r_n)$$

where $r_1, r_2, r_3, \ldots, r_n$ are either real or nonreal complex numbers and a_n is the lead coefficient of $P(x)$.

Proof   Let $P(x)$ be a polynomial of degree n ($n > 0$). Because of the fundamental theorem of algebra, we know that $P(x)$ has a zero r_1 and that the equation $P(x) = 0$ has r_1 for a root. By the factor theorem, we know that $x - r_1$ is a factor of $P(x)$. Thus,

$$P(x) = (x - r_1)Q_1(x)$$

If the lead coefficient of the nth-degree polynomial $P(x)$ is a_n, then $Q_1(x)$ is a polynomial of degree $n - 1$ whose lead coefficient is also a_n.

By the fundamental theorem of algebra, we know that $Q_1(x)$ also has a zero, r_2. By the factor theorem, $x - r_2$ is a factor of $Q_1(x)$, and

$$P(x) = (x - r_1)(x - r_2)Q_2(x)$$

where $Q_2(x)$ is a polynomial of degree $n - 2$ with lead coefficient a_n.

This process can continue only to n factors of the form $x - r_i$ until the final quotient $Q_n(x)$ is a polynomial of degree $n - n$, or degree 0. Thus, the polynomial $P(x)$ factors completely as

1. $P(x) = a_n(x - r_1)(x - r_2)(x - r_3) \cdots (x - r_n)$ □

If we substitute any one of the numbers $r_1, r_2, r_3, \ldots, r_n$ for x in Equation 1, $P(x)$ will equal 0. Thus, each value of r is a zero of $P(x)$ and a root of the equation $P(x) = 0$. There can be no other roots, because no single factor in Equation 1 is 0 for any value of x not included in the list $r_1, r_2, r_3, \ldots, r_n$.

The values of r in the previous list need not be distinct. Any number r_i that occurs k times as a root of a polynomial equation is called a **root of multiplicity k.** The following theorem summarizes the previous discussion.

> **Theorem**
> If multiple roots are counted individually, the polynomial equation $P(x) = 0$ with degree n $(n > 0)$ has exactly n roots among the complex numbers.

■ CONJUGATE PAIRS THEOREM

Recall that the complex numbers $a + bi$ and $a - bi$ are called **complex conjugates** of each other. The next theorem points out that *complex roots of polynomial equations with real coefficients occur in complex conjugate pairs.*

> **Conjugate Pairs Theorem**
> If a polynomial equation $P(x) = 0$ with real number coefficients has a complex root $a + bi$ with $b \neq 0$, then its conjugate $a - bi$ is also a root.

EXAMPLE 1   Find a second-degree polynomial equation with a root of $2 + i$.

Solution   Because $2 + i$ is a root, its complex conjugate $2 - i$ is also a root. The equation is

$$[x - (2 + i)][x - (2 - i)] = 0$$
$$(x - 2 - i)(x - 2 + i) = 0$$
$$x^2 - 4x + 5 = 0 \qquad \text{Multiply.}$$

The equation $x^2 - 4x + 5 = 0$ will have roots of $2 + i$ and $2 - i$.   ■

Self Check   Find a polynomial equation with a root of $1 - i$.
Answer $x^2 - 2x + 2 = 0$

EXAMPLE 2   Find a fourth-degree polynomial equation with real coefficients and i as a root of multiplicity 2.

Solution   Because i is a root twice and a fourth-degree polynomial equation has four roots, we must find the other two roots. According to the conjugate pairs theorem, the missing roots are the conjugates of the given roots. Thus, the complete solution set is

$$\{i, i, -i, -i\}$$

The equation is

$$(x - i)(x - i)[x - (-i)][x - (-i)] = 0$$
$$(x - i)(x + i)(x - i)(x + i) = 0$$
$$(x^2 + 1)(x^2 + 1) = 0$$
$$x^4 + 2x^2 + 1 = 0$$ ∎

___

Self Check   Find a polynomial equation with $-i$ as a root of multiplicity 2.
Answer $x^4 + 2x^2 + 1 = 0$

___

EXAMPLE 3   Find a quadratic equation with a double root of i.

Solution   If i is a root twice in a quadratic equation, the equation is

$$(x - i)(x - i) = 0$$
$$x^2 - 2ix - 1 = 0$$

In this equation, the roots are not in complex conjugate pairs. This is not surprising, because the coefficient $2i$ is a not a real number.   ∎

___

Self Check   Find a quadratic equation with a double root of $-i$.
Answer $x^2 + 2ix - 1 = 0$

___

 WARNING!   The theorem "Complex roots of real polynomial equations occur in complex conjugate pairs" applies only to polynomial equations with real coefficients.

■ DESCARTES' RULE OF SIGNS

René Descartes is credited with a theorem known as **Descartes' rule of signs,** which enables us to estimate the number of positive, negative, and nonreal roots of a polynomial equation.

If a polynomial is written in descending powers of x and we scan it from left to right, we say that a variation in sign occurs whenever successive terms have opposite signs. For example, the polynomial

$$+ \text{to} - \quad\quad - \text{to} + + \text{to} -$$
$$P(x) = \overbrace{3x^5} - 2x^4 - \overbrace{5x^3 + x^2} - x - 9$$

has three variations in sign, and the polynomial

$$P(-x) = 3(-x)^5 - 2(-x)^4 - 5(-x)^3 + (-x)^2 - (-x) - 9$$
$$- \text{to} + \quad\quad\quad + \text{to} -$$
$$= -3x^5 - \overbrace{2x^4} + 5x^3 + x^2 + \overbrace{x} - 9$$

has two variations in sign.

Descartes' Rule of Signs

If $P(x)$ is a polynomial with real coefficients, the number of positive roots of $P(x) = 0$ is either equal to the number of variations in sign of $P(x)$ or less than that by an even number.

The number of negative roots of $P(x) = 0$ is either equal to the number of variations in sign of $P(-x)$ or less than that by an even number.

To illustrate Descartes' rule of signs, we consider the equation

$$P(x) = x^8 + x^6 + x^4 + x^2 + 1 = 0$$

Since $P(x)$ is of eighth degree, the equation has 8 roots. Since there are no variations in sign of $P(x)$, there are 0 positive roots.
Because

$$P(-x) = (-x)^8 + (-x)^6 + (-x)^4 + (-x)^2 + 1 = 0$$
$$P(-x) = x^8 + x^6 + x^4 + x^2 + 1 = 0$$

has no variations in sign either, there are 0 negative roots. Thus, all 8 roots are non-real complex numbers, and they will occur in complex conjugate pairs

EXAMPLE 4

Discuss the possibilities for the roots of $P(x) = 3x^3 - 2x^2 + x - 5 = 0$.

Solution Since there are three variations of sign in $P(x) = 3x^3 - 2x^2 + x - 5 = 0$, there can be either 3 positive roots or only 1 (1 is less than 3 by the even number 2). Because

$$P(-x) = 3(-x)^3 - 2(-x)^2 + (-x) - 5$$
$$= -3x^3 - 2x^2 - x - 5$$

has no variations in sign, there are 0 negative roots. Furthermore, 0 is not a root, because the terms of the polynomial do not have a common factor of x.

If there are 3 positive roots, then all of the roots are accounted for. If there is 1 positive root, the remaining two roots must be nonreal complex numbers. The following chart shows these possibilities.

Number of positive roots	Number of negative roots	Number of nonreal roots
3	0	0
1	0	2

The number of nonreal complex roots is the number needed to bring the total number of roots up to 3. ∎

Self Check Discuss the possibilities for the roots of $5x^3 + 2x^2 - x + 3 = 0$.

Answer

Num. of pos. roots	Num. of neg. roots	Num. of nonreal roots
2	1	0
0	1	2

EXAMPLE 5 Discuss the possibilities for the roots of $P(x) = 5x^5 - 3x^3 - 2x^2 + x - 1 = 0$.

Solution Since there are three variations of sign in $P(x)$, there are either 3 or 1 positive roots. Because $P(-x) = -5x^5 + 3x^3 - 2x^2 - x - 1$ has two variations in sign, there are 2 or 0 negative roots. The possibilities are shown as follows:

Number of positive roots	Number of negative roots	Number of nonreal roots
1	0	4
3	0	2
1	2	2
3	2	0

In each case, the number of nonreal complex roots is an even number. This is expected, because this polynomial has real coefficients, and its nonreal complex roots will occur in conjugate pairs. ■

Self Check Discuss the possibilities for the roots of $5x^5 - 2x^2 - x - 1 = 0$.

Answer

Num. of pos. roots	Num. of neg. roots	Num. of nonreal roots
1	2	2
1	0	4

■ BOUNDS ON ROOTS

A final theorem provides a way to find **bounds** on the roots of a polynomial equation, enabling us to look for roots where they can be found.

Upper and Lower Bounds on Roots

Let the lead coefficient of a polynomial $P(x)$ with real coefficients be positive, and do a synthetic division of the coefficients by a positive number c. If each term in the last row of the division is nonnegative, then no number greater than c can be a root of $P(x) = 0$. (c is called an **upper bound** of the real roots.)

If $P(x)$ is synthetically divided by a negative number d and the signs in the last row alternate,* then no value less than d can be a root of $P(x) = 0$. (d is called a **lower bound** of the real roots.)

*If 0 appears in the third row, that 0 can be assigned either a + or − sign to help the signs alternate.

EXAMPLE 6   Establish integer bounds for the roots of $2x^3 + 3x^2 - 5x - 7 = 0$.

Solution   We will do several synthetic divisions by positive integers, looking for nonnegative values in the last row. Then we will divide by several negative integers, looking for alternating signs in the last row. Trying 1 first gives

$$
\begin{array}{r|rrrr}
1 & 2 & 3 & -5 & -7 \\
  &   & 2 & 5 & 0 \\
\hline
  & +2 & +5 & 0 & -7
\end{array}
$$

Because one of the signs in the last row is negative, we cannot claim that 1 is an upper bound. We now try 2.

$$
\begin{array}{r|rrrr}
2 & 2 & 3 & -5 & -7 \\
  &   & 4 & 14 & 18 \\
\hline
  & +2 & +7 & +9 & +11
\end{array}
$$

Because the last row is entirely nonnegative, 2 is an upper bound. That is, no number greater than 2 can be a root of the equation.

Now we try some negative divisors, beginning with −3.

$$
\begin{array}{r|rrrr}
-3 & 2 & 3 & -5 & -7 \\
   &   & -6 & 9 & -12 \\
\hline
   & +2 & -3 & +4 & -19
\end{array}
$$

Since the signs in the last row alternate, −3 is a lower bound. That is, no number less than −3 can be a root. To see whether there is a greater lower bound, we try −2.

$$
\begin{array}{r|rrrr}
-2 & 2 & 3 & -5 & -7 \\
   &   & -4 & 2 & 6 \\
\hline
   & +2 & -1 & -3 & -1
\end{array}
$$

Since the signs in the last row do not alternate, we cannot say that −2 is a lower bound.

All of the real roots are in the interval $(-3, 2)$. ∎

Self Check Establish integer bounds for the roots of $2x^3 + 3x^2 - 11x - 7 = 0$.

Answer Roots are in $(-4, 3)$.

It is important to understand what the theorem on the bounds of roots says and what it doesn't say. If we divide synthetically by a positive number c and the last row of the synthetic division is entirely nonnegative, the theorem guarantees that c is an upper bound of the roots. However, if the last row contains some negative values, c could still be an upper bound.

If we divide by a negative number d and the signs in the last row alternate, the theorem guarantees that d is a lower bound of the roots. However, if the signs in the last row do not alternate, d could still be a lower bound. This is illustrated in Example 6. It can be shown that the smallest negative root of the equation is approximately -1.81. Thus, -2 is a lower bound for the roots of the equation. However, when we checked -2, the last row of the synthetic division did not have alternating signs. Unfortunately, the theorem does not always determine the best bounds for the roots of the equation.

EXERCISE 6.2

VOCABULARY AND CONCEPTS *In Exercises 1–10, fill in the blank to make a true statement.*

1. If $P(x)$ is a polynomial with positive degree, then $P(x)$ has at least one _____.

2. The statement in Exercise 1 is called the _____.

3. The _____ of $a + bi$ is $a - bi$.

4. The polynomial $6x^4 + 5x^3 - 2x^2 + 3$ has __ variations in sign.

5. The polynomial $(-x)^3 - (-x)^2 - 4$ has __ variations in sign.

6. The equation $7x^4 + 5x^3 - 2x + 1 = 0$ can have at most __ positive roots.

7. The equation $7x^4 + 5x^3 - 2x + 1$ can have at most __ negative roots.

8. Complex roots occur in complex _____ pairs. (Assume that the equation has real coefficients.)

9. If no number less than d can be a root of $P(x) = 0$, then d is called a(n) _____.

10. If no number greater than c can be a root of $P(x) = 0$, then c is called a(n) _____.

PRACTICE *In Exercises 11–14, tell how many roots each equation has.*

11. $x^{10} = 1$

12. $x^{40} = 1$

13. $3x^4 - 4x^2 - 2x = -7$

14. $-32x^{111} - x^5 = 1$

15. One root of $x(3x^4 - 2) = 12x$ is 0. How many other roots are there?

16. One root of $3x^2(x^7 - 14x + 3) = 0$ is 0. How many other roots are there?

In Exercises 17–20, write a third-degree polynomial equation with real coefficients and the given roots.

17. $3, -i$

18. $1, i$

19. $2, 2 + i$

20. $-2, 3 - i$

In Exercises 21–24, write a fourth-degree polynomial equation with real coefficients and the given roots.

21. $3, 2, i$

22. $1, 2, 1 + i$

23. $i, 1 - i$

24. $i, 2 - i$

In Exercises 25–38, use Descartes' rule of signs to find the number of possible positive, negative, and nonreal roots of each equation. **Do not try to find the roots.**

25. $3x^3 + 5x^2 - 4x + 3 = 0$

26. $3x^3 - 5x^2 - 4x - 3 = 0$

27. $2x^3 + 7x^2 + 5x + 5 = 0$

28. $-2x^3 - 7x^2 - 5x - 4 = 0$

29. $8x^4 = -5$

30. $-3x^3 = -5$

31. $x^4 + 8x^2 - 5x - 10 = 0$

32. $5x^7 + 3x^6 - 2x^5 + 3x^4 + 9x^3 + x^2 + 1 = 0$

33. $-x^{10} - x^8 - x^6 - x^4 - x^2 - 1 = 0$

34. $x^{10} + x^8 + x^6 + x^4 + x^2 + 1 = 0$

35. $x^9 + x^7 + x^5 + x^3 + x = 0$ (Is 0 a root?)

36. $-x^9 - x^7 - x^5 - x^3 - x = 0$ (Is 0 a root?)

37. $-2x^4 - 3x^2 + 2x + 3 = 0$

38. $-7x^5 - 6x^4 + 3x^3 - 2x^2 + 7x - 4 = 0$

In Exercises 39–48, find integer bounds for the roots of each equation.

39. $x^2 - 2x - 4 = 0$

40. $9x^2 - 6x - 1 = 0$

41. $18x^2 - 6x - 1 = 0$

42. $2x^2 - 10x - 9 = 0$

43. $6x^3 - 13x^2 - 110x = 0$

44. $12x^3 + 20x^2 - x - 6 = 0$

45. $x^5 + x^4 - 8x^3 - 8x^2 + 15x + 15 = 0$

46. $3x^4 - 5x^3 - 9x^2 + 15x = 0$

47. $3x^5 - 11x^4 - 2x^3 + 38x^2 - 21x - 15 = 0$

48. $3x^6 - 4x^5 - 21x^4 + 4x^3 + 8x^2 + 8x + 32 = 0$

DISCOVERY AND WRITING

49. Explain why the fundamental theorem of algebra guarantees that every polynomial equation of positive degree has at least one root.

50. Explain why the fundamental theorem of algebra and the factor theorem guarantee that an nth-degree polynomial equation has n roots.

51. Prove that any odd-degree polynomial equation with real coefficients must have at least one real root.

52. If a, b, c, and d are positive numbers, prove that $ax^4 + bx^2 + cx - d = 0$ has exactly two nonreal roots.

REVIEW *Assume that k represents a positive real number. Complete each sentence.*

53. The graph of $y = f(x - k)$ looks like the graph of $y = f(x)$, except that it has been translated _____.

54. The graph of $y = f(x) - k$ looks like the graph of $y = f(x)$, except that it has been translated _____.

55. The graph of $y = f(-x)$ looks like the graph of $y = f(x)$, except that it has been _____.

56. The graph of $y = -f(x)$ looks like the graph of $y = f(x)$, except that it has been _____.

57. If $a > 0$, the graph of $y = af(x)$ looks like the graph of $y = f(x)$, except that it has been

_____ .

58. If $a > 0$, the graph of $y = f(ax)$ looks like the graph of $y = f(x)$, except that it has been

_____ .

6.3 Rational Roots of Polynomial Equations

■ FINDING POSSIBLE RATIONAL ROOTS ■ FINDING RATIONAL ROOTS ■ AN APPLICATION

In this section, we will find rational roots of polynomial equations with integer co-efficients. The following theorem enables us to list the possible rational roots of such equations.

Rational Root Theorem

Let the polynomial equation

$$P(x) = a_n x^n + a_{n-1} x^{n-1} + a_{n-2} x^{n-2} + \cdots + a_1 x + a_0 = 0$$

have integer coefficients. If the rational number $\frac{p}{q}$ (written in lowest terms) is a root of $P(x) = 0$, then p is a factor of the constant a_0, and q is a factor of the lead coefficient a_n.

Proof Let $\frac{p}{q}$ (written in lowest terms) be a rational root of $P(x) = 0$. Then the equation is satisfied by $\frac{p}{q}$:

1. $a_n \left(\dfrac{p}{q} \right)^n + a_{n-1} \left(\dfrac{p}{q} \right)^{n-1} + a_{n-2} \left(\dfrac{p}{q} \right)^{n-2} + \cdots + a_1 \left(\dfrac{p}{q} \right) + a_0 = 0$

We can clear Equation 1 of fractions by multiplying both sides by q^n.

2. $a_n p^n + a_{n-1} p^{n-1} q + a_{n-2} p^{n-2} q^2 + \cdots + a_1 p q^{n-1} + a_0 q^n = 0$

We can factor p from all but the last term and subtract $a_0 q^n$ from both sides to get

$$p(a_n p^{n-1} + a_{n-1} p^{n-2} q + a_{n-2} p^{n-3} q^2 + \cdots + a_1 q^{n-1}) = -a_0 q^n$$

Since p is a factor of the left-hand side, it is also a factor of the right-hand side. So p is a factor of $-a_0 q^n$, but because $\frac{p}{q}$ is written in lowest terms, p cannot be a factor of q^n. Therefore, p is a factor of a_0.

We can factor q from all but the first term of Equation 2 and subtract $a_n p^n$ from both sides to get

$$q(a_{n-1} p^{n-1} + a_{n-2} p^{n-2} q + a_{n-3} p^{n-3} q^2 + \cdots + a_0 q^{n-1}) = -a_n p^n$$

Since q is a factor of the left-hand side, it is also a factor of the right-hand side. Because q is not a factor of p^n, it must be a factor of a_n. □

■ FINDING POSSIBLE RATIONAL ROOTS

To illustrate the previous theorem, we consider the equation

$$\frac{1}{2}x^4 + \frac{2}{3}x^3 + 3x^2 - \frac{3}{2}x + 3 = 0$$

Because the theorem requires integer coefficients, we multiply both sides of the equation by 6 to clear it of fractions.

$$3x^4 + 4x^3 + 18x^2 - 9x + 18 = 0$$

By the previous theorem, the only possible numerators for the rational roots of the equation are the factors of the constant term 18:

$$\pm 1, \pm 2, \pm 3, \pm 6, \pm 9, \quad \text{and} \quad \pm 18$$

The only possible denominators are the factors of the lead coefficient 3:

$$\pm 1 \quad \text{and} \quad \pm 3$$

We can form a list of all possible rational solutions by listing the combinations of possible numerators and denominators:

$$\pm\frac{1}{1}, \pm\frac{2}{1}, \pm\frac{3}{1}, \pm\frac{6}{1}, \pm\frac{9}{1}, \pm\frac{18}{1}, \pm\frac{1}{3}, \pm\frac{2}{3}, \pm\frac{3}{3}, \pm\frac{6}{3}, \pm\frac{9}{3}, \pm\frac{18}{3}$$

Since several of these possibilities are duplicates, we can condense the list to get

Possible Rational Roots

$$\pm 1, \pm 2, \pm 3, \pm 6, \pm 9, \pm 18, \pm\frac{1}{3}, \pm\frac{2}{3}$$

EXAMPLE 1   Prove that $\sqrt{2}$ is an irrational number.

Solution   We know that $\sqrt{2}$ is a real root of the equation $x^2 - 2 = 0$. Since any rational root of this equation will have the form $\frac{\text{factor of the constant 2}}{\text{factor of the lead coefficient 1}}$, the possible rational roots are

$$\pm\frac{1}{1} \quad \text{and} \quad \pm\frac{2}{1}$$

Since none of the numbers 1, −1, 2, or −2 satisfy the equation, $\sqrt{2}$ must be irrational.   ■

Self Check   Prove that $\sqrt{3}$ is irrational.

Answer $\pm\frac{1}{1}$ or $\pm\frac{3}{1}$ do not satisfy $x^2 - 3 = 0$.

■ FINDING RATIONAL ROOTS

EXAMPLE 2   Solve $P(x) = 2x^3 + 3x^2 - 8x + 3 = 0$.

Solution   Since the equation is of third degree, it has 3 roots. By Descartes' rule of signs, there are two possible combinations of positive, negative, and nonreal roots. They are summarized as follows:

Number of positive roots	Number of negative roots	Number of nonreal roots
2	1	0
0	1	2

Since any rational roots will have the form $\frac{\text{factor of the constant 3}}{\text{factor of the lead coefficient 2}}$, the possible rational roots are

$$\pm\frac{3}{1}, \pm\frac{1}{1}, \pm\frac{3}{2}, \pm\frac{1}{2}$$

or, written in order of increasing size,

$$-3, -\frac{3}{2}, -1, -\frac{1}{2}, \frac{1}{2}, 1, \frac{3}{2}, 3$$

We check each possibility to see whether it is a root. We can start with $\frac{3}{2}$.

$$\begin{array}{r|rrrr} \tfrac{3}{2} & 2 & 3 & -8 & 3 \\ & & 3 & 9 & \tfrac{3}{2} \\ \hline & 2 & 6 & 1 & \tfrac{9}{2} \end{array}$$

Since the remainder is not 0, $\frac{3}{2}$ is not a root and can be crossed off the list. Since every number in the last row of the synthetic division is positive, $\frac{3}{2}$ is an upper bound. So 3 cannot be a root and can be crossed off the list as well.

$$-3, -\frac{3}{2}, -1, -\frac{1}{2}, \frac{1}{2}, 1, \cancel{\frac{3}{2}}, \cancel{3}$$

We now try $\frac{1}{2}$:

$$\begin{array}{r|rrrr} \tfrac{1}{2} & 2 & 3 & -8 & 3 \\ & & 1 & 2 & -3 \\ \hline & 2 & 4 & -6 & 0 \end{array}$$

Since the remainder is 0, $\frac{1}{2}$ is a root. The binomial $x - \frac{1}{2}$ is a factor of $P(x)$, and any remaining roots must be supplied by the remaining factor, which is the quotient $2x^2 + 4x - 6$. We can find the other roots by solving the equation $2x^2 + 4x - 6 = 0$, called the **depressed equation.**

$$2x^2 + 4x - 6 = 0$$
$$x^2 + 2x - 3 = 0 \qquad \text{Divide both sides by 2.}$$
$$(x - 1)(x + 3) = 0 \qquad \text{Factor } x^2 + 2x - 3 = 0.$$
$$x - 1 = 0 \quad \text{or} \quad x + 3 = 0$$
$$x = 1 \qquad\qquad x = -3$$

The solution set of the equation is $\left\{\frac{1}{2}, 1, -3\right\}$. Note that two positive roots and one negative root is a predicted possibility. ■

Self Check Solve $3x^3 - 10x^2 + 9x - 2 = 0$.

Answer $1, 2, \frac{1}{3}$

EXAMPLE 3 Solve $P(x) = x^7 - 2x^6 - 5x^5 + 6x^4 - x^3 + 2x^2 + 5x - 6 = 0$.

Solution Because the equation is of seventh degree, it has 7 roots. By Descartes' rule of signs, there are six possible combinations of positive, negative, and nonreal roots.

Number of positive roots	Number of negative roots	Number of nonreal roots
5	2	0
3	2	2
1	2	4
5	0	2
3	0	4
1	0	6

The possible rational roots are

$$-6, -3, -2, -1, 1, 2, 3, 6$$

We check each one, eliminating those that don't satisfy the equation. We begin with -3.

$$
\begin{array}{r|rrrrrrrr}
-3 & 1 & -2 & -5 & 6 & -1 & 2 & 5 & -6 \\
   &   & -3 & 15 & -30 & 72 & -213 & 633 & -1914 \\
\hline
   & 1 & -5 & 10 & -24 & 71 & -211 & 638 & -1920
\end{array}
$$

Since the last number in the synthetic division is not 0, -3 is not a root and can be crossed off the list. Since the signs in the last row alternate, -3 is a lower bound, and we can cross off -6 as well.

$$-\cancel{6}, -\cancel{3}, -2, -1, 1, 2, 3, 6,$$

We now try -2:

$$
\begin{array}{r|rrrrrrr}
-2 & 1 & -2 & -5 & 6 & -1 & 2 & 5 & -6 \\
   &   & -2 & 8 & -6 & 0 & 2 & -8 & 6 \\
\hline
   & 1 & -4 & 3 & 0 & -1 & 4 & -3 & 0
\end{array}
$$

Since the remainder is 0, -2 is a root.

Because this root is negative, we can revise the chart of possibilities to eliminate the possibility that there are 0 negative roots.

Number of positive roots	Number of negative roots	Number of nonreal roots
5	2	0
3	2	2
1	2	4

Because -2 is a root, the factor theorem states that $x - (-2)$, or $x + 2$, is a factor of $P(x)$. Any remaining roots can be found by solving the depressed equation

$$x^6 - 4x^5 + 3x^4 - x^2 + 4x - 3 = 0$$

Because the constant term of this equation is different from the constant term of the original equation, we can cross off other possible rational roots. The number -2 cannot be a root a second time, because it is not a factor of -3. The numbers 2 and 6 are no longer possible roots, because neither is a factor of -3.

The list of possible roots is now

$$-\cancel{6}, \; -\cancel{3}, \; -\cancel{2}, \; -1, \; 1, \; \cancel{2}, \; 3, \; \cancel{6}$$

Since we know that there is one more negative root, we will synthetically divide the coefficients of the depressed equation by -1.

$$
\begin{array}{r|rrrrrr}
-1 & 1 & -4 & 3 & 0 & -1 & 4 & -3 \\
   &   & -1 & 5 & -8 & 8 & -7 & 3 \\
\hline
   & 1 & -5 & 8 & -8 & 7 & -3 & 0
\end{array}
$$

Since the remainder is 0, -1 is a root and the solution set so far is $\{-2, -1, \ldots\}$. The number -1 cannot be a root again, because we have found both negative roots. When we cross off -1, we have only two possibilities left.

$$-\cancel{6}, \; -\cancel{3}, \; -\cancel{2}, \; -\cancel{1}, \; 1, \; \cancel{2}, \; 3, \; \cancel{6}$$

The depressed equation is now $x^5 - 5x^4 + 8x^3 - 8x^2 + 7x - 3 = 0$. We can synthetically divide the coefficients of this equation by 1 to get

$$
\begin{array}{r|rrrrrr}
1 & 1 & -5 & 8 & -8 & 7 & -3 \\
  &   & 1 & -4 & 4 & -4 & 3 \\
\hline
  & 1 & -4 & 4 & -4 & 3 & 0
\end{array}
$$

The depressed equation is now $x^4 - 4x^3 + 4x^2 - 4x + 3 = 0$.

Thus, 1 joins the solution set $\{-2, -1, 1, \ldots\}$. To see whether 1 is a root a second time, we synthetically divide the coefficients of the new depressed equation by 1.

$$
\begin{array}{r|rrrrr}
1 & 1 & -4 & 4 & -4 & 3 \\
  &   & 1 & -3 & 1 & -3 \\
\hline
  & 1 & -3 & 1 & -3 & 0
\end{array}
$$

The depressed equation is now
$x^3 - 3x^2 + x - 3 = 0$.

Again, 1 is a root, and the solution set is now $\{-2, -1, 1, 1, \ldots\}$ To see whether 1 is a root a third time, we synthetically divide the coefficients of the new depressed equation by 1.

$$
\begin{array}{r|rrrr}
1 & 1 & -3 & 1 & -3 \\
  &   & 1 & -2 & -1 \\
\hline
  & 1 & -2 & -1 & -4
\end{array}
$$

Since the remainder is not 0, the number 1 is not a root for a third time, and we can cross 1 off the list of possibilities, leaving only 3.

$$-\cancel{6}, -\cancel{3}, -\cancel{2}, -\cancel{1}, \cancel{1}, \cancel{2}, 3, \cancel{6}$$

To see whether 3 is a root, we synthetically divide the coefficients of $x^3 - 3x^2 + x - 3 = 0$ by 3.

$$
\begin{array}{r|rrrr}
3 & 1 & 3 & 1 & -3 \\
  &   & 3 & 0 & 3 \\
\hline
  & 1 & 0 & 1 & 0
\end{array}
$$

Since the remainder is 0, 3 joins the solution set, which is now $\{-2, -1, 1, 1, 3 \ldots\}$.

The depressed equation is now $x^2 + 1 = 0$, which can be solved as a quadratic equation.

$$x^2 + 1 = 0$$
$$x^2 = -1$$
$$x = i \quad \text{or} \quad x = -i$$

The complete solution set is $\{-2, -1, 1, 1, 3, i, -i\}$. The solution set contains 3 positive roots, 2 negative roots, and 2 nonreal roots that are complex conjugates. This combination was one of the predicted possibilities. ∎

■ AN APPLICATION

EXAMPLE 4

Growing cranberries To protect cranberry crops from the damage of early freezes, growers flood the cranberry bogs. Three irrigation sources, used together, can flood a cranberry bog in one day. If the sources are used one at a time, the second source requires one day longer to flood the bog than the first, and the third requires four days longer than the first. If the bog must be flooded before a freeze that is predicted in three days, can the water in the last two sources be diverted to other bogs?

Solution   Let x represent the number of days it would take the first irrigation source to flood the bog. Then $x + 1$ and $x + 4$ represent the number of days it would take the second and third sources, respectively, to flood the bog.

Because the first source, alone, requires x days to flood the bog, that source could fill $\frac{1}{x}$ of the bog in one day. In one day's time, the remaining sources could flood $\frac{1}{x+1}$ and $\frac{1}{x+4}$ of the bog. This gives the equation

The part of the bog the first source can flood in one day	+	the part the second source can flood in one day	+	the part the third source can flood in one day	=	one bog.
$\dfrac{1}{x}$	+	$\dfrac{1}{x+1}$	+	$\dfrac{1}{x+4}$	=	1

We multiply both sides of the equation by $x(x + 1)(x + 4)$ to clear it of fractions and then simplify to get

$$x(x + 1)(x + 4)\left(\frac{1}{x} + \frac{1}{x + 1} + \frac{1}{x + 4}\right) = 1 \cdot x(x + 1)(x + 4)$$
$$(x + 1)(x + 4) + x(x + 4) + x(x + 1) = x(x + 1)(x + 4)$$
$$x^2 + 5x + 4 + x^2 + 4x + x^2 + x = x^3 + 5x^2 + 4x$$
$$0 = x^3 + 2x^2 - 6x - 4$$

To solve the equation $x^3 + 2x^2 - 6x - 4 = 0$, we first list its possible rational roots, which are the factors of the constant term, -4.

$$-4, -2, -1, 1, 2, \quad \text{and} \quad 4$$

One solution of this equation is $x = 2$, because when we synthetically divide by 2, the remainder is 0:

$$
\begin{array}{r|rrrr}
2 & 1 & 2 & -6 & -4 \\
  &   & 2 & 8 & 4 \\
\hline
  & 1 & 4 & 2 & 0
\end{array}
$$

We find the remaining solutions by using the quadratic formula to solve the depressed equation, which is $x^2 + 4x + 2 = 0$. The two solutions are $-2 + \sqrt{2}$ and $-2 - \sqrt{2}$. Since these numbers are negative and the time it takes to flood the bog cannot be negative, these roots must be discarded. The only meaningful solution is 2.

Since the first source, alone, can flood the bog in two days, and it is three days until the freeze, the other two water sources can be diverted to flood other bogs. ∎

EXERCISE 6.3

VOCABULARY AND CONCEPTS *In Exercises 1–4, fill in the blank to make a true statement.*

1. The rational roots of the equation $3x^3 + 4x - 7 = 0$ will have the form $\frac{p}{q}$, where p is a factor of ___ and q is a factor of 3.

2. The rational roots of the equation $5x^3 + 3x^2 - 4 = 0$ will have the form $\frac{p}{q}$, where p is a factor of -4 and q is a factor of ___.

3. Consider the synthetic division of $5x^3 - 7x^2 - 3x - 63 = 0$ by 3.

$$\begin{array}{r|rrrr} 3 & 5 & -7 & -3 & -63 \\ & & 15 & 24 & 63 \\ \hline & 5 & 8 & 21 & 0 \end{array}$$

Since the remainder is 0, 3 is a ___ of the equation.

4. In Exercise 3, the depressed equation is _____.

PRACTICE *In Exercises 5–32, find all rational roots of each equation.*

5. $x^3 - 5x^2 - x + 5 = 0$

6. $x^3 + 7x^2 - x - 7 = 0$

7. $x^3 - 2x^2 - 9x + 18 = 0$

8. $x^3 + 3x^2 - 4x - 12 = 0$

9. $x^3 - 2x^2 - x + 2 = 0$

10. $x^3 + 2x^2 - x - 2 = 0$

11. $x^4 - 10x^3 + 35x^2 - 50x + 24 = 0$

12. $x^4 + 4x^3 + 6x^2 + 4x + 1 = 0$

13. $x^4 + 3x^3 - 13x^2 - 9x + 30 = 0$

14. $x^4 - 8x^3 + 14x^2 + 8x - 15 = 0$

15. $x^5 + 3x^4 - 5x^3 - 15x^2 + 4x + 12 = 0$

16. $x^5 - 3x^4 - 5x^3 + 15x^2 + 4x - 12 = 0$

17. $x^7 - 12x^5 + 48x^3 - 64x = 0$

18. $x^7 + 7x^6 + 21x^5 + 35x^4 + 35x^3 + 21x^2 + 7x + 1 = 0$

19. $3x^3 - 2x^2 + 12x - 8 = 0$

20. $4x^4 - 8x^3 - x^2 + 8x - 3 = 0$

21. $3x^4 - 14x^3 + 11x^2 + 16x - 12 = 0$

22. $2x^4 - x^3 - 2x^2 - 4x - 40 = 0$

23. $12x^4 + 20x^3 - 41x^2 + 20x - 3 = 0$

24. $4x^5 - 12x^4 + 15x^3 - 45x^2 - 4x + 12 = 0$

25. $6x^5 - 7x^4 - 48x^3 + 81x^2 - 4x - 12 = 0$

26. $36x^4 - x^2 + 2x - 1 = 0$

27. $30x^3 - 47x^2 - 9x + 18 = 0$

28. $20x^3 - 53x^2 - 27x + 18 = 0$

29. $15x^3 - 61x^2 - 2x + 24 = 0$

30. $12x^4 + x^3 + 42x^2 + 4x - 24 = 0$

31. $20x^3 - 44x^2 + 9x + 18 = 0$

32. $24x^3 - 82x^2 + 89x - 30 = 0$

In Exercises 33–36, $1 + i$ is a root of each equation. Find the other roots.

33. $x^3 - 5x^2 + 8x - 6 = 0$

34. $x^3 - 2x + 4 = 0$

35. $x^4 - 2x^3 - 7x^2 + 18x - 18 = 0$

36. $x^4 - 2x^3 - 2x^2 + 8x - 8 = 0$

In Exercises 37–40, solve each equation.

37. $x^3 - \dfrac{4}{3}x^2 - \dfrac{13}{3}x - 2 = 0$

38. $x^3 - \dfrac{19}{6}x^2 + \dfrac{1}{6}x + 1 = 0$

39. $x^{-5} - 8x^{-4} + 25x^{-3} - 38x^{-2} + 28x^{-1} - 8 = 0$

40. $1 - x^{-1} - x^{-2} - 2x^{-3} = 0$

APPLICATIONS

41. Parallel resistance   If three resistors with resistances of R_1, R_2, and R_3 are wired in parallel, their combined resistance R is given by the formula

$$\frac{1}{R} = \frac{1}{R_1} + \frac{1}{R_2} + \frac{1}{R_3}$$

The design of a voltmeter requires that the resistance R_2 be 10 ohms greater than the resistance R_1, that the resistance R_3 be 50 ohms greater than R_1, and that their combined resistance be 6 ohms. Find the value of each resistance.

42. Fabricating sheet metal   The open tray in Illustration 1 is to be manufactured from a 12-by-14-inch rectangular sheet of metal by cutting squares from each corner and folding up the sides. If the volume of the tray is to be 160 cubic inches, what size squares should be cut from each corner?

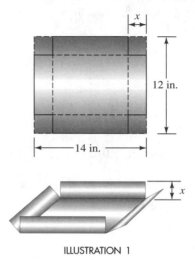

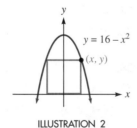

ILLUSTRATION 1

DISCOVERY AND WRITING

43. If n is an even integer and c is a positive constant, show that $x^n + c = 0$ has no real roots.

44. If n is an even positive integer and c is a positive constant, show that $x^n - c = 0$ has two real roots.

45. Precalculus   A rectangle is inscribed in the parabola $y = 16 - x^2$, as shown in Illustration 2. Find the point (x, y) if the area of the rectangle is 42 square units.

$y = 16 - x^2$

(x, y)

ILLUSTRATION 2

46. Precalculus   One corner of the rectangle in Illustration 3 is at the origin, and the opposite corner (x, y) lies in the first quadrant on the curve $y = x^3 - 2x^2$. Find the point (x, y) if the area of the rectangle is 27 square units.

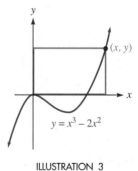

(x, y)

$y = x^3 - 2x^2$

ILLUSTRATION 3

REVIEW   *Simplify each radical expression. Assume that all variables represent nonnegative real numbers.*

47. $\sqrt{72a^3b^5c}$

48. $\dfrac{5a}{\sqrt{5a}}$

49. $\sqrt{18a^2b} + a\sqrt{50b}$

50. $\left(\sqrt{3} + \sqrt{5b}\right)\left(\sqrt{12} - \sqrt{45b}\right)$

51. $\dfrac{2}{\sqrt{3} - 1}$

52. $\dfrac{\sqrt{11} + \sqrt{x}}{\sqrt{11} - \sqrt{x}}$

6.4 Irrational Roots of Polynomial Equations

■ THE INTERMEDIATE VALUE THEOREM ■ THE BISECTION METHOD ■ FINDING ROOTS WITH A GRAPHING CALCULATOR

First-degree equations are easy to solve, and all quadratic equations can be solved with the quadratic formula. There are formulas for solving third- and fourth-degree polynomial equations, although they are complicated. However, there are no formulas for solving polynomial equations of fifth degree or greater. This fact was proved by the Norwegian mathematician Niels Henrik Abel (1802–1829) and (for equations of degree greater than 5) by the French mathematician Evariste Galois (1811–1832).

To solve a high-degree polynomial equation with integer coefficients, we can use the methods of the previous section to find its rational roots. Once we find them, the final depressed equation would have to be a first- or second-degree equation to enable us to complete the solution. The purpose of this section is to discuss ways of approximating irrational roots of these higher-degree polynomial equations.

■ THE INTERMEDIATE VALUE THEOREM

The following theorem, called the **intermediate value theorem,** will lead to a way of locating an interval that contains a root.

Intermediate Value Theorem
Let $P(x)$ be a polynomial with real coefficients. If $P(a) \neq P(b)$ for $a < b$, then $P(x)$ takes on all values between $P(a)$ and $P(b)$ on the closed interval $[a, b]$.

Justification This theorem becomes clear when we consider the graph of the polynomial $y = P(x)$, shown in Figure 6-1. We have seen that graphs of polynomials are *continuous* curves, a technical term that means, roughly, that they can be drawn without lifting the pencil from the paper. If $P(a) \neq P(b)$, then the continuous curve joining the points $A(a, P(a))$ and $B(b, P(b))$ must take on all values between $P(a)$ and $P(b)$ in the interval $[a, b]$, because the curve has no gaps in it.

The next theorem follows from the intermediate value theorem.

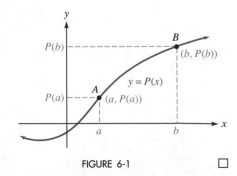

FIGURE 6-1   □

> **Location Theorem**
> Let $P(x)$ be a polynomial with real coefficients. If $P(a)$ and $P(b)$ have opposite signs, there is at least one number r in the interval (a, b) for which $P(r) = 0$.

Proof   See Figure 6-2. By the intermediate value theorem, $P(x)$ takes on all values between $P(a)$ and $P(b)$. Since $P(a)$ and $P(b)$ have opposite signs, the number 0 lies between them. Thus, there is a number r between a and b for which $P(r) = 0$. This number r is a zero of $P(x)$, and a root of the equation $P(x) = 0$.

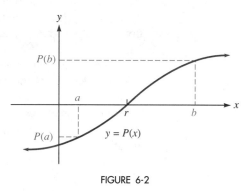

FIGURE 6-2   □

■ THE BISECTION METHOD

The previous theorem provides a method (called the **bisection method**) for finding the roots of $P(x) = 0$ to any desired degree of accuracy.

Suppose we find, by trial and error, that the numbers x_l and x_r (for left and right) straddle a root; that is, $x_l < x_r$ and $P(x_l)$ and $P(x_r)$ have opposite signs. Further suppose that $P(x_l) < 0$ and $P(x_r) > 0$. We can compute the number c that is halfway between x_l and x_r and then compute $P(c)$. If $P(c) = 0$, we've found a root. However, if $P(c)$ is not 0, we proceed in one of two ways:

1. If $P(c) < 0$, the root r lies between c and x_r, as shown in Figure 6-3(a). In this case, we let c become a new x_l and repeat the procedure.

2. If $P(c) > 0$, the root r lies between x_l and c, as shown in Figure 6-3(b). In this case, we let c become a new x_r and repeat the procedure.

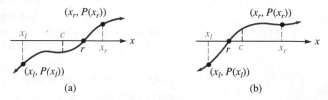

FIGURE 6-3

At any stage in this procedure, the root is contained between the current values of x_l and x_r. If the original bounds were 1 unit apart, after 10 repetitions of this procedure, the root would be between bounds that were 2^{-10} units apart. This means that the zero of $P(x)$ would be within 0.0001 units of either x_l or x_r.

After 20 repetitions, the root would be between bounds that were 2^{-20} units apart. This means that the zero of $P(x)$ would be within 0.000001 of either x_l or x_r.

EXAMPLE 1   Solve $x^2 - 2 = 0$ and express the positive root to the nearest tenth.

Solution   Since $P(1) = -1$ and $P(2) = 2$ have opposite signs, there is a root between 1 and 2. We can let $x_l = 1$ and $x_r = 2$ and compute the midpoint c:

$$c = \frac{x_l + x_r}{2} = \frac{1 + 2}{2} = 1.5$$

Because $P(c) = P(1.5) = 0.25$ is a positive number, we let c become a new x_r and calculate a new midpoint, which we will call c_1.

$$c_1 = \frac{x_l + x_r}{2} = \frac{1 + 1.5}{2} = 1.25$$

Because $P(c_1) = P(1.25) = -0.4375$ is a negative number, we let c_1 become a new x_l and calculate a new midpoint, which we will call c_2.

$$c_2 = \frac{x_l + x_r}{2} = \frac{1.25 + 1.5}{2} = 1.375$$

Because $P(c_2) = P(1.375) = -0.109375$ is a negative number, we let c_2 become a new x_l and calculate a new midpoint, which we will call c_3.

$$c_3 = \frac{x_l + x_r}{2} = \frac{1.375 + 1.5}{2} = 1.4375$$

Because $P(c_3) = P(1.4375) = 0.066406$ is a positive number, we let c_3 become a new x_r and calculate a new midpoint, which we will call c_4.

$$c_4 = \frac{x_l + x_r}{2} = \frac{1.375 + 1.4375}{2} = 1.40625$$

From here on, the first two digits of the midpoints will remain 1.4. The root r of the equation (to the nearest tenth) is 1.4. ∎

Self Check   Find the negative root of $x^2 - 2 = 0$ to the nearest tenth.

Answer -1.4

■ FINDING ROOTS WITH A GRAPHING CALCULATOR

We can approximate the roots of an equation by using the TRACE and ZOOM capabilities of a graphing calculator.

■ ■ ■ ■ ■ ■ ■ ■ ■

ACCENT ON TECHNOLOGY

To use a graphing calculator to solve polynomial equation $x^4 - 6x^2 + 9 = 0$, we graph the function $y = x^4 - 6x^2 + 9$ and then trace and zoom to find the x-intercepts.

We begin by observing that the graph of $y = x^4 - 6x^2 + 9$ will be symmetric about the y-axis, so that if $(r, 0)$ is a positive x-intercept, then $(-r, 0)$ will also be an x-intercept.

If we use window settings of $[-8, 8]$ for x and $[-8, 8]$ for y and graph the function, we will obtain the graph shown in Figure 6-4(a). If we trace and move the cursor near the positive x-intercept, we will obtain Figure 6-4(b). If we zoom in, we will obtain Figure 6-4(c), which shows that the x-intercept is approximately $(1.7021277, 0)$.

To get better results, we can zoom in and trace again to obtain Figure 6-4(d), which shows that the x-coordinate of the x-intercept is approximately 1.7340426. After zooming and tracing a few more times, we will see that to three decimal places, $x = 1.732$. By symmetry, we know that the negative solution is -1.732.

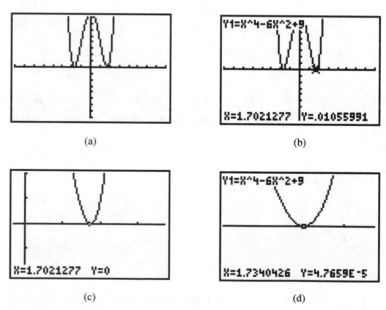

(a)

(b)

(c)

(d)

FIGURE 6-4

These graphs illustrate a situation in which the bisection method would not work. Since the graph doesn't cross the axis, the values of the function on opposite sides of the intercepts do not have opposite signs. In this case, the bisection method would be useless.

■ ■ ■ ■ ■ ■ ■ ■ ■

EXERCISE 6.4

VOCABULARY AND CONCEPTS *In Exercises 1–4, fill in the blank to make a true statement.*

1. If $P(x)$ is a polynomial with real coefficients and $P(a) \neq P(b)$ for $a < b$, then $P(x)$ takes on all values between _____ in the interval $[a, b]$.

2. If $P(x)$ has real coefficients and $P(a)$ and $P(b)$ have opposite signs, there is at least one number r in (a, b) for which _____.

3. In the bisection method, we find two numbers x_l and x_r that straddle a root. We then compute a new guess for the root (a number c) by finding the average of _____.

4. A _____ curve has no gaps in it.

PRACTICE   *In Exercises 5–14, show that each equation has at least one real root between the specified numbers.*

5. $2x^2 + x - 3 = 0$; -2 and -1

6. $2x^3 + 17x^2 + 31x - 20 = 0$; -1 and 2

7. $3x^3 - 11x^2 - 14x = 0$; 4 and 5

8. $2x^3 - 3x^2 + 2x - 3 = 0$; 1 and 2

9. $x^4 - 8x^2 + 15 = 0$; 1 and 2

10. $x^4 - 8x^2 + 15 = 0$; 2 and 3

11. $30x^3 + 10 = 61x^2 + 39x$; 2 and 3

12. $30x^3 + 10 = 61x^2 + 39x$; -1 and 0

13. $30x^3 + 10 = 61x^2 + 39x$; 0 and 1

14. $5x^3 - 9x^2 - 4x + 9 = 0$; -1 and 2

In Exercises 15–22, use the bisection method to find the following values to the nearest tenth.

15. The positive root of $x^2 - 3 = 0$.

16. The negative root of $x^2 - 3 = 0$.

17. The negative root of $x^2 - 5 = 0$.

18. The positive root of $x^2 - 5 = 0$.

19. The positive root of $x^3 - x^2 - 2 = 0$.

20. The negative root of $x^3 - x + 2 = 0$.

21. The negative root of $3x^4 + 3x^3 - x^2 - 4x - 4 = 0$.

22. The positive root of $x^5 + x^4 - 4x^3 - 4x^2 - 5x - 5 = 0$.

In Exercises 23–26, use a graphing calculator to find the distinct real solutions of each equation to the nearest tenth. Which roots, if any, would the bisection method fail to find?

23. $x^2 - 5 = 0$

24. $x^2 - 10x + 25 = 0$

25. $x^3 - 5x^2 + 8x - 4 = 0$

26. $x^3 - 5x^2 - 2x + 10 = 0$

APPLICATIONS

27. Precalculus Use the bisection method or a graphing calculator to find the coordinates of the two points on the graph of $y = x^3$ that lie 1 unit from the origin. (See Illustration 1.) Give the result to the nearest hundredth.

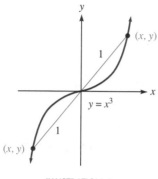

ILLUSTRATION 1

28. Building a crate The length of the shipping crate shown in Illustration 2 is to be 2 feet greater than its height and 5 feet longer than its width, and its volume is to be 170 cubic feet. Use the bisection method or a graphing calculator to find the height of the crate to the nearest tenth of a foot.

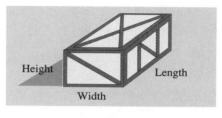

ILLUSTRATION 2

DISCOVERY AND WRITING

29. Explain when the bisection method will not work.

30. Describe how you could tell from a graph that a polynomial equation has no real roots.

REVIEW *Give the equations of all of the asymptotes of each rational function.* ***Don't graph the functions.***

31. $y = \dfrac{5x - 3}{x^2 - 4}$

32. $y = \dfrac{5x^2 - 3}{x^2 - 25}$

33. $y = \dfrac{x^2}{x - 2}$

34. $y = \dfrac{x + 1}{x^2 + 1}$

■ ■ ■ ■ ■ ■ ■ ■ ■ **PROBLEMS AND PROJECTS**

1. If $x^2 - 8x + 15$ is a factor of $P(x)$, find $P(3) + P(5)$.

2. $x^2 - 6kx + 15k = 0$ has two positive solutions, one 5 times the other. Find k and solve the equation.

3. Using Descartes' rule of signs, construct **a.** a fifth-degree equation with only one positive real root and **b.** a fourth-degree equation with no real roots.

4. For what real value(s) of k does

$$x^3 + k^2x^2 + 2kx + 1 = 0$$

have rational solutions?

PROJECT 1   Find the dimensions of the right-triangular brace in Illustration 1.

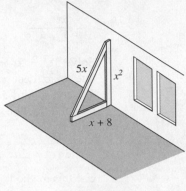

ILLUSTRATION 1

PROJECT 2   The bisection method is one of several techniques for finding approximate solutions of equations. Another is **Newton's method,** discovered by Sir Isaac Newton (1642–1727).

The process begins with a guess of a solution and then transforms that guess into a better estimate of the solution. When Newton's method is applied to that estimate, a third solution results—one more accurate than any previous. As the method is applied repeatedly, the results converge on the actual solution.

To solve the quadratic equation $x^2 + x - 3 = 0$, for example, Newton's method uses the fraction

$$\frac{x^2 + 3}{2x + 1}$$

to generate better solutions. We begin with a guess (3, for example) and substitute 3 for x in the fraction.

$$\frac{x^2 + 3}{2x + 1} = \frac{3^2 + 3}{2(3) + 1} \approx 1.714286$$

The number 1.714286 is a better estimate of the solution than the original guess, 3. We apply Newton's method again and substitute 1.714286 for x.

$$\frac{x^2 + 3}{2x + 1} = \frac{1.714286^2 + 3}{2(1.714286) + 1} \approx 1.341014$$

The estimate 1.341014 is better yet. More passes produce even better accuracy.

- Continue with Newton's method until your results remain unchanged to six decimal places.

■ ■ ■ ■ ■ ■ ■ ■ ■ ■ **PROBLEMS AND PROJECTS** *(continued)*

- Solve the equation using the quadratic formula. Is Newton's method correct?
- To solve the equation $ax^2 + bx + c = 0$, Newton's method uses the fraction

$$\frac{ax^2 - c}{2ax + b}$$

Solve $3x^2 - 5x - 2 = 0$ using Newton's method. Find two solutions, accurate to six decimal places.

- To solve a third-degree equation $ax^3 + bx^2 + cx + d = 0$, Newton's method uses the fraction

$$\frac{2ax^3 + bx^2 - d}{3ax^2 + 2bx + c}$$

Use this method to find three solutions of $x^3 - x^2 - 5x + 5 = 0$, accurate to six decimal places.

CHAPTER SUMMARY

CONCEPT

REVIEW EXERCISES

The Remainder and Factor Theorems; Synthetic Division

The remainder theorem:
If $P(x)$ is a polynomial, r is any number, and $P(x)$ is divided by $x - r$, the remainder is $P(r)$.

1. Let $P(x) = 4x^4 + 2x^3 - 3x^2 - 2$. Find the remainder when $P(x)$ is divided by each binomial.
    a. $x - 1$ **b.** $x - 2$ **c.** $x + 3$ **d.** $x + 2$

The factor theorem:
Let $P(x)$ be any polynomial and r any number. Then $P(r) = 0$ if and only if $x - r$ is a factor of $P(x)$.

2. Use the factor theorem to decide whether each statement is true.
    a. $x - 2$ is a factor of $x^3 + 4x^2 - 2x + 4$.

    b. $x + 3$ is a factor of $2x^4 + 10x^3 + 4x^2 + 7x + 21$.

    c. $x - 5$ is a factor of $x^5 - 3125$.

    d. $x - 6$ is a factor of $x^5 - 6x^4 - 4x + 24$.

3. Find the polynomial of lowest degree with integer coefficients and the given zeros
    a. $1, 2,$ and $\dfrac{3}{2}$ **b.** $1, -4,$ and $\dfrac{1}{2}$
    c. $2, -5, i$ and $-i$ **d.** $-3, 2, i,$ and $-i$

4. Use synthetic division to find the quotient and remainder when each polynomial is divided by the given divisor.
    a. $3x^4 + 2x^2 + 3x + 7; x - 3$

    b. $2x^4 - 3x^2 + 3x - 1; x - 2$

    c. $5x^5 - 4x^4 + 3x^3 - 2x^2 + x - 1; x + 2$

    d. $4x^5 + 2x^4 - x^3 + 3x^2 + 2x + 1; x + 1$

Descartes' Rule of Signs and Bounds on Roots

Fundamental theorem of algebra:
If $P(x)$ is a polynomial with positive degree, then $P(x)$ has at least one zero.

5. How many roots does each equation have?
    a. $3x^6 - 4x^5 + 3x + 2 = 0$

    b. $2x^6 - 5x^4 + 5x^3 - 4x + x - 12 = 0$

    c. $3x^{65} - 4x^{50} + 3x^{17} + 2x = 0$

The polynomial equation $P(x) = 0$ with degree n ($n > 0$) has exactly n roots among the complex numbers.

Complex roots of polynomial equations with real coefficients occur in complex conjugate pairs.

Descartes' rule of signs:
If $P(x)$ has real coefficients, the number of positive roots of $P(x) = 0$ is equal to the number of variations in sign of $P(x)$, or is less than that by an even number.

   The number of negative roots is equal to the number of variations in sign of $P(-x)$, or is less than that by an even number.

Upper bounds:
Let the lead coefficient of the polynomial $P(x)$ with real coefficients be positive, and do a synthetic division of the coefficients of $P(x)$ by the positive number c. If none of the terms in the bottom row is negative, then c is an upper bound for the real roots of $P(x) = 0$.

Lower bounds:
Let the lead coefficient of the polynomial $P(x)$ with real coefficients be positive, and do a synthetic division of the coefficients of $P(x)$ by the negative number d. If the signs in the bottom row alternate, then d is a lower bound for the real roots of $P(x) = 0$.

d. $x^{1984} - 12 = 0$

6. Find another root of a polynomial equation if the given quantity is one root.
   a. $2 + i$ **b.** $-i$

7. Find the number of possible positive, negative, and nonreal roots for each equation. **Do not attempt to solve the equation.**
   a. $3x^4 + 2x^3 - 4x + 2 = 0$

   b. $2x^4 - 3x^3 + 5x^2 + x - 5 = 0$

   c. $4x^5 + 3x^4 + 2x^3 + x^2 + x = 7$

   d. $3x^7 - 4x^5 + 3x^3 + x - 4 = 0$

   e. $x^4 + x^2 + 24{,}567 = 0$

   f. $-x^7 - 5 = 0$

8. Find integer bounds for the roots of each equation.
   a. $5x^3 - 4x^2 - 2x + 4$

   b. $x^4 + 3x^3 - 5x^2 - 9x + 1$

| SECTION 6.3 | *Rational Roots of Polynomial Equations* |

If $P(x) = 0$ has integer coefficients and $\frac{p}{q}$ (written in lowest terms) is a root, then p is a factor of the constant, and q is a factor of the lead coefficient.

9. Find all rational roots of each equation.
 a. $2x^3 + 17x^2 + 41x + 30 = 0$
 b. $3x^3 + 2x^2 + 2x - 1 = 0$
 c. $4x^4 - 25x^2 + 36 = 0$
 d. $2x^4 - 11x^3 - 6x^2 + 64x + 32 = 0$

| SECTION 6.4 | *Irrational Roots of Polynomial Equations* |

Intermediate value theorem: Let $P(x)$ be a polynomial with real coefficients. If $P(a) \neq P(b)$ for $a < b$, then $P(x)$ takes on all values between $P(a)$ and $P(b)$ on the closed interval $[a, b]$.

Let $P(x)$ be a polynomial with real coefficients. If $P(a)$ and $P(b)$ have opposite signs, then there is at least one number r between a and b for which $P(r) = 0$.

10. Show that each polynomial has a zero between the two given numbers.

 a. $5x^3 + 37x^2 + 59x + 18 = 0$; -1 and 0
 b. $6x^3 - x^2 - 10x - 3 = 0$; 1 and 2

11. Use the bisection method to find the positive root of each equation to the nearest tenth.
 a. $x^3 - 2x^2 - 9x - 2 = 0$
 b. $6x^2 - 13x - 5 = 0$

12. Use a graphing calculator to find the positive root of each equation to the nearest hundredth.
 a. $6x^2 - 7x - 5 = 0$
 b. $3x^2 + x - 2 = 0$

13. Designing solar collectors The space available for the installation of three solar collecting panels requires that their lengths differ by the amounts shown in Illustration 1, and that the total of their widths be 15 meters. To be equally effective, each panel must measure exactly 60 square meters. Find the dimensions of each panel.

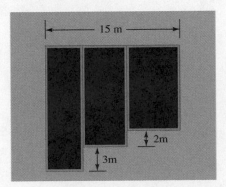

ILLUSTRATION 1

14. Designing a storage tank The design specifications for the cylindrical storage tank in Illustration 2 require that its height be 3 feet greater than the radius of its circular base, and that the volume of the tank be 19,000 cubic feet. Use the bisection method or a graphing calculator to find the radius of the tank to the nearest hundredth of a foot.

ILLUSTRATION 2

■ Chapter 6 Test

In Questions 1–2, use the remainder theorem to find each value.

1. $P(x) = 3x^3 - 9x - 5$; $P(2)$

2. $P(x) = x^5 + 2$; $P(-2)$

In Questions 3–4, find a polynomial with the given zeros.

3. $5, -1, 0$

4. $i, -i, \sqrt{3}, -\sqrt{3}$

In Questions 5–8, let $P(x) = 3x^3 - 2x^2 + 4$. Use synthetic division to find each value.

5. $P(1)$

6. $P(-2)$

7. $P\left(-\dfrac{1}{3}\right)$

8. $P(i)$

In Questions 9–10, use synthetic division to express $P(x) = 2x^3 - 3x^2 - 4x - 1$ in the form (divisor)(quotient) + remainder for each divisor.

9. $x - 2$

10. $x + 1$

In Questions 11–12, use synthetic division to do each division.

11. $\dfrac{2x^2 - 7x - 15}{x - 5}$

12. $\dfrac{3x^3 + 7x^2 + 2x}{x + 2}$

In Questions 13–14, write a third-degree polynomial equation with real coefficients and the given roots.

13. $2, i$

14. $1, 2 + i$

*In Questions 15–16, use Descartes' rule of signs to find the number of possible positive, negative, and nonreal roots of each equation. **Do not find the roots.***

15. $3x^5 - 2x^4 + 2x^2 - x - 3 = 0$

16. $2x^3 - 5x^2 - 2x - 1 = 0$

In Questions 17–18, find integer bounds for the roots of each equation.

17. $x^5 - x^4 - 5x^3 + 5x^2 + 4x - 5 = 0$

18. $2x^3 - 11x^2 + 10x + 3 = 0$

In Question 19, find all roots of the equation.

19. $2x^3 + 3x^2 - 11x - 6 = 0$

In Question 20, use the bisection method to find the positive root of the equation to the nearest tenth.

20. $x^2 - 11 = 0$

■ Cumulative Review Exercises

In Exercises 1–4, graph the function defined by each equation.

1. $y = 3^x - 2$

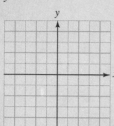

2. $y = 2e^x$

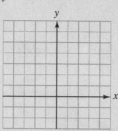

3. $y = \log_3 x$

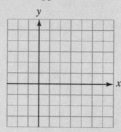

4. $y = \ln(x - 2)$

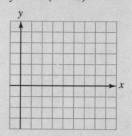

In Exercises 5–8, find each value.

5. $\log_2 64$

6. $\log_{1/2} 8$

7. $\ln e^3$

8. $2^{\log_2 2}$

In Exercises 9–12, write each expression in terms of the logarithms of a, b, and c.

9. $\log abc$

10. $\log \dfrac{a^2 b}{c}$

11. $\log \sqrt{\dfrac{ab}{c^3}}$

12. $\log \dfrac{\sqrt{ab^2}}{c}$

In Exercises 13–14, write each expression as the logarithm of a single quantity.

13. $3 \log a - 3 \log b$

14. $\dfrac{1}{2} \log a + 3 \log b - \dfrac{2}{3} \log c$

In Exercises 15–18, solve each equation.

15. $3^{x+1} = 8$

16. $3^{x-1} = 3^{2x}$

17. $\log x + \log 2 = 3$

18. $\log(x + 1) + \log(x - 1) = 1$

In Exercises 19–22, let $P(x) = 4x^3 + 3x + 2$. Use synthetic division to find each value.

19. $P(1)$

20. $P(-2)$

21. $P\left(\dfrac{1}{2}\right)$

22. $P(i)$

In Exercises 23–26, tell whether each binomial is a factor of $P(x) = x^3 + 2x^2 - x - 2$. Use synthetic division.

23. $x + 1$

24. $x - 2$

25. $x - 1$

26. $x + 2$

In Exercises 27–28, tell how many roots each equation has.

27. $x^{12} - 4x^8 + 2x^4 + 12 = 0$

28. $x^{2000} - 1 = 0$

In Exercises 29–30, tell the number of possible positive, negative, and nonreal roots of each equation.

29. $x^4 + 2x^3 - 3x^2 + x + 2 = 0$

30. $x^4 - 3x^3 - 2x^2 - 3x - 5 = 0$

In Exercises 31–32, solve each equation.

31. $x^3 + x^2 - 9x - 9 = 0$

32. $x^3 - 2x^2 - x + 2 = 0$

7 Linear Systems

IN THIS CHAPTER, WE LEARN TO SOLVE SYSTEMS OF EQUATIONS—SETS OF SEVERAL EQUATIONS, EACH WITH MORE THAN ONE VARIABLE. ONE METHOD OF SOLUTION USES MATRICES, AN IMPORTANT TOOL IN MATHEMATICS AND ITS APPLICATIONS.

7.1 Systems of Linear Equations

Equations with two variables have infinitely many solutions. For example, the following tables give a few of the solutions of $x + y = 5$ and $x - y = 1$.

$$x + y = 5 \qquad x - y = 1$$

x	y	x	y
1	4	8	7
2	3	3	2
3	2	1	0
5	0	0	-1

Only the pair $x = 3$ and $y = 2$ satisfies both equations. The pair of equations

$$\begin{cases} x + y = 5 \\ x - y = 1 \end{cases}$$

is called a **system of equations,** and the solution $x = 3$, $y = 2$ is called its **simultaneous solution,** or just its **solution.** The process of finding the solution of a system of equations is called **solving the system.**

The graph of an equation in two variables displays the equation's infinitely many solutions. For example, the graph of one of the lines in Figure 7-1 represents the infinitely many solutions of the equation $5x - 2y = 1$. The other line in the figure is the graph of the infinitely many solutions of $2x + 3y = 8$.

Because only point $P(1, 2)$ lies on both lines, only its coordinates satisfy both equations. Thus, the simultaneous solution of the system of equations

$$\begin{cases} 5x - 2y = 1 \\ 2x + 3y = 8 \end{cases}$$

is the pair of numbers $x = 1$ and $y = 2$, or simply the pair $(1, 2)$.

This discussion suggests a graphical method of solving systems of equations in two variables.

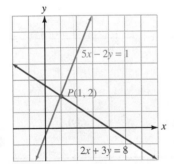

FIGURE 7-1

■ THE GRAPHING METHOD

We use the following steps to solve a system of two equations in two variables by graphing.

Strategy for Using the Graphing Method

1. On one coordinate grid, graph each equation.
2. Find the coordinates of the point or points where all of the graphs intersect. These coordinates give the solutions of the system.
3. If the graphs have no point in common, the system has no solution.

EXAMPLE 1   Use the graphing method to solve each system. **a.** $\begin{cases} 3x + y = 1 \\ -x + 2y = 9 \end{cases}$

b. $\begin{cases} 2x - 3y = 4 \\ 4x = -4 + 6y \end{cases}$ **c.** $\begin{cases} y = 4 - x \\ 2x + 2y = 8 \end{cases}$

Solution   **a.** The graphs of the equations are the lines shown in Figure 7-2(a). The solution of this system is given by the coordinates of the point $(-1, 4)$, where the lines intersect. By checking both values in both equations, we can verify that the solution is $x = -1$ and $y = 4$.

b. The graphs of the equations are the parallel lines shown in Figure 7-2(b). Since parallel lines do not intersect, the system has no solutions.

c. The graphs of the equations are the lines shown in Figure 7-2(c). In this case, the lines are the same. Because the two lines have infinitely many points in common, the system has infinitely many solutions. All ordered pairs whose coordinates satisfy one of the equations satisfy the other also.

To find some solutions, we substitute numbers for x in the first equation and solve for y. If $x = 3$, for example, then $y = 1$. One solution is the pair $(3, 1)$. Other solutions are $(0, 4)$ and $(5, -1)$.

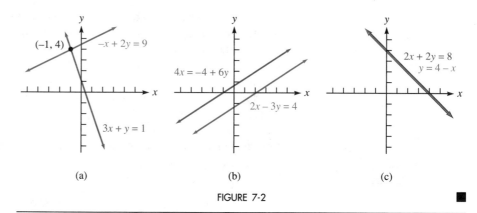

(a)   (b)   (c)

FIGURE 7-2

Self Check   Solve $\begin{cases} y = 2x \\ x + y = 3 \end{cases}$.

Answer $(1, 2)$

Example 1 illustrates three situations that can occur when we solve systems of equations. If a system of equations has at least one solution, as in parts **a** and **c** of Example 1, the system is called **consistent**. If it has no solutions, as in part **b**, it is called **inconsistent**.

If a system of two equations in two variables has exactly one solution as in part **a**, or no solution as in part **b**, the equations in the system are called **independent**. If a system of linear equations has infinitely many solutions, as in part **c**, the equations of the system are called **dependent**.

The following table shows the three possibilities that can occur when two equations, each with two variables, are graphed.

Possible figure	If the	then
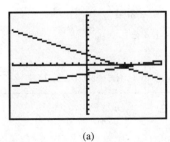 lines are distinct and intersect,	lines are distinct and intersect,	the equations are independent, and the system is consistent. The system has one solution.
lines are distinct and parallel,	lines are distinct and parallel,	the equations are independent, and the system is inconsistent. The system has no solutions.
lines coincide,	lines coincide,	the equations are dependent, and the system is consistent. The system has infinitely many solutions.

■ ■ ■ ■ ■ ■ ■ ■ ■ ■

ACCENT ON TECHNOLOGY

To use a graphing calculator to solve the system

$$\begin{cases} x - 4y = 7 \\ x + 2y = 4 \end{cases}$$

we must solve both equations for y. We do so to get the equivalent system

$$\begin{cases} y = \dfrac{x - 7}{4} \\ y = \dfrac{4 - x}{2} \end{cases}$$

Depending on the viewing window, the two graphs will be similar to those in Figure 7-3(a). We zoom in on the intersection as in Figure 7-3(b) and trace to find the solution: $x = 5$, $y = -\frac{1}{2}$.

(a)

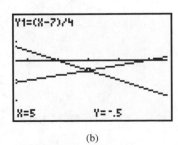

(b)

FIGURE 7-3

We now consider two algebraic methods to find solutions of systems containing any number of variables.

■ THE SUBSTITUTION METHOD

We use the following steps to solve a system of two equations in two variables by substitution.

> **Strategy for Using the Substitution Method**
> 1. Solve one equation for a variable—say, y.
> 2. Substitute the expression obtained for y for every y in the second equation.
> 3. Solve the equation that results.
> 4. Substitute the solution found in Step 3 into the equation found in Step 1 and solve for y.

EXAMPLE 2   Use the substitution method to solve $\begin{cases} 3x + y = 1 \\ -x + 2y = 9 \end{cases}$.

Solution   We can solve the first equation for y and then substitute that result for y in the second equation.

$$\begin{cases} 3x + y = 1 \rightarrow y = \boxed{1 - 3x} \\ -x + 2y = 9 \end{cases}$$

The substitution gives one linear equation with one variable, which we can solve for x:

$$-x + 2(1 - 3x) = 9$$
$$-x + 2 - 6x = 9$$
$$-7x = 7 \qquad \text{Combine like terms and subtract 2 from both sides.}$$
$$x = -1 \qquad \text{Divide both sides by } -7.$$

To find y, we substitute -1 for x in the equation $y = 1 - 3x$ and simplify:

$$y = 1 - 3x$$
$$= 1 - 3(-1)$$
$$= 1 + 3$$
$$= 4$$

The solution is the pair $(-1, 4)$. ■

Self Check   Solve $\begin{cases} 2x + 3y = 1 \\ x - y = 3 \end{cases}$.

Answer $(2, -1)$

■ THE ADDITION METHOD

As with substitution, the addition method combines the equations of a system to eliminate terms involving one of the variables. To do so, we follow these steps.

Strategy for Using the Addition Method

1. Write the equations of the system in general form so that terms with the same variable are aligned vertically.
2. Multiply all the terms of one or both of the equations by constants chosen to make the coefficients of x (or y) differ only in sign.
3. Add the equations. Solve the equation that results, if possible.
4. Substitute the value obtained in Step 3 into either of the original equations, and solve for the remaining variable.
5. The results obtained in Steps 3 and 4 are the solution of the system.

EXAMPLE 3 Use the addition method to solve $\begin{cases} 3x + 2y = 8 \\ 2x - 5y = 18 \end{cases}$.

Solution To eliminate y, we multiply the terms of the first equation by 5 and the terms of the second equation by 2, to obtain an equivalent system in which the coefficients of y differ only in sign. Then we add the equations to eliminate y.

$$\begin{cases} 3x + 2y = \;\;8 \rightarrow 15x + 10y = 40 \quad &\text{Multiply by 5.} \\ 2x - 5y = 18 \rightarrow \underline{\;\;4x - 10y = 36} \quad &\text{Multiply by 2.} \\ \qquad\qquad\qquad\quad 19x \qquad\quad = 76 \quad &\text{Add the equations.} \\ \qquad\qquad\qquad\qquad\quad x = 4 \quad &\text{Divide both sides by 19.} \end{cases}$$

We now substitute 4 for x into either of the original equations and solve for y. If we use the first equation,

$$3x + 2y = 8$$
$$3(4) + 2y = 8 \qquad \text{Substitute 4 for } x.$$
$$12 + 2y = 8 \qquad \text{Simplify.}$$
$$2y = -4 \qquad \text{Subtract 12 from both sides.}$$
$$y = -2 \qquad \text{Divide both sides by 2.}$$

The solution is $(4, -2)$. ■

Self Check Solve $\begin{cases} 2x + 5y = 8 \\ 2x - 4y = -10 \end{cases}$.

Answer $(-1, 2)$

■ A SYSTEM WITH INFINITELY MANY SOLUTIONS

EXAMPLE 4   Use the addition method to solve $\begin{cases} x + 2y = 3 \\ 2x + 4y = 6 \end{cases}$.

Solution   To eliminate the variable y, we can multiply both sides of the first equation by -2 and add the result to the second equation to get

$$-2x - 4y = -6$$
$$\underline{2x + 4y = 6}$$
$$0 = 0$$

Although the result $0 = 0$ is true, it does not give the value of y.

Since the second equation in the original system is twice the first, the equations are equivalent. If we were to solve this system by graphing, the two lines would coincide. The (x, y) coordinates of points on that one line form the infinite set of solutions to the given system. The system is consistent, but the equations are dependent.

To find some solutions, we can substitute values for x into either equation and solve for y. If $x = 5$, for example, then $y = -1$, and the pair $(5, -1)$ is a solution. Other solutions are $(3, 0)$, $\left(0, \frac{3}{2}\right)$, and $(-7, 5)$.   ■

Self Check   Solve $\begin{cases} 3x - y = 2 \\ 6x - 2y = 4 \end{cases}$.

Answer   Some solutions are $(1, 1)$, $(0, -2)$, and $(3, 7)$.

■ AN INCONSISTENT SYSTEM

EXAMPLE 5   Solve $\begin{cases} x + y = 3 \\ x + y = 2 \end{cases}$.

Solution   We can multiply both sides of the second equation by -1 and add the results to the first equation to get

$$x + y = 3$$
$$\underline{-x - y = -2}$$
$$0 = 1$$

Because $0 \neq 1$, the system has no solutions. If we graph each equation in this system, the graphs will be parallel lines. This system is inconsistent, and the equations are independent.   ■

Self Check   Solve $\begin{cases} 3x + 2y = 2 \\ 3x + 2y = 3 \end{cases}$.

Answer   The system is inconsistent—no solutions.

EXAMPLE 6 Solve $\begin{cases} \dfrac{x+2y}{4} + \dfrac{x-y}{5} = \dfrac{6}{5} \\ \dfrac{x+y}{7} - \dfrac{x-y}{3} = -\dfrac{12}{7} \end{cases}.$

Solution To clear the first equation of fractions, we multiply both sides by the lowest common denominator, 20. Similarly, we multiply both sides of the second equation by 21.

$$\begin{cases} 20\left(\dfrac{x+2y}{4} + \dfrac{x-y}{5}\right) = 20\left(\dfrac{6}{5}\right) \\ 21\left(\dfrac{x+y}{7} - \dfrac{x-y}{3}\right) = 21\left(-\dfrac{12}{7}\right) \end{cases}$$

$$\begin{cases} 5(x+2y) + 4(x-y) = 24 \\ 3(x+y) - 7(x-y) = -36 \end{cases}$$

$$\begin{cases} 9x + 6y = 24 \\ -4x + 10y = -36 \end{cases} \qquad \text{Remove parentheses and combine like terms.}$$

$$\begin{cases} 3x + 2y = 8 \\ 2x - 5y = 18 \end{cases} \qquad \begin{array}{l} \text{Divide both sides by 3.} \\ \text{Divide both sides by } -2. \end{array}$$

In Example 3, we saw that the solution to this final system is $(4, -2)$. ∎

Self Check Solve $\begin{cases} \dfrac{x+y}{3} + \dfrac{x-y}{4} = \dfrac{3}{4} \\ \dfrac{x-y}{3} - \dfrac{2x-y}{2} = -\dfrac{1}{3} \end{cases}.$

Answer $(1, 2)$

■ THREE EQUATIONS IN THREE VARIABLES

To solve three equations in three variables, we use addition to eliminate one variable. This will produce a system of two equations in two variables, which we can solve using the methods previously discussed.

EXAMPLE 7 Solve the system

$$\begin{array}{ll} \textbf{1.} & \\ \textbf{2.} & \begin{cases} x + 2y + z = 8 \\ 2x + y - z = 1 \\ x + y - 2z = -3 \end{cases} \\ \textbf{3.} & \end{array}$$

Solution We can add Equations 1 and 2 to eliminate z

$$\begin{array}{ll} \textbf{1.} & \quad x + 2y + z = 8 \\ \textbf{2.} & \underline{2x + y - z = 1} \\ \textbf{4.} & \quad 3x + 3y \phantom{{}- z} = 9 \end{array}$$

and divide both sides of Equation 4 by 3 to get

$$\textbf{5.} \quad x + y = 3$$

We now choose a different pair of equations—say, Equations 1 and 3—and eliminate z again. If we multiply both sides of Equation 1 by 2 and add the result to Equation 3, we get

$$
\begin{array}{rl}
 & 2x + 4y + 2z = 16 \\
\textbf{3.} & \underline{x + y - 2z = -3} \\
\textbf{6.} & 3x + 5y = 13
\end{array}
$$

Equations 5 and 6 form a system of two equations in two variables, which we can solve by substitution.

5. $x + y = 3 \rightarrow y = 3 - x$

6. $3x + 5y = 13$

$3x + 5(3 - x) = 13$ Substitute $3 - x$ for y.

$3x + 15 - 5x = 13$ Remove parentheses.

$-2x = -2$ Combine like terms and subtract 15 from both sides.

$x = 1$ Divide both sides by -2.

To find y, we substitute 1 for x in the equation $y = 3 - x$.

$y = 3 - 1$

$y = 2$

To find z, we substitute 1 for x and 2 for y in any one of the original equations, and find that $z = 3$. The solution is the triple

$$(x, y, z) = (1, 2, 3)$$

Because there is one solution, the system is consistent, and its equations are independent.   ∎

■ APPLICATIONS

EXAMPLE 8   An airplane flies 600 miles with the wind for 2 hours and returns against the wind in 3 hours. Find the speed of the wind and the air speed of the plane.

Solution   If a represents the air speed and w represents wind speed, the ground speed of the plane with the wind is the combined speed $a + w$. On the return trip, against the wind, the ground speed is $a - w$. The information in this problem is organized in Figure 7-4 and can be used to give a system of two equations in the variables a and w.

	d	$=$	r	$\cdot$	t
Outbound trip	600		$a + w$		2
Return trip	600		$a - w$		3

FIGURE 7-4

Since $d = rt$, we have $\begin{cases} 600 = 2(a + w) \\ 600 = 3(a - w) \end{cases}$, which can be written as

7. $\begin{cases} 300 = a + w \\ 200 = a - w \end{cases}$

We can add these equations to get

$$500 = 2a$$
$$a = 250 \qquad \text{Divide both sides by 2.}$$

To find w, we substitute 250 for a into either of the previous equations (we'll use Equation 7) and solve for w:

7. $300 = a + w$
$$300 = 250 + w$$
$$w = 50 \qquad \text{Subtract 250 from both sides.}$$

The air speed of the plane is 250 mph. With a 50-mph tailwind, the ground speed is 250 + 50, or 300 mph. At 300 mph, the 600-mile trip will take 2 hours.

With a 50-mph headwind, the ground speed is 250 − 50, or 200 mph. At 200 mph, the 600-mile trip will take 3 hours. The answers check. ■

EXERCISE 7.1

VOCABULARY AND CONCEPTS *In Exercises 1–12, fill in the blank to make a true statement.*

1. A set of several equations with several variables is called a _____ of equations.

2. Any set of numbers that satisfies each equation of a system is called a _____ of the system.

3. If a system of equations has a solution, the system is _____.

4. If a system of equations has no solution, the system is _____.

5. If a system of equations has only one solution, the equations of the system are _____.

6. If a system of equations has infinitely many solutions, the equations of the system are _____.

7. The system $\begin{cases} x + y = 5 \\ x - y = 1 \end{cases}$ is _____ (consistent, inconsistent).

8. The system $\begin{cases} x + y = 5 \\ x + y = 1 \end{cases}$ is _____ (consistent, inconsistent).

9. The equations of the system $\begin{cases} x + y = 5 \\ 2x + 2y = 10 \end{cases}$ are _____ (dependent, independent).

10. The equations of the system $\begin{cases} x + y = 5 \\ x - y = 1 \end{cases}$ are _____ (dependent, independent).

11. The pair $(1, 3)$ ___ (is, is not) a solution of the system $\begin{cases} x + 2y = 7 \\ 2x - y = -1 \end{cases}$

12. The pair $(1, 3)$ ___ (is, is not) a solution of the system $\begin{cases} 3x + y = 6 \\ x - 3y = -8 \end{cases}$

PRACTICE *In Exercises 13–16, solve each system of equations by graphing.*

13. $\begin{cases} y = -3x + 5 \\ x - 2y = -3 \end{cases}$

14. $\begin{cases} x - 2y = -3 \\ 3x + y = -9 \end{cases}$

15. $\begin{cases} 3x + 2y = 2 \\ -2x + 3y = 16 \end{cases}$

16. $\begin{cases} x + y = 0 \\ 5x - 2y = 14 \end{cases}$

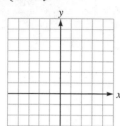

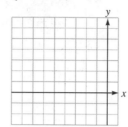

In Exercises 17–20, use a graphing calculator to approximate the solutions of each system. Give answers to the nearest tenth.

17. $\begin{cases} y = -5.7x + 7.8 \\ y = 37.2 - 19.1x \end{cases}$

18. $\begin{cases} y = 3.4x - 1 \\ y = -7.1x + 3.1 \end{cases}$

19. $\begin{cases} y = \dfrac{5.5 - 2.7x}{3.5} \\ 5.3x - 9.2y = 6.0 \end{cases}$

20. $\begin{cases} 29x + 17y = 7 \\ -17x + 23y = 19 \end{cases}$

In Exercises 21–28, solve each system by substitution, if possible.

21. $\begin{cases} y = x - 1 \\ y = 2x \end{cases}$

22. $\begin{cases} y = 2x - 1 \\ x + y = 5 \end{cases}$

23. $\begin{cases} 2x + 3y = 0 \\ y = 3x - 11 \end{cases}$

24. $\begin{cases} 2x + y = 3 \\ y = 5x - 11 \end{cases}$

25. $\begin{cases} 4x + 3y = 3 \\ 2x - 6y = -1 \end{cases}$

26. $\begin{cases} 4x + 5y = 4 \\ 8x - 15y = 3 \end{cases}$

27. $\begin{cases} x + 3y = 1 \\ 2x + 6y = 3 \end{cases}$

28. $\begin{cases} x - 3y = 14 \\ 3(x - 12) = 9y \end{cases}$

In Exercises 29–44, solve each system by the addition method, if possible.

29. $\begin{cases} 5x - 3y = 12 \\ 2x - 3y = 3 \end{cases}$

30. $\begin{cases} 2x + 3y = 8 \\ -5x + y = -3 \end{cases}$

31. $\begin{cases} x - 7y = -11 \\ 8x + 2y = 28 \end{cases}$

32. $\begin{cases} 3x + 9y = 9 \\ -x + 5y = -3 \end{cases}$

33. $\begin{cases} 3(x - y) = y - 9 \\ 5(x + y) = -15 \end{cases}$

34. $\begin{cases} 2(x + y) = y + 1 \\ 3(x + 1) = y - 3 \end{cases}$

35. $\begin{cases} 2 = \dfrac{1}{x + y} \\ 2 = \dfrac{3}{x - y} \end{cases}$

36. $\begin{cases} \dfrac{1}{x + y} = 12 \\ \dfrac{3x}{y} = -4 \end{cases}$

37. $\begin{cases} 0.5x = 0.1y \\ y - 0.5x = 1.5 \end{cases}$

38. $\begin{cases} -0.3x + 0.1y = -0.1 \\ 6x - 2y = 2 \end{cases}$

39. $\begin{cases} x + 2(x - y) = 2 \\ 3(y - x) - y = 5 \end{cases}$

40. $\begin{cases} x + \dfrac{y}{3} = \dfrac{5}{3} \\ \dfrac{x + y}{3} = 3 - x \end{cases}$

41. $\begin{cases} \dfrac{3}{2}x + \dfrac{1}{3}y = 2 \\ \dfrac{2}{3}x + \dfrac{1}{9}y = 1 \end{cases}$

42. $\begin{cases} \dfrac{x + y}{2} + \dfrac{x - y}{5} = 2 \\ x = \dfrac{y}{2} + 1 \end{cases}$

43. $\begin{cases} \dfrac{x - y}{5} + \dfrac{x + y}{2} = 6 \\ \dfrac{x - y}{2} - \dfrac{x + y}{4} = 3 \end{cases}$

44. $\begin{cases} \dfrac{x - 2}{5} + \dfrac{y + 3}{2} = 5 \\ \dfrac{x + 3}{2} + \dfrac{y - 2}{3} = 6 \end{cases}$

In Exercises 45–62, solve each system, if possible.

45. $\begin{cases} x + y + z = 3 \\ 2x + y + z = 4 \\ 3x + y - z = 5 \end{cases}$

46. $\begin{cases} x - y - z = 0 \\ x + y - z = 0 \\ x - y + z = 2 \end{cases}$

47. $\begin{cases} x - y + z = 0 \\ x + y + 2z = -1 \\ -x - y + z = 0 \end{cases}$

48. $\begin{cases} 2x + y - z = 7 \\ x - y + z = 2 \\ x + y - 3z = 2 \end{cases}$

49. $\begin{cases} 2x + y = 4 \\ x \quad - z = 2 \\ y + z = 1 \end{cases}$

50. $\begin{cases} 3x + y + z = 0 \\ 2x - y + z = 0 \\ 2x + y + z = 0 \end{cases}$

51. $\begin{cases} x + y + z = 6 \\ 2x + y + 3z = 17 \\ x + y + 2z = 11 \end{cases}$

52. $\begin{cases} x + y + z = 3 \\ 2x + y + z = 6 \\ x + 2y + 3z = 2 \end{cases}$

53. $\begin{cases} x + y + z = 3 \\ x + z = 2 \\ 2x + 2y + 2z = 3 \end{cases}$

54. $\begin{cases} x + y + z = 3 \\ x + z = 2 \\ 2x + y + 2z = 5 \end{cases}$

55. $\begin{cases} x + 2y - z = 2 \\ 2x - y = -1 \\ 3x + y + z = 1 \end{cases}$

56. $\begin{cases} x + y = 2 \\ y + z = 2 \\ 3x + 3y = 2 \end{cases}$

57. $\begin{cases} 3x + 4y + 2z = 4 \\ 6x - 2y + z = 4 \\ 3x - 8y - 6z = -3 \end{cases}$

58. $\begin{cases} x + y = 2 \\ y + z = 2 \\ x - z = 0 \end{cases}$

59. $\begin{cases} 2x - y - z = 0 \\ x - 2y - z = -1 \\ x - y - 2z = -1 \end{cases}$

60. $\begin{cases} x + 3y - z = 5 \\ 3x - y + z = 2 \\ 2x + y = 1 \end{cases}$

61. $\begin{cases} (x + y) + (y + z) + (z + x) = 6 \\ (x - y) + (y - z) + (z - x) = 0 \\ x + y + 2z = 4 \end{cases}$

62. $\begin{cases} (x + y) + (y + z) = 1 \\ (x + z) + (x + z) = 3 \\ (x - y) - (x - z) = -1 \end{cases}$

APPLICATIONS *In Exercises 63–72, use systems of equations to solve each problem.*

63. Planning for harvest A farmer raises corn and soybeans on 350 acres of land. Because of expected prices at harvest time, he thinks it would be wise to plant 100 more acres of corn than of soybeans. How many acres of each does he plant?

64. **Country club membership** There is an initiation fee to join the Pine River Country Club, as well as monthly dues. The total cost after 7 months' membership will be \$3025, and after $1\frac{1}{2}$ years, \$3850. Find both the initiation fee and the monthly dues.

65. **Framing pictures** A rectangular picture frame has a perimeter of 1900 centimeters and a width that is 250 centimeters less than its length. Find the area of the picture.

66. **Boating** A Mississippi riverboat can travel 30 kilometers downstream in three hours and can make the return trip in five hours. Find the speed of the boat in still water.

67. **Making an alloy** A metallurgist wants to make 60 grams of an alloy that is to be 34% copper. She has samples that are 9% copper and 84% copper. How many grams of each must she use?

68. **Archimedes' law of the lever** The two weights shown in Illustration 1 will be in balance if the product of one weight and its distance from the fulcrum is equal to the product of the other weight and its distance from the fulcrum. Two weights are in balance when one is 2 meters and the other 3 meters from the fulcrum. If the fulcrum remained in the same spot and the weights were interchanged, the closer weight would need to be increased by 5 pounds to maintain balance. Find the weights.

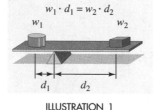

ILLUSTRATION 1

69. **Lifting a weight** A 112-pound force can lift the 448-pound load in Illustration 2. If the fulcrum is moved 1 foot away from the load, a 192-pound force is required. Find the length of the lever.

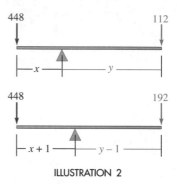

ILLUSTRATION 2

70. **Writing test questions** For a test question, a mathematics teacher wants to find two constants a and b such that the test item "Simplify $a(x + 2y) - b(2x - y)$" will have an answer of $-3x + 9y$. What constants a and b should the teacher use?

71. **Break-even point** Rollowheel, Inc. can manufacture a pair of rollerblades for \$43.53. Daily fixed costs attributable to their manufacture amount to \$742.72. A pair of rollerblades can be sold for \$89.95. Find equations expressing the expenses, E, and the revenue, R, as functions of x, the number of pairs manufactured and sold. At what production level will expenses equal revenues?

72. **Choosing salary options** For its sales staff, a company offers two salary options. One is \$326 per week and a commission of $3\frac{1}{2}$% of sales. The other is \$200 per week and $4\frac{1}{4}$% of sales. Find equations that express incomes as functions of sales, and find the weekly sales level that produces equal salaries.

In Exercises 73–78, use systems of three equations in three variables to solve each problem.

73. **Work schedule** A college student earns \$198.50 per week working three part-time jobs. Half of his 30-hour work week is spent cooking hamburgers at a fast-food chain, earning \$5.70 per hour. In addition, the student earns \$6.30 per hour working at a gas station and \$10 per hour doing janitorial work. How many hours per week does the student work at each job?

74. **Investment income** A woman invested a \$22,000 rollover IRA account in three banks paying 5%, 6%, and 7% annual interest. She invested \$2000 more at 6% than at 5%. The total annual interest she earned was \$1370. How much did she invest at each rate?

75. Age distribution Approximately 3 million people live in Costa Rica. 2.61 million are less than 50 years old, and 1.95 million are older than 14. How many people are in each of the categories 0–14 years, 15–49 years, and 50 years and older?

76. Designing an arch The engineer designing a parabolic arch knows that its equation has the form $y = ax^2 + bx + c$. Use the information in Illustration

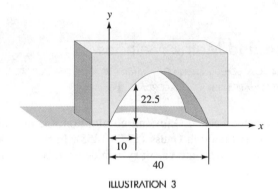

22.5

10

40

ILLUSTRATION 3

3 to find a, b, and c. Assume that the distances are given in feet. (*Hint:* The coordinates of points on the parabola satisfy its equation.)

77. Geometry The sum of the angles of a triangle is 180°. The largest angle is 20° greater than the sum of the other two and is 10° greater than 3 times the smallest. How large is each angle?

78. Ballistics The path of a thrown object is a parabola with the equation $y = ax^2 + bx + c$. Use the information in Illustration 4 to find a, b, and c. (Distances are in feet.)

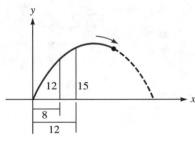

12 15

8

12

ILLUSTRATION 4

DISCOVERY AND WRITING

79. Use a graphing calculator to attempt to find the solution of the system

$$\begin{cases} x - 8y = -51 \\ 3x - 25y = -160 \end{cases}$$

80. Solve the system of Exercise 79 algebraically. Which method is easier, and why?

81. Use a graphing calculator to attempt to find the solution of the system

$$\begin{cases} 17x - 23y = -76 \\ 29x + 19y = 278 \end{cases}$$

82. Solve the system of Exercise 81 algebraically. Which method is easier, and why?

83. Invent a system of two equations with the solution $x = 1$, $y = 2$, that is difficult to solve with a graphing calculator (as in Exercise 81).

84. Invent a system of two equations with the solutions $x = 1$, $y = 2$, that is easier to solve graphically than algebraically.

REVIEW *Graph each function.*

85. $y = 3^x$

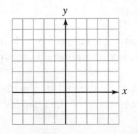

86. $y = \log_2 x$

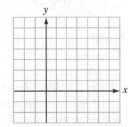

Solve for x:

87. $3 = \log_2 x$

88. $x = \log_5 \frac{1}{25}$

Use the properties of logarithms to write each expression in terms of the logarithms of x, y, and z.

89. $\log \dfrac{x}{y^2 z}$

90. $\log x(y^z)$

Use the properties of logarithms to write each expression as a single logarithm.

91. $\log x + 3 \log y - \dfrac{1}{2} \log z$

92. $\dfrac{1}{2}(\log A - 3 \log B)$

7.2 Gaussian Elimination and Matrix Methods

■ ELEMENTARY ROW OPERATIONS ■ ROW ECHELON FORM OF A MATRIX ■ AN INCONSISTENT SYSTEM ■ A SYSTEM OF DEPENDENT EQUATIONS ■ GAUSS–JORDAN ELIMINATION

Carl Friedrich Gauss
(1777–1855)

Many people consider Gauss to be the greatest mathematician of all time. He made contributions in the areas of number theory, solutions of equations, the geometry of curved surfaces, and statistics. For his efforts, he earned the title "Prince of the Mathematicians."

In this section, we introduce a method called **Gaussian elimination,** the invention of the German mathematician Carl Friedrich Gauss (1777–1855). In this method, we transform a system of equations into an equivalent system that can be solved by a process called **back substitution.**

Using Gaussian elimination, we solve systems of equations by working only with the coefficients of the variables. From the system of equations

$$\begin{cases} x + 2y + z = 8 \\ 2x + y - z = 1 \\ x + y - 2z = -3 \end{cases}$$

for example, we can form three rectangular arrays of numbers. Each is called a **matrix.** The first is the **coefficient matrix,** which contains the coefficients of the variables of the system. The second matrix contains the constants from the right-hand side of the system. We can write the third matrix, called the **augmented matrix** or the **system matrix,** by joining the two.

Coefficient Matrix   *Constants*   *System Matrix*

$$\begin{bmatrix} 1 & 2 & 1 \\ 2 & 1 & -1 \\ 1 & 1 & -2 \end{bmatrix} \qquad \begin{bmatrix} 8 \\ 1 \\ -3 \end{bmatrix} \qquad \begin{bmatrix} 1 & 2 & 1 & 8 \\ 2 & 1 & -1 & 1 \\ 1 & 1 & -2 & -3 \end{bmatrix}$$

Each row of the system matrix represents one equation of the system.

- The first row represents the equation $x + 2y + z = 8$.
- The second row represents the equation $2x + y - z = 1$.
- The third row represents the equation $x + y - 2z = -3$.

Because the system matrix has three rows and four columns, it is called a 3×4 matrix (read as "3 by 4"). The coefficient matrix is 3×3, and the constants form a 3×1 matrix.

To illustrate Gaussian elimination, we solve this system by the addition method of the previous section. But in the second column, we keep track of the changes to the system matrix.

EXAMPLE 1

Use Gaussian elimination to solve

1. $\quad\begin{cases} x + 2y + z = 8 \\ 2x + y - z = 1 \\ x + y - 2z = -3 \end{cases}$
2.
3.

Solution   We multiply each term in Equation 1 by -2 and add the result to Equation 2 to obtain Equation 4. Then we multiply each term of Equation 1 by -1 and add the result to Equation 3 to obtain Equation 5. This gives the equivalent system

1. $\quad\begin{cases} x + 2y + z = 8 \\ -3y - 3z = -15 \\ -y - 3z = -11 \end{cases}$
4.
5.

$$\begin{bmatrix} 1 & 2 & 1 & 8 \\ 0 & -3 & -3 & -15 \\ 0 & -1 & -3 & -11 \end{bmatrix}$$

Now we divide both sides of Equation 4 by -3 to obtain Equation 6

1. $\quad\begin{cases} x + 2y + z = 8 \\ y + z = 5 \\ -y - 3z = -11 \end{cases}$
6.
5.

$$\begin{bmatrix} 1 & 2 & 1 & 8 \\ 0 & 1 & 1 & 5 \\ 0 & -1 & -3 & -11 \end{bmatrix}$$

and add Equation 6 to Equation 5 to obtain Equation 7.

1. $\quad\begin{cases} x + 2y + z = 8 \\ y + z = 5 \\ -2z = -6 \end{cases}$
6.
7.

$$\begin{bmatrix} 1 & 2 & 1 & 8 \\ 0 & 1 & 1 & 5 \\ 0 & 0 & -2 & -6 \end{bmatrix}$$

Finally, we divide both sides of Equation 7 by -2 to obtain the system

1. $\quad\begin{cases} x + 2y + z = 8 \\ y + z = 5 \\ z = 3 \end{cases}$
6.

$$\begin{bmatrix} 1 & 2 & 1 & 8 \\ 0 & 1 & 1 & 5 \\ 0 & 0 & 1 & 3 \end{bmatrix}$$

The system can now be solved by *back substitution*. Because $z = 3$, we can substitute 3 for z in Equation 6 and solve for y:

6. $y + z = 5$
$y + 3 = 5$
$y = 2$

We can now substitute 2 for y and 3 for z in Equation 1 and solve for x:

1. $x + 2y + z = 8$
$x + 2(2) + 3 = 8$
$x + 7 = 8$
$x = 1$

The solution is the ordered triple $(1, 2, 3)$. ∎

Self Check   Solve $\begin{cases} x - 3y = 1 \\ 2x + y = 9 \end{cases}$.

Answer $(4, 1)$

■ ELEMENTARY ROW OPERATIONS

In Example 1, we performed operations on the equations as well as on the corresponding matrices. These matrix operations are called **elementary row operations.**

Elementary Row Operations

If any of the following operations are performed on the rows of a system matrix, the matrix of an equivalent system results:

Type 1 row operation: Two rows of a matrix can be interchanged.

Type 2 row operation: The elements of a row of a matrix can be multiplied by a nonzero constant.

Type 3 row operation: Any row can be changed by adding a multiple of another row to it.

- A type 1 row operation is equivalent to writing the equations of a system in a different order.
- A type 2 row operation is equivalent to multiplying both sides of an equation by a nonzero constant.
- A type 3 row operation is equivalent to adding a multiple of one equation to another.

Any matrix that can be obtained from another matrix by a sequence of elementary row operations is **row equivalent** to the original matrix. This means that row equivalent matrices represent equivalent systems of equations.

■ ROW ECHELON FORM OF A MATRIX

The process of Example 1 changed a system matrix into a special row equivalent form, called **row echelon form.**

Row Echelon Form of a Matrix

A matrix is in **row echelon form** if it has these three properties:

1. The first nonzero entry in each row (called the **lead entry**) is 1.
2. Lead entries appear farther to the right as we move down the rows of the matrix.
3. Any rows containing only 0's are at the bottom of the matrix.

The matrices

$$\begin{bmatrix} 1 & 3 & 0 & 7 \\ 0 & 0 & 1 & 8 \\ 0 & 0 & 0 & 0 \end{bmatrix}, \begin{bmatrix} 0 & 1 & 2 \\ 0 & 0 & 1 \end{bmatrix}, \text{ and } \begin{bmatrix} 1 & 1 \\ 0 & 0 \\ 0 & 0 \\ 0 & 0 \end{bmatrix}$$

are in row echelon form, because the first nonzero entry in each row is 1, the lead entries appear farther to the right as we move down the rows, and any rows that consist entirely of 0's are at the bottom of the matrix.

The next three matrices are not in row echelon form, for the reasons given:

$$\begin{bmatrix} 1 & 3 & 0 & 7 \\ 0 & 0 & 1 & 8 \\ 0 & 0 & 1 & 0 \end{bmatrix}$$ The lead entry in the last row is not to the right of the lead entry of the middle row.

$$\begin{bmatrix} 0 & 3 & 0 \\ 0 & 0 & 1 \end{bmatrix}$$ The lead entry of the first row is not 1.

$$\begin{bmatrix} 0 & 0 \\ 1 & 0 \\ 0 & 1 \end{bmatrix}$$ The row of 0's is not last.

EXAMPLE 2   Solve the following system by transforming its system matrix into row echelon form and back substituting.

$$\begin{cases} x + 2y + 3z = 4 \\ 2x - y - 2z = 0 \\ x - 3y - 3z = -2 \end{cases}$$

Solution   This system is represented by the system matrix

$$\begin{bmatrix} 1 & 2 & 3 & 4 \\ 2 & -1 & -2 & 0 \\ 1 & -3 & -3 & -2 \end{bmatrix}$$

We will reduce the system matrix to row echelon form. We can use the 1 in the upper left-hand corner to zero out the rest of the first column. To do so, we multiply the first row by -2 and add the result to the second row. We indicate this operation with the notation $(-2)R1 + R2$.

$$\begin{bmatrix} 1 & 2 & 3 & 4 \\ 2 & -1 & -2 & 0 \\ 1 & -3 & -3 & -2 \end{bmatrix} (-2)R1 + R2 \rightarrow \begin{bmatrix} 1 & 2 & 3 & 4 \\ 0 & -5 & -8 & -8 \\ 1 & -3 & -3 & -2 \end{bmatrix}$$

Next, we multiply row one by -1 and add the result to row three to get a new row three. This fills the rest of the first column with 0's.

$$(-1)R1 + R3 \rightarrow \begin{bmatrix} 1 & 2 & 3 & 4 \\ 0 & -5 & -8 & -8 \\ 0 & -5 & -6 & -6 \end{bmatrix}$$

We multiply row two by -1 and add the result to row three to get

$$(-1)R2 + R3 \rightarrow \begin{bmatrix} 1 & 2 & 3 & 4 \\ 0 & -5 & -8 & -8 \\ 0 & 0 & 2 & 2 \end{bmatrix}$$

Finally, we multiply row two by $-\frac{1}{5}$ and row three by $\frac{1}{2}$.

$$\begin{matrix} \left(-\frac{1}{5}\right) R2 \rightarrow \\ \left(\frac{1}{2}\right) R3 \rightarrow \end{matrix} \begin{bmatrix} 1 & 2 & 3 & 4 \\ 0 & 1 & \frac{8}{5} & \frac{8}{5} \\ 0 & 0 & 1 & 1 \end{bmatrix}$$

The final matrix, now in row echelon form, represents the equivalent system

$$\begin{cases} x + 2y + 3z = 4 \\ y + \dfrac{8}{5}z = \dfrac{8}{5} \\ z = 1 \end{cases}$$

We can now use back substitution to solve the system. To find y, we substitute 1 for z in the second equation.

$$y + \frac{8}{5}z = \frac{8}{5}$$

$$y + \frac{8}{5}(\mathbf{1}) = \frac{8}{5}$$

$$y = 0$$

To solve for x, we substitute 0 for y and 1 for z in the first equation.

$$x + 2y + 3\mathbf{z} = 4$$
$$x + 2(\mathbf{0}) + 3(\mathbf{1}) = 4$$
$$x + 3 = 4$$
$$x = 1$$

The solution of the original system is the triple $(x, y, z) = (1, 0, 1)$.   ∎

Self Check   Solve $\begin{cases} x + y + 2z = 5 \\ y - 2z = 0 \\ x - z = 0 \end{cases}$.

Answer $(1, 2, 1)$

Since the systems of Examples 1 and 2 have solutions, each system is consistent. The next two examples illustrate a system that is inconsistent and a system whose equations are dependent.

■ AN INCONSISTENT SYSTEM

EXAMPLE 3 Use matrix methods to solve

$$\begin{cases} x + y + z = 3 \\ 2x - y + z = 2 \\ 3y + z = 1 \end{cases}$$

Solution We form the system matrix and use row operations to write it in row echelon form. We use the 1 in the top left position to zero out the rest of the first column.

$$\begin{bmatrix} 1 & 1 & 1 & 3 \\ 2 & -1 & 1 & 2 \\ 0 & 3 & 1 & 1 \end{bmatrix} \begin{array}{c} \\ (-2)R1 + R2 \rightarrow \\ \\ \end{array} \begin{bmatrix} 1 & 1 & 1 & 3 \\ 0 & -3 & -1 & -4 \\ 0 & 3 & 1 & 1 \end{bmatrix}$$

$$\begin{array}{c} \\ \\ (1)R2 + R3 \rightarrow \end{array} \begin{bmatrix} 1 & 1 & 1 & 3 \\ 0 & 3 & 1 & -4 \\ 0 & 0 & 0 & -3 \end{bmatrix}$$

Since the last row of the matrix represents the equation

$$0x + 0y + 0z = -3$$

and no values of x, y, and z could make $0 = -3$, there is no point in continuing. The given system has no solution and is inconsistent. ■

Self Check Solve $\begin{cases} x + y + z = 5 \\ y - z = 1 \\ x + 2y = 3 \end{cases}$.

Answer The system is inconsistent.

■ A SYSTEM OF DEPENDENT EQUATIONS

EXAMPLE 4 Use matrices to solve

$$\begin{cases} x + 2y + z = 8 \\ 2x + y - z = 1 \\ x - y - 2z = -7 \end{cases}$$

Solution We can set up the system matrix and use row operations to reduce it to row echelon form:

$$
\begin{bmatrix} 1 & 2 & 1 & 8 \\ 2 & 1 & -1 & 1 \\ 1 & -1 & -2 & -7 \end{bmatrix}
\begin{matrix} \\ (-2)R1 + R2 \to \\ (-1)R1 + R3 \to \end{matrix}
\begin{bmatrix} 1 & 2 & 1 & 8 \\ 0 & -3 & -3 & -15 \\ 0 & -3 & -3 & -15 \end{bmatrix}
$$

$$
\begin{matrix} \\ (-\tfrac{1}{3})R2 \to \\ (-\tfrac{1}{3})R3 \to \end{matrix}
\begin{bmatrix} 1 & 2 & 1 & 8 \\ 0 & 1 & 1 & 5 \\ 0 & 1 & 1 & 5 \end{bmatrix}
$$

$$
\begin{matrix} \\ \\ (-1)R2 + R3 \to \end{matrix}
\begin{bmatrix} 1 & 2 & 1 & 8 \\ 0 & 1 & 1 & 5 \\ 0 & 0 & 0 & 0 \end{bmatrix}
$$

The final matrix is in echelon form and represents the system

1. $x + 2y + z = 8$
2. $y + z = 5$
3. $0x + 0y + 0z = 0$

Since all coefficients in Equation 3 are 0, it can be ignored. To solve this system by back substitution, we solve Equation 2 for the variable y,

$$y = 5 - z$$

and then substitute $5 - z$ for y in Equation 1.

1. $x + 2y + z = 8$
    $x + 2(5 - z) + z = 8$
    $x + 10 - 2z + z = 8$ Remove parentheses.
    $x + 10 - z = 8$ Combine like terms.
    $x = -2 + z$ Solve for x.

The solution of this system is $(x, y, z) = (-2 + z, 5 - z, z)$.

There are infinitely many solutions to this system. We can choose any real number for z, but once it is chosen, x and y are determined. For example,

- If $z = 3$, then $x = 1$ and $y = 2$. One possible solution of this system is $x = 1$, $y = 2$, and $z = 3$.
- If $z = 2$, then $x = 0$, and $y = 3$. Another solution is $x = 0$, $y = 3$, and $z = 2$.

Because this system has solutions, it is consistent. Because there are infinitely many solutions, the equations are dependent. ∎

Self Check Solve $\begin{cases} x + 2y + z = 2 \\ x - z = 2 \\ 2x + 2y = 4 \end{cases}$.

Answer $(x, y, z) = (2 + z, -z, z)$

■ GAUSS–JORDAN ELIMINATION

In Gaussian elimination, we perform row operations until the system matrix is in row echelon form. Then we convert the matrix back into equation form and solve by back substitution. A modification of this method, called **Gauss–Jordan elimination,** uses row operations to produce a matrix in *reduced* **row echelon** form. With this method, we can obtain the solution of the system from that matrix directly, and back substitution is not needed.

> **Reduced Row Echelon Form of a Matrix**
> A matrix is in **reduced row echelon form** if
>
> **1.** It is in row echelon form.
> **2.** Entries *above* each lead entry are also zero.

In the next example, we solve a system by Gauss–Jordan elimination.

EXAMPLE 5   Use Gauss–Jordan elimination to solve $\begin{cases} w + 2x + 3y + z = 4 \\ x + 4y - z = 0 \\ w - x - y + 2z = 2 \end{cases}$.

Solution   We row reduce the augmented matrix as follows. We first use the 1 in the upper left corner to zero out the rest of the first column:

$$\begin{bmatrix} 1 & 2 & 3 & 1 & 4 \\ 0 & 1 & 4 & -1 & 0 \\ 1 & -1 & -1 & 2 & 2 \end{bmatrix} \begin{matrix} \\ \\ (-1)R1 + R3 \rightarrow \end{matrix} \begin{bmatrix} 1 & 2 & 3 & 1 & 4 \\ 0 & 1 & 4 & -1 & 0 \\ 0 & -3 & -4 & 1 & -2 \end{bmatrix}$$

The lead entry in the second row is already 1. We use it to zero out the rest of the second column.

$$\begin{matrix} (-2)R2 + R1 \rightarrow \\ \\ (3)R2 + R3 \rightarrow \end{matrix} \begin{bmatrix} 1 & 0 & -5 & 3 & 4 \\ 0 & 1 & 4 & -1 & 0 \\ 0 & 0 & 8 & -2 & -2 \end{bmatrix}$$

To make the lead entry in the third row equal to 1, we multiply the third row by $\frac{1}{8}$ and use it to zero out the rest of the third column.

$$\begin{matrix} \\ \\ \left(\tfrac{1}{8}\right)R3 \rightarrow \end{matrix} \begin{bmatrix} 1 & 0 & -5 & 3 & 4 \\ 0 & 1 & 4 & -1 & 0 \\ 0 & 0 & 1 & -\tfrac{1}{4} & -\tfrac{1}{4} \end{bmatrix}$$

$$\begin{matrix} (5)R3 + R2 \rightarrow \\ (-4)R3 + R2 \rightarrow \\ \\ \end{matrix} \begin{bmatrix} 1 & 0 & 0 & \tfrac{7}{4} & \tfrac{11}{4} \\ 0 & 1 & 0 & 0 & 1 \\ 0 & 0 & 1 & -\tfrac{1}{4} & -\tfrac{1}{4} \end{bmatrix}$$

The final matrix is in reduced row echelon form. Note that each lead entry is 1, and each is alone in its column. The matrix represents the system of equations

$$\begin{cases} w + \dfrac{7}{4}z = \dfrac{11}{4} \\ x = 1 \\ y - \dfrac{1}{4}z = -\dfrac{1}{4} \end{cases} \quad \text{or} \quad \begin{aligned} w &= \dfrac{11}{4} - \dfrac{7}{4}z \\ x &= 1 \\ y &= -\dfrac{1}{4} + \dfrac{1}{4}z \end{aligned}$$

The system has infinitely many solutions. To find some of them, we choose values for z, and the corresponding values of w, x, and y will be determined. For example, if $z = 1$, then $w = 1$, $x = 1$, and $y = 0$. Thus, $(w, x, y, z) = (1, 1, 0, 1)$ is a solution. If $z = -1$, another solution is $(w, x, y, z) = \left(\frac{9}{2}, 1, -\frac{1}{2}, -1\right)$. ∎

In Example 5, there were more variables than equations. In the next example, there are more equations than variables.

EXAMPLE 6   Use Gauss–Jordan elimination to solve $\begin{cases} 2x + y = 4 \\ x - 3y = 9 \\ x + 4y = -5 \end{cases}$.

Solution   To get a 1 in the top left corner, we could multiply the first row by $\frac{1}{2}$, but that would introduce fractions. Instead, we can exchange the first two rows and proceed as follows:

$$\begin{bmatrix} 2 & 1 & 4 \\ 1 & -3 & 9 \\ 1 & 4 & -5 \end{bmatrix} \xrightarrow{R1 \leftrightarrow R2} \begin{bmatrix} 1 & -3 & 9 \\ 2 & 1 & 4 \\ 1 & 4 & -5 \end{bmatrix}$$

$$\begin{array}{c} \\ (-2)R1 + R2 \rightarrow \\ (-1)R1 + R3 \rightarrow \end{array} \begin{bmatrix} 1 & -3 & 9 \\ 0 & 7 & -14 \\ 0 & 7 & -14 \end{bmatrix}$$

$$\begin{array}{c} \\ \left(\frac{1}{7}\right)R2 \rightarrow \\ \left(\frac{1}{7}\right)R3 \rightarrow \end{array} \begin{bmatrix} 1 & -3 & 9 \\ 0 & 1 & -2 \\ 0 & 1 & -2 \end{bmatrix}$$

$$\begin{array}{c} (3)R2 + R1 \rightarrow \\ \\ (-1)R2 + R3 \rightarrow \end{array} \begin{bmatrix} 1 & 0 & 3 \\ 0 & 1 & -2 \\ 0 & 0 & 0 \end{bmatrix}$$

This final matrix is in reduced echelon form. It represents the system

$$\begin{cases} x = 3 \\ y = -2 \end{cases}$$

Verify that $x = 3$ and $y = -2$ satisfy the equations of the original system. ∎

Self Check   Solve $\begin{cases} x + y = 5 \\ x - y = 1 \\ x + 2y = 7 \end{cases}$.

Answer $x = 3$, $y = 2$

EXERCISE 7.2

VOCABULARY AND CONCEPTS   *In Exercises 1–12, fill in the blank to make a true statement.*

1. A rectangular array of numbers is called a _____.

2. A 3×5 matrix has __ rows and __ columns.

3. The matrix containing the coefficients of the variables is called the _____ matrix.

4. The coefficient matrix joined to the column of constants is called the _____ matrix or the _____ matrix.

5. Each row of a system matrix represents one _____.

6. The rows of the system matrix are changed using elementary _____.

7. If one system matrix is changed to another using row operations, the matrices are _____.

8. If two system matrices are row equivalent, then the systems have the _____.

9. In a type 1 row operation, two rows of a matrix can be _____.

10. In a type 2 row operation, one entire row can be _____ by a nonzero constant.

11. In a type 3 row operation, any row can be changed by _____ to it any _____ of another row.

12. The first nonzero entry in a row is called that row's _____.

PRACTICE   *In Exercises 13–20, use Gaussian elimination to solve each system.* **Do not use matrices.**

13. $\begin{cases} x + y = 7 \\ x - 2y = -1 \end{cases}$

14. $\begin{cases} x + 3y = 8 \\ 2x - 5y = 5 \end{cases}$

15. $\begin{cases} x - y = 1 \\ 2x - y = 8 \end{cases}$

16. $\begin{cases} x - 5y = 4 \\ 2x + 3y = 21 \end{cases}$

17. $\begin{cases} x + 2y - z = 2 \\ x - 3y + 2z = 1 \\ x + y - 3z = -6 \end{cases}$

18. $\begin{cases} x + 5y - z = 2 \\ x + 2y + z = 3 \\ x + y + z = 2 \end{cases}$

19. $\begin{cases} x - y - z = -3 \\ 5x + y = 6 \\ y + z = 4 \end{cases}$

20. $\begin{cases} x + y = 1 \\ x + z = 3 \\ y + z = 2 \end{cases}$

In Exercises 21–24, tell whether each matrix is in row echelon form, reduced row echelon form, or neither.

21. $\begin{bmatrix} 1 & 3 & 0 & 5 \\ 0 & 1 & 2 & 7 \\ 0 & 0 & 1 & 0 \end{bmatrix}$

22. $\begin{bmatrix} 1 & 3 & 0 & 5 \\ 0 & 1 & 2 & 7 \\ 0 & 0 & 0 & 0 \end{bmatrix}$

23. $\begin{bmatrix} 1 & 0 & 1 \\ 0 & 1 & 5 \\ 0 & 0 & 0 \\ 0 & 0 & 0 \end{bmatrix}$

24. $\begin{bmatrix} 1 & 0 & 1 \\ 0 & 1 & 5 \\ 0 & 0 & 1 \\ 0 & 0 & 0 \end{bmatrix}$

In Exercises 25–40, write each system as a matrix and solve it by Gaussian elimination.

25. $\begin{cases} 2x + y = 3 \\ x - 3y = 5 \end{cases}$

26. $\begin{cases} x + 2y = -1 \\ 3x - 5y = 19 \end{cases}$

27. $\begin{cases} x - 7y = -2 \\ 5x - 2y = -10 \end{cases}$

28. $\begin{cases} 3x - y = 3 \\ 2x + y = -3 \end{cases}$

29. $\begin{cases} 2x - y = 5 \\ x + 3y = 6 \end{cases}$

30. $\begin{cases} 3x - 5y = -25 \\ 2x + y = 5 \end{cases}$

31. $\begin{cases} x - 2y = 3 \\ -2x + 4y = 6 \end{cases}$

32. $\begin{cases} 2x - y = 7 \\ -x + \dfrac{1}{3}y = -\dfrac{7}{3} \end{cases}$

33. $\begin{cases} x - y + z = 3 \\ 2x - y + z = 4 \\ x + 2y - z = -1 \end{cases}$

34. $\begin{cases} 2x + y - z = 1 \\ x + y - z = 0 \\ 3x + y + 2z = 2 \end{cases}$

35. $\begin{cases} x + y - z = -1 \\ 3x + y = 4 \\ y - 2z = -4 \end{cases}$

36. $\begin{cases} 3x + y = 7 \\ x - z = 0 \\ y - 2z = -8 \end{cases}$

37. $\begin{cases} x - y + z = 2 \\ 2x + y + z = 5 \\ 3x - 4z = -5 \end{cases}$

38. $\begin{cases} x + z = -1 \\ 3x + y = 2 \\ 2x + y + 5z = 3 \end{cases}$

39. $\begin{cases} x + y + 2z = 4 \\ -x - y - 3z = -5 \\ 2x + y + z = 2 \end{cases}$

40. $\begin{cases} 2x - y + z = 6 \\ 3x + y - z = 2 \\ -x + 3y - 3z = 8 \end{cases}$

In Exercises 41–60, write each system as a matrix and solve it by Gauss–Jordan elimination. If a system has infinitely many solutions, show how individual solutions are determined.

41. $\begin{cases} x - 2y = 7 \\ y = 3 \end{cases}$

42. $\begin{cases} x - 2y = 7 \\ y = 8 \end{cases}$

43. $\begin{cases} x + 2y - z = 3 \\ y + 3z = 1 \\ z = -2 \end{cases}$

44. $\begin{cases} x - 3y + 2z = -1 \\ y - 2z = 3 \\ z = 5 \end{cases}$

45. $\begin{cases} x - y = 7 \\ x + y = 13 \end{cases}$

46. $\begin{cases} x + 2y = 7 \\ 2x - y = -1 \end{cases}$

47. $\begin{cases} x - \dfrac{1}{2}y = 0 \\ x + 2y = 0 \end{cases}$

48. $\begin{cases} x - y = 5 \\ -x + \dfrac{1}{5}y = -9 \end{cases}$

49. $\begin{cases} x + y + 2z = 0 \\ x + y + z = 2 \\ x + z = 1 \end{cases}$

50. $\begin{cases} x + 2y = -3 \\ x + 4y = -2 \\ 2x + z = -8 \end{cases}$

51. $\begin{cases} 2x + y - 2z = 1 \\ -x + y - 3z = 0 \\ 4x + 3y = 4 \end{cases}$

52. $\begin{cases} 3x + y = 3 \\ 3x + y - z = 2 \\ 6x + z = 5 \end{cases}$

53. $\begin{cases} 2x - 2y + 3z + t = 2 \\ x + y + z + t = 5 \\ -x + 2y - 3z + 2t = 2 \\ x + y + 2z - t = 4 \end{cases}$

54. $\begin{cases} x + y + 2z + t = 1 \\ x + 2y + z + t = 2 \\ 2x + y + z + t = 4 \\ x + y + z + 2t = 3 \end{cases}$

55. $\begin{cases} x + y + t = 4 \\ x + z + t = 2 \\ 2x + 2y + z + 2t = 8 \\ x - y + z - t = -2 \end{cases}$

56. $\begin{cases} x - y + 2z + t = 3 \\ 3x - 2y - z - t = 4 \\ 2x + y + 2z - t = 10 \\ x + 2y + z - 3t = 8 \end{cases}$

57. $\begin{cases} \dfrac{1}{3}x + \dfrac{3}{4}y - \dfrac{2}{3}z = -2 \\ x + \dfrac{1}{2}y + \dfrac{1}{3}z = 1 \\ \dfrac{1}{6}x - \dfrac{1}{8}y - z = 0 \end{cases}$

58. $\begin{cases} \dfrac{1}{4}x + y + 3z = 1 \\ \dfrac{1}{2}x - 4y + 6z = -1 \\ \dfrac{1}{3}x - 2y - 2z = -1 \end{cases}$

59. $\begin{cases} \dfrac{1}{2}x + \dfrac{1}{4}y - z = 2 \\ \dfrac{2}{3}x + \dfrac{1}{4}y + \dfrac{1}{2}z = \dfrac{3}{2} \\ \dfrac{2}{3}x + z = -\dfrac{1}{3} \end{cases}$

60. $\begin{cases} \dfrac{5}{7}x - \dfrac{1}{3}y + z = 0 \\ \dfrac{2}{7}x + y + \dfrac{1}{8}z = 9 \\ 6x + 4y - \dfrac{27}{4}z = 20 \end{cases}$

In Exercises 61–68, each system does not contain the same number of equations as variables. Solve each system using Gauss–Jordan elimination. If a system has infinitely many solutions, show a general solution.

61. $\begin{cases} x + y = -2 \\ 3x - y = 6 \\ 2x + 2y = -4 \\ x - y = 4 \end{cases}$

62. $\begin{cases} x - y = -3 \\ 2x + y = -3 \\ 3x - y = -7 \\ 4x + y = -7 \end{cases}$

63. $\begin{cases} x + 2y + z = 4 \\ 3x - y - z = 2 \end{cases}$

64. $\begin{cases} x + 2y - 3z = -5 \\ 5x + y - z = -11 \end{cases}$

65. $\begin{cases} w + x = 1 \\ w + y = 0 \\ x + z = 0 \end{cases}$

66. $\begin{cases} w + x - y + z = 2 \\ 2w - x - 2y + z = 0 \\ w - 2x - y + z = -1 \end{cases}$

67. $\begin{cases} x + y = 3 \\ 2x + y = 1 \\ 3x + 2y = 2 \end{cases}$

68. $\begin{cases} x + 2y + z = 4 \\ x - y + z = 1 \\ 2x + y + 2z = 2 \\ 3x + 3z = 6 \end{cases}$

APPLICATIONS *Use matrix methods to solve each problem.*

69. Flight range The speed of an airplane with a tailwind is 300 miles per hour, and with a headwind 220 miles per hour. On a day with no wind, how far could the plane travel on a 5-hour fuel supply?

70. Resource allocation 120,000 gallons of fuel are to be divided between two airlines. Triple A Airways requires twice as much as UnityAir. How much fuel should be allocated to Triple A?

71. Library shelving To use space effectively, librarians like to fill shelves completely. One 35-inch shelf can hold 3 dictionaries, 5 atlases, and 1 thesaurus; or 6 dictionaries and 2 thesauruses; or 2 dictionaries, 4 atlases, and 3 thesauruses. How wide is one copy of each book?

72. Copying machine productivity When both copying machines A and B are working, secretaries can make 100 copies in one minute. In one minute's time, copiers A and C together produce 140 copies, and all three working together produce 180 copies. How many copies per minute can each machine produce separately?

DISCOVERY AND WRITING

73. Explain the difference between the row echelon form and the reduced row echelon form of a matrix.

74. If the upper-left corner entry of a matrix is zero, what row operation might you do first?

75. What characteristic of a row reduced matrix would let you conclude that the system is inconsistent?

76. Explain the differences between Gaussian elimination and Gauss–Jordan elimination.

Use matrix methods to solve each system.

77. $\begin{cases} x^2 + y^2 + z^2 = 14 \\ 2x^2 + 3y^2 - 2z^2 = -7 \\ x^2 - 5y^2 + z^2 = 8 \end{cases}$
(*Hint:* Solve first as a system in x^2, y^2, and z^2.)

78. $\begin{cases} 5\sqrt{x} + 2\sqrt{x} + \sqrt{z} = 22 \\ \sqrt{x} + \sqrt{y} - \sqrt{z} = 5 \\ 3\sqrt{x} - 2\sqrt{y} - 3\sqrt{z} = 10 \end{cases}$

REVIEW *In Exercises 77–80, fill in the blank to make a true statement.*

79. The slope–intercept form of the equation of a line is _____.

80. The point–slope form of the equation of a line is _____.

81. The slopes of parallel lines are _____.

82. The slopes of _____ lines are negative reciprocals.

Write the equation of a line with the given properties.

83. The line has a slope of 2 and a *y*-intercept of 7.

84. The line has a slope of -3 and passes through the point $(2, -3)$.

85. The line is vertical and passes through $(2, -3)$.

86. The line is horizontal and passes through $(2, -3)$.

7.3 Matrix Algebra

■ MULTIPLYING A MATRIX BY A CONSTANT ■ ADDING AND SUBTRACTING MATRICES
■ MULTIPLYING MATRICES ■ AN APPLICATION OF MATRICES ■ THE IDENTITY MATRIX

Suppose there are 66 security officers employed at two locations:

Downtown Office		
	Male	**Female**
Day shift	12	18
Night shift	3	0

Suburban Office		
	Male	**Female**
Day shift	14	12
Night shift	5	2

The information about the security force is contained in the following matrices.

$$D = \begin{bmatrix} 12 & 18 \\ 3 & 0 \end{bmatrix} \quad \text{and} \quad S = \begin{bmatrix} 14 & 12 \\ 5 & 2 \end{bmatrix}$$

The entry 12 in matrix D gives the information that 12 males work the day shift at the downtown office. Company management can add the corresponding terms of matrices D and S to find corporate-wide totals:

$$D + S = \begin{bmatrix} 12 & 18 \\ 3 & 0 \end{bmatrix} + \begin{bmatrix} 14 & 12 \\ 5 & 2 \end{bmatrix} = \begin{bmatrix} 26 & 30 \\ 8 & 2 \end{bmatrix}$$

We interpret the total to mean:

	Male	Female
Day shift	26	30
Night shift	8	2

If one-third of the force at the downtown location retires, the downtown staff would be reduced to $\frac{2}{3}D$ people. We can compute $\frac{2}{3}D$ by multiplying each entry by $\frac{2}{3}$.

	After retirements, downtown staff would be	
	Male	Female
Day shift	8	12
Night shift	2	0

$$\frac{2}{3}D = \frac{2}{3}\begin{bmatrix} 12 & 18 \\ 3 & 0 \end{bmatrix} = \begin{bmatrix} 8 & 12 \\ 2 & 0 \end{bmatrix}$$

These examples illustrate two calculations used in the algebra of matrices, which is the topic of this section.

Matrices

An $m \times n$ *matrix* is a rectangular array of mn numbers arranged in m rows and n columns. We say that the matrix is of **order** $m \times n$.

Matrices are often denoted by letters such as A, B, and C. To denote the entries in an $m \times n$ matrix A, we use double-subscript notation: The entry in the first row, third column is a_{13}, and the entry in the ith row, jth column is a_{ij}. We can use any of the following notations to denote the $m \times n$ matrix A:

$$A, \qquad [a_{ij}], \qquad \left.\begin{bmatrix} a_{11} & a_{12} & a_{13} & \cdots & a_{1n} \\ a_{21} & a_{22} & a_{23} & \cdots & a_{2n} \\ \vdots & \vdots & \vdots & \ddots & \vdots \\ a_{m1} & a_{m2} & a_{m3} & \cdots & a_{mn} \end{bmatrix}\right\} m \text{ rows}$$
$$\underbrace{\phantom{a_{11} \quad a_{12} \quad a_{13} \quad \cdots \quad a_{1n}}}_{n \text{ columns}}$$

Two matrices are equal if they are the same size, with the same entries in corresponding positions.

Equality of Matrices

If $A = [a_{ij}]$ and $B = [b_{ij}]$ are both $m \times n$ matrices, then

$A = B$ if and only if $a_{ij} = b_{ij}$ for all i and j

The following matrices are equal, because they are the same size and corresponding entries are equal.

$$\begin{bmatrix} \sqrt{9} & 0.5 \\ 1 & 4 \end{bmatrix} = \begin{bmatrix} 3 & \frac{1}{2} \\ 1 & 2^2 \end{bmatrix}$$

The following matrices are not equal, because they are not the same size.

$$\begin{bmatrix} 1 & 2 & 3 \\ 1 & 2 & 3 \end{bmatrix} \neq \begin{bmatrix} 1 & 2 & 3 \\ 1 & 2 & 3 \\ 1 & 2 & 3 \end{bmatrix}$$

The first matrix is 2×3 and the second is 3×3.

■ MULTIPLYING A MATRIX BY A CONSTANT

We can multiply a matrix by a constant by multiplying each of its entries by that constant. If

$$A = \begin{bmatrix} 1 & 2 \\ 3 & 4 \end{bmatrix}, \quad \text{then} \quad 5A = 5\begin{bmatrix} 1 & 2 \\ 3 & 4 \end{bmatrix} = \begin{bmatrix} 5 & 10 \\ 15 & 20 \end{bmatrix}$$

If A is a matrix, the real number k in the product kA is called a **scalar.**

Multiplying a Matrix by a Scalar
If $A = [a_{ij}]$ is an $m \times n$ matrix and k is a scalar, then
$$kA = k[a_{ij}] = [ka_{ij}]$$

EXAMPLE 1 Let $\begin{bmatrix} 5 & y \\ 15 & z \end{bmatrix} = 5\begin{bmatrix} x & 3 \\ 3 & y \end{bmatrix}$. Find y and z.

Solution We simplify the right-hand side of the expression by multiplying each entry of the matrix by 5.

$$\begin{bmatrix} 5 & y \\ 15 & z \end{bmatrix} = \begin{bmatrix} 5x & 15 \\ 15 & 5y \end{bmatrix}$$

Because the matrices are equal, their corresponding entries are equal. So $y = 15$, and $z = 5y$. We conclude that $y = 15$ and $z = 75$. ■

Self Check Find x.
Answer 1

■ ADDING AND SUBTRACTING MATRICES

We can add matrices of the same size by adding the entries in corresponding positions.

Sum of Two Matrices
Let $A = [a_{ij}]$ and $B = [b_{ij}]$ be two $m \times n$ matrices. The sum, $A + B$, is the $m \times n$ matrix found by adding the corresponding entries of matrices A and B:
$$A + B = [a_{ij} + b_{ij}] \text{ for all } i \text{ and } j$$

EXAMPLE 2 Add the matrices $\begin{bmatrix} 2 & 1 & 3 \\ 1 & -1 & 0 \end{bmatrix} + \begin{bmatrix} 1 & -1 & 2 \\ -1 & 1 & 5 \end{bmatrix}$.

Solution   Since each matrix is 2×3, we find their sum by adding their corresponding entries.

$$\begin{bmatrix} 2 & 1 & 3 \\ 1 & -1 & 0 \end{bmatrix} + \begin{bmatrix} 1 & -1 & 2 \\ -1 & 1 & 5 \end{bmatrix} = \begin{bmatrix} 2+1 & 1-1 & 3+2 \\ 1-1 & -1+1 & 0+5 \end{bmatrix}$$

$$= \begin{bmatrix} 3 & 0 & 5 \\ 0 & 0 & 5 \end{bmatrix} \qquad \blacksquare$$

Self Check   Add $\begin{bmatrix} 3 & -5 \\ 2 & 0 \\ -6 & 5 \end{bmatrix} + \begin{bmatrix} -3 & 4 \\ -1 & -3 \\ 7 & 0 \end{bmatrix}$.

Answer $\begin{bmatrix} 0 & -1 \\ 1 & -3 \\ 1 & 5 \end{bmatrix}$

WARNING!   Matrices that are not the same size cannot be added.

In arithmetic, 0 is the **additive identity,** because $a + 0 = 0 + a = a$ for any real number a. In matrix algebra, the matrix $0 = \begin{bmatrix} 0 & 0 \\ 0 & 0 \end{bmatrix}$ is called an additive identity, because $A + 0 = 0 + A = A$. For example,

$$\begin{bmatrix} 1 & 2 \\ 3 & 4 \end{bmatrix} + \begin{bmatrix} 0 & 0 \\ 0 & 0 \end{bmatrix} = \begin{bmatrix} 0 & 0 \\ 0 & 0 \end{bmatrix} + \begin{bmatrix} 1 & 2 \\ 3 & 4 \end{bmatrix} = \begin{bmatrix} 1 & 2 \\ 3 & 4 \end{bmatrix}$$

The Additive Identity Matrix
Let A be any $m \times n$ matrix. There is an $m \times n$ matrix **0**, called the **zero matrix** or the **additive identity matrix,** for which

   $A + 0 = 0 + A = A$

The matrix **0** consists of m rows and n columns of 0's.

Every matrix also has an additive inverse.

The Additive Inverse of a Matrix
Any $m \times n$ matrix A has an **additive inverse,** an $m \times n$ matrix $-A$ with the property that the sum of A and $-A$ is the zero matrix:

   $A + (-A) = (-A) + A = 0$

The entries of $-A$ are the negatives of the corresponding entries of A:

   $-A = (-1)A$

Arthur Cayley
(1821–1895)

Cayley taught mathematics at Cambridge University. When he refused to take religious vows, he was fired and became a lawyer. After 14 years, he returned to mathematics and to Cambridge. Cayley was a major force in developing the theory of matrices.

The additive inverse of $A = \begin{bmatrix} 1 & -3 & 2 \\ 0 & 1 & -5 \end{bmatrix}$ is the matrix

$$-A = (-1)A = \begin{bmatrix} -1 & 3 & -2 \\ 0 & -1 & 5 \end{bmatrix}$$

because their sum is the zero matrix:

$$A + (-A) = \begin{bmatrix} 1 & -3 & 2 \\ 0 & 1 & -5 \end{bmatrix} + \begin{bmatrix} -1 & 3 & -2 \\ 0 & -1 & 5 \end{bmatrix}$$

$$= \begin{bmatrix} 1-1 & -3+3 & 2-2 \\ 0+0 & 1-1 & -5+5 \end{bmatrix}$$

$$= \begin{bmatrix} 0 & 0 & 0 \\ 0 & 0 & 0 \end{bmatrix}$$

Subtraction of matrices is similar to the subtraction of real numbers.

Difference of Two Matrices

If A and B are $m \times n$ matrices, their difference, $A - B$, is the sum of A and the additive inverse of B:

$$A - B = A + (-B)$$

For example,

$$\begin{bmatrix} 3 & 7 \\ -4 & 0 \end{bmatrix} - \begin{bmatrix} -1 & 4 \\ -5 & 1 \end{bmatrix} = \begin{bmatrix} 3 & 7 \\ -4 & 0 \end{bmatrix} + \begin{bmatrix} 1 & -4 \\ 5 & -1 \end{bmatrix} = \begin{bmatrix} 4 & 3 \\ 1 & -1 \end{bmatrix}$$

■ MULTIPLYING MATRICES

We will consider how to find the product of two matrices by finding the product of a 2×3 matrix A and a 3×3 matrix B. The result will be matrix C.

$$AB = \begin{bmatrix} 1 & 2 & 3 \\ 4 & 5 & 6 \end{bmatrix} \begin{bmatrix} a & b & c \\ d & e & f \\ g & h & i \end{bmatrix} = C$$

Each entry of matrix C is the result of multiplying the numbers in a row of A and the corresponding numbers in a column of B. For example, the first-row, third-column entry of matrix C is the sum of the products of corresponding entries of the first row of A and the third column of B:

$$\begin{bmatrix} \mathbf{1} & \mathbf{2} & \mathbf{3} \\ 4 & 5 & 6 \end{bmatrix} \begin{bmatrix} a & b & \mathbf{c} \\ d & e & \mathbf{f} \\ g & h & \mathbf{i} \end{bmatrix} = \begin{bmatrix} ? & ? & \mathbf{1c + 2f + 3i} \\ ? & ? & ? \end{bmatrix}$$

The second-row, second-column entry of matrix C is the sum of the products of the second row of A and the second column of B.

$$\begin{bmatrix} 1 & 2 & 3 \\ 4 & 5 & 6 \end{bmatrix} \begin{bmatrix} a & b & c \\ d & e & f \\ g & h & i \end{bmatrix} = \begin{bmatrix} ? & ? & 1c + 2f + 3i \\ ? & 4b + 5e + 6h & ? \end{bmatrix}$$

The other entries of the product are computed similarly.

For the product AB to exist, the number of columns of A must equal the number of rows of B. If the product exists, it will have as many rows as A and as many columns as B:

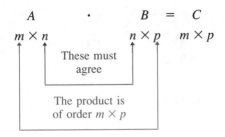

Product of Two Matrices

Let $A = [a_{ij}]$ be an $m \times n$ and $B = [b_{ij}]$ be an $n \times p$ matrix. The product, AB, is the $m \times p$ matrix C, found as follows:

$$AB = C = [c_{ij}]$$

where c_{ij} is the sum of the products of the corresponding entries in the ith row of A and the jth column of B, where $i = 1, 2, 3, \cdots, m$ and $j = 1, 2, 3, \cdots, p$.

EXAMPLE 3   Find $C = AB$ if $A = \begin{bmatrix} 1 & 2 & 4 \\ -2 & 1 & -1 \end{bmatrix}$ and $B = \begin{bmatrix} 1 & 5 \\ -2 & 4 \\ 1 & -3 \end{bmatrix}$.

Solution   Because matrix A is 2×3 and B is 3×2, the product C is defined and is 2×2. To find entry c_{11} of C, we find the total of the products of the entries in the first row of A and the first column of B:

$$c_{11} = 1 \cdot 1 + 2 \cdot (-2) + 4 \cdot (1) = 1$$

$$\begin{bmatrix} 1 & 2 & 4 \\ -2 & 1 & -1 \end{bmatrix} \begin{bmatrix} 1 & 5 \\ -2 & 4 \\ 1 & -3 \end{bmatrix} = \begin{bmatrix} 1 & ? \\ ? & ? \end{bmatrix}$$

To find entry c_{12}, we move across the first row of A and down the second column of B:

$$c_{12} = 1 \cdot 5 + 2 \cdot 4 + 4 \cdot (-3) = 1$$

$$\begin{bmatrix} 1 & 2 & 4 \\ -2 & 1 & -1 \end{bmatrix} \begin{bmatrix} 1 & 5 \\ -2 & 4 \\ 1 & -3 \end{bmatrix} = \begin{bmatrix} 1 & 1 \\ ? & ? \end{bmatrix}$$

To find entry c_{21}, we move across the second row of A and down the first column of B:

$$c_{21} = (-2) \cdot 1 + 1 \cdot (-2) + (-1) \cdot 1 = -5$$

$$\begin{bmatrix} 1 & 2 & 4 \\ -2 & 1 & -1 \end{bmatrix} \begin{bmatrix} 1 & 5 \\ -2 & 4 \\ 1 & -3 \end{bmatrix} = \begin{bmatrix} 1 & 1 \\ -5 & ? \end{bmatrix}$$

Finally, we find entry c_{22}:

$$c_{22} = (-2) \cdot 5 + 1 \cdot 4 + (-1) \cdot (-3) = -3$$

$$\begin{bmatrix} 1 & 2 & 4 \\ -2 & 1 & -1 \end{bmatrix} \begin{bmatrix} 1 & 5 \\ -2 & 4 \\ 1 & -3 \end{bmatrix} = \begin{bmatrix} 1 & 1 \\ -5 & -3 \end{bmatrix}$$ ∎

Self Check Find $D = EF$ if $E = \begin{bmatrix} 1 & -2 \\ 2 & 0 \end{bmatrix}$ and $F = \begin{bmatrix} 2 & 1 \\ 3 & 5 \end{bmatrix}$.

Answer $\begin{bmatrix} -4 & -9 \\ 4 & 2 \end{bmatrix}$

EXAMPLE 4 Find the product $\begin{bmatrix} 1 & -1 & 2 \\ 1 & 3 & 0 \\ 0 & 1 & 1 \end{bmatrix} \begin{bmatrix} 2 & 1 \\ 1 & 3 \\ 0 & 1 \end{bmatrix}$.

Solution Because the matrices are 3×3 and 3×2, the product is a 3×2 matrix.

$$\begin{bmatrix} 1 & -1 & 2 \\ 1 & 3 & 0 \\ 0 & 1 & 1 \end{bmatrix} \begin{bmatrix} 2 & 1 \\ 1 & 3 \\ 0 & 1 \end{bmatrix} = \begin{bmatrix} 1 \cdot 2 + (-1) \cdot 1 + 2 \cdot 0 & 1 \cdot 1 + (-1) \cdot 3 + 2 \cdot 1 \\ 1 \cdot 2 + 3 \cdot 1 + 0 \cdot 0 & 1 \cdot 1 + 3 \cdot 3 + 0 \cdot 1 \\ 0 \cdot 2 + 1 \cdot 1 + 1 \cdot 0 & 0 \cdot 1 + 1 \cdot 3 + 1 \cdot 1 \end{bmatrix}$$

$$= \begin{bmatrix} 1 & 0 \\ 5 & 10 \\ 1 & 4 \end{bmatrix}$$ ∎

Self Check   Find the product $\begin{bmatrix} 1 & -2 \\ 3 & 1 \\ -4 & 0 \end{bmatrix} \begin{bmatrix} 4 & -3 \\ 1 & -1 \end{bmatrix}$.

Answer $\begin{bmatrix} 2 & -1 \\ 13 & -10 \\ -16 & 12 \end{bmatrix}$

EXAMPLE 5   Find the product $\begin{bmatrix} 1 & 2 & 3 \end{bmatrix} \begin{bmatrix} 4 \\ 5 \\ 6 \end{bmatrix}$.

Solution   Because the first matrix is 1×3 and the second matrix is 3×1, the product is a 1×1 matrix:

$$\begin{bmatrix} 1 & 2 & 3 \end{bmatrix} \begin{bmatrix} 4 \\ 5 \\ 6 \end{bmatrix} = [1 \cdot 4 + 2 \cdot 5 + 3 \cdot 6]$$

$$[32]$$ ∎

Self Check   Find the product $\begin{bmatrix} 1 & 3 & 5 \end{bmatrix} \begin{bmatrix} 1 \\ 0 \\ 1 \end{bmatrix}$.

Answer   [6]

EXAMPLE 6   Find the product $\begin{bmatrix} 1 \\ 2 \\ 3 \end{bmatrix} \begin{bmatrix} 4 & 5 & 6 \end{bmatrix}$.

Solution   Because the first matrix is 3×1 and the second matrix is 1×3, the product is a 3×3 matrix:

$$\begin{bmatrix} 1 \\ 2 \\ 3 \end{bmatrix} \begin{bmatrix} 4 & 5 & 6 \end{bmatrix} = \begin{bmatrix} 1 \cdot 4 & 1 \cdot 5 & 1 \cdot 6 \\ 2 \cdot 4 & 2 \cdot 5 & 2 \cdot 6 \\ 3 \cdot 4 & 3 \cdot 5 & 3 \cdot 6 \end{bmatrix}$$

$$= \begin{bmatrix} 4 & 5 & 6 \\ 8 & 10 & 12 \\ 12 & 15 & 18 \end{bmatrix}$$ ∎

Self Check Find the product $\begin{bmatrix} 1 \\ 2 \end{bmatrix}$ [3   4].

Answer $\begin{bmatrix} 3 & 4 \\ 6 & 8 \end{bmatrix}$

The multiplication of matrices is not commutative. To show this, we compute AB and BA, where $A = \begin{bmatrix} 1 & 1 \\ 0 & 0 \end{bmatrix}$ and $B = \begin{bmatrix} 0 & 1 \\ 0 & 1 \end{bmatrix}$.

$$AB = \begin{bmatrix} 1 & 1 \\ 0 & 0 \end{bmatrix} \begin{bmatrix} 0 & 1 \\ 0 & 1 \end{bmatrix} = \begin{bmatrix} 0 & 2 \\ 0 & 0 \end{bmatrix}$$

$$BA = \begin{bmatrix} 0 & 1 \\ 0 & 1 \end{bmatrix} \begin{bmatrix} 1 & 1 \\ 0 & 0 \end{bmatrix} = \begin{bmatrix} 0 & 0 \\ 0 & 0 \end{bmatrix}$$

Since the products are not equal, matrix multiplication is not commutative.

■ ■ ■ ■ ■ ■ ■ ■ ■

ACCENT ON TECHNOLOGY

Several models of graphing calculators are able to do matrix arithmetic. For example, to find the sum and the product of the matrices

$$A = \begin{bmatrix} 2 & 3.7 \\ -2.1 & 3 \end{bmatrix} \quad \text{and} \quad B = \begin{bmatrix} 2 & -1 \\ 0 & 0.3 \end{bmatrix}$$

on one model, we press MATRIX , select EDIT, and enter the size and entries of matrix A, as shown in Figure 7-5(a). Similarly, we enter matrix B as in Figure 7-5(b). Figure 7-5(c) displays the result $A + B$, as well as the product AB.

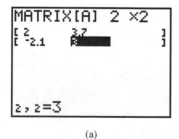

(a)

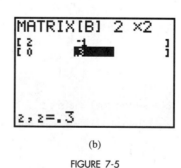

(b)

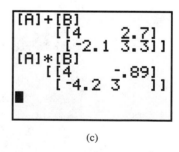

(c)

FIGURE 7-5

■ ■ ■ ■ ■ ■ ■ ■ ■

■ AN APPLICATION OF MATRICES

EXAMPLE 7

Suppose that supplies must be purchased for the security officers discussed at the beginning of the section. The quantities and prices of each item required for each shift are as follows:

	Quantities			Unit Prices (in $)	
	Uniforms	**Badges**	**Whistles**	**Uniforms**	47
Day shift	17	13	19	**Badges**	7
Night shift	14	24	27	**Whistles**	5

Find the cost of supplies for each shift.

Solution   We can write the quantities and prices in matrix form and multiply them to get a cost matrix.

$$C = QP$$

$$= \begin{bmatrix} 17 & 13 & 19 \\ 14 & 24 & 27 \end{bmatrix} \begin{bmatrix} 47 \\ 7 \\ 5 \end{bmatrix}$$

$$= \begin{bmatrix} 17 \cdot 47 + 13 \cdot 7 + 19 \cdot 5 \\ 14 \cdot 47 + 24 \cdot 7 + 27 \cdot 5 \end{bmatrix}$$

(17 uniforms)($47) + (13 badges)($7) +
(19 whistles)($5).
(14 uniforms)($47) + (24 badges)($7) +
(27 whistles)($5).

$$\begin{bmatrix} 985 \\ 961 \end{bmatrix}$$

It will cost $985 to buy supplies for the day shift and $961 to buy supplies for the night shift.   ■

■ THE IDENTITY MATRIX

The number 1 is called the **identity for multiplication,** because multiplying a number by 1 does not change the number: $a \cdot 1 = 1 \cdot a = a$. There is a **multiplicative identity matrix** with a similar property.

The Identity Matrix
Let A be an $n \times n$ matrix. There is an $n \times n$ identity matrix I for which

$$AI = IA = A$$

It is the matrix I consisting of 1's on its diagonal and 0's elsewhere.

$$I = \begin{bmatrix} 1 & 0 & 0 & \cdots & 0 \\ 0 & 1 & 0 & \cdots & 0 \\ 0 & 0 & 1 & \cdots & 0 \\ \vdots & \vdots & \vdots & \ddots & \vdots \\ 0 & 0 & 0 & \cdots & 1 \end{bmatrix}$$

WARNING! An identity matrix is always a square matrix—a matrix that has the same number of rows and columns.

We illustrate the previous definition for the 3×3 identity matrix.

$$\begin{bmatrix} 1 & 0 & 0 \\ 0 & 1 & 0 \\ 0 & 0 & 1 \end{bmatrix} \begin{bmatrix} 1 & 2 & 3 \\ 4 & 5 & 6 \\ 7 & 8 & 9 \end{bmatrix} = \begin{bmatrix} 1 \cdot 1 + 0 \cdot 4 + 0 \cdot 7 & 1 \cdot 2 + 0 \cdot 5 + 0 \cdot 8 & 1 \cdot 3 + 0 \cdot 6 + 0 \cdot 9 \\ 0 \cdot 1 + 1 \cdot 4 + 0 \cdot 7 & 0 \cdot 2 + 1 \cdot 5 + 0 \cdot 8 & 0 \cdot 3 + 1 \cdot 6 + 0 \cdot 9 \\ 0 \cdot 1 + 0 \cdot 4 + 1 \cdot 7 & 0 \cdot 2 + 0 \cdot 5 + 1 \cdot 8 & 0 \cdot 3 + 0 \cdot 6 + 1 \cdot 9 \end{bmatrix}$$

$$= \begin{bmatrix} 1 & 2 & 3 \\ 4 & 5 & 6 \\ 7 & 8 & 9 \end{bmatrix}$$

$$\begin{bmatrix} 1 & 2 & 3 \\ 4 & 5 & 6 \\ 7 & 8 & 9 \end{bmatrix} \begin{bmatrix} 1 & 0 & 0 \\ 0 & 1 & 0 \\ 0 & 0 & 1 \end{bmatrix} = \begin{bmatrix} 1 & 2 & 3 \\ 4 & 5 & 6 \\ 7 & 8 & 9 \end{bmatrix}$$

The following properties of real numbers carry over to matrices.

Properties of Matrices

Let A, B, and C be matrices and a and b be scalars.

The Commutative Property of Addition: $A + B = B + A$

The Associative Property of Addition: $A + (B + C) = (A + B) + C$

The Associative Properties of Scalar Multiplication: $\begin{cases} a(bA) = (ab)A \\ a(AB) = (aA)B \end{cases}$

Distributive Properties of Scalar Multiplication: $(a + b)A = aA + bA$

$a(A + B) = aA + aB$

The Associative Property of Multiplication: $A(BC) = (AB)C$

The Distributive Properties of Matrix Multiplication: $\begin{cases} A(B + C) = AB + AC \\ (A + B)C = AC + BC \end{cases}$

EXAMPLE 8   Verify the distributive property $A(B + C) = AB + AC$ using the matrices

$$A = \begin{bmatrix} 2 & 1 \\ 3 & 1 \end{bmatrix}, \quad B = \begin{bmatrix} 3 & -4 \\ 1 & -1 \end{bmatrix}, \quad \text{and} \quad C = \begin{bmatrix} -1 & 3 \\ 0 & 1 \end{bmatrix}$$

Solution   We do the operations on the left-hand side and the right-hand side separately and compare the results.

$$\begin{bmatrix} 2 & 1 \\ 3 & 1 \end{bmatrix} \left(\begin{bmatrix} 3 & -4 \\ 1 & -1 \end{bmatrix} + \begin{bmatrix} -1 & 3 \\ 0 & 1 \end{bmatrix} \right) = \begin{bmatrix} 2 & 1 \\ 3 & 1 \end{bmatrix} \left(\begin{bmatrix} 2 & -1 \\ 1 & 0 \end{bmatrix} \right) \qquad \text{Do the addition within the parentheses.}$$

$$= \begin{bmatrix} 5 & -2 \\ 7 & -3 \end{bmatrix} \qquad \text{Multiply.}$$

$$\begin{bmatrix} 2 & 1 \\ 3 & 1 \end{bmatrix} \begin{bmatrix} 3 & -4 \\ 1 & -1 \end{bmatrix} + \begin{bmatrix} 2 & 1 \\ 3 & 1 \end{bmatrix} \begin{bmatrix} -1 & 3 \\ 0 & 1 \end{bmatrix} = \begin{bmatrix} 7 & -9 \\ 10 & -13 \end{bmatrix} + \begin{bmatrix} -2 & 7 \\ -3 & 10 \end{bmatrix} \qquad \text{Do the multiplications.}$$

$$= \begin{bmatrix} 5 & -2 \\ 7 & -3 \end{bmatrix} \qquad \text{Add.}$$

Because the left- and right-hand sides agree, this example illustrates the distributive property. ∎

EXERCISE 7.3

VOCABULARY AND CONCEPTS *In Exercises 1–6, fill in the blank to make a true statement.*

1. In a matrix A, the symbol a_{ij} is the entry in row ___ and column ___.

2. For matrices A and B to be equal, they must _____, and corresponding entries must be ___.

3. To multiply a matrix by a scalar, we multiply _____ by that scalar.

4. To find the sum of matrices A and B, we add the _____ entries.

5. Among 2×2 matrices, $\begin{bmatrix} 0 & 0 \\ 0 & 0 \end{bmatrix}$ is the _____ matrix.

6. Among 2×2 matrices, $\begin{bmatrix} 1 & 0 \\ 0 & 1 \end{bmatrix}$ is the _____ matrix.

PRACTICE *In Exercises 7–10, find values of x and y, if any, that will make the two matrices equal.*

7. $\begin{bmatrix} x & y \\ 1 & 3 \end{bmatrix} = \begin{bmatrix} 2 & 5 \\ 1 & 3 \end{bmatrix}$

8. $\begin{bmatrix} x & 5 \\ 3 & y \end{bmatrix} = \begin{bmatrix} 0 & 5 \\ 3 & 2 \end{bmatrix}$

9. $\begin{bmatrix} x + y & 3 + x \\ -2 & 5y \end{bmatrix} = \begin{bmatrix} 3 & 4 \\ -2 & 10 \end{bmatrix}$

10. $\begin{bmatrix} x + y & x - y \\ 2x & 3y \end{bmatrix} = \begin{bmatrix} -x & x - 2 \\ -y & 8 - y \end{bmatrix}$

In Exercises 11–14, find 5A.

11. $A = \begin{bmatrix} 3 & -3 \\ 0 & -2 \end{bmatrix}$

12. $A = \begin{bmatrix} 3 & \frac{3}{5} \\ 0 & -1 \end{bmatrix}$

13. $A = \begin{bmatrix} 5 & 15 & -2 \\ -2 & -5 & 1 \end{bmatrix}$

14. $A = \begin{bmatrix} -3 & 1 & 2 \\ -8 & -2 & -5 \end{bmatrix}$

In Exercises 15–16, find $A + B$.

15. $A = \begin{bmatrix} 2 & 1 & -1 \\ -3 & 2 & 5 \end{bmatrix}, B = \begin{bmatrix} -3 & 1 & 2 \\ -3 & -2 & -5 \end{bmatrix}$

16. $A = \begin{bmatrix} 3 & 2 & 1 \\ -2 & 3 & -3 \\ -4 & -2 & -1 \end{bmatrix}, B = \begin{bmatrix} -2 & 6 & -2 \\ 5 & 7 & -1 \\ -4 & -6 & 7 \end{bmatrix}$

In Exercises 17–18, find $A - B$.

17. $A = \begin{bmatrix} -3 & 2 & -2 \\ -1 & 4 & -5 \end{bmatrix}, B = \begin{bmatrix} 3 & -3 & -2 \\ -2 & 5 & -5 \end{bmatrix}$

18. $A = \begin{bmatrix} 2 & 2 & 0 \\ -2 & 8 & 1 \\ 3 & -3 & -8 \end{bmatrix}, B = \begin{bmatrix} -4 & 3 & 7 \\ -1 & 2 & 0 \\ 1 & 4 & -1 \end{bmatrix}$

In Exercises 19–20, find 5A + 3B.

19. $A = \begin{bmatrix} 3 & 1 & -2 \\ -4 & 3 & -2 \end{bmatrix}, B = \begin{bmatrix} 1 & -2 & 2 \\ -5 & -5 & 3 \end{bmatrix}$ **20.** $A = \begin{bmatrix} 2 & -5 \\ -5 & 2 \end{bmatrix}, B = \begin{bmatrix} 5 & -2 \\ 2 & -5 \end{bmatrix}$

In Exercises 21–22, find the additive inverse of each matrix.

21. $A = \begin{bmatrix} 5 & -2 & 7 \\ -5 & 0 & 3 \\ -2 & 3 & -5 \end{bmatrix}$ **22.** $A = \begin{bmatrix} 3 & -\frac{2}{3} & -5 & \frac{1}{2} \end{bmatrix}$

In Exercises 23–34, find each product, if possible.

23. $\begin{bmatrix} 2 & 3 \\ 3 & -2 \end{bmatrix} \begin{bmatrix} 1 & 2 \\ 0 & -2 \end{bmatrix}$ **24.** $\begin{bmatrix} -2 & 3 \\ 3 & -2 \end{bmatrix} \begin{bmatrix} 2 & 4 \\ -5 & 7 \end{bmatrix}$

25. $\begin{bmatrix} -4 & -2 \\ 21 & 0 \end{bmatrix} \begin{bmatrix} -5 & 6 \\ 21 & -1 \end{bmatrix}$ **26.** $\begin{bmatrix} -5 & 4 \\ 4 & -5 \end{bmatrix} \begin{bmatrix} 6 & -2 \\ 1 & 3 \end{bmatrix}$

27. $\begin{bmatrix} 2 & 1 & 3 \\ 1 & 2 & -1 \\ 0 & 1 & 0 \end{bmatrix} \begin{bmatrix} 1 & 2 & 3 \\ 2 & -2 & 1 \\ 0 & 0 & 1 \end{bmatrix}$ **28.** $\begin{bmatrix} 2 & 1 & 1 \\ 1 & 1 & 2 \\ 1 & -2 & -1 \end{bmatrix} \begin{bmatrix} 1 & 2 & 3 \\ 1 & 2 & -3 \\ -1 & -1 & 3 \end{bmatrix}$

29. $\begin{bmatrix} 1 & -2 & -3 \\ 2 & 0 & 1 \end{bmatrix} \begin{bmatrix} 4 \\ -5 \\ -6 \end{bmatrix}$ **30.** $\begin{bmatrix} 1 \\ -2 \\ -3 \end{bmatrix} \begin{bmatrix} 4 & -5 & -6 \end{bmatrix}$

31. $\begin{bmatrix} 1 & 2 & 3 \end{bmatrix} \begin{bmatrix} 4 & 5 & 6 \\ 7 & 8 & 9 \end{bmatrix}$ **32.** $\begin{bmatrix} 2 & 3 & 4 \\ 1 & 2 & 3 \\ -2 & 2 & 2 \end{bmatrix} \begin{bmatrix} -1 \\ 2 \\ 3 \end{bmatrix}$

33. $\begin{bmatrix} 2 & 5 \\ -3 & 1 \\ 0 & -2 \\ 1 & -5 \end{bmatrix} \begin{bmatrix} 3 & -2 & 4 \\ -2 & -3 & 1 \end{bmatrix}$ **34.** $\begin{bmatrix} 1 & 4 & 0 & 0 \\ -4 & 1 & 0 & -2 \\ 0 & 0 & 1 & 0 \\ 0 & 2 & 0 & 1 \end{bmatrix} \begin{bmatrix} 1 \\ 2 \\ -2 \\ -1 \end{bmatrix}$

In Exercises 35–38, let $A = \begin{bmatrix} 2.3 & -1.7 & 3.1 \\ -2 & 3.5 & 1 \\ -8 & 4.7 & 9.1 \end{bmatrix}$, $B = \begin{bmatrix} -2.5 \\ 5.2 \\ -7 \end{bmatrix}$, and $C = \begin{bmatrix} -5.8 \\ 2.9 \\ 4.1 \end{bmatrix}$. Use a graphing calculator to find each result.

35. AB

36. $B + C$

37. A^2

38. $AB + C$

In Exercises 39–42, $A = \begin{bmatrix} 2 & 3 \\ 1 & 3 \end{bmatrix}$, $B = \begin{bmatrix} 2 & 1 & -5 \\ 1 & 1 & 2 \end{bmatrix}$, $C = \begin{bmatrix} -2 & -1 & 6 \\ 0 & -1 & -1 \end{bmatrix}$, $D = \begin{bmatrix} 1 & 2 \\ 1 & 3 \end{bmatrix}$, and $E = \begin{bmatrix} 1 & -2 \\ 2 & 3 \end{bmatrix}$. Verify each property by doing the operations on each side of the equation and comparing the results.

39. Distributive property:

$$A(B + C) = AB + AC$$

40. Associative property of scalar multiplication:

$$5(6A) = (5 \cdot 6)A$$

41. Associative property of scalar multiplication:

$$3(AB) = (3A)B$$

42. Associative property of multiplication:

$$A(DE) = (AD)E$$

In Exercises 43–50, let $A = \begin{bmatrix} 1 & 3 \\ 2 & 5 \end{bmatrix}$, $B = \begin{bmatrix} -1 \\ 3 \end{bmatrix}$, and $C = [\,3 \quad 2\,]$. Do the operations, if possible.

43. $A - BC$

44. $AB + B$

45. $CB - AB$

46. CAB

47. ABC

48. $CA + C$

49. A^2B

50. $(BC)^2$

APPLICATIONS Use a graphing calculator to help solve each problem.

51. Beverage sales Beverages were sold to parents and children at a school basketball game in the quantities and prices given in Illustrations 1 and 2. Find matrices Q and P that represent the quantities and prices, find the product QP, and interpret the result.

Price	
Coffee	$ .75
Milk	$1.00
Cola	$1.25

ILLUSTRATION 2

	Quantities		
	Coffee	Milk	Cola
Adult males	217	23	319
Adult females	347	24	340
Children	3	97	750

ILLUSTRATION 1

52. Production costs Each of four factories manufactures three products in the daily quantities and unit costs given in Illustrations 3 and 4. Find a suitable matrix product to represent production costs.

	Production quantities		
Factory	**Product A**	**Product B**	**Product C**
Ashtabula	19	23	27
Boston	17	21	22
Chicago	21	18	20
Denver	27	25	22

ILLUSTRATION 3

	Unit Production Costs	
	Day shift	**Night shift**
Product A	$1.20	$1.35
Product B	$ .75	$ .85
Product C	$3.50	$3.70

ILLUSTRATION 4

DISCOVERY AND WRITING

55. Routing telephone calls A long distance telephone carrier has established several direct microwave links among four cities. In the following connectivity matrix, entries a_{ij} and a_{ji} indicate the number of direct links between cities i and j. For example, cities 2 and 4 are not linked directly but could be connected through city 3. Find and interpret matrix A^2.

$$A = \begin{bmatrix} 0 & 2 & 1 & 0 \\ 2 & 0 & 1 & 0 \\ 1 & 1 & 0 & 2 \\ 0 & 0 & 2 & 0 \end{bmatrix}$$

53. Connectivity matrix An entry of 1 in the following connectivity matrix A indicates that the person associated with that row knows the address of the person associated with that column. For example, the 1 in Bill's row and Al's column indicates that Bill can write to Al. The 0 in Bill's row and Carl's column indicates that Bill cannot write to Carl. However, Bill could ask Al to forward his letter to Carl. The matrix A^2 indicates the number of ways that one person can write to another with a letter that is forwarded exactly once. Find A^2.

$$\begin{array}{c} \\ \text{Al} \\ \text{Bill} \\ \text{Carl} \end{array} \begin{array}{ccc} \text{Al} & \text{Bill} & \text{Carl} \end{array} \\ \begin{bmatrix} 0 & 1 & 1 \\ 1 & 0 & 0 \\ 0 & 1 & 0 \end{bmatrix} = A$$

54. Communication routing Refer to Exercise 53. Find and interpret the matrix $A + A^2$. Can everyone receive a letter from everyone else with at most one forwarding?

56. Communication on one-way channels Three communication centers are linked as indicated in Illustration 5, with communication only in the direction of the arrows. Thus, location 1 can send a message directly to location 2 along two paths, but location 2 can return a message directly on only one path. Entry c_{ij} of matrix C indicates the number of channels from i to j. Find and interpret C^2.

$$C = \begin{bmatrix} 0 & 2 & 1 \\ 1 & 0 & 1 \\ 2 & 0 & 0 \end{bmatrix}$$

ILLUSTRATION 5

where $A \neq 0$, to show that such a law does not hold for all matrices.

57. If A and B are 2×2 matrices, is $(AB)^2$ equal to A^2B^2? Support your answer.

59. Another property of the real numbers is that if $ab = 0$, then either $a = 0$ or $b = 0$. To show that this property is not true for matrices, find two nonzero 2×2 matrices A and B, such that $AB = 0$.

58. Let a, b, and c be real numbers. If $ab = ac$ and $a \neq 0$, then $b = c$. Find 2×2 matrices A, B, and C,

60. Find 2×2 matrices to show that $(A + B)(A - B) \neq A^2 - B^2$.

REVIEW *Do the operations and simplify.*

61. $(3x + 2)(2x - 3) - (2 - x)$

62. $\dfrac{x^2 + 3x - 4}{2x + 5 - (x + 1)}$

63. $\dfrac{1 + \dfrac{1}{x}}{1 - \dfrac{1}{x}}$

64. $\dfrac{1 - x^{-1}}{1 + x^{-1}}$

65. Solve the formula for a.

$$s = \frac{n(a + l)}{2}$$

66. Solve the formula for x_1.

$$y - y_1 = m(x - x_1)$$

7.4 Matrix Inversion

■ FINDING AN INVERSE BY ROW OPERATIONS ■ SOLVING A SYSTEM OF EQUATIONS
■ AN APPLICATION OF MATRIX INVERSION

Two real numbers are called **multiplicative inverses** if their product is the multiplicative identity 1. Some matrices have multiplicative inverses also.

Inverse of a Matrix

If A and B are $n \times n$ matrices, I is the $n \times n$ identity matrix, and

$$AB = BA = I$$

then A and B are called **multiplicative inverses**. Matrix A is the **inverse** of B, and B is the **inverse** of A.

It can be shown that if a matrix A has an inverse, it only has one inverse. The inverse of A is written as A^{-1}.

$$AA^{-1} = A^{-1}A = I$$

EXAMPLE 1 Show that A and B are inverses.

$$A = \begin{bmatrix} 1 & 1 & 0 \\ 4 & 3 & 0 \\ 2 & 1 & -1 \end{bmatrix} \qquad B = \begin{bmatrix} -3 & 1 & 0 \\ 4 & -1 & 0 \\ -2 & 1 & -1 \end{bmatrix}$$

Solution We must show that both AB and BA are equal to I.

$$AB = \begin{bmatrix} 1 & 1 & 0 \\ 4 & 3 & 0 \\ 2 & 1 & -1 \end{bmatrix}\begin{bmatrix} -3 & 1 & 0 \\ 4 & -1 & 0 \\ -2 & 1 & -1 \end{bmatrix}$$

$$= \begin{bmatrix} -3+4+0 & 1-1+0 & 0+0+0 \\ -12+12+0 & 4-3+0 & 0+0+0 \\ -6+4+2 & 2-1-1 & 0+0+1 \end{bmatrix} = \begin{bmatrix} 1 & 0 & 0 \\ 0 & 1 & 0 \\ 0 & 0 & 1 \end{bmatrix}$$

$$BA = \begin{bmatrix} -3 & 1 & 0 \\ 4 & -1 & 0 \\ -2 & 1 & -1 \end{bmatrix}\begin{bmatrix} 1 & 1 & 0 \\ 4 & 3 & 0 \\ 2 & 1 & -1 \end{bmatrix} = \begin{bmatrix} 1 & 0 & 0 \\ 0 & 1 & 0 \\ 0 & 0 & 1 \end{bmatrix} \blacksquare$$

Self Check Are C and D inverses? $C = \begin{bmatrix} 2 & 5 \\ 1 & 3 \end{bmatrix}$ $D = \begin{bmatrix} 3 & -5 \\ -1 & 2 \end{bmatrix}$

Answer yes

■ FINDING AN INVERSE BY ROW OPERATIONS

A matrix that has an inverse is called a **nonsingular matrix** and is said to be **invertible.** If it does not have an inverse, it is called a **singular matrix** and is not invertible. The following theorem, stated without proof, provides a way to find the inverse of an invertible matrix.

Finding a Matrix Inverse
If a sequence of row operations performed on the $n \times n$ matrix A reduces A to the $n \times n$ identity matrix I, then those same row operations, performed in the same order on I, will transform I into A^{-1}.

If no sequence of row operations will reduce A to I, then A is not invertible.

To use this theorem, we perform row operations on matrix A to change it to the identity matrix I. At the same time, we perform the same row operations on I. This changes I into A^{-1}.

A notation for this process uses an n-row-by-$2n$-column matrix, with matrix A as the left half and matrix I as the right half. If A is invertible, the proper row operations performed on $[A|I]$ will transform it into $[I|A^{-1}]$.

EXAMPLE 2   Find the inverse of matrix A if $A = \begin{bmatrix} 2 & -4 \\ 4 & -7 \end{bmatrix}$.

Solution   We can set up a 2×4 matrix with A on the left and I on the right of the broken line:

$$[A|I] = \begin{bmatrix} 2 & -4 & | & 1 & 0 \\ 4 & -7 & | & 0 & 1 \end{bmatrix}$$

We perform row operations on the entire matrix to transform the left half into I:

$$\begin{bmatrix} 2 & -4 & | & 1 & 0 \\ 4 & -7 & | & 0 & 1 \end{bmatrix} \begin{array}{c} (\frac{1}{2})R1 \rightarrow \\ (-2)R1 + R2 \rightarrow \end{array} \begin{bmatrix} 1 & -2 & | & \frac{1}{2} & 0 \\ 0 & 1 & | & -2 & 1 \end{bmatrix}$$

$$(2)R2 + R1 \rightarrow \begin{bmatrix} 1 & 0 & | & -\frac{7}{2} & 2 \\ 0 & 1 & | & -2 & 1 \end{bmatrix}$$

Since matrix A has been transformed into I, the right side of the previous matrix is A^{-1}. We can verify this by finding AA^{-1} and $A^{-1}A$ and showing that each product is I:

$$AA^{-1} = \begin{bmatrix} 2 & -4 \\ 4 & -7 \end{bmatrix} \begin{bmatrix} -\frac{7}{2} & 2 \\ -2 & 1 \end{bmatrix} = \begin{bmatrix} 1 & 0 \\ 0 & 1 \end{bmatrix}$$

$$A^{-1}A = \begin{bmatrix} -\frac{7}{2} & 2 \\ -2 & 1 \end{bmatrix} \begin{bmatrix} 2 & -4 \\ 4 & -7 \end{bmatrix} = \begin{bmatrix} 1 & 0 \\ 0 & 1 \end{bmatrix}$$ ∎

Self Check   Find the inverse of $A = \begin{bmatrix} 3 & 2 \\ 4 & 3 \end{bmatrix}$.

Answer $\begin{bmatrix} 3 & -2 \\ -4 & 3 \end{bmatrix}$

EXAMPLE 3   Find the inverse of matrix A.

$$A = \begin{bmatrix} 1 & 1 & 0 \\ 1 & 2 & 1 \\ 2 & 3 & 2 \end{bmatrix}$$

Solution   We set up a 3×6 matrix with A on the left and I on the right of the broken line.

$$[A|I] = \begin{bmatrix} 1 & 1 & 0 & | & 1 & 0 & 0 \\ 1 & 2 & 1 & | & 0 & 1 & 0 \\ 2 & 3 & 2 & | & 0 & 0 & 1 \end{bmatrix}$$

We then perform row operations on the matrix to transform the left half into I.

$$\begin{bmatrix} 1 & 1 & 0 & \vdots & 1 & 0 & 0 \\ 1 & 2 & 1 & \vdots & 0 & 1 & 0 \\ 2 & 3 & 2 & \vdots & 0 & 0 & 1 \end{bmatrix} \begin{matrix} \\ (-1)R1 + R2 \rightarrow \\ (-2)R1 + R3 \rightarrow \end{matrix} \begin{bmatrix} 1 & 1 & 0 & \vdots & 1 & 0 & 0 \\ 0 & 1 & 1 & \vdots & -1 & 1 & 0 \\ 0 & 1 & 2 & \vdots & -2 & 0 & 1 \end{bmatrix}$$

$$\begin{matrix} (-1)R2 + R1 \rightarrow \\ \\ (-1)R2 + R3 \rightarrow \end{matrix} \begin{bmatrix} 1 & 0 & -1 & \vdots & 2 & -1 & 0 \\ 0 & 1 & 1 & \vdots & -1 & 1 & 0 \\ 0 & 0 & 1 & \vdots & -1 & -1 & 1 \end{bmatrix}$$

$$\begin{matrix} R3 + R1 \rightarrow \\ (-1)R3 + R2 \rightarrow \\ \\ \end{matrix} \begin{bmatrix} 1 & 0 & 0 & \vdots & 1 & -2 & 1 \\ 0 & 1 & 0 & \vdots & 0 & 2 & -1 \\ 0 & 0 & 1 & \vdots & -1 & -1 & 1 \end{bmatrix}$$

Since the left half has been transformed into the identity matrix, the right half has become A^{-1}, and

$$A^{-1} = \begin{bmatrix} 1 & -2 & 1 \\ 0 & 2 & -1 \\ -1 & -1 & 1 \end{bmatrix}$$ ■

Self Check   Find the inverse of $B = \begin{bmatrix} 1 & 1 & 1 \\ 2 & 1 & 4 \\ 2 & 2 & 3 \end{bmatrix}$.

Answer $\begin{bmatrix} 5 & 1 & -3 \\ -2 & -1 & 2 \\ -2 & 0 & 1 \end{bmatrix}$

EXAMPLE 4   Find the inverse of $A = \begin{bmatrix} 1 & 2 \\ 2 & 4 \end{bmatrix}$, if possible.

Solution   We form the 2×4 matrix

$$[A \,|\, I] = \begin{bmatrix} 1 & 2 & \vdots & 1 & 0 \\ 2 & 4 & \vdots & 0 & 1 \end{bmatrix}$$

and begin to transform the left side of the matrix into the identity matrix I:

$$\begin{bmatrix} 1 & 2 & \vdots & 1 & 0 \\ 2 & 4 & \vdots & 0 & 1 \end{bmatrix} (-2)R1 + R2 \rightarrow \begin{bmatrix} 1 & 2 & \vdots & 1 & 0 \\ 0 & 0 & \vdots & -2 & 1 \end{bmatrix}$$

In obtaining the second-row, first-column position of A, the entire second row of A is zeroed out. Since we cannot transform matrix A to the identity, A is not invertible.   ■

Self Check   Find the inverse, if possible: $B = \begin{bmatrix} 1 & -2 \\ -3 & 6 \end{bmatrix}$.

Answer   not possible

■ ■ ■ ■ ■ ■ ■ ■ ■

ACCENT ON TECHNOLOGY

We can use a graphing calculator that performs matrix operations to find the inverse of a matrix. For example, to find the inverse of

$$A = \begin{bmatrix} 2 & 2 & 3 \\ 1 & 2 & 3 \\ 1 & 0 & 1 \end{bmatrix}$$

on one model, we press MATRIX , select EDIT, and enter matrix A. We exit entry mode and display A^{-1} by pressing A x^{-1} ENTER . The display appears in Figure 7-6(a). To verify that $AA^{-1} = I$, we press A   A x^{-1} ENTER to obtain the result shown in Figure 7-6(b).

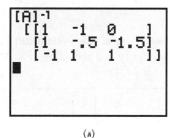

(a)

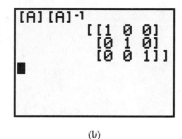

(b)

FIGURE 7-6

■ ■ ■ ■ ■ ■ ■ ■ ■

■ SOLVING A SYSTEM OF EQUATIONS

If we multiply the matrices on the left-hand side of the equation

$$\begin{bmatrix} 1 & 1 & 0 \\ 1 & 2 & 1 \\ 2 & 3 & 2 \end{bmatrix} \begin{bmatrix} x \\ y \\ z \end{bmatrix} = \begin{bmatrix} 20 \\ 30 \\ 55 \end{bmatrix}$$

and set the corresponding entries equal, we get the following system of equations:

$$\begin{cases} x + y = 20 \\ x + 2y + z = 30 \\ 2x + 3y + 2z = 55 \end{cases}$$

A system of equations can always be written as a matrix equation $AX = B$, where A is the coefficient matrix of the system, X is a column matrix of variables, and B is the column matrix of constants. If matrix A is invertible, the matrix equation $AX = B$ is easy to solve.

EXAMPLE 5

Solve $\begin{bmatrix} 1 & 1 & 0 \\ 1 & 2 & 1 \\ 2 & 3 & 2 \end{bmatrix} \begin{bmatrix} x \\ y \\ z \end{bmatrix} = \begin{bmatrix} 20 \\ 30 \\ 55 \end{bmatrix}$.

Solution The 3×3 matrix on the left is the matrix whose inverse was found in Example 3. We multiply each side of the equation on the left by this inverse to obtain an equivalent system of equations.

$$\begin{bmatrix} 1 & -2 & 1 \\ 0 & 2 & -1 \\ -1 & -1 & 1 \end{bmatrix} \begin{bmatrix} 1 & 1 & 0 \\ 1 & 2 & 1 \\ 2 & 3 & 2 \end{bmatrix} \begin{bmatrix} x \\ y \\ z \end{bmatrix} = \begin{bmatrix} 1 & -2 & 1 \\ 0 & 2 & -1 \\ -1 & -1 & 1 \end{bmatrix} \begin{bmatrix} 20 \\ 30 \\ 55 \end{bmatrix}$$

$$\begin{bmatrix} 1 & 0 & 0 \\ 0 & 1 & 0 \\ 0 & 0 & 1 \end{bmatrix} \begin{bmatrix} x \\ y \\ z \end{bmatrix} = \begin{bmatrix} 15 \\ 5 \\ 5 \end{bmatrix}$$ Multiply the matrices. On the left, remember that $A^{-1}A = I$.

$$\begin{bmatrix} x \\ y \\ z \end{bmatrix} = \begin{bmatrix} 15 \\ 5 \\ 5 \end{bmatrix}$$ $IX = X$.

The solution of this system can be read directly from the matrix on the right-hand side. Verify that the values $x = 15$, $y = 5$, $z = 5$ satisfy the original equations. ■

Self Check Solve $\begin{bmatrix} 2 & -4 \\ 4 & -7 \end{bmatrix} \begin{bmatrix} x \\ y \end{bmatrix} = \begin{bmatrix} -4 \\ 2 \end{bmatrix}$. (The inverse was found in Example 2.)

Answer $\begin{bmatrix} x \\ y \end{bmatrix} = \begin{bmatrix} 18 \\ 10 \end{bmatrix}$

Example 5 suggests the following result.

Solving Systems of Equations
If A is invertible, the solution of the matrix equation $AX = B$ is
$$X = A^{-1}B$$

This method is especially useful for finding solutions of several systems of equations that differ from each other only in the column matrix B. If the coefficient matrix A remains unchanged from one system of equations to the next, then A^{-1} needs to be found only once. The solution of each system is found by a single matrix multiplication, $A^{-1}B$.

■ AN APPLICATION OF MATRIX INVERSION

EXAMPLE 6 A company that manufactures medical equipment spends time on paperwork, manufacture, and testing for each of three versions of a circuit board. The times spent on each and the total time available are given in the following tables.

	Hours Required per Unit		
	Product A	**Product B**	**Product C**
Paperwork	1	1	0
Manufacture	1	2	1
Testing	2	3	2

Hours Available	
Paperwork	20
Manufacture	30
Testing	55

Solution   We can let x, y, and z represent the number of units of products A, B, and C to be manufactured, respectively. We can then set up the following system of equations:

Paperwork: $x + y = 20$ One hour is needed for every A, and one hour for every B.

Manufacture: $x + 2y + z = 30$ One hour is needed for every A and C, and two hours for every B.

Testing: $2x + 3y + 2z = 55$ Two hours are needed for every A and C, and three hours for every B.

In matrix form, the system becomes

$$\begin{bmatrix} 1 & 1 & 0 \\ 1 & 2 & 1 \\ 2 & 3 & 2 \end{bmatrix} \begin{bmatrix} x \\ y \\ z \end{bmatrix} = \begin{bmatrix} 20 \\ 30 \\ 55 \end{bmatrix}$$

We solved this equation in Example 5 to get $x = 15$, $y = 5$, and $z = 5$. To use all of the available time, the company should manufacture 15 units of product A and 5 units each of products B and C.   ∎

EXERCISE 7.4

VOCABULARY AND CONCEPTS   *In Exercises 1–4, fill in the blank to make a true statement.*

1. Matrices A and B are multiplicative inverses if _____.

2. A nonsingular matrix ___ (is, is not) invertible.

3. If A is invertible, elementary row operations can change $[A\,|\,I]$ into _____.

4. If A is invertible, the solution of $AX = B$ is _____.

PRACTICE   *In Exercises 5–20, find the inverse of each matrix, if possible.*

5. $\begin{bmatrix} 3 & -4 \\ -2 & 3 \end{bmatrix}$

6. $\begin{bmatrix} 2 & 3 \\ 3 & 5 \end{bmatrix}$

7. $\begin{bmatrix} 3 & 7 \\ 2 & 5 \end{bmatrix}$

8. $\begin{bmatrix} 1 & -2 \\ 2 & -5 \end{bmatrix}$

9. $\begin{bmatrix} 1 & 0 & 3 \\ -1 & 1 & 3 \\ -2 & 1 & 1 \end{bmatrix}$ **10.** $\begin{bmatrix} 2 & 1 & -1 \\ 2 & 2 & -1 \\ -1 & -1 & 1 \end{bmatrix}$ **11.** $\begin{bmatrix} 3 & 2 & 1 \\ 1 & 1 & -1 \\ 4 & 3 & 1 \end{bmatrix}$ **12.** $\begin{bmatrix} -2 & 1 & -3 \\ 2 & 3 & 0 \\ 1 & 0 & 1 \end{bmatrix}$

13. $\begin{bmatrix} 1 & 3 & 5 \\ 0 & 1 & 6 \\ 1 & 4 & 11 \end{bmatrix}$ **14.** $\begin{bmatrix} 1 & 2 & 3 \\ 4 & 5 & 6 \\ 7 & 8 & 9 \end{bmatrix}$ **15.** $\begin{bmatrix} 1 & 2 & 3 \\ 0 & 1 & 2 \\ 0 & 0 & 1 \end{bmatrix}$ **16.** $\begin{bmatrix} 1 & 2 & 3 \\ 0 & 1 & 1 \\ 0 & -1 & 0 \end{bmatrix}$

17. $\begin{bmatrix} 1 & 6 & 4 \\ 1 & -2 & -5 \\ 2 & 4 & -1 \end{bmatrix}$ **18.** $\begin{bmatrix} 1 & 1 & 1 \\ 1 & 0 & -1 \\ 1 & 2 & 3 \end{bmatrix}$

19. $\begin{bmatrix} 1 & 2 & 3 & 4 \\ 0 & 1 & 2 & 3 \\ 0 & 0 & 1 & 2 \\ 0 & 0 & 0 & 1 \end{bmatrix}$ **20.** $\begin{bmatrix} 1 & 0 & 0 & 0 \\ 1 & 1 & 0 & 0 \\ 1 & 1 & 1 & 0 \\ 1 & 2 & 2 & 1 \end{bmatrix}$

In Exercises 21–24, use a graphing calculator to find the inverse of each matrix.

21. $\begin{bmatrix} 1 & 1 & -1 \\ 0.5 & 1 & 0.5 \\ 1 & 1 & -1.5 \end{bmatrix}$ **22.** $\begin{bmatrix} -2 & -1 & 1 \\ 0.5 & -1.5 & -0.5 \\ 0 & 1 & 0.5 \end{bmatrix}$

23. $\begin{bmatrix} 3 & 3 & -3 & 2 \\ 1 & -4 & 3 & -5 \\ 3 & 0 & -2 & -1 \\ -1 & 5 & -3 & 6 \end{bmatrix}$ **24.** $\begin{bmatrix} 1 & 0 & 0 & 0 \\ 2 & 1 & 0 & 0 \\ 3 & 2 & 1 & 0 \\ 4 & 3 & 2 & 1 \end{bmatrix}$

In Exercises 25–32, use the method of Example 5 to solve each system of equations. Note that several systems have the same coefficient matrix.

25. $\begin{cases} 3x - 4y = 1 \\ -2x + 3y = 5 \end{cases}$ **26.** $\begin{cases} 3x - 4y = -1 \\ -2x + 3y = 3 \end{cases}$ **27.** $\begin{cases} 3x - 4y = 0 \\ -2x + 3y = 0 \end{cases}$ **28.** $\begin{cases} 3x - 4y = -3 \\ -2x + 3y = -2 \end{cases}$

29. $\begin{cases} 2x + y - z = 2 \\ 2x + 2y - z = 4 \\ -x - y + z = -1 \end{cases}$ **30.** $\begin{cases} 2x + y - z = 3 \\ 2x + 2y - z = -1 \\ -x - y + z = 4 \end{cases}$ **31.** $\begin{cases} -2x + y - 3z = 2 \\ 2x + 3y = -3 \\ x + z = 5 \end{cases}$ **32.** $\begin{cases} -2x + y - 3z = 5 \\ 2x + 3y = 1 \\ x + z = -2 \end{cases}$

In Exercises 33–36, use a graphing calculator to solve each system of equations. Use the method of Example 5.

33. $\begin{cases} 5x + 3y = 13 \\ -7x + 5y = -9 \end{cases}$ **34.** $\begin{cases} 8x - 3y = 7 \\ -3x + 2y = 0 \end{cases}$ **35.** $\begin{cases} 5x + 2y + 3z = 12 \\ 2x + 5z = 7 \\ 3x + z = 4 \end{cases}$ **36.** $\begin{cases} 3x + 2y - z = 0 \\ 5x - 2y = 5 \\ 3x + y + z = 6 \end{cases}$

APPLICATIONS

37. Manufacturing and testing   The numbers of hours required to manufacture and test each of two models of heart monitor are given in Illustration 1, and the number of hours available each week for manufacturing and testing is given in Illustration 2.

	Hours Required per Unit	
	Model A	**Model B**
Manufacturing	23	27
Testing	21	22

ILLUSTRATION 1

	Hours Available
Manufacturing	127
Testing	108

ILLUSTRATION 2

How many of each model can be manufactured each week?

38. Cryptography   The letters of a message, called **plaintext,** are assigned values 1–26 (for a-z) and are written in groups of 3 as 3×1 matrices. To write the message in **cyphertext,** each 3×1 matrix is multiplied by matrix A, where

$$A = \begin{bmatrix} 1 & 1 & 0 \\ 2 & 3 & 3 \\ 1 & 1 & 1 \end{bmatrix}$$

The cyphertext of one message is

$$AY = \begin{bmatrix} 30 \\ 122 \\ 49 \end{bmatrix}$$

Find the plaintext.

DISCOVERY AND WRITING   *In Exercises 39–40, use examples chosen from 2×2 matrices to support each answer.*

39. Does $(AB)^{-1} = A^{-1}B^{-1}$?

40. Does $(AB)^{-1} = B^{-1}A^{-1}$?

In Exercises 41–42, let $A = \begin{bmatrix} -1 & -1 \\ 1 & 1 \end{bmatrix}$.

41. Show that $A^2 = 0$.

42. Show that the inverse of $I - A$ is $I + A$.

In Exercises 43–44, let $A = \begin{bmatrix} 3 & 0 & 0 \\ -2 & -1 & -2 \\ 3 & 6 & 3 \end{bmatrix}$ and $X = \begin{bmatrix} x \\ y \\ z \end{bmatrix}$. Solve each equation. Each solution is called an eigenvector of the matrix A.

43. $(A - 2I)X = 0$

44. $(A - 3I)X = 0$

45. Suppose that A, B, and C are $n \times n$ matrices and A is invertible. If $AB = AC$, prove that $B = C$.

46. Prove that $\begin{bmatrix} a & b \\ c & d \end{bmatrix}$ has an inverse if and only if $ad - bc \neq 0$. (*Hint:* Try to find the inverse and see what happens.)

47. Suppose that B is any matrix for which $B^2 = 0$. Show that $I - B$ is invertible by showing that the inverse of $I - B$ is $I + B$.

48. Suppose that C is any matrix for which $C^3 = 0$. Show that $I - C$ is invertible by showing that the inverse of $I - C$ is $I + C + C^2$.

REVIEW *Find the domain of each function.*

49. $y = \dfrac{3x - 5}{x^2 - 4}$

50. $y = \dfrac{3x - 5}{x^2 + 4}$

51. $y = \dfrac{3x - 5}{\sqrt{x^2 + 4}}$

52. $y = \dfrac{3x - 5}{\sqrt{x^2 - 4}}$

Find the range of each function.

53. $y = x^2$

54. $y = x^3$

55. $x = \log x$

56. $y = 2^x$

7.5 Determinants

■ EVALUATING DETERMINANTS OF HIGHER-ORDER MATRICES ■ PROPERTIES OF DETERMINANTS
■ USING DETERMINANTS TO SOLVE SYSTEMS OF EQUATIONS ■ WRITING EQUATIONS OF LINES
■ FINDING AREAS OF TRIANGLES

The **determinant function** associates a number with any square matrix. The function is written as $\det(A)$ or as $|A|$.

Determinants

If a, b, c, and d are numbers, the determinant of $A = \begin{bmatrix} a & b \\ c & d \end{bmatrix}$ is

$$\det(A) = \begin{vmatrix} a & b \\ c & d \end{vmatrix} = ad - bc$$

WARNING! Do not confuse the notation $|A|$ with absolute value symbols.

EXAMPLE 1

a. $\begin{vmatrix} 1 & 2 \\ 3 & 4 \end{vmatrix} = 1 \cdot 4 - 2 \cdot 3$

$= 4 - 6$

$= -2$

b. $\begin{vmatrix} -2 & 3 \\ -\pi & \frac{1}{2} \end{vmatrix} = (-2) \cdot \left(\frac{1}{2} \right) - (3) \cdot (-\pi)$

$= -1 + 3\pi$ ∎

Self Check Evaluate $\begin{vmatrix} 3 & -2 \\ 5 & -4 \end{vmatrix}$.

Answer -2

■ EVALUATING DETERMINANTS OF HIGHER-ORDER MATRICES

To evaluate determinants of higher-order matrices, we must define the **minor** and the **cofactor** of an element in a matrix.

> **Minor and Cofactor of a Matrix**
> Let $A = [a_{ij}]$ be a square matrix of order $n \geq 2$.
> **1.** The **minor** of a_{ij}, denoted as M_{ij}, is the determinant of the $n - 1 \times n - 1$ matrix formed by deleting the ith row and jth column of A.
> **2.** The **cofactor** of a_{ij}, denoted as C_{ij}, is $\begin{cases} M_{ij} \text{ when } i + j \text{ is even} \\ -M_{ij} \text{ when } i + j \text{ is odd} \end{cases}$.

EXAMPLE 2

In matrix $A = \begin{bmatrix} 1 & 2 & 3 \\ 4 & 5 & 6 \\ 7 & 8 & 9 \end{bmatrix}$, find the minor and cofactor of **a.** a_{31} and **b.** a_{12}.

Solution

a. The minor M_{31} is the minor of $a_{31} = 7$ appearing in row 3, column 1. It is found by deleting row 3 and column 1:

$$M_{31} = \begin{bmatrix} 1 & 2 & 3 \\ 4 & 5 & 6 \\ 7 & 8 & 9 \end{bmatrix} = \begin{vmatrix} 2 & 3 \\ 5 & 6 \end{vmatrix}$$

Because $i + j$ is even $(3 + 1 = 4)$, the cofactor of the minor M_{31} is M_{31}:

$$C_{31} = M_{31} = \begin{vmatrix} 2 & 3 \\ 5 & 6 \end{vmatrix} = 2 \cdot 6 - 3 \cdot 5 = 12 - 15 = -3$$

b. The minor M_{12} is the minor of $a_{12} = 2$ appearing in row 1, column 2. It is found by deleting row 1 and column 2.

$$M_{12} = \begin{bmatrix} 1 & 2 & 3 \\ 4 & 5 & 6 \\ 7 & 8 & 9 \end{bmatrix} = \begin{vmatrix} 4 & 6 \\ 7 & 9 \end{vmatrix}$$

Because $i + j$ is odd ($1 + 2 = 3$), the cofactor of the minor M_{12} is $-M_{12}$:

$$C_{12} = -M_{12} = -\begin{vmatrix} 4 & 6 \\ 7 & 9 \end{vmatrix} = -(4 \cdot 9 - 6 \cdot 7) = -(36 - 42) = -(-6) = 6 \quad ■$$

Self Check Find the cofactor of a_{23}.

Answer 6

We are now ready to evaluate determinants of higher-order matrices.

Value of a Determinant

If A is a square matrix of order $n \geq 2$, $|A|$ is the sum of the products of the elements in any row (or column) and the cofactors of those elements.

In Example 3, we evaluate a 3×3 determinant by expanding the determinant in three ways: along two different rows and along a column. This method is called **expanding a determinant by minors.**

EXAMPLE 3 Evaluate $\begin{vmatrix} 1 & 2 & -3 \\ -1 & 0 & 1 \\ -2 & 2 & 1 \end{vmatrix}$ by expanding along the designated row or column.

Expanding on row 1:

$$\begin{vmatrix} 1 & 2 & -3 \\ -1 & 0 & 1 \\ -2 & 2 & 1 \end{vmatrix} = a_{11}C_{11} + a_{12}C_{12} + a_{13}C_{13}$$

$$= 1\begin{vmatrix} 0 & 1 \\ 2 & 1 \end{vmatrix} + 2\left[-\begin{vmatrix} -1 & 1 \\ -2 & 1 \end{vmatrix} \right] + (-3)\begin{vmatrix} -1 & 0 \\ -2 & 2 \end{vmatrix}$$

$$= 1(-2) + 2(-1) - 3(-2)$$

$$= -2 - 2 + 6$$

$$= 2$$

Expanding on row 3:

$$\begin{vmatrix} 1 & 2 & -3 \\ -1 & 0 & 1 \\ -2 & 2 & 1 \end{vmatrix} = a_{31}C_{31} + a_{32}C_{32} + a_{33}C_{33}$$

$$= -2\begin{vmatrix} 2 & -3 \\ 0 & 1 \end{vmatrix} + 2\left[-\begin{vmatrix} 1 & -3 \\ -1 & 1 \end{vmatrix}\right] + 1\begin{vmatrix} 1 & 2 \\ -1 & 0 \end{vmatrix}$$

$$= -2(2) + 2(+2) + 1(2)$$

$$= -4 + 4 + 2$$

$$= 2$$

Expanding on column 2:

$$\begin{vmatrix} 1 & 2 & -3 \\ -1 & 0 & 1 \\ -2 & 2 & 1 \end{vmatrix} = a_{12}C_{12} + a_{22}C_{22} + a_{32}C_{32}$$

$$= 2\left[-\begin{vmatrix} -1 & 1 \\ -2 & 1 \end{vmatrix}\right] + 0\begin{vmatrix} 1 & -3 \\ -2 & 1 \end{vmatrix} + 2\left[-\begin{vmatrix} 1 & -3 \\ -1 & 1 \end{vmatrix}\right]$$

$$= 2(-1) + 0(-5) + 2(+2)$$

$$= -2 + 4$$

$$= 2$$

In each case, the result is 2. ∎

Self Check Expand $\begin{vmatrix} 1 & 0 & 1 \\ 2 & -1 & 0 \\ 3 & 1 & -1 \end{vmatrix}$ on its first row.

Answer 6

EXAMPLE 4 Evaluate $\begin{vmatrix} 0 & 0 & 2 & 0 \\ 1 & 2 & 17 & -3 \\ -1 & 0 & 28 & 1 \\ -2 & 2 & -37 & 1 \end{vmatrix}$.

Solution Because row 1 contains three 0's, we expand the determinant along row 1. Then only one cofactor needs to be evaluated.

$$\begin{vmatrix} 0 & 0 & 2 & 0 \\ 1 & 2 & 17 & -3 \\ -1 & 0 & 28 & 1 \\ -2 & 2 & -37 & 1 \end{vmatrix} = 0|?| - 0|?| + 2\begin{vmatrix} 1 & 2 & -3 \\ -1 & 0 & 1 \\ -2 & 2 & 1 \end{vmatrix} - 0|?|$$

$$= 2(2) \qquad \text{See Example 3.}$$

$$= 4 \qquad\qquad\qquad\qquad ∎$$

Self Check   Evaluate $\begin{vmatrix} 0 & 1 & 0 \\ 2 & 11 & -2 \\ 1 & 13 & 1 \end{vmatrix}$.

Answer -4

Example 4 suggests the following theorem.

Zero Row or Column Theorem

If every entry in a row or column of a square matrix A is 0, then $|A| = 0$.

Many graphing calculators are able to evaluate determinants. Simply enter a square matrix (say, A) and press det(and A. For details, consult the owner's manual.

■ PROPERTIES OF DETERMINANTS

We have seen that there are row operations for transforming matrices. There are similar row and column operations for transforming determinants.

Row and Column Operations for Determinants

Let A be a square matrix and k be a real number.

1. If a matrix B is obtained from matrix A by interchanging two rows (or columns), then $|B| = -|A|$.

2. If B is obtained from A by multiplying every element in a row (or column) of A by k, then $|B| = k|A|$.

3. If B is obtained from A by adding k times any row (or column) of A to another row (or column) of A, then $|B| = |A|$.

We illustrate each row operation by showing that it is true for the 2×2 determinant

$$|A| = \begin{vmatrix} a & b \\ c & d \end{vmatrix}$$

Interchanging two rows: Let B be obtained from A by interchanging its two rows. Then

$$|B| = \begin{vmatrix} c & d \\ a & b \end{vmatrix} = cb - da = -(ad - bc) = -|A|$$

Multiplying every element in a row by k: Let B be obtained from A by multiplying its second row by k. Then

$$|B| = \begin{vmatrix} a & b \\ kc & kd \end{vmatrix} = akd - bkc = k(ad - bc) = k|A|$$

Adding k times any row of A to another row of A: Let B be obtained from A by adding k times its first row to its second row. Then

$$|B| = \begin{vmatrix} a & b \\ ka + c & kb + d \end{vmatrix} = a(kb + d) - b(ka + c) = akb + ad - bka - bc = ad - bc = |A|$$

We will use several row and column operations in the next example.

EXAMPLE 5   Evaluate $\begin{vmatrix} 10 & 20 & -10 & 20 \\ 2 & 1 & 1 & 1 \\ 1 & 2 & -3 & 2 \\ 2 & -1 & -1 & 1 \end{vmatrix}$.

Solution   To get smaller numbers in the first row, we can use a type 2 row operation and multiply each entry in the first row by $\frac{1}{10}$. However, we must then multiply the resulting determinant by 10 to retain its original value.

1. $\begin{vmatrix} 10 & 20 & -10 & 20 \\ 2 & 1 & 1 & 1 \\ 1 & 2 & -3 & 2 \\ 2 & -1 & -1 & 1 \end{vmatrix} \begin{array}{c} (\frac{1}{10})R1 \to \\ = 10 \end{array} \begin{vmatrix} 1 & 2 & -1 & 2 \\ 2 & 1 & 1 & 1 \\ 1 & 2 & -3 & 2 \\ 2 & -1 & -1 & 1 \end{vmatrix}$

To get three 0's in the first row of the second determinant in Equation 1, we perform a type 3 row operation and expand the new determinant along its first row.

$10 \begin{vmatrix} 1 & 2 & -1 & 2 \\ 2 & 1 & 1 & 1 \\ 1 & 2 & -3 & 2 \\ 2 & -1 & -1 & 1 \end{vmatrix} \begin{array}{c} (-1)R3 + R1 \to \\ = 10 \end{array} \begin{vmatrix} 0 & 0 & 2 & 0 \\ 2 & 1 & 1 & 1 \\ 1 & 2 & -3 & 2 \\ 2 & -1 & -1 & 1 \end{vmatrix}$

2. $= 10(2) \begin{vmatrix} 2 & 1 & 1 \\ 1 & 2 & 2 \\ 2 & -1 & 1 \end{vmatrix}$

To introduce 0's into the 3×3 determinant in Equation 2, we perform a column operation on the 3×3 determinant and expand the result on its first column.

$$(-2)C3 + C1$$
$$\downarrow$$

$$20\begin{vmatrix} 2 & 1 & 1 \\ 1 & 2 & 2 \\ 2 & -1 & 1 \end{vmatrix} = 20\begin{vmatrix} 0 & 1 & 1 \\ -3 & 2 & 2 \\ 0 & -1 & 1 \end{vmatrix}$$

$$= 20\left[-(-3)\begin{vmatrix} 1 & 1 \\ -1 & 1 \end{vmatrix}\right]$$

$$= 20(3)[1 - (-1)]$$
$$= 60(2)$$
$$= 120 \qquad\blacksquare$$

Self Check   Evaluate $\begin{vmatrix} 15 & 20 & 5 & 10 \\ 1 & 2 & 2 & 3 \\ 0 & 3 & -1 & 1 \\ 1 & 0 & 0 & -1 \end{vmatrix}$.

Answer   30

We will consider one final theorem.

Theorem

If A is a square matrix with two identical rows (or columns), then $|A| = 0$.

Proof   If the square matrix A has two identical rows (or columns), we can apply a type 3 row (or column) operation to zero out one of those rows (or columns). Since the matrix would then have an all-zero row (or column), its determinant would be 0, by the zero row or column theorem. $\square$

■ USING DETERMINANTS TO SOLVE SYSTEMS OF EQUATIONS

We can solve the system $\begin{cases} ax + bx = e \\ cx + dy = f \end{cases}$ by multiplying the first equation by d, multiplying the second equation by $-b$, and adding to get

$$adx + bdy = ed$$
$$\underline{-bcx - bdy = -bf}$$
$$adx - bcx = ed - bf$$

If $ad \neq bc$, we can solve the resulting equation for x:

$$adx - bcx = ed - bf$$

$$(ad - bc)x = ed - bf \qquad \text{Factor out } x.$$

3. $$x = \frac{ed - bf}{ad - bc} \qquad \text{Divide both sides by } ad - bc.$$

If $ad \neq bc$, we can also solve the system for y to get

4. $$y = \frac{af - ec}{ad - bc}$$

We can write the values of x and y in Equations 3 and 4 using determinants.

$$x = \frac{\begin{vmatrix} e & b \\ f & d \end{vmatrix}}{\begin{vmatrix} a & b \\ c & d \end{vmatrix}} = \frac{ed - bf}{ad - bc} \qquad y = \frac{\begin{vmatrix} a & e \\ c & f \end{vmatrix}}{\begin{vmatrix} a & b \\ c & d \end{vmatrix}} = \frac{af - ec}{ad - bc}$$

If we compare these formulas with the original system,

$$\begin{cases} ax + by = e \\ cx + dy = f \end{cases}$$

we see that the denominators are the determinant of the coefficient matrix:

$$\text{denominator determinant} = \begin{vmatrix} a & b \\ c & d \end{vmatrix}$$

To find the numerator determinant for x, we replace the a and c in the first column of the denominator determinant with the constants e and f.

To find the numerator determinant for y, we replace the b and d in the second column of the denominator determinant with the constants e and f.

$$x = \frac{\begin{vmatrix} e & b \\ f & d \end{vmatrix}}{\begin{vmatrix} a & b \\ c & d \end{vmatrix}} \qquad y = \frac{\begin{vmatrix} a & e \\ b & f \end{vmatrix}}{\begin{vmatrix} a & b \\ c & d \end{vmatrix}}$$

This method of using determinants to solve systems of equations is called **Cramer's rule.**

Cramer's Rule for Two Equations in Two Variables

If the system $\begin{cases} ax + by = e \\ cx + dy = f \end{cases}$ has a single solution, it is given by

$$x = \frac{D_x}{D} \qquad \text{and} \qquad y = \frac{D_y}{D}$$

where $D = \begin{vmatrix} a & b \\ c & d \end{vmatrix}$, $D_x = \begin{vmatrix} e & b \\ f & d \end{vmatrix}$, and $D_y = \begin{vmatrix} a & e \\ c & f \end{vmatrix}$.

If D, D_x, and D_y are all 0, the system is consistent, but the equations are dependent. If $D = 0$ and $D_x \neq 0$ or $D_y \neq 0$, the system is inconsistent.

EXAMPLE 6   Use Cramer's rule to solve $\begin{cases} 3x + 2y = 7 \\ -x + 5y = 9 \end{cases}$.

Solution

$$x = \frac{\begin{vmatrix} 7 & 2 \\ 9 & 5 \end{vmatrix}}{\begin{vmatrix} 3 & 2 \\ -1 & 5 \end{vmatrix}} = \frac{7 \cdot 5 - 2 \cdot 9}{3 \cdot 5 - 2(-1)} = \frac{35 - 18}{15 + 2} = \frac{17}{17} = 1$$

$$y = \frac{\begin{vmatrix} 3 & 7 \\ -1 & 9 \end{vmatrix}}{\begin{vmatrix} 3 & 2 \\ -1 & 5 \end{vmatrix}} = \frac{3 \cdot 9 - 7(-1)}{3 \cdot 5 - 2(-1)} = \frac{27 + 7}{15 + 2} = \frac{34}{17} = 2$$

Verify that the pair $(1, 2)$ satisfies both of the equations in the system.   ■

Self Check   Solve $\begin{cases} 2x + 5y = 9 \\ 3x + 7y = 13 \end{cases}$.

Answer $(2, 1)$

We can use Cramer's rule to solve systems of n equations in n variables where each equation has the form

$$a_1 x_1 + a_2 x_2 + \cdots + a_n x_n = c$$

To do so, we let D be the determinant of the coefficient matrix of the system and let D_{x_i} be the determinant formed by replacing the ith column of D by the column of constants from the right of the equal signs. If $D \neq 0$, Cramer's rule provides the following solution:

$$x_1 = \frac{D_{x_1}}{D}, \quad x_2 = \frac{D_{x_2}}{D}, \ldots, x_n = \frac{D_{x_n}}{D}$$

EXAMPLE 7   Use Cramer's rule to solve the system $\begin{cases} 2x - y + 2z = 3 \\ x - y + z = 2 \\ x + y + 2z = 3 \end{cases}$.

Solution   Each of the values x, y, and z is the quotient of two 3×3 determinants. The denominator of each quotient is the determinant consisting of the nine coefficients of the variables. The numerators for x, y, and z are modified copies of this denominator determinant. We substitute the column of constants for the coefficients of the variable for which we are solving.

$$\begin{cases} 2x - y + 2z = 3 \\ x - y + z = 2 \\ x + y + 2z = 3 \end{cases}$$

$$x = \frac{\begin{vmatrix} 3 & -1 & 2 \\ 2 & -1 & 1 \\ 3 & 1 & 2 \end{vmatrix}}{\begin{vmatrix} 2 & -1 & 2 \\ 1 & -1 & 1 \\ 1 & 1 & 2 \end{vmatrix}} = \frac{3\begin{vmatrix} -1 & 1 \\ 1 & 2 \end{vmatrix} - (-1)\begin{vmatrix} 2 & 1 \\ 3 & 2 \end{vmatrix} + 2\begin{vmatrix} 2 & -1 \\ 3 & 1 \end{vmatrix}}{2\begin{vmatrix} -1 & 1 \\ 1 & 2 \end{vmatrix} - (-1)\begin{vmatrix} 1 & 1 \\ 1 & 2 \end{vmatrix} + 2\begin{vmatrix} 1 & -1 \\ 1 & 1 \end{vmatrix}} = \frac{2}{-1} = -2$$

$$y = \frac{\begin{vmatrix} 2 & 3 & 2 \\ 1 & 2 & 1 \\ 1 & 3 & 2 \end{vmatrix}}{\begin{vmatrix} 2 & -1 & 2 \\ 1 & -1 & 1 \\ 1 & 1 & 2 \end{vmatrix}} = \frac{2\begin{vmatrix} 2 & 1 \\ 3 & 2 \end{vmatrix} - 3\begin{vmatrix} 1 & 1 \\ 1 & 2 \end{vmatrix} + 2\begin{vmatrix} 1 & 2 \\ 1 & 3 \end{vmatrix}}{-1} = \frac{1}{-1} = -1$$

$$z = \frac{\begin{vmatrix} 2 & -1 & 3 \\ 1 & -1 & 2 \\ 1 & 1 & 3 \end{vmatrix}}{\begin{vmatrix} 2 & -1 & 2 \\ 1 & -1 & 1 \\ 1 & 1 & 2 \end{vmatrix}} = \frac{2\begin{vmatrix} -1 & 1 \\ 1 & 3 \end{vmatrix} - (-1)\begin{vmatrix} 1 & 2 \\ 1 & 3 \end{vmatrix} + 3\begin{vmatrix} 1 & -1 \\ 1 & 1 \end{vmatrix}}{-1} = \frac{-3}{-1} = 3$$

Verify that the triple $(-2, -1, 3)$ satisfies each equation in the system. ∎

■ WRITING EQUATIONS OF LINES

If we are given the coordinates of two points in the xy-plane, we can use determinants to write the equation of the line passing through those points.

> **Two-Point Form of the Equation of a Line**
> The equation of the line passing through points $P(x_1, y_1)$ and $Q(x_2, y_2)$ is given by
> $$\begin{vmatrix} x & y & 1 \\ x_1 & y_1 & 1 \\ x_2 & y_2 & 1 \end{vmatrix} = 0$$

EXAMPLE 8   Write the equation of the line passing through $P(-2, 3)$ and $Q(4, -5)$.

Solution   We set up the equation $\begin{vmatrix} x & y & 1 \\ -2 & 3 & 1 \\ 4 & -5 & 1 \end{vmatrix} = 0$ and expand along the first row to get

$$[3(1) - 1(-5)]x - [-2(1) - 1(4)]y + [(-2)(-5) - 3(4)]1 = 0$$
$$8x + 6y - 2 = 0$$
$$4x + 3y = 1 \qquad \text{Add 2 to both sides and divide both sides by 2.}$$

The equation of the line is $4x + 3y = 1$.   ∎

Self Check   Find the equation of the line passing through $(1, 3)$ and $(3, 5)$.

Answer $x - y = -2$

■ FINDING AREAS OF TRIANGLES

> **Area of a Triangle**
> If points $P(x_1, y_1)$, $Q(x_2, y_2)$, and $R(x_3, y_3)$ are the vertices of a triangle, then the area of the triangle is given by
> $$A = \pm\frac{1}{2}\begin{vmatrix} x_1 & y_1 & 1 \\ x_2 & y_2 & 1 \\ x_3 & y_3 & 1 \end{vmatrix} \qquad \text{Pick either + or − to make the area positive.}$$

EXAMPLE 9   Find the area of the triangle shown in Figure 7-7.

Solution   We set up the equation

$$A = \pm\frac{1}{2}\begin{vmatrix} 0 & 0 & 1 \\ 5 & 0 & 1 \\ 5 & 12 & 1 \end{vmatrix}$$

and expand the determinant along the first row to get

$$= \pm\frac{1}{2}\left[1\begin{vmatrix} 5 & 0 \\ 5 & 12 \end{vmatrix}\right]$$

$$= \pm\frac{1}{2}(60 - 0)$$

$$= 30$$

The area of the triangle is 30 square units.   ∎

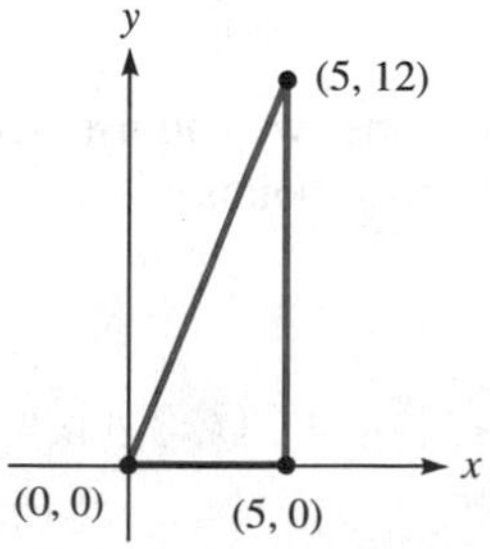

FIGURE 7-7

Self Check   Find the area of the triangle with vertices $(1, 2)$, $(2, 3)$, and $(-1, 4)$.

Answer   2

EXERCISE 7.5

VOCABULARY AND CONCEPTS   *In Exercises 1–6, fill in the blank to make a true statement.*

1. The determinant of a square matrix A is written _____ or as _____.

2. $\begin{vmatrix} a & b \\ c & d \end{vmatrix} =$ _____.

3. If every entry in one row or one column of A is zero, then $|A| =$ __.

4. If a matrix B is obtained from matrix A by adding one row to another, then $|B| =$ _____.

5. If two columns of A are identical, then $|A| =$ __.

6. In Cramer's rule, the denominator is the determinant of the _____.

PRACTICE   *In Exercises 7–10, evaluate each determinant.*

7. $\begin{vmatrix} 2 & 1 \\ -2 & 3 \end{vmatrix}$

8. $\begin{vmatrix} -3 & -6 \\ 2 & -5 \end{vmatrix}$

9. $\begin{vmatrix} 2 & -3 \\ -3 & 5 \end{vmatrix}$

10. $\begin{vmatrix} 5 & 8 \\ -6 & -2 \end{vmatrix}$

In Exercises 11–18, $A = \begin{vmatrix} 1 & -2 & 3 \\ 4 & 5 & -6 \\ -7 & 8 & 9 \end{vmatrix}$. *Find each minor or cofactor.*

11. M_{21}

12. M_{13}

13. M_{33}

14. M_{32}

15. C_{21}

16. C_{13}

17. C_{33}

18. C_{32}

In Exercises 19–30, evaluate each determinant.

19. $\begin{vmatrix} 2 & -3 & 5 \\ -2 & 1 & 3 \\ 1 & 3 & -2 \end{vmatrix}$

20. $\begin{vmatrix} 1 & 3 & 1 \\ -2 & 5 & 3 \\ 3 & -2 & -2 \end{vmatrix}$

21. $\begin{vmatrix} 1 & -1 & 2 \\ 2 & 1 & 3 \\ 1 & 1 & -1 \end{vmatrix}$

22. $\begin{vmatrix} 1 & 3 & 1 \\ 2 & 1 & -1 \\ 2 & -1 & 1 \end{vmatrix}$

23. $\begin{vmatrix} 2 & 1 & -1 \\ 1 & 3 & 5 \\ 2 & -5 & 3 \end{vmatrix}$

24. $\begin{vmatrix} 3 & 1 & -2 \\ -3 & 2 & 1 \\ 1 & 3 & 0 \end{vmatrix}$

25. $\begin{vmatrix} 0 & 1 & -3 \\ -3 & 5 & 2 \\ 2 & -5 & 3 \end{vmatrix}$

26. $\begin{vmatrix} 1 & -7 & -2 \\ -2 & 0 & 3 \\ -1 & 7 & 1 \end{vmatrix}$

27. $\begin{vmatrix} 0 & 0 & 1 & 0 \\ -2 & 1 & 0 & 1 \\ 1 & 0 & 1 & 2 \\ 2 & 0 & 1 & 2 \end{vmatrix}$

28. $\begin{vmatrix} 1 & 0 & -2 & 1 \\ 0 & 1 & 0 & 1 \\ 0 & 3 & -1 & 2 \\ 0 & -1 & 0 & 1 \end{vmatrix}$

29. $\begin{vmatrix} 1 & 2 & 1 & 3 \\ -2 & 1 & -3 & 1 \\ -1 & 0 & 1 & -2 \\ 2 & -1 & -1 & 3 \end{vmatrix}$

30. $\begin{vmatrix} -1 & 3 & -2 & 5 \\ 2 & 1 & 0 & 1 \\ 1 & 3 & -2 & 5 \\ 2 & -1 & 0 & -1 \end{vmatrix}$

In Exercises 31–34, decide whether each statement is true. **Do not evaluate the determinants.**

31. $\begin{vmatrix} 1 & 3 & -4 \\ -2 & 1 & 3 \\ 1 & 3 & 2 \end{vmatrix} = -\begin{vmatrix} -2 & 1 & 3 \\ 1 & 3 & -4 \\ 1 & 3 & 2 \end{vmatrix}$

32. $\begin{vmatrix} 4 & 6 & 8 \\ 10 & 5 & 15 \\ 20 & 5 & 10 \end{vmatrix} = \begin{vmatrix} 2 & 3 & 4 \\ 10 & 5 & 15 \\ 20 & 5 & 10 \end{vmatrix}$

33. $\begin{vmatrix} -2 & -3 & -4 \\ 5 & -1 & 2 \\ 1 & 2 & 3 \end{vmatrix} = -\begin{vmatrix} 2 & 3 & 4 \\ -5 & 1 & -2 \\ 1 & 2 & 3 \end{vmatrix}$

34. $\begin{vmatrix} 1 & 2 & 3 \\ 4 & 5 & 6 \\ 7 & 8 & 9 \end{vmatrix} = \begin{vmatrix} 5 & 7 & 9 \\ 4 & 5 & 6 \\ 7 & 8 & 9 \end{vmatrix}$

In Exercises 35–38, assume that $\begin{vmatrix} a & b & c \\ d & e & f \\ g & h & i \end{vmatrix} = 3$ *and find the value of each determinant.*

35. $\begin{vmatrix} d & e & f \\ a & b & c \\ -g & -h & -i \end{vmatrix}$

36. $\begin{vmatrix} 5a & 5b & 5c \\ -d & -e & -f \\ 3g & 3h & 3i \end{vmatrix}$

37. $\begin{vmatrix} a+g & b+h & c+i \\ d & e & f \\ g & h & i \end{vmatrix}$

38. $\begin{vmatrix} g & h & i \\ a & b & c \\ d & e & f \end{vmatrix}$

In Exercises 39–50, use Cramer's rule to find the solution of each system, if possible.

39. $\begin{cases} 3x + 2y = 7 \\ 2x - 3y = -4 \end{cases}$

40. $\begin{cases} x - 5y = -6 \\ 3x + 2y = -1 \end{cases}$

41. $\begin{cases} x - y = 3 \\ 3x - 7y = 9 \end{cases}$

42. $\begin{cases} 2x - y = -6 \\ x + y = 0 \end{cases}$

43. $\begin{cases} x + 2y + z = 2 \\ x - y + z = 2 \\ x + y + 3z = 4 \end{cases}$

44. $\begin{cases} x + 2y - z = -1 \\ 2x + y - z = 1 \\ x - 3y - 5z = 17 \end{cases}$

45. $\begin{cases} 2x - y + z = 5 \\ 3x - 3y + 2z = 10 \\ x + 3y + z = 0 \end{cases}$

46. $\begin{cases} x - y - z = 2 \\ x + y + z = 2 \\ -x - y + z = -4 \end{cases}$

47. $\begin{cases} \dfrac{x}{2} + \dfrac{y}{3} + \dfrac{z}{2} = 11 \\ \dfrac{x}{3} + y - \dfrac{z}{6} = 6 \\ \dfrac{x}{2} + \dfrac{y}{6} + z = 16 \end{cases}$

48. $\begin{cases} \dfrac{x}{2} + \dfrac{y}{5} + \dfrac{z}{3} = 17 \\ \dfrac{x}{5} + \dfrac{y}{2} + \dfrac{z}{5} = 32 \\ x + \dfrac{y}{3} + \dfrac{z}{2} = 30 \end{cases}$

49. $\begin{cases} 2p - q + 3r - s = 0 \\ p + q - s = -1 \\ 3p - r = 2 \\ p - 2q + 3s = 7 \end{cases}$

50. $\begin{cases} a + b + c + d = 8 \\ a + b + c + 2d = 7 \\ a + b + 2c + 3d = 3 \\ a + 2b + 3c + 4d = 4 \end{cases}$

In Exercises 51–54, write the equation of the line that passes through the given points.

51. $P(0, 0)$, $Q(4, 6)$

52. $P(2, 3)$, $Q(6, 8)$

53. $P(-2, 3)$, $Q(5, -3)$

54. $P(1, -2)$, $Q(-4, 3)$

In Exercises 55–58, find the area of each triangle with vertices at the given points.

55. $P(0, 0)$, $Q(12, 0)$, $R(12, 5)$

56. $P(0, 0)$, $Q(0, 5)$, $R(12, 5)$

57. $P(2, 3)$, $Q(10, 8)$, $R(0, 20)$

58. $P(1, 1)$, $Q(6, 6)$, $R(2, 10)$

In Exercises 59–61, illustrate each column operation by showing that it is true for the determinant $\begin{vmatrix} a & b \\ c & d \end{vmatrix}$.

59. Interchanging two columns

60. Multiplying each element in a column by k.

61. Adding k times any column of A to another column of A.

62. Use the method of addition to solve $\begin{cases} ax + by = e \\ cx + dy = f \end{cases}$ for y, and thereby show that $y = \dfrac{af - ec}{ad - bc}$.

In Exercises 63–66, expand the determinants and solve for x.

63. $\begin{vmatrix} 3 & x \\ 1 & 2 \end{vmatrix} = \begin{vmatrix} 2 & -1 \\ x & -5 \end{vmatrix}$

64. $\begin{vmatrix} 4 & x^2 \\ 1 & -1 \end{vmatrix} = \begin{vmatrix} x & 4 \\ 2 & 3 \end{vmatrix}$

65. $\begin{vmatrix} 3 & x & 1 \\ x & 0 & -2 \\ 4 & 0 & 1 \end{vmatrix} = \begin{vmatrix} 2 & x \\ x & 4 \end{vmatrix}$

66. $\begin{vmatrix} x & -1 & 2 \\ -2 & x & 3 \\ 4 & -3 & -1 \end{vmatrix} = \begin{vmatrix} 2 & 2 \\ 5 & x \end{vmatrix}$

DISCOVERY AND WRITING In Exercises 67–70, find each determinant. What do you discover?

67. $\begin{vmatrix} 1 & 3 & 4 \\ 0 & 5 & 3 \\ 0 & 0 & 2 \end{vmatrix}$

68. $\begin{vmatrix} 2 & 1 & -2 \\ 0 & 3 & 4 \\ 0 & 0 & -1 \end{vmatrix}$

69. $\begin{vmatrix} 1 & 2 & 4 & 3 \\ 0 & 2 & 2 & 1 \\ 0 & 0 & 3 & 2 \\ 0 & 0 & 0 & 4 \end{vmatrix}$

70. $\begin{vmatrix} 2 & 1 & -2 & 1 \\ 0 & 2 & 2 & -1 \\ 0 & 0 & 3 & 1 \\ 0 & 0 & 0 & 2 \end{vmatrix}$

71. Use an example chosen from 2×2 matrices to show that the determinant of the product of two matrices is the product of the determinants of those two matrices.

72. Find an example among 2×2 matrices to show that the determinant of a sum of two matrices is not equal to the sum of the determinants of those matrices.

73. A determinant is a function that associates a number with every square matrix. Give the domain and the range of that function.

74. Use an example chosen from 2×2 matrices to show that for $n \times n$ matrices A and B, $AB \neq BA$ but $|AB| = |BA|$.

75. If A and B are matrices and $|AB| = 0$, must $|A| = 0$ or $|B| = 0$? Explain.

76. If A and B are matrices and $|AB| = 0$, must $A = 0$ or $B = 0$? Explain.

In Exercises 77–78, use a graphing calculator to evaluate each determinant.

77. $\begin{vmatrix} 2.3 & 5.7 & 6.1 \\ 3.4 & 6.2 & 8.3 \\ 5.8 & 8.2 & 9.2 \end{vmatrix}$

78. $\begin{vmatrix} .32 & -7.4 & -6.7 \\ 3.3 & 5.5 & -.27 \\ -8 & -.13 & 5.47 \end{vmatrix}$

REVIEW *Factor each polynomial, if possible.*

79. $x^2 + 3x - 4$ **80.** $2x^2 - 5x - 12$ **81.** $9x^3 - x$ **82.** $x^2 + 3x + 5$

Add the fractions.

83. $\dfrac{1}{x-2} + \dfrac{2}{2x-1}$ **84.** $\dfrac{2}{x} + \dfrac{3}{x^2} - \dfrac{1}{x-1}$

85. $\dfrac{2}{x^2+1} + \dfrac{1}{x}$ **86.** $\dfrac{2x}{x^2+1} + \dfrac{1}{x}$

7.6 Partial Fractions

■ WHEN THE DENOMINATOR HAS DISTINCT LINEAR FACTORS ■ WHEN THE DENOMINATOR HAS DISTINCT QUADRATIC FACTORS ■ WHEN THE DENOMINATOR HAS REPEATED LINEAR FACTORS ■ WHEN THE DENOMINATOR HAS REPEATED QUADRATIC FACTORS ■ WHEN THE DEGREE OF *P(x)* IS EQUAL TO OR GREATER THAN THE DEGREE OF *Q(x)*

Johann Bernoulli
(1667–1748)

A Swiss mathematician and teacher of Euler. His method of decomposition by partial fractions was a major contribution to the calculus.

In this section, we discuss how to write complicated fractions as sums of simpler fractions. We begin by reviewing how to add fractions. For example, to find the sum

$$\frac{2}{x} + \frac{6}{x+1} + \frac{-1}{(x+1)^2}$$

we write each fraction with an LCD of $x(x+1)^2$, add the fractions by adding their numerators and keeping the common denominator, and simplify.

$$\frac{2}{x} + \frac{6}{x+1} + \frac{-1}{(x+1)^2} = \frac{2(x+1)^2}{x(x+1)^2} + \frac{6x(x+1)}{(x+1)x(x+1)} + \frac{-1x}{(x+1)^2 x}$$

$$= \frac{2x^2 + 4x + 2 + 6x^2 + 6x - x}{x(x+1)^2}$$

$$= \frac{8x^2 + 9x + 2}{x(x+1)^2}$$

To reverse the addition process and write a fraction as the sum of simpler fractions with denominators of smallest possible degree, we must **decompose a fraction into partial fractions.** For example, to decompose the fraction

$$\frac{8x^2 + 9x + 2}{x(x+1)^2}$$

into partial fractions, we will assume that there are constants A, B, and C such that

$$\frac{8x^2 + 9x + 2}{x(x+1)^2} = \frac{A}{x} + \frac{B}{x+1} + \frac{C}{(x+1)^2}$$

After writing the terms on the right-hand side as fractions with an LCD of $x(x + 1)^2$, we add the fractions to get

$$\frac{8x^2 + 9x + 2}{x(x + 1)^2} = \frac{A(x + 1)^2}{x(x + 1)^2} + \frac{Bx(x + 1)}{x(x + 1)(x + 1)} + \frac{Cx}{(x + 1)^2 x}$$

$$= \frac{Ax^2 + 2Ax + A + Bx^2 + Bx + Cx}{x(x + 1)^2}$$

1. $\dfrac{8x^2 + 9x + 2}{x(x + 1)^2} = \dfrac{(A + B)x^2 + (2A + B + C)x + A}{x(x + 1)^2}$

Factor out x^2 from $Ax^2 + Bx^2$.
Factor x from $2Ax + Bx + Cx$.

Since the fractions on the left- and right-hand sides of Equation 1 are equal, the coefficients of their polynomial numerators are equal.

$$\begin{cases} A + B & = 8 \\ 2A + B + C = 9 \\ A & = 2 \end{cases}$$

The coefficients of x^2.
The coefficients of x.
The constants.

We can solve this system to find that $A = 2$, $B = 6$, and $C = -1$. Then we know that

$$\frac{8x^2 + 9x + 2}{x(x + 1)^2} = \frac{2}{x} + \frac{6}{x + 1} + \frac{-1}{(x + 1)^2}$$

The method of partial fractions uses the following theorem.

Polynomial Factorization Theorem
The factorization of any polynomial $Q(x)$ with real coefficients is the product of polynomials of the forms

$$(ax + b)^n \qquad \text{and} \qquad (ax^2 + bx + c)^n$$

where n is a positive integer and $ax^2 + bx + c$ is irreducible over the real numbers.

The example and this theorem illustrate the topic of this section. We begin with an algebraic fraction—like $\frac{P(x)}{Q(x)}$, the quotient of two polynomials—and write it as the sum of two or more fractions with simpler denominators. By the theorem, we know that they will be either first-degree or irreducible second-degree polynomials, or powers of those. We consider each case separately.

We suppose that $P(x)$ and $Q(x)$ have real coefficients, that the degree of $P(x)$ is less than the degree of $Q(x)$, and that the fraction $\frac{P(x)}{Q(x)}$ is in lowest terms.

■ WHEN THE DENOMINATOR HAS DISTINCT LINEAR FACTORS

EXAMPLE 1   Decompose $\dfrac{9x + 2}{(x + 2)(3x - 2)}$ into partial fractions.

Solution   Since each denominator is linear, there are constants A and B such that

$$\frac{9x + 2}{(x + 2)(3x - 2)} = \frac{A}{x + 2} + \frac{B}{3x - 2}$$

After writing the terms on the right-hand side as fractions with an LCD of $(x + 2)(3x - 2)$, we add the fractions to get

$$\frac{9x + 2}{(x + 2)(3x - 2)} = \frac{A(3x - 2)}{(x + 2)(3x - 2)} + \frac{B(x + 2)}{(x + 2)(3x - 2)}$$

$$\frac{9x + 2}{(x + 2)(3x - 2)} = \frac{3Ax - 2A + Bx + 2B}{(x + 2)(3x - 2)}$$

2. $\quad\dfrac{9x + 2}{(x + 2)(3x - 2)} = \dfrac{(3A + B)x - 2A + 2B}{(x + 2)(3x - 2)}$ Factor x from $3Ax + Bx$.

Since the fractions in Equation 2 are equal, the coefficients of their polynomial numerators are equal, and we have

$$\begin{cases} 9 = 3A + B & \text{The coefficients of } x. \\ 2 = -2A + 2B & \text{The constants.} \end{cases}$$

After solving this system to find that $A = 2$ and $B = 3$, we have

$$\frac{9x + 2}{(x + 2)(3x - 2)} = \frac{2}{x + 2} + \frac{3}{3x - 2} \qquad\qquad ■$$

Self Check   Decompose $\dfrac{2x + 2}{x(x + 2)}$ into partial fractions.

Answer $\frac{1}{x} + \frac{1}{x + 2}$

■ WHEN THE DENOMINATOR HAS DISTINCT QUADRATIC FACTORS

If $Q(x)$ has prime quadratic factor of the form $ax^2 + bx + c$, the partial fraction decomposition of $\frac{P(x)}{Q(x)}$ will have a corresponding term of the form

$$\frac{Bx + C}{ax^2 + bx + c}$$

 WARNING!   Be sure that the quadratic factor really is prime—check it with the discriminant. If $b^2 - 4ac$ is negative, the factor is prime.

EXAMPLE 2 Decompose $\dfrac{2x^2 + x + 1}{x^3 + x}$ into partial fractions.

Solution Since the denominator can be written as a product of a linear factor and prime quadratic factor, the partial fractions have the form

$$\frac{2x^2 + x + 1}{x(x^2 + 1)} = \frac{A}{x} + \frac{Bx + C}{x^2 + 1}$$

$$= \frac{A(x^2 + 1)}{x(x^2 + 1)} + \frac{(Bx + C)x}{x(x^2 + 1)}$$

$$= \frac{Ax^2 + A + Bx^2 + Cx}{x(x^2 + 1)}$$

$$= \frac{(A + B)x^2 + Cx + A}{x(x^2 + 1)} \qquad \text{Factor } x^2 \text{ from } Ax^2 + Bx^2.$$

We can equate the corresponding coefficients of the numerators $2x^2 + x + 1$ and $(A + B)x^2 + Cx + A$ to get the system

$$\begin{cases} A + B = 2 & \text{The coefficients of } x^2. \\ C = 1 & \text{The coefficients of } x. \\ A = 1 & \text{The constants} \end{cases}$$

with solutions $A = 1$, $B = 1$, $C = 1$. The partial fraction decomposition is

$$\frac{2x^2 + x + 1}{x(x^2 + 1)} = \frac{A}{x} + \frac{Bx + C}{x^2 + 1}$$

$$= \frac{1}{x} + \frac{1x + 1}{x^2 + 1}$$

$$= \frac{1}{x} + \frac{x + 1}{x^2 + 1}$$

∎

Self Check Decompose $\dfrac{2x^2 + 1}{x^3 + x}$ into partial fractions.

Answer $\dfrac{1}{x} + \dfrac{x}{x^2 + 1}$

■ **WHEN THE DENOMINATOR HAS REPEATED LINEAR FACTORS**

If $Q(x)$ has n linear factors of $ax + b$, then $(ax + b)^n$ is a factor of $Q(x)$. *Each* factor of the form $(ax + b)^n$ generates the following sum of n partial fractions:

$$\frac{A}{ax + b} + \frac{B}{(ax + b)^2} + \frac{C}{(ax + b)^3} + \cdots + \frac{D}{(ax + b)^n}$$

EXAMPLE 3 Decompose $\dfrac{3x^2 - x + 1}{x(x - 1)^2}$ into partial fractions.

Solution Here, each factor in the denominator is linear. The linear factor x appears once, and the linear factor $x - 1$ appears twice. Thus, there are constants A, B, and C such that

$$\frac{3x^2 - x + 1}{x(x - 1)^2} = \frac{A}{x} + \frac{B}{x - 1} + \frac{C}{(x - 1)^2}$$

After writing the terms on the right-hand side as fractions with an LCD of $x(x - 1)^2$, we combine them to get

$$\frac{3x^2 - x + 1}{x(x - 1)^2} = \frac{A(x - 1)^2}{x(x - 1)^2} + \frac{Bx(x - 1)}{x(x - 1)(x - 1)} + \frac{Cx}{(x - 1)^2 x}$$

$$= \frac{Ax^2 - 2Ax + A + Bx^2 - Bx + Cx}{x(x - 1)^2}$$

$$= \frac{(A + B)x^2 + (-2A - B + C)x + A}{x(x - 1)^2} \qquad \begin{array}{l}\text{Factor } x^2 \text{ from } Ax^2 + Bx^2. \\ \text{Factor } x \text{ from } -2Ax - Bx + Cx.\end{array}$$

Since the fractions are equal, the coefficients of the polynomial numerators are equal, and we have

$$\begin{cases} A + B = 3 & \text{The coefficients of } x^2. \\ -2A - B + C = -1 & \text{The coefficients of } x. \\ A = 1 & \text{The constants.} \end{cases}$$

We can solve this system to find that $A = 1$, $B = 2$, and $C = 3$. So, we have

$$\frac{3x^2 - x + 1}{x(x - 1)^2} = \frac{A}{x} + \frac{B}{x - 1} + \frac{C}{(x - 1)^2}$$

$$= \frac{1}{x} + \frac{2}{x - 1} + \frac{3}{(x - 1)^2} \qquad \blacksquare$$

Self Check Decompose $\dfrac{3x^2 + 7x + 1}{x(x + 1)^2}$ into partial fractions.

Answer $\dfrac{1}{x} + \dfrac{2}{x + 1} + \dfrac{3}{(x + 1)^2}$

■ **WHEN THE DENOMINATOR HAS REPEATED QUADRATIC FACTORS**

If $Q(x)$ has n prime factors of $ax^2 + bx + c$, then $(ax^2 + bx + c)^n$ is a factor. Each factor of the form $(ax^2 + bx + c)^n$ generates a sum of n partial fractions of the form

$$\frac{Ax + B}{ax^2 + bx + c} + \frac{Cx + D}{(ax^2 + bx + c)^2} + \cdots + \frac{Ex + F}{(ax^2 + bx + c)^n}$$

EXAMPLE 4   Decompose $\dfrac{3x^2 + 5x + 5}{(x^2 + 1)^2}$ into partial fractions.

Solution   Since the quadratic factor $x^2 + 1$ is used twice, we must find constants A, B, C, and D such that

$$\frac{3x^2 + 5x + 5}{(x^2 + 1)^2} = \frac{Ax + B}{x^2 + 1} + \frac{Cx + D}{(x^2 + 1)^2}$$

We add the fractions on the right-hand side to get

$$\frac{3x^2 + 5x + 5}{(x^2 + 1)^2} = \frac{(Ax + B)(x^2 + 1)}{(x^2 + 1)(x^2 + 1)} + \frac{Cx + D}{(x^2 + 1)^2}$$

$$= \frac{Ax^3 + Ax + Bx^2 + B + Cx + D}{(x^2 + 1)^2}$$

$$= \frac{Ax^3 + Bx^2 + (A + C)x + B + D}{(x^2 + 1)^2} \qquad \text{Factor } x \text{ from } Ax + Cx.$$

If we add the term $0x^3$ to the numerator on the left-hand side, we can equate the corresponding coefficients in the numerators to get

$$A = 0 \qquad\qquad\qquad \text{The coefficients of } x^3.$$
$$D = 3 \qquad\qquad\qquad \text{The coefficients of } x^2.$$
$$A + C = 5 \qquad\qquad\quad \text{The coefficients of } x.$$
$$B + D = 5 \qquad\qquad\quad \text{The constants.}$$

with a solution of $A = 0$, $B = 3$, $C = 5$, and $D = 2$. So we have

$$\frac{3x^2 + 5x + 5}{(x^2 + 1)^2} = \frac{Ax + B}{x^2 + 1} + \frac{Cx + D}{(x^2 + 1)^2}$$

$$= \frac{0x + 3}{x^2 + 1} + \frac{5x + 2}{(x^2 + 1)^2}$$

$$= \frac{3}{x^2 + 1} + \frac{5x + 2}{(x^2 + 1)^2} \qquad\qquad ■$$

Self Check   Decompose $\dfrac{x^3 + 2x + 3}{(x^2 + 2)^2}$ into partial fractions.

Answer $\dfrac{x}{x^2 + 2} + \dfrac{3}{(x^2 + 2)^2}$

■ WHEN THE DEGREE OF $P(x)$ IS EQUAL TO OR GREATER THAN THE DEGREE OF $Q(x)$

When the degree of $P(x)$ is equal to or greater than the degree of $Q(x)$ in the fraction $\frac{P(x)}{Q(x)}$, we do a long division before decomposing the fraction into partial fractions.

EXAMPLE 5   Decompose $\dfrac{x^2 + 4x + 2}{x^2 + x}$ into partial fractions.

Solution   Because the degree of the numerator and denominator are the same, we must do a long division and express the fraction in quotient $+ \frac{\text{remainder}}{\text{divisor}}$ form:

$$\begin{array}{r} 1 \\ x^2 + x\overline{)x^2 + 4x + 2} \\ \underline{x^2 + x} \\ 3x + 2 \end{array}$$

So we can write

3. $\dfrac{x^2 + 4x + 2}{x^2 + x} = 1 + \dfrac{3x + 2}{x^2 + x}$

Because the degree of the numerator of the fraction on the right-hand side of Equation 1 is less than the degree of the denominator, we can find its partial fraction decomposition:

$$\begin{aligned} \frac{3x + 2}{x^2 + x} &= \frac{3x + 2}{x(x + 1)} \\ &= \frac{A}{x} + \frac{B}{x + 1} \\ &= \frac{A(x + 1) + Bx}{x(x + 1)} \\ &= \frac{(A + B)x + A}{x(x + 1)} \end{aligned}$$

We equate the corresponding coefficients in the numerator and solve the resulting system of equations to find that the solution is $A = 2$, $B = 1$. So we have

$$\frac{x^2 + 4x + 2}{x^2 + x} = 1 + \frac{2}{x} + \frac{1}{x + 1} \qquad \blacksquare$$

Self Check   Decompose $\dfrac{x^2 + x - 1}{x^2 - x}$ into partial fractions.

Answer $1 + \dfrac{1}{x} + \dfrac{1}{x - 1}$

EXERCISE 7.6

VOCABULARY AND CONCEPTS   *In Exercises 1–2, fill in the blank to make a true statement.*

1. A polynomial with real coefficients factors as the product of _____ and _____ factors or powers of those.

2. The second-degree factors of a polynomial with real coefficients are _____, which means they don't factor further.

PRACTICE *In Exercises 3–46, decompose each fraction into partial fractions.*

3. $\dfrac{3x - 1}{x(x - 1)}$

4. $\dfrac{4x + 6}{x(x + 2)}$

5. $\dfrac{2x - 15}{x(x - 3)}$

6. $\dfrac{5x + 21}{x(x + 7)}$

7. $\dfrac{3x + 1}{(x + 1)(x - 1)}$

8. $\dfrac{9x - 3}{(x + 1)(x - 2)}$

9. $\dfrac{-2x + 11}{x^2 - x - 6}$

10. $\dfrac{7x + 2}{x^2 + x - 2}$

11. $\dfrac{3x - 23}{x^2 + 2x - 3}$

12. $\dfrac{-x - 17}{x^2 - x - 6}$

13. $\dfrac{9x - 31}{2x^2 - 13x + 15}$

14. $\dfrac{-2x - 6}{3x^2 - 7x + 2}$

15. $\dfrac{4x^2 + 4x - 2}{x(x^2 - 1)}$

16. $\dfrac{x^2 - 6x - 13}{(x + 2)(x^2 - 1)}$

17. $\dfrac{x^2 + x + 3}{x(x^2 + 3)}$

18. $\dfrac{5x^2 + 2x + 2}{x^3 + x}$

19. $\dfrac{3x^2 + 8x + 11}{(x + 1)(x^2 + 2x + 3)}$

20. $\dfrac{-3x^2 + x - 5}{(x + 1)(x^2 + 2)}$

21. $\dfrac{5x^2 + 9x + 3}{x(x + 1)^2}$

22. $\dfrac{2x^2 - 7x + 2}{x(x - 1)^2}$

23. $\dfrac{-2x^2 + x - 2}{x^2(x - 1)}$

24. $\dfrac{x^2 + x + 1}{x^3}$

25. $\dfrac{3x^2 - 13x + 18}{x^3 - 6x^2 + 9x}$

26. $\dfrac{3x^2 + 13x + 20}{x^3 + 4x^2 + 4x}$

27. $\dfrac{x^2 - 2x - 3}{(x - 1)^3}$

28. $\dfrac{x^2 + 8x + 18}{(x + 3)^3}$

29. $\dfrac{x^3 + 4x^2 + 2x + 1}{x^4 + x^3 + x^2}$

30. $\dfrac{3x^3 + 5x^2 + 3x + 1}{x^2(x^2 + x + 1)}$

31. $\dfrac{4x^3 + 5x^2 + 3x + 4}{x^2(x^2 + 1)}$

32. $\dfrac{2x^2 + 1}{x^4 + x^2}$

33. $\dfrac{-x^2 - 3x - 5}{x^3 + x^2 + 2x + 2}$

34. $\dfrac{-2x^3 + 7x^2 + 6}{x^2(x^2 + 2)}$

35. $\dfrac{x^3 + 4x^2 + 3x + 6}{(x^2 + 2)(x^2 + x + 2)}$

36. $\dfrac{x^3 + 3x^2 + 2x + 4}{(x^2 + 1)(x^2 + x + 2)}$

37. $\dfrac{2x^4 + 6x^3 + 20x^2 + 22x + 25}{x(x^2 + 2x + 5)^2}$

38. $\dfrac{x^3 + 3x^2 + 6x + 6}{(x^2 + x + 5)(x^2 + 1)}$

39. $\dfrac{x^3}{x^2 + 3x + 2}$

40. $\dfrac{2x^3 + 6x^2 + 3x + 2}{x^3 + x^2}$

41. $\dfrac{3x^3 + 3x^2 + 6x + 4}{3x^3 + x^2 + 3x + 1}$

42. $\dfrac{x^4 + x^3 + 3x^2 + x + 4}{(x^2 + 1)^2}$

43. $\dfrac{x^3 + 3x^2 + 2x + 1}{x^3 + x^2 + x}$ **44.** $\dfrac{x^4 - x^3 + x^2 - x + 1}{(x^2 + 1)^2}$ **45.** $\dfrac{2x^4 + 2x^3 + 3x^2 - 1}{(x^2 - x)(x^2 + 1)}$ **46.** $\dfrac{x^4 - x^3 + 5x^2 + x + 6}{(x^2 + 3)(x^2 + 1)}$

DISCOVERY AND WRITING

47. Is the polynomial $x^3 + 1$ prime?

48. Decompose $\dfrac{1}{x^3 + 1}$ into partial fractions.

REVIEW *Simplify each radical expression. Use absolute value symbols if necessary.*

49. $\sqrt{8a^3b}$

50. $\sqrt{x^2 + 6x + 9}$

51. $\sqrt{18x^5} + x^2\sqrt{50x} - 5x^2\sqrt{2x}$

52. $\dfrac{\sqrt{x^2y^3}}{\sqrt{xy^5}}$

Solve each equation.

53. $\sqrt{x - 5} = x - 7$

54. $x - 5 = (x - 7)^2$

7.7 Graphs of Linear Inequalities

■ GRAPHS OF INEQUALITIES ■ GRAPHS OF SYSTEMS OF INEQUALITIES

■ GRAPHS OF INEQUALITIES

The **graph of an inequality** in x and y is the graph of all ordered pairs (x, y) that satisfy the inequality. In this section, we will consider graphs of **linear inequalities**—inequalities that can be expressed in a form such as

$$Ax + By < C, \quad Ax + By > C, \quad Ax + By \le C, \quad \text{or} \quad Ax + By \ge C$$

To graph the inequality $y > 3x + 2$, we note that one of the following statements is true:

$$y = 3x + 2, \quad y < 3x + 2, \quad \text{or} \quad y > 3x + 2$$

The graph of $y = 3x + 2$ is a line, as shown in Figure 7-8(a). The graphs of the inequalities are half-planes, one on each side of that line. We can think of the graph of $y = 3x + 2$ as a boundary separating the two half-planes. The graph of $y = 3x + 2$ is drawn with a broken line to show that it is not part of the graph of $y > 3x + 2$.

To find which half-plane is the graph of $y > 3x + 2$, we can substitute the coordinates of any point on one side of the line—say, the origin $(0, 0)$—into the inequality and simplify:

$$y > 3x + 2$$
$$0 > 3(0) + 2$$
$$0 > 2$$

Since $0 > 2$ is false, the coordinates $(0, 0)$ do not satisfy the inequality, and the origin is not in the half-plane that is the graph of $y > 3x + 2$. Thus, the graph is the half-plane located on the other side of the broken line. The graph of the inequality $y > 3x + 2$ is shown in Figure 7-8(b).

$$y = 3x + 2$$

x	y	(x, y)
-1	-1	$(-1, -1)$
0	2	$(0, 2)$
1	5	$(1, 5)$

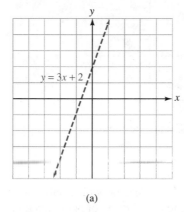

(a)

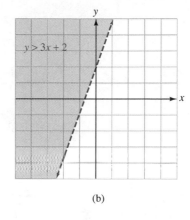

(b)

FIGURE 7-8

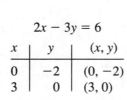

EXAMPLE 1   Graph the inequality $2x - 3y \leq 6$.

Solution   This inequality is the combination of $2x - 3y < 6$ and $2x - 3y = 6$. We start by graphing $2x - 3y = 6$ to establish the boundary that separates two half-planes. This time we draw a solid line, because equality is permitted. See Figure 7-9(a).

$$2x - 3y = 6$$

x	y	(x, y)
0	-2	$(0, -2)$
3	0	$(3, 0)$

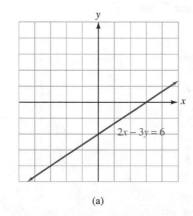

(a)

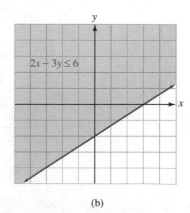

(b)

FIGURE 7-9

To decide which half-plane represents $2x - 3y \leq 6$, we check whether the coordinates of the origin satisfy the inequality:

$$2x - 3y \leq 6$$
$$2(0) - 3(0) \leq 6$$
$$0 \leq 6$$

Because $0 \leq 6$ is true, the origin lies in the graph of $2x - 3y \leq 6$. The graph is shown in Figure 7-9(b). ∎

Self Check   Graph $3x + 2y \leq 6$.

Answer

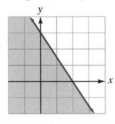

EXAMPLE 2   Graph the inequality $y < 2x$.

Solution   Since the graph of $y = 2x$ is not part of the graph of the inequality, we graph the boundary with a broken line, as in Figure 7-10(a).

To decide which half-plane represents the graph of $y < 2x$, we check whether the coordinates of some fixed point satisfy the inequality. This time, we cannot use the origin as a test point, because the boundary line passes through the origin. So we choose some other point—say, $(3, 1)$—for a test point:

$$y < 2x$$
$$1 < 2(3)$$
$$1 < 6$$

Since $1 < 6$, the point $(3, 1)$ lies in the graph, and we have the graph shown in Figure 7-10(b).

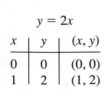

$y = 2x$

x	y	(x, y)
0	0	$(0, 0)$
1	2	$(1, 2)$

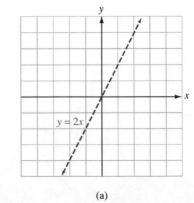

(a)

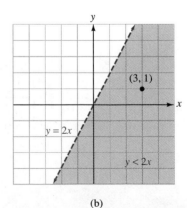

(b)

FIGURE 7-10   ∎

Self Check   Graph $y > 3x$.
   Answer

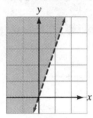

■ GRAPHS OF SYSTEMS OF INEQUALITIES

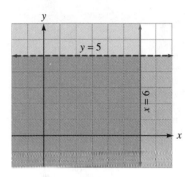

FIGURE 7-11

We now consider systems of inequalities. To graph the solution set of the system

$$\begin{cases} y < 5 \\ x \le 6 \end{cases}$$

we graph each inequality on the same set of coordinate axes, as in Figure 7-11. The graph of the inequality $y < 5$ is the half-plane that lies below the line $y = 5$. The graph of the inequality $x \le 6$ includes the half-plane that lies to the left of the line $x = 6$ together with the line $x = 6$.

The portion of the xy-plane where the two graphs intersect is the graph of the system. Any point that lies in the doubly shaded region has coordinates that satisfy both inequalities in the system.

    EXAMPLE 3   Graph the solution set of $\begin{cases} x + y \le 1 \\ 2x - y > 2 \end{cases}$.

   Solution   On the same set of coordinate axes, we graph each inequality, as in Figure 7-12. The graph of $x + y \le 1$ includes the graph of $x + y = 1$ and all points below it. Because the boundary line is included, we draw it as a solid line.

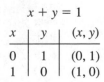

$x + y = 1$		
x	y	(x, y)
0	1	$(0, 1)$
1	0	$(1, 0)$

$2x - y = 2$		
x	y	(x, y)
0	-2	$(0, -2)$
1	0	$(1, 0)$

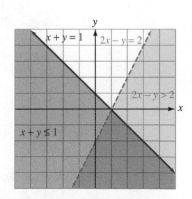

FIGURE 7-12

The graph of $2x - y > 2$ contains only those points below the line graph of $2x - y = 2$. Because the boundary line is not included, we draw it as a broken line.

The area that is shaded twice represents the solution of the system of inequalities. Any point in the doubly shaded region has coordinates that satisfy both inequalities. ∎

Self Check   Graph the solution set of $\begin{cases} x + y < 2 \\ x - 2y \geq 2 \end{cases}$.

Answer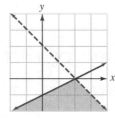

EXAMPLE 4   Graph the solution set of $\begin{cases} y < x^2 \\ y > \dfrac{x^2}{4} - 2 \end{cases}$.

Solution   The graph of $y = x^2$ is a parabola opening upward with vertex at the origin, as shown in Figure 7-13. The points with coordinates that satisfy the inequality are the points below the parabola.

The graph of $y = \frac{x^2}{4} - 2$ is also a parabola opening upward. This time, the points that satisfy the inequality are the points above the parabola. The graph of the solution set is the shaded area between the two parabolas.

$y = x^2$

x	y	(x, y)
0	0	$(0, 0)$
1	1	$(1, 1)$
-1	1	$(-1, 1)$
2	4	$(2, 4)$
-2	4	$(-2, 4)$

$y = \dfrac{x^2}{4} - 2$

x	y	(x, y)
0	-2	$(0, -2)$
2	-1	$(2, -1)$
-2	-1	$(-2, -1)$
4	2	$(4, 2)$
-4	2	$(-4, 2)$

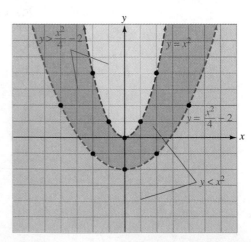

FIGURE 7-13

∎

Self Check   Graph the solution set of $\begin{cases} y \le 4 - x^2 \\ y > \dfrac{x^2}{4} \end{cases}$.

Answer

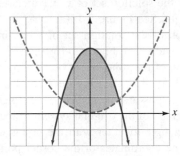

EXAMPLE 5   Graph the solution set of $\begin{cases} x + y \le 4 \\ x - y \le 6. \\ x \ge 0 \end{cases}$

Solution   We graph each inequality, as in Figure 7-14. The graph of $x + y \le 4$ includes the line $x + y = 4$ and all points below it. Because the boundary line is included, we draw it as a solid line. The graph of $x - y \le 6$ contains the line $x - y = 6$ and all points above it.

The graph of the inequality $x \ge 0$ contains the y-axis and all points to the right of the y-axis. The solution of the system of inequalities is the shaded area in the figure.

The coordinates of the corner points of the shaded area are $(0, 4)$, $(0, -6)$, and $(5, -1)$.

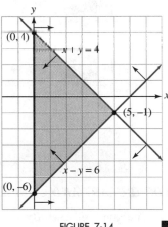

FIGURE 7-14

Self Check   Graph the solution set of $\begin{cases} x + y \le 5 \\ x - 3y \le -3. \\ x \ge 0 \end{cases}$

Answer

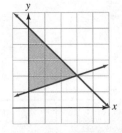

EXAMPLE 6   Graph the solution set of the system $\begin{cases} x \geq 1 \\ y \geq x \\ 4x + 5y < 20 \end{cases}$.

Solution   The graph of the solution set of $x \geq 1$ includes those points on the graph of $x = 1$ and to the right. See Figure 7-15(a).

The graph of the solution set of $y \geq x$ includes those points on the graph of $y = x$ and above it. See Figure 7-15(b).

The graph of the solution set of $4x + 5y < 20$ includes those points below the graph of $4x + 5y = 20$. See Figure 7-15(c).

If these graphs are merged onto a single set of coordinate axes, as in Figure 7-15(d), the graph of the original system of inequalities includes those points within the shaded triangle together with the points on the sides of the triangle drawn as solid lines. The coordinates of the corner points are $\left(1, \frac{16}{5}\right)$, $(1, 1)$, and $\left(\frac{20}{9}, \frac{20}{9}\right)$.

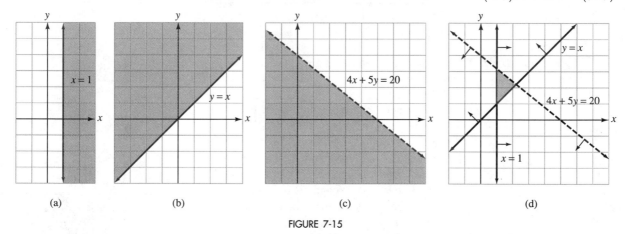

(a)   (b)   (c)   (d)

FIGURE 7-15

EXERCISE 7.7

VOCABULARY AND CONCEPTS   *In Exercises 1–4, fill in the blank to make a true statement.*

1. The graph of $Ax + By = C$ is a line. The graph of $Ax + By \leq C$ is a _____. The line is its _____.

2. The boundary of the graph of $Ax + By < C$ is _____ (included, excluded) from the graph.

3. The origin _____ (is, is not) included in the graph of $3x - 4y > 4$.

4. The origin _____ (is, is not) included in the graph of $4x + 3y \leq 5$.

PRACTICE   *In Exercises 5–16, graph each linear inequality.*

5. $2x + 3y < 12$

6. $4x - 3y > 6$

7. $x < 3$

8. $y > -1$

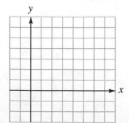
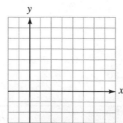
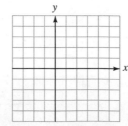
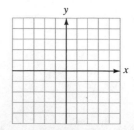

9. $4x - y > 4$

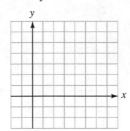

10. $x - 2y < 5$

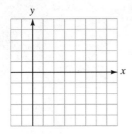

11. $y > 2x$

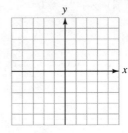

12. $y < 3x$

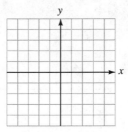

13. $y \le \dfrac{1}{2}x + 1$

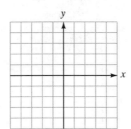

14. $y \ge \dfrac{1}{3}x - 1$

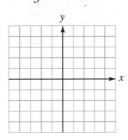

15. $2y \ge 3x - 2$

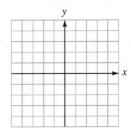

16. $3y \le 2x + 3$

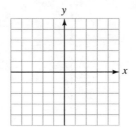

In Exercises 17–36, graph the solution set of each system.

17. $\begin{cases} y < 3 \\ x \ge 2 \end{cases}$

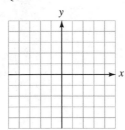

18. $\begin{cases} y \ge -2 \\ x < 0 \end{cases}$

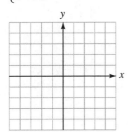

19. $\begin{cases} y \ge 1 \\ x < 2 \end{cases}$

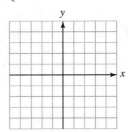

20. $\begin{cases} y \le -1 \\ x > -1 \end{cases}$

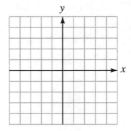

21. $\begin{cases} y \le x - 2 \\ y \ge 2x + 1 \end{cases}$

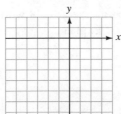

22. $\begin{cases} y < 3x + 2 \\ y < -2x + 3 \end{cases}$

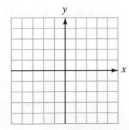

23. $\begin{cases} x + y < 2 \\ x + y \le 1 \end{cases}$

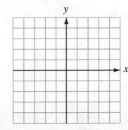

24. $\begin{cases} 3x + 2y \ge 6 \\ x + 3y \le 2 \end{cases}$

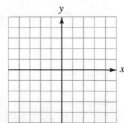

25. $\begin{cases} x + 2y < 3 \\ 2x - 4y < 8 \end{cases}$

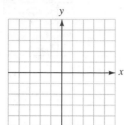

26. $\begin{cases} 3x + y \le 1 \\ -x + 2y \ge 9 \end{cases}$

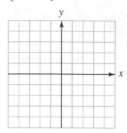

27. $\begin{cases} 2x - 3y \ge 6 \\ 3x + 2y < 6 \end{cases}$

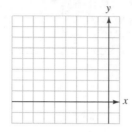

28. $\begin{cases} 4x + 2y \le 6 \\ 2x - 4y \ge 10 \end{cases}$

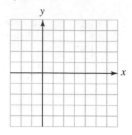

29. $\begin{cases} 2x - y \le 0 \\ x + 2y \le 10 \\ y \ge 0 \end{cases}$

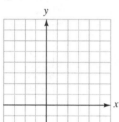

30. $\begin{cases} 3x - 2y \ge 5 \\ 2x + y \ge 8 \\ x \le 5 \end{cases}$

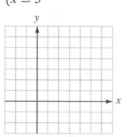

31. $\begin{cases} x - 2y \ge 0 \\ x - y \le 2 \\ x \ge 0 \end{cases}$

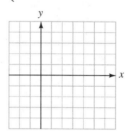

32. $\begin{cases} 2x + 3y \le 6 \\ x - y \ge 4 \\ y \ge -4 \end{cases}$

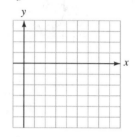

33. $\begin{cases} x + y \le 4 \\ x - y \le 4 \\ x \ge 0 \\ y \ge 0 \end{cases}$

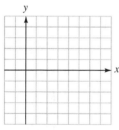

34. $\begin{cases} 2x + 3y \ge 12 \\ 2x - 3y \le 6 \\ x \ge 0 \\ y \le 4 \end{cases}$

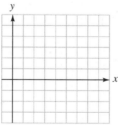

35. $\begin{cases} 3x - 2y \le 6 \\ x + 2y \le 10 \\ x \ge 0 \\ y \ge 0 \end{cases}$

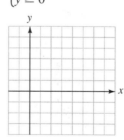

36. $\begin{cases} 3x + 2y \ge 12 \\ 5x - y \le 15 \\ x \ge 0 \\ y \le 4 \end{cases}$

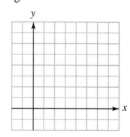

DISCOVERY AND WRITING

37. When graphing a linear inequality, explain how to determine the boundary.

38. When graphing a linear inequality, explain how to decide whether the boundary is included.

39. When graphing a linear inequality, explain how to decide which side of the boundary to shade.

40. Does the method you describe in Exercise 39 work for the inequality $3x - 2y \le 0$? Explain.

REVIEW

41. State the remainder theorem.

42. State the factor theorem.

In Exercises 43–46, fill in the blank to make a true statement.

43. According to Descartes' rule of signs, $53x^4 + 13x^2 - 12 = 0$ has ____ positive solution(s) and ____ negative solution(s).

44. A root of the polynomial equation $P(x) = 0$ is a ____ of the polynomial $P(x)$.

45. If $x^{37} + 3x^{19} - 4$ is divided by $x - 1$, the remainder is __.

46. Is $x - 2$ a factor of $x^4 - 7x - 2$? ____.

7.8 Linear Programming

■ APPLICATIONS OF LINEAR PROGRAMMING

Charles Babbage
(1792–1871)

In 1823, Babbage built a steam-powered digital calculator, which he called a *difference engine.* Thought to be a crackpot by his London neighbors, Babbage was a visionary. His machine embodied principles still used in modern computers.

Systems of inequalities are the basis for an area of applied mathematics called **linear programming.** Linear programming helps answer questions such as "How can a business make as much money as possible?" or "How can a school cafeteria plan a nutritious menu at the least cost?"

In such problems, the solution depends on certain **constraints:** The business has limited resources, and the menu must contain sufficient vitamins, minerals, and so on. Any solution that satisfies the constraints is called a **feasible solution.** In linear programming, the constraints are expressed as a system of linear inequalities, and the quantity that is to be maximized (or minimized) is expressed as a linear function of several variables.

To solve a linear programming problem, we will maximize or minimize a function subject to constraints. For example, suppose that the profit P to be earned by a company is given by the function $P = y + 2x$, subject to the constraints

$$\begin{cases} x + y \geq 1 \\ x - y \leq 1 \\ x - y \geq 0 \\ x \leq 2 \end{cases}$$

We can find the solution to this system of inequalities and then find the coordinates of each corner of the region R shown in Figure 7-16(a). We then write the equation

$$P = y + 2x$$

in the equivalent form

$$y = -2x + P$$

This is the equation for a set of parallel lines, each with a slope of -2 and a y-intercept of P. Many such lines pass through the region R. To decide which of these provides the maximum value of P, we refer to Figure 7-16(b) and locate the line with the greatest y-intercept. Since line l has the greatest y-intercept and crosses region R at the corner $(2, 2)$, the maximum value of P (subject to the given constraints) is

$$P = y + 2x = 2 + 2(2) = 6$$

Thus, the profit P has a maximum value of 6, subject to the given constraints. This profit occurs when $x = 2$ and $y = 2$.

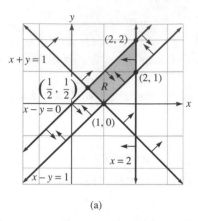

(a)

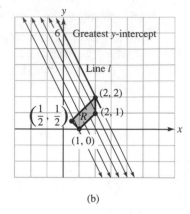

(b)

FIGURE 7-16

In a linear programming problem, the function to be maximized (or minimized) is called the **objective function.** The solution region R for the set of constraints is called the **feasible region.**

The previous example illustrates the following important theorem.

Maximum or Minimum Values of an Objective Function

If a linear function, subject to the constraints of a system of linear inequalities in two variables, attains a maximum or a minimum value, that value will occur at a corner or along an entire edge of the region R that represents the solution of the system.

EXAMPLE 1   If $P = 2x + 3y$, find the maximum value of P subject to the given constraints:

$$\begin{cases} x \geq 0 \\ y \geq 0 \\ x + y \leq 4 \\ 2x + y \leq 6 \end{cases}$$

Solution   We solve the system of inequalities to find the feasible region R shown in Figure 7-17. The coordinates of its corners are $(0, 0)$, $(3, 0)$, $(0, 4)$, and $(2, 2)$.

Because the maximum value of P will occur at a corner of R, we substitute the coordinates of each corner point into the objective function

$$P = 2x + 3y$$

and find the one that gives the maximum value of P.

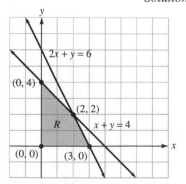

FIGURE 7-17

Point	$P = 2x + 3y$
$(0, 0)$	$P = 2(0) + 3(0) = 0$
$(3, 0)$	$P = 2(3) + 3(0) = 6$
$(2, 2)$	$P = 2(2) + 3(2) = 10$
$(0, 4)$	$P = 2(0) + 3(4) = 12$

The maximum value $P = 12$ occurs when $x = 0$ and $y = 4$.   ■

Self Check   Find the maximum value of $P = 3x + y$, subject to the same constraints as in Example 1.

Answer   9

EXAMPLE 2   If $P = 3x + 2y$, find the minimum value of P subject to the given constraints:

$$\begin{cases} x + y \geq 1 \\ x - y \leq 1 \\ x - y \geq 0 \\ x \leq 2 \end{cases}$$

Solution   We refer to the feasible region shown in Figure 7-18, with corners at $\left(\frac{1}{2}, \frac{1}{2}\right)$, $(2, 2)$, $(2, 1)$, and $(1, 0)$. Because the minimum value of P occurs at a corner point of R, we substitute the coordinates of each corner point into the objective function

$$P = 3x + 2y$$

and find which one gives the minimum value of P.

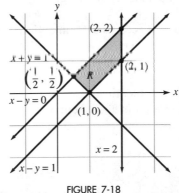

FIGURE 7-18

Point	$P = 3x + 2y$
$\left(\dfrac{1}{2}, \dfrac{1}{2}\right)$	$P = 3\left(\dfrac{1}{2}\right) + 2\left(\dfrac{1}{2}\right) = \dfrac{5}{2}$
$(2, 2)$	$P = 3(2) + 2(2) = 10$
$(2, 1)$	$P = 3(2) + 2(1) = 8$
$(1, 0)$	$P = 3(1) + 2(0) = 3$

The minimum value $P = \frac{5}{2}$ occurs when $x = \frac{1}{2}$ and $y = \frac{1}{2}$.   ■

Self Check   Find the minimum value of $P = x + 3y$, subject to the same constraints as in Example 2.

Answer   1

EXAMPLE 3   An accountant earns $80 for preparing an individual tax return and $150 for preparing a commercial tax return. On average, an individual return requires 3 hours of the accountant's time, of which 1 hour is spent working on a computer. Each commercial return requires 4 hours of time, of which 2 hours is spent working on a computer. If the accountant's time is limited to 240 hours and the available computer time is limited to 100 hours, how should he divide his time between individual and commercial returns to maximize his income?

Solution   Suppose that x represents the number of individual returns completed and y represents the number of commercial returns completed. Because each of the x individual returns brings in $80 and each of the y commercial returns brings in $150, the profit function P is given by the equation

$$P = 80x + 150y$$

The following table provides the information about time requirements:

	Individual return	Commercial return	Time available
The accountant's time	3	4	240
Computer time	1	2	100

The profit is subject to the following constraints:

$$\begin{cases} x \geq 0 \\ y \geq 0 \\ 3x + 4y \leq 240 \\ x + 2y \leq 100 \end{cases}$$

The inequalities $x \geq 0$ and $y \geq 0$ indicate that the number of individual and commercial returns cannot be negative.

The inequality $3x + 4y \leq 240$ is a constraint on the accountant's time. Each of the x individual returns takes 3 hours of his time, and each of the y commercial returns takes 4 hours. The sum of these two amounts of time must be less than or equal to his available time, which is 240 hours.

The inequality $x + 2y \leq 100$ is a constraint on computer time. Each of the x individual returns takes 1 hour of computer time, and each of the y commercial returns takes 2 hours of computer time. The sum of x and $2y$ must be less than or equal to the available time, which is 100 hours.

We graph each of the constraints to find the feasible region R, as in Figure 7-19. The four corners of region R have coordinates of $(0, 0)$, $(80, 0)$, $(40, 30)$, and $(0, 50)$.

We substitute the coordinates of each corner point into the profit function

$$P = 80x + 150y$$

to find the maximum profit P.

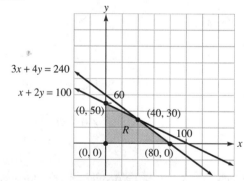

FIGURE 7-19

Point	$P = 80x + 150y$
$(0, 0)$	$P = 80(0) + 150(0) = 0$
$(80, 0)$	$P = 80(80) + 150(0) = 6{,}400$
$(40, 30)$	$P = 80(40) + 150(30) = 7{,}700$
$(0, 50)$	$P = 80(0) + 150(50) = 7{,}500$

The accountant will maximize his income at $7700 if he prepares 40 individual returns and 30 commercial returns. ∎

■ APPLICATIONS OF LINEAR PROGRAMMING

Linear programming was developed during World War II to move huge quantities of men, materials, and supplies as efficiently and economically as possible. Subject to constraints on time, energy, money, or other resources, linear programming provides *optimal* (best) solutions to problems such as finding the best production or transportation schedules, efficiently routing telephone calls, or maximizing corporate profits.

EXAMPLE 4

A diet problem Vitamix and Megamin are two diet supplements. Each Vitamix tablet contains 3 units of calcium, 20 units of vitamin C, and 40 units of iron; it costs $0.50. Each Megamin tablet contains 4 units of calcium, 40 units of vitamin C, and 30 units of iron; it costs $0.60. At least 24 units of calcium, 200 units of vitamin C, and 120 units of iron are required for the daily needs of a particular patient. How many tablets of each supplement should be taken daily for a minimum cost? Find the daily minimum cost.

Solution

We can let x represent the number of Vitamix tablets to be taken daily and y the number of Megamin tablets to be taken daily. We then construct a table displaying the calcium, vitamin C, iron, and cost information.

	Vitamix	**Megamin**	**Amount required**
Calcium	3	4	24
Vitamin C	20	40	200
Iron	40	30	120
Cost	$ 0.50	$ 0.60	

Since we are minimizing cost, the cost function is the objective function

$$C = 0.50x + 0.60y$$

We then write the constraints. Since there are requirements for calcium, vitamin C, and iron, there is a constraint for each. Note that neither x nor y can be negative.

Calcium	$3x + 4y \geq 24$
Vitamin C	$20x + 40y \geq 200$
Iron	$40x + 30y \geq 120$
Nonnegative constraints	$x \geq 0, y \geq 0$

We graph the inequalities to find the feasible region and the coordinates of its corners, as in Figure 7-20.

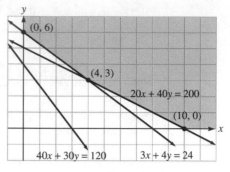

FIGURE 7-20

In this case, the feasible region is not bounded on all sides. The coordinates of the corner points are $(0, 6)$, $(4, 3)$, and $(10, 0)$. To find the minimum cost, we substitute each pair of coordinates into the objective function.

Point	$C = 0.50x + 0.60y$
$(0, 6)$	$C = 0.50(0) + 0.60(6) = 3.60$
$(4, 3)$	$C = 0.50(4) + 0.60(3) = 3.80$
$(10, 0)$	$C = 0.50(10) + 0.60(0) = 5.00$

A minimum cost will occur if no Vitamix and 6 Megamin tablets are taken daily. The minimum daily cost is $3.60.   ■

EXAMPLE 5

A production-schedule problem   A program director must schedule comedy skits and musical numbers for television variety shows. Each skit requires 2 hours of rehearsal time, costs $3000, and brings in $20,000 from sponsors. Each musical number requires 1 hour of rehearsal time, costs $6000, and generates $12,000. If 250 hours are available for rehearsal and $600,000 is budgeted for comedy and music, how many of each type of segment should be produced to maximize income? Find the maximum income.

Solution   We let x represent the number of skits and y the number of musical numbers to be scheduled. We then construct a table with information about rehearsal time, production cost, and income generated.

	Comedy	Musical	Available
Rehearsal time (hours)	2	1	250
Cost (in $1000s)	3	6	600
Generated income (in $1000s)	20	12	

Since each of the x skits generates $20 thousand, the income generated by the skits is $20x$ thousand. The musical numbers produce $12y$ thousand. The objective function to be maximized is

$$V = 20x + 12y$$

Then we write the constraints. Since there are limits on rehearsal time and budget, there is a constraint for each. Note that neither x nor y can be negative.

Constraint on rehearsal time	$2x + y \leq 250$
Constraint on cost	$3x + 6y \leq 600$
Nonnegative constraints	$x \geq 0, y \geq 0$

We graph the inequalities to find the feasible region shown in Figure 7-21 and find the coordinates of each corner point.

The coordinates of the corner points of the feasible region are $(0, 0)$, $(0, 100)$, $(100, 50)$, and $(125, 0)$. To find the maximum income, we substitute each pair of coordinates into the objective function.

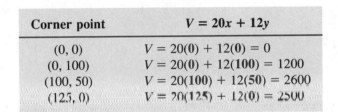

Corner point	$V = 20x + 12y$
$(0, 0)$	$V = 20(0) + 12(0) = 0$
$(0, 100)$	$V = 20(0) + 12(100) = 1200$
$(100, 50)$	$V = 20(100) + 12(50) = 2600$
$(125, 0)$	$V = 20(125) + 12(0) = 2500$

FIGURE 7-21

Maximum income occurs if 100 comedy skits and 50 musical numbers are scheduled. The maximum income will be 2600 thousand dollars, or $2,600,000. ■

EXERCISE 7.8

VOCABULARY AND CONCEPTS   *In Exercises 1–4, fill in the blank to make a true statement.*

1. In linear programming, the inequalities are called _____.

2. A linear programming solution that satisfies the inequalities is called a _____ solution.

3. The function to be maximized or minimized is called the _____ function.

4. A linear function attains a maximum or a minimum value, subject to the constraints, at a _____ or along an _____ of the feasibility region.

PRACTICE   *In Exercises 5–12, maximize P subject to the following constraints.*

5. $P = 2x + 3y$
$$\begin{cases} x \geq 0 \\ y \geq 0 \\ x + y \leq 4 \end{cases}$$

6. $P = 3x + 2y$
$$\begin{cases} x \geq 0 \\ y \geq 0 \\ x + y \leq 4 \end{cases}$$

7. $P = y + \dfrac{1}{2}x$
$$\begin{cases} x \geq 0 \\ y \geq 0 \\ 2y - x \leq 1 \\ y - 2x \geq -2 \end{cases}$$

8. $P = 4y - x$
$$\begin{cases} x \leq 2 \\ y \geq 0 \\ x + y \geq 1 \\ 2y - x \leq 1 \end{cases}$$

9. $P = 2x + y$
$$\begin{cases} y \geq 0 \\ y - x \leq 2 \\ 2x + 3y \leq 6 \\ 3x + y \leq 3 \end{cases}$$

10. $P = x - 2y$
$$\begin{cases} x + y \leq 5 \\ y \leq 3 \\ x \leq 2 \\ x \geq 0 \\ y \geq 0 \end{cases}$$

11. $P = 3x - 2y$
$$\begin{cases} x \leq 1 \\ x \geq -1 \\ y - x \leq 1 \\ x - y \leq 1 \end{cases}$$

12. $P = x - y$
$$\begin{cases} 5x + 4y \leq 20 \\ y \leq 5 \\ x \geq 0 \\ y \geq 0 \end{cases}$$

In Exercises 13–20, minimize P subject to the following constraints.

13. $P = 5x + 12y$
$$\begin{cases} x \geq 0 \\ y \geq 0 \\ x + y \leq 4 \end{cases}$$

14. $P = 3x + 6y$
$$\begin{cases} x \geq 0 \\ y \geq 0 \\ x + y \leq 4 \end{cases}$$

15. $P = 3y + x$
$$\begin{cases} x \geq 0 \\ y \geq 0 \\ 2y - x \leq 1 \\ y - 2x \geq -2 \end{cases}$$

16. $P = 5y + x$
$$\begin{cases} x \leq 2 \\ y \geq 0 \\ x + y \geq 1 \\ 2y - x \leq 1 \end{cases}$$

17. $P = 6x + 2y$
$$\begin{cases} y \geq 0 \\ x \geq 0 \\ 2x + y \geq 3 \\ x + 2y \geq 3 \end{cases}$$

18. $P = 2y - x$
$$\begin{cases} x \geq 0 \\ y \geq 0 \\ x + y \leq 5 \\ x + 2y \geq 2 \end{cases}$$

19. $P = 2x - 2y$
$$\begin{cases} x \leq 1 \\ x \geq -1 \\ y - x \leq 1 \\ x - y \leq 1 \end{cases}$$

20. $P = 2x + 2y$
$$\begin{cases} x \geq 1 \\ y \geq 0 \\ 2x + y \geq 5 \\ x + 2y \geq 4 \end{cases}$$

APPLICATIONS *In Exercises 21–32, write the objective function and the inequalities that describe the constraints in each problem. Graph the feasible region, showing corner points. Then find the maximum or minimum value of the objective function.*

21. Making furniture Two woodworkers, Tom and Carlos, earn a profit of $100 for making a table and $80 for making a chair. On average, Tom must work 3 hours and Carlos 2 hours to make a chair. Tom must work 2 hours and Carlos 6 hours to make a table. If neither wishes to work more than 42 hours per week, how many tables and how many chairs should they make each week to maximize their profit? The information is summarized in Illustration 1.

22. Making crafts Two artists make winter yard ornaments. They receive a profit of $80 for each wooden snowman they make and $64 for each wooden Santa Claus. On average, Nina must work 4 hours and Roberta 2 hours to make a snowman. Nina must work 3 hours and Roberta 4 hours to make a Santa Claus. If neither wishes to work more than 20 hours per week, how many of each should they make each week to maximize their profits? The information is summarized in Illustration 2.

	Table	Chair	Time available
Profit (dollars)	100	80	
Tom's time (hours)	2	3	42
Carlos' time (hours)	6	2	42

ILLUSTRATION 1

	Snowman	Santa Claus	Time available
Profit (dollars)	80	64	
Nina's time (hours)	4	3	20
Roberta's time (hours)	2	4	20

ILLUSTRATION 2

23. Inventories The manager of an electronics store maintains an inventory of IBM-compatible computers and Macintosh computers in the quantities given in Illustration 3. There is room in the store to stock at most 60 computers. The manager receives commissions on the sale of each computer in the amounts given in the table. If the manager can sell all of the computers, how many of each should she stock to maximize her commissions?

Inventory	IBM	Macintosh
Minimum	20	30
Maximum	30	50
Commission	$50	$40

ILLUSTRATION 3

24. Food for a trip Bill packs two foods for his camping trip. The cost and nutrient content of each are shown in Illustration 4, along with Bill's daily requirements. However, Bill does not want to carry more than 60 ounces of food for each day of his trip. How much of each type of food should he carry per day to minimize cost? Find the minimum cost per day.

	Food x	Food y	Bill's daily requirements
Cost	35¢	27¢	
Calories	150	60	3000
Vitamins	21	42	1260

ILLUSTRATION 4

25. Nutrition The owner of Burger Bar purchases hamburger from suppliers A and B, with protein, carbohydrate, and fat contents as given in Illustration 5. He will make a mix that must provide the minimums given in the table. How many ounces from each supplier should he use to minimize fat? Find the minimum fat content.

	Supplier A	Supplier B	Minimum requirements
Protein (units per ounce)	2	8	360
Carbohydrate (units per ounce)	20	40	2000
Fat content	10%	30%	

ILLUSTRATION 5

26. Diet problem A diet requires at least 16 units of vitamin C and at least 34 units of vitamin B complex. Two food supplements are available that provide these nutrients, in the amounts and costs shown in Illustration 6. How much of each should be used to minimize the cost?

Supplement	Vitamin C	Vitamin B	Cost
A	3 units per gram	2 units per gram	3¢ per gram
B	2 units per gram	6 units per gram	4¢ per gram

ILLUSTRATION 6

27. Production problem Machines A and B are available 24 hours a day, 7 days a week (168 hours per week) to manufacture chewing gum and bubble gum. To make a case of chewing gum, machine A must run 2 hours and machine B must run 8 hours. To make a case of bubble gum, machine A must run 4 hours and machine B 2 hours. Machine times and profit information are given in Illustration 7. How many cases of each should be manufactured weekly to maximize profit? Find the maximum profit.

	Times to make 1 case (in hours)		
	Chewing gum	Bubble gum	Maximum time available
Machine A	2	4	168 hr
Machine B	8	2	168 hr
Profit (per case)	$200	$160	

ILLUSTRATION 7

28. Production problem   Manufacturing VCRs and TVs requires the use of the electronics, assembly, and finishing departments of a factory, according to the schedule in Illustration 8. VCRs have a profit of $40 each, and TVs have a profit of $32. How many VCRs and TVs should be manufactured weekly to maximize profit? Find the maximum profit.

	Hours for VCR	Hours for TV	Hours available per week
Electronics	3	4	180
Assembly	2	3	120
Finishing	2	1	60

ILLUSTRATION 8

29. Financial planning   A stockbroker has $200,000 to invest in stocks and bonds. She wants to invest at least $100,000 in stocks and at least $50,000 in bonds. If stocks have an annual yield of 9% and bonds have an annual yield of 7%, how much should she invest in each to maximize her return? Find the maximum return.

30. Agriculture   A small country exports soybeans and flowers. Soybeans require 8 workers per acre, flowers require 12 workers per acre, and 100,000 workers are available. Government contracts demand that there be at least 3 times as many acres planted in soybeans as in flowers. It costs $250 per acre to plant soybeans and $300 per acre to plant flowers, and there is a budget of $3 million. If the profit from soybeans is $1600 per acre and the profit from flowers is $2000 per acre, how many acres of each crop should be planted to maximize profit? Find the maximum profit.

31. Band trip   A high school band trip will require renting buses and trucks to transport no fewer than 100 students and 18 or more large instruments. Each bus can accommodate 40 students plus three large instruments; it costs $350 to rent. Each truck can accommodate 10 students plus 6 large instruments and costs $200 to rent. How many of each type of vehicle should be rented for the cost to be minimum? Find the minimum cost.

32. Making ice cream   An ice cream store sells two new flavors: Fantasy and Excess. Each barrel of Fantasy requires 4 pounds of nuts and 3 pounds of chocolate and has a profit of $500. Each barrel of Excess requires 4 pounds of nuts and 2 pounds of chocolate and has a profit of $400. There are 18 pounds of nuts and 16 pounds of chocolate in stock, and the owner does not want to buy more for this batch. How many barrels of each should be made for a maximum profit? Find the maximum profit.

DISCOVERY AND WRITING

33. Does the objective function attain a maximum at the corners of a region defined by nonlinear inequalities? Attempt to maximize $P(x) = x + y$ on the region
$$\begin{cases} x \geq 0 \\ y \geq 0 \\ y \leq 4 - x^2 \end{cases}$$
and write a paragraph on your findings.

34. Attempt to minimize the objective function of Exercise 33.

REVIEW   *In Exercises 35–36, write each matrix in reduced row echelon form.*

35. $\begin{bmatrix} 1 & 2 & 3 \\ 1 & -2 & 3 \\ 0 & 2 & -3 \\ 2 & 0 & 6 \end{bmatrix}$

36. $\begin{bmatrix} 1 & 3 & -2 & 1 \\ 3 & 9 & -3 & 2 \end{bmatrix}$

37. The matrix in Exercise 36 is the system matrix of a system of equations. Find the general solution of the system.

38. Find the inverse of $\begin{bmatrix} 7 & 2 & 5 \\ 3 & 1 & 2 \\ 3 & 1 & 3 \end{bmatrix}$.

■ ■ ■ ■ ■ ■ ■ ■ ■   **PROBLEMS AND PROJECTS**

1. Solve $\begin{cases} 16x + 21y = 14 \\ 19x + 25y = 14 \end{cases}$ both graphically and algebraically. Which method is easier, and why?

2. Solve $\begin{cases} \dfrac{2}{x} + \dfrac{5}{y} = -1 \\ \dfrac{3}{x} - \dfrac{1}{y} = 7 \end{cases}$.

3. Solve $\begin{cases} 3.7x + 2.9y = 6.61 \\ 4.2x + 3.3y = 7.49 \end{cases}$.

4. Solve $\begin{cases} 3.7x + 2.9y = 6.6 \\ 4.2x + 3.3y = 7.5 \end{cases}$, which is the system of Problem 3, with the constants on the right-hand side rounded to the nearest tenth. Does this solution approximate the solution to Problem 3?

5. Let $A = \begin{bmatrix} 1 & 1 & 0 \\ 0 & 1 & 1 \end{bmatrix}$ and $B = \begin{bmatrix} 1 & -1 \\ 0 & 1 \\ 0 & 0 \end{bmatrix}$.

Verify that the product AB is the identity matrix. Is A invertible? Explain.

6. If $A = \begin{bmatrix} 0 & 0 & 1 \\ 0 & 1 & 0 \\ 1 & 0 & 0 \end{bmatrix}$, find $A^2, A^3, \ldots$.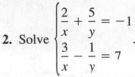

Describe the pattern.

PROJECT 1   A 13-foot and a 15-foot ladder lean against opposite walls of a 12-foot-wide alley, as in Illustration 1. How far above the ground do they cross? (*Hint:* Set up a coordinate system.)

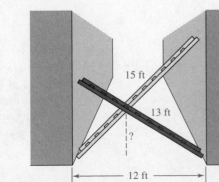

15 ft

13 ft

?

12 ft

ILLUSTRATION 1

(continued)

PROJECT 2

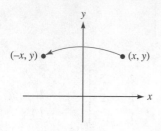

ILLUSTRATION 2

The appearance of motion in computer graphics is accomplished, point-by-point, by matrix multiplication. For example, to reflect the point (x, y) in the y-axis, as in Illustration 2, the graphics software multiplies the point's **coordinate matrix,**
$\begin{bmatrix} x \\ y \end{bmatrix}$, by $A = \begin{bmatrix} -1 & 0 \\ 0 & 1 \end{bmatrix}$, as follows:

$$A\begin{bmatrix} x \\ y \end{bmatrix} = \begin{bmatrix} -1 & 0 \\ 0 & 1 \end{bmatrix}\begin{bmatrix} x \\ y \end{bmatrix} = \begin{bmatrix} -x \\ y \end{bmatrix}$$

1. Describe geometrically the effect on a point (x, y) that results from multiplying its coordinate matrix by A.

a. $A = \begin{bmatrix} 1 & 0 \\ 0 & -1 \end{bmatrix}$ **b.** $A = \begin{bmatrix} -1 & 0 \\ 0 & -1 \end{bmatrix}$ **c.** $A = \begin{bmatrix} 3 & 0 \\ 0 & 3 \end{bmatrix}$

d. $A = \begin{bmatrix} 0 & 1 \\ 1 & 0 \end{bmatrix}$

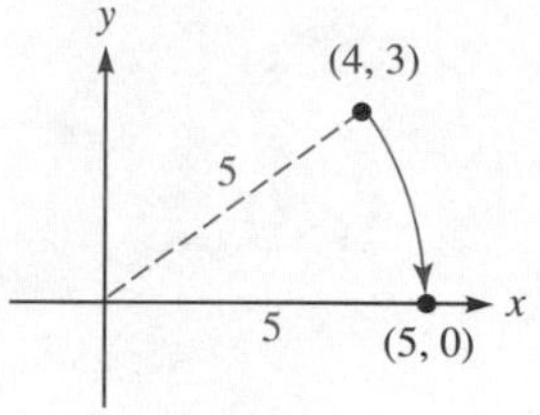

2. To rotate the point $(4, 3)$ clockwise until it reaches the x-axis at $(5, 0)$, the coordinate matrix $\begin{bmatrix} 4 \\ 3 \end{bmatrix}$ is multiplied by a rotation matrix of the form $A = \begin{bmatrix} a & -b \\ b & a \end{bmatrix}$. See Illustration 3. Find the matrix A such that $A\begin{bmatrix} 4 \\ 3 \end{bmatrix} = \begin{bmatrix} 5 \\ 0 \end{bmatrix}$.

ILLUSTRATION 3

PROJECT 3

The temperatures of the edges of a steel plate are shown in Illustration 4. The temperature at each other point is the average of the temperatures of four surrounding points. Find the temperatures of the six points on the plate in the illustration.

$\left(Hint: T_1 = \dfrac{0° + 10° + T_2 + T_3}{4}.\right)$

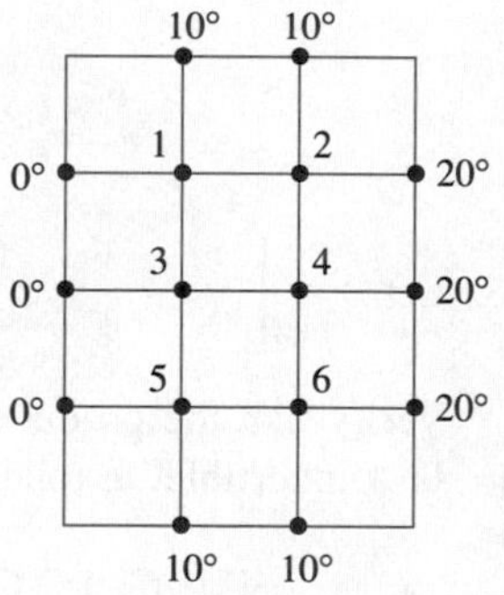

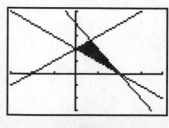

ILLUSTRATION 4

PROJECT 4

Some graphing calculators can display the graphs of linear inequalities. Explore the capabilities of your calculator, and see if you can produce the display in Illustration 5. It is the graph of the system

$$\begin{cases} y \le x + 2 \\ y \ge 2 - x \\ y \le 4 - 2x \end{cases}$$

ILLUSTRATION 5

CHAPTER SUMMARY

CONCEPT   REVIEW EXERCISES

| SECTION 7.1 | *Systems of Linear Equations* |

To solve a system of equations by graphing, find the coordinates of the point where all graphs intersect.

1. Solve each system by graphing.

a. $\begin{cases} 2x - y = -1 \\ x + y = 7 \end{cases}$

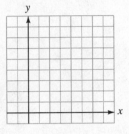

b. $\begin{cases} 5x + 2y = 1 \\ 2x - y = -5 \end{cases}$

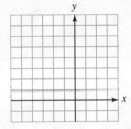

c. $\begin{cases} y = 5x + 7 \\ x = y - 7 \end{cases}$

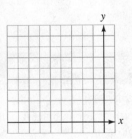

To solve a system by substitution, solve one equation for one variable, and substitute that result into the other equation.

2. Solve by substitution:

a. $\begin{cases} 2y + x = 0 \\ x = y + 3 \end{cases}$

b. $\begin{cases} 2x + y = -3 \\ x - y = 3 \end{cases}$

c. $\begin{cases} \dfrac{x + y}{2} + \dfrac{x - y}{3} = 1 \\ y = 3x - 2 \end{cases}$

To solve a system by addition, multiply one or both equations by suitable constants so that when the results are added, one variable will drop out.

3. Solve by addition:

a. $\begin{cases} x + 5y = 7 \\ 3x + y = -7 \end{cases}$

b. $\begin{cases} 2x + 3y = 11 \\ 3x - 7y = -41 \end{cases}$

c. $\begin{cases} 2(x + y) - x = 0 \\ 3(x + y) + 2y = 1 \end{cases}$

4. Solve by any method:

a. $\begin{cases} 3x + 2y - z = 2 \\ x + y - z = 0 \\ 2x + 3y - z = 1 \end{cases}$

b. $\begin{cases} 5x - y + z = 3 \\ 3x + y + 2z = 2 \\ x + y = 2 \end{cases}$

c. $\begin{cases} 2x - y + z = 1 \\ x - y + 2z = 3 \\ x - y + z = 1 \end{cases}$

5. The buyer for a large department store must order 40 coats, some fake fur and some leather. He is unsure of the expected sales. He can buy 25 fur coats and the rest leather for $9300, or 10 fur coats and the rest leather for $12,600. How much does he pay if he decides to split the order evenly?

6. Adult tickets for the championship game are usually $5, but on Seniors' Day, seniors paid $4. Children's tickets were $2.50. Sales of 1800 tickets netted $7425, and children and seniors accounted for one-half of the tickets sold. How many of each were sold?

SECTION 7.2   *Gaussian Elimination and Matrix Methods*

The three elementary row operations for matrices:

1. In a type 1 row operation, any two rows of a matrix are interchanged.

2. In a type 2 row operation, the elements of any row of a matrix are multiplied by any nonzero constant.

3. In a type 3 row operation, any row of a matrix is altered by adding to it any multiple of another row.

7. Solve by matrix methods, if possible:

a. $\begin{cases} 2x + 5y = 7 \\ 3x - y = 2 \end{cases}$

b. $\begin{cases} x + 3y - z = 8 \\ 2x + y - 2z = 11 \\ x - y + 5z = -8 \end{cases}$

c. $\begin{cases} x + 3y + z = 3 \\ 2x - y + z = -11 \\ 3x + 2y + 3z = 2 \end{cases}$

d. $\begin{cases} x + y + z = 4 \\ 3x - 2y - 2z = -3 \\ 4x - y - z = 0 \end{cases}$

SECTION 7.3	*Matrix Algebra*

Matrices are equal if and only if they are the same size and have the same corresponding entries.

8. Solve for x and y. $\begin{bmatrix} 1 & -4 \\ x & 2 \\ 0 & x+7 \end{bmatrix} = \begin{bmatrix} 1 & x \\ -4 & 2 \\ x+4 & y \end{bmatrix}$

9. Do the matrix operations, if possible:

Two $m \times n$ matrices are added by adding the corresponding elements of those matrices.

a. $\begin{bmatrix} 3 & 2 & 1 \\ 3 & 2 & 1 \end{bmatrix} + \begin{bmatrix} -2 & 1 & 3 \\ 1 & -2 & 1 \end{bmatrix}$

b. $\begin{bmatrix} 2 & 3 & 5 \\ 1 & -2 & 4 \\ 2 & 1 & -2 \end{bmatrix} - \begin{bmatrix} 0 & -2 & 1 \\ 3 & 4 & -2 \\ 6 & -4 & 1 \end{bmatrix}$

Two $m \times n$ matrices are subtracted by the rule $A - B = A + (-B)$.

c. $\begin{bmatrix} 1 & -2 \\ -3 & 1 \end{bmatrix} \begin{bmatrix} 2 & 3 \\ -1 & 2 \end{bmatrix}$

The product AB of the $m \times n$ matrix A and the $n \times p$ matrix B is the $m \times p$ matrix C. The ith-row, jth-column entry of C is found by keeping a running total of the products of the elements in the ith row of A with the corresponding elements in the jth column of B.

d. $\begin{bmatrix} -2 & 3 & 5 \\ 1 & -2 & -3 \end{bmatrix} \begin{bmatrix} 2 & 1 \\ -1 & 2 \\ -2 & 3 \end{bmatrix}$

e. $\begin{bmatrix} 1 & -3 & 2 \end{bmatrix} \begin{bmatrix} 2 \\ 1 \\ 3 \end{bmatrix}$

f. $\begin{bmatrix} 1 \\ 2 \\ 1 \\ 5 \end{bmatrix} \begin{bmatrix} 2 & -1 & 1 & 3 \end{bmatrix}$

g. $\begin{bmatrix} 1 & -5 & 3 \\ 2 & 1 & -1 \end{bmatrix} \begin{bmatrix} 2 \\ -2 \\ 3 \end{bmatrix} \begin{bmatrix} 1 & -1 \\ -1 & 3 \end{bmatrix} \begin{bmatrix} 1 \\ -2 \end{bmatrix}$

h. $\begin{bmatrix} 1 & -3 & 2 \end{bmatrix} \begin{bmatrix} 2 \\ 1 \\ -5 \end{bmatrix} + \begin{bmatrix} 1 & -3 \end{bmatrix} \begin{bmatrix} 2 \\ 5 \end{bmatrix}$

i. $\left(\begin{bmatrix} 1 & -3 \\ 3 & 1 \end{bmatrix} + \begin{bmatrix} -1 & 3 \\ 1 & 1 \end{bmatrix} \right) \begin{bmatrix} 1 \\ -5 \end{bmatrix}$

| **SECTION 7.4** | *Matrix Inversion* |

The inverse of an $n \times n$ matrix A is A^{-1}, where
$$AA^{-1} = A^{-1}A = I.$$

Use elementary row operations to transform $[A\,|\,I]$ into $[I\,|\,A^{-1}]$, where I is an identity matrix. If A cannot be transformed into I, then A is singular.

10. Find the inverse of each matrix, if possible.

a. $\begin{bmatrix} 2 & 3 \\ 3 & 5 \end{bmatrix}$

b. $\begin{bmatrix} 1 & 0 & 0 \\ 2 & 0 & -2 \\ 1 & 2 & 2 \end{bmatrix}$

c. $\begin{bmatrix} 1 & 0 & 8 \\ 3 & 7 & 6 \\ 1 & 2 & 3 \end{bmatrix}$

d. $\begin{bmatrix} 4 & 4 & 1 \\ 1 & 1 & 1 \\ -1 & -1 & 0 \end{bmatrix}$

If A is invertible, then the solution of $AX = B$ is $X = A^{-1}B$.

11. Use the inverse of the coefficient matrix to solve each system of equations.

a. $\begin{cases} 4x - y + 2z = 0 \\ x + y + 2z = 1 \\ x + z = 0 \end{cases}$

b. $\begin{cases} w + 3x + y + 3z = 1 \\ w + 4x + y + 3z = 2 \\ x + y = 1 \\ w + 2x - y + 2z = 1 \end{cases}$

| **SECTION 7.5** | *Determinants* |

The determinant:
$$\det(A) = |A| =$$
$$\begin{vmatrix} a & b \\ c & d \end{vmatrix} = ad - bc$$

The determinant of an $n \times n$ matrix A is the sum of the products of the elements of any row (or column) and the cofactors of those elements.

12. Evaluate each determinant.

a. $\begin{vmatrix} 3 & -2 \\ 1 & -3 \end{vmatrix}$

b. $\begin{vmatrix} 1 & -2 & 3 \\ 2 & -1 & 3 \\ 1 & -1 & 0 \end{vmatrix}$

c. $\begin{vmatrix} 1 & 3 & -1 \\ 1 & 2 & 1 \\ 1 & 0 & 2 \end{vmatrix}$

d. $\begin{vmatrix} 1 & 2 & 3 & 4 \\ -1 & 3 & -3 & 2 \\ 0 & 0 & 0 & -1 \\ 3 & 3 & 4 & 3 \end{vmatrix}$

Cramer's rule:
Form quotients of two determinants. The denominator is the determinant of the coefficient matrix, A. The numerator is the determinant of a modified coefficient matrix; when solving for the ith variable, replace the ith column of A with a column of constants, B.

13. Use Cramer's rule to solve each system.

a. $\begin{cases} x + 3y = -5 \\ -2x + y = -4 \end{cases}$

b. $\begin{cases} x - y + z = -1 \\ 2x - y + 3z = -4 \\ x - 3y + z = -1 \end{cases}$

c. $\begin{cases} x - 3y + z = 7 \\ x + y - 3z = -9 \\ x + y + z = 3 \end{cases}$

d. $\begin{cases} w + x - y + z = 4 \\ 2w + x + z = 4 \\ x + 2y + z = 0 \\ w + y + z = 2 \end{cases}$

Properties of Determinants
If a row (or column) of a matrix is multiplied by a constant k, the value of its determinant is multiplied by k.

If a row (or column) of a matrix is altered by adding to it a multiple of another row (or column), the value of its determinant is unchanged.

14. If $\begin{vmatrix} a & b & c \\ d & e & f \\ g & h & i \end{vmatrix} = 7$, evaluate each determinant.

a. $\begin{vmatrix} 3a & 3b & 3c \\ d & e & f \\ g & h & i \end{vmatrix}$

b. $\begin{vmatrix} a & b & c \\ d+g & e+h & f+i \\ g & h & i \end{vmatrix}$

SECTION 7.6   *Partial Fractions*

The fraction $\frac{P(x)}{Q(x)}$ can be written as the sum of simpler fractions with denominators determined by the prime factors of $Q(x)$.

15. Decompose into partial fractions.

a. $\dfrac{7x + 3}{x^2 + x}$

b. $\dfrac{4x^3 + 3x + x^2 + 2}{x^4 + x^2}$

c. $\dfrac{x^2 + 5}{x^3 + x^2 + 5x}$

d. $\dfrac{x^2 + 1}{(x + 1)^3}$

SECTION 7.7	*Graphs of Linear Inequalities*

Systems of inequalities in two variables can be solved by graphing.

The solution is represented by a plane region with boundaries determined by graphing the inequalities as if they were equations.

16. Solve each system by graphing

a. $\begin{cases} 3x + 2y \le 6 \\ x - y > 3 \end{cases}$

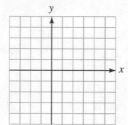

b. $\begin{cases} y \le x^2 + 1 \\ y \ge x^2 - 1 \end{cases}$

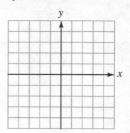

SECTION 7.8	*Linear Programming*

The maximum and the minimum values of a linear function in two variables, subject to the constraints of a system of linear inequalities, are attained at a corner or along an entire edge of the region determined by the system of inequalities.

17. Maximize P subject to the given conditions.

a. $P = 2x + y$
$\begin{cases} x \ge 0 \\ y \ge 0 \\ x + y \le 3 \end{cases}$

b. $P = 2x - 3y$
$\begin{cases} x \ge 0 \\ y \le 3 \\ x - y \le 4 \end{cases}$

c. $P = 3x - y$
$\begin{cases} y \ge 1 \\ y \le 2 \\ y \le 3x + 1 \\ x \le 1 \end{cases}$

d. $P = y - 2x$
$\begin{cases} x + y \ge 1 \\ x \le 1 \\ y \le \dfrac{x}{2} + 2 \\ x + y \le 2 \end{cases}$

18. A company manufactures two fertilizers, x and y. Each 50-pound bag of fertilizer requires three ingredients, which are available in the limited quantities shown in Illustration 1. The profit on each bag of fertilizer x is \$6 and on each bag of y, \$5. How many bags of each product should be produced to maximize the profit?

Ingredient	Number of pounds in fertilizer x	Number of pounds in fertilizer y	Total number of pounds available
Nitrogen	6	10	20,000
Phosphorus	8	6	16,400
Potash	6	4	12,000

ILLUSTRATION 1

■ Chapter 7 Test

In Questions 1–2, solve each system of equations by the graphing method.

1. $\begin{cases} x - 3y = -5 \\ 2x - y = 0 \end{cases}$

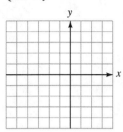

2. $\begin{cases} x = 2y + 5 \\ y = 2x - 4 \end{cases}$

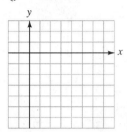

In Questions 3–4, solve each system of equations by the substitution or addition method.

3. $\begin{cases} 3x + y = 0 \\ 2x - 5y = 17 \end{cases}$

4. $\begin{cases} \dfrac{x + y}{2} + x = 7 \\ \dfrac{x - y}{2} - y = -6 \end{cases}$

5. Mixing solutions A chemist has two solutions, one 20% concentration and the other 45% concentration. How many liters of each must she mix to obtain 10 liters of 30% concentration?

6. Wholesale distribution Ace Electronics, Hi-Fi Stereo, and CD World buy a total of 175 cassette tape decks from the same distributor each month. Because CD World buys 25 more tape decks than the other two stores combined, CDW's cost is only $160 per unit. The decks cost Hi-Fi $165 each and Ace $170 each. How many decks does each retailer buy each month if the distributor receives $28,500 each month from the sale of tape decks to the three stores?

In Questions 7–8, write each system of equations as a matrix and solve it by Gaussian elimination.

7. $\begin{cases} 3x - 2y = 4 \\ 2x + 3y = 7 \end{cases}$

8. $\begin{cases} x + 3y - z = 6 \\ 2x - y - 2z = -2 \\ x + 2y + z = 6 \end{cases}$

In Questions 9–10, write each system of equations as a matrix and solve it by Gauss–Jordan elimination. If the system has infinitely many solutions, show how each can be determined.

9. $\begin{cases} x + 2y + 3z = -5 \\ 3x + y - 2z = 7 \\ y - z = 2 \end{cases}$

10. $\begin{cases} x + 2y + z = 0 \\ 3x - 2y - 2z = 7 \\ 4x - z = 7 \end{cases}$

In Questions 11–12, do the operations.

11. $3\begin{bmatrix} 2 & -3 & 5 \\ 0 & 3 & -1 \end{bmatrix} - 5\begin{bmatrix} -2 & 1 & -1 \\ 0 & 3 & 2 \end{bmatrix}$

12. $\begin{bmatrix} 1 & 2 & 3 \end{bmatrix} \begin{bmatrix} 2 & -2 \\ -2 & 2 \\ 1 & 0 \end{bmatrix} \begin{bmatrix} 3 \\ -2 \end{bmatrix}$

In Questions 13–14, find the inverse of each matrix, if possible.

13. $\begin{bmatrix} 5 & 19 \\ 2 & 7 \end{bmatrix}$

14. $\begin{bmatrix} -1 & 3 & -2 \\ 4 & 1 & 4 \\ 0 & 3 & -1 \end{bmatrix}$

In Questions 15–16, use the inverses found in Questions 13–14 to solve each system.

15. $\begin{cases} 5x + 19y = 3 \\ 2x + 7y = 2 \end{cases}$

16. $\begin{cases} -x + 3y - 2z = 1 \\ 4x + y + 4z = 3 \\ 3y - z = -1 \end{cases}$

In Questions 17–18, evaluate each determinant.

17. $\begin{vmatrix} 3 & -5 \\ -3 & 1 \end{vmatrix}$

18. $\begin{vmatrix} 3 & 5 & -1 \\ -2 & 3 & -2 \\ 1 & 5 & -3 \end{vmatrix}$

In Questions 19–20, use Cramer's rule to solve each system.

19. $\begin{cases} 3x - 5y = 3 \\ -3x + y = 2 \end{cases}$

20. $\begin{cases} 3x + 5y - z = 2 \\ -2x + 3y - 2z = 1 \\ x + 5y - 3z = 0 \end{cases}$

In Questions 21–22, decompose each fraction into partial fractions.

21. $\dfrac{5x}{2x^2 - x - 3}$

22. $\dfrac{3x^2 + x + 2}{x^3 + 2x}$

In Questions 23–24, graph the solution set of each system.

23. $\begin{cases} x - 3y \geq 3 \\ x + 3y \leq 3 \end{cases}$

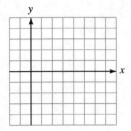

24. $\begin{cases} 3x + 4y \leq 12 \\ 3x + 4y \geq 6 \\ x \geq 0 \\ y \geq 0 \end{cases}$

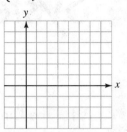

25. Maximize $P = 3x + 2y$ subject to

$$\begin{cases} y \geq 0 \\ x \geq 0 \\ 2x + y \leq 4 \\ y \leq 2 \end{cases}$$

26. Minimize $P = y - x$ subject to

$$\begin{cases} x \geq 0 \\ y \geq 0 \\ x + y \leq 8 \\ 2x + y \geq 2 \end{cases}$$

8
Conic Sections and Quadratic Systems

In this chapter, we will study second-degree equations in x and y. They have many applications in such fields as navigation, astronomy, and satellite communications.

8.1 The Circle and the Parabola

■ THE CIRCLE ■ THE PARABOLA ■ GRAPHING EQUATIONS OF PARABOLAS

Second-degree equations in x and y have the form

$$Ax^2 + Bxy + Cy^2 + Dx + Ey + F = 0$$

where at least one of the coefficients A, B, and C is not zero. The graphs of the equations fall into one of several categories: a single point, a pair of lines, a circle, a parabola, an ellipse, a hyperbola, or no graph at all. These graphs are called **conic sections,** because each is the intersection of a plane and a right-circular cone, as shown in Figure 8-1.

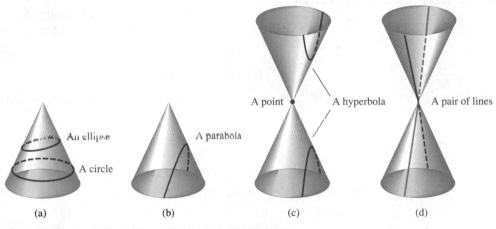

An ellipse

A circle

A parabola

A point   A hyperbola   A pair of lines

(a)   (b)   (c)   (d)

FIGURE 8-1

Although these shapes have been known since the time of the ancient Greeks, it wasn't until the 17th century that René Descartes (1596–1650) and Blaise Pascal (1623–1662) developed the mathematics needed to study them in detail.

■ THE CIRCLE

In Section 3.4, we saw that the graph of any equation that can be written in the form

$$(x - h)^2 + (y - k)^2 = r^2$$

is a circle with radius r and center at point (h, k). This is called *the standard equation of the circle.*

We have also seen that the graph of $x^2 + y^2 = r^2$ is a circle with radius r and center at the origin. Both circles appear in Figure 8-2.

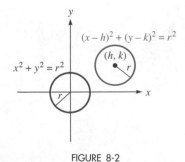

FIGURE 8-2

EXAMPLE 1   Find the equation of the circle shown in Figure 8-3.

Solution   To find the radius of the circle, we substitute the points' coordinates in the distance formula and simplify:

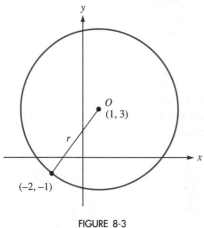

$$r = \sqrt{(x_2 - x_1)^2 + (y_2 - y_1)^2}$$

$$r = \sqrt{(-2 - 1)^2 + (-1 - 3)^2}$$ Substitute 1 for x_1, 3 for y_1, -2 for x_2, and -1 for y_2.

$$= \sqrt{(-3)^2 + (-4)^2}$$
$$= \sqrt{9 + 16}$$
$$= \sqrt{25}$$
$$= 5$$

To find the equation of a circle with radius 5 and center at $(1, 3)$, we substitute 1 for h, 3 for k, and 5 for r in the standard equation of the circle.

$$(x - h)^2 + (y - k)^2 = r^2$$
$$(x - 1)^2 + (y - 3)^2 = 5^2$$

FIGURE 8-3

To write the equation in general form, we square the binomials and simplify.

$$x^2 - 2x + 1 + y^2 - 6y + 9 = 25$$
$$x^2 + y^2 - 2x - 6y - 15 = 0$$ Subtract 25 from both sides and simplify.

∎

Self Check   Find the equation of a circle with center at $O(-2, 1)$ and radius of 4.
Answer $x^2 + y^2 + 4x - 2y - 11 = 0$

The final equation in Example 1 can be written as

$$1x^2 + 0xy + 1y^2 - 2x - 6y - 15 = 0$$

which illustrates that the graph of

$$Ax^2 + Bxy + Cy^2 + Dx + Ey + F = 0$$

is a circle whenever $B = 0$ and $A = C$.

EXAMPLE 2   Graph the circle $2x^2 + 2y^2 + 4x + 3y = 3$.

Solution   To find the coordinates of the center and the radius, we complete the square on x and y. We begin by dividing both sides of the equation by 2 and rearranging terms to get

$$x^2 + 2x + y^2 + \frac{3}{2}y = \frac{3}{2}$$

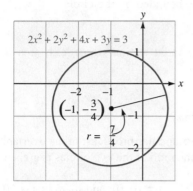

FIGURE 8-4

We add 1 and $\frac{9}{16}$ to both sides to complete the square to get

$$x^2 + 2x + 1 + y^2 + \frac{3}{2}y + \frac{9}{16} = \frac{3}{2} + 1 + \frac{9}{16}$$

Then we factor $x^2 + 2x + 1$ and $y^2 + \frac{3}{2}y + \frac{9}{16}$.

$$(x + 1)^2 + \left(y + \frac{3}{4}\right)^2 = \frac{49}{16}$$

$$[x - (-1)]^2 + \left[y - \left(-\frac{3}{4}\right)\right]^2 = \left(\frac{7}{4}\right)^2$$

From the equation, we see that the coordinates of the center of the circle are $h = -1$ and $k = -\frac{3}{4}$ and that the radius of the circle is $\frac{7}{4}$. The graph is shown in Figure 8-4.

■

Self Check Graph the circle $x^2 + y^2 - 6x - 2y = -9$.

Answer

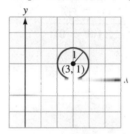

ACCENT ON TECHNOLOGY

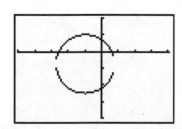

FIGURE 8-5

To graph the result of Example 2 on a graphing calculator, we first solve $(x + 1)^2 + \left(y + \frac{3}{4}\right)^2 = \frac{49}{16}$ for y:

$$\left(y + \frac{3}{4}\right)^2 = \frac{49}{16} - (x + 1)^2 \qquad \text{Subtract } (x + 1)^2 \text{ from both sides.}$$

$$y + \frac{3}{4} = \pm\sqrt{\frac{49}{16} - (x + 1)^2} \qquad \begin{array}{l}\text{Take the square root of both sides.}\\ \text{There are two possibilities, one}\\ \text{positive and one negative.}\end{array}$$

$$y = -\frac{3}{4} \pm \sqrt{\frac{49}{16} - (x + 1)^2} \qquad \text{Subtract } \frac{3}{4} \text{ from both sides.}$$

This expression represents the two functions

$$y = -\frac{3}{4} + \sqrt{\frac{49}{16} - (x + 1)^2} \quad \text{and} \quad y = -\frac{3}{4} - \sqrt{\frac{49}{16} - (x + 1)^2}$$

Their graphs appear in Figure 8-5.

EXAMPLE 3   The effective broadcast area of a radio station is bounded by the circle

$$x^2 + y^2 = 2500$$

where x and y are measured in miles. Another radio station's broadcast area is bounded by the circle

$$(x - 100)^2 + (y - 100)^2 = 900$$

Is there any location that can receive both stations?

Solution   It is possible to receive both stations only if their circular broadcast areas overlap. (See Figure 8-6.) This happens when the sum of the radii of the two circles is greater than the distance between their centers.

The center of the circle $x^2 + y^2 = 2500$ is $(x_1, y_1) = (0, 0)$, and its radius is 50 miles. The center of the circle $(x - 100)^2 + (y - 100)^2 = 900$ is $(x_2, y_2) = (100, 100)$, and its radius is 30 miles.

We can use the distance formula to find the distance between the centers.

$$d = \sqrt{(x_2 - x_1)^2 + (y_2 - y_1)^2}$$
$$d = \sqrt{(\mathbf{100} - \mathbf{0})^2 + (\mathbf{100} - \mathbf{0})^2}$$
$$= \sqrt{100^2 + 100^2}$$
$$= \sqrt{100^2 \cdot 2}$$
$$= 100\sqrt{2}$$
$$\approx 141 \text{ miles}$$

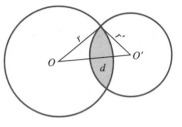

FIGURE 8-6

The sum of the radii of the two circles is $(50 + 30)$ miles, or 80 miles. Since this is less than the distance between their centers (141 miles), there is no location where both stations can be received. ∎

Self Check   In Example 3, if the station at the origin boosted its power to cover the area within

$$x^2 + y^2 = 9000$$

would the coverage overlap?

Answer   no

■ THE PARABOLA

In Chapter 4, we saw that the graphs of some functions are parabolas that open upward or downward. We now discuss parabolas that open to the left and to the right and examine the properties of all parabolas in greater detail.

> **The Parabola**
> A **parabola** is the set of all points in a plane equidistant from a line *l*, called the **directrix,** and a fixed point *F*, called the **focus.**

Sonya Kovalevskaya
(1850–1891)

This talented young Russian woman hoped to study mathematics at the University of Berlin, but strict rules prohibited women from attending lectures. Undaunted, Sonya studied privately with the great mathematician Karl Weierstrauss and published several important papers.

The point on the parabola that is closest to the directrix is called the **vertex,** and the line passing through the vertex and the focus is called the **axis.** In this section, we consider parabolas that have a vertex at point (h, k) and open to the left, to the right, upward, and downward.

The parabola shown in Figure 8-7 opens to the right and passes through its vertex $V(h, k)$ and some point $P(x, y)$. Since each point is equidistant from the focus (point F) and the directrix, we can let $d(DV) = d(VF) = p$, where p is a positive constant.

Because of the geometry of the figure,

$$d(MP) = p - h + x$$

and by the distance formula,

$$d(PF) = \sqrt{[x - (p + h)]^2 + (y - k)^2}$$

FIGURE 8-7

By the definition of the parabola, $d(MP) = d(PF)$. Thus,

$$p - h + x = \sqrt{[x - (p + h)]^2 + (y - k)^2}$$
$$(p - h + x)^2 = [x - (p + h)]^2 + (y - k)^2 \qquad \text{Square both sides.}$$

Finally, we expand the expression on each side of the equation and simplify:

$$p^2 - ph + px - ph + h^2 - hx + px - hx + x^2 = x^2 - 2px - 2hx + p^2 + 2ph + h^2 + (y - k)^2$$
$$-2ph + 2px = -2px + 2ph + (y - k)^2$$
$$4px - 4ph = (y - k)^2$$

1.
$$4p(x - h) = (y - k)^2$$

Equation 1 is one of four **standard equations of a parabola.**

Standard Equation of a Parabola with Vertex at (h, k)

The standard equation of a parabola with vertex at $V(h, k)$ and opening to the right is

$$(y - k)^2 = 4p(x - h)$$

where p is the distance from the vertex to the focus.

If the parabola has its vertex at the origin, both h and k are 0, and we have the following result.

Standard Equation of a Parabola with Vertex at the Origin

The standard equation of a parabola with vertex at the origin and opening to the right is

$$y^2 = 4px$$

where p is the distance from the vertex to the focus.

Equations of parabolas that open to the right, left, upward, and downward are summarized as follows.

Standard Equations of the Parabola

If $p > 0$, then

Parabola opening	Vertex at origin	Vertex at $V(h, k)$
Right	$y^2 = 4px$	$(y - k)^2 = 4p(x - h)$
Left	$y^2 = -4px$	$(y - k)^2 = -4p(x - h)$
Upward	$x^2 = 4py$	$(x - h)^2 = 4p(y - k)$
Downward	$x^2 = -4py$	$(x - h)^2 = -4p(y - k)$

EXAMPLE 4   Find the equation of the parabola with vertex at the origin and focus at $(3, 0)$.

Solution   A sketch of the parabola is shown in Figure 8-8. Because the focus is to the right of the vertex, the parabola opens to the right, and because the vertex is the origin, the standard equation is $y^2 = 4px$. The distance between the focus and the vertex is $p = 3$. We can substitute 3 for p into the standard equation to get

$$y^2 = 4px$$
$$y^2 = 4(\mathbf{3})x$$
$$y^2 = 12x$$

The equation of the parabola is $y^2 = 12x$.

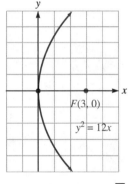

FIGURE 8-8   ∎

Self Check   Find the equation of the parabola with vertex at the origin and focus at $(-3, 0)$.
Answer $y^2 = -12x$

EXAMPLE 5   Find the equation of the parabola that opens upward, has vertex at the point $(4, 5)$, and passes through the point $(0, 7)$.

Solution   Because the parabola opens upward, we use the equation

$$(x - h)^2 = 4p(y - k)$$

Since $(h, k) = (4, 5)$ and the point $(0, 7)$ is on the curve, we can substitute 4 for h, 5 for k, 0 for x, and 7 for y into the standard equation and solve for p.

$$(x - h)^2 = 4p(y - k)$$
$$(0 - 4)^2 = 4p(7 - 5)$$
$$16 = 8p$$
$$2 = p$$

To find the equation of the parabola, we substitute 4 for h, 5 for k, and 2 for p in the standard equation and simplify:

$$(x - h)^2 = 4p(y - k)$$
$$(x - 4)^2 = 4 \cdot 2(y - 5)$$
$$(x - 4)^2 = 8(y - 5)$$

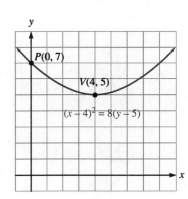

FIGURE 8-9

The graph of the equation appears in Figure 8-9.   ■

Self Check   Find the equation of the parabola that opens upward, has vertex at $(4, 5)$, and passes through $(0, 9)$.

Answer $(x - 4)^2 = 4(y - 5)$

EXAMPLE 6   Find the equations of two parabolas with a vertex at $(2, 4)$ that pass through $(0, 0)$.

Solution   The two parabolas are shown in Figure 8-10.

Part 1: To find the parabola that opens to the left, we use the equation $(y - k)^2 = -4p(x - h)$. Since the curve passes through the point $(x, y) = (0, 0)$ and the vertex $(h, k) = (2, 4)$, we substitute 0 for x, 0 for y, 2 for h, and 4 for k into the standard equation and solve for p:

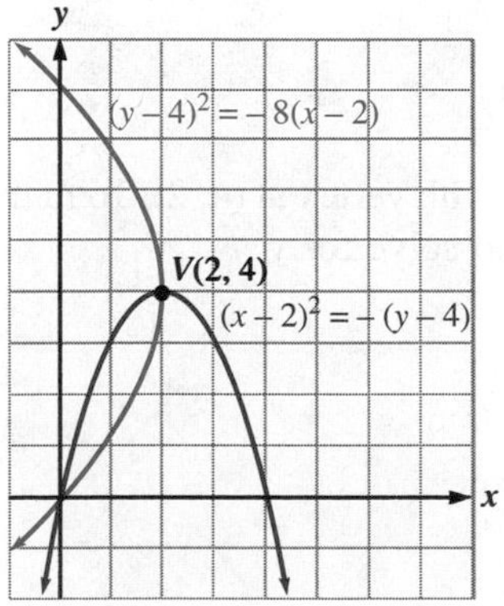

$$(y - k)^2 = -4p(x - h)$$
$$(0 - 4)^2 = -4p(0 - 2)$$
$$16 = 8p$$
$$2 = p$$

Since $h = 2$, $k = 4$, $p = 2$, and the parabola opens to the left, its equation is

$$(y - k)^2 = -4p(x - h)$$
$$(y - 4)^2 = -4(2)(x - 2)$$
$$(y - 4)^2 = -8(x - 2)$$

FIGURE 8-10

Part 2: To find the equation of the parabola that opens downward, we use the equation $(x - h)^2 = -4p(h - k)$ and substitute 2 for h, 4 for k, 0 for x, and 0 for y and solve for p:

$$(x - h)^2 = -4p(y - k)$$
$$(0 - 2)^2 = -4p(0 - 4)$$
$$4 = 16p$$
$$p = \frac{1}{4}$$

Since $h = 2$, $k = 4$, $p = \frac{1}{4}$, and the parabola opens downward, its equation is

$$(x - h)^2 = -4p(y - k)$$
$$(x - 2)^2 = -4\left(\frac{1}{4}\right)(y - 4) \qquad \text{Substitute 2 for } h, \tfrac{1}{4} \text{ for } p, \text{ and 4 for } k.$$
$$(x - 2)^2 = -(y - 4)$$ ∎

■ GRAPHING EQUATIONS OF PARABOLAS

EXAMPLE 7   Find the vertex and y-intercepts of the parabola $y^2 + 8x - 4y = 28$ and graph it.

Solution   We can complete the square on y to write the equation in standard form:

$$y^2 + 8x - 4y = 28$$
$$y^2 - 4y = -8x + 28 \qquad \text{Subtract } 8x \text{ from both sides.}$$
$$y^2 - 4y + 4 = -8x + 28 + 4 \qquad \text{Add 4 to both sides.}$$
2. $$(y - 2)^2 = -8(x - 4) \qquad \text{Factor both sides.}$$

Equation 2 represents a parabola opening to the left with vertex at $(4, 2)$. To find the y-intercepts, we substitute 0 for x in Equation 2 and solve for y.

$$(y - 2)^2 = -8(x - 4)$$
$$(y - 2)^2 = -8(0 - 4) \qquad \text{Substitute 0 for } x.$$
$$y^2 - 4y + 4 = 32 \qquad \text{Remove parentheses.}$$
3. $$y^2 - 4y - 28 = 0$$

We can use the quadratic formula to find that the roots of Equation 3 are $y \approx 7.7$ and $y \approx -3.7$. So, the points with coordinates of approximately $(0, 7.7)$ and $(0, -3.7)$ lie on the graph of the parabola. We can use this information and the knowledge that the graph opens to the left and has a vertex at $(4, 2)$ to draw the graph as shown in Figure 8-11. ∎

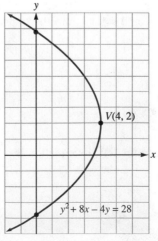

FIGURE 8-11

Self Check   Find the vertex and y-intercepts of the parabola $y^2 - 8x + 6y = 7$. Then graph it.

Answers   vertex $(-2, -3)$; y-intercepts $(0, 1)$, $(0, -7)$

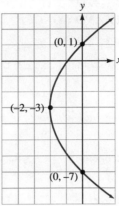

EXAMPLE 8   If a stone is thrown straight up, the equation $s = 128t - 16t^2$ expresses its height in feet t seconds after it was thrown. Find the maximum height reached by the stone.

Solution   The graph of $s = 128t - 16t^2$, expressing the height of the stone t seconds after it was thrown, is the parabola shown in Figure 8-12.

To find the maximum height reached by the stone, we find the x-coordinate k of the vertex of the parabola. To find k, we write the equation of the parabola in standard form by completing the square on t.

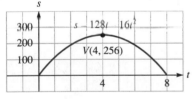

FIGURE 8-12

$$s = 128t - 16t^2$$

$$16t^2 - 128t = -s \qquad \text{Multiply both sides by } -1.$$

$$t^2 - 8t = \frac{-s}{16} \qquad \text{Divide both sides by 16.}$$

$$t^2 - 8t + \mathbf{16} = \frac{-s}{16} + \mathbf{16} \qquad \text{Add 16 to both sides to complete the square.}$$

$$(t - 4)^2 = \frac{-s + 256}{16} \qquad \text{Factor } t^2 - 8t + 16 \text{ and combine like terms.}$$

$$(t - 4)^2 = -\frac{1}{16}(s - 256) \qquad \text{Factor out } -\frac{1}{16}.$$

This equation indicates that the maximum height of 256 feet is reached in 4 seconds.

   WARNING!   The parabola shown in Figure 8-12 is not the path of the stone. The stone goes straight up and straight down.   ■

Self Check   At what time will the stone strike the ground? (*Hint:* find the t-intercept.)

Answer $t = 8$ seconds

EXERCISE 8.1

VOCABULARY AND CONCEPTS *In Exercises 1–4, give the coordinates of the circle's center and its radius.*

1. $(x - 2)^2 + (y + 5)^2 = 9$: center (__, __); radius __

2. $x^2 + y^2 - 36 = 0$: center (__, __); radius __

3. $x^2 + y^2 = 5$: center (__, __); radius ____

4. $2(x - 9)^2 + 2y^2 = 7$: center (__, __); radius ____

In Exercises 5–8, tell whether the parabolic graph of the equation opens upward, downward, to the left, or to the right.

5. $y^2 = -4x$: opens _____

6. $y^2 = 10x$: opens _____

7. $x^2 = -8(y - 3)$: opens _____

8. $(x - 2)^2 = (y + 3)$: opens _____

In Exercises 9–10, fill in the blank to make a true statement.

9. A parabola is the set of all points in a plane equidistant from a line, called the _____, and a fixed point, called the _____.

10. The general form of a second-degree equation in the variables x and y is
Ax^2 _____ $= 0$.

PRACTICE *In Exercises 11–16, write the equation of each circle.*

11.

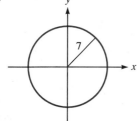

12.

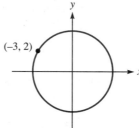

13.

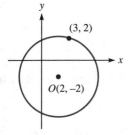

14.

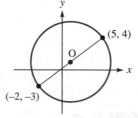

15. Radius of 6; center at the intersection of $3x + y = 1$ and $-2x - 3y = 4$

16. Radius of 8; center at the intersection of $x + 2y = 8$ and $2x - 3y = -5$

In Exercises 17–20, graph each equation.

17. $x^2 + y^2 = 4$

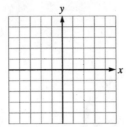

18. $x^2 - 2x + y^2 = 15$

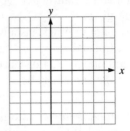

19. $3x^2 + 3y^2 - 12x - 6y = 12$

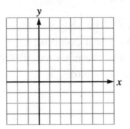

20. $2x^2 + 2y^2 + 4x - 8y + 2 = 0$

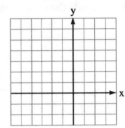

APPLICATIONS

21. Broadcast ranges   A television tower broadcasts a signal with a circular range, as shown in Illustration 1. Can a city 50 miles east and 70 miles north of the tower receive the signal?

22. Warning siren   A tornado warning siren can be heard in the circular range shown in Illustration 2. Can a person 4 miles west and 5 miles south of the siren hear its sound?

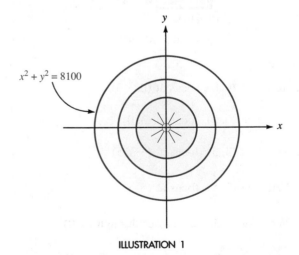

$x^2 + y^2 = 8100$

ILLUSTRATION 1

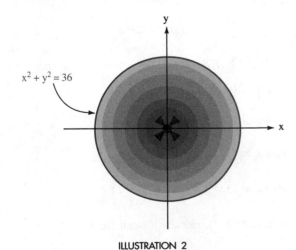

$x^2 + y^2 = 36$

ILLUSTRATION 2

23. Radio translators Some radio stations extend their broadcast range by installing a translator—a remote device that receives the signal and retransmits it. A station with a broadcast range given by $x^2 + y^2 = 1600$, where x and y are in miles, installs a translator with a broadcast area bounded by $x^2 + y^2 - 70y + 600 = 0$. Find the greatest distance from the main transmitter that the signal can be received.

24. Ripples in a pond When a stone is thrown into the center of a pond, the ripples spread out in a circular pattern, moving at a rate of 3 feet per second. If the stone is dropped at the point $(0, 0)$ in Illustration 3, when will the ripple reach the seagull floating at the point $(15, 36)$?

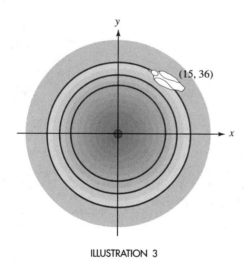

ILLUSTRATION 3

25. Writing equations of circles Find the equation of the outer rim of the circular arch shown in Illustration 4.

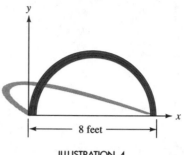

ILLUSTRATION 4

26. Writing equations of circles The shape of the window shown in Illustration 5 is a combination of a rectangle and a semicircle. Find the equation of the circle of which the semicircle is a part.

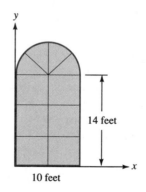

ILLUSTRATION 5

PRACTICE *In Exercises 27–42, find the equation of each parabola.*

27. Vertex at $(0, 0)$; focus at $(0, 3)$

28. Vertex at $(0, 0)$; focus at $(0, -3)$

29. Vertex at $(0, 0)$; focus at $(-3, 0)$

30. Vertex at $(0, 0)$; focus at $(3, 0)$

31. Vertex at $(3, 5)$; focus at $(3, 2)$

32. Vertex at $(3, 5)$; focus at $(-3, 5)$

33. Vertex at $(3, 5)$; focus at $(3, -2)$

34. Vertex at $(3, 5)$; focus at $(6, 5)$

35. Vertex at $(2, 2)$; passes through $(0, 0)$

36. Vertex at $(-2, -2)$; passes through $(0, 0)$

37. Vertex at $(-4, 6)$; passes through $(0, 3)$

38. Vertex at $(-2, 3)$; passes through $(0, -3)$

39. Vertex at $(6, 8)$; passes through $(5, 10)$ and $(5, 6)$

40. Vertex at $(2, 3)$; passes through $\left(1, \dfrac{13}{4}\right)$ and $\left(-1, \dfrac{21}{4}\right)$

41. Vertex at $(3, 1)$; passes through $(4, 3)$ and $(2, 3)$

42. Vertex at $(-4, -2)$; passes through $(-3, 0)$ and $\left(\dfrac{9}{4}, 3\right)$

In Exercises 43–52, change each equation to standard form and graph it.

43. $y = x^2 + 4x + 5$

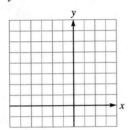

44. $2x^2 - 12x - 7y = 10$

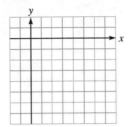

45. $y^2 + 4x - 6y = -1$

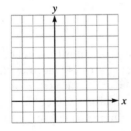

46. $x^2 - 2y - 2x = -7$

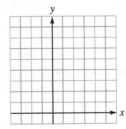

47. $y^2 + 2x - 2y = 5$

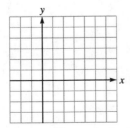

48. $y^2 - 4y = -8x + 20$

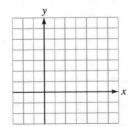

49. $x^2 - 6y + 22 = -4x$

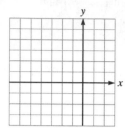

50. $4y^2 - 4y + 16x = 7$

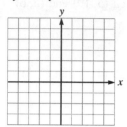

51. $4x^2 - 4x + 32y = 47$

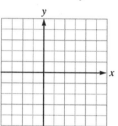

52. $4y^2 - 16x + 17 = 20y$

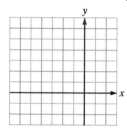

APPLICATIONS

53. Solar furnaces   A parabolic mirror collects rays of the sun and concentrates them at its focus. In Illustration 6, how far from the vertex of the parabolic mirror will it get the hottest? (Assume that all measurements are in feet.)

54. Searchlight reflectors   A parabolic mirror reflects light in a beam when the light source is placed at its focus. In Illustration 7, how far from the vertex of the parabolic reflector should the light source be placed? (Assume that all measurements are in feet.)

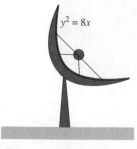

ILLUSTRATION 6

ILLUSTRATION 7

55. Writing equations of parabolas   Derive the equation of the parabolic arch shown in Illustration 8.

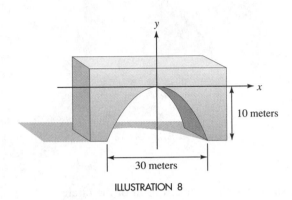

ILLUSTRATION 8

56. Design of satellite antennas   The cross section of the satellite antenna shown in Illustration 9 is a parabola with the pickup at its focus. Find the distance d from the pickup to the center of the dish.

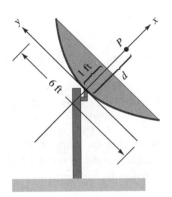

ILLUSTRATION 9

57. Operating a resort   A resort owner plans to build and rent n cabins for d dollars per week. The price, d, that she can charge for each cabin depends on the number of cabins she builds, where $d = -45\left(\frac{n}{32} - \frac{1}{2}\right)$. Find the number of cabins that she should build to maximize her weekly income.

58. Toy rockets   A toy rocket is s feet above the earth at the end of t seconds, where $s = -16t + 80\sqrt{3}t$. Find the maximum height of the rocket.

59. Design of parabolic reflectors   Find the outer diameter (the length $\overline{AB}$) of the parabolic reflector shown in Illustration 10.

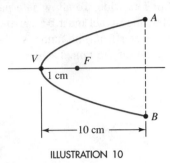

ILLUSTRATION 10

60. Design of suspension bridges   The cable between the towers of the suspension bridge shown in Illustration 11 has the shape of a parabola with a vertex 15 feet above the roadway. Find the equation of the parabola.

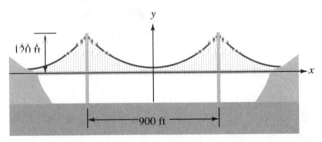

ILLUSTRATION 11

61. Gateway Arch   The Gateway Arch in St. Louis has a shape that approximates a parabola. See Illustration 12. Find the width w of the arch 200 feet above the ground.

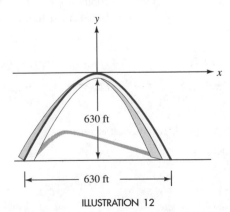

ILLUSTRATION 12

62. Building tunnels A construction firm plans to build a tunnel whose arch is in the shape of a parabola. (See Illustration 13.) The tunnel will span a two-lane highway 8 meters wide. To allow safe passage for most vehicles, the tunnel must be 5 meters high at a distance of 1 meter from the tunnel's edge. Find the maximum height of the tunnel.

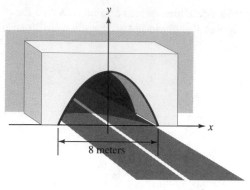

ILLUSTRATION 13

DISCOVERY AND WRITING

63. Show that the standard form of the parabola $(y - 2)^2 = 8(x - 1)$ is a special case of the general form of a second-degree equation in two variables.

64. Show that the standard form of the circle $(x + 2)^2 + (y - 5)^2 = 36$ is a special case of the general form of a second-degree equation in two variables.

In Exercises 65-66, find the equation, in the form $(x - h)^2 + (y - k)^2 = r^2$, of the circle passing through the given points.

65. $(0, 8)$, $(5, 3)$, and $(4, 6)$

66. $(-2, 0)$, $(2, 8)$, and $(5, -1)$

In Exercises 67–68, find the equation of the parabola passing through the given points. Give the equation in the form $y = ax^2 + bx + c$.

67. $(1, 8)$, $(-2, -1)$, and $(2, 15)$

68. $(1, -3)$, $(-2, 12)$, and $(-1, 3)$

69. Ballistics A stone tossed upward is s feet above the earth after t seconds, where $s = -16t^2 + 128t$. Show that the stone's height x seconds after it is thrown is equal to its height x seconds before it hits the ground.

70. Ballistics Show that the stone in Exercise 69 reaches its greatest height in one-half of the time it takes until it strikes the ground.

REVIEW *In Exercises 71–74, find the number that must be added to make a trinomial perfect square.*

71. $x^2 + 4x +$ ___

72. $y^2 - 12y +$ ___

73. $x^2 - 7x +$ ___

74. $y^2 + 11y +$ ___

In Exercises 75–78, solve each equation.

75. $x^2 + 4x = 5$

76. $y^2 - 12y = 13$

77. $x^2 - 7x - 18 = 0$

78. $y^2 + 11y = -18$

8.2 The Ellipse

■ THE ELLIPSE ■ GRAPHING EQUATIONS OF ELLIPSES

A third important conic is the ellipse.

■ THE ELLIPSE

The Ellipse

An **ellipse** is the set of all points P in a plane such that the sum of the distances from P to two other fixed points F and F' is a positive constant.

In the ellipse shown in Figure 8-13, the fixed points F and F' are called **foci** (each is a **focus**), the midpoint of the chord FF' is called the **center,** and the chord VV' is called the **major axis.** Each of the endpoints V and V' of the major axis is called a **vertex.** The chord BB', perpendicular to the major axis and passing through the center C of the ellipse, is called the **minor axis.**

To derive the equation of the ellipse shown in Figure 8-14, we note that point O is the midpoint of chord FF' and let $d(OF) = d(OF') - c$, where $c > 0$. Then the coordinates of point F are $(c, 0)$, and the coordinates of F' are $(-c, 0)$. We also let $P(x, y)$ be any point on the ellipse.

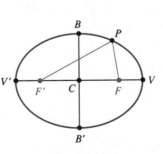

FIGURE 8-13

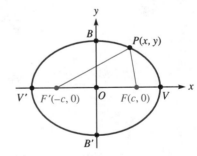

FIGURE 8-14

By the definition of an ellipse, $d(F'P) + d(PF)$ must be a positive constant, which we will call $2a$. Thus,

1. $d(F'P) + d(PF) = 2a$

We can use the distance formula to compute the lengths of $F'P$ and PF,

$$d(F'P) = \sqrt{[x - (-c)]^2 + y^2} \qquad \text{and} \qquad d(PF) = \sqrt{(x - c)^2 + y^2}$$

and substitute these values into Equation 1 to obtain

$$\sqrt{[x - (-c)]^2 + y^2} + \sqrt{(x - c)^2 + y^2} = 2a$$

or

2. $\sqrt{[x + c]^2 + y^2} = 2a - \sqrt{(x - c)^2 + y^2}$ Subtract $\sqrt{(x - c)^2 + y^2}$ from both sides.

We can square both sides of Equation 2 and simplify to get

$$(x + c)^2 + y^2 = 4a^2 - 4a\sqrt{(x - c)^2 + y^2} + [(x - c)^2 + y^2]$$

$$x^2 + 2cx + c^2 + y^2 = 4a^2 - 4a\sqrt{(x - c)^2 + y^2} + x^2 - 2cx + c^2 + y^2$$

$$4cx = 4a^2 - 4a\sqrt{(x - c)^2 + y^2}$$

$$cx = a^2 - a\sqrt{(x - c)^2 + y^2}$$

$$cx - a^2 = -a\sqrt{(x - c)^2 + y^2}$$

We square both sides again and simplify to get

$$c^2x^2 - 2a^2cx + a^4 = a^2[(x - c)^2 + y^2]$$

$$c^2x^2 - 2a^2cx + a^4 = a^2(x^2 - 2cx + c^2 + y^2)$$

$$c^2x^2 - 2a^2cx + a^4 = a^2x^2 - 2a^2cx + a^2c^2 + a^2y^2$$

$$c^2x^2 + a^4 = a^2x^2 + a^2c^2 + a^2y^2$$

$$a^4 - a^2c^2 = a^2x^2 - c^2x^2 + a^2y^2$$

3. $\quad a^2(a^2 - c^2) = (a^2 - c^2)x^2 + a^2y^2$

Because the shortest distance between two points is a line segment, $d(F'P) + d(PF) > d(F'F)$. Therefore, $2a > 2c$. Thus, $a > c$ and $a^2 - c^2$ is a positive number, which we will call b^2. Letting $b^2 = a^2 - c^2$ and substituting into Equation 3, we have

4. $\quad a^2b^2 = b^2x^2 + a^2y^2$

Dividing both sides of Equation 4 by a^2b^2 gives the equation

$$\frac{x^2}{a^2} + \frac{y^2}{b^2} = 1 \quad \text{where } a > b > 0$$

To find the coordinates of the vertices V and V', we substitute 0 for y and solve for x:

$$\frac{x^2}{a^2} + \frac{y^2}{b^2} = 1$$

$$\frac{x^2}{a^2} + \frac{0^2}{b^2} = 1$$

$$\frac{x^2}{a^2} = 1$$

$$x^2 = a^2$$

$$x = a \quad \text{or} \quad x = -a$$

Since the coordinates of V are $(a, 0)$ and the coordinates of V' are $(-a, 0)$, a is the distance between the center of the ellipse and either of its vertices. Thus, the center of the ellipse is the midpoint of the major axis.

To find the coordinates of B and B', we substitute 0 for x and solve for y:

$$\frac{x^2}{a^2} + \frac{y^2}{b^2} = 1$$

$$\frac{0^2}{a^2} + \frac{y^2}{b^2} = 1$$

$$y^2 = b^2$$

$$y = b \quad \text{or} \quad y = -b$$

Since the coordinates of B are $(0, b)$ and the coordinates of B' are $(0, -b)$, the distance between the center of the ellipse and either endpoint of the minor axis is b. We have the following results.

The Ellipse: Major Axis on x-Axis, Center at (0, 0)

The standard equation of an ellipse with center at the origin and major axis on the x-axis is

$$\frac{x^2}{a^2} + \frac{y^2}{b^2} = 1 \qquad \text{where } a > b > 0$$

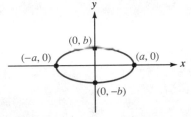

Vertices: $\qquad V(a, 0)$ and $V'(-a, 0)$
Ends of the minor axis: $B(0, b)$ and $B'(0, -b)$

The Ellipse: Major Axis on y-Axis, Center at (0, 0)

The standard equation of an ellipse with center at the origin and major axis on the y-axis is

$$\frac{y^2}{a^2} + \frac{x^2}{b^2} = 1 \qquad \text{where } a > b > 0$$

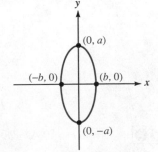

Vertices: $\qquad V(0, a)$ and $V'(0, -a)$
Ends of the minor axis: $B(b, 0)$ and $B'(-b, 0)$

To translate the ellipse to a new position centered at the point (h, k) instead of the origin, we replace x and y in the equations with $x - h$ and $y - k$, respectively, to get these results:

The Ellipse: Major Axis Horizontal, Center at (h, k)

The standard equation of an ellipse with center at the origin and major axis horizontal is

$$\frac{(x - h)^2}{a^2} + \frac{(y - k)^2}{b^2} = 1 \qquad \text{where } a > b > 0$$

Vertices: $V(a + h, k)$ and $V'(-a + h, k)$
Ends of the minor axis: $B(h, b + k)$ and $B'(h, -b + k)$

The Ellipse: Major Axis Vertical, Center at (h, k)

The standard equation of an ellipse with center at the origin and major axis vertical is

$$\frac{(y - k)^2}{a^2} + \frac{(x - h)^2}{b^2} = 1 \qquad \text{where } a > b > 0$$

Vertices: $V(h, a + k)$ and $V'(h, -a + k)$
Ends of the minor axis: $B(b + h, k)$ and $B'(-b + h, k)$

In all cases, the length of the major axis is $2a$, and the length of the minor axis is $2b$.

EXAMPLE 1   Find the equation of the ellipse with center at the origin, major axis of length 6 units located on the x-axis, and minor axis of length 4 units.

Solution   Since the center is the origin and the length of the major axis is 6, $a = 3$ and the coordinates of the vertices are $(3, 0)$ and $(-3, 0)$, as shown in Figure 8-15.

Since the length of the minor axis is 4, $b = 2$ and the coordinates of B and B' are $(0, 2)$ and $(0, -2)$. To find the equation, we substitute 3 for a and 2 for b in the standard equation of an ellipse with center at the origin and major axis on the x-axis.

$$\frac{x^2}{a^2} + \frac{y^2}{b^2} = 1$$

$$\frac{x^2}{3^2} + \frac{y^2}{2^2} = 1$$

$$\frac{x^2}{9} + \frac{y^2}{4} = 1$$

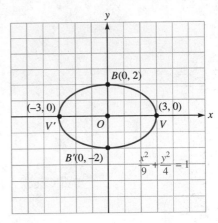

FIGURE 8-15 ∎

Self Check   Find the equation of the ellipse with center at the origin, major axis of length 10 on the x-axis, and minor axis of length 8.

Answer $\dfrac{x^2}{25} + \dfrac{y^2}{16} = 1$

EXAMPLE 2   Find the equation of the ellipse with focus at $(0, 3)$ and vertices V and V' at $(3, 3)$ and $(-5, 3)$.

Solution   Since the midpoint of the major axis is the center of the ellipse, the coordinates of the center are $(-1, 3)$, as in Figure 8-16. Because the major axis is parallel to the x-axis, the standard equation to use is

$$\frac{(x - h)^2}{a^2} + \frac{(y - k)^2}{b^2} = 1$$

where $a > b > 0$

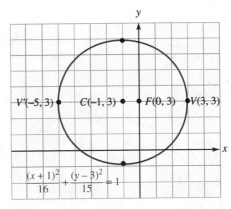

FIGURE 8-16

From Figure 8-16, we see that the distance between the center of the ellipse and either vertex is $a = 4$. We also see that the distance between the focus and the center is $c = 1$.

Since $b^2 = a^2 - c^2$ in an ellipse, we can substitute 4 for a and 1 for c and find b^2:

$$b^2 = a^2 - c^2$$
$$= 4^2 - 1^2$$
$$= 15$$

To find the equation of the ellipse, we substitute -1 for h, 3 for k, 16 for a^2, and 15 for b^2 in the standard equation and simplify:

$$\frac{(x - h)^2}{a^2} + \frac{(y - k)^2}{b^2} = 1$$

$$\frac{[x - (-1)]^2}{16} + \frac{(y - 3)^2}{15} = 1$$

$$\frac{(x + 1)^2}{16} + \frac{(y - 3)^2}{15} = 1 \qquad \blacksquare$$

Self Check   Find the equation of the ellipse with focus at $(3, 1)$ and vertices at $V(5, 1)$ and $V'(-5, 1)$.

Answer $\dfrac{x^2}{25} + \dfrac{(y - 1)^2}{16} = 1$

EXAMPLE 3   The orbit of the earth is approximately an ellipse, with the sun at one focus. The ratio of c to a (called the **eccentricity** of the ellipse) is about $\frac{1}{62}$, and the length of the major axis is approximately 186,000,000 miles. How close does the earth get to the sun?

Solution   We will assume that the ellipse has its center at the origin and vertices V' and V at $(-93,000,000, 0)$ and $(93,000,000, 0)$, as shown in Figure 8-17. Because the eccentricity $\frac{c}{a}$ is given to be $\frac{1}{62}$ and $a = 93,000,000$, we have

$$\frac{c}{a} = \frac{1}{62}$$

$$c = \frac{1}{62}a$$

$$c = \frac{1}{62}(93,000,000)$$

$$= 1,500,000$$

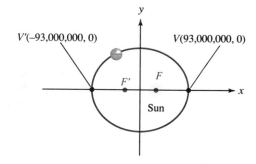

FIGURE 8-17

The distance $d(FV)$ is the shortest distance between the earth and the sun. (You'll be asked to prove this in the exercises.) Thus,

$$d(FV) = a - c = 93,000,000 - 1,500,000 = 91,500,000 \text{ miles}$$

The earth's point of closest approach to the sun (called the **perihelion**) is approximately 91.5 million miles. $\qquad \blacksquare$

We can use the eccentricity of an ellipse to judge its shape. If the eccentricity is close to 1, the ellipse is relatively flat, as in Figure 8-18(a). If the eccentricity is close to 0, the ellipse more circular, as in Figure 8-18(b). Since the eccentricity of the earth's orbit is $\frac{1}{62}$, the earth's orbit is almost a circle.

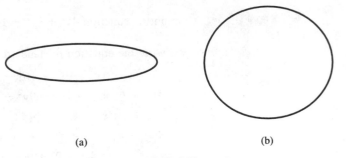

(a)   (b)

FIGURE 8-18

■ GRAPHING EQUATIONS OF ELLIPSES

EXAMPLE 4 Graph the ellipse $\dfrac{(x+2)^2}{4} + \dfrac{(y-2)^2}{9} = 1$.

Solution The center is at $(-2, 2)$, and the major axis is parallel to the y-axis. Because $a = 3$, the vertices are 3 units above and below the center at points $(-2, 5)$ and $(-2, -1)$. Because $b = 2$, the endpoints of the minor axis are 2 units to the right and left of the center at points $(0, 2)$ and $(-4, 2)$. Using these points, we can sketch the ellipse shown in Figure 8-19.

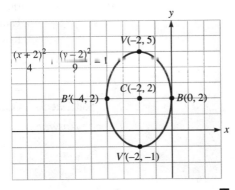

FIGURE 8-19 ■

Self Check Graph the ellipse $\dfrac{(x-2)^2}{9} + \dfrac{(y+1)^2}{25} = 1$.

Answer

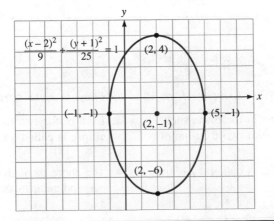

EXAMPLE 5 Graph the equation $4x^2 + 9y^2 - 16x - 18y = 11$.

Solution We write the equation in standard form by completing the square on x and y:

$$4x^2 + 9y^2 - 16x - 18y = 11$$
$$4x^2 - 16x + 9y^2 - 18y = 11$$
$$4(x^2 - 4x) + 9(y^2 - 2y) = 11$$
$$4(x^2 - 4x + 4) + 9(y^2 - 2y + 1) = 11 + 16\ + 9$$
$$4(x - 2) + 9(y - 1)^2 = 36$$
$$\frac{(x - 2)^2}{9} + \frac{(y - 1)^2}{4} = 1$$

We can now see that the graph is an ellipse with center at $(2, 1)$ and major axis parallel to the x-axis. Because $a = 3$, the vertices are at $(-1, 1)$ and $(5, 1)$. Because $b = 2$, the endpoints of the minor axis are at $(2, -1)$ and $(2, 3)$. Using these points, we can sketch the ellipse shown in Figure 8-20.

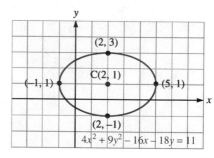

FIGURE 8-20 ∎

Self Check Graph the equation $4x^2 + 9y^2 - 8x + 36y = -4$.

Answer

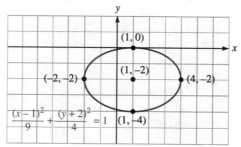

EXERCISE 8.2

VOCABULARY AND CONCEPTS *In Exercises 1–6, fill in the blank to make a true statement.*

1. An ellipse is the set of all points in the plane such that the _____ of the distances from two fixed points is a positive _____.

2. Each of the two fixed points in the definition of an ellipse is called a _____ of the ellipse.

3. The chord that joins the _____ is called the major axis of the ellipse.

4. The chord through the center of an ellipse and perpendicular to the major axis is called the _____ axis.

5. In the ellipse $\frac{x^2}{a^2} + \frac{y^2}{b^2} = 1$ $(a > b)$, the vertices are $V(\underline{\quad}, \underline{\quad})$ and $V'(\underline{\quad}, \underline{\quad})$

6. In an ellipse, the relation between a, b, and c is $\underline{\hspace{3cm}}$.

PRACTICE   *In Exercises 7–12, write the equation of the ellipse that has its center at the origin.*

7. Focus at $(3, 0)$; vertex at $(5, 0)$

8. Focus at $(0, 4)$; vertex at $(0, 7)$

9. Focus at $(0, 1)$; $\frac{4}{3}$ is one-half the length of the minor axis.

10. Focus at $(1, 0)$; $\frac{4}{3}$ is one-half the length of the minor axis.

11. Focus at $(0, 3)$; major axis equal to 8

12. Focus at $(5, 0)$; major axis equal to 12

In Exercises 9–22, write the equation of each ellipse.

13. Center at $(3, 4)$; $a = 3$, $b = 2$; major axis parallel to the y-axis

14. Center at $(3, 4)$; passes through $(3, 10)$ and $(3, -2)$; $b = 2$

15. Center at $(3, 4)$; $a = 3$, $b = 2$; major axis parallel to the x-axis

16. Center at $(3, 4)$; passes through $(8, 4)$ and $(-2, 4)$; $b = 2$

17. Foci at $(-2, 4)$ and $(8, 4)$; $b = 4$

18. Foci at $(8, 5)$ and $(4, 5)$; $b = 3$

19. Vertex at $(6, 4)$; foci at $(-4, 4)$ and $(4, 4)$

20. Center at $(-4, 5)$; $\frac{c}{a} = \frac{1}{3}$; vertex at $(-4, -1)$

21. Foci at $(6, 0)$ and $(-6, 0)$; $\frac{c}{a} = \frac{3}{5}$

22. Vertices at $(2, 0)$ and $(-2, 0)$; $\frac{2b^2}{a} = 2$

In Exercises 23–30, graph each ellipse.

23. $\frac{x^2}{25} + \frac{y^2}{49} = 1$

24. $4x^2 + y^2 = 4$

25. $\frac{x^2}{16} + \frac{(y + 2)^2}{36} = 1$

26. $(x - 1)^2 + \frac{4y^2}{25} = 4$

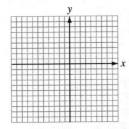

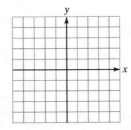

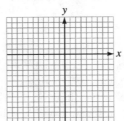

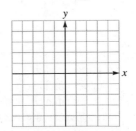

27. $x^2 + 4y^2 - 4x + 8y + 4 = 0$

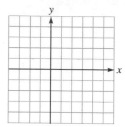

28. $x^2 + 4y^2 - 2x - 16y = -13$

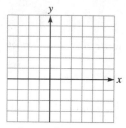

29. $16x^2 + 25y^2 - 160x - 200y + 400 = 0$

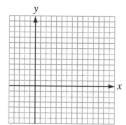

30. $3x^2 + 2y^2 + 7x - 6y = -1$

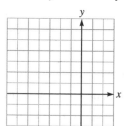

APPLICATIONS

31. Astronomy   The moon has an orbit that is an ellipse, with the earth at one focus. If the major axis of the orbit is 378,000 miles and the ratio of c to a is approximately $\frac{11}{200}$, how far does the moon get from the earth? (This farthest point in an orbit is called the **apogee.**)

32. Equation of an arch   An arch is a semiellipse 10 meters wide and 5 meters high. Write the equation of the ellipse if the ellipse is centered at the origin.

33. Design of a track   A track is built in the shape of an ellipse with a maximum length of 100 meters and a maximum width of 60 meters. Write the equation of the ellipse and find its **focal width.** That is, find the length of a chord that is perpendicular to the major axis and passes through either focus of the ellipse.

34. Whispering galleries   Any sound from one focus of an ellipse reflects off the ellipse directly back to the other focus. This property explains whispering galleries such as Statuary Hall in Washington, DC. The whispering gallery shown in Illustration 1 has the shape of a semiellipse. Find the distance sound travels as it leaves focus F and returns to focus F'.

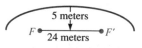

ILLUSTRATION 1

35. Finding the width of a mirror   The oval mirror shown in Illustration 2 is in the shape of an ellipse. Find the width of the mirror 12 inches above its base.

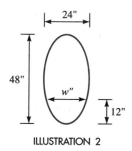

ILLUSTRATION 2

36. Finding the height of a window   The window shown in Illustration 3 has the shape of an ellipse. Find the height of the window 20 inches from one end.

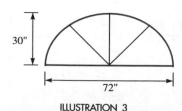

ILLUSTRATION 3

DISCOVERY AND WRITING

37. If F is a focus of the ellipse shown in Illustration 4 and B is an endpoint of the minor axis, use the distance formula to prove that the length of segment FB is a. (*Hint:* In an ellipse, $a^2 - c^2 = b^2$.)

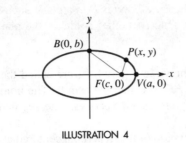

ILLUSTRATION 4

38. If F is a focus of the ellipse shown in Illustration 4, and P is any point on the ellipse, use the distance formula to show that the length of FP is $a - \frac{c}{a}x$. (*Hint:* In an ellipse, $a^2 - c^2 = b^2$.)

39. Finding the focal width In the ellipse shown in Illustration 5, chord AA' passes through the focus F and is perpendicular to the major axis. Show that the length of AA' (called the **focal width**) is $\frac{2b^2}{a}$.

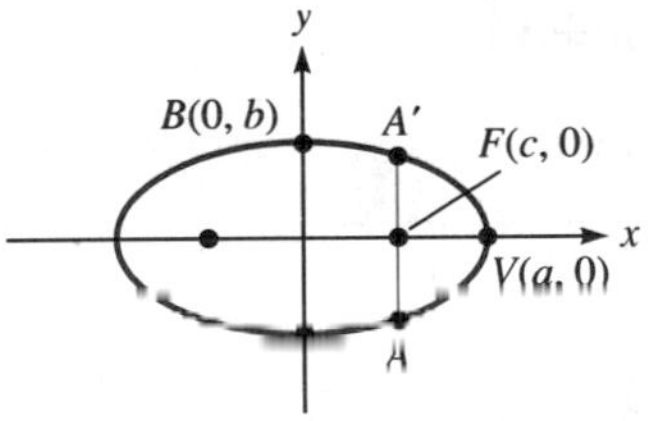

ILLUSTRATION 5

40. Prove that segment FV in Example 3 is the shortest distance between the earth and the sun. (*Hint:* Refer to Exercise 39.)

41. Constructing an ellipse The ends of a piece of string 6 meters long are attached to two thumbtacks that are 2 meters apart. A pencil catches the loop and draws it tight. As the pencil is moved about the thumbtacks (always keeping the tension), an ellipse is produced, with the thumbtacks as foci. Write the equation of the ellipse. (*Hint:* You'll have to establish a coordinate system.)

42. The distance between point $P(x, y)$ and the point $(0, 2)$ is $\frac{1}{3}$ of the distance of point P from the line $y = 18$. Find the equation of the curve on which point P lies.

43. Prove that $a > b$ in the development of the standard equation of an ellipse.

44. Show that the expansion of the standard equation of an ellipse is a special case of the general second-degree equation in two variables.

REVIEW *In Exercises 45–50, let* $A = \begin{bmatrix} 3 & -1 & 2 \\ 0 & 2 & -1 \\ 3 & 1 & 1 \end{bmatrix}$, $B = \begin{bmatrix} 1 & 2 \\ 2 & 0 \\ -1 & 1 \end{bmatrix}$, *and* $C = \begin{bmatrix} 0 & 1 \\ -1 & 1 \\ -2 & 0 \end{bmatrix}$. *Find each of the following, if possible.*

45. AB

46. $B + C$

47. $5B - 2C$

48. $AC + B$

49. the additive inverse of B

50. the multiplicative inverse of A.

8.3 The Hyperbola

■ ASYMPTOTES OF A HYPERBOLA ■ GRAPHING EQUATIONS OF HYPERBOLAS

The definition of a hyperbola is similar to the definition of an ellipse, except that we require a constant difference of $2a$ instead of a constant sum.

The Hyperbola

A **hyperbola** is the set of all points P in a plane such that the difference of the distances from point P to two other points in the plane is a positive constant.

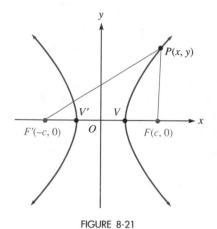

FIGURE 8-21

Points F and F' shown in Figure 8-21 are called the **foci** of the hyperbola, and the midpoint of chord FF' is called the **center.** The points V and V', where the hyperbola intersects FF', are called **vertices.** The segment VV' is called the **transverse axis.**

To develop the equation of the hyperbola centered at the origin, we note that the origin is the midpoint of chord FF', and we let $d(F'O) = d(OF) = c$, where $c > 0$. Then F is at $(c, 0)$, and F' is at $(-c, 0)$. The definition of a hyperbola requires that $d(F'P) - d(PF) = 2a$, where $2a$ is a positive constant. We use the distance formula to compute the lengths of $F'P$ and PF:

$$d(F'P) = \sqrt{[x - (-c)]^2 + y^2}$$
$$d(PF) = \sqrt{(x - c)^2 + y^2}$$

Substituting these values into the equation $d(F'P) - d(PF) = 2a$ gives

$$\sqrt{(x + c)^2 + y^2} - \sqrt{(x - c)^2 + y^2} = 2a$$

or

$$\sqrt{(x + c)^2 + y^2} = 2a + \sqrt{(x - c)^2 + y^2}$$

After squaring, we have

$$(x + c)^2 + y^2 = 4a^2 + 4a\sqrt{(x - c)^2 + y^2} + (x - c)^2 + y^2$$
$$x^2 + 2cx + c^2 + y^2 = 4a^2 + 4a\sqrt{(x - c)^2 + y^2} + x^2 - 2cx + c^2 + y^2$$
$$4cx = 4a^2 + 4a\sqrt{(x - c)^2 + y^2}$$
$$cx - a^2 = a\sqrt{(x - c)^2 + y^2}$$

Squaring both sides again and simplifying gives

$$c^2x^2 - 2a^2cx + a^4 = a^2(x^2 - 2cx + c^2 + y^2)$$
$$c^2x^2 - 2a^2cx + a^4 = a^2x^2 - 2a^2cx + a^2c^2 + a^2y^2$$
$$c^2x^2 + a^4 = a^2x^2 + a^2c^2 + a^2y^2$$

1. $(c^2 - a^2)x^2 - a^2y^2 = a^2(c^2 - a^2)$

Because $c > a$ (you will be asked to prove this in the exercises), $c^2 - a^2$ is a positive number. So we can let $b^2 = c^2 - a^2$ and substitute b^2 for $c^2 - a^2$ in Equation 1 to get

$$b^2x^2 - a^2y^2 = a^2b^2 \qquad (b^2 = c^2 - a^2)$$

We divide both sides of the previous equation by a^2b^2 to get the standard equation for a hyperbola with center at the origin and foci on the x-axis:

$$\frac{x^2}{a^2} - \frac{y^2}{b^2} = 1$$

To find the x-intercepts of the graph, we let $y = 0$ and solve for x. We get

$$\frac{x^2}{a^2} = 1$$
$$x^2 = a^2$$
$$x = a \quad \text{or} \quad x = -a$$

We now know that the x-intercepts are the vertices $V(a, 0)$ and $V'(-a, 0)$. The distance between the center of the hyperbola and either vertex is a, and the center of the hyperbola is the midpoint of the segment $V'V$ as well as of the segment FF'.

We attempt to find the y-intercepts by letting $x = 0$. Then the equation becomes

$$\frac{-y^2}{b^2} = 1 \quad \text{or} \quad y^2 = -b^2$$

Because $-b^2$ represents a negative number, and y^2 cannot be negative, the equation has no real solutions. Since there are no y-values corresponding to $x = 0$, the hyperbola does not intersect the y-axis.

This discussion suggests the following results.

Hyperbola: Foci on x-Axis, Center at (0, 0)

The standard equation of a hyperbola with center at the origin and foci on the x-axis is

$$\frac{x^2}{a^2} - \frac{y^2}{b^2} = 1$$

where $a^2 b^2 = c^2$.

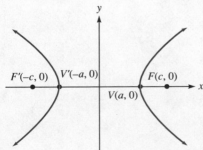

Vertices: $V(a, 0)$ and $V'(-a, 0)$
Foci: $F(c, 0)$ and $F'(-c, 0)$

If the foci are on the *y*-axis, a similar equation results.

Hyperbola: Foci on *y*-Axis, Center at (0, 0)

The standard equation of a hyperbola with center at the origin and foci on the *y*-axis is

$$\frac{y^2}{a^2} - \frac{x^2}{b^2} = 1$$

where $a^2 + b^2 = c^2$.

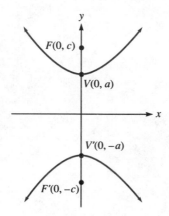

Vertices: $V(0, a)$ and $V'(0, -a)$
Foci: $F(0, c)$ and $F'(0, -c)$

To translate the hyperbola to a new position centered at the point (h, k) instead of the origin, we replace *x* and *y* with $x - h$ and $y - k$, respectively. We get the following results.

Hyperbola: Transverse Axis Horizontal, Center at (*h*, *k*)

The standard equation of a hyperbola with center at (h, k) and foci on a line parallel to the *x*-axis is

$$\frac{(x - h)^2}{a^2} - \frac{(y - k)^2}{b^2} = 1$$

where $a^2 + b^2 = c^2$.

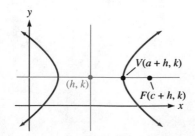

Vertices: $V(a + h, k)$ and $V'(-a + h, k)$
Foci: $F(c + h, k)$ and $F'(-c + h, k)$

Hyperbola: Transverse Axis Vertical, Center at (**h, k**)

The standard equation of a hyperbola with center at (h, k) and foci on a line parallel to the y-axis is

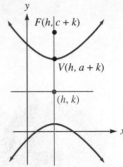

$$\frac{(y - k)^2}{a^2} - \frac{(x - h)^2}{b^2} = 1$$

where $a^2 + b^2 = c^2$.

Vertices: $V(h, a + k)$ and $V'(h, -a + k)$
Foci: $F(h, c + k)$ and $F'(h, -c + k)$

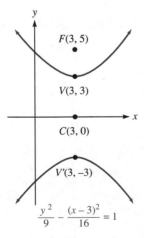

FIGURE 8-22

EXAMPLE 1

Write the equation of the hyperbola with vertices $(3, -3)$ and $(3, 3)$ and a focus at $(3, 5)$.

Solution   First, we plot the vertices and focus, as shown in Figure 8-22. Because the foci lie on a vertical line, the equation to use is

$$\frac{(y - k)^2}{a^2} - \frac{(x - h)^2}{b^2} = 1$$

The center of the hyperbola is midway between the vertices V and V'. Thus, the center is point $(3, 0)$; $h = 3$; and $k = 0$. The distance between the vertex and the center is $a = 3$, and the distance between the focus and the center is $c = 5$. We can find b^2 by substituting 3 for a and 5 for c in the following equation to get

$$b^2 = c^2 - a^2 \qquad \text{In a hyperbola, } b^2 = c^2 - a^2.$$
$$b^2 = 5^2 - 3^2$$
$$b^2 = 16$$

Substituting the values for h, k, a^2, and b^2 into the standard equation gives the equation of the hyperbola:

$$\frac{(y - k)^2}{a^2} - \frac{(x - h)^2}{b^2} = 1$$
$$\frac{(y - 0)^2}{9} - \frac{(x - 3)^2}{16} = 1$$
$$\frac{y^2}{9} - \frac{(x - 3)^2}{16} = 1$$

∎

Self Check Write the equation of the hyperbola with vertices $(3, 1)$ and $(-3, 1)$ and a focus at $(5, 1)$.

Answer $\dfrac{x^2}{9} - \dfrac{(y-1)^2}{16} = 1$

■ ASYMPTOTES OF A HYPERBOLA

The values of a and b are important aids in graphing hyperbolas. To see their value, we consider the hyperbola

$$\frac{x^2}{a^2} - \frac{y^2}{b^2} = 1$$

with center at the origin and vertices at $V(a, 0)$ and $V'(-a, 0)$. We can plot points V, V', $B(0, b)$, and $B'(0, -b)$ and form rectangle $RSQP$, called the **fundamental rectangle,** as shown in Figure 8-23. We can show that the extended diagonals of this rectangle are asymptotes of the hyperbola. In the exercises, you will be asked to show that the equations of these two lines are

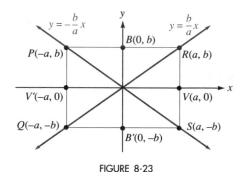

FIGURE 8-23

$$y = \frac{b}{a}x \qquad \text{and} \qquad y = -\frac{b}{a}x$$

To show that the extended diagonals are asymptotes of the hyperbola, we solve the equation

$$\frac{x^2}{a^2} - \frac{y^2}{b^2} = 1$$

for y and modify its form as follows:

$$\frac{x^2}{a^2} - \frac{y^2}{b^2} = 1$$

$$b^2x^2 - a^2y^2 = a^2b^2 \qquad \text{Multiply both sides by } a^2b^2.$$

$$y^2 = \frac{b^2x^2 - a^2b^2}{a^2} \qquad \text{Subtract } b^2x^2 \text{ from both sides and divide both sides by } -a^2.$$

$$y^2 = \frac{b^2x^2}{a^2}\left(1 - \frac{a^2}{x^2}\right) \qquad \text{Factor out } b^2x^2 \text{ in the numerator.}$$

2. $\qquad y = \pm\dfrac{bx}{a}\sqrt{1 - \dfrac{a^2}{x^2}} \qquad \text{Take the square root of both sides.}$

Pierre de Fermat (1601–1665)

Pierre de Fermat shares the honor of discovering analytic geometry (with Descartes) and of developing the theory of probability (with Pascal). But to Fermat alone goes credit for founding number theory. He is probably most famous for a theorem called Fermat's last theorem. It states that if n represents a number greater than 2, there are no whole numbers a, b, and c that satisfy the equation $a^n + b^n = c^n$. Only recently have mathematicians made significant progress toward proving it.

In Equation 2, if a is constant and $|x| \to \infty$, then $\dfrac{a^2}{x^2} \to 0$, and $\sqrt{1 - \dfrac{a^2}{x^2}}$ approaches 1. Thus, the hyperbola approaches the lines

$$y = \frac{b}{a}x \qquad \text{and} \qquad y = -\frac{b}{a}x$$

Knowing the asymptotes makes it easy to sketch a hyperbola. We simply convert its equation into standard form, find the coordinates of its vertices, and construct the fundamental rectangle with its extended diagonals. Using the vertices and the asymptotes as guides, we can sketch the hyperbola shown in Figure 8-24. The segment BB' is called the **conjugate axis** of the hyperbola.

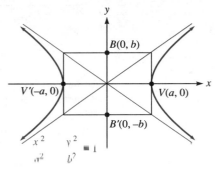

FIGURE 8-24

■ GRAPHING EQUATIONS OF HYPERBOLAS

EXAMPLE 2   Graph the hyperbola $x^2 - y^2 - 2x + 4y = 12$.

Solution   We complete the square on x and y to write the equation into standard form:

$$x^2 - 2x - y^2 + 4y = 12$$
$$x^2 - 2x - (y^2 - 4y) = 12$$
$$x^2 - 2x + 1 - (y^2 - 4y + 4) = 12 + 1 - 4$$
$$(x - 1)^2 - (y - 2)^2 = 9$$
$$\frac{(x - 1)^2}{9} - \frac{(y - 2)^2}{9} = 1$$

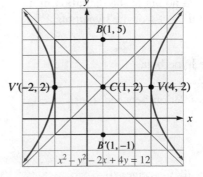

FIGURE 8-25

From the standard equation of a hyperbola, we see that the center is $(1, 2)$, that $a = 3$ and $b = 3$, and that the vertices are on a line segment parallel to the x-axis, as shown in Figure 8-25. The vertices V and V' are 3 units to the right and left of the center and have coordinates of $(4, 2)$ and $(-2, 2)$. Points B and B', 3 units above and below the center, have coordinates of $(1, 5)$ and $(1, -1)$. After using points V, V', B, and B' to construct the fundamental rectangle and its extended diagonals, we sketch the graph. ■

Self Check

Graph the hyperbola $x^2 - y^2 + 6x + 2y = 8$.

Answer

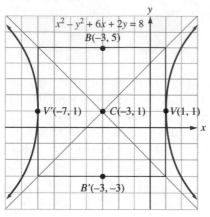

We have considered only those hyperbolas with a major axis that is horizontal or vertical. However, some hyperbolas have nonhorizontal or nonvertical major axes. For example, the graph of the equation $xy = 4$ is a hyperbola with vertices at $(2, 2)$ and $(-2, -2)$, as shown in Figure 8-26.

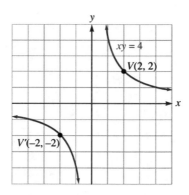

FIGURE 8-26

EXERCISE 8.3

VOCABULARY AND CONCEPTS *In Exercises 1–6, fill in the blank to make a true statement.*

1. A hyperbola is the set of all points in the plane such that the _____ of the distances from two fixed points is a positive _____.

2. Each of the two fixed points in the definition of a hyperbola is called a _____ of the hyperbola.

3. The vertices of the hyperbola $\dfrac{x^2}{a^2} - \dfrac{y^2}{b^2} = 1$ are $V(\text{\_\_}, \text{\_\_})$ and $V'(\text{\_\_\_\_}, \text{\_\_})$.

4. The vertices of the hyperbola $\dfrac{y^2}{a^2} - \dfrac{x^2}{b^2} = 1$ are $V(\text{\_\_}, \text{\_\_})$ and $V'(\text{\_\_}, \text{\_\_\_\_})$.

5. The chord that joins the vertices is called the _____ of the hyperbola.

6. In a hyperbola, the relation between a, b, and c is _____.

PRACTICE *In Exercises 7–18, write the equation of each hyperbola.*

7. Vertices $(5, 0)$ and $(-5, 0)$; focus $(7, 0)$

8. Focus $(3, 0)$; vertex $(2, 0)$; center $(0, 0)$

9. Center $(2, 4)$; $a = 2$, $b = 3$; transverse axis is horizontal

10. Center $(-1, 3)$; vertex $(1, 3)$; focus $(2, 3)$

11. Center $(5, 3)$; vertex $(5, 6)$; passes through $(1, 8)$

12. Foci $(0, 10)$ and $(0, -10)$; $\dfrac{c}{a} = \dfrac{5}{4}$

13. Vertices $(0, 3)$ and $(0, -3)$; $\dfrac{c}{a} = \dfrac{5}{3}$

14. Focus $(4, 0)$; vertex $(2, 0)$; center $(0, 0)$

15. Center $(1, -3)$; $a^2 = 4$; $b^2 = 16$

16. Center $(1, 4)$; focus $(7, 4)$; vertex $(3, 4)$

17. Center at the origin; passes through $(4, 2)$ and $(8, -6)$

18. Center $(3, -1)$; y-intercept -1; x-intercept $3 + \dfrac{3\sqrt{5}}{2}$

In Exercises 19–22, find the area of the fundamental rectangle of each hyperbola.

19. $4(x - 1)^2 - 9(y + 2)^2 = 36$

20. $x^2 - y^2 - 4x - 6y = 6$

21. $x^2 + 6x - y^2 + 2y = -11$

22. $9x^2 - 4y^2 = 18x + 24y + 63$

In Exercises 23–26, write the equation of each hyperbola.

23. Center $(-2, -4)$; $a = 2$; area of fundamental rectangle is 36 square units

24. Center $(3, -5)$; $b = 6$; area of fundamental rectangle is 24 square units

25. Vertex $(6, 0)$; one end of conjugate axis at $\left(0, \dfrac{5}{4}\right)$

26. Vertex $(3, 0)$; focus $(-5, 0)$; center $(0, 0)$

In Exercises 27–36, graph each hyperbola.

27. $\dfrac{x^2}{9} - \dfrac{y^2}{4} = 1$

28. $\dfrac{y^2}{4} - \dfrac{x^2}{9} = 1$

29. $4x^2 - 3y^2 = 36$

30. $3x^2 - 4y^2 = 36$

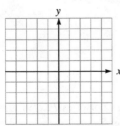
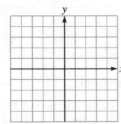
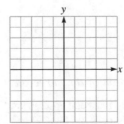
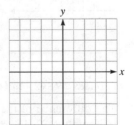

31. $y^2 - x^2 = 1$

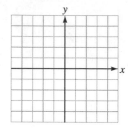

32. $9(y + 2)^2 - 4(x - 1)^2 = 36$

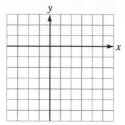

33. $4x^2 - 2y^2 + 8x - 8y = 8$

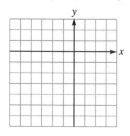

34. $x^2 - y^2 - 4x - 6y = 6$

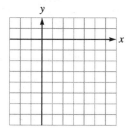

35. $y^2 - 4x^2 + 6y + 32x = 59$

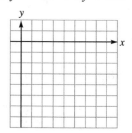

36. $x^2 + 6x - y^2 + 2y = -11$

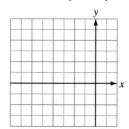

In Exercises 37–38, graph each hyperbola.

37. $-xy = 6$

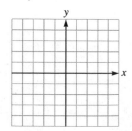

38. $xy = 20$

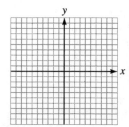

In Exercises 39–42, find the equation of the curve on which point P lies.

39. The difference of the distances between $P(x, y)$ and the points $(-2, 1)$ and $(8, 1)$ is 6.

40. The difference of the distances between $P(x, y)$ and the points $(3, -1)$ and $(3, 5)$ is 5.

41. The distance between point $P(x, y)$ and the point $(0, 3)$ is $\frac{3}{2}$ of the distance between P and the line $y = -2$.

42. The distance between point $P(x, y)$ and the point $(5, 4)$ is $\frac{5}{3}$ of the distance between P and the line $y = -3$.

APPLICATIONS

43. Astronomy   Some comets may have a hyperbolic orbit, with the sun as one focus. When the comet shown in Illustration 1 is far away from earth, it appears to be approaching earth along the line $y = 2x$. Find the equation of its orbit if the comet comes within 100 million miles of the earth.

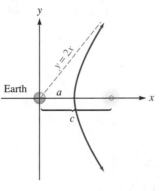

ILLUSTRATION 1

44. Physics   Parallel beams of similarly charged particles are shot from two atomic accelerators 20 meters apart, as shown in Illustration 2. If the particles were not deflected, the beams would be 2.0×10^{-4} meters apart. However, because the charged particles repel each other, the beams follow the hyperbolic path $y = \frac{k}{x}$, for some k. Find k.

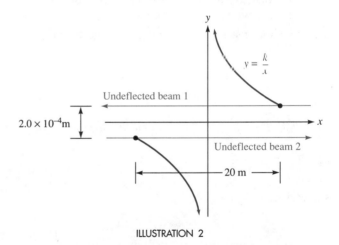

ILLUSTRATION 2

45. Navigation   The **LORAN** system (LOng RAnge Navigation) in Illustration 3 uses two radio transmitters 26 miles apart to send simultaneous signals. The navigator on a ship at $P(x, y)$ receives the closer signal first, and determines that the difference of the distances between the ship and each transmitter is 24 miles. That places the ship on a certain curve. Identify the curve and find its equation.

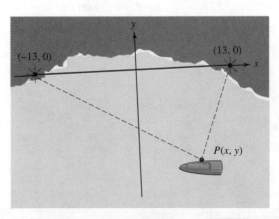

ILLUSTRATION 3

46. Wave propagation Stones dropped into a calm pond at points A and B create ripples that propagate in widening circles. In Illustration 4, points A and B are 20 feet apart, and the radii of the circles differ by 12 feet. The point $P(x, y)$ where the circles intersect moves along a curve. Identify the curve and find its equation.

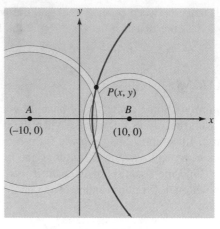

ILLUSTRATION 4

DISCOVERY AND WRITING

47. Prove that $c > a$ for a hyperbola with center at $(0, 0)$ and line segment FF' on the x-axis.

48. Show that the extended diagonals of the fundamental rectangle of the hyperbola
$$\frac{x^2}{a^2} - \frac{y^2}{b^2} = 1 \text{ are } y = \frac{b}{a}x \text{ and } y = -\frac{b}{a}x.$$

49. Show that the expansion of the standard equation of a hyperbola is special case of the general equation of second degree with $B = 0$.

50. Write a paragraph describing how you can tell from the equation of a hyperbola whether the transverse axis is vertical or horizontal.

REVIEW *In Exercises 51–54, find the inverse of each function. Write the answer in $y = f^{-1}(x)$ form.*

51. $f(x) = 3x - 2$

52. $f(x) = \dfrac{x + 1}{x}$

53. $f(x) = \dfrac{5x}{x + 2}$

54. $f(x) = x$

In Exercises 55–58, let $f(x) = x^2 + 1$ and $g(x) = (x + 1)^2$. Find each composite function.

55. $f(g(x))$

56. $g(f(x))$

57. $f(f(x))$

58. $g(g(x))$

8.4 Solving Simultaneous Second-Degree Equations

■ SOLVING SYSTEMS BY GRAPHING ■ SOLVING SYSTEMS ALGEBRAICALLY

We now discuss techniques for solving systems of two equations in two variables, where at least one of the equations is of second degree.

■ SOLVING SYSTEMS BY GRAPHING

EXAMPLE 1   Solve $\begin{cases} x^2 + y^2 = 25 \\ 2x + y = 10 \end{cases}$ by graphing.

Solution   The graph of $x^2 + y^2 = 25$ is a circle with center at the origin and radius of 5. The graph of $2x + y = 10$ is a straight line. Depending on whether the line is a secant (intersecting the circle at two points) or a tangent (intersecting the circle at one point) or does not intersect the circle at all, there are two, one, or no solutions to the system, respectively.

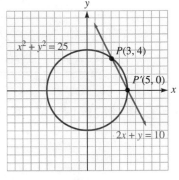

After graphing the circle and the line, as shown in Figure 8-27, we see that there are two intersection points, $P(3, 4)$ and $P'(5, 0)$. The solutions to the system are

$$\begin{cases} x = 3 \\ y = 4 \end{cases} \quad \text{and} \quad \begin{cases} x = 5 \\ y = 0 \end{cases}$$

Verify that these are exact solutions.

FIGURE 8-27   ■

Self Check   Solve $\begin{cases} x^2 + y^2 = 25 \\ 2x - y = 5 \end{cases}$ graphically.

Answer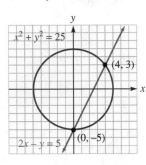

$$\begin{cases} x = 4 \\ y = 3 \end{cases} \quad \text{and} \quad \begin{cases} x = 0 \\ y = -5 \end{cases}$$

■ ■ ■ ■ ■ ■ ■ ■

ACCENT ON TECHNOLOGY

To solve the system of equations in Example 1 on a graphing calculator, we must first solve each of the equations for y. From the first equation, $x^2 + y^2 = 25$, we get two equations to graph:

$$y_1 = \sqrt{25 - x^2}$$
$$y_2 = -\sqrt{25 - x^2}$$

From the other equation, $2x + y = 10$, we get a third equation:

$$y_3 = 10 - 2x$$

The graphs of these equations will be similar to those shown in Figure 8-28(a). To find the top solution, we use ZOOM and TRACE to read the coordinates of the point of intersection, as shown in Figure 8-28(b): $x \approx 3$ and $y \approx 4$. Because the graph is not complete at the other point of intersection shown in Figure 8-28(c), we must use our best judgment. We read the coordinates to be $x \approx 5$ and $y \approx 0$.

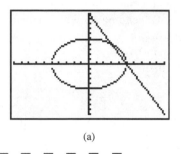

(a)

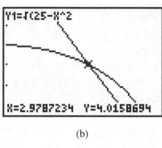

(b)

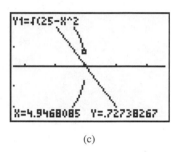

(c)

FIGURE 8-28

■ SOLVING SYSTEMS ALGEBRAICALLY

Algebraic methods can be used to find exact solutions.

EXAMPLE 2   Solve $\begin{cases} x^2 + y^2 = 25 \\ 2x + y = 10 \end{cases}$ algebraically.

Solution   If a system contains one equation of second degree and another of first degree, we can solve it by substitution. Solving the linear equation for y gives

$$2x + y = 10$$
$$y = -2x + 10$$

We can now substitute $-2x + 10$ for y in the second-degree equation and solve the resulting quadratic equation for x:

$$x^2 + y^2 = 25$$
$$x^2 + (-2x + 10)^2 = 25$$
$$x^2 + 4x^2 - 40x + 100 = 25 \qquad \text{Remove parentheses.}$$
$$5x^2 - 40x + 75 = 0 \qquad \text{Subtract 25 from both sides and combine like terms.}$$
$$x^2 - 8x + 15 = 0 \qquad \text{Divide both sides by 5.}$$
$$(x - 5)(x - 3) = 0 \qquad \text{Factor } x^2 - 8x + 15 = 0.$$
$$x - 5 = 0 \quad \text{or} \quad x - 3 = 0$$
$$x = 5 \qquad\qquad x = 3$$

Because $y = -2x - 10$, if $x = 5$, then $y = 0$; and if $x = 3$, then $y = 4$. The two solutions are

$$\begin{cases} x = 5 \\ y = 0 \end{cases} \quad \text{or} \quad \begin{cases} x = 3 \\ y = 4 \end{cases}$$ ∎

Self Check   Solve $\begin{cases} x^2 + y^2 = 25 \\ 2x - y = 5 \end{cases}$ algebraically.

Answer $\begin{cases} x = 4 \\ y = 3 \end{cases}$ and $\begin{cases} x = 0 \\ y = -5 \end{cases}$

EXAMPLE 3   Solve $\begin{cases} 4x^2 + 9y^2 = 5 \\ y = x^2 \end{cases}$ algebraically.

Solution   We can solve this system by substitution.

$$4x^2 + 9y^2 = 5$$
$$4y + 9y^2 = 5 \qquad\qquad \text{Substitute } y \text{ for } x^2.$$
$$9y^2 + 4y - 5 = 0 \qquad\qquad \text{Add } -5 \text{ to both sides.}$$
$$(9y - 5)(y + 1) = 0 \qquad\qquad \text{Factor } 9y^2 + 4y - 5.$$
$$9y - 5 = 0 \quad \text{or} \quad y + 1 = 0$$
$$y = \frac{5}{9} \qquad\qquad y = -1$$

Because $y = x^2$, we can find x by solving the equations

$$x^2 = \frac{5}{9} \quad \text{and} \quad x^2 = -1$$

The solutions of $x^2 = \frac{5}{9}$ are

$$x = \frac{\sqrt{5}}{3} \quad \text{and} \quad x = -\frac{\sqrt{5}}{3}$$

The equation $x^2 = -1$ has no real solutions. Thus, the solutions of the system are

$$\left(\frac{\sqrt{5}}{3}, \frac{5}{9} \right) \quad \text{and} \quad \left(-\frac{\sqrt{5}}{3}, \frac{5}{9} \right)$$ ∎

Self Check   Solve $\begin{cases} x^2 + 3y^2 = 13 \\ x = y^2 - 1 \end{cases}$.

Answer $\left(2, \sqrt{3}\right)$ and $\left(2, -\sqrt{3}\right)$

EXAMPLE 4   Solve $\begin{cases} 3x^2 + 2y^2 = 36 \\ 4x^2 - y^2 = 4 \end{cases}$ algebraically.

Solution   When we have two second-degree equations of the form $ax^2 + by^2 = c$, we can solve the system by eliminating one of the variables by addition. To eliminate the terms involving y^2, we copy the first equation and multiply the second equation by 2 to obtain the following equivalent system.

$$\begin{cases} 3x^2 + 2y^2 = 36 \\ 8x^2 - 2y^2 = 8 \end{cases}$$

We can then add the equations and solve the resulting equation for x:

$$11x^2 = 44$$
$$x^2 = 4$$
$$x = 2 \qquad \text{or} \qquad x = -2$$

To find y, we substitute 2 for x and then -2 for x in the first equation.

For $x = 2$	**For $x = -2$**
$3x^2 + 2y^2 = 36$	$3x^2 + 2y^2 = 36$
$3(2)^2 + 2y^2 = 36$	$3(-2)^2 + 2y^2 = 36$
$12 + 2y^2 = 36$	$12 + 2y^2 = 36$
$2y^2 = 24$	$2y^2 = 24$
$y^2 = 12$	$y^2 = 12$
$y = \sqrt{12}$ or $y = -\sqrt{12}$	$y = \sqrt{12}$ or $y = -\sqrt{12}$
$y = 2\sqrt{3}$ $\qquad$ $y = -2\sqrt{3}$	$y = 2\sqrt{3}$ $\qquad$ $y = -2\sqrt{3}$

The four solutions of this system are

$$\left(2, 2\sqrt{3}\right), \qquad \left(2, -2\sqrt{3}\right), \qquad \left(-2, 2\sqrt{3}\right), \quad \text{and} \quad \left(-2, -2\sqrt{3}\right) \qquad \blacksquare$$

Self Check   Solve $\begin{cases} 2x^2 + y^2 = 23 \\ 3x^2 - 2y^2 = 17 \end{cases}$.

Answer $\left(3, \sqrt{5}\right), \left(3, -\sqrt{5}\right), \left(-3, \sqrt{5}\right), \left(-3, -\sqrt{5}\right)$

EXERCISE 8.4

VOCABULARY AND CONCEPTS   *In Exercises 1–2, fill in the blank to make a true statement.*

1. Solutions of systems of second-degree equations are points of intersection of _____ sections.

2. Approximate solutions can be found _____, and exact solutions are found _____.

PRACTICE   *In Exercises 3–12, solve each system of equations by graphing.*

3. $\begin{cases} 8x^2 + 32y^2 = 256 \\ x = 2y \end{cases}$

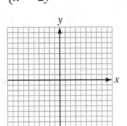

4. $\begin{cases} x^2 + y^2 = 2 \\ x + y = 2 \end{cases}$

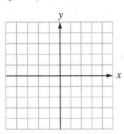

5. $\begin{cases} x^2 + y^2 = 90 \\ y = x^2 \end{cases}$

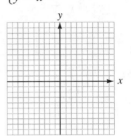

6. $\begin{cases} x^2 + y^2 = 5 \\ x + y = 3 \end{cases}$

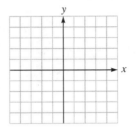

7. $\begin{cases} x^2 + y^2 = 25 \\ 12x^2 + 64y^2 = 768 \end{cases}$

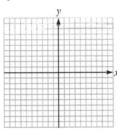

8. $\begin{cases} x^2 + y^2 = 13 \\ y = x^2 - 1 \end{cases}$

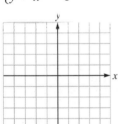

9. $\begin{cases} x^2 - 13 = -y^2 \\ y = 2x - 4 \end{cases}$

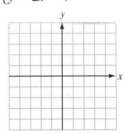

10. $\begin{cases} x^2 + y^2 = 20 \\ y = x^2 \end{cases}$

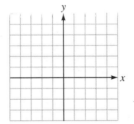

11. $\begin{cases} x^2 - 6x - y = -5 \\ x^2 - 6x + y = -5 \end{cases}$

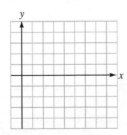

12. $\begin{cases} x^2 - y^2 = -5 \\ 3x^2 + 2y^2 = 30 \end{cases}$

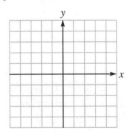

In Exercises 13–16, use a graphing calculator to solve each system of equations.

13. $\begin{cases} y = x + 1 \\ y = x^2 + x \end{cases}$

14. $\begin{cases} y = 6 - x^2 \\ y = x^2 - x \end{cases}$

15. $\begin{cases} 6x^2 + 9y^2 = 10 \\ 3y - 2x = 0 \end{cases}$

16. $\begin{cases} x^2 + y^2 = 68 \\ y^2 - 3x^2 = 4 \end{cases}$

In Exercises 17–42, solve each system of equations algebraically for real values of x and y.

17. $\begin{cases} 25x^2 + 9y^2 = 225 \\ 5x + 3y = 15 \end{cases}$

18. $\begin{cases} x^2 + y^2 = 20 \\ y = x^2 \end{cases}$

19. $\begin{cases} x^2 + y^2 = 2 \\ x + y = 2 \end{cases}$

20. $\begin{cases} x^2 + y^2 = 36 \\ 49x^2 + 36y^2 = 1764 \end{cases}$

21. $\begin{cases} x^2 + y^2 = 5 \\ x + y = 3 \end{cases}$

22. $\begin{cases} x^2 - x - y = 2 \\ 4x - 3y = 0 \end{cases}$

23. $\begin{cases} x^2 + y^2 = 13 \\ y = x^2 - 1 \end{cases}$

24. $\begin{cases} x^2 + y^2 = 25 \\ 2x^2 - 3y^2 = 5 \end{cases}$

25. $\begin{cases} x^2 + y^2 = 30 \\ y = x^2 \end{cases}$

26. $\begin{cases} 9x^2 - 7y^2 = 81 \\ x^2 + y^2 = 9 \end{cases}$

27. $\begin{cases} x^2 + y^2 = 13 \\ x^2 - y^2 = 5 \end{cases}$

28. $\begin{cases} 2x^2 + y^2 = 6 \\ x^2 - y^2 = 3 \end{cases}$

29. $\begin{cases} x^2 + y^2 = 20 \\ x^2 - y^2 = -12 \end{cases}$

30. $\begin{cases} xy = -\dfrac{9}{2} \\ 3x + 2y = 6 \end{cases}$

31. $\begin{cases} y^2 = 40 - x^2 \\ y = x^2 - 10 \end{cases}$

32. $\begin{cases} x^2 - 6x - y = -5 \\ x^2 - 6x + y = -5 \end{cases}$

33. $\begin{cases} y = x^2 - 4 \\ x^2 - y^2 = -16 \end{cases}$

34. $\begin{cases} 6x^2 + 8y^2 = 182 \\ 8x^2 - 3y^2 = 24 \end{cases}$

35. $\begin{cases} x^2 - y^2 = -5 \\ 3x^2 + 2y^2 = 30 \end{cases}$

36. $\begin{cases} \dfrac{1}{x} + \dfrac{1}{y} = 5 \\ \dfrac{1}{x} - \dfrac{1}{y} = -3 \end{cases}$

37. $\begin{cases} \dfrac{1}{x} + \dfrac{2}{y} = 1 \\ \dfrac{2}{x} - \dfrac{1}{y} = \dfrac{1}{3} \end{cases}$

38. $\begin{cases} \dfrac{1}{x} + \dfrac{3}{y} = 4 \\ \dfrac{2}{x} - \dfrac{1}{y} = 7 \end{cases}$

39. $\begin{cases} 3y^2 = xy \\ 2x^2 + xy - 84 = 0 \end{cases}$

40. $\begin{cases} x^2 + y^2 = 10 \\ 2x^2 - 3y^2 = 5 \end{cases}$

41. $\begin{cases} xy = \dfrac{1}{6} \\ y + x = 5xy \end{cases}$

42. $\begin{cases} xy = \dfrac{1}{12} \\ y + x = 7xy \end{cases}$

APPLICATIONS

43. Geometry   The area of a rectangle is 63 square centimeters, and its perimeter is 32 centimeters. Find the dimensions of the rectangle.

44. Investments   Grant receives $225 annual income from one investment. Jeff invested $500 more than Grant, but at an annual rate of 1% less. Jeff's annual income is $240. Find the amount and rate of Grant's investment.

45. Investments   Carol receives $67.50 annual income from one investment. John invested $150 more than Carol at an annual rate of $1\frac{1}{2}$% more. John's annual income is $94.50. Find the amount and rate of Carol's investment. (*Hint:* There are two answers.)

46. Finding rates and time   Jim drove 306 miles. Jim's brother made the same trip at a speed 17 miles per hour slower than Jim did and required an extra $1\frac{1}{2}$ hours. Find Jim's rate and time.

47. 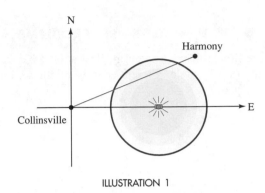 **Radio reception** A radio station located 120 miles due east of Collinsville has a listening radius of 100 miles. A straight road joins Collinsville with Harmony, a town 200 miles to the east and 100 miles north. See Illustration 1. How far from Collinsville will a driver first pick up the station?

48. **Listening range** For how many miles will a driver in Exercise 47 continue to receive the signal?

ILLUSTRATION 1

DISCOVERY AND WRITING

49. Is it possible for a system of second-degree equations to have no common solution? If so, sketch the graphs of the equations of such a system.

50. Can a system have exactly one solution? If so, sketch the graphs of the equations of such a system.

51. Can a system have exactly two solutions? If so, sketch possible graphs.

52. Can a system have three solutions? If so, sketch the graphs.

53. Can it have exactly four solutions? If so, sketch the graphs.

54. Can a system have more than four solutions? If so, sketch the graphs.

REVIEW *In Exercises 55–58, find the vertical, horizontal, or slant asymptotes of each of the following rational functions.*

55. $y = \dfrac{3x + 1}{x - 1}$

56. $y = \dfrac{x^2 + 3}{x - 1}$

57. $y = \dfrac{3x + 1}{x^2 - 1}$

58. $y = \dfrac{x^2 + 3}{x^2 + 1}$

In Exercises 59–62, tell whether the graph has x-axis, y-axis, or origin symmetry, or none of these symmetries.

59. $y = \dfrac{3x^2 + 5}{5x^2 + 3}$

60. $y = \dfrac{3x^3 + 5}{5x^2 + 3}$

61. $y = \dfrac{3x^3}{5x^2 + 3}$

62. $y^2 = \dfrac{3x^3 + 5}{5x^2 + 3}$

■ ■ ■ ■ ■ ■ ■ ■ ■ PROBLEMS AND PROJECTS

1. In the equation of a hyperbola,

$$\frac{x^2}{a^2} - \frac{y^2}{b^2} = 1$$

replace the 1 on the right-hand side with 0. Solve the resulting equation for y by factoring the left-hand side (it is the difference of two squares) and setting each factor equal to 0. What are the graphs of these two equations, and what is their relation to the hyperbola?

(continued)

■ ■ ■ ■ ■ ■ ■ ■ ■ ■ **PROBLEMS AND PROJECTS** *(continued)*

2. The 2-inch-diameter pipe in Illustration 1 has been cut at a 45° angle. The cut edge is an ellipse. Find its equation.

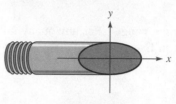

ILLUSTRATION 1

In the equations in Problems 3–6, each different value of k produces a different graph. The collection of all these graphs is called a **family of curves.** By choosing several values of k, graph a few members of each family. What common characteristic do members of the family share?

3. A family of circles $(x - k)^2 + (y - k)^2 = k^2$

4. A family of parabolas $y = x^2 + k$

5. A family of ellipses $x^2 + \dfrac{y^2}{k^2} = 1$

6. A family of hyperbolas $xy = k$

PROJECT 1

To draw a hyperbola, tie a small loop in a long piece of string, large enough to hold the point of a pencil. As in Illustration 2, arrange the string around nails at A and B and hold a pencil in the loop at P. While keeping the string taut, carefully pull both ends together. Explain why the pencil traces a hyperbola. From the measured distance between A and B and the *difference* of lengths PA and PB, find the equation of the curve. How would you draw the other branch of the hyperbola?

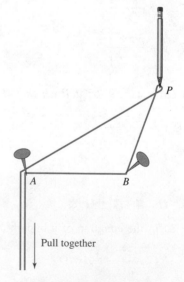

ILLUSTRATION 2

PROJECT 2   Draw a line on a sheet of paper and mark a point P about 2 inches away, as in Illustration 3(a). (Use wax paper, and draw the line and point with a fine-tip marker.) Fold the paper over to place the point on the line and make a sharp crease, as in Illustration 3(b). Fold the paper several times, placing P at different locations on the line. What curve do these creases seem to describe? What is the relation of the point and the line to that curve? What happens if point P is closer to the line? Write a paragraph to describe your findings.

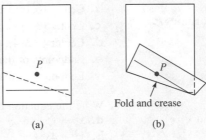

(a)   (b)

ILLUSTRATION 3

PROJECT 3   Draw a circle on a sheet of paper and mark point P inside. Fold and crease several times, placing P at several locations on the circle, as in Illustration 4. What curve do the creases describe? What is the significance of point P?

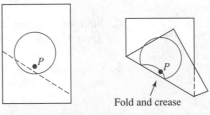

ILLUSTRATION 4

PROJECT 4   This time, draw a circle and mark point P on the outside, as in Illustration 5. Fold and crease as before. What curve is described by these creases, and what is the significance of P? Do you get two branches of the curve?

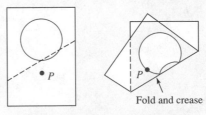

ILLUSTRATION 5

CHAPTER SUMMARY

CONCEPT REVIEW EXERCISES

SECTION 8.1 | *The Circle and the Parabola*

Circle, center at $(0, 0)$, radius r:
$x^2 + y^2 = r^2$

Circle, center at (h, k), radius r:
$(x - h)^2 + (y - k)^2 = r^2$

General form of a second-degree equation in x and y:
$Ax^2 + Bxy + Cy^2 + Dx + Ey + F = 0$

Parabola

Parabola opening	Vertex at origin
Right	$y^2 = 4px$
Left	$y^2 = -4px$
Upward	$x^2 = 4py$
Downward	$x^2 = -4py$

Parabola opening	Vertex at $V(h, k)$
Right	$(y - k)^2 = 4p(x - h)$
Left	$(y - k)^2 = -4p(x - h)$
Upward	$(x - h)^2 = 4p(y - k)$
Downward	$(x - h)^2 = -4p(y - k)$

1. Write the equation of each circle.
 a. Center $(0, 0)$; radius 4
 b. Center $(0, 0)$; passes through $(6, 8)$
 c. Center $(3, -2)$; radius 5
 d. Center $(-2, 4)$; passes through $(1, 0)$
 e. Endpoints of diameter $(-2, 4)$ and $(12, 16)$
 f. Endpoints of diameter $(-3, -6)$ and $(7, 10)$

2. Write the equation of each circle in standard form.
 a. $x^2 + y^2 - 6x + 4y = 3$
 b. $x^2 + 4x + y^2 - 10y = -13$

3. Write the equation of each parabola.
 a. Vertex $(0, 0)$; passes through $(-8, 4)$ and $(-8, -4)$
 b. Vertex $(0, 0)$; passes through $(-8, 4)$ and $(8, 4)$

4. Find the equation of the parabola with vertex at $(-2, 3)$, curve passing through point $(-4, -8)$, and opening downward.

5. Graph each equation.
 a. $x^2 - 4y - 2x + 9 = 0$

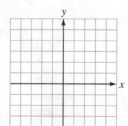

 b. $y^2 - 6y = 4x - 13$

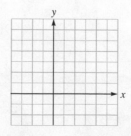

SECTION 8.2 — *The Ellipse*

Ellipse, center at (0, 0):
Major axis is $2a$, minor axis is $2b$ ($a > b > 0$)

Center-to-focus distance is c, where $a^2 = b^2 + c^2$

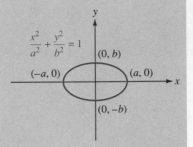

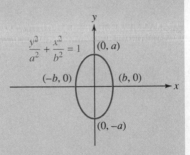

Ellipse, center at (h, k):
Major axis is $2a$, minor axis is $2b$ ($a > b > 0$)

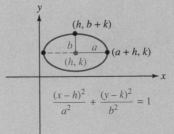

6. Write the equation of the ellipse with center at the origin, major axis that is horizontal and 12 units long, and minor axis 8 units long.

7. Write the equation of the ellipse with center at the origin, major axis that is vertical and 10 units long, and minor axis 4 units long.

8. Write the equation of the ellipse with center at point $(-2, 3)$ and curve passing through points $(-2, 0)$ and $(2, 3)$.

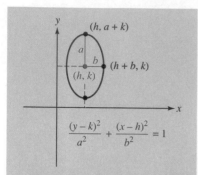

$$\frac{(y-k)^2}{a^2} + \frac{(x-h)^2}{b^2} = 1$$

9. Write the equation in standard form and graph it.

$$4x^2 + y^2 - 16x + 2y = -13$$

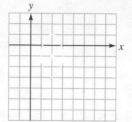

SECTION 8.3

The Hyperbola

Hyperbola:
Center at $(0, 0)$,
center-to-focus distance is c,
where $a^2 + b^2 = c^2$

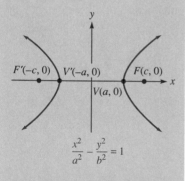

$$\frac{x^2}{a^2} - \frac{y^2}{b^2} = 1$$

10. Write the equation of the hyperbola with center at the origin, passing through points $(-2, 0)$ and $(2, 0)$, and having a focus at $(4, 0)$.

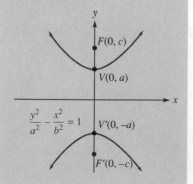

$$\frac{y^2}{a^2} - \frac{x^2}{b^2} = 1$$

11. Write the equation of the hyperbola with center at the origin, one focus at $(0, 5)$, and one vertex at $(0, 3)$.

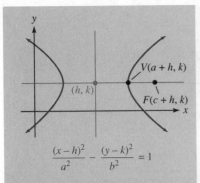

$$\frac{(x-h)^2}{a^2} - \frac{(y-k)^2}{b^2} = 1$$

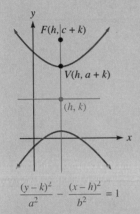

$$\frac{(y-k)^2}{a^2} - \frac{(x-h)^2}{b^2} = 1$$

The extended diagonals of the fundamental rectangle are asymptotes of the graph of a hyperbola.

12. Write the equation of the hyperbola with vertices at point $(-3, 3)$ and $(3, 3)$ and a focus at point $(5, 3)$.

13. Write the equation of the hyperbola with vertices at point $(3, -3)$ and $(3, 3)$ and a focus at point $(3, 5)$.

14. Write the equation of the asymptotes of the hyperbola

$$\frac{x^2}{25} - \frac{y^2}{16} = 1$$

15. Write the equation in standard form and graph it.

$$9x^2 - 4y^2 - 16y - 18x = 43$$

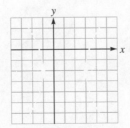

16. Graph the equation $4xy = 1$.

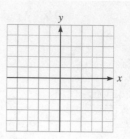

SECTION 8.4   *Solving Simultaneous Second-Degree Equations*

Good estimates for solutions of systems of simultaneous second-degree equations can be found by graphing.

17. Solve by graphing: $\begin{cases} 3x^2 + y^2 = 52 \\ x^2 - y^2 = 12 \end{cases}$.

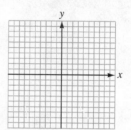

18. Solve by graphing: $\begin{cases} x^2 + y^2 = 16 \\ -\sqrt{3}y + 4\sqrt{3} = 3x \end{cases}$

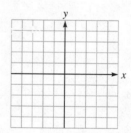

19. Solve by graphing: $\begin{cases} \dfrac{x^2}{16} + \dfrac{y^2}{12} = 1 \\ x^2 - \dfrac{y^2}{3} = 1 \end{cases}$.

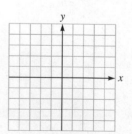

Exact solutions to systems of simultaneous second-degree equations can often be found by algebraic techniques.

20. Solve algebraically: $\begin{cases} 3x^2 + y^2 = 52 \\ x^2 - y^2 = 12 \end{cases}$.

21. Solve algebraically: $\begin{cases} x^2 + y^2 = 16 \\ -\sqrt{3}y + 4\sqrt{3} = 3x \end{cases}$.

22. Solve algebraically: $\begin{cases} \dfrac{x^2}{16} + \dfrac{y^2}{12} = 1 \\ x^2 - \dfrac{y^2}{3} = 1 \end{cases}$.

■ Chapter 8 Test

In Questions 1–3, write the equation of each circle.

1. Center $(2, 3)$; $r = 3$

2. Ends of diameter at $(-2, -2)$ and $(6, 8)$

3. Center $(2, -5)$, passes through $(7, 7)$

4. Change the equation of the circle $x^2 + y^2 - 4x + 6y + 4 = 0$ to standard form and graph it.

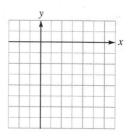

In Questions 5–7, find the equation of each parabola.

5. Vertex $(3, 2)$; focus at $(3, 6)$

6. Vertex $(4, -6)$; passes through $(3, -8)$ and $(3, -4)$

7. Vertex $(2, -3)$; passes through $(0, 0)$

8. Change the equation of the parabola $x^2 - 6x - 8y = 7$ into standard form and graph it.

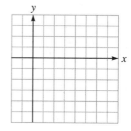

In Questions 9–11, find the equation of each ellipse.

9. Vertex $(10, 0)$, center at the origin, focus at $(6, 0)$

10. Minor axis 24, center at the origin, focus at $(5, 0)$

11. Center $(2, 3)$; passes through $(2, 9)$ and $(0, 3)$

12. Change the equation of the ellipse $9x^2 + 4y^2 - 18x - 16y - 11 = 0$ into standard form and graph it.

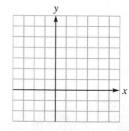

In Questions 13–15, find the equation of each hyperbola.

13. Center at the origin, focus at $(13, 0)$, vertex at $(5, 0)$

14. Vertices $(6, 0)$ and $(-6, 0)$; $\dfrac{c}{a} = \dfrac{13}{12}$

15. Center $(2, -1)$, major axis horizontal and of length 16, distance of 20 between foci

16. Change the equation of the hyperbola $x^2 - 4y^2 + 16y = 8$ into standard form and graph it.

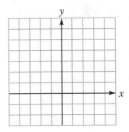

In Questions 17–18, solve each system algebraically.

17. $\begin{cases} x^2 + y^2 = 23 \\ y = x^2 - 3 \end{cases}$

18. $\begin{cases} 2x^2 - 3y^2 = 9 \\ x^2 + y^2 = 27 \end{cases}$

In Questions 19–20, complete the square to write each equation in standard form, and identify the curve.

19. $y^2 - 4y - 6x - 14 = 0$

20. $2x^2 + 3y^2 - 4x + 12y + 8 = 0$

■ Cumulative Review Exercises

In Exercises 1–16, simplify each expression. Assume that all variables represent positive numbers, and write answers without using negative exponents.

1. $64^{2/3}$

2. $8^{1/3}$

3. $\dfrac{y^{2/3} y^{5/3}}{y^{1/3}}$

4. $\dfrac{x^{5/3} x^{1/2}}{x^{3/4}}$

5. $(x^{2/3} - x^{1/3})(x^{2/3} + x^{1/3})$

6. $(x^{1/2} + x^{1/2})^2$

7. $\sqrt[3]{-27x^3}$

8. $\sqrt{48t^3}$

9. $\sqrt[3]{\dfrac{128x^4}{2x}}$

10. $\sqrt{x^2 + 6x + 9}$

11. $\sqrt{50} - \sqrt{8} + \sqrt{32}$

12. $-3\sqrt[4]{32} - 2\sqrt[4]{162} + 5\sqrt[4]{48}$

13. $3\sqrt{2}(2\sqrt{3} - 4\sqrt{12})$

14. $\dfrac{5}{\sqrt[3]{x}}$

15. $\dfrac{\sqrt{x}+2}{\sqrt{x}-1}$

16. $\sqrt[6]{x^3 y^3}$

In Exercises 17–18, solve each equation.

17. $5\sqrt{x+2} = x + 8$

18. $\sqrt{x} + \sqrt{x+2} = 2$

19. Use the method of completing the square to solve the equation $2x^2 + x - 3 = 0$.

20. Use the quadratic formula to solve the equation $3x^2 + 4x - 1 = 0$.

In Exercises 21–26, write each complex number in $a + bi$ form.

21. $(3 + 5i) + (4 - 3i)$

22. $(7 - 4i) - (12 + 3i)$

23. $(2 - 3i)(2 + 3i)$

24. $(3 + i)(3 - 3i)$

25. $(3 - 2i) - (4 + i)^2$

26. $\dfrac{5}{3 - i}$

In Exercises 27–28, find each value.

27. $|3 + 2i|$

28. $|5 - 6i|$

29. For what values of k will the solutions of $2x^2 + 4x = k$ be equal?

30. Find the coordinates of the vertex of the graph of the equation $y = \dfrac{1}{2}x^2 - x + 1$.

In Exercises 31–32, give the result in interval notation.

31. Solve $x^2 - x - 6 > 0$.

32. Solve $x^2 - x - 6 \le 0$.

In Exercises 33–36, $f(x) = 3x^2 + 2$ and $g(x) = 2x - 1$. Find each value or function.

33. $f(-1)$

34. $(g \circ f)(2)$

35. $(f \circ g)(x)$

36. $(g \circ f)(x)$

37. Write $y = \log_2 x$ in exponential equation.

38. Write $3^b = a$ in logarithmic notation.

In Exercises 39–42, find x.

39. $\log_x 25 = 2$

40. $\log_5 125 = x$

41. $\log_3 x = -3$

42. $\log_5 x = 0$

43. Find the inverse of $y = \log_2 x$.

44. If $\log_{10} 10^x = y$, then y equals what quantity?

In Exercises 45–48, $\log 7 = 0.8451$ and $\log 14 = 1.1461$. Evaluate each expression without using a calculator or tables.

45. $\log 98$

46. $\log 2$

47. $\log 49$

48. $\log \dfrac{7}{5}$ (*Hint*: $\log 10 = 1$.)

49. Solve $2^{x+2} = 3^x$.

50. Solve $2 \log 5 + \log x - \log 4 = 2$.

🔲 *In Exercises 51–52, use a calculator.*

51. Boat depreciation How much will a $9000 boat be worth after 9 years if it depreciates 12% per year?

52. Find $\log_6 8$.

53. Use graphing to solve $\begin{cases} 2x + y = 5 \\ x - 2y = 0 \end{cases}$.

54. Use substitution to solve $\begin{cases} 3x + y = 4 \\ 2x - 3y = -1 \end{cases}$.

55. Use addition to solve $\begin{cases} x + 2y = -2 \\ 2x - y = 6 \end{cases}$.

56. Use any method to solve $\begin{cases} \dfrac{x}{10} + \dfrac{y}{5} = \dfrac{1}{2} \\ \dfrac{x}{2} - \dfrac{y}{5} = \dfrac{13}{10} \end{cases}$.

57. Evaluate $\begin{vmatrix} 3 & -2 \\ 1 & -1 \end{vmatrix}$.

58. Use Cramer's rule and solve for y only:
$$\begin{cases} 4x - 3y = -1 \\ 3x + 4y = -7 \end{cases}$$

59. Solve $\begin{cases} x + y + z = 1 \\ 2x - y - z = -4 \\ x - 2y + z = 4 \end{cases}$.

60. Solve $\begin{cases} x + 2y + 3z = 6 \\ 3x + 2y + z = 6 \\ 2x + 3y + z = 6 \end{cases}$.

9 Natural Number Functions and Probability

IN THIS CHAPTER, WE INTRODUCE A WAY TO EXPAND POWERS OF BINOMIALS OF THE FORM $(x + y)^n$. THIS METHOD LEADS TO IDEAS NEEDED WHEN WORKING WITH PROBABILITY AND STATISTICS.

9.1 The Binomial Theorem

■ PASCAL'S TRIANGLE ■ FACTORIAL NOTATION ■ THE BINOMIAL THEOREM ■ FINDING A PARTICULAR TERM OF A BINOMIAL EXPANSION

**Blaise Pascal
(1623–1662)**

Pascal was torn between religion and mathematics. Each surfaced at times in his life to dominate his interest. In mathematics, Pascal made contributions to the study of conic sections, probability, and differential calculus. At the age of 19, he invented a calculating machine. He is best known for a triangular array of numbers that bears his name.

We begin this chapter by introducing a way to expand binomials of the form $(x + y)^n$. This method, called the **binomial theorem,** is important when working with probability and statistics. Consider the following expansions.

$$(a + b)^0 = 1$$
$$(a + b)^1 = a + b$$
$$(a + b)^2 = a^2 + 2ab + b^2$$
$$(a + b)^3 = a^3 + 3a^2b + 3ab^2 + b^3$$
$$(a + b)^4 = a^4 + 4a^3b + 6a^2b^2 + 4ab^3 + b^4$$
$$(a + b)^5 = a^5 + 5a^4b + 10a^3b^2 + 10a^2b^3 + 5ab^4 + b^5$$
$$(a + b)^6 = a^6 + 6a^5b + 15a^4b^2 + 20a^3b^3 + 15a^2b^4 + 6ab^5 + b^6$$

Four patterns are apparent in the above expansions:

1. Each expansion has one more term than the power of the binomial.
2. The degree of each term in each expansion equals the exponent of the binomial.
3. The first term in each expansion is a raised to the power of the binomial.
4. The exponents of a decrease by 1 in each successive term, and the exponents of b, beginning with b^0 in the first term, increase by 1 in each successive term.

■ PASCAL'S TRIANGLE

To see another pattern, we write the coefficients in a triangular array:

$$
\begin{array}{rccccccccccccc}
(a + b)^0 = & & & & & & & 1 & & & & & & \\
(a + b)^1 = & & & & & & 1 & & 1 & & & & & \\
(a + b)^2 = & & & & & 1 & & 2 & & 1 & & & & \\
(a + b)^3 = & & & & 1 & & 3 & & 3 & & 1 & & & \\
(a + b)^4 = & & & 1 & & 4 & & 6 & & 4 & & 1 & & \\
(a + b)^5 = & & 1 & & 5 & & 10 & & 10 & & 5 & & 1 & \\
(a + b)^6 = & 1 & & 6 & & 15 & & 20 & & 15 & & 6 & & 1
\end{array}
$$

In this array, each entry other than the 1's is the sum of the closest pair of numbers in the line above it. For example, the 20 in the bottom row is the sum of the 10's above it. The first 3 in the fourth row is the sum of the 1 and 2 above it.

This array, called **Pascal's triangle** after Blaise Pascal (1623–1662), continues with the same pattern forever. The next two lines are

$$
\begin{array}{rccccccccccccccc}
(a + b)^7 = & & 1 & & 7 & & 21 & & 35 & & 35 & & 21 & & 7 & & 1 \\
(a + b)^8 = & 1 & & 8 & & 28 & & 56 & & 70 & & 56 & & 28 & & 8 & & 1
\end{array}
$$

EXAMPLE 1 Expand $(x + y)^6$.

Solution The first term is x^6, and the exponents of x will decrease by 1 in each successive term. A y will appear in the second term, and the exponents of y will increase by 1 in each successive term, concluding when the term y^6 is reached. The variables in the expansion are

$$x^6 \quad x^5y \quad x^4y^2 \quad x^3y^3 \quad x^2y^4 \quad xy^5 \quad y^6$$

We can use Pascal's triangle to find the coefficients of the variables. Because the binomial is raised to the sixth power, we choose the row where the second entry is 6. The coefficients of the variables are the numbers in that row.

$$1 \quad 6 \quad 15 \quad 20 \quad 15 \quad 6 \quad 1$$

Putting this information together gives the expansion:

$$(x + y)^6 = x^6 + 6x^5y + 15x^4y^2 + 20x^3y^3 + 15x^2y^4 + 6xy^5 + y^6 \qquad ■$$

Self Check Expand $(p + q)^3$.
Answer $p^3 + 3p^2q + 3pq^2 + q^3$

EXAMPLE 2 Expand $(x - y)^6$.

Solution We write the binomial as $[x + (-y)]^6$ and substitute $-y$ for y in the result of Example 1.

$$[x + (-y)]^6$$
$$= x^6 + 6x^5(-y) + 15x^4(-y)^2 + 20x^3(-y)^3 + 15x^2(-y)^4 + 6x(-y)^5 + (-y)^6$$
$$= x^6 - 6x^5y + 15x^4y^2 - 20x^3y^3 + 15x^2y^4 - 6xy^5 + y^6$$

In general, the signs in the expansion of $(x - y)^n$ alternate. The sign of the first term is $+$, the sign of the second term is $-$, and so on. ■

Self Check Expand $(p - q)^3$.
Answer $p^3 - 3p^2q + 3pq^2 - q^3$

■ **FACTORIAL NOTATION**

To use the binomial theorem to expand a binomial, we will need **factorial notation**.

Factorial Notation
If n is a natural number, the symbol $n!$ (read either as **"n factorial"** or as **"factorial n"**) is defined as

$$n! = n(n - 1)(n - 2)(n - 3) \cdots (3)(2)(1)$$

EXAMPLE 3   Evaluate **a.** 3!, **b.** 6!, and **c.** 10!.

Solution   **a.** $3! = 3 \cdot 2 \cdot 1 = 6$

b. $6! = 6 \cdot 5 \cdot 4 \cdot 3 \cdot 2 \cdot 1 = 720$

c. $10! = 10 \cdot 9 \cdot 8 \cdot 7 \cdot 6 \cdot 5 \cdot 4 \cdot 3 \cdot 2 \cdot 1 = 3,628,800$ ■

Self Check   Evaluate **a.** 4! and **b.** 7!

Answer   **a.** 24, **b.** 5,040

There are two fundamental properties of factorials.

Properties of Factorials
1. By definition, $0! = 1$.
2. If n is a natural number, $(n - 1)! = n!$.

EXAMPLE 4   Show that **a.** $6 \cdot 5! = 6!$ and **b.** $8 \cdot 7! = 8!$.

Solution   **a.** $6 \cdot 5! = 6(5 \cdot 4 \cdot 3 \cdot 2 \cdot 1)$ **b.** $8 \cdot 7! = 8(7 \cdot 6 \cdot 5 \cdot 4 \cdot 3 \cdot 2 \cdot 1)$

$\qquad\qquad = 6 \cdot 5 \cdot 4 \cdot 3 \cdot 2 \cdot 1$ $\qquad\qquad = 8 \cdot 7 \cdot 6 \cdot 5 \cdot 4 \cdot 3 \cdot 2 \cdot 1$

$\qquad\qquad = 6!$ $\qquad\qquad\qquad = 8!$ ■

Self Check   Show that $4 \cdot 3! = 4!$.

Answer $4 \cdot (3 \cdot 2 \cdot 1) = 4!$

■ THE BINOMIAL THEOREM

We can now state the binomial theorem.

The Binomial Theorem
If n is any positive number, then

$$(a + b)^n = a^n + \frac{n!}{1!(n-1)!}a^{n-1}b + \frac{n!}{2!(n-2)!}a^{n-2}b^2$$
$$+ \frac{n!}{3!(n-3)!}a^{n-3}b^3 + \cdots + b^n$$

A proof of the binomial theorem appears in Appendix I.

In the binomial theorem, the exponents of the variables in each term on the right-hand side follow the familiar patterns:

1. The sum of the exponents of a and b in each term is n.

2. The exponents on a decrease by 1 in each successive term.

3. The exponents on b increase by 1 in each successive term.

However, the method of finding the coefficients is different. Except for the first and last terms, $n!$ is the numerator of each fractional coefficient. When the exponent on b is 2, the factors in the denominator of the fractional coefficient are 2! and $(n - 2)!$. When the exponent on b is 3, the factors in the denominator are 3! and $(n - 3)!$, and so on.

EXAMPLE 5   Expand $(a + b)^5$.

Solution   We substitute directly into the binomial theorem.

$$(a + b)^5 = a^5 + \frac{5!}{1!(5 - 1)!}a^4b + \frac{5!}{2!(5 - 2)!}a^3b^2 + \frac{5!}{3!(5 - 3)!}a^2b^3 + \frac{5!}{4!(5 - 4)!}ab^4 + b^5$$

$$= a^5 + \frac{5 \cdot 4!}{1 \cdot 4!}a^4b + \frac{5 \cdot 4 \cdot 3!}{2 \cdot 1 \cdot 3!}a^3b^2 + \frac{5 \cdot 4 \cdot 3!}{3! \cdot 2 \cdot 1}a^2b^3 + \frac{5 \cdot 4!}{4! \cdot 1}ab^4 + b^5$$

$$= a^5 + 5a^4b + 10a^3b^2 + 10a^2b^3 + 5ab^4 + b^5$$

We note that the coefficients are the same numbers as in the sixth row of Pascal's triangle. ■

Self Check   Expand $(p + q)^4$.
Answer $p^4 + 4p^3q + 6p^2q^2 + 4pq^3 + q^4$

EXAMPLE 6   Expand $(2x - 3y)^4$.

Solution   We first find the expansion of $(a + b)^4$.

$$(a + b)^4 = a^4 + \frac{4!}{1!(4 - 1)!}a^3b + \frac{4!}{2!(4 - 2)!}a^2b^2 + \frac{4!}{3!(4 - 3)!}ab^3 + b^4$$

$$= a^4 + \frac{4 \cdot 3!}{1 \cdot 3!}a^3b + \frac{4 \cdot 3 \cdot 2!}{2 \cdot 1 \cdot 2!}a^2b^2 + \frac{4 \cdot 3!}{3!1}ab^3 + b^4$$

$$= a^4 + 4a^3b + 6a^2b^2 + 4ab^3 + b^4$$

We then substitute $2x$ for a and $-3y$ for b in the result.

$$(a + b)^4 = a^4 + 4a^3b + 6a^2b^2 + 4ab^3 + b^4$$

$$[2x + (-3y)]^4 = (2x)^4 + 4(2x)^3(-3y) + 6(2x)^2(-3y)^2 + 4(2x)(-3y)^3 + (-3y)^4$$

$$(2x - 3y)^4 = 16x^4 - 96x^3y + 216x^2y^2 - 216xy^3 + 81y^4$$ ■

Self Check   Expand $(3x - 2y)^4$.
Answer $81x^4 - 216x^3y + 216x^2y^2 - 96xy^3 + 16y^4$

■ FINDING A PARTICULAR TERM OF A BINOMIAL EXPANSION

Suppose that we wish to find the fifth term of the expansion of $(a + b)^{11}$. It would be tedious to raise the binomial to the 11th power and then look at the fifth term. The binomial theorem provides an easier way.

EXAMPLE 7   Find the fifth term of the expansion of $(a + b)^{11}$.

Solution   In the fifth term, the exponent on b is 4 (the exponent on b is always 1 less than the number of the term). Since the exponent on b added to the exponent on a equals 11, the exponent on a is 7. The variables of the fifth term are a^7b^4.

The number in the numerator of the fractional coefficient is $n!$, which in this case is 11!. The factors in the denominator are 4! and $(11 - 4)!$. The complete fifth term is

$$\frac{11!}{4!(11 - 4)!} a^7b^4 - \frac{11!}{4!7!} a^7b^4$$

$$= \frac{11 \cdot 10 \cdot 9 \cdot 8 \cdot 7!}{4 \cdot 3 \cdot 2 \cdot 1 \cdot 7!} a^7b^4$$

$$= 330a^7b^4 \qquad ■$$

Self Check   Find the sixth term of the expansion in Example 7.
Answer $462a^6b^5$

EXAMPLE 8   Find the sixth term of the expansion of $(a + b)^9$.

Solution   In the sixth term, the exponent on b is 5, and the exponent on a is $9 - 5$, or 4. The numerator of the fractional coefficient is 9!, and the factors in the denominator are 5! and $(9 - 5)!$. The sixth term of the expansion is

$$\frac{9!}{5!(9 - 5)!} a^4b^5 = \frac{9 \cdot 8 \cdot 7 \cdot 6 \cdot 5!}{5! \cdot 4!} a^4b^5$$

$$= \frac{9 \cdot 8 \cdot 7 \cdot 6}{4 \cdot 3 \cdot 2 \cdot 1} a^4b^5$$

$$= 126a^4b^5 \qquad ■$$

Self Check   Find the fifth term of the expansion in Example 8.
Answer $126a^5b^4$

EXAMPLE 9 Find the third term of the expansion of $(3x - 2y)^6$.

Solution We begin by finding the third term of the expansion of $(a + b)^6$.

$$2. \quad \frac{6!}{2!(6-2)!} a^4 b^2 = \frac{6 \cdot 5 \cdot 4!}{2 \cdot 1 \cdot 4!} a^4 b^2 = 15 a^4 b^2$$

We can then substitute $3x$ for a and $-2y$ for b in Equation 2 to obtain the third term of the expansion of $(3x - 2y)^6$.

$$\begin{aligned} 15 a^4 b^2 &= 15(3x)^4 (-2y)^2 \\ &= 15(3)^4 (-2)^2 x^4 y^2 \\ &= 4860 x^4 y^2 \end{aligned}$$ ∎

Self Check Find the fourth term of the expansion in Example 9.
Answer $-4320 x^3 y^3$

EXERCISE 9.1

VOCABULARY AND CONCEPTS *In Exercises 1–8, fill in the blank to make a true statement.*

1. In the expansion of a binomial, there will be one more term than the _____ of the binomial.

2. The _____ of each term in a binomial expansion is the same as the exponent of the binomial.

3. The ____ term in a binomial expansion is the first term raised to the power of the binomial.

4. In the expansion of $(p + q)^n$, the _____ of p decrease by 1 in each successive term.

5. $7! = $ _____.

6. $0! = $ ___.

7. $n \cdot$ _____ $= n!$

8. In the seventh term of $(a + b)^{11}$, the exponent on a is ___.

PRACTICE *In Exercises 9–20, evaluate each expression.*

9. $4!$

10. $-5!$

11. $3! \cdot 6!$

12. $0! \cdot 7!$

13. $6! + 6!$

14. $5! - 2!$

15. $\dfrac{9!}{12!}$

16. $\dfrac{8!}{5!}$

17. $\dfrac{5! \cdot 7!}{9!}$

18. $\dfrac{3! \cdot 5! \cdot 7!}{1!8!}$

19. $\dfrac{18!}{6!(18-6)!}$

20. $\dfrac{15!}{9!(15-9)!}$

In Exercises 21–36, use the binomial theorem to expand each binomial.

21. $(a + b)^3$

22. $(a + b)^4$

23. $(a - b)^5$

24. $(x - y)^4$

25. $(2x + y)^3$

26. $(x + 2y)^3$

27. $(x - 2y)^3$

28. $(2x - y)^3$

29. $(2x + 3y)^4$

30. $(2x - 3y)^4$

31. $(x - 2y)^4$

32. $(x + 2y)^4$

33. $(x - 3y)^5$

34. $(3x - y)^5$

35. $\left(\dfrac{x}{2} + y\right)^4$

36. $\left(x + \dfrac{y}{2}\right)^4$

In Exercises 37–56, find the required term in each binomial expansion.

37. $(a + b)^4$; 3rd term

38. $(a - b)^4$; 2nd term

39. $(a + b)^7$; 5th term

40. $(a + b)^5$; 4th term

41. $(a - b)^5$; 6th term

42. $(a - b)^8$; 7th term

43. $(a + b)^{17}$; 5th term

44. $(a - b)^{12}$; 3rd term

45. $\left(a - \sqrt{2}\right)^4$; 2nd term

46. $\left(a - \sqrt{3}\right)^8$; 3rd term

47. $\left(a + \sqrt{3}b\right)^9$; 5th term

48. $\left(\sqrt{2}a - b\right)^7$; 4th term

49. $\left(\dfrac{x}{2} + y\right)^4$; 3rd term

50. $\left(m + \dfrac{n}{2}\right)^8$; 3rd term

51. $\left(\dfrac{r}{2} - \dfrac{s}{2}\right)^{11}$; 10th term

52. $\left(\dfrac{p}{2} - \dfrac{q}{2}\right)^9$; 6th term

53. $(a + b)^n$; 4th term

54. $(a - b)^n$; 5th term

55. $(a + b)^n$; rth term

56. $(a + b)^n$; $(r + 1)$th term

DISCOVERY AND WRITING

57. Find the sum of the numbers in each row of the first ten rows of Pascal's triangle. Do see you a pattern?

58. Show that the sum of the coefficients in the binomial expansion of $(x + y)^n$ is 2^n. (*Hint:* Let $x = y = 1$.)

59. Find the constant term in the expansion of $\left(a - \dfrac{1}{a}\right)^{10}$.

60. Find the coefficient of x^5 in the expansion of $\left(x + \dfrac{1}{x}\right)^9$.

61. If we apply the pattern of coefficients to the coefficient of the first term in the binomial theorem, it would be $\frac{n!}{0!(n - 0)!}$. Show that this expression equals 1.

62. If we apply the pattern of coefficients to the coefficient of the last term in the binomial theorem, it would be $\frac{n!}{n!(n - n)!}$. Show that this expression equals 1.

63. Define factorial notation and explain how to evaluate 10!.

64. With a calculator, evaluate 69!. Explain why you cannot find 70! with a calculator.

65. Explain how the rth term of a binomial expansion is constructed.

66. Explain the four patterns apparent in a binomial expansion.

REVIEW *In Exercises 67–70, factor each expression.*

67. $3x^3y^2z^4 - 6xyz^5 + 15x^2yz^2$

68. $3z^2 - 15tz + 12t^2$

69. $a^4 - b^4$

70. $3r^4 - 36r^3 - 135r^2$

In Exercises 71–72, simplify each complex fraction.

71. $\dfrac{\dfrac{1}{x} + \dfrac{1}{3}}{\dfrac{1}{x} - \dfrac{1}{3}}$

72. $\dfrac{x - \dfrac{1}{y}}{y - \dfrac{1}{x}}$

9.2 Sequences, Series, and Summation Notation

■ SEQUENCES ■ RECURSIVE DEFINITION OF A SEQUENCE ■ SERIES ■ SUMMATION NOTATION

In this section, we will introduce a function whose domain is the set of natural numbers. This function, called a **sequence,** is a list of numbers in a specific order.

■ SEQUENCES

Sequences

A **sequence** is a function whose domain is the set of natural numbers.

Since a sequence is a function whose domain is the set of natural numbers, we can write its terms as a list of numbers. For example, if n is a natural number, the function defined by $f(n) = 2n - 1$ generates the sequence

$$1, 3, 5, \ldots, 2n - 1, \ldots$$

The number 1 is the first term, 3 is the second term, and $2n - 1$ is the **general,** or **nth term.** If n is a natural number, the function $f(n) = 3n^2 + 1$ generates the sequence

$$4, 13, 28, \ldots, 3n^2 + 1, \ldots$$

The number 4 is the first term, 13 is the second term, 28 is the third term, and $3n^2 + 1$ is the general term.

A constant function such as $g(n) = 1$ is a sequence, because it generates the sequence

$$1, 1, 1, \ldots$$

We seldom use function notation to denote a sequence, because it is often difficult or even impossible to write the general term. In such cases, if there is a pattern that is assumed to be continued, we simply list several terms of the sequence. Some examples of sequences follow:

$$1^2, 2^2, 3^2, \ldots, n^2, \ldots$$

$$3, 9, 19, 33, \ldots, 2n^2 + 1, \ldots$$

$$1, 3, 6, 10, 15, 21, \ldots, \frac{n(n + 1)}{2}, \ldots$$

$$1, 1, 2, 3, 5, 8, 13, 21, \ldots \qquad \textbf{(Fibonacci sequence)}$$

$$2, 3, 5, 7, 11, 13, 17, 19, 23, \ldots \qquad \textbf{(prime numbers)}$$

The **Fibonacci sequence** is named after the 12th-century mathematician Leonardo of Pisa, also known as Fibonacci. After the two 1's in the Fibonacci sequence, each term is the sum of the two terms that immediately precede it. The Fibonacci sequence occurs in many fields, such as the growth patterns of plants, the reproductive habits of bees, and music.

Leonardo Fibonacci (late 12th and early 13th centuries)

Fibonacci, an Italian mathematician, is also known as Leonardo da Pisa. In his work *Liber abaci,* he advocated the adoption of Arabic numerals, the numerals that we use today.

■ RECURSIVE DEFINITION OF A SEQUENCE

A sequence can be defined **recursively** by giving its first term and a rule showing how to obtain the $(n + 1)$th term from the nth term. For example, the information

$$a_1 = 5 \qquad \text{(the first term)} \qquad \text{and} \qquad a_{n+1} = 3a_n - 2 \qquad \begin{array}{l}\text{(the rule showing}\\ \text{how to get the}\\ (n + 1)\text{th term}\\ \text{from the }n\text{th term)}\end{array}$$

defines a sequence recursively. To find the first five terms of this sequence, we proceed as follows:

$$a_1 = 5$$
$$a_2 = 3(a_1) - 2 = 3(5) - 2 = 13 \qquad \text{Substitute 5 for } a_1 \text{ and simplify to get } a_2.$$
$$a_3 = 3(a_2) - 2 = 3(13) - 2 = 37 \qquad \text{Substitute 13 for } a_2 \text{ and simplify to get } a_3.$$
$$a_4 = 3(a_3) - 2 = 3(37) - 2 = 109 \qquad \text{Substitute 37 for } a_3 \text{ and simplify to get } a_4.$$
$$a_5 = 3(a_4) - 2 = 3(109) - 2 = 325 \qquad \text{Substitute 109 for } a_4 \text{ and simplify to get } a_5.$$

■ SERIES

To add the terms of a sequence, we replace each comma between its terms with a + sign to form a **series.** Because each sequence is infinite, the number of terms in the series associated with it is infinite also. Two examples of infinite series are

$$1^2 + 2^2 + 3^2 + \cdots + n^2 + \cdots$$

and

$$1 + 2 + 3 + 5 + 8 + 13 + 21 + \cdots$$

If the signs between successive terms of an infinite series alternate, the series is called an **alternating infinite series.** Two examples of alternating infinite series are

$$-3 + 6 - 9 + 12 - \cdots + (-1)^n 3n + \cdots$$

and

$$2 - 4 + 8 - 16 + \cdots + (-1)^{n+1} 2^n + \cdots$$

■ SUMMATION NOTATION

Summation notation is a shorthand way to indicate the sum of the first n terms, or the **nth partial sum,** of a sequence. For example, the expression

$$\sum_{n=1}^{3} (2n^2 + 1) \qquad \text{The symbol } \Sigma \text{ is the capital letter sigma in the Greek alphabet.}$$

indicates the sum of the three terms obtained if we successively substitute 1, 2, and 3 for n in the expression $2n^2 + 1$.

$$\sum_{n=1}^{3} (2n^2 + 1) = [2(1)^2 + 1] + [2(2)^2 + 1] + [2(3)^2 + 1]$$
$$= 3 + 9 + 19$$
$$= 31$$

EXAMPLE 1   Evaluate $\displaystyle\sum_{n=1}^{4} (n^2 - 1)$.

Solution   Since n runs from 1 to 4, we substitute 1, 2, 3, and 4 for n in the expression $n^2 - 1$ and find the sum of the resulting terms:

$$\sum_{n=1}^{4} (n^2 - 1) = (1^2 - 1) + (2^2 - 1) + (3^2 - 1) + (4^2 - 1)$$
$$= 0 + 3 + 8 + 15$$
$$= 26 \qquad\qquad\qquad ■$$

Self Check   Evaluate $\displaystyle\sum_{n=1}^{5} (n^2 - 1)$.

Answer   50

EXAMPLE 2   Evaluate $\displaystyle\sum_{n=3}^{5} (3n + 2)$.

Solution Since n runs from 3 to 5, we substitute 3, 4, and 5 for n in the expression $3n + 2$ and find the sum of the resulting terms:

$$\sum_{n=3}^{5}(3n + 2) = [3(3) + 2] + [3(4) + 2] + [3(5) + 2]$$
$$= 11 + 14 + 17$$
$$= 42$$ ■

Self Check Evaluate $\sum_{n=2}^{5}(3n + 2)$.

Answer 50

There are three basic properties of summations. The first states that *the summation of a constant as k runs from 1 to n is n times the constant.*

Summation of a Constant

If c is a constant, then $\sum_{k=1}^{n} c = nc$.

Proof Because c is a constant, each term is c for each value of k as k runs from 1 to n.

$$\sum_{k=1}^{n} c = \overbrace{c + c + c + c + \cdots + c}^{n \text{ number of } c\text{'s}} = nc$$ □

EXAMPLE 3 Evaluate $\sum_{n=1}^{5} 13$.

Solution
$$\sum_{n=1}^{5} 13 = 13 + 13 + 13 + 13 + 13$$
$$= 5(13)$$
$$= 65$$ ■

Self Check Evaluate $\sum_{n=1}^{6} 12$.

Answer 72

A second property states that *a constant factor can be brought outside a summation sign.*

Summation of a Product

If c is a constant, then $\displaystyle\sum_{k=1}^{n} cf(k) = c\sum_{k=1}^{n} f(k)$.

Proof

$$\sum_{k=1}^{n} cf(k) = cf(1) + cf(2) + cf(3) + \cdots + cf(n)$$
$$= c[f(1) + f(2) + f(3) + \cdots + f(n)] \qquad \text{Factor out } c.$$
$$= c\sum_{k=1}^{n} f(k)$$

□

EXAMPLE 4 Show that $\displaystyle\sum_{k=1}^{3} 5k^2 = 5\sum_{k=1}^{3} k^2$.

Solution

$$\sum_{k=1}^{3} 5k^2 = 5(1)^2 + 5(2)^2 + 5(3)^2 \qquad 5\sum_{k=1}^{3} k^2 = 5[(1)^2 + (2)^2 + (3)^2]$$
$$= 5 + 20 + 45 \qquad\qquad\qquad = 5[1 + 4 + 9]$$
$$= 70 \qquad\qquad\qquad\qquad = 5(14)$$
$$\qquad\qquad\qquad\qquad\qquad\qquad = 70$$

The quantities are equal. ■

Self Check Evaluate $\displaystyle\sum_{k=1}^{4} 3k$.

Answer 30

The third property states that *the summation of a sum is equal to the sum of the summations.*

Summation of a Sum

$$\sum_{k=1}^{n} [f(k) + g(k)] = \sum_{k=1}^{n} f(k) + \sum_{k=1}^{n} g(k)$$

Proof

$$\sum_{k=1}^{n} [f(k) + g(k)] = [f(1) + g(1)] + [f(2) + g(2)] + [f(3) + g(3)] + \cdots + [f(n) + g(n)]$$
$$= [f(1) + f(2) + f(3) + \cdots + f(n)] + [g(1) + g(2) + g(3) + \cdots + g(n)]$$
$$= \sum_{k=1}^{n} f(k) + \sum_{k=1}^{n} g(k)$$

□

EXAMPLE 5   Show that $\displaystyle\sum_{k=1}^{3}(k + k^2) = \sum_{k=1}^{3}k + \sum_{k=1}^{3}k^2$.

Solution
$$\sum_{k=1}^{3}(k + k^2) = (1 + 1^2) + (2 + 2^2) + (3 + 3^2)$$
$$= 2 + 6 + 12$$
$$= 20$$

$$\sum_{k=1}^{3}k + \sum_{k=1}^{3}k^2 = (1 + 2 + 3) + (1^2 + 2^2 + 3^2)$$
$$= 6 + 14$$
$$= 20$$ ∎

Self Check   Evaluate $\displaystyle\sum_{k=1}^{3}(k^2 + 2k)$.

Answer   26

EXAMPLE 6   Evaluate $\displaystyle\sum_{k=1}^{5}(2k - 1)^2$ directly. Then expand the binomial, apply the previous properties, and evaluate the expression again.

Solution   Part 1: $\displaystyle\sum_{k=1}^{5}(2k - 1)^2 = 1 + 9 + 25 + 49 + 81 = 165$

Part 2: $\displaystyle\sum_{k=1}^{5}(2k - 1)^2 = \sum_{k=1}^{5}(4k^2 - 4k + 1)$

$$= \sum_{k=1}^{5}4k^2 + \sum_{k=1}^{5}(-4k) + \sum_{k=1}^{5}1$$ The summation of a sum is the sum of the summations.

$$= 4\sum_{k=1}^{5}k^2 - 4\sum_{k=1}^{5}k + \sum_{k=1}^{5}1$$ Bring the constant factors outside the summation signs.

$$= 4\sum_{k=1}^{5}k^2 - 4\sum_{k=1}^{5}k + 5$$ The summation of a constant as k runs from 1 to 5 is 5 times that constant.

$$= 4(1 + 4 + 9 + 16 + 25) - 4(1 + 2 + 3 + 4 + 5) + 5$$
$$= 4(55) - 4(15) + 5$$
$$= 220 - 60 + 5$$
$$= 165$$

Either way, the sum is 165. ∎

Self Check   Evaluate $\displaystyle\sum_{k=1}^{4} (2k - 1)^2$.

Answer   84

EXERCISE 9.2

VOCABULARY AND CONCEPTS   *In Exercises 1–8, fill in the blank to make a true statement.*

1. A sequence is a function whose _____ is the set of natural numbers.

2. A _____ is formed when we add the terms of a sequence.

3. _____ is a shorthand way to indicate the sum of the first n terms of a sequence.

4. The symbol $\displaystyle\sum_{k=1}^{5} (k^2 - 3)$ indicates the ____ of the five terms obtained when we successively substitute 1, 2, 3, 4, and 5 for n.

5. $\displaystyle\sum_{k=1}^{5} 6k^2 = \underline{} \sum_{k=1}^{5} k^2$

6. $\displaystyle\sum_{k=1}^{5} (k^2 + 3k) = \sum_{k=1}^{5} k^2 + \underline{}$

7. $\displaystyle\sum_{k=1}^{5} c$, where c is a constant, equals ___.

8. The summation of a sum is equal to the ____ of the summations.

PRACTICE   *In Exercises 9–10, write the first six terms of the sequence defined by each function.*

9. $f(n) = 5n(n - 1)$

10. $f(n) = n\left(\dfrac{n-1}{2}\right)\left(\dfrac{n-2}{3}\right)$

In Exercises 11–16, find the next term of each sequence.

11. 1, 6, 11, 16, . . .

12. 1, 8, 27, 64, . . .

13. $a, a + d, a + 2d, a + 3d, \ldots$

14. $a, ar, ar^2, ar^3, \ldots$

15. 1, 3, 6, 10, . . .

16. 20, 17, 13, 8, . . .

In Exercises 17–24, find the sum of the first five terms of the sequence with the given general term.

17. n

18. $2k$

19. 3

20. $4k^0$

21. $2\left(\dfrac{1}{3}\right)^n$

22. $(-1)^n$

23. $3n - 2$

24. $2k + 1$

In Exercises 25–32, a sequence is defined recursively. Find the first four terms of each sequence.

25. $a_1 = 3$ and $a_{n+1} = 2a_n + 1$

26. $a_1 = -5$ and $a_{n+1} = -a_n - 3$

27. $a_1 = -4$ and $a_{n+1} = \dfrac{a_n}{2}$

28. $a_1 = 0$ and $a_{n+1} = 2a_n^2$

29. $a_1 = k$ and $a_{n+1} = a_n^2$

30. $a_1 = 3$ and $a_{n+1} = ka_n$

31. $a_1 = 8$ and $a_{n+1} = \dfrac{2a_n}{k}$

32. $a_1 = m$ and $a_{n+1} = \dfrac{a_n^2}{m}$

In Exercises 33–36, tell whether each series is an alternating infinite series.

33. $-1 + 2 - 3 + \cdots + (-1)^n n + \cdots$

34. $a + \dfrac{a}{b} + \dfrac{a}{b^2} + \cdots + a\left(\dfrac{1}{b}\right)^{n-1} + \cdots ; b = 4$

35. $a + a^2 + a^3 + \cdots + a^n + \cdots ; a = 3$

36. $a + a^2 + a^3 + \cdots + a^n + \cdots ; a = -2$

In Exercises 37–50, evaluate each sum.

37. $\displaystyle\sum_{k=1}^{5} 2k$

38. $\displaystyle\sum_{k=3}^{6} 3k$

39. $\displaystyle\sum_{k=3}^{4} (-2k^2)$

40. $\displaystyle\sum_{k=1}^{100} 5$

41. $\displaystyle\sum_{k=1}^{5} (3k - 1)$

42. $\displaystyle\sum_{n=2}^{5} (n^2 + 3n)$

43. $\displaystyle\sum_{k=1}^{1000} \dfrac{1}{2}$

44. $\displaystyle\sum_{x=4}^{5} \dfrac{2}{x}$

45. $\displaystyle\sum_{x=3}^{4} \dfrac{1}{x}$

46. $\displaystyle\sum_{x=2}^{6} (3x^2 + 2x) - 3\sum_{x=2}^{6} x^2$

47. $\displaystyle\sum_{x=1}^{4} (4x + 1)^2 - \sum_{x=1}^{4} (4x - 1)^2$

48. $\displaystyle\sum_{x=0}^{10} (2x - 1)^2 + 4\sum_{x=0}^{10} x(1 - x)$

49. $\displaystyle\sum_{x=6}^{8} (5x - 1)^2 + \sum_{x=6}^{8} (10x - 1)$

50. $\displaystyle\sum_{x=2}^{7} (3x + 1)^2 - 3\sum_{x=2}^{7} x(3x + 2)$

DISCOVERY AND WRITING

51. Find a counterexample to disprove the proposition that the summation of a product is the product of the summations. In other words, prove that

$$\sum_{k=1}^{n} f(k)g(k) \neq \sum_{k=1}^{n} f(k) \sum_{k=1}^{n} g(k)$$

52. Find a counterexample to disprove the proposition that the summation of a quotient is the quotient of the summations. In other words, prove that

$$\sum_{k=1}^{n} \frac{f(k)}{g(k)} \neq \frac{\displaystyle\sum_{k=1}^{n} f(k)}{\displaystyle\sum_{k=1}^{n} g(k)}$$

53. Explain what it means to define something recursively.

54. Define the concept of *n*th partial sums.

REVIEW *In Exercises 55–56, the triangles are similar. Find the value of x.*

55.

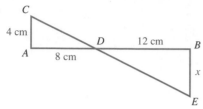

56.

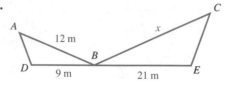

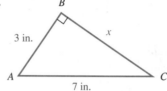

In Exercises 57–58, find the third side of each right triangle.

57.

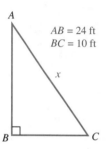

$AB = 24$ ft
$BC = 10$ ft

58.

3 in.

x

7 in.

9.3 Arithmetic Sequences

■ ARITHMETIC SEQUENCES ■ ARITHMETIC MEANS ■ SUM OF THE FIRST *N* TERMS OF AN ARITHMETIC SEQUENCE ■ APPLICATIONS

The German mathematician Carl Friedrich Gauss (1777–1855) was once a student in the class of a strict teacher. One day, the teacher asked the students to add together all of the natural numbers from 1 through 100. Gauss immediately knew the sum.

He recognized that in the sum

$$1 + 2 + 3 + \cdots + 98 + 99 + 100$$

the first number (1) added to the last number (100) is 101, the second number (2) added to the second from the last number (99) is 101, and the third number (3) added to the third from the last number (98) is 101. Gauss reasoned that there would be fifty pairs of such numbers, and that there would be fifty sums of 101. He multiplied 101 by 50 to get the correct answer of 5050.

This story illustrates a problem involving the sum of the terms of a sequence, called an **arithmetic sequence,** in which each term except the first is found by adding a constant to the preceding term.

■ ARITHMETIC SEQUENCES

> **Arithmetic Sequences**
> An **arithmetic sequence** is a sequence of the form
> $$a, \quad a + d, \quad a + 2d, \quad a + 3d, \ldots, a + (n - 1)d, \ldots$$
> where a is the **first term,** $a + (n - 1)d$ is the **nth term,** and d is the **common difference.**

In this definition, the second term has an addend of d, the third term has an addend of $2d$, the fourth term has an addend of $3d$, and so on. This is why the nth term has an addend of $(n - 1)d$.

EXAMPLE 1   Write the first six terms and the 21st term of an arithmetic sequence with a first term of 7 and a common difference of 5.

Solution   Since the first term a is 7 and the common difference d is 5, the first six terms are

$$7, 7 + 5, 7 + 2(5), 7 + 3(5), 7 + 4(5), 7 + 5(5)$$

or

$$7, 12, 17, 22, 27, 32$$

To find the 21st term, we substitute 21 for n in the formula for the nth term:

$$n\text{th term} = a + (n - 1)d$$
$$21\text{st term} = 7 + (21 - 1)5$$
$$= 7 + (20)5$$
$$= 107$$

The 21st term is 107.   ■

Self Check   Write the first five terms and the 18th term of an arithmetic sequence with a first term of 3 and a common difference of 6.

Answer   3, 9, 15, 21, 27; 105

EXAMPLE 2   Find the 98th term of an arithmetic sequence whose first three terms are 2, 6, and 10.

Solution   Here $a = 2$, $n = 98$, and $d = 6 - 2 = 10 - 6 = 4$. Because we want to find the 98th term, we substitute these numbers into the formula for the nth term:

$$n\text{th term} = a + (n - 1)d$$
$$98\text{th term} = 2 + (98 - 1)4$$
$$= 2 + (97)4$$
$$= 390$$

■

Self Check | Write the 50th term of the arithmetic sequence whose first three terms are 3, 8, and 13.

Answer | 248

■ ARITHMETIC MEANS

Numbers inserted between a first and last term to form a segment of an arithmetic sequence are called **arithmetic means.** When finding arithmetic means, we consider the last term, l, to be the nth term:

$$l = a + (n - 1)d$$

EXAMPLE 3 | Insert three arithmetic means between -3 and 12.

Solution | Since we are inserting three terms between -3 and 12, the total number of terms is five. Thus, $a = -3$, $l = 12$, and $n = 5$. To find the common difference, we substitute -3 for a, 12 for l, and 5 for n in the formula for the last term and solve for d:

$$l = a + (n - 1)d$$
$$12 = -3 + (5 - 1)d$$
$$15 = 4d \qquad \text{Add 3 to both sides and simplify.}$$
$$\frac{15}{4} = d \qquad \text{Divide both sides by 4.}$$

Once we know d, we can find the other terms of the sequence:

$$a + d = -3 + \frac{15}{4} = \frac{3}{4}$$
$$a + 2d = -3 + 2\left(\frac{15}{4}\right) = -3 + \frac{30}{4} = 4\frac{1}{2}$$
$$a + 3d = -3 + 3\left(\frac{15}{4}\right) = -3 + \frac{45}{4} = 8\frac{1}{4}$$

The three arithmetic means are $\frac{3}{4}$, $4\frac{1}{2}$, and $8\frac{1}{4}$. ■

Self Check | Find three arithmetic means between -5 and 23.
Answer | 2, 9, 16

■ SUM OF THE FIRST N TERMS OF AN ARITHMETIC SEQUENCE

To find the sum of the first n terms of an arithmetic sequence, we use the following formula.

Sum of the First *n* Terms of an Arithmetic Sequence
The formula

$$S_n = \frac{n(a + l)}{2}$$

gives the sum of the first *n* terms of an arithmetic sequence. In this formula, *a* is the first term, *l* is the last (or *n*th) term, and *n* is the number of terms.

Proof   We write the first *n* terms of an arithmetic sequence (letting S_n represent their sum). Then we write the same sum in reverse order and add the equations term by term:

$$S_n = \quad\quad a \quad\quad + \quad\quad (a + d) \quad\quad + \cdots + \ [a + (n - 2)d] \ + \ [a + (n - 1)d]$$
$$\underline{S_n = [a + (n - 1)d] \ + \ [a + (n - 2)d] \ + \cdots + \quad\quad (a + d) \quad\quad + \quad\quad a \quad\quad}$$
$$2S_n = [2a + (n - 1)d] + [2a + (n - 1)d] + \cdots + [2a + (n - 1)d] + [2a + (n - 1)d]$$

Because there are *n* equal terms on the right-hand side of the previous equation,

$$2S_n = n[2a + (n - 1)d]$$

or

1. $2S_n = n\{a + [a + (n - 1)d]\}$ Write $2a$ as $a + a$.

We can substitute *l* for $a + (n - 1)d$ in the right-hand side of Equation 1 and divide both sides by 2 to get

$$S_n = \frac{n(a + l)}{2}$$

$\square$

EXAMPLE 4   Find the sum of the first 30 terms of the arithmetic sequence 5, 8, 11, ...

Solution   Here $a = 5$, $n = 30$, $d = 3$, and $l = 5 + 29(3) = 92$. Substituting these values into the formula for the sum of an arithmetic sequence gives

$$S_n = \frac{n(a + l)}{2}$$
$$S_{30} = \frac{30(5 + 92)}{2}$$
$$= 15(97)$$
$$= 1455$$

The sum of the first 30 terms is 1455.   ■

Self Check   Find the sum of the first 50 terms of the arithmetic sequence −2, 5, 12, ...
Answer   8475

■ APPLICATIONS

EXAMPLE 5   A student deposits $50 in a non-interest-bearing account and plans to add $7 a week. How much will she have in the account one year after her first deposit?

Solution   Her weekly balances form an arithmetic sequence:

50, 57, 64, 71, 78, . . .

with a first term of 50 and a common difference of 7. To find her balance in one year (52 weeks), we substitute 50 for a, 7 for d, and 52 for n in the formula for the last term.

$$l = a + (n - 1)d$$
$$l = 50 + (52 - 1)7$$
$$l = 50 + (51)7$$
$$l = 407$$

After one year, the balance will be $407.   ■

Self Check   How much will the student in Example 5 have in the account after 60 weeks?
Answer $463

EXAMPLE 6   The equation $s = 16t^2$ represents the distance s (in feet) that an object will fall in t seconds.

- In 1 second, the object will fall 16 feet.
- In 2 seconds, the object will fall 64 feet.
- In 3 seconds, the object will fall 144 feet.

The object fell 16 feet during the first second, 48 feet during the next second, and 80 feet during the third second. Find the distance the object will fall during the 12th second.

Solution   The sequence 16, 48, 80, . . . is an arithmetic sequence with $a = 16$ and $d = 32$. To find the 12th term, we substitute these values into the formula for the last term.

$$l = a + (n - 1)d$$
$$l = 16 + (12 - 1)32$$
$$l = 16 + 11(32)$$
$$l = 368$$

During the 12th second, the object falls 368 feet.   ■

Self Check   How far will the object in Example 6 fall during the 20th second?
Answer   624 ft

EXERCISE 9.3

VOCABULARY AND CONCEPTS   *In Exercises 1–6, fill in the blank to make a true statement.*

1. An arithmetic sequence is a sequence of the form

$$a, a + d, a + 2d, a + 3d, \ldots a + \underline{\hspace{1cm}} d$$

2. In an arithmetic sequence, a is the _____ term, d is the common _____, and n is the _____ of terms.

3. The last term of an arithmetic sequence is given by the formula

_____

4. The formula for the sum of the first n terms of an arithmetic sequence is given by the formula

_____

5. _____ are numbers inserted between a first and last term to form an arithmetic sequence.

6. The formula _____ gives the distance (in feet) that an object will fall in t seconds.

PRACTICE   *In Exercises 7–12, write the first six terms of the arithmetic sequences with the given properties.*

7. $a = 1$; $d = 2$

8. $a = -12$; $d = -5$

9. $a = 5$; 3rd term is 2

10. $a = 4$; 5th term is 12

11. 7th term is 24; common difference is $\dfrac{5}{2}$

12. 20th term is -49; common difference is -3

In Exercises 13–20, find the missing term in each arithmetic sequence.

13. Find the 40th term of an arithmetic sequence with a first term of 6 and a common difference of 8.

14. Find the 35th term of an arithmetic sequence with a first term of 50 and a common difference of -6.

15. The 6th term of an arithmetic sequence is 28, and the first term is -2. Find the common difference.

16. The 7th term of an arithmetic sequence is -42, and the common difference is -6. Find the first term.

17. Find the 55th term of an arithmetic sequence whose first three terms are -8, -1, and 6.

18. Find the 37th term of an arithmetic sequence whose second and third terms are -4 and 6.

19. If the fifth term of an arithmetic sequence is 14 and the second term is 5, find the 15th term.

20. If the fourth term of an arithmetic sequence is 13 and the second term is 3, find the 24th term.

In Exercises 21–24, find the required means.

21. Insert three arithmetic means between 10 and 20.

22. Insert five arithmetic means between 5 and 15.

23. Insert four arithmetic means between -7 and $\dfrac{2}{3}$.

24. Insert three arithmetic means between -11 and -2.

In Exercises 25–28, find the sum of the first n terms of each arithmetic sequence.

25. $5 + 7 + 9 + \cdots$ (to 15 terms)

26. $-3 + (-4) + (-5) + \cdots$ (to 10 terms)

27. $\displaystyle\sum_{n=1}^{20}\left(\frac{3}{2}n + 12\right)$

28. $\displaystyle\sum_{n=1}^{10}\left(\frac{2}{3}n + \frac{1}{3}\right)$

In Exercises 29–32, solve each problem.

29. Find the sum of the first 30 terms of an arithmetic sequence with 25th term of 10 and a common difference of $\frac{1}{2}$.

30. Find the sum of the first 100 terms of an arithmetic sequence with 15th term of 86 and first term of 2.

31. Find the sum of the first 200 natural numbers.

32. Find the sum of the first 1000 natural numbers

APPLICATIONS

33. Borrowing money   To pay for college, a student borrows $5000 interest-free from his father. If he pays his father back at the rate of $200 per month, how much will he still owe after 12 months?

34. Borrowing money   If Juanita borrows $5500 interest-free from her mother to buy a new car and agrees to pay her mother back at the rate of $105 per month, how much will she still owe after four years?

35. Jogging   One day, some students jogged $\frac{1}{2}$ mile. Because it was fun, they decided to increase the jogging distance each day by a certain amount. If they jogged $6\frac{3}{4}$ miles on the 51st day, how much was the distance increased each day?

36. Sales   The year it incorporated, a company had sales of $237,500. Its sales were expected to increase by $150,000 annually for the next several years. If the forecast was correct, what will sales be in 10 years?

37. Falling objects   Find how many feet a brick will travel during the 10th second of its fall.

38. Falling objects   If a rock is dropped from the Golden Gate Bridge, how far will it fall in the third second?

39. Pile of logs   Several logs are stored in a pile with 20 logs on the bottom layer, 19 on the second layer, 18 on the third layer, and so on. If the top layer has one log, how many logs are in the pile?

40. Theater seating   The first row in a movie theater contains 24 seats. As you move toward the back, each row has 1 additional seat. If there are 30 rows, what is the capacity of the theater?

DISCOVERY AND WRITING

41. Can an arithmetic sequence have a first term of 4, a 25th term of 126, and a common difference of $4\frac{1}{4}$? Explain.

42. In an arithmetic sequence, can a and d be negative, but l positive?

43. Can an arithmetic sequence be an alternating sequence? Explain.

44. Between 5 and $10\frac{1}{3}$ are three arithmetic means. One of them is 9. Find the other two.

REVIEW   *In Exercises 45–50, solve each equation.*

45. $x + \sqrt{x + 3} = 9$

46. $\dfrac{x + 3}{x} + \dfrac{x - 3}{5} = 2$

47. $\dfrac{1}{x} + \dfrac{2}{x^2} + \dfrac{1}{x^3} = 0$

48. $x + 3\sqrt{x} - 10 = 0$

49. $x^4 - 1 = 0$

50. $x^4 - 29x^2 + 100 = 0$

9.4 Geometric Sequences

■ GEOMETRIC SEQUENCES ■ GEOMETRIC MEANS ■ SUM OF THE FIRST N TERMS OF A GEOMETRIC SEQUENCE ■ INFINITE GEOMETRIC SEQUENCES ■ APPLICATIONS

Another common sequence is the *geometric sequence.*

■ GEOMETRIC SEQUENCES

A **geometric sequence** is a sequence in which each term, except the first, is found by multiplying the preceding term by a constant.

> **Geometric Sequences**
> A **geometric sequence** is a sequence of the form
> $$a, \quad ar, \quad ar^2, \quad ar^3, \ldots, ar^{n-1}, \ldots$$
> where a is the first term, ar^{n-1} is the nth term, and r is the common ratio.

In this definition, the second term of the sequence has a factor of r^1, the third term has a factor of r^2, the fourth term has a factor of r^3, and so on. This explains why the nth term has a factor of r^{n-1}.

EXAMPLE 1   Write the first six terms and the 15th term of the geometric sequence whose first term is 3 and whose common ratio is 2.

Solution   We first write the first six terms of the geometric sequence:

$$3, \ 3(2), \ 3(2)^2, \ 3(2)^3, \ 3(2)^4, \ 3(2)^5$$

or

$$3, \ 6, \ 12, \ 24, \ 48, \ 96$$

To find the 15th term, we substitute 15 for n, 3 for a, and 2 for r in the formula for the nth term:

$$
\begin{aligned}
n\text{th term} &= ar^{n-1} \\
\mathbf{15}\text{th term} &= 3(2)^{15-1} \\
&= 3(2)^{14} \\
&= 3(16{,}384) \\
&= 49{,}152
\end{aligned}
$$ ■

Self Check   Write the first five terms of the geometric sequence whose first term is 2 and whose common ratio is 3. Find the 10th term.

Answers   2, 6, 18, 54, 162; 39,366

EXAMPLE 2   Find the eighth term of a geometric sequence whose first three terms are 9, 3, and 1.

Solution   Here $a = 9$, $r = \frac{1}{3}$, and $n = 8$. To find the eighth term, we substitute these values into the formula for the nth term.

$$n\text{th term} = ar^{n-1}$$
$$8\text{th term} = 9\left(\frac{1}{3}\right)^{8-1}$$
$$= 9\left(\frac{1}{3}\right)^{7}$$
$$= \frac{1}{243}$$
∎

Self Check   Find the eighth term of a geometric sequence whose first three terms are $\frac{1}{3}$, 1, and 3.

Answer   729

■ GEOMETRIC MEANS

Numbers inserted between a first and last term to form a segment of a geometric sequence are called **geometric means.** When finding geometric means, we consider the last term, l, to be the nth term.

EXAMPLE 3   Insert two geometric means between 4 and 256.

Solution   The first term is $a = 4$, and because 256 is the fourth term, $n = 4$ and $l = 256$. To find the common ratio, we substitute these values into the formula for the nth term and solve for r:

$$ar^{n-1} = l$$
$$4r^{4-1} = 256$$
$$r^3 = 64$$
$$r = 4$$

The common ratio is 4. The two geometric means are the second and third terms of the geometric sequence:

$$ar = 4 \cdot 4 = 16$$
$$ar^2 = 4 \cdot 4^2 = 4 \cdot 16 = 64$$

The first four terms of the geometric sequence are 4, 16, 64, and 256. The two geometric means between 4 and 256 are 16 and 64.   ∎

Self Check   Insert two geometric means between −3 and 192.

Answer   12, −48

■ SUM OF THE FIRST N TERMS OF A GEOMETRIC SEQUENCE

To find the sum of the first n terms of a geometric sequence, we use the following formula.

> **Sum of the First n Terms of a Geometric Sequence**
> The formula
> $$S_n = \frac{a - ar^n}{1 - r} \quad (r \neq 1)$$
> gives the sum of the first n terms of a geometric sequence. In the formula, S_n is the sum, a is the first term, r is the common ratio, and n is the number of terms.

Proof We write the sum of the first n terms of the geometric sequence:

2. $S_n = a + ar + ar^2 + \cdots + ar^{n-3} + ar^{n-2} + ar^{n-1}$

Multiplying both sides of this equation by r gives

3. $S_n r = ar + ar^2 + \cdots + ar^{n-2} + ar^{n-1} + ar^n$

We now subtract Equation 3 from Equation 2 and solve for S_n:

$$S_n - S_n r = a - ar^n$$
$$S_n(1 - r) = a - ar^n \qquad \text{Factor out } S_n.$$
$$S_n = \frac{a - ar^n}{1 - r} \qquad \text{Divide both sides by } 1 - r.$$

□

EXAMPLE 4 Find the sum of the first six terms of the geometric sequence 8, 4, 2, ...

Solution Here $a = 8$, $n = 6$, and $r = \frac{1}{2}$. Substituting these values into the formula for the sum of the first n terms of a geometric sequence gives

$$S_n = \frac{a - ar^n}{1 - r}$$

$$S_6 = \frac{8 - 8\left(\dfrac{1}{2}\right)^6}{1 - \dfrac{1}{2}}$$

$$= 2\left(\frac{63}{8}\right)$$

$$= \frac{63}{4}$$

The sum of the first six terms is $\dfrac{63}{4}$.

■

Self Check   Find the sum of the first eight terms of the geometric sequence $81, 27, 9, \ldots$.

Answer $\dfrac{3280}{27}$

■ INFINITE GEOMETRIC SEQUENCES

Under certain conditions, we can find the sum of all of the terms in an **infinite geometric sequence**. To define this sum, we consider the geometric sequence

$$a, ar, ar^2, \ldots$$

- The first partial sum, S_1, of the sequence is $S_1 = a$.
- The second partial sum, S_2, of the sequence is $S_2 = a + ar$.
- The nth partial sum, S_n, of the sequence is $S_n = a + ar + ar^2 + \cdots + ar^{n-1}$.

If the nth partial sum, S_n, of an infinite geometric sequence approaches some number S as $n \to \infty$, then S is called the **sum of the infinite geometric sequence.** The following symbol denotes the sum, S, of an infinite geometric sequence, provided the sum exists.

$$S = \sum_{n=1}^{\infty} ar^{n-1}$$

To develop a formula for finding the sum of all the terms in an infinite geometric sequence, we consider the formula

4. $S_n = \dfrac{a - ar^n}{1 - r}$ $(r \neq 1)$

If $|r| < 1$ and a is a constant, then as $n \to \infty$, $ar^n \to 0$, and the term ar^n in Equation 4 can be dropped. This argument gives the following formula.

Sum of the Terms of an Infinite Geometric Sequence
If $|r| < 1$, the sum of the terms of an infinite geometric sequence is given by

$$S = \dfrac{a}{1 - r}$$

where a is the first term and r is the common ratio.

WARNING!   If $r \geq 1$, the terms get larger and larger, and the sum does not approach a limit. In this case, the previous theorem does not apply.

EXAMPLE 5 Change $0.\overline{4}$ to a common fraction.

Solution We write the decimal as an infinite geometric series and find its sum:

$$S = \frac{4}{10} + \frac{4}{100} + \frac{4}{1000} + \frac{4}{10,000} + \cdots$$

$$S = \frac{4}{10} + \frac{4}{10}\left(\frac{1}{10}\right) + \frac{4}{10}\left(\frac{1}{10}\right)^2 + \frac{4}{10}\left(\frac{1}{10}\right)^3 + \cdots$$

Since the common ratio r equals $\frac{1}{10}$ and $|\frac{1}{10}| < 1$, we can use the formula for the sum of an infinite geometric series:

$$S = \frac{a}{1 - r} = \frac{\dfrac{4}{10}}{1 - \dfrac{1}{10}} = \frac{\dfrac{4}{10}}{\dfrac{9}{10}} = \frac{4}{9}$$

Long division will verify that $\frac{4}{9} = 0.\overline{4}$. ∎

Self Check Change $0.\overline{7}$ to a common fraction.

Answer $\frac{7}{9}$

■ APPLICATIONS

Many of the exponential growth problems discussed in Chapter 5 can be solved using the concepts of geometric sequences.

EXAMPLE 6 A town with a population of 3500 people has a predicted growth rate of 6% per year for the next 20 years. How many people are expected to live in the town 20 years from now?

Solution Let p_0 be the initial population of the town. After 1 year, the population p_1 will be the initial population (p_0) plus the growth (the product of p_0 and the rate of growth, r).

$$p_1 = p_0 + p_0 r$$
$$ = p_0(1 + r) \qquad \text{Factor out } p_0.$$

The population p_2 at the end of 2 years will be

$$p_2 = p_1 + p_1 r$$
$$p_2 = p_1(1 + r) \qquad \text{Factor out } p_1.$$
$$p_2 = p_0(1 + r)(1 + r) \qquad \text{Substitute } p_0(1 + r) \text{ for } p_1.$$
$$p_2 = p_0(1 + r)^2$$

The population at the end of the third year will be $p_3 = p_0(1 + r)^3$. Writing the terms in a sequence gives

$$p_0, \quad p_0(1 + r), \quad p_0(1 + r)^2, \quad p_0(1 + r)^3, \quad p_0(1 + r)^4, \ldots$$

This is a geometric sequence with p_0 as the first term and $1 + r$ as the common ratio. In this example, $p_0 = 3500$, $1 + r = 1.06$, and (since the population after 20 years will be the value of the 21st term of the geometric sequence) $n = 21$. We can substitute these values into the formula for the last term of a geometric sequence to get

$$l = ar^{n-1}$$
$$l = 3500(1.06)^{21-1}$$
$$l = 3500(1.06)^{20}$$
$$l \approx 11{,}224.97415 \qquad \text{Use a calculator.}$$

The population after 20 years will be approximately 11,225. ∎

EXAMPLE 7   A woman deposits \$2500 in a bank at 7% annual interest, compounded daily. If the investment is left untouched for 60 years, how much money will be in the account?

Solution   We let the initial amount in the account be a_0 and r be the rate. At the end of the first day, the account is worth

$$a_1 = a_0 + a_0\left(\frac{r}{365}\right) = a_0\left(1 + \frac{r}{365}\right)$$

After the second day, the account is worth

$$a_2 = a_1 + a_1\left(\frac{r}{365}\right) = a_1\left(1 + \frac{r}{365}\right) = a_0\left(1 + \frac{r}{365}\right)^2$$

The daily amounts form the following geometric sequence

$$a_0, \, a_0\left(1 + \frac{r}{365}\right), \quad a_0\left(1 + \frac{r}{365}\right)^2, \quad a_0\left(1 + \frac{r}{365}\right)^3, \ldots$$

where a_0 is the initial deposit and r is the annual rate of interest.

Because interest is compounded daily for 60 years (21,900 days), the amount at the end of 60 years will be the 21,901th term of the sequence.

$$a_{21{,}901} = 2500\left(1 + \frac{0.07}{365}\right)^{21{,}900}$$

We can use a calculator to find that $a_{21{,}901} \approx \$166{,}648.70$. ∎

EXAMPLE 8   A pump can remove 20% of the gas in a container with each stroke. Find the percentage of gas that remains in the container after six strokes.

Solution   We let V represent the volume of the container. Because each stroke of the pump removes 20% of the gas, 80% remains after each stroke, and we have the geometric sequence

$$V, \quad 0.80V, \quad 0.80(0.80V), \quad 0.80[0.80(0.80V)], \ldots$$

or

$$V, \quad 0.80V, \quad (0.8)^2V, \quad (0.8)^3V, \quad (0.8)^4V, \ldots$$

The amount of gas remaining after six strokes is the seventh term, l, of the sequence:

$$l = ar^{n-1}$$
$$l = V(0.8)^{7-1}$$
$$l = V(0.8)^6$$

We can use a calculator to find that approximately 26% of the gas remains after six strokes. ∎

EXERCISE 9.4

VOCABULARY AND CONCEPTS *In Exercises 1–6, fill in the blank to make a true statement.*

1. A geometric sequence is a sequence of the form a, ar, ar^2, ar^3, ... The nth term is $a(\underline{\quad})$.

2. In a geometric sequence, a is the _____ term, r is the common _____, and n is the _____ of terms.

3. The last term of a geometric sequence is given by the formula

_____

4. The formula for the sum of the first n terms of a geometric sequence is given by

_____

5. _____ are numbers inserted between a first and a last term to form a geometric sequence.

6. If $|r| < 1$, the formula _____ gives the sum of the terms of an infinite geometric sequence.

PRACTICE *In Exercises 7–14, write the first four terms of each geometric sequence with the given properties.*

7. $a = 10$; $r = 2$

8. $a = -3$; $r = 2$

9. $a = -2$ and $r = 3$

10. $a = 64$; $r = \dfrac{1}{2}$

11. $a = 3$; $r = \sqrt{2}$

12. $a = 2$; $r = \sqrt{3}$

13. $a = 2$; 4th term is 54

14. 3rd term is 4; $r = \dfrac{1}{2}$

In Exercises 15–18, find the requested term of each geometric sequence.

15. Find the sixth term of the geometric sequence whose first three terms are $\frac{1}{4}$, 1, and 4.

16. Find the eighth term of the geometric sequence whose second and fourth terms are 0.2 and 5.

17. Find the fifth term of a geometric sequence whose second term is 6 and whose third term is -18.

18. Find the sixth term of a geometric sequence whose second term is 3 and whose fourth term is $\frac{1}{3}$.

In Exercises 19–22, solve each problem.

19. Insert three positive geometric means between 10 and 20.

20. Insert five geometric means between -5 and 5, if possible.

21. Insert four geometric means between 2 and 2048.

22. Insert three geometric means between 162 and 2. (There are two possibilities.)

In Exercises 23–28, find the sum of the indicated terms of each geometric sequence.

23. 4, 8, 16, . . . (to 5 terms)

24. 9, 27, 81, . . . (to 6 terms)

25. 2, −6, 18, . . . (to 10 terms)

26. $\dfrac{1}{8}, \dfrac{1}{4}, \dfrac{1}{2}, \ldots$ (to 12 terms)

27. $\displaystyle\sum_{n=1}^{6} 3\left(\dfrac{3}{2}\right)^{n-1}$

28. $\displaystyle\sum_{n=1}^{6} 12\left(-\dfrac{1}{2}\right)^{n-1}$

In Exercises 29–32, find the sum of each infinite geometric sequence.

29. $6 + 4 + \dfrac{8}{3} + \cdots$

30. $8 + 4 + 2 + 1 + \cdots$

31. $\displaystyle\sum_{n=1}^{\infty} 12\left(-\dfrac{1}{2}\right)^{n-1}$

32. $\displaystyle\sum_{n=1}^{\infty} \left(\dfrac{1}{3}\right)^{n-1}$

In Exercises 33–36, change each decimal to a common fraction.

33. $0.\overline{5}$

34. $0.\overline{6}$

35. $0.\overline{25}$

36. $0.\overline{37}$

APPLICATIONS   ⊞   *In Exercises 37–54, use a calculator to help solve each problem.*

37. Staffing a department The number of students studying algebra at State College is 623. The Department Chair expects enrollment to increase 10% each year. How many professors will be needed in 8 years to teach algebra if one professor can handle 60 students?

38. Bouncing balls A Super Ball rebounds to approximately 95% of the height from which it is dropped. If the ball is dropped from a height of 10 meters, how high will it rebound after the 13th bounce?

39. Investing money If a married couple invests $1000 in a 1-year certificate of deposit at $6\frac{3}{4}\%$ annual interest, compounded daily, how much interest will be earned during the year?

40. Biology If a single cell divides into two cells every 30 minutes, how many cells will there be at the end of ten hours?

41. Depreciation A lawn tractor, costing c dollars when new, depreciates 20% of the previous year's value each year. How much is the lawn tractor worth after 5 years?

42. Financial planning Maria can invest $1000 at $7\frac{1}{2}\%$, compounded annually, or at $7\frac{1}{4}\%$, compounded daily. If she invests the money for a year, which is the better investment?

43. Population study If the population of the earth were to double every 30 years, approximately how many people would there be in the year 3010? (Consider the population in 1990 to be 5 billion and use 1990 as the base year.)

44. Investing money If Linda deposits $1300 in a bank at 7% interest, compounded annually, how much will be in the bank 17 years later? (Assume that there are no other transactions on the account.)

45. Real estate appreciation If a house purchased for $50,000 in 1988 appreciates in value by 6% each year, how much will the house be worth in the year 2010?

46. Compound interest Find the value of $1000 left on deposit for 10 years at an annual rate of 7%, compounded annually.

47. Compound interest   Find the value of $1000 left on deposit for 10 years at an annual rate of 7%, compounded quarterly.

48. Compound interest   Find the value of $1000 left on deposit for 10 years at an annual rate of 7%, compounded monthly.

49. Compound interest   Find the value of $1000 left on deposit for 10 years at an annual rate of 7%, compounded daily.

50. Compound interest   Find the value of $1000 left on deposit for 10 years at an annual rate of 7%, compounded hourly.

51. Saving for retirement   When John was 20 years old, he opened an individual retirement account by investing $2000 at 11% interest, compounded quarterly. How much will his investment be worth when he is 65 years old?

52. Biology   One bacterium divides into two bacteria every 5 minutes. If two bacteria multiply enough to completely fill a petri dish in 2 hours, how long will it take one bacterium to fill the dish?

53. Mathematical myths   A legend tells of a king who offered to grant the inventor of the game of chess any request. The inventor said, "Simply place one grain of wheat on the first square of a chessboard, two grains on the second, four on the third, and so on, until the board is full. Then give me the wheat." The king agreed. How many grains did the king need to fill the chessboard?

54. Mathematical myths   Estimate the size of the wheat pile in Exercise 53. (*Hint:* There are about one-half million grains of wheat in a bushel.)

DISCOVERY AND WRITING   *Write a paragraph using your own words.*

55. Does 0.999999 = 1? Explain.

56. Does 0.999 . . . = 1? Explain.

REVIEW   *In Exercises 57–64, do each operation.* $\left(i = \sqrt{-1}.\right)$

57. $(3 + 2i) + (2 - 5i)$

58. $(7 + 8i) - (2 - 5i)$

59. $(3 + i)(3 - i)$

60. $\left(3 + \sqrt{3}i\right)\left(3 - \sqrt{3}i\right)$

61. $\dfrac{2 + 3i}{2 - i}$

62. $(7 + 3i)^2$

63. i^{127}

64. i^{-127}

9.5 Mathematical Induction

■ THE METHOD OF MATHEMATICAL INDUCTION

In Section 9.4, we developed the following formula for the sum of the first n terms of an arithmetic sequence:

$$S_n = \frac{n(a + l)}{2}$$

If we apply this formula to the arithmetic series $1 + 2 + 3 + \cdots + n$, we have

$$S_n = 1 + 2 + 3 + \cdots + n = \frac{n(1 + n)}{2}$$

To see that this formula is true, we can check it for some positive numbers n:

For $n = 1$: $1 = \dfrac{1(1 + 1)}{2}$ is a true statement, because $1 = 1$.

For $n = 2$: $1 + 2 = \dfrac{2(1 + 2)}{2}$ is a true statement, because $3 = 3$.

For $n = 3$: $1 + 2 + 3 = \dfrac{3(1 + 3)}{2}$ is a true statement, because $6 = 6$.

For $n = 6$: $1 + 2 + 3 + 4 + 5 + 6 = \dfrac{6(1 + 6)}{2}$ is a true statement, because $21 = 21$.

However, because the set of positive numbers is infinite, it is impossible to prove the formula by verifying it for all positive numbers. To verify this sequence formula for all positive numbers n, we must have a method of proof called **mathematical induction,** a method first used extensively by Giuseppe Peano (1858–1932).

■ THE METHOD OF MATHEMATICAL INDUCTION

Suppose that we are standing in line for a movie and are worrying about whether we will be admitted. Even if the first person gets in, our worries would still be justified. Perhaps there is only room in the theater for a few people.

It would be good news to hear the theater manager say, "If anyone gets in, the next person in line will get in also." However, this promise does not guarantee that anyone will be admitted. Perhaps the theater is already full, and no one will get in.

However, when we see the first person in line walk in, we can conclude that everyone will be admitted, because we know two things:

- The first person was admitted.

- Because of the promise, if the first person is admitted, then so is the second, and when the second person is admitted, then so is the third, and so on until everyone gets in.

This situation is similar to a game played with dominoes. Suppose that some dominoes are placed on end, as in Figure 9-1. When the first domino is knocked over, it knocks over the second. The second domino, in turn, knocks over the third, which knocks over the fourth, and so on until all of the dominoes fall. Two things must happen to guarantee that all of the dominoes fall:

FIGURE 9-1

- The first domino must be knocked over.

- Every domino that falls must knock over the next one.

When both conditions are met, it is certain that all of the dominoes will fall.

The preceding examples illustrate the basic principle of mathematical induction.

Mathematical Induction

If a statement involving the natural number n has the following two properties

1. The statement is true for $n = 1$, and

2. If the statement is true for $n = k$, then it is true for $n = k + 1$,

then the statement is true for all natural numbers.

Mathematical induction provides a way to prove many formulas. Any proof by induction involves two parts. First, we must show that the formula is true for the number 1. Second, we must show that, if the formula is true for any natural number k, then it also is true for the natural number $k + 1$. A proof by induction is complete only when both of these properties are established.

EXAMPLE 1 Use mathematical induction to prove that the following formula is true for every natural number n.

$$1 + 2 + 3 + \cdots + n = \frac{n(n + 1)}{2}$$

Solution *Part 1:* Verify that the formula is true for $n = 1$. When $n = 1$, there is a single term, the number 1, on the left-hand side of the equation. Substituting 1 for n on the right-hand side, we have

$$1 = \frac{n(n + 1)}{2}$$

$$1 = \frac{(1)(1 + 1)}{2}$$

$$1 = 1$$

The formula is true when $n = 1$. Part 1 of the proof is complete.

Part 2: We assume that the given formula is true when $n = k$. By this assumption, called the **induction hypothesis,** we accept that

1. $1 + 2 + 3 + \cdots + k = \dfrac{k(k + 1)}{2}$

is a true statement. We must show that the induction hypothesis forces the given formula to be true when $n = k + 1$. We can show this by verifying the statement

2. $1 + 2 + 3 + \cdots + k + (k + 1) = \dfrac{(k + 1)[(k + 1) + 1]}{2}$

obtained from the given formula by replacing n with $k + 1$.

Comparing the left-hand sides of Equations 1 and 2 shows that the left-hand side of Equation 2 contains an extra term of $k + 1$. Thus, we add $k + 1$ to both sides of Equation 1 (which was assumed to be true) to obtain the equation

$$1 + 2 + 3 + \cdots + k + (k + 1) = \frac{k(k + 1)}{2} + (k + 1)$$

Because both terms on the right-hand side of this equation have a common factor of $k + 1$, the right-hand side factors, and the equation can be written as follows:

$$1 + 2 + 3 + \cdots + k + (k + 1) = (k + 1)\left(\frac{k}{2} + 1\right)$$
$$= (k + 1)\left(\frac{k + 2}{2}\right)$$
$$= \frac{(k + 1)(k + 2)}{2}$$
$$= \frac{(k + 1)[(k + 1) + 1]}{2}$$

This final result is Equation 2. Because the truth of Equation 1 implies the truth of Equation 2, Part 2 of the proof is complete. Parts 1 and 2 together establish that the formula is true for any natural number n. ∎

EXAMPLE 2   Use mathematical induction to prove the following formula for all natural numbers n.

$$1 + 5 + 9 + \cdots + (4n - 3) = n(2n - 1)$$

Solution   *Part 1:* First we verify the formula for $n = 1$. When $n = 1$, there is a single term, the number 1, on the left-hand side of the equation. Substituting 1 for n on the right-hand side, we have

$$1 = \mathbf{1}[2(\mathbf{1}) - 1]$$
$$1 = 1$$

The formula is true for $n = 1$. Part 1 of the proof is complete.

Part 2: We assume that the formula is true for $n = k$. Hence,

3. $1 + 5 + 9 + \cdots + (4k - 3) = k(2k - 1)$

is a true statement. To show that the induction hypothesis guarantees the truth of the formula for $k + 1$ terms, we add the $(k + 1)$th term to both sides of Equation 3. Because the terms on the left-hand side increase by 4, the $(k + 1)$th term is $(4k - 3) + 4$, or $4k + 1$. Adding $4k + 1$ to both sides of Equation 3 gives

$$1 + 5 + 9 + \cdots + (4k - 3) + (4k + 1) = k(2k - 2) + (4k + 1)$$

We can simplify the right-hand side and write the previous equation as follows:

$$1 + 5 + 9 + \cdots + (4k - 3) + [4(k + 1) - 3] = 2k^2 + 3k + 1$$
$$= (k + 1)(2k + 1)$$
$$= (k + 1)[2(k + 1) - 1]$$

Since this result has the same form as the given formula, except that $k + 1$ replaces n, the truth of the formula for $n = k$ implies the truth of the formula for $n = k + 1$. Part 2 of the proof is complete.

Because both of the induction requirements are true, the formula is true for all natural numbers n.   ∎

EXAMPLE 3   Prove that $\dfrac{1}{2} + \dfrac{1}{4} + \dfrac{1}{8} + \cdots + \dfrac{1}{2^n} < 1$.

Solution   *Part 1:* We verify the formula for $n = 1$. When $n = 1$, there is a single term, the fraction $\frac{1}{2}$, on the left-hand side of the equation. Substituting 1 for n on the right-hand side, we have the following true statement:

$$\frac{1}{2} < 1$$

The formula is true for $n = 1$. Part 1 of the proof is complete.

Part 2: We assume that the inequality is true for $n = k$. Thus,

$$\frac{1}{2} + \frac{1}{4} + \frac{1}{8} + \cdots + \frac{1}{2^k} < 1$$

We can multiply both sides of the above inequality by $\frac{1}{2}$ to get

$$\frac{1}{2}\left(\frac{1}{2} + \frac{1}{4} + \frac{1}{8} + \cdots + \frac{1}{2^k}\right) < 1\left(\frac{1}{2}\right)$$

or

$$\frac{1}{4} + \frac{1}{8} + \frac{1}{16} + \cdots + \frac{1}{2^{k+1}} < \frac{1}{2}$$

We now add $\frac{1}{2}$ to both sides of this inequality to get

$$\frac{1}{2} + \frac{1}{4} + \frac{1}{8} + \frac{1}{16} + \cdots + \frac{1}{2^{k+1}} < \frac{1}{2} + \frac{1}{2}$$

or

$$\frac{1}{2} + \frac{1}{4} + \frac{1}{8} + \frac{1}{16} + \cdots + \frac{1}{2^{k+1}} < 1$$

The resulting inequality is the same as the original inequality, except that $k + 1$ appears in place of n. Thus, the truth of the inequality for $n = k$ implies the truth of the inequality for $n = k + 1$. Part 2 of the proof is complete.

Because both of the induction requirements have been verified, this inequality is true for all natural numbers.   ∎

Some statements are not true when $n = 1$, but are true for all natural numbers equal to or greater than some given natural number (say, q). In these cases, we verify the given statements for $n = q$ in Part 1 of the induction proof. After establishing Part 2 of the induction proof, the given statement is proved for all natural numbers that are greater than q.

EXERCISE 9.5

VOCABULARY AND CONCEPTS *In Exercises 1–4, fill in the blank to make a true statement.*

1. Any proof by induction requires _____ parts.

2. Part 1 is to show that the statement is true for _____.

3. Part 2 is to show that the statement is true for _____ whenever it is true for $n = k$.

4. When we assume that a formula is true for $n = k$, we call the assumption the induction _____.

PRACTICE *In Exercises 5–8, verify each formula for $n = 1, 2, 3,$ and 4.*

5. $5 + 10 + 15 + \cdots + 5n = \dfrac{5n(n + 1)}{2}$

6. $1^2 + 2^2 + 3^2 + \cdots + n^2 = \dfrac{n(n + 1)(2n + 1)}{6}$

7. $7 + 10 + 13 + \cdots + (3n + 4) = \dfrac{n(3n + 11)}{2}$

8. $1(3) + 2(4) + 3(5) + \cdots + n(n + 2) =$
$\dfrac{n}{6}(n + 1)(2n + 7)$

In Exercises 9–34, prove each formula by mathematical induction, if possible.

9. $2 + 4 + 6 + \cdots + 2n = n(n + 1)$

10. $1 + 3 + 5 + \cdots + (2n - 1) = n^2$

11. $3 + 7 + 11 + \cdots + (4n - 1) = n(2n + 1)$

12. $4 + 8 + 12 + \cdots + 4n = 2n(n + 1)$

13. $10 + 6 + 2 + \cdots + (14 - 4n) = 12n - 2n^2$

14. $8 + 6 + 4 + \cdots + (10 - 2n) = 9n - n^2$

15. $2 + 5 + 8 + \cdots + (3n - 1) = \dfrac{n(3n + 1)}{2}$

16. $3 + 6 + 9 + \cdots + 3n = \dfrac{3n(n + 1)}{2}$

17. $1^2 + 2^2 + 3^2 + \cdots + n^2 = \dfrac{n(n + 1)(2n + 1)}{6}$

18. $1 + 2 + 3 + \cdots + (n - 1) + n + (n - 1) + \cdots + 3 + 2 + 1 = n^2$

19. $\dfrac{1}{3} + 2 + \dfrac{11}{3} + \cdots + \left(\dfrac{5}{3}n - \dfrac{4}{3}\right) = n\left(\dfrac{5}{6}n - \dfrac{1}{2}\right)$

20. $\dfrac{1}{1 \cdot 2} + \dfrac{1}{2 \cdot 3} + \dfrac{1}{3 \cdot 4} + \cdots + \dfrac{1}{n(n + 1)} = \dfrac{n}{n + 1}$

21. $\dfrac{1}{2} + \dfrac{1}{4} + \dfrac{1}{8} + \cdots + \left(\dfrac{1}{2}\right)^n = 1 - \left(\dfrac{1}{2}\right)^n$

22. $\dfrac{1}{3} + \dfrac{2}{9} + \dfrac{4}{27} + \cdots + \dfrac{1}{3}\left(\dfrac{2}{3}\right)^{n-1} = 1 - \left(\dfrac{2}{3}\right)^n$

23. $2^0 + 2^1 + 2^2 + 2^3 + \cdots + 2^{n-1} = 2^n - 1$

24. $1^3 + 2^3 + 3^3 + \cdots + n^3 = \left[\dfrac{n(n + 1)}{2}\right]^2$

25. Prove that $x - y$ is a factor of $x^n - y^n$. (*Hint:* Consider subtracting and adding xy^k to the binomial $x^{k+1} - y^{k+1}$).

26. Prove that $n < 2^n$.

27. There are $180°$ in the sum of the angles of any triangle. Prove that $(n - 2)180°$ is the sum of the angles of any simple polygon when n is its number of sides. (*Hint:* If a polygon has $k + 1$ sides, it has $k - 2$ sides plus three more sides.)

28. Consider the equation

$$1 + 3 + 5 = \cdots + 2n - 1 = 3n - 2$$

a. Is the equation true for $n = 1$?
b. Is the equation true for $n = 2$?
c. Is the equation true for all natural numbers n?

29. If $1 + 2 + 3 + \cdots + n = \frac{n}{2}(n + 1) + 1$ were true for $n = k$, show that it would be true for $n = k + 1$. Is it true for $n = 1$?

30. Prove that $n + 1 = 1 + n$ for each natural number n.

31. If n is any natural number, prove that $7^n - 1$ is divisible by 6.

32. Prove that $1 + 2n < 3^n$ for $n > 1$.

33. Prove that, if r is a real number where $r \neq 1$, then

$$1 + r + r^2 + \cdots + r^n = \frac{1 - r^{n+1}}{1 - r}$$

34. Prove the formula for the sum of the first n terms of an arithmetic sequence:

$$a + [a + d] + [a + 2d] + [a + (n - 1)d] = \frac{n(a + l)}{2}$$

where $l = a + (n - 1)d$.

DISCOVERY AND WRITING

35. The expression a^m, where m is a natural number, was defined in Section 1.3. An alternative definition of a^m is (Part 1) $a^1 = a$ and (Part 2) $a^{m+1} = a^m \cdot a$. Use induction on n to prove the law of exponents, $a^m a^n = a^{m+n}$.

36. Use induction on n to prove the law of exponents, $(a^m)^n = a^{mn}$. (See Exercise 35.)

REVIEW *In Exercises 37–40, graph each inequality.*

37. $3x + 4y \leq 12$

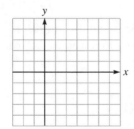

38. $4x - 3y < 12$

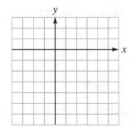

39. $2x + y > 5$

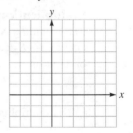

40. $x + 3y \geq 7$

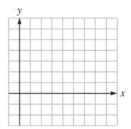

9.6 Permutations and Combinations

■ THE MULTIPLICATION PRINCIPLE FOR EVENTS ■ PERMUTATIONS ■ FORMULAS FOR PERMUTATIONS
■ COMBINATIONS ■ FORMULAS FOR COMBINATIONS ■ THE BINOMIAL THEOREM
■ DISTINGUISHABLE WORDS

■ THE MULTIPLICATION PRINCIPLE FOR EVENTS

Lydia plans to go to dinner and attend a movie. If she has a choice of four restaurants and three movies, in how many ways can she spend her evening? There are four choices of restaurants and, for any one of these choices, there are three choices of movies, as shown in the tree diagram in Figure 9-2.

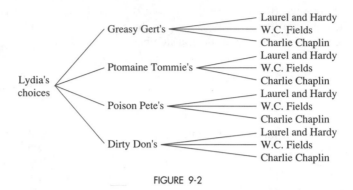

Lydia's choices

Greasy Gert's — Laurel and Hardy / W.C. Fields / Charlie Chaplin

Ptomaine Tommie's — Laurel and Hardy / W.C. Fields / Charlie Chaplin

Poison Pete's — Laurel and Hardy / W.C. Fields / Charlie Chaplin

Dirty Don's — Laurel and Hardy / W.C. Fields / Charlie Chaplin

FIGURE 9-2

The diagram shows that Lydia has 12 ways to spend her evening. One possibility is to eat at Ptomaine Tommie's and watch W. C. Fields. Another is to eat at Dirty Don's and watch Laurel and Hardy.

Any situation that has several outcomes is called an **event.** Lydia's first event (choosing a restaurant) can occur in 4 ways. Her second event (choosing a movie) can occur in 3 ways. Thus, she has $4 \cdot 3$, or 12, ways to spend her evening. This example illustrates the **multiplication principle for events.**

Multiplication Principle for Events

Let E_1 and E_2 be two events. If E_1 can be done in a_1 ways, and if—after E_1 has occurred—E_2 can be done in a_2 ways, then the event "E_1 followed by E_2" can be done in $a_1 \cdot a_2$ ways.

The multiplication principle can be extended to n events.

EXAMPLE 1   If a traveler has four ways to go from New York to Chicago, three ways to go from Chicago to Denver, and six ways to go from Denver to San Francisco, in how many ways can she go from New York to San Francisco?

Solution   We can let E_1 be the event "going from New York to Chicago," E_2 be the event "going from Chicago to Denver," and E_3 be the event "going from Denver to San Francisco." Since there are 4 ways to accomplish E_1, 3 ways to accomplish E_2, and 6 ways to accomplish E_3, the number of routes available is

$$4 \cdot 3 \cdot 6 = 72$$ ∎

Self Check   If a person has four sweaters and 5 pairs of slacks, how many outfits can he wear?

Answer   20

■ PERMUTATIONS

Suppose we want to arrange 7 books on a shelf. We can fill the first space with any one of the 7 books, the second space with any of the remaining 6 books, the third space with any of the remaining 5 books, and so on, until there is only one space left to fill with the last book. According to the multiplication principle, the number of ways that we can arrange the books is

$$7 \cdot 6 \cdot 5 \cdot 4 \cdot 3 \cdot 2 \cdot 1 = 5040$$

When finding the number of possible arrangements of books on a shelf, we are finding the number of **permutations.** The number of permutations of 7 books, using all the books, is 5040. The symbol $P(n, r)$ is read as "the number of permutations of n things r at a time." Thus, $P(7, 7) = 5040$.

EXAMPLE 2   Assume that there are 7 signal flags of 7 different colors to hang on a mast. How many different signals can be sent when 3 flags are used?

Solution   We are asked to find $P(7, 3)$, the number of permutations of 7 things using 3 of them. Any one of the 7 flags can hang in the top position on the mast. Any one of the 6 remaining flags can hang in the middle position, and any one of the remaining 5 flags can hang in the bottom position. By the multiplication principle, we have

$$P(7, 3) = 7 \cdot 6 \cdot 5 = 210$$

It is possible to send 210 different signals.   ■

Self Check   In Example 2, how many signals can be sent if two of the seven flags are missing?

Answer   60

■ FORMULAS FOR PERMUTATIONS

Although it is correct to write $P(7, 3) = 7 \cdot 6 \cdot 5$, we will change the form of the answer to obtain a convenient formula. To derive this formula, we proceed as follows:

$$P(7, 3) = 7 \cdot 6 \cdot 5$$

$$= \frac{7 \cdot 6 \cdot 5 \cdot 4 \cdot 3 \cdot 2 \cdot 1}{4 \cdot 3 \cdot 2 \cdot 1} \qquad \text{Multiply numerator and denominator by } 4 \cdot 3 \cdot 2 \cdot 1.$$

$$= \frac{7!}{4!}$$

$$= \frac{7!}{(7 - 3)!}$$

The generalization of this idea gives the following formula.

Formula for P(n, r)

The number of permutations of n things r at a time is given by

$$P(n, r) = \frac{n!}{(n - r)!}$$

EXAMPLE 3   Find   **a.** $(8, 4)$,   **b.** $P(n, n)$,   and   **c.** $P(n, 0)$.

Solution   **a.** $P(8, 4) = \dfrac{8!}{(8 - 4)!} = \dfrac{8 \cdot 7 \cdot 6 \cdot 5 \cdot 4!}{4!} = 1680$

b. $P(n, n) = \dfrac{n!}{(n - n)!} = \dfrac{n!}{0!} = n!$

c. $P(n, 0) = \dfrac{n!}{(n - 0)!} = \dfrac{n!}{n!} = 1$ ∎

Self Check   Find   **a.** $P(7, 5)$ and   **b.** $P(6, 0)$.
Answers   **a.** 2520,   **b.** 1

Parts b and c of Example 3 establish the following formulas.

Formulas for P(n, n) and P(n, 0)

The number of permutations of n things n at a time and n things 0 at a time are given by the formulas

$$P(n, n) = n! \qquad \text{and} \qquad P(n, 0) = 1$$

EXAMPLE 4   In how many ways can a baseball manager arrange a batting order of 9 players if there are 25 players on the team?

Solution   To find the number of permutations of 25 things 9 at a time, we substitute 25 for n and 9 for r in the formula for finding $P(n, r)$.

$$P(n, r) = \frac{n!}{(n - r)!}$$

$$P(25, 9) = \frac{25!}{(25 - 9)!}$$

$$= \frac{25!}{16!}$$

$$= \frac{25 \cdot 24 \cdot 23 \cdot 22 \cdot 21 \cdot 20 \cdot 19 \cdot 18 \cdot 17 \cdot 16!}{16!}$$

$$\approx 741,354,768,000$$

The number of permutations is approximately 741,354,768,000.   ∎

Self Check | In how many ways can the baseball manager arrange a batting order if two players can't play?

Answer | approximately 296,541,907,200

EXAMPLE 5 | In how many ways can 5 people stand in a line if 2 people refuse to stand next to each other?

Solution | The total number of ways that 5 people can stand in line is

$$P(5, 5) = 5! = 5 \cdot 4 \cdot 3 \cdot 2 \cdot 1 = 120$$

To find the number of ways that 5 people can stand in line if 2 people insist on standing together, we consider the two people as one person. Then there are 4 people to stand in line, and this can be done in $P(4, 4) = 4! = 24$ ways. However, because either of the two who are paired could be first, there are two arrangements for the pair who insist on standing together. Thus, there are $2 \cdot 4!$, or 48 ways that 5 people can stand in line if 2 people insist on standing together.

The number of ways that 5 people can stand in line if two people refuse to stand together is $5! = 120$ (the total number of ways to line up 5 people) minus $2 \cdot 4! = 48$ (the number of ways to line up the 5 people if 2 do stand together):

$$120 - 48 = 72$$

There are 72 ways to line up the people. ∎

Self Check | In how many ways can 5 people stand in a line if one person demands to be first?

Answer | 24

EXAMPLE 6 | In how many ways can 5 people be seated at a round table?

Solution | If we were to seat 5 people in a row, there would be 5! possible arrangements. However, at a round table, each person has a neighbor to the left and to the right. If each person moves one, two, three, four, or five places to the left, everyone has the same neighbors and the arrangement has not changed. Thus, we must divide 5! by 5 to get rid of these duplications. The number of ways that 5 people can be seated at a round table is

$$\frac{5!}{5} = 4! = 4 \cdot 3 \cdot 2 \cdot 1 = 24$$ ∎

Self Check | In how many ways can 6 people be seated at a round table?

Answer | 120

The results of Example 6 suggest the following fact.

Circular Arrangements
There are $(n - 1)!$ ways to arrange n things in a circle.

■ COMBINATIONS

Suppose that a class of 12 students selects a committee of 3 persons to plan a party. With committees, order is not important. A committee of John, Maria, and Raul is the same as a committee of Maria, Raul, and John. However, if we assume for the moment that order is important, we can find the number of permutations of 12 things 3 at a time.

$$P(12, 3) = \frac{12!}{(12 - 3)!} = \frac{12 \cdot 11 \cdot 10 \cdot 9!}{9!} = 1320$$

However, since we do not care about order, this result of 1320 ways is too large. Because there are 6 ways ($3! = 6$) of ordering every committee of 3 students, the result of $P(12, 3) = 1320$ is exactly 6 times too big. To get the correct number of committees, we must divide $P(12, 3)$ by 6:

$$\frac{P(12, 3)}{6} = \frac{1320}{6} = 220$$

In cases of selection where order is not important, we are interested in **combinations,** not permutations. The symbols $C(n, r)$ and $\binom{n}{r}$ both mean the number of combinations of n things r at a time.

■ FORMULAS FOR COMBINATIONS

If a committee of r people is chosen from a total of n people, the number of possible committees is $C(n, r)$, and there will be $r!$ arrangements of each committee. If we consider the committee as an ordered grouping, the number of orderings is $P(n, r)$. Thus, we have

1. $r!C(n, r) = P(n, r)$

We can divide both sides of Equation 1 by $r!$ to obtain the formula for finding $C(n, r)$.

$$C(n, r) = \binom{n}{r} = \frac{P(n, r)}{r!} = \frac{n!}{r!(n - r)!}$$

Formula for $C(n, r)$
The number of combinations of n things r at a time is given by

$$C(n, r) = \binom{n}{r} = \frac{n!}{r!(n - r)!}$$

In the exercises, you will be asked to prove the following formulas.

> **Formulas for C(n, n) and C(n, 0)**
> If n is a whole number, then
> $$C(n, n) = 1 \quad \text{and} \quad C(n, 0) = 1$$

EXAMPLE 7   If Carla must read 4 books from a reading list of 10 books, how many choices does she have?

Solution   Because the order in which the books are read is not important, we find the number of combinations of 10 things 4 at a time:

$$C(10, 4) = \frac{10!}{4!(10 - 4)!} = \frac{10 \cdot 9 \cdot 8 \cdot 7 \cdot 6!}{4 \cdot 3 \cdot 2 \cdot 1 \cdot 6!}$$

$$= \frac{10 \cdot 9 \cdot 8 \cdot 7}{4 \cdot 3 \cdot 2}$$

$$= 210$$

Carla has 210 choices. ■

Self Check   How many choices would Carla have if she had to read 5 books?

Answer   252

EXAMPLE 8   A class consists of 15 men and 8 women. In how many ways can a debate team be chosen with 3 men and 3 women?

Solution   There are $C(15, 3)$ ways of choosing 3 men and $C(8, 3)$ ways of choosing 3 women. By the multiplication principle, there are $C(15, 3) \cdot C(8, 3)$ ways of choosing members of the debate team:

$$C(15, 3) \cdot C(8, 3) = \frac{15!}{3!(15 - 3)!} \cdot \frac{8!}{3!(8 - 3)!}$$

$$= \frac{15 \cdot 14 \cdot 13}{6} \cdot \frac{8 \cdot 7 \cdot 6}{6}$$

$$= 25,480$$

There are 25,480 ways to choose the debate team. ■

Self Check   In Example 7, in how many ways can the debate team be chosen if it is to be 4 men and 2 women?

Answer   38,220

■ THE BINOMIAL THEOREM

The formula

$$C(n, r) = \frac{n!}{r!(n-r)!}$$

gives the coefficient of the $(r + 1)$th term of the binomial expansion of $(a + b)^n$. This implies that the coefficients of a binomial expansion can be used to solve problems involving combinations. The binomial theorem is restated below—this time listing the $(r + 1)$th term and using combination notation.

Binomial Theorem
If n is any positive integer, then

$$(a + b)^n = \binom{n}{0}a^n + \binom{n}{1}a^{n-1}b + \binom{n}{2}a^{n-2}b^2 + \cdots$$

$$+ \binom{n}{r}a^{n-r}b^r + \cdots + \binom{n}{n}b^n$$

EXAMPLE 9   Use Pascal's triangle to compute $C(7, 5)$.

Solution   Consider the eighth row of Pascal's triangle and the corresponding combinations:

$$\begin{array}{cccccccc}
1 & 7 & 21 & 35 & 35 & 21 & 7 & 1 \\
\binom{7}{0} & \binom{7}{1} & \binom{7}{2} & \binom{7}{3} & \binom{7}{4} & \binom{7}{5} & \binom{7}{6} & \binom{7}{7}
\end{array}$$

$C(7, 5) = \binom{7}{5} = 21.$ ■

Self Check   Use Pascal's triangle to compute $C(6, 5)$.
Answer   6

■ DISTINGUISHABLE WORDS

A "word" is a distinguishable arrangement of letters. For example, six words can be formed with the letters a, b, and c if each letter is used exactly once. The six words are

$$abc, \quad acb, \quad bac, \quad bca, \quad cab, \quad \text{and} \quad cba$$

If there are n distinct letters and each letter is used once, the number of distinct words that can be formed is $n! = P(n, n)$. It is more complicated to compute the number of distinguishable words that can be formed with n letters when some of the letters are duplicates.

EXAMPLE 10   Find the number of "words" that can be formed if each of the 6 letters of the word *little* is used once.

Solution   For the moment, assume that the letters of the word *little* are distinguishable: "LitTle." The number of words that can be formed using each letter once is $6! = P(6, 6)$. However, in reality we cannot tell the *l*'s or the *t*'s apart. Therefore, we must divide by a number to get rid of these duplications. Because there are 2! orderings of the two *l*'s and 2! orderings of the two *t*'s, we divide by $2! \cdot 2!$. The number of words that can be formed using each letter of the word *little* is

$$\frac{P(6, 6)}{2! \cdot 2!} = \frac{6!}{2! \cdot 2!} = \frac{6 \cdot 5 \cdot 4 \cdot 3 \cdot 2 \cdot 1}{2 \cdot 1 \cdot 2 \cdot 1} = 180$$ ■

Self Check   How many words can be formed if each letter of the word *balloon* is used once?

Answer   1260

Example 10 illustrates the following general principle.

Distinguishable Words

If a word with n letters has a of one letter, b of another letter, and so on, the number of distinguishable words that can be formed using each letter of the n-letter word exactly once is

$$\frac{n!}{a!b! \cdots}$$

EXERCISE 9.6

VOCABULARY AND CONCEPTS   *In Exercises 1–12, fill in the blank to make a true statement.*

1. If E_1 and E_2 are two events and E_1 can be done in 4 ways and E_2 can be done in 6 ways, then the event E_1 followed by E_2 can be done in ___ ways.

2. An arrangement of n objects is called a _____.

3. $P(n, r) =$ _____

4. $P(n, n) =$ ___

5. $P(n, 0) =$ ___

6. There are _____ ways to arrange n things in a circle.

7. Another notation for $C(n, r)$ is___.

8. $C(n, r) =$ _____

9. $C(n, n) =$ ___

10. $C(n, 0) =$ ___

11. If a word with n letters has a of one letter, b of another letter, and so on, the number of different words that can be formed is

_____

12. Where the order of selection is not important, we are interested in _____, not _____.

PRACTICE   In Exercises 13–28, evaluate each expression.

13. $P(7, 4)$

14. $P(8, 3)$

15. $C(7, 4)$

16. $C(8, 3)$

17. $P(5, 5)$

18. $P(5, 0)$

19. $\binom{5}{4}$

20. $\binom{8}{4}$

21. $\binom{5}{0}$

22. $\binom{5}{5}$

23. $P(5, 4) \cdot C(5, 3)$

24. $P(3, 2) \cdot C(4, 3)$

25. $\binom{5}{3}\binom{4}{3}\binom{3}{3}$

26. $\binom{5}{5}\binom{6}{6}\binom{7}{7}\binom{8}{8}$

27. $\binom{68}{66}$

28. $\binom{100}{99}$

APPLICATIONS

29. Choosing lunch   A lunchroom has a machine with eight kinds of sandwiches, a machine with four kinds of soda, a machine with both white and chocolate milk, and a machine with three kinds of ice cream. How many different lunches can be chosen? (Consider a lunch to be one sandwich, one drink, and one ice cream.)

30. Manufacturing license plates   How many six-digit license plates can be manufactured if no license plate number begins with 0?

31. Available phone numbers   How many different seven-digit phone numbers can be used in one area code if no phone number begins with 0 or 1?

32. Arranging letters   In how many ways can the letters of the word *number* be arranged?

33. Arranging letters with restrictions   In how many ways can the letters of the word *number* be arranged if the e and r must remain next to each other?

34. Arranging letters with restrictions   In how many ways can the letters of the word *number* be arranged if the e and r cannot be side by side?

35. Arranging letters with repetitions   How many ways can five Scrabble tiles bearing the letters, F, F, F, L, and U be arranged to spell the word *fluff*?

36. Arranging letters with repetitions   How many ways can six Scrabble tiles bearing the letters B, E, E, E, F, and L be arranged to spell the word *feeble*?

37. Placing people in line   In how many arrangements can 8 women be placed in a line?

38. Placing people in line   In how many arrangements can 5 women and 5 men be placed in a line if the women and men alternate?

39. Placing people in line   In how many arrangements can 5 women and 5 men be placed in a line if all the men line up first?

40. Placing people in line   In how many arrangements can 5 women and 5 men be placed in a line if all the women line up first?

41. Combination locks   How many permutations does a combination lock have if each combination has 3 numbers, no two numbers of the combination are the same, and the lock dial has 30 notches?

42. Combination locks   How many permutations does a combination lock have if each combination has 3 numbers, no two numbers of the combination are the same, and the lock dial has 100 notches?

43. Seating at a table   In how many ways can 8 people be seated at a round table?

44. Seating at a table   In how many ways can 7 people be seated at a round table?

45. Seating at a table   In how many ways can 6 people be seated at a round table if 2 of the people insist on sitting together?

46. Seating arrangements with conditions In how many ways can 6 people be seated at a round table if 2 of the people refuse to sit together?

47. Arrangements in a circle In how many ways can 7 children be arranged in a circle if Sally and John want to sit together and Martha and Peter want to sit together?

48. Arrangements in a circle In how many ways can 8 children be arranged in a circle if Laura, Scott, and Paula want to sit together?

49. Selecting candy bars In how many ways can 4 candy bars be selected from 10 different candy bars?

50. Selecting birthday cards In how many ways can 6 birthday cards be selected from 24 different cards?

51. Circuit wiring A wiring harness containing a red, a green, a white, and a black wire must be attached to a control panel. In how many different orders can the wires be attached?

52. Grading homework A professor grades homework by randomly checking 7 of the 20 problems assigned. In how many different ways can this be done?

53. Forming words with distinct letters How many words can be formed from the letters of the word *plastic* if each letter is to be used once?

54. Forming words with repeated letters How many words can be formed from the letters of the word *banana* if each letter is to be used once?

55. Manufacturing license plates How many license plates can be made using two different letters followed by four different digits if the first digit cannot be 0 and the letter O is not used?

56. Planning class schedules If there are seven class periods in a school day, and a typical student takes 5 classes, how many different time patterns are possible for the student?

57. Selecting golf balls From a bucket containing 6 red and 8 white golf balls, in how many ways can we draw 6 golf balls of which 3 are red and 3 are white?

58. Selecting committees In how many ways can you select a committee of 3 Republicans and 3 Democrats from a group containing 18 Democrats and 11 Republicans?

59. Selecting committees In how many ways can you select a committee of 4 Democrats and 3 Republicans from a group containing 12 Democrats and 10 Republicans?

60. Drawing cards In how many ways can you select a group of 5 red cards and 2 black cards from a deck containing 10 red cards and 8 black cards?

61. Planning dinner In how many ways can a husband and wife choose 2 different dinners from a menu of 17 dinners?

62. Placing people in line In how many ways can 7 people stand in a row if 2 of the people refuse to stand together?

63. Geometry How many lines are determined by 8 points if no 3 points lie on a straight line?

64. Geometry How many lines are determined by 10 points if no 3 points lie on a straight line?

65. Coaching basketball How many different teams can a basketball coach start if the entire squad consists of 10 players? (Assume that a starting team has 5 players and each player can play all positions.)

66. Managing baseball How many different teams can a manager start if the entire squad consists of 25 players? (Assume that a starting team has 9 players and each player can play all positions.)

67. Selecting job applicants There are 30 qualified applicants for 5 openings in the sales department. In how many different ways can the group of 5 be selected?

68. Sales promotion If a customer purchases a new stereo system during the spring sale, he may choose any six CDs from 20 classical and 30 jazz selections. In how many ways can the customer choose three of each?

69. Guessing on matching questions Ten words are to be paired with the correct 10 out of 12 possible definitions. How many ways are there of guessing?

70. Guessing on true–false exams How many possible ways are there of guessing on a 10-question true–false exam, if it is known that the instructor will have 5 true and 5 false responses?

DISCOVERY AND WRITING

71. Prove that $C(n, n) = 1$.

73. Prove that $\binom{n}{r} = \binom{n}{n-r}$.

72. Prove that $C(n, 0) = 1$.

74. Show that the binomial theorem can be expressed in the form

$$(a + b)^n = \sum_{k=0}^{n}\binom{n}{k}a^{n-k}b^k$$

75. Explain how to use Pascal's triangle to find $C(8, 5)$.

76. Explain how to use Pascal's triangle to find $C(10, 8)$.

REVIEW   *In Exercises 77–80, find the value of x.*

77. $\log_x 16 = 4$

78. $\log_\pi x = \dfrac{1}{2}$

79. $\log_{\sqrt{7}} 49 = x$

80. $\log_x \dfrac{1}{2} = -\dfrac{1}{3}$

In Exercises 81–84, tell whether the statement is true.

81. $\log_{17} 1 = 0$

82. $\log_5 0 = 1$

83. $\log_b b^b = b$

84. $\dfrac{\log_7 A}{\log_7 B} = \log_7 \dfrac{A}{B}$

9.7 Probability

■ PROBABILITY   ■ MULTIPLICATION PROPERTY OF PROBABILITIES

■ PROBABILITY

The probability that an event will occur is a measure of the likelihood of that event. A tossed coin, for example, can land in two ways, either heads or tails. Because one of these two equally likely outcomes is heads, we expect that out of several tosses, about half will be heads. We say that the probability of obtaining heads in a single toss of the coin is $\frac{1}{2}$.

If records show that out of 100 days with weather conditions like today's, 30 have received rain, the weather service will report, "There is a $\frac{30}{100}$ or 30% probability of rain today."

An **experiment** is a process for which the outcome is uncertain. Tossing a coin, rolling a die, drawing a card, and predicting rain are examples of experiments. For any experiment, the set of all possible outcomes is called a **sample space.**

The sample space, S, for the experiment of tossing two coins is the set

$$S = \{(H, H), (H, T), (T, H), (T, T)\}$$

where the ordered pair (H, T) represents the outcome "heads on the first coin and tails on the second coin." Because there are two possible outcomes for the first coin and two for the second coin, we know (by the multiplication principle for events) that there are $2 \cdot 2 = 4$ possible outcomes. Since there are 4 elements in the sample space S, we write

$$n(S) = 4 \qquad \text{Read as "The number of elements in set } S \text{ is 4."}$$

An **event** associated with an experiment is any subset of the sample space of that experiment. For example, if E is the event "getting at least one heads" in the experiment of tossing two coins, then

$$E = \{(H, H), (H, T), (T, H)\}$$

and $n(E) = 3$. Because the outcome of getting at least one heads can occur in 3 out of 4 possible ways, we say that the **probability** of a favorable outcome is $\frac{3}{4}$.

$$P(E) = P(\text{at least one heads}) = \frac{3}{4}$$

We define the probability of an event as follows.

Probability of an Event

If S is the sample space of an experiment with n distinct and equally likely outcomes, and E is an event that occurs in s of those ways, then the **probability of E** is

$$P(E) = \frac{n(E)}{n(S)} = \frac{s}{n}$$

Because $0 \le s \le n$, it follows that $0 \le \frac{s}{n} \le 1$. This implies that all probabilities have values from 0 to 1. An event that cannot happen has probability 0. An event that is certain to happen has probability 1.

To say that the probability of tossing heads on one toss of a coin is $\frac{1}{2}$ means that if a fair coin is tossed a large number of times, the ratio of the number of heads to the total number of tosses is nearly $\frac{1}{2}$.

To say that the probability of rolling 5 on one roll of a die is $\frac{1}{6}$ means that as the number of rolls approaches infinity, the ratio of the number of favorable outcomes (rolling a 5) to the total number of outcomes (rolling a 1, 2, 3, 4, 5, or 6) approaches $\frac{1}{6}$.

EXAMPLE 1   Show the sample space of the experiment "rolling two dice a single time."

Solution   We can list ordered pairs and let the first number be the result on the first die and the second number the result on the second die. The sample space, S, is the set with the following elements:

$$
\begin{array}{cccccc}
(1, 1) & (1, 2) & (1, 3) & (1, 4) & (1, 5) & (1, 6) \\
(2, 1) & (2, 2) & (2, 3) & (2, 4) & (2, 5) & (2, 6) \\
(3, 1) & (3, 2) & (3, 3) & (3, 4) & (3, 5) & (3, 6) \\
(4, 1) & (4, 2) & (4, 3) & (4, 4) & (4, 5) & (4, 6) \\
(5, 1) & (5, 2) & (5, 3) & (5, 4) & (5, 5) & (5, 6) \\
(6, 1) & (6, 2) & (6, 3) & (6, 4) & (6, 5) & (6, 6)
\end{array}
$$

Since there are 6 possible outcomes with the first die and 6 possible outcomes with the second die, we expect $6 \cdot 6 = 36$ equally likely possible outcomes, and we have $n(S) = 36$. ■

Self Check   In Example 1, how many pairs in the sample space have a sum of 7?
Answer   6

EXAMPLE 2   Find the probability of the event "tossing a sum of 7 on one toss of two dice."

Solution   The sample space is listed in Example 1. We let E be the set of favorable outcomes, those that give a sum of 7:

$$E = \{(1, 6), (2, 5), (3, 4), (4, 3), (5, 2), (6, 1)\}$$

Since there are 6 favorable outcomes among the 36 equally likely outcomes, $n(E) = 6$, and

$$P(E) = P(\text{tossing a 7}) = \frac{n(E)}{n(S)} = \frac{6}{36} = \frac{1}{6}$$ ■

Self Check   In Example 2, find the probability of tossing a sum of 4.
Answer $\frac{1}{12}$

A standard playing deck of 52 cards has two red suits, hearts and diamonds, and two black suits, clubs and spades. Each suit has 13 cards, including an ace, a king, a queen, a jack, and cards numbered from 2 to 10. We will refer to a standard deck of cards in many examples and exercises.

EXAMPLE 3   Find the probability of drawing 5 cards, all hearts, from a standard deck of cards.

Solution   Since the number of ways to draw 5 hearts from the 13 hearts is $C(13, 5)$, we have $n(E) = C(13, 5)$. Since the number of ways to draw 5 cards from the deck is $C(52, 5)$, we have $n(S) = C(52, 5)$. The probability of drawing 5 hearts is the ratio of the number of favorable outcomes to the number of possible outcomes.

$$P(5 \text{ hearts}) = \frac{C(13, 5)}{C(52, 5)}$$

$$P(5 \text{ hearts}) = \frac{\dfrac{13!}{5!8!}}{\dfrac{52!}{5!47!}}$$

$$= \frac{13!}{5!8!} \cdot \frac{5!47!}{52!}$$

$$= \frac{13 \cdot 12 \cdot 11 \cdot 10 \cdot 9 \cdot 8!}{8!} \cdot \frac{47!}{52 \cdot 51 \cdot 50 \cdot 49 \cdot 48 \cdot 47!}$$

$$= \frac{13 \cdot 12 \cdot 11 \cdot 10 \cdot 9}{52 \cdot 51 \cdot 50 \cdot 49 \cdot 48}$$

$$= \frac{33}{66,640}$$

The probability of drawing 5 hearts is $\frac{33}{66,640}$. ■

Self Check Find the probability of drawing 6 cards, all diamonds, from the deck.

Answer $\frac{33}{391,510}$

■ MULTIPLICATION PROPERTY OF PROBABILITIES

There is a property of probabilities that is similar to the multiplication principle for events. In the following theorem, we read $P(A \cap B)$ as "the probability of A and B" and $P(B|A)$ as "the probability of B given A." If A and B are events, the set $A \cap B$ contains the outcomes that are in both A and B.

> **Multiplication Property of Probabilities**
> If $P(A)$ represents the probability of event A, and $P(B|A)$ represents the probability that event B will occur after event A, then
> $$P(A \cap B) = P(A) \cdot P(B|A)$$

EXAMPLE 4 A box contains 40 cubes of the same size. Of these cubes, 17 are red, 13 are blue, and the rest are yellow. If 2 cubes are drawn at random, without replacement, find the probability that 2 yellow cubes will be drawn.

Solution Of the 40 cubes in the box, 10 are yellow. The probability of getting a yellow cube on the first draw is

$$P(\text{yellow cube on the first draw}) = \frac{10}{40} = \frac{1}{4}$$

Because there is no replacement after the first draw, 39 cubes remain in the box, and 9 of these are yellow. The probability of drawing a yellow cube on the second draw is

$$P(\text{yellow cube on the second draw}) = \frac{9}{39} = \frac{3}{13}$$

The probability of drawing 2 yellow cubes in succession is the product of the probability of drawing a yellow cube on the first draw and the probability of drawing a yellow cube on the second draw.

$$P(\text{drawing two yellow cubes}) = \frac{1}{4} \cdot \frac{3}{13} = \frac{3}{52}$$ ■

Self Check   In Example 4, find the probability that 2 blue cubes will be drawn.

Answer $\dfrac{1}{10}$

EXAMPLE 5   Repeat Example 3 using the multiplication property of probabilities.

Solution   The probability of drawing a heart on the first draw is $\frac{13}{52}$. The probability of drawing a heart on the second draw *given that we got a heart on the first draw* is $\frac{12}{51}$. The probability is $\frac{11}{50}$ on the third draw, $\frac{10}{49}$ on the fourth draw, and $\frac{9}{48}$ on the fifth draw. By the multiplication property of probabilities,

$$P(5 \text{ hearts in a row}) = \frac{13}{52} \cdot \frac{12}{51} \cdot \frac{11}{50} \cdot \frac{10}{49} \cdot \frac{9}{48}$$

$$= \frac{33}{66{,}640}$$

Self Check   Find the probability of drawing 6 cards, all diamonds, from the deck.

Answer $\dfrac{33}{391{,}510}$

EXAMPLE 6   In a school, 30% of the students are gifted in mathematics and 10% are gifted in art and mathematics. If a student is gifted in mathematics, find the probability that the student is also gifted in art.

Solution   Let $P(M)$ be the probability that a randomly chosen student is gifted in mathematics, and let $P(M \cap A)$ be the probability that the student is gifted in both art and mathematics. We must find $P(A|M)$, the probability that the student is gifted in art, given that he or she is gifted in mathematics. To do so, we substitute the given values

$$P(M) = 0.3 \quad \text{and} \quad P(M \cap A) = 0.1$$

in the formula for multiplication of probabilities and solve for $P(A|M)$:

$$P(M \cap A) = P(M) \cdot P(A|M)$$
$$0.1 = (0.3)P(A|M)$$
$$P(A|M) = \frac{0.1}{0.3}$$
$$= \frac{1}{3}$$

If a student is gifted in mathematics, there is a probability of $\frac{1}{3}$ that he or she is also gifted in art.

Self Check   If 40% of the students in Example 6 are gifted in art, find the probability that a student gifted in art is also gifted in mathematics.

Answer $\dfrac{1}{4}$

EXERCISES 9.7

VOCABULARY AND CONCEPTS   *In Exercises 1–4, fill in the blank to make a true statement.*

1. An _____ is any process for which the outcome is uncertain.

2. A list of all possible outcomes for an experiment is called a _____.

3. The probability of an event E is defined as $P(E) =$ _____ $= \dfrac{s}{n}$.

4. $P(A \cap B) =$ _____.

PRACTICE   *In Exercises 4–8, list the sample space of each experiment.*

5. Rolling one die and tossing one coin

6. Tossing three coins

7. Selecting a letter of the alphabet

8. Picking a one-digit number

In Exercises 9–12, an ordinary die is tossed. Find the probability of each event.

9. Tossing a 2

10. Tossing a number greater than 4

11. Tossing a number larger than 1 but less than 6

12. Tossing an odd number

In Exercises 13–16, balls numbered from 1 to 42 are placed in a container and stirred. If one is drawn at random, find the probability of each result.

13. The number is less than 20.

14. The number is less than 50.

15. The number is a prime number.

16. The number is less than 10 or greater than 40.

In Exercises 17–20, refer to the spinner in Illustration 1. If the spinner is spun, find the probability of each event. Assume that the spinner never stops on a line.

17. The spinner stops on red.

18. The spinner stops on green.

19. The spinner stops on orange.

20. The spinner stops on yellow.

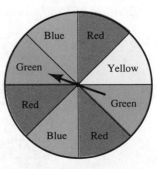

ILLUSTRATION 1

In Exercises 21–38, find the probability of each given event.

21. Rolling a sum of 4 on one roll of two dice.

22. Drawing a diamond on one draw from a card deck.

23. Drawing two aces in succession from a card deck if the card is replaced and the deck is shuffled after the first draw.

24. Drawing two aces from a card deck without replacing the card after the first draw.

25. Drawing a red egg from a basket containing 5 red eggs and 7 blue eggs.

26. Getting 2 red eggs in a single scoop from a bucket containing 5 red eggs and 7 yellow eggs.

27. Drawing a bridge hand of 13 cards, all of one suit.

28. Drawing 6 diamonds from a card deck without replacing the cards after each draw.

29. Drawing 5 aces from a card deck without replacing the cards after each draw.

30. Drawing 5 clubs from the black cards in a card deck.

31. Drawing a face card (king, queen, or jack) from a card deck.

32. Drawing 6 face cards in a row from a card deck without replacing the cards after each draw.

33. Drawing 5 orange cubes from a bowl containing 5 orange cubes and 1 beige cube.

34. Rolling a sum of 4 with one roll of three dice.

35. Rolling a sum of 11 with one roll of three dice.

36. Picking, at random, 5 Republicans from a group containing 8 Republicans and 10 Democrats.

37. Tossing 3 heads in 5 tosses of a fair coin.

38. Tossing 5 heads in 5 tosses of a fair coin.

In Exercises 39–44, assume that the probability that an airplane engine will fail during a torture test is $\frac{1}{2}$ and that the aircraft in question has 4 engines.

39. Construct a sample space for the torture test.

40. Find the probability that all engines will survive the test.

41. Find the probability that exactly 1 engine will survive.

42. Find the probability that exactly 2 engines will survive.

43. Find the probability that exactly 3 engines will survive.

44. Find the probability that no engines will survive.

45. Find the sum of the probabilities in Exercises 40 through 44.

In Exercises 46–48, assume that a survey of 282 people is taken to determine the opinions of doctors, teachers, and lawyers on a proposed piece of legislation, with the following results. A person is chosen at random from those surveyed. Refer to Illustration 2 to find each probability.

	Number that favor	Number that oppose	Number with no opinion	Total
Doctors	70	32	17	119
Teachers	83	24	10	117
Lawyers	23	15	8	46
Total	176	71	35	282

ILLUSTRATION 2

46. The person favors the legislation.

47. A doctor opposes the legislation.

48. A person who opposes the legislation is a lawyer.

49. Quality control In a batch of 10 tires, 2 are known to be defective. If 4 tires are chosen at random, find the probability that all 4 tires are good.

50. Medicine Out of a group of 9 patients treated with a new drug, 4 suffered a relapse. Find the probability that 3 patients of this group, chosen at random, will remain disease-free.

In Exercises 51–58, use the multiplication property of probabilities.

51. If $P(A) = 0.3$ and $P(B|A) = 0.6$, find $P(A \cap B)$.

52. If $P(A \cap B) = 0.3$ and $P(B|A) = 0.6$, find $P(A)$.

53. Conditional probability The probability that a person owns a luxury car is 0.2, and the probability that the owner of such a car also owns a personal computer is 0.7. Find the probability that a person, chosen at random, owns both a luxury car and a computer.

54. Conditional probability If 40% of the population have completed college, and 85% of college graduates are registered to vote, what percent of the

population are college graduates and registered voters?

55. Conditional probability About 25% of the population watches the evening television news coverage as well as the morning soap operas. If 75% of the population watches the news, what percent of those who watch the news also watch the soaps?

56. Conditional probability The probability of rain today is 0.40. If it rains, the probability that Bill will forget his raincoat is 0.70. Find the probability that Bill will get wet.

DISCOVERY AND WRITING

57. If $P(A \cap B) = 0.7$, is it possible that $P(B|A) = 0.6$? Explain.

58. Is it possible that $P(A \cap B) = P(A)$? Explain.

REVIEW *In Exercises 59–60, solve each equation.*

59. $|x + 3| = 7$

60. $|3x - 2| = |2x - 3|$

In Exercises 61–62, graph each inequality.

61. $|x - 3| < 7$

62. $|3x - 2| \geq |2x - 3|$

9.8 Computation of Compound Probabilities

■ COMPOUND EVENTS ■ INDEPENDENT EVENTS

■ COMPOUND EVENTS

Sometimes we must find the probability of one event *or* another, or the probability of one event *and* another. Such events are called **compound events.** As we have seen, if A and B are two events, the probability that A and B will both occur is

$P(A \cap B)$. In this section, we also discuss $P(A \cup B)$, the probability that either A or B will occur.

Suppose we want to find the probability of drawing a king or a heart from a standard card deck. If K is the event "drawing a king" and H is the event "drawing a heart," then $P(K) = \frac{4}{52}$, and $P(H) = \frac{13}{52}$. However, the probability of drawing a king *or* a heart is not the sum of these two probabilities. Because the king of hearts was counted twice, once as a king and once as a heart, and because the probability of drawing the king of hearts is $\frac{1}{52}$, we must subtract $\frac{1}{52}$ from the sum of $\frac{4}{52}$ and $\frac{13}{52}$ to get the correct probability.

$$P(\text{king } or \text{ heart}) = P(\text{king}) + P(\text{heart}) - P(\text{king of hearts})$$
$$P(K \cup H) = P(K) + P(H) - P(K \cap H)$$
$$= \frac{4}{52} + \frac{13}{52} - \frac{1}{52}$$
$$= \frac{16}{52}$$
$$= \frac{4}{13}$$

In general, we have the following rule.

$P(A \cup B)$

If A and B are two events, then

$$P(A \cup B) = P(A) + P(B) - P(A \cap B)$$

If events A and B have no outcomes in common, then $A \cap B = \emptyset$ and $P(A \cap B) = P(\emptyset) = 0$. Such events are **mutually exclusive** (if one event occurs, the other cannot).

$P(A \cup B)$

If A and B cannot occur simultaneously, then

$$P(A \cup B) = P(A) + P(B)$$

The event $\overline{A}$ (read as "not A") contains all outcomes of the sample space that are not elements of event A. Because the events A and $\overline{A}$ are mutually exclusive,

$$P(A \cup \overline{A}) = P(A) + P(\overline{A})$$

Because either event A or event $\overline{A}$ must happen, $P(A \cup \overline{A}) = 1$. Thus,

$$P(A \cup \overline{A}) = 1$$
$$P(A) + P(\overline{A}) = 1$$
$$P(\overline{A}) = 1 - P(A) \qquad \text{Add } -P(A) \text{ to both sides.}$$

This result gives another property of compound probabilities.

> **P($\overline{A}$)**
> If A is any event, then
> $$P(\overline{A}) = 1 - P(A)$$

EXAMPLE 1   A counselor tells a student that his probability of earning a grade of D in algebra is $\frac{1}{5}$ and his probability of earning an F is $\frac{1}{25}$. Find the probability that the student earns a C or better.

Solution   Because "earning a D" and "earning an F" are mutually exclusive, the probability of earning a D or F is given by

$$P(D \cup F) = P(D) + P(F) \qquad \text{Note that } P(D \text{ and } F) = 0.$$

$$= \frac{1}{5} + \frac{1}{25}$$

$$= \frac{6}{25}$$

The probability that the student will receive a C or better is

$$P(C \text{ or better}) = 1 - P(D \cup F)$$

$$= 1 - \frac{6}{25}$$

$$= \frac{19}{25}$$

The probability of earning a C or better is $\frac{19}{25}$. ∎

Self Check   Find the probability that the student in Example 1 earns less than a C.

Answer $\dfrac{6}{25}$

■ INDEPENDENT EVENTS

If two events do not influence each other, they are called **independent events.**

> **Independent Events**
> The events A and B are said to be **independent events** if and only if $P(B) = P(B \mid A)$.

Substituting $P(B)$ for $P(B|A)$ in the multiplication property for probabilities gives a formula for computing probabilities of compound independent events.

Formula for $P(A \cap B)$

If A and B are independent events, then

$$P(A \cap B) = P(A) \cdot P(B)$$

The event A of "drawing an ace from a standard deck of cards" and the event B of "tossing heads" on one toss of a coin are independent events, because neither event influences the other. Consequently,

$$P(A \cap B) = P(A) \cdot P(B)$$

$$P(\text{drawing an ace and tossing heads}) = \textbf{P(drawing an ace)} \cdot \textbf{P(tossing heads)}$$

$$= \frac{4}{52} \cdot \frac{1}{2}$$

$$= \frac{1}{26}$$

EXAMPLE 2   The probability that a baseball player can get a hit is $\frac{1}{3}$. Find the probability that she will get three hits in a row.

Solution   Assume that the three times at bat are independent events: One time at bat does not influence her chances of getting a hit on another turn at bat. Because $P(E_1) = \frac{1}{3}$, $P(E_2) = \frac{1}{3}$, and $P(E_3) = \frac{1}{3}$,

$$P(E_1 \cap E_2 \cap E_3) = \frac{1}{3} \cdot \frac{1}{3} \cdot \frac{1}{3} = \frac{1}{27}$$

The probability that she gets three hits in a row is $\frac{1}{27}$.   ■

Self Check   See Example 2. The probability that another player can get a hit is $\frac{1}{4}$. Find the probability that she gets four hits in a row.

Answer $\frac{1}{256}$

EXAMPLE 3   A die is tossed three times. Find the probability that the outcome is 6 on the first toss, an even number on the second toss, and an odd prime number on the third toss.

Solution   The probability of a 6 on any toss is $P(6) = \frac{1}{6}$. Because there are three even integers represented on a die, the probability of tossing an even number is $P(\text{even number}) = P(E) = \frac{3}{6} = \frac{1}{2}$. Since 3 and 5 are the only odd prime numbers on a die, the probability of tossing an odd prime is $P(\text{odd prime}) = P(O) = \frac{2}{6} = \frac{1}{3}$.

Because these three events are independent, the probability of the events happening in succession is the product of the probabilities:

$$P(\text{six and even number and odd prime}) = P(6 \cap E \cap O)$$
$$= P(6) + P(E) + P(O)$$
$$= \frac{1}{6} \cdot \frac{1}{2} \cdot \frac{1}{3}$$
$$= \frac{1}{36} \qquad \blacksquare$$

Self Check   See Example 3. Find the probability that the outcome is five on the first toss, an odd number on the second toss, and two on the third toss.

Answer $\frac{1}{72}$

EXAMPLE 4   The probability that a drug will cure dandruff is $\frac{1}{8}$. However, if the drug is used, the probability that it will cause side effects is $\frac{1}{6}$. Find the probability that a patient who uses the drug will be cured and will suffer no side effects.

Solution   The probability that the drug will cure dandruff is $P(C) = \frac{1}{8}$. The probability of having side effects is $P(E) = \frac{1}{6}$. The probability that the patient will have no side effects is

$$P(\overline{E}) = 1 - P(E) = 1 - \frac{1}{6} = \frac{5}{6}$$

Since these events are independent,

$$P(\text{cure and no side effects}) = P(C \cap \overline{E})$$
$$= P(C) \cdot P(\overline{E})$$
$$= \frac{1}{8} \cdot \frac{5}{6}$$
$$= \frac{5}{48} \qquad \blacksquare$$

Self Check   See Example 4. Find the probability that the patient will be cured and suffer side effects.

Answer $\frac{1}{48}$

EXERCISE 9.8

VOCABULARY AND CONCEPTS   *In Exercises 1–8, fill in the blank to make a true statement.*

1. A _____ event is one event *or* another or one event *followed by* another.

2. $P(A \cup B) = $ _____.

3. If A and B are _____, then $P(A \cup B) = P(A) + P(B)$.

4. The event $\overline{A}$ is read as "_____."

5. $P(\overline{A}) =$ _____

6. Two events, A and B, are called independent events when _____.

7. If A and B are independent events, then $P(A \cap B) =$ _____.

8. If two events do not influence each other, they are called _____ events.

PRACTICE   *In Exercises 9–12, assume that you draw one card from a card deck. Find the probability of each event.*

9. Drawing a black card

10. Drawing a jack

11. Drawing a black card or an ace

12. Drawing a red card or a face card

In Exercises 13–16, assume that you draw two cards from a card deck, without replacement. Find the probability of each event.

13. Drawing two aces

14. Drawing three aces

15. Drawing a club and then another black card

16. Drawing a heart and then a spade

In Exercises 17–20, assume that you roll two dice once. Find the probability of each result.

17. Rolling a sum of 7 or 6

18. Rolling a sum of 5 or an even sum

19. Rolling a sum of 10 or an odd sum

20. Rolling a sum of 12 or 1

In Exercises 21–24, assume that you have a bucket containing 7 beige capsules, 3 blue capsules, and 6 green capsules. You make a single draw from the bucket, taking one capsule. Find the probability of each result.

21. Drawing a beige or a blue capsule

22. Drawing a green capsule

23. Not drawing a blue capsule

24. Not drawing either a beige or a blue capsule

In Exercises 25–28, assume that you are using the same bucket of capsules as in Exercises 21–24.

25. On two draws from the bucket, find the probability of drawing a beige capsule followed by a green capsule. (Assume that the capsule is returned to the bucket after the first draw.)

26. On two draws from the bucket, find the probability of drawing one blue capsule and one green capsule. (Assume that the capsule is not returned to the bucket after the first draw.)

27. On three successive draws from the bucket (without replacement), find the probability of failing to draw a beige capsule.

28. Jeff rolls a die and draws one card from a card deck. Find the probability of his rolling a 4 and drawing a four.

29. Birthday problem   Three people are in an elevator together. Find the probability that all three were born on the same day of the week.

30. Birthday problem   Three people are on a bus together. Find the probability that at least one was born on a different day of the week from the others.

31. Birthday problem   Five people are in a room together. Find the probability that all five were born on a different day of the year.

32. Birthday problem   Five people are on a bus together. Find the probability that at least two of them were born on the same day of the year.

33. **Sharing homework**   If the probability that Rick will solve a problem is $\frac{1}{4}$ and the probability that Dinah will solve it is $\frac{2}{5}$, find the probability that at least one of them will solve it.

34. **Signaling**   A bugle is used for communication at camp. The call for dinner is based on four pitches and is five notes long. If a child can play these four pitches on a bugle, find the probability that the first five notes that the child plays will call the camp for food. (Assume that the child is equally likely to play any of the four pitches each time a note is blown.)

In Exercises 35–38, assume that a woman visits her cabin in Canada. The probability that her lawnmower will start is $\frac{1}{2}$, the probability that her gas-powered saw will start is $\frac{1}{3}$, and the probability that her outboard motor will start is $\frac{3}{4}$. Find each probability.

35. That all three will start

36. That none will start

37. That exactly one will start

38. That exactly two will start

APPLICATIONS

39. **Immigration**   Three children leave Thailand to start a new life in either the United States or France. The probability that May Xao will go to France is $\frac{1}{3}$, that Tou Lia will go to France is $\frac{1}{2}$, and that May Moua will go to France is $\frac{1}{6}$. Find the probability that exactly two of them will end up in the United States.

40. **Preparing for the GED**   The administrators of a program to prepare people for the high school equivalency exam have found that 80% of the students require tutoring in math, 60% need help in English, and 45% need work in both math and English. Find the probability that a student selected at random needs help with either math or English.

41. **Insurance losses**   The insurance underwriters have determined that in any one year, the probability that George will have a car accident is 0.05, and that if he has an accident, the probability that he will be hospitalized is 0.40. Find the probability that George will have an accident but not be hospitalized.

42. **Grading homework**   One instructor grades homework by randomly choosing 3 out of the 15 problems assigned. Bill did only 8 problems. What is the probability that he won't get caught?

DISCOVERY AND WRITING

43. Explain the difference between dependent and independent events, and give examples of each.

44. Explain why   **a.** $P(A|A) = 1$ and   **b.** $P(A|\overline{A}) = 0$.

REVIEW   *In Exercises 45–46, maximize P subject to the given constraints.*

45. $P = 2x + y$

$$\begin{cases} x \geq 0 \\ y \geq 0 \\ x \leq 4 \\ x + 2y \leq 8 \end{cases}$$

46. $P = 3x - y$

$$\begin{cases} x \geq 0 \\ y \geq 0 \\ y \leq 2 \\ x \leq 2 \\ x + y \leq 3 \end{cases}$$

9.9 Odds And Mathematical Expectation

■ ODDS ■ MATHEMATICAL EXPECTATION

A concept closely related to probability is **mathematical odds.**

■ ODDS

Odds
The **odds for an event** is the probability of a favorable outcome divided by the probability of an unfavorable outcome.

The **odds against an event** is the probability of an unfavorable outcome divided by the probability of a favorable outcome.

EXAMPLE 1   The probability that a horse will win a race is $\frac{1}{4}$. Find the odds for and the odds against the horse.

Solution   Because the probability that the horse will win is $\frac{1}{4}$, the probability that the horse will not win is $\frac{3}{4}$. The odds for the horse are

$$\frac{\text{probability of a win}}{\text{probability of a loss}} = \frac{\frac{1}{4}}{\frac{3}{4}} = \frac{1}{3}$$

or 1 to 3. The odds against the horse are

$$\frac{\text{probability of a loss}}{\text{probability of a win}} = \frac{\frac{3}{4}}{\frac{1}{4}} = 3$$

or 3 to 1. The odds for an event is the reciprocal of the odds against the event. ■

Self Check   The probability that a horse will lose a race is $\frac{1}{5}$. Find the odds for and the odds against the horse.

Answer   4 to 1; 1 to 4

■ MATHEMATICAL EXPECTATION

Suppose we have a chance to play a simple game with the following rules:

1. Roll a single die once.

2. If a six appears, win $3.

3. If a five appears, win $1.

4. If any other number appears, win 50¢.

5. The cost to play (one roll of the die) is $1.

In this game, the probability of any one of the six outcomes—rolling a 6, 5, 4, 3, 2, or 1—is $\frac{1}{6}$, and the winnings are $3, $1, and 50¢. The expected winnings can be found by using the following equation and simplifying the right-hand side:

$$E = \frac{1}{6}(3) + \frac{1}{6}(1) + \frac{1}{6}(0.50) + \frac{1}{6}(0.50) + \frac{1}{6}(0.50) + \frac{1}{6}(0.50)$$

$$= \frac{1}{6}(3 + 1 + 0.50 + 0.50 + 0.50 + 0.50)$$

$$= \frac{1}{6}(6)$$

$$= 1$$

Over the long run, we could expect to win $1 with every play of the game. Since it costs $1 to play the game, the expected gain or loss is 0. Because the expected winnings are equal to the admission price, the game is fair.

> **Mathematical Expectation**
>
> If a certain event has n different outcomes with probabilities $P_1, P_2, \ldots, P_n$ and the winnings assigned to each outcome are $x_1, x_2, \ldots x_n$, the expected winnings, or **mathematical expectation,** E is given by
>
> $$E = P_1x_1 + P_2x_2 + \cdots + P_nx_n$$

EXAMPLE 2   It costs $1 to play the following game: Roll two dice, collect $5 if we roll a sum of 7, and collect $2 if we roll a sum of 11. All other numbers pay nothing. Is it wise to play the game?

Solution   The probability of rolling a 7 on a single roll of two dice is $\frac{6}{36}$, the probability of rolling an 11 is $\frac{2}{36}$, and the probability of rolling something else is $\frac{28}{36}$. The mathematical expectation is

$$E = \frac{6}{36}(5) + \frac{2}{36}(2) + \frac{28}{36}(0) = \frac{17}{18} \approx \$0.944$$

By playing the game for a long period of time, we can expect to get back about 95¢ for every dollar spent. For the fun of playing, the cost is about 5¢ a game. If the game is enjoyable, it might be worth the expected loss. However, the game is slightly unfair. ∎

Self Check   Is it wise to play the game in Example 2 if you win $4 when you roll a sum of 7 and $6 when you roll a sum of 11?

Answer   The game is even.

EXERCISES 9.9

VOCABULARY AND CONCEPTS   *In Exercises 1–6, fill in the blank to make a true statement.*

1. The _____ is the probability of a favorable outcome divided by the probability of an unfavorable outcome.

2. The _____ is the probability of an unfavorable outcome divided by the probability of a favorable outcome.

3. If the odds for an event are 1 to 4, the probability of winning is __.

4. Mathematical expectation E is given by $E =$ _____.

PRACTICE   *In Exercises 5–10, assume a single roll of a die.*

5. Find the probability of rolling a 6.

6. Find the odds in favor of rolling a 6.

7. Find the odds against rolling a 6.

8. Find the probability of rolling an even number.

9. Find the odds in favor of rolling an even number.

10. Find the odds against rolling an even number.

In Exercises 11–16, assume a single roll of two dice.

11. Find the probability of rolling a sum of 6.

12. Find the odds in favor of rolling a sum of 6.

13. Find the odds against rolling a sum of 6.

14. Find the probability of rolling an even sum.

15. Find the odds in favor of rolling an even sum.

16. Find the odds against rolling an even sum.

In Exercises 17–20, assume that you are drawing one card from a card deck.

17. Find the odds in favor of drawing a queen.

18. Find the odds against drawing a black card.

19. Find the odds in favor of drawing a face card.

20. Find the odds against drawing a diamond.

21. If the odds in favor of victory are 5 to 2, find the probability of victory.

22. If the odds in favor of victory are 5 to 2, find the odds against victory.

23. If the odds against winning are 90 to 1, find the odds in favor of winning.

24. Find the odds in favor of rolling a 7 on a single toss of two dice.

25. Find the odds against tossing four heads in a row with a fair nickel.

26. Find the odds in favor of a couple having four girl babies in succession. $\left(\text{Assume } P(\text{boy}) = \frac{1}{2}.\right)$

27. The odds against a horse are 8 to 1. Find the probability that the horse will win.

28. The odds against a horse are 1 to 1. Find the probability that the horse will lose.

29. It costs $2 to play the following game:
    a. Draw one card from a card deck.
    b. Collect $5 if an ace is drawn.
    c. Collect $4 if a king is drawn.
    d. Collect nothing for all other cards drawn.
    Is it wise to play this game? Explain.

30. **Lottery tickets**   One thousand tickets are sold for a lottery with two grand prizes of $800. Find a fair price for the tickets.

31. Find the odds against a couple having three baby boys in a row. $\left(\text{Assume } P(\text{boy}) = \frac{1}{2}.\right)$

32. Find the odds in favor of tossing at least three heads in five tosses of a fair coin.

33. Suppose you toss a coin five times and collect $5 if you toss five heads, $4 if you toss four heads, $3 if you toss three heads, and no money for any other combination. How much should you pay to play the game if the game is to be fair?

34. If you toss two dice one time and collect $10 for double 6's and $1 for double 1's, what is a fair price for playing the game?

35. Counting an ace as 1, a face card as 10, and all others at their numerical value, find the expected value if you draw one card from a card deck.

36. Find the expected sum of one roll of two dice.

37. A multiple-choice test of eight questions gives five possible answers for each question. Only one of the answers for each question is right. Find the probability of getting seven right answers by simple guessing.

38. In the situation described in Exercise 37, find the odds in favor of getting seven answers right.

DISCOVERY AND WRITING

39. If "the odds are against an event," what can be said about its probability?

40. To disguise the unlikely chance of winning, a contest promoter publishes the odds in favor of winning as 0.0000000372 to 1. What are the odds against winning?

REVIEW

41. Find the equation of the line perpendicular to $3x - 2y = 9$ and passing through the point $(1, 1)$.

42. Find the equations of the parabolas with vertex at the point $(3, -2)$ and passing through the origin.

43. Find the equation of the circle with center at $(3, 5)$ and tangent to the x-axis.

44. Write the equation $2x^2 + 4y^2 + 12x - 24y + 53 = 0$ in standard form, and identify the curve.

■ ■ ■ ■ ■ ■ ■ ■ ■ **PROBLEMS AND PROJECTS**

1. Solve for n: $\dfrac{n!}{(n-2)!} = 9900$.

2. Subtract: $\dfrac{1}{(n-1)!} - \dfrac{1}{n!}$.

3. Use mathematical induction to prove that
$$3^0 + 3^1 + 3^2 + \cdots + 3^{n-1} = \frac{3^n - 1}{2}$$

4. What is the sum of all possible three-digit numbers that can be formed from the digits 2, 5, and 7?

5. If $P(n, r) = 5040$ and $C(n, r) = 210$, find r. (*Hint:* $P(n, r) = r!C(n, r)$.)

6. In a bus accident, 3 out of the 20 passengers were injured. Four of the 20 were politicians. What is the probability that all the injured were politicians?

PROJECT 1 By the binomial theorem,
$$(1 + x)^m = 1 + mx + \frac{m(m-1)}{2!}x^2 + \frac{m(m-1)(m-2)}{3!}x^3 + \cdots$$
$$+ \frac{m(m-1)\cdots(m-n+1)}{n!}x^n + \cdots$$

(continued)

■ ■ ■ ■ ■ ■ ■ ■ ■ ■ **PROBLEMS AND PROJECTS** *(continued)*

- If m is not a positive integer or zero, explain why the coefficients of the terms of the binomial theorem never become zero. In this case, the formula is called the **binomial series.**

- In the binomial series, let $x = -\frac{1}{2}$ and $m = \frac{1}{2}$, so that the left-hand side becomes

$$\left(1 - \frac{1}{2}\right)^{1/2} = \left(\frac{1}{2}\right)^{1/2} = \sqrt{\frac{1}{2}} = \frac{\sqrt{2}}{2}$$

  Use a calculator to evaluate this expression.

- On the right-hand side of the equation, also let $x = -\frac{1}{2}$ and $m = \frac{1}{2}$. Calculate the first 4 terms of the series. How close is their sum to the value in the previous question? Is the sum of the first 5 terms more accurate?

PROJECT 2   If n, a, b, and c are nonnegative integers and $a + b + c = n$, then the **multinomial coefficient** $\binom{n}{a, b, c}$ is defined as $\frac{n!}{a!b!c!}$.

- Expand $(x + y + z)^3$.

- In this expansion, verify that the coefficient of x^2y is $\binom{3}{2, 1, 0}$ and that the coefficient of xyz is $\binom{3}{1, 1, 1}$. What is the multinomial coefficient of x^3? (*Hint:* $x^3 = x^3y^0z^0$.) What is the multinomial coefficient of xz^2?

- In the expansion of $(w + x + y + z)^7$, what is the coefficient of w^2x^3yz?

- In the expansion of $(x + y)^n$, explain why the multinomial coefficients are just the binomial coefficients.

PROJECT 3   - Find the hockey-stick pattern in the numbers of Pascal's triangle in Illustration 1. What is the missing number in the rightmost hockey stick? Does the pattern work with larger hockey sticks? Experiment and report your conclusions.

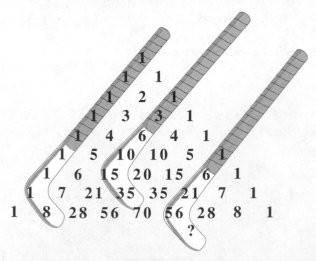

ILLUSTRATION 1

• In Illustration 2, find the pattern in the sums of increasingly larger portions of Pascal's triangle. Find the sum of all of the numbers up to and including the row that begins 1   10   45 . . . .

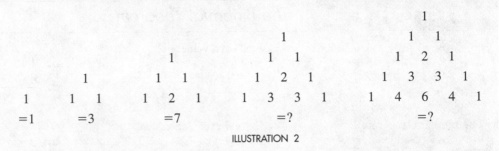

```
                                                                    1
                                                 1               1   1
                              1               1   1           1   2   1
          1               1   1           1   2   1       1   3   3   1
1       1   1       1   2   1       1   3   3   1       1   4   6   4   1
=1        =3            =7                =?                    =?
```

ILLUSTRATION 2

• There are many other patterns hidden in Pascal's triangle. Find some more and share them with your class. Illustration 3 is an idea to get you started.

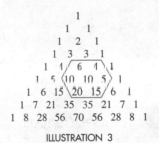

```
                    1
                  1   1
                1   2   1
              1   3   3   1
            1   1 / 6   4 \ 1
          1   5 ⟨10  10  5⟩ 1
        1   6  15 \20  15/ 6   1
      1   7  21  35  35  21  7   1
    1   8  28  56  70  56  28  8   1
```

ILLUSTRATION 3

C H A P T E R   S U M M A R Y

CONCEPT

REVIEW EXERCISES

SECTION 9.1 — The Binomial Theorem

$n! = n(n-1)(n-2) \cdots 3 \cdot 2 \cdot 1$

$0! = 1 \quad n(n-1)! = n!$

The binomial theorem:
If n is any positive integer, then

$$(a+b)^n = a^n + \frac{n!}{1!(n-1)!}a^{n-1}b$$

$$+ \frac{n!}{2!(n-2)!}a^{n-2}b^2 +$$

$$\frac{n!}{3!(n-3)!}a^{n-3}b^3 + \cdots + b^n$$

1. Find each value.
 a. $6!$
 b. $7! \cdot 0! \cdot 1! \cdot 3!$
 c. $\dfrac{8!}{7!}$
 d. $\dfrac{5! \cdot 7! \cdot 8!}{6! \cdot 9!}$

2. Expand each expression.
 a. $(x+y)^3$
 b. $(p+q)^4$
 c. $(a-b)^5$
 d. $(2a-b)^3$

3. Find the required term of each expansion.
 a. $(a+b)^8$; 4th term
 b. $(2x-y)^5$; 3rd term
 c. $(x-y)^9$; 7th term
 d. $(4x+7)^6$; 4th term

SECTION 9.2 — Sequences, Series, and Summation Notation

A **sequence** is a function whose domain is the set of natural numbers.

4. Write the fourth term in each sequence.
 a. $0, 7, 26, \ldots, n^3 - 1, \ldots$
 b. $\dfrac{3}{2}, 3, \dfrac{11}{2}, \ldots, \dfrac{n^2+2}{2}, \ldots$

5. Find the first four terms of each sequence.
 a. $a_1 = 5$ and $a_{n+1} = 3a_n + 2$
 b. $a_1 = -2$ and $a_{n+1} = 2a_n^2$

If c is a constant, then

$$\sum_{k=1}^{n} c = nc$$

If c is a constant, then

$$\sum_{k=1}^{n} cf(k) = c\sum_{k=1}^{n} f(k)$$

$$\sum_{k=1}^{n} [f(k) + g(k)] =$$

$$\sum_{k=1}^{n} f(k) + \sum_{k=1}^{n} g(k)$$

6. Evaluate each expression.
 a. $\displaystyle\sum_{k=1}^{4} 3k^2$
 b. $\displaystyle\sum_{k=1}^{10} 6$
 c. $\displaystyle\sum_{k=5}^{8} (k^3 + 3k^2)$
 d. $\displaystyle\sum_{k=1}^{30}\left(\frac{3}{2}k - 12\right) - \frac{3}{2}\sum_{k=1}^{30} k$

| **SECTION 9.3** | *Arithmetic Sequences* |

If a is the first term, n is the number of terms, and d is the common difference, then the formula $l = a + (n - 1)d$ gives the last (or nth) term of an arithmetic sequence.

7. Find the required term of each arithmetic sequence.
   a. 5, 9, 13, . . .; 29th term
   b. 8, 15, 22, . . .; 40th term

   c. 6, −1, −8, . . .; 15th term
   d. $\frac{1}{2}, -\frac{3}{2}, -\frac{7}{2}, \ldots$; 35th term

8. Find three arithmetic means between 2 and 8.

9. Find five arithmetic means between 10 and 100.

The formula
$$S_n = \frac{n(a + 1)}{2}$$

gives the sum of the first n terms of an arithmetic sequence, where S_n is the sum, a is the first term, l is the last (or nth) term, and n is the number of terms.

10. Find the sum of the first 40 terms in each sequence.
   a. 5, 9, 13, . . .
   b. 8, 15, 22, . . .

   c. 6, −1, −8, . . .
   d. $\frac{1}{2}, -\frac{3}{2}, -\frac{7}{2}, \ldots$

| **SECTION 9.4** | *Geometric Sequences* |

If a is the first term and r is the common ratio, then the formula $l = ar^{n-1}$ gives the nth (or last) term of a geometric sequence.

11. Find the required term of each geometric sequence.
   a. 81, 27, 9, . . .; 11th term
   b. 2, 6, 18, . . .; 9th term

   c. $9, \frac{9}{2}, \frac{9}{4}, \ldots$ 15th term
   d. $8, -\frac{8}{5}, \frac{8}{25}, \ldots$; 7th term

12. Find three geometric means between 2 and 8.

13. Find four geometric means between −2 and 64.

14. Find the positive geometric mean between 4 and 64.

The formula
$$S_n = \frac{a - ar^n}{1 - r} \quad (r \neq 1)$$

gives the sum of the first n terms of a geometric sequence, where S_n is the sum, a is the first term, r is the common ratio, and n is the number of terms.

15. Find the sum of the first 8 terms in each sequence.
   a. 81, 27, 9, . . .
   b. 2, 6, 18, . . .

   c. $9, \frac{9}{2}, \frac{9}{4}, \ldots$
   d. $8, -\frac{8}{5}, \frac{8}{25}, \ldots$

16. Find the sum of the first eight terms of the sequence $\frac{1}{3}, 1, 3, \ldots$.

17. Find the seventh term of the sequence $2\sqrt{2}, 4, 4\sqrt{2}, \ldots$.

If $|r| < 1$, the formula

$$S = \frac{a}{1 - r}$$

gives the sum of the terms of an infinite geometric sequence, where S is the sum, a is the first term, and r is the common ratio.

18. Find the sum of each infinite sequence, if possible.

a. $\dfrac{1}{3}, \dfrac{1}{6}, \dfrac{1}{12}, \ldots$ **b.** $\dfrac{1}{5}, -\dfrac{2}{15}, \dfrac{4}{45}, \ldots$

c. $1, \dfrac{3}{2}, \dfrac{9}{4}, \ldots$ **d.** $0.5, 0.25, 0.125, \ldots$

19. Change each decimal into a common fraction.
   a. $0.\overline{3}$ **b.** $0.\overline{9}$ **c.** $0.\overline{17}$ **d.** $0.\overline{45}$

20. **Investment problem** If Leonard invests \$3000 in a 6-year certificate of deposit at the annual rate of 7.75%, compounded daily, how much money will be in the account when it matures?

21. **College enrollments** The enrollment at Hometown College is growing at the rate of 5% over each previous year's enrollment. If the enrollment is currently 4000 students, what will it be 10 years from now? What was it 5 years ago?

22. **House trailer depreciation** A house trailer that originally cost \$10,000 depreciates in value at the rate of 10% per year. How much will the trailer be worth after 10 years?

SECTION 9.5   *Mathematical Induction*

Mathematical induction: If a statement involving the natural number n has the two properties that
1. the statement is true for $n = 1$, and
2. if the statement is true for $n = k$, then it is true for $n = k + 1$,
the statement is true for all natural numbers.

23. Verify the following formula for $n = 1$, $n = 2$, $n = 3$, and $n = 4$:

$$1^3 + 2^3 + 3^3 + \cdots + n^3 = \frac{n^2(n + 1)^2}{4}$$

24. Prove the formula given in Exercise 23 by mathematical induction.

SECTION 9.6   *Permutations and Combinations*

$$P(n, r) = \frac{n!}{(n - r)!}$$

$$P(n, n) = n!$$

$$P(n, 0) = 1$$

$$C(n, r) = \binom{n}{r} = \frac{n!}{r!(n - r)!}$$

$$C(n, n) = 1$$

$$C(n, 0) = 1$$

25. Evaluate each expression.
   a. $P(8, 5)$ **b.** $C(7, 4)$
   c. $0! \cdot 1!$ **d.** $P(10, 2) \cdot C(10, 2)$
   e. $P(8, 6) \cdot C(8, 6)$ **f.** $C(8, 5) \cdot C(6, 2)$
   g. $C(7, 5) \cdot P(4, 0)$ **h.** $C(12, 10) \cdot C(11, 0)$
   i. $\dfrac{P(8, 5)}{C(8, 5)}$ **j.** $\dfrac{C(8, 5)}{C(13, 5)}$
   k. $\dfrac{C(6, 3)}{C(10, 3)}$ **l.** $\dfrac{C(13, 5)}{C(52, 5)}$

There are $(n-1)!$ ways to place n things in a circle.

If an n-letter word has a of one letter, b of another letter, and so on, the number of distinguishable words that can be formed using each letter exactly once is

$$\frac{n!}{a!b!\cdots}$$

26. In how many ways can 10 teenagers be seated at a round table if 2 girls wish to sit with their boyfriends?

27. How many distinguishable words can be formed from the letters of the word *casserole* if each letter is used exactly once?

Probability

An event that cannot happen has a probability of 0. An event that is certain to happen has a probability of 1. All other events have probabilities between 0 and 1.

If S is the sample space of an experiment with n distinct and equally likely outcomes, and if E is an event that occurs in s of those ways, then the probability of E is

$$P(E) = \frac{n(E)}{n(S)} = \frac{s}{n}$$

$$P(A \cap B) = P(A) \cdot P(B \mid A)$$

28. Make a tree diagram to illustrate the possible results of tossing a coin four times.

29. In how many ways can you draw a five-card poker hand of 3 aces and 2 kings?

30. Find the probability of drawing the hand described in Exercise 29.

31. Find the probability of not drawing the hand described in Exercise 29.

32. Find the probability of having a 13-card bridge hand consisting of 4 aces, 4 kings, 4 queens, and 1 jack.

33. Find the probability of choosing a committee of 3 men and 2 women from a group of 8 men and 6 women.

34. Find the probability of drawing a club or a spade on one draw from a card deck.

35. Find the probability of drawing a black card or a king on one draw from a card deck.

36. Find the probability of getting an ace-high royal flush in hearts (ace, king, queen, jack, and ten of hearts) in poker.

37. Find the probability of being dealt 13 cards of one suit in a bridge hand.

38. Find the probability of getting 3 heads or fewer on 4 tosses of a fair coin.

| SECTION 9.8 | Computation of Compound Probabilities |

If A and B are two events, then

$$P(A \cup B) = .$$
$$P(A) + P(B) - P(A \cap B)$$

If A and B cannot occur simultaneously, then

$$P(A \cup B) = P(A) + P(B)$$

If A is any event, then

$$P(\overline{A}) = 1 - P(A)$$

If A and B are independent events, then

$$P(A \cap B) = P(A) \cdot P(B)$$

39. If the probability that a drug cures a certain disease is 0.83 and we give the drug to 800 people with the disease, find the number of people that we can expect to be cured.

| SECTION 9.9 | Odds and Mathematical Expectation |

The concepts of probabilities and odds are related.

40. Find the odds against a horse if the probability that the horse will win is $\frac{7}{8}$.

41. Find the odds in favor of a couple having 4 baby girls in a row.

42. If the probability that Joe will marry is $\frac{5}{6}$ and the probability that John will marry is $\frac{3}{4}$, find the odds against either one becoming a husband.

43. If the odds against Priscilla's graduation from college are $\frac{10}{11}$, find the probability that she will graduate.

A game is fair if its cost to play equals the expected winnings.

44. Find the expected earnings if you collect $1 for every heads you get when you toss a fair coin 4 times.

45. If the total number of subsets that a set with n elements can have is 2^n, explain why

$$\binom{n}{0} + \binom{n}{1} + \binom{n}{2} + \cdots + \binom{n}{n} = 2^n$$

■ Chapter 9 Test

In Questions 1–2, find each value.

1. $3! \cdot 0! \cdot 4! \cdot 1!$

2. $\dfrac{2! \cdot 4! \cdot 6! \cdot 8!}{3! \cdot 5! \cdot 7!}$

In Questions 3–4, find the required term in each expansion.

3. $(x + 2y)^5$; 2nd term

4. $(2a - b)^8$; 7th term

In Questions 5–6, find each sum.

5. $\displaystyle\sum_{k=1}^{3} (4k + 1)$

6. $\displaystyle\sum_{k=2}^{4} (3k - 21)$

In Questions 7–8, find the sum of the first ten terms of each sequence.

7. 2, 5, 8, . . .

8. 5, 1, −3, . . .

9. Find three arithmetic means between 4 and 24.

10. Find two geometric means between −2 and −54.

In Questions 11–12, find the sum of the first ten terms of each sequence.

11. $\dfrac{1}{4}, \dfrac{1}{2}, 1, \ldots$

12. $6, 2, \dfrac{2}{3}, \ldots$

13. A car costing \$$c$ when new depreciates 25% of the previous year's value each year. How much is the car worth after 3 years?

14. A house costing \$$c$ when new appreciates 10% of the previous year's value each year. How much will the house be worth after 4 years?

15. Prove by induction:

$$1^3 + 2^3 + 3^3 + \cdots + n^3 = \frac{n^2(n + 1)^2}{4}$$

16. How many six-digit license plates can be made if no plate begins with 0 or 1?

In Questions 17–20, find each value.

17. $P(7, 2)$

18. $P(4, 4)$

19. $C(8, 2)$

20. $C(12, 0)$

21. How many ways can 4 men and 4 women stand in line if all the women are first?

22. How many different ways can 6 people be seated at a round table?

23. How many different words can be formed from the letters of the word *bluff* if each letter is used once?

24. Show the sample space of the experiment: toss a fair coin three times.

In Questions 25–28, find each probability.

25. Rolling a 5 on one roll of a die.

26. Drawing a jack or a queen from a standard card deck.

27. Receiving 5 hearts for a 5-card poker hand.

28. Tossing 2 heads in 5 tosses of a fair coin.

29. If 30% of the population enjoy jazz, and 80% enjoy popular music, and 30% don't like either, what percent of the people like both types of music?

30. If the probability that a person is sick is 0.1, find the probability that the person is well.

In Questions 31–32, find each probability.

31. Drawing a black card or a face card on one draw from a standard card deck.

32. Rolling a sum of 3 or an even sum with one roll of two dice.

33. If the odds for a horse are 3 to 1, find the probability that the horse will win.

34. Assume a game with the following rules:

Roll a single die once.

If 6 appears, win $4.

If 5 appears, win $2.

Find the fair price to play the game.

10 The Mathematics of Finance

IN THIS CHAPTER, WE DISCUSS THE MATHEMATICS OF FINANCE—THE RULES
THAT GOVERN INVESTING AND BORROWING MONEY.

10.1 Compound Interest

■ COMPOUND INTEREST ■ BANK NOTES ■ EFFECTIVE RATE OF INTEREST ■ PRESENT VALUE

In this section, we revisit the concept of compound interest, as preparation for more complicated business formulas.

■ COMPOUND INTEREST

A deposit in a savings account earns **interest,** which the bank pays for the privilege of using someone else's money. The amount deposited is called the **principal,** and the interest, paid at regular intervals, is usually a percentage of the amount currently on deposit. Any interest left in the account also earns interest. In that case, the account earns **compound interest.**

EXAMPLE 1   Yolanda deposits $10,000 into a savings account paying 6% interest, compounded annually. Find the balance in her account after each of the first three years.

Solution   At the end of the first year, the interest earned is 6% of the principal, or

$$0.06(\$10,000) = \$600$$

This interest is added to the principal; after one year, the balance is $10,600. The second year's earned interest is 6% of the current balance, or

$$0.06(\$10,600) = \$636$$

This interest is also added to the principal, giving a second-year balance of $11,236. The interest earned during the third year is

$$0.06(\$11,236) = \$674.16$$

giving Yolanda a balance of $11,236 + $674.16, or $11,910.16, after three years.

■

Self Check   In Example 1, what will be the balance in Yolanda's account after two more years?
Answer $13,382.26

We can generalize this example to find a formula for compound interest calculations. Suppose that the original deposit, the **principal,** is P dollars, that interest is paid at an **annual rate** r, and that the **accumulated amount** or the **future value** in the account at the end of the first year is A_1. Then the interest earned that year is Pr, and

The amount after one year	=	the original deposit	+	the interest earned.

$$A_1 = P + Pr$$
$$= P(1 + r) \qquad \text{Factor out the common factor, } P.$$

The amount at the end of the first year is also the deposit at the beginning of the second year. The amount at the end of the second year, A_2, is

The amount after two years	=	the initial amount	+	the interest earned.

$$A_2 = A_1 + A_1 r$$
$$= A_1(1 + r) \qquad \text{Factor out the common factor, } A_1.$$
$$= P(1 + r)(1 + r) \qquad \text{Substitute } P(1 + r) \text{ for } A_1.$$
$$= P(1 + r)^2 \qquad \text{Simplify.}$$

By the end of the third year, the amount will be

$$A_3 = P(1 + r)^3$$

The pattern continues, and we have the following result.

Compound Interest (Annual Compounding)

A single deposit, P, called the **principal**, earning compound interest for n years at an annual rate r, will grow to a **future value** FV according to the formula

$$FV = P(1 + r)^n$$

EXAMPLE 2 For their newborn child, parents deposit $10,000 in a college account that pays 8% interest, compounded annually. What sum will be available on the child's 17th birthday?

Solution Substitute $P = 10,000$, $r = 0.08$, and $n = 17$ into the compound interest formula and use a calculator to find the future value.

$$FV = P(1 + r)^n$$
$$FV = \mathbf{10,000}(1 + \mathbf{0.08})^{17}$$
$$= 10,000(1.08)^{17}$$
$$= 37,000.18054801$$
$$= 37,000.18 \qquad \text{Round to the nearest cent.}$$

$37,000.18 will be available on the child's 17th birthday. ∎

Self Check If the parents in Example 2 leave the money on deposit for two more years, what amount will be available?

Answer $43,157.01

Interest compounded once each year is **compounded annually.** Many financial institutions compound interest more often. For example, instead of paying an annual rate of 8% once a year, they might pay 4% twice each year, or 2% four times each year. The annual rate, 8%, is also called the **nominal rate,** and the time between interest calculations is the **conversion period.** If there are k periods each year, interest is paid at the **periodic rate** given by the following formula.

Periodic Rate

$$\text{Periodic rate} = \frac{\text{annual rate}}{\text{number of periods per year}}$$

This formula is often written as

$$i = \frac{r}{k}$$

where i is the periodic interest rate, r is the annual rate, and k is the number of times interest is paid each year.

If interest is calculated k times during each year, in n years there will be kn conversions. Each conversion is at the periodic rate i. This leads to another form of the compound interest formula.

Compound Interest Formula

A principal P, earning interest compounded k times a year for n years at an annual rate r, will grow to the future value FV according to the formula

$$FV = P(1 + i)^{kn}$$

where $i = \frac{r}{k}$ is the periodic interest rate.

Interest paid twice each year is called **semiannual** compounding, four times each year **quarterly** compounding, twelve times each year **monthly** compounding, and 360 or 365 times **daily** compounding.

EXAMPLE 3

If the parents of Example 2 invested that $10,000 in an account paying 8%, but compounded quarterly, how much more money would be available after 17 years?

Solution First calculate the periodic rate, i.

$$i = \frac{r}{k}$$

$$i = \frac{0.08}{4} \qquad \text{Substitute } r = 0.08 \text{ and } k = 4.$$

$$i = 0.02$$

Substitute $P = 10,000$, $i = 0.02$, $k = 4$, and $n = 17$ into the compound interest formula, and use a calculator.

$$FV = P(1 + i)^{kn}$$
$$FV = \mathbf{10,000}(1 + \mathbf{0.02})^{4 \cdot 17}$$
$$= 10,000(1.02)^{68}$$
$$= 38442.50502546$$
$$= 38,442.51 \qquad \text{Round to the nearest cent.}$$

$38,442.51 will be available, an increase of $1,442.33 over annual compounding. ∎

Self Check   **a.** What would $10,000 become in 17 years if compounded monthly at a nominal rate of 8%?   **b.** How does this compare with quarterly compounding?

Answers   **a.** $38,786.48,   **b.** $343.97 more than quarterly compounding

ACCENT ON TECHNOLOGY

We can use a graphing calculator to find the time it would take a $10,000 investment to triple, assuming an 8% annual rate, compounded quarterly.

In n years, $10,000 earning 8% interest, compounded quarterly, will become the future value

$$10,000(1.02)^{4n}$$

To watch this value grow, we enter the function

$$Y_1 = 10000 * 1.02^{\wedge}(4 * X)$$

in a graphing calculator, and set the window to $0 \le X \le 10$ (for 10 years) and $0 \le Y \le 40000$ (for the dollar amount).

The graph appears in Figure 10-1(a). To find the time it would take for the investment to triple, we use TRACE to move to the point with a Y-value close to 30,000. The X-value in Figure 10-1(b) shows that the investment would triple in about 13.9 years.

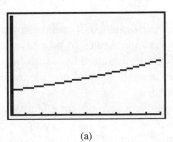

(a)

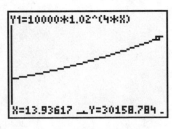

(b)

FIGURE 10-1

■ BANK NOTES

When a customer borrows money from a bank, the bank is making an investment in that person. The amount of the loan is the bank's deposit, and the bank expects to be repaid with interest in a single **balloon payment** at a later date. These loans, called **notes,** are based on a 360-day year, and they are usually written for 30 days, 90 days, or 180 days. We use the formula for compound interest to calculate the terms of the loan.

EXAMPLE 4

A student needs $4000 for tuition. His bank writes a 9%, 180-day note, with interest compounded daily. What will she owe?

Solution In making the loan, the bank invests the principal, $4000. The amount to be repaid is the expected future value,

$$FV = P(1 + i)^{kn}$$

where the principal, P, is $4000; the frequency of compounding, k, is 360; the periodic rate, i, is $\frac{0.09}{360} = 0.00025$; and the term, n, is 0.5 (180 days is one-half of 360 days.)

$$FV = P(1 + i)^{kn}$$
$$FV = 4000(1 + 0.00025)^{360 \cdot 0.5}$$
$$= 4000(1.00025)^{180}$$
$$= 4184.087907996$$
$$= 4184.09 \qquad \text{Round to the nearest cent.}$$

The student must repay $4184.09. ■

Self Check A man borrows $7500 for 90 days at 12%. If interest is compounded daily, how much will he owe?

Answer $7728.37

■ EFFECTIVE RATE OF INTEREST

The true performance of an investment depends on the compounding frequency as well as the annual rate. To help investors compare different savings plans, financial institutions are required by law to provide the **effective rate**—the rate that, if compounded annually, would provide the same yield as the plan that is more frequently compounded.

To derive a formula for effective rate, we assume that P dollars are invested for n years at an annual rate r, compounded k times per year. That same principal P, compounded annually at the effective rate R, would produce the same accumulated value. We have the equation

| Accumulated amount at effective rate R, compounded annually | $=$ | accumulated amount at annual rate r, compounded k times per year |

$$P(1 + R)^n = P(1 + i)^{kn} \qquad i \text{ is the periodic rate, } i = \tfrac{r}{k}.$$

We can solve this equation for R.

$$P(1 + R)^n = P(1 + i)^{kn}$$
$$(1 + R)^n = (1 + i)^{kn} \qquad \text{Divide both sides by } P.$$
$$[(1 + R)^n]^{1/n} = [(1 + i)^{kn}]^{1/n} \qquad \text{Raise both sides to the } 1/n \text{ power.}$$
$$1 + R = (1 + i)^k \qquad \text{Multiply the exponents.}$$
$$R = (1 + i)^k - 1 \qquad \text{Subtract 1 from both sides.}$$

Effective Rate of Interest

The **effective rate of interest** R for an account paying a nominal rate r, compounded k times per year, is

$$R = (1 + i)^k - 1$$

where i is the periodic rate, $i = \dfrac{r}{k}$.

EXAMPLE 5   A bank offers the savings plans shown in the table. Calculate the effective interest rates.

Savings plan	**a.** Money Market Fund	**b.** Certificate of Deposit
Annual rate	6.5%	7%
Compounding	quarterly	monthly
Effective rate		

Solution   **a.** For the money market fund, $r = 0.065$ and $k = 4$, so $i = \tfrac{r}{k} = \tfrac{0.065}{4} = 0.01625$. Substitute $k = 4$ and $i = 0.01625$ into the formula for effective rate.

$$R = (1 + r)^k - 1$$
$$R = (1 + \mathbf{0.01625})^4 - 1$$
$$= 0.0666016087915$$

As a percent, the effective rate is 6.66%, or $6\tfrac{2}{3}\%$.

b. For the certificate of deposit, $r = 0.07$ and $k = 12$, so

$$i = \frac{0.07}{12} = 0.00583333$$

$$i = (1 + 0.00583333)^{12} - 1$$
$$= 0.0722900808562$$

As a percent, the effective rate is 7.23%.   ■

Self Check   A bank's passbook savings account offers daily compounding (365 days per year) at an annual rate of 6%. Find the effective rate.

Answer   6.18%

■ PRESENT VALUE

From the initial deposit (the principal P), the compound interest formula predicts the future value n years later. This is the situation suggested by Figure 10-2, where we know the beginning amount and need to find the future value.

Known principal P Unknown future value FV

FIGURE 10-2

Often, the situation is reversed: We need to make a deposit now, because several years from now we will need a specific amount—perhaps to buy a car or pay tuition. As Figure 10-3 suggests, we need to know what single deposit *now* will accomplish that goal: What **present value** PV will yield a *specific* future value?

Unknown present value PV Known future value FV

FIGURE 10-3

To derive the formula for present value, we rename the principal PV and solve the compound interest formula for PV.

$$FV = PV(1 + i)^{kn}$$ The principal is now the present value, PV.

$$\frac{FV}{(1 + i)^{kn}} = \frac{PV(1 + i)^{kn}}{(1 + i)^{kn}}$$ Divide both sides by $(1 + i)^{kn}$.

$$\frac{FV}{(1 + i)^{kn}} = PV \qquad \text{Simplify.}$$

$$FV(1 + i)^{-kn} = PV \qquad \text{Use the definition}$$

of negative exponent: $\dfrac{1}{a^x} = a^{-x}$.

Present Value

The **present value** PV that must be deposited now to provide a **future value** FV n years from now is given by the formula

$$PV = FV(1 + i)^{-kn}$$

where interest is compounded k times per year at an annual rate r (i is the periodic rate, $\frac{r}{k}$).

EXAMPLE 6

When a medical student graduates in eight years, she will need $25,700 to buy medical equipment. What amount must she deposit now (at 8%, compounded twice per year) to meet this future obligation?

Solution   Use the annual rate ($r = 0.08$) and the frequency of compounding ($k = 2$) to find the periodic rate:

$$i = \frac{r}{k} = \frac{0.08}{2} = 0.04$$

Into the present value formula, we substitute

the number of years, $n = 8$,
the periodic rate, $i = 0.04$,
the frequency of compounding, $k = 2$, and
the future value, $FV = 25,700$.

$$PV = FV(1 + i)^{-kn}$$
$$PV = 25,700\left(1 + \frac{0.08}{2}\right)^{-2 \cdot 8}$$
$$= 25,700(1.04)^{-16}$$
$$= 13,721.44$$

$13,721.44, deposited now, will earn enough to meet the student's obligations 8 years from now. ∎

Self Check   If the student decides to take two extra years to complete medical school, her obligation will be $27,000. What present value will meet her goal?

Answer $12,322.45

EXERCISE 10.1

VOCABULARY AND CONCEPTS

1. A bank pays _____ for the privilege of using your money.

2. If interest is left on deposit to earn more interest, the account earns _____ interest.

3. Interest compounded once each year is called _____ compounding.

4. The initial deposit is called the _____ or the _____ value.

5. After a specific time, the principal grows to a _____ value.

6. Interest is calculated as a _____ of the amount on deposit.

7. Future value = principal + _____ earned.

8. The annual rate is also called the _____ rate.

9. _____ = $\dfrac{\text{annual rate}}{\text{number of periods per year}}$

10. The time between interest calculations is the _____ period.

11. In the future value formula $FV = P(1 + i)^{kn}$,
P is the _____,
i is the _____,
k is the _____,
and n is the _____.

12. In the present value formula $PV = FV(1 + i)^{-kn}$,
FV is the _____,
i is the _____,
k is the _____,
and n is the _____.

13. To help consumers compare savings plans, banks advertise the _____ rate of interest.

14. If after one year, $100 grows to $110, the effective rate is ___%.

PRACTICE
In Exercises 15–18, assume that $1200 is deposited in an account that compounds interest annually at a rate of 8%. Find the accumulated amount after the given number of years.

15. 1 year **16.** 3 years **17.** 5 years **18.** 20 years

In Exercises 19–22, assume that $1200 is deposited in an account that compounds interest annually at the given rate. Find the accumulated amount after 10 years.

19. 3% **20.** 5% **21.** 9% **22.** 12%

In Exercises 23–26, assume that $1200 is deposited in an account that compounds interest at the given frequency, at an annual rate of 6%. Find the accumulated amount after 15 years.

23. $k = 2$ **24.** $k = 4$ **25.** $k = 12$ **26.** $k = 365$

In Exercises 27–30, find the effective interest rate with the given annual rate r and compounding frequency k.

27. $r = 6\%, k = 4$ **28.** $r = 8\%, k = 12$ **29.** $r = 9\frac{1}{2}\%, k = 2$ **30.** $r = 10\%, k = 360$

In Exercises 31–34, find the present value of $20,000 due in 6 years, at the given annual rate and compounding frequency.

31. 6%, semiannually **32.** 8%, quarterly **33.** 9%, monthly **34.** 7%, daily (360 days/year)

APPLICATIONS

35. Saving for college At the birth of their child, the Fieldsons deposited $7000 into an account paying 6% interest, compounded quarterly. How much will be available when the child turns 18?

36. Planning a celebration When the Fernandez family made reservations at the end of 1993 for the December 1999 New Year's celebration in Paris, they placed $5700 into an account paying 8% interest, compounded monthly. What amount will be available for the celebration?

37. Planning for retirement When Jim retires in 12 years, he expects to live lavishly on the money in a retirement account that is earning $7\frac{1}{2}\%$ interest, compounded semiannually. If the account now contains $147,500, how much will be available at retirement?

38. Pension fund management The managers of a pension fund invested $3 million in government bonds paying 8.73% annual interest, compounded semiannually. After 8 years, what will the investment be worth?

39. Real estate investing Property values in the suburbs have been appreciating about 11% annually. If this trend continues, what will a $137,000 home be worth in four years? Give the result to the nearest dollar.

40. Real estate investing Property in suburbs closer to the city is appreciating about 8.5% annually. If this trend continues, what will a $47,000 one-acre lot be worth in five years? Give the result to the nearest dollar.

41. Gas consumption The gas utilities expect natural gas consumption to increase at 7.2% per year for the next decade. Monthly consumption for one county is currently 4.3 million cubic feet. What monthly demand for gas is expected in ten years?

42. Comparing banks Bank One offers a passbook account with 4.35% annual rate, compounded quarterly. Bank Two offers a money market account at 4.3%, compounded monthly. Which account provides the better growth? (*Hint:* Find the effective rates.)

43. Comparing accounts A savings and loan offers the two accounts shown in Illustration 1. Find the effective rates.

	Annual rate	Compounding	Effective rate
NOW account	7.2%	quarterly	
Money Market	6.9%	monthly	

ILLUSTRATION 1

44. Comparing accounts A credit union offers the two accounts shown in Illustration 2. Find the effective rates.

	Annual rate	Compounding	Effective rate
Certificate of deposit	6.2%	semiannual	
Passbook	5.25%	quarterly	

ILLUSTRATION 2

45. Car repair Craig borrows $1230 for unexpected car repair costs. His bank writes a 90-day note at 12%, with interest compounded daily. What will Craig owe?

46. Fly now, pay later For a 7-day Hawaii vacation, Beth borrowed $2570 for 9 months at an annual rate of 11.4%, compounded monthly. What did she owe?

47. Buying a computer A man estimates that the computer he plans to buy in 18 months will cost $4200. To meet this goal, how much should he deposit in an account paying 5.75%, compounded monthly?

48. Buying a copier An accounting firm plans to deposit enough money now into an account paying 7.6% interest, compounded quarterly, to finance the purchase of a $2780 copier in 18 months. What should be the amount of that deposit?

DISCOVERY AND WRITING

49. Adding to an investment To prepare for his retirement in 14 years, Jay deposited $12,000 into an account paying 7.5% annual interest, compounded monthly. Ten years later, he deposited another $12,000. How much will be available at retirement?

50. Changing rates Ten years ago, a man deposited $1100 into a five-year certificate of deposit paying 10%, compounded monthly. When the CD matured, he invested the proceeds in another five-year CD, paying 8%, compounded semiannually. How much is available now?

51. The power of time "A young person's most powerful money-making scheme," said an investment advisor, "is *time*." Write a paragraph explaining what the advisor meant.

52. Watching money grow $10,000 is invested at 10%, compounded annually. Use a graphing calculator to find how long it will take for the accumulated value to exceed $1 million.

REVIEW *Simplify each expression. Assume that all variables represent positive numbers.*

53. $\dfrac{x^2 - 2x - 15}{2x^2 - 9x - 5}$

54. $3x(x^2 - 5) - (x^3 - 2x)$

55. $\dfrac{(3 - x)(x + 3)}{-x^2 + 9}$

56. $-\sqrt{x^2 - 6x + 9}\ (x \geq 3)$

10.2 Annuities and Future Value

■ FUTURE VALUE OF AN ANNUITY ■ SINKING FUNDS

■ FUTURE VALUE OF AN ANNUITY

Financial plans that involve a series of payments are called **annuities.** Monthly mortgage payments, for example, are part of an annuity, as are regular contributions to a retirement plan.

Annuity
A plan involving periodic payments made at regular intervals is called an **annuity.**

Future Value
The **future value** of an annuity is the sum of all the payments and the interest those payments earn.

Term
The time over which the payments are made is called the **term** of the annuity.

Ordinary Annuity
In an **ordinary annuity,** the payments are made at the *end* of each time interval.

In this book, we consider only ordinary annuities with equal periodic payments made for a fixed term.

To understand how an annuity works, imagine a savings account that pays 12% annual interest, compounded monthly. Its periodic rate is $\frac{12\%}{12}$, or 1%—each month the bank pays, as interest, 1% of the current balance. Imagine also that each month for the next year, we deposit $100 into that account.

To determine the future value of this annuity, think of each monthly payment as a one-time initial contribution to a compound-interest savings account. The future value of the annuity is the sum of the accumulated values of 12 individual accounts. As Figure 10-4 suggests, the first deposit earns interest for 11 months, the second for 10 months, and so on. Because the last deposit is made at the end of the year, it earns no interest. A few minute's work with a calculator finds the total to be $1268.25.

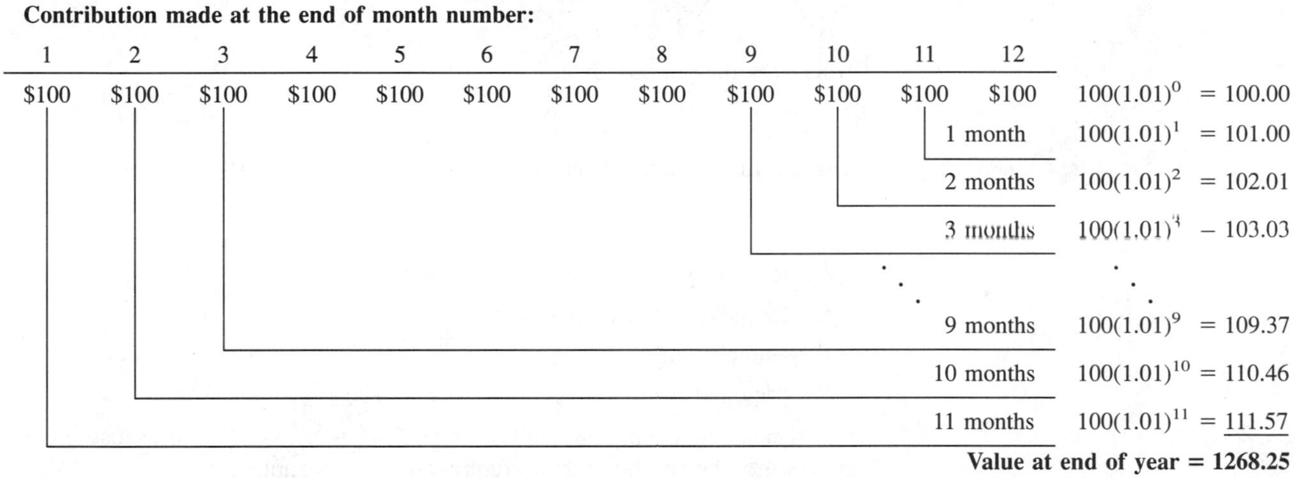

Contribution made at the end of month number:

1	2	3	4	5	6	7	8	9	10	11	12		
$100	$100	$100	$100	$100	$100	$100	$100	$100	$100	$100	$100	$100(1.01)^0$	$= 100.00$
											1 month	$100(1.01)^1$	$= 101.00$
											2 months	$100(1.01)^2$	$= 102.01$
											3 months	$100(1.01)^4$	$- 103.03$
											9 months	$100(1.01)^9$	$= 109.37$
											10 months	$100(1.01)^{10}$	$= 110.46$
											11 months	$100(1.01)^{11}$	$= \underline{111.57}$

Value at end of year = 1268.25

FIGURE 10-4

We will now generalize this example to derive a formula for future value. Consider an annuity with regular deposits of P dollars and with interest compounded k times per year for n years, at an annual rate r $\left(\text{periodic rate } i = \frac{r}{k}\right)$. During n years, there will be kn periods and kn deposits.

The last deposit, made at the end of the last period, earns no interest. The next-to-last deposit earns interest for one period, and so on. The first deposit, made at the end of the first period, earns compound interest for $kn - 1$ periods.

Future value of the annuity	=	future value of the last deposit	+	future value of the next-to-last deposit	$+ \cdots +$	future value of the first deposit.

$$FV = P + P(1 + i)^1 + P(1 + i)^2 + P(1 + i)^3 + \cdots + P(1 + i)^{kn-1}$$

This is a geometric sequence with first term P and common ratio $(1 + i)$. The sum is given by

$$FV = \frac{P[(1 + i)^{kn} - 1]}{i}$$

> Recall that the sum of the terms of the geometric sequence
> $$S = a + ar + ar^2 + ar^3 + \cdots + ar^{n-1} \text{ is}$$
> $$S = \frac{a(r^n - 1)}{r - 1}$$

We summarize this result.

Future Value of an Annuity

The future value FV of an ordinary annuity with deposits of P dollars made regularly k times each year for n years, with interest compounded k times per year at an annual rate r, is

$$FV = \frac{P[(1 + i)^{kn} - 1]}{i}$$

where i is the periodic rate, $i = \frac{r}{k}$.

EXAMPLE 1 Verify the future value of the annuity outlined in Figure 10-4.

Solution From Figure 10-4, we find

the term, in years:	$n = 1$
the frequency of compounding:	$k = 12$
the annual rate:	$r = 0.12$
the principal:	$P = 100$

We then calculate the periodic interest rate: $i = \frac{r}{k} = \frac{0.12}{12} = 0.01$, and substitute these numbers into the formula for the future value of an annuity.

$$FV = \frac{P[(1 + i)^{kn} - 1]}{i}$$

$$FV = \frac{100[(1 + 0.01)^{12 \cdot 1} - 1]}{0.01}$$

$$= \frac{100[(1.01)^{12} - 1]}{0.01}$$

$$= 1268.25030132 \qquad \text{Use a calculator.}$$

$$= 1268.25 \qquad \text{Round to the nearest cent.}$$

The annuity of Figure 10-4 will provide \$1268.25 by the end of the year. ∎

Self Check Under the payroll savings plan, Hogan contributes \$50 a month to an ordinary annuity paying $7\frac{1}{2}\%$ annual interest, compounded monthly. How much will he have in 5 years?

Answer \$3626.36

■ ■ ■ ■ ■ ■ ■ ■ ■

ACCENT ON
TECHNOLOGY

The value of an annuity accumulates rapidly. To watch the accumulated amount in the annuity of Example 1 grow, we graph the function

$$FV = \frac{100[(1.01)^{12n} - 1]}{0.01}$$

$$= 10,000(1.01^{12n} - 1) \qquad \frac{100}{0.01} = 10,000.$$

We enter the function

$$Y_1 = 10000*(1.01^\wedge(12*X) - 1)$$

into a graphing calculator, set the window values to be $0 \le X \le 50$ and $0 \le Y \le 1,000,000$, and graph it, as in Figure 10-5(a). Note how the amount increases more rapidly as the years go by. Using TRACE, as in Figure 10-5(b), we see that over \$1 million accumulates in just 39 years.

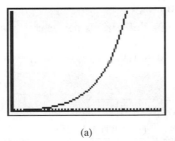
(a)

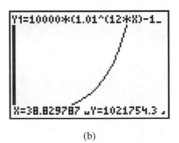
(b)

FIGURE 10-5

■ ■ ■ ■ ■ ■ ■ ■ ■

■ SINKING FUNDS

If we know the amount of each deposit, we can calculate the future value of an annuity using the formula on page 724. The situation is often reversed: What regular deposits, made now, will provide a specific future amount? An annuity created to produce a fixed future value is called a **sinking fund.** To determine the required periodic payment P, we solve the future value formula for P.

$$FV = \frac{P[(1 + i)^{kn} - 1]}{i}$$

$$FV\frac{i}{(1 + i)^{kn} - 1} = \frac{P[(1 + i)^{kn} - 1]}{i} \cdot \frac{i}{(1 + i)^{kn} - 1} \qquad \text{To isolate } P, \text{ multiply both sides by } \frac{i}{(1 + i)^{kn} - 1}.$$

$$\frac{FVi}{(1 + i)^{kn} - 1} = P \qquad \text{Simplify.}$$

Sinking Fund Payment

For an annuity to provide a future value FV, regular deposits P are made k times per year for n years, with interest compounded k times per year at an annual rate r. The payment P is given by

$$P = \frac{FVi}{(1 + i)^{kn} - 1}$$

where i is the periodic rate, $i = \dfrac{r}{k}$.

EXAMPLE 2   An accounting firm will need $17,000 in five years to replace its computer system. What periodic deposits to an annuity paying quarterly interest at a 9% annual rate will achieve that goal?

Solution   Substitute the given values into the formula to find the sinking fund payment:

future value:	$FV = \$17{,}000$
annual rate:	$r = 0.09$
term:	$n = 5$
number of periods per year:	$k = 4$
periodic rate:	$i = \dfrac{r}{k} = \dfrac{0.09}{4} = 0.0225$

$$P = \frac{FVi}{(1 + i)^{kn} - 1}$$

$$= \frac{(17{,}000)(0.0225)}{(1 + 0.0225)^{4 \cdot 5} - 1}$$

$= 682.415203056$ Use a calculator.

$= 682.42$ Round to the nearest cent.

Quarterly payments of $682.42 will raise $17,000 in five years.   ∎

Self Check   What quarterly deposits to the account in Example 2 are required to raise the $50,000 start-up cost of a branch office in seven years?

Answer $1301.26

EXERCISE 10.2

VOCABULARY AND CONCEPTS

1. Plans involving periodic payments made at regular intervals are called _____.

2. In an ordinary annuity, payments are made at the ____ of each time period.

3. The future value of an annuity is the sum of all the _____ and _____.

4. The time over which the payments are made is the ____ of the annuity.

5. In the future value formula,
$$FV = \frac{P[(1 + i)^{kn} - 1]}{i},$$
P is the _____,
i is the _____,
k is the _____,
and n is the _____.

6. An annuity created to fund a specific future obligation is a _____ fund.

PRACTICE *In Exercises 7–10, assume that $100 is deposited at the end of each year into an account that compounds interest annually at a rate of 6%. Find the accumulated amount after the given number of years.*

7. 10 years   **8.** 5 years   **9.** 3 years   **10.** 20 years

In Exercises 11–14, assume that $100 is deposited at the end of each year into an account that compounds interest annually at the given rate. Find the accumulated amount after 10 years.

11. 4%   **12.** 7%   **13.** 9.5%   **14.** 8.5%

In Exercises 15–18, assume that $100 is deposited at the end of each period into an account that compounds interest at the given frequency, at an annual rate of 8%. Find the accumulated amount after 15 years.

15. $k = 2$ **16.** $k = 4$ **17.** $k = 12$ **18.** $k = 1$

In Exercises 19–22, find the amount of each regular payment to provide $20,000 in 10 years, at the given annual rate and compounding frequency.

19. 4%, annually   **20.** 6%, quarterly   **21.** 9%, semiannually   **22.** 8%, monthly

APPLICATIONS

23. Saving for a vacation   For next year's vacation, the Phelps family is saving $200 each month in an account paying 6% annual interest, compounded monthly. How much will be available a year from now?

24. Planning for retirement   Hank's regular $1300 quarterly contributions to his retirement account have earned 6.5% annual interest, compounded quarterly, since he started 21 years ago. How much is in his account now?

25. Pension fund management   The managers of a company's pension fund invest the monthly employee contributions of $135,000 into a government fund paying 8.7%, compounded monthly. To what value will the fund grow in 20 years?

26. Saving for college   A mother has been saving regularly for her daughter's college—$25 each month for 11 years. The money has been earning $7\frac{1}{2}\%$ annual interest, compounded monthly. How much is now in the account?

27. Buying office machines   A company's new corporate headquarters will be completed in $2\frac{1}{2}$ years. At that time, $750,000 will be needed for office equipment. How much should be invested monthly to fund that expense? Assume 9.75% interest, compounded monthly.

28. Retirement lifestyle   A woman would like to receive a $500,000 lump-sum distribution from her retirement account when she retires in 25 years. She begins making monthly contributions now to an annuity paying 8.5%, compounded monthly. Find the amount of that contribution.

29. Comparing accounts Which account will require the lower annual contributions to fund a $10,000 obligation in 20 years? (*Hint:* Compare the yearly total contributions.)

Bank A	5.5%; annually
Bank B	5.35%; monthly

30. Avoiding a balloon payment The last payment of a home mortgage is a balloon payment of $47,000, which the owner is scheduled to pay in 12 years. How much *extra* should he start including in each monthly payment to eliminate the balloon payment? His mortgage is at 10.2%, compounded monthly.

DISCOVERY AND WRITING

31. Retirement strategy Jim will retire in 30 years. He will invest $100 each month for 15 years and then let the accumulated value continue to grow for the next 15 years. How much will be available at retirement? Assume 8%, compounded monthly.

32. Retirement strategy Jim's brother Jack will also retire in 30 years. He plans on doing nothing during the first 15 years, then contributing twice as much—$200 monthly—to "catch up." How much will be available at retirement? Assume 8%, compounded monthly.

33. Changing plans A woman needs $13,500 in 10 years. She would like to make regular annual contributions for the first five years and then let the amount grow at compound interest for the next five years. What should her contributions be? Assume 9%, compounded annually.

34. Talking financial sense How would you explain to a friend who has just been hired for her first job that now is the time to start thinking about retirement?

REVIEW *Solve each equation.*

35. $\dfrac{2(5x - 12)}{x} = 8$

36. $\dfrac{2(5x - 12)}{x} = x$

37. $\sqrt{2x + 3} = 3$

38. $\sqrt{2x + 3} = x$

10.3 Present Value of an Annuity; Amortization

■ PRESENT VALUE OF AN ANNUITY ■ AMORTIZATION

■ PRESENT VALUE OF AN ANNUITY

Instead of using an annuity to create a future value *A*, we might ask, "What *single* deposit made *now* would create that same future value?" The one deposit that gives the same final result as an annuity is called the **present value** of that annuity.

To find a formula for the present value of an annuity, we combine two previous formulas. A series of regular payments of *P* dollars for *n* years will grow to a future value *FV* given by

1. $\quad FV = \dfrac{P[(1 + i)^{kn} - 1]}{i}$

From a formula in Section 10.1, the present value of a future asset is given by

2. $\quad PV = FV(1 + i)^{-kn}$

We find the present value of a series of future payments by substituting the right-hand side of Equation 1 into Equation 2.

$$PV = FV(1 + i)^{-kn}$$

Equation 2.

$$PV = \frac{P[(1 + i)^{kn} - 1]}{i}(1 + i)^{-kn}$$

Substitute $\dfrac{P[(1 + i)^{kn} - 1]}{i}$

for FV in Equation 2.

$$= \frac{P[(1 + i)^{-kn}(1 + i)^{kn} - 1(1 + i)^{-kn}]}{i}$$

Use the distributive property.

$$= \frac{P[1 - (1 + i)^{-kn}]}{i}$$

Simplify.

Present Value of an Annuity

The present value PV of an annuity with payments of P dollars made k times per year for n years, with interest compounded k times per year at an annual rate r, is

$$PV = \frac{P[1 - (1 + i)^{-kn}]}{i}$$

where i is the periodic rate, $i = \frac{r}{k}$.

EXAMPLE 1

To buy a jet ski in 2 years, the Higgins family plans to save $200 a month in an account that pays 12% interest, compounded monthly.   **a.** What will be the total of the payments?   **b.** What will the value of the account be in 2 years?   **c.** What single deposit in that account now would do as well?

Solution   **a.** At $200 per month for 24 months, the total amount contributed is

$200(24) = $4800

b. To find the value after 2 years, we use the formula for future value of an annuity:

$$FV = \frac{P[(1 + i)^{kn} - 1]}{i}$$

$$FV = \frac{200[1.01^{12 \cdot 2} - 1]}{0.01}$$

$$= 5394.69$$

c. To find the present value of the annuity, we substitute

the term, in years:	$n = 2$
the frequency of compounding:	$k = 12$
the annual rate:	$r = 0.12$
the payment:	$P = 200$
the periodic interest rate:	$i = \frac{r}{k} = \frac{0.12}{12} = 0.01$

into the present value formula.

$$PV = \frac{P[1 - (1 + i)^{-kn}]}{i}$$

$$PV = \frac{200[1 - (1 + 0.01)^{-12 \cdot 2}]}{0.01}$$

$$= \frac{200[1 - (1.01)^{-24}]}{0.01} \qquad \text{Simplify.}$$

$$= 4248.677451 \qquad \text{Use a calculator.}$$

$$= 4248.68 \qquad \text{Round to the nearest cent.}$$

$4248.68 is the present value of the annuity: That one deposit now will provide the same final amount, $5394.69, as the annuity. ∎

Self Check For his retirement in 30 years, a man plans to make monthly contributions of $25 to an ordinary annuity paying $8\frac{1}{2}\%$ annually, compounded monthly. **a.** What will be the total of his contributions? **b.** What single deposit now will provide the same retirement benefit?

Answers **a.** $9000, **b.** $3251.34

State lottery winnings are usually paid as a 20-year annuity. That is to the state's advantage, because it can fund the annuity with a single amount that is much smaller than the total prize.

EXAMPLE 2 Britta won the lottery. She will receive $75,000 per month for the next 20 years—a total of $18 million. What single deposit should the lottery commission make now to fund Britta's annuity? Assume 8.4% annual interest, compounded monthly.

Solution The lottery commission finds the present value of the annuity, with

the payment: $P = 75,000$

the annual rate: $r = 0.084$

the frequency of compounding: $k = 12$

the periodic rate: $i = \frac{r}{k} = \frac{0.084}{12} = 0.007$

the term, in years: $n = 20$

These values are used in the formula for the present value of an annuity.

$$PV = \frac{P[1 - (1 + i)^{-kn}]}{i}$$

$$PV = \frac{75,000[1 - (1.007)^{-12 \cdot 20}]}{0.007}$$

$$= 8705700.365$$

$$= 8,705,700.37$$

To fund the $18 million prize, the lottery commission must deposit a bit more than $8.7 million. ∎

Self Check The lottery pays a total prize of $120,000 in monthly installments, as a 10-year annuity. Assuming 8.4% interest, compounded monthly, what current deposit is needed to fund the annuity?

Answer about $81,000

EXAMPLE 3 As a settlement in an automobile injury lawsuit, Robyn will receive $30,000 each year for the next 25 years, for a total of $750,000. The insurance company is offering a one-payment settlement of $300,000, now. Should she accept? Assume that the money can be invested at 9% annual interest.

Solution Robyn should calculate the present value of an annuity with

the payment: $P = 30,000$

the annual rate: $r = 0.09$

the frequency of compounding: $k = 1$ (annual)

the periodic rate: $i = \dfrac{r}{k} = \dfrac{0.09}{1} = 0.09$

the term, in years: $n = 25$

She should use these values in the formula for the present value of an annuity.

$$PV = \frac{P[1 - (1 + i)^{-kn}]}{i}$$

$$PV = \frac{30,000[1 - (1.09)^{-1 \cdot 25}]}{0.09}$$

$$= 294,677$$

The annuity is worth only $294,677, and the company is offering $300,000. Robyn should accept the $300,000. ∎

Self Check If Robyn could invest the settlement at 8% interest, should she still accept the lump-sum offer?

Answer No; the annuity is now worth more than $320,243.

When a worker is employed, regular contributions are usually made to a retirement fund. After retirement, those funds are given back, either as an annuity or as a lump-sum distribution.

EXAMPLE 4 Pete wants to fund an annuity to supplement his retirement income. How much should he deposit now to generate retirement income of $1000 a month for the next 20 years? Assume that he can get $9\frac{3}{4}\%$ interest, compounded monthly.

Solution   Pete must calculate the present value of a future stream of income, with

the payment:	$P = 1000$
the annual rate:	$r = 0.0975$
the frequency of compounding:	$k = 12$
the periodic rate:	$i = \dfrac{r}{k} = \dfrac{0.0975}{12} = 0.008125$
the term, in years:	$n = 20$

He should use these values in the formula for the present value of an annuity.

$$PV = \frac{P[1 - (1 + i)^{-kn}]}{i}$$

$$PV = \frac{1000[1 - (1.008125)^{-12 \cdot 20}]}{0.008125}$$

$$= 105,428$$

$105,428 put on deposit now will supplement Pete's retirement income for 20 years.

■

Self Check   If Pete can invest at $8\frac{3}{4}\%$, what deposit is needed now?

Answer $113,159

■ AMORTIZATION

Before a bank will allow you to borrow money, you must sign a **promissory note** indicating that you promise to pay the money back. We discussed one-payment notes in Section 10.1. Most loans, however, are repaid in installments instead of all at once. Spreading the repayment over several equal payments is called **amortization.**

When a loan is made, the lending institution is buying an annuity from the borrower—the lender pays the borrower an agreed-upon amount and expects regular payments in return. To calculate the amount of these regular installment payments, we solve the present value formula for P to get

$$P = \frac{PV\, i}{1 - (1 + i)^{-kn}}$$

In this context, the present value is the amount of the loan. Using the letter A (for *amount*) instead of PV, we have the following formula.

Installment payments

The periodic payment P required to repay an amount A is given by

$$P = \frac{Ai}{1 - (1 + i)^{-kn}}$$

where

r is the annual rate

k is the frequency of compounding (usually monthly)

i is the periodic rate, $i = \frac{r}{k}$, and

n is the term of the loan.

EXAMPLE 5 The Almondi family took a 15-year mortgage of $200,000 for their new home, at 10.8%, compounded monthly. **a.** What is the amount of their monthly payment? **b.** What is the total of their payments over the full term?

Solution **a.** Into the formula for calculating the installment payment, we substitute

the amount:	$A = 200{,}000$
the annual rate:	$r = 0.108$
the frequency of compounding:	$k = 12$
the periodic rate:	$i = \dfrac{r}{k} = \dfrac{0.108}{12} = 0.009$
the term, in years:	$n = 15$

$$P = \frac{Ai}{1 - (1 + i)^{-kn}}$$

$$P = \frac{(200{,}000)(0.009)}{1 - (1.009)^{-12 \cdot 15}}$$

$$= 2248.14$$

Each monthly mortgage payment is $2,248.14.

b. There are $12 \cdot 15 = 180$ payments of $2248.14 each, for a total of $404,665— over twice the amount borrowed! ∎

Self Check Instead of a 15-year mortgage, the Almondis considered a 30-year mortgage. Answer the two questions again.

Answers **a.** $1874.48, **b.** $674,814

EXERCISE 10.3

VOCABULARY AND CONCEPTS

1. The current worth of a future stream of income is the _____ of an annuity.

2. The amount required now to produce a future stream of income is the _____ of an annuity.

3. A loan is called a _____ because you promise to repay it.

4. Often, loan repayment is spread out over several _____.

5. Spreading repayment of a loan over several *equal* payments is called _____ the loan.

6. An amortized loan is also called a _____.

PRACTICE *In Exercises 7–8, find the present value of an annuity with the given terms.*

7. Annual payments of $3500 at 5.25%, compounded annually for 25 years.

8. Semiannual payments of $375 at a 4.92% annual rate, compounded semiannually for 10 years.

In Exercises 9–10, find the periodic payment required to repay a loan with the given terms.

9. $25,000 repaid over 15 years with monthly payments at a 12% annual rate.

10. $1750 repaid in 18 monthly installments, at an annual rate of 19%.

11. Funding retirement Instead of making quarterly contributions of $700 to a retirement fund for the next 15 years, Jason would rather make only one contribution, now. How much should that be? Assume $6\frac{1}{4}$% annual interest, compounded quarterly.

12. Funding a lottery To fund Jamie's lottery winnings of $15,000 per month for the next 20 years, the lottery commission needs to make a single deposit now. Assuming 9.2% compounded monthly, what should the deposit be?

13. Money up front Instead of receiving an annuity of $12,000 each year for the next 15 years, Manuel would like a one-time payment, now. Assuming Manuel could invest the proceeds at $8\frac{1}{2}$%, what would be a fair amount?

14. Funding retirement What single amount deposited now into an account paying $7\frac{2}{3}$% annual interest,

compounded quarterly, would fund an annuity paying $5000 quarterly for the next 25 years?

15. Buying a car The Jepsens are buying a $21,700 car and financing it over the next four years. They secure an 8.4% loan. What will their monthly payments be?

16. Total cost of buying a car What will be the total amount the Jepsens will pay over the life of the loan? (See Exercise 15.)

17. Choosing a mortgage One lender offers two mortgages—a 15-year mortgage at 12%, and a 20-year mortgage at 11%. For each, find the monthly payment to repay $130,000.

18. Total cost of a mortgage For each of the mortgages in Exercise 17, find the total of the monthly payments.

DISCOVERY AND WRITING

19. Getting an early start As Jorge starts working now at the age of 20, he decides to make regular contributions to a savings account. He wants to accumulate enough by age 55 to fund an annuity of $5000 per month until age 80. What should his monthly contributions be? Assume that both accounts pay 8.75%, compounded monthly.

20. Comparing annuities Which of these 20-year plans is best, and why? All are at 8% annually.
   a. $1000 each year for 10 years, and then let the accumulated amount grow for 10 years.
   b. $500 each year for 20 years.

   c. Do nothing for 10 years, and then contribute $2000 each year for 10 years.
   d. One payment of $8000 now, and let it grow.

21. Changing the payments A woman contributed $500 per quarter for the first 10 years of an annuity, but changed to quarterly payments of $1,500 for the last 10 years. Assuming $7\frac{1}{4}$% annual interest compounded quarterly, what is her accumulated value?

22. Changing the rates Amy contributed $150 per month for ten years to an account that paid 5% for the first five years, but 6.5% for the last five years. How much has she saved?

REVIEW Simplify each expression. Assume that all variables represent positive numbers.

23. $\dfrac{6\sqrt{30}}{3\sqrt{5}}$ **24.** $\dfrac{6}{\sqrt{7}-2}$ **25.** $3\sqrt{5x}+5\sqrt{20x}$ **26.** $\sqrt{\dfrac{x^3y^5}{x^5y^6}}$

■ ■ ■ ■ ■ ■ ■ ■ ■ **PROBLEMS AND PROJECTS**

In Problems 1–6, write a brief paragraph explaining the given terms.

1. *interest* and *interest rate*

2. *present value* and *future value*

3. *compound interest* and *simple interest*

4. *annual rate* and *effective rate*

5. *annual rate* and *periodic rate*

6. *annuity* and *sinking fund*

7. The United States gross federal debt in 1997 was approximately $5.5 trillion. If the interest on that debt is 5%, how much does each of the country's 260 million people owe in interest each year?

8. With an annual inflation rate of 5%, $100 today will have the purchasing power (in today's dollars) of $100(1.05)^{-1}$, or $95.24, a year from now. At a 5% inflation rate, what monthly income would be required in 20 years to provide the purchasing power of $2000 per month today?

PROJECT 1   In an installment loan, the borrower agrees to pay back principal and interest in a series of installments for a specified period of time. For a mortgage, the interest is calculated as compound interest. Another type of installment loan uses simple interest, which is added on to the amount of the loan. This method, called **add-on interest,** is often used for car loans.

For example, you need to borrow $1000 for two years at an add-on interest rate of 12%. The simple interest is $240, because

$$I = Prt$$
$$I = (1000)(0.12)(2) \qquad P = 1000,\ r = 0.12,\ t = 2.$$
$$= 240$$

The amount to be repaid is $1240, the sum of the interest and the principal. The required monthly payment A is

$$A = \dfrac{1240}{24} \qquad \text{In two years, there are 24 monthly payments.}$$
$$= 51.67$$

Each month's payment is $51.67.

- If this had been a mortgage loan, what would the payment have been?

- Notice that you have paid two years' interest on the full amount, even though you will not owe the full amount for two years. Explain the effect this has on the real interest rate.

(continued)

■ ■ ■ ■ ■ ■ ■ ■ ■ ■ **PROBLEMS AND PROJECTS** (continued)

• To help consumers compare interest rates, the Truth-in-Lending Act requires that all lenders state the true annual interest, called the **Annual Percentage Rate (APR)**. It is calculated by the formula

$$\text{APR} = \frac{2Nr}{N + 1}$$

where N is the total number of payments and r is the add-on interest rate. Find the APR for this loan, if satisfied in two years.

• By doubling each monthly payment, this loan would be satisfied in one year. Is it wise to pay off this loan early? Calculate the APR and explain. (*Hint:* $240 interest on $1000 for 1 year is no longer 12%.)

PROJECT 2   Another method for calculating the terms of an installment loan is called the **rule of 78**. It is based on the fact that the sum of the fractions

$$\frac{12}{78} + \frac{11}{78} + \frac{10}{78} + \frac{9}{78} + \cdots + \frac{1}{78}$$

is equal to 1. For a one-year loan, the interest charged is simple interest, but $\frac{12}{78}$ of that interest is repaid the first month, $\frac{11}{78}$ of the interest the second month, and so on.

• Show that $\frac{12}{78} + \frac{11}{78} + \frac{10}{78} + \frac{9}{78} + \cdots + \frac{1}{78} = 1$.

• For a loan of $1000 for one year at 12%, simple interest of $120 would be added to the principal, and each month's payment would be $\frac{1}{12}$ of $1120, or $99.33. Of this first month's payment, $\frac{12}{78}$ of $120, or $18.46, would be interest. Of the *last* month's payment, $\frac{1}{78}$ of $120, or only $1.54, would be interest.

  Make a table showing how the monthly payment is divided between interest and principal for each of the 12 months.

• If a "rule of 78" loan is paid back early, the borrower is entitled to a return of the unpaid interest. Suppose the previous loan ($1000 at 12% for one year) is repaid in 6 months. What interest will be refunded? Why is this not one-half of $120?

C H A P T E R   S U M M A R Y

CONCEPT

REVIEW EXERCISES

SECTION 10.1   *Compound Interest*

Funds in a savings account earn **interest** at an **annual rate** *r*.

The amount deposited is the **principal, *P*.**

Interest left in the account earns **compound interest.**

The **accumulated amount** or the **future value** *FV* after the **term** of *n* years is found by the formula

$$FV = P(1 + i)^n$$
(annual compounding)

1. $2000 is deposited in an account that compounds interest annually at 9%. Find the value in 5 years.

2. Brian borrows $2350 for medical bills. The bank writes a 60-day note at 14%, with interest compounded daily. What will Brian owe?

If interest is calculated *k* times each year, then the **periodic rate** is $i = \frac{r}{k}$, and the future value is

$$FV = P(1 + i)^{kn}$$
(compounding *k* times per year)

3. $2000 earns interest, compounded quarterly, at an annual rate of 7.6% for 16 years. Find the future value.

The **effective rate** *R* is used to compare different savings plans.

$$R = (1 + i)^k - 1$$

4. BigBank advertises a savings account at a 6.3% rate, compounded quarterly. BestBank offers 6.21%, compounded daily. Calculate each effective rate and choose the better account.

The **present value** *PV* is the single deposit *now* that will yield a specific future value, *FV*.

$$PV = FV(1 + i)^{-kn}$$

5. What amount deposited now in an account paying 5.75% interest, compounded semiannually, will yield $7900 in 6 years?

| **SECTION 10.2** | *Annuities and Future Value* |

An **annuity** is a series of periodic payments made at regular intervals.

The **future value** *FV* of an annuity is the sum of all the payments and the interest those payments earn.

The time over which the payments are made is called the **term** of the annuity.

In an **ordinary annuity,** the payments are made at the *end* of each time interval.

The future value *FV* of an ordinary annuity with deposits of *P* dollars made regularly *k* times each year for *n* years, with interest compounded *k* times per year at an annual rate *r* is

$$FV = \frac{P[(1 + i)^{kn} - 1]}{i}$$

where *i* is the periodic rate, $i = \frac{r}{k}$.

An annuity with the purpose of funding a future obligation is a **sinking fund.**

To yield a specific future value *FV*, regular deposits *P* are made *k* times per year for *n* years, with interest compounded *k* times per year at an annual rate *r*. The payment *P* is

$$P = \frac{FV\,i}{(1 + i)^{kn} - 1}$$

where *i* is the periodic rate, $i = \frac{r}{k}$.

6. $500 is deposited at the end of each year into an annuity that compounds interest annually at 5%. Find the accumulated amount after 13 years.

7. $150 is deposited monthly into an account that pays 8% annual interest, compounded monthly. Find the future value after 20 years.

8. The owners of a small dry cleaning shop will need $40,700 to open a second shop in 7 years. What monthly payments to a sinking fund earning 7.5% interest, compounded monthly, will meet that obligation?

SECTION 10.3 — *Present Value of an Annuity; Amortization*

The **present value** of an annuity is the current worth of a future stream of income.

The present value PV of an annuity with payments of P dollars made k times per year for n years, with interest compounded k times per year at an annual rate r, is

$$PV = \frac{P[1 - (1 + i)^{-kn}]}{i}$$

where i is the periodic rate, $i = \frac{r}{k}$.

Loans are often paid off in **installments.** If equal installments are paid over a fixed time, the payments are **amortized.**

The periodic payment P required to repay an amount A is given by

$$P = \frac{Ai}{1 - (1 + i)^{-kn}}$$

where

r is the annual rate,

k is the frequency of compounding,

i is the periodic rate, $i = \frac{r}{k}$, and

n is the term of the loan.

9. An annuity pays $250 semiannually for 20 years. At a semiannually compounded rate of 6.5%, what is the present value?

10. The lottery must fund a 20-year annuity of $50,000 per year. At 9.6% compounded annually, what must be invested now?

11. What are the monthly payments for a $150,500, 15-year, 10.75% mortgage? What is the total amount paid?

12. Answer the previous question, but for a 30-year mortgage.

■ Chapter 10 Test

In Questions 1–8, fill each blank with an appropriate answer.

1. When interest is left on deposit to earn more interest, the account earns _____ interest.

2. The annual rate of interest divided by the number of periods is called the _____ interest rate.

3. To compare different savings plans, compare the _____ rates of interest.

4. The nominal rate of interest is also called the _____ rate.

5. Plans involving regular periodic payments are called _____.

6. An annuity to fund a specific future obligation is a _____.

7. The current value of a series of future payments is the _____ of an annuity.

8. Repaying a loan over several regular, equal payments is called _____ the loan.

9. $1300 is deposited into an account that earns 5% interest, compounded annually. What will it be worth in 10 years?

10. $1300 is deposited in an account that earns 5% interest, compounded monthly. What will it be worth in 10 years?

11. What is the effective rate of the savings plan in Question 10?

12. What single deposit now will yield $5000 in 10 years? Assume 7% annual interest, compounded quarterly.

13. Each month for 5 years, Bob made $700 payments to an account paying 7.3% annual interest, compounded monthly. Find the accumulated amount.

14. What monthly payment to a sinking fund will raise $8000 in 5 years? Assume 6.5% annual interest, compounded monthly.

15. Find the present value of an annuity that pays $1000 each month for 15 years. Assume 6.8% annual interest, compounded monthly.

16. What are the monthly payments for a 15-year, $90,000 mortgage at 8.95%?

■ Cumulative Review Exercises

In Exercises 1–2, solve each system by graphing.

1. $\begin{cases} 2x + y = 8 \\ x - 2y = -1 \end{cases}$

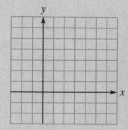

2. $\begin{cases} 3x = -y + 2 \\ y + x - 4 = -2x \end{cases}$

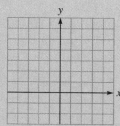

In Exercises 3–4, solve each system.

3. $\begin{cases} 5x = 3y + 12 \\ 2x - 3y = 3 \end{cases}$

4. $\begin{cases} 2x + y - z = 7 \\ x - y + z = 2 \\ x + y - 3z = 2 \end{cases}$

In Exercises 5–6, solve each system using matrices.

5. $\begin{cases} 2x + y - z = 0 \\ x - y + z = 3 \\ x + y - 3y = -5 \end{cases}$

6. $\begin{cases} 2x - 2y + 3z + t = 2 \\ x + y + z + t = 5 \\ -x + 2y - 3z + 2t = 2 \\ x + y + 2z - t = 4 \end{cases}$

In Exercises 7–10, let $A = \begin{bmatrix} 2 & 1 \\ 1 & 4 \end{bmatrix}$, $B = \begin{bmatrix} -1 & 2 \\ 2 & 3 \end{bmatrix}$, *and* $C = \begin{bmatrix} 2 & 0 & -1 \\ -1 & 2 & 2 \end{bmatrix}$. *Find each matrix.*

7. $A + B$ **8.** $B - A$ **9.** AC **10.** $B^2 + 2A$

In Exercises 11–12, find the inverse of each matrix, if possible.

11. $\begin{bmatrix} 2 & 6 \\ 2 & 4 \end{bmatrix}$

12. $\begin{bmatrix} 1 & -1 & 1 \\ 1 & 4 & 0 \\ 2 & 4 & 1 \end{bmatrix}$

In Exercises 13–14, find each determinant.

13. $\begin{vmatrix} -3 & 5 \\ 4 & 7 \end{vmatrix}$

14. $\begin{vmatrix} 2 & -3 & 2 \\ 0 & 1 & -1 \\ 1 & -2 & 1 \end{vmatrix}$

In Exercises 15–16, set up the determinants to find x and y in the system $\begin{cases} 4x + 3y = 11 \\ -2x + 5y = 24 \end{cases}$.

15. $x =$ **16.** $y =$

In Exercises 17–18, decompose each fraction into partial fractions.

17. $\dfrac{-x + 1}{(x + 1)(x + 2)}$

18. $\dfrac{x - 4}{(2x - 5)^2}$

In Exercises 19–20, find each solution by graphing.

19. $y \le 2x + 6$

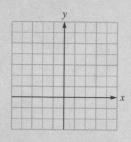

20. $\begin{cases} 2x + 3y \ge 6 \\ 2x - 3y \le 6 \end{cases}$

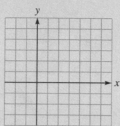

In Exercises 21–22, write the equation of each circle with the given center O and radius r.

21. $O(0, 0)$; $r = 4$

22. $O(2, -3)$; $r = 11$

In Exercises 23–26, complete the square on x and/or y and graph each equation.

23. $x^2 + y^2 - 4y = 12$ **24.** $x^2 - 2y - 2x = -7$ **25.** $x^2 + 4y^2 + 2x = 3$ **26.** $x^2 - 9y^2 - 4x = 5$

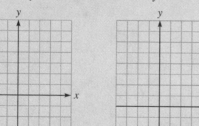
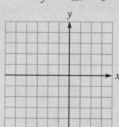
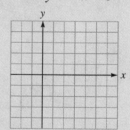

In Exercises 27–28, write the equation of each ellipse.

27. Center $(0, 0)$; horizontal major axis of 12; minor axis of 8

28. Center $(2, 3)$; $a = 5$; $c = 2$, major axis vertical

In Exercises 29–30, write the equation of each hyperbola.

29. Center $(0, 0)$; focus $(3, 0)$; vertex $(2, 0)$

30. Center $(2, 4)$; area of fundamental rectangle is 36 square units; $a = b$; transverse axis parallel to y-axis

In Exercises 31–32, find the required term of the expansion of $(x + 2y)^8$.

31. 2nd term

32. 6th term

In Exercises 33–34, find each sum.

33. $\displaystyle\sum_{k=1}^{5} 2$

34. $\displaystyle\sum_{k=2}^{6} (3x + 1)$

In Exercises 35–36, find the sum of the first six terms of each sequence.

35. $-2, 1, 4, \ldots$

36. $\dfrac{1}{9}, \dfrac{1}{3}, 1, \ldots$

In Exercises 37–40, find each value.

37. $P(8, 4)$

38. $P(24, 0)$

39. $C(12, 10)$

40. $P(4, 4) \cdot C(6, 6)$

41. In how many ways can 6 men and 4 women be placed in a line if the women line up first?

42. In how many ways can a committee of 4 people be selected from a group of 12 people?

In Exercises 43–46, find each probability.

43. Rolling 11 on one roll of two dice

44. Being dealt an all-red 5-card poker hand from a standard deck

45. If the probability that a person is married is 0.6 and the probability that a married person has children is 0.8, find the probability that a randomly chosen person is married with children.

46. The probability that a student earns an A in algebra is 0.3, and the probability that a student earns a B is 0.4. Find the probability that a student earns a C or less.

47. Prove the formula by induction:

$$4 + 7 + 10 + \cdots + (3n + 1) = \frac{n(3n + 5)}{2}$$

48. Compute the fair cost to play the following game.
   1. Draw a card from a standard card deck.
   2. If you draw an ace, you receive $100.
   3. If you draw a king, you receive $10.
   4. All other cards pay nothing.

49. What single deposit made now into an account that pays $8\frac{1}{2}\%$ interest, compounded annually, will grow to $10,000 in 12 years?

50. A bank offers a $110,000, 20-year mortgage at 8.75%. Find the monthly payment.

A PROOF OF THE BINOMIAL THEOREM

The binomial theorem can be proved for positive integral exponents by using mathematical induction.

The Binomial Theorem

If n is a positive integer, then

$$(a + b)^n = a^n + \frac{n!}{1!(n-1)!}a^{n-1}b + \frac{n!}{2!(n-2)!}a^{n-2}b^2 + \cdots$$

$$+ \frac{n!}{r!(n-r)!}a^{n-r}b^r + \cdots + b^n$$

Proof   As in all induction proofs, there are two parts.

Part 1: Substituting the number 1 for n on both sides of the equation, we have

$$(a + b)^1 = a^1 + \frac{1!}{1!(1-1)!}a^{1-1}b^1$$

$$a + b = a + a^0 b$$

$$a + b = a + b$$

and the theorem is true when $n = 1$. Part 1 is complete.

Part 2: We write expressions for two general terms in the statement of the induction hypothesis. We assume that the theorem is true for $n = k$:

$$(a + b)^k = a^k + \frac{k!}{1!(k-1)!}a^{k-1}b + \frac{k!}{2!(k-2)!}a^{k-2}b^2 + \cdots$$

$$+ \frac{k!}{(r-1)!(k-r+1)!}a^{k-r+1}b^{r-1}$$

$$+ \frac{k!}{r!(k-r)!}a^{k-r}b^r + \cdots + b^k$$

We multiply both sides of this equation by $a + b$ and hope to obtain a similar equation in which the quantity $k + 1$ replaces all of the n values in the binomial theorem:

$$(a + b)^k(a + b)$$

$$= (a + b)\left[a^k + \frac{k!}{1!(k - 1)!}a^{k-1}b + \frac{k!}{2!(k - 2)!}a^{k-2}b^2 + \cdots \right.$$

$$\left. + \frac{k!}{(r - 1)!(k - r + 1)!}a^{k-r+1}b^{r-1} + \frac{k!}{r!(k - r)!}a^{k-r}b^r + \cdots + b^k \right]$$

We distribute the multiplication first by a and then by b:

$$(a + b)^{k+1} = \left[a^{k+1} + \frac{k!}{1!(k - 1)!}a^k b + \frac{k!}{2!(k - 2)!}a^{k-1}b^2 + \cdots \right.$$

$$\left. + \frac{k!}{(r - 1)!(k - r + 1)!}a^{k-r+2}b^{r-1} + \frac{k!}{r!(k - r)!}a^{k-r+1}b^r + \cdots + ab^k \right]$$

$$+ \left[a^k b + \frac{k!}{1!(k - 1)!}a^{k-1}b^2 + \frac{k!}{2!(k - 2)!}a^{k-2}b^3 + \cdots \right.$$

$$\left. + \frac{k!}{(r - 1)!(k - r + 1)!}a^{k-r+1}b^r + \frac{k!}{r!(k - r)!}a^{k-r}b^{r+1} + \cdots + b^{k+1} \right]$$

Combining like terms, we have

$$(a + b)^{k+1} = a^{k+1} + \left[\frac{k!}{1!(k - 1)!} + 1 \right]a^k b$$

$$+ \left[\frac{k!}{2!(k - 2)!} + \frac{k!}{1!(k - 1)!} \right]a^{k-1}b^2 + \cdots$$

$$+ \left[\frac{k!}{r!(k - r)!} + \frac{k!}{(r - 1)!(k - r + 1)!} \right]a^{k-r+1}b^r + \cdots + b^{k+1}$$

These results may be written as

$$(a + b)^{k+1} = a^{k+1} + \frac{(k + 1)!}{1!(k + 1 - 1)!}a^{(k+1)-1}b + \frac{(k + 1)!}{2!(k + 1 - 2)!}a^{(k+1)-2}b^2$$

$$+ \cdots + \frac{(k + 1)!}{r!(k + 1 - r)!}a^{(k+1)-r}b^r + \cdots + b^{k+1}$$

This formula has precisely the same form as the binomial theorem, with the quantity $k + 1$ replacing all of the original n values. Therefore, the truth of the theorem for $n = k$ implies the truth of the theorem for $n = k + 1$. Because both parts of the axiom of mathematical induction are verified, the theorem is proved.   □

TABLES

Table A Powers and Roots

n	n^2	$\sqrt{n}$	n^3	$\sqrt[3]{n}$	n	n^2	$\sqrt{n}$	n^3	$\sqrt[3]{n}$
1	1	1.000	1	1.000	51	2,601	7.141	132,651	3.708
2	4	1.414	8	1.260	52	2,704	7.211	140,608	3.733
3	9	1.732	27	1.442	53	2,809	7.280	148,877	3.756
4	16	2.000	64	1.587	54	2,916	7.348	157,464	3.780
5	25	2.236	125	1.710	55	3,025	7.416	166,375	3.803
6	36	2.449	216	1.817	56	3,136	7.483	175,616	3.826
7	49	2.646	343	1.913	57	3,249	7.550	185,193	3.849
8	64	2.828	512	2.000	58	3,364	7.616	195,112	3.871
9	81	3.000	729	2.080	59	3,481	7.681	205,379	3.893
10	100	3.162	1,000	2.154	60	3,600	7.746	216,000	3.915
11	121	3.317	1,331	2.224	61	3,721	7.810	226,981	3.936
12	144	3.464	1,728	2.289	62	3,844	7.874	238,328	3.958
13	169	3.606	2,197	2.351	63	3,969	7.937	250,047	3.979
14	196	3.742	2,744	2.410	64	4,096	8.000	262,144	4.000
15	225	3.873	3,375	2.466	65	4,225	8.062	274,625	4.021
16	256	4.000	4,096	2.520	66	4,356	8.124	287,496	4.041
17	289	4.123	4,913	2.571	67	4,489	8.185	300,763	4.062
18	324	4.243	5,832	2.621	68	4,624	8.246	314,432	4.082
19	361	4.359	6,859	2.668	69	4,761	8.307	328,509	4.102
20	400	4.472	8,000	2.714	70	4,900	8.367	343,000	4.121
21	441	4.583	9,261	2.759	71	5,041	8.426	357,911	4.141
22	484	4.690	10,648	2.802	72	5,184	8.485	373,248	4.160
23	529	4.796	12,167	2.844	73	5,329	8.544	389,017	4.179
24	576	4.899	13,824	2.884	74	5,476	8.602	405,224	4.198
25	625	5.000	15,625	2.924	75	5,625	8.660	421,875	4.217
26	676	5.099	17,576	2.962	76	5,776	8.718	438,976	4.236
27	729	5.196	19,683	3.000	77	5,929	8.775	456,533	4.254
28	784	5.292	21,952	3.037	78	6,084	8.832	474,552	4.273
29	841	5.385	24,389	3.072	79	6,241	8.888	493,039	4.291
30	900	5.477	27,000	3.107	80	6,400	8.944	512,000	4.309
31	961	5.568	29,791	3.141	81	6,561	9.000	531,441	4.327
32	1,024	5.657	32,768	3.175	82	6,724	9.055	551,368	4.344
33	1,089	5.745	35,937	3.208	83	6,889	9.110	571,787	4.362
34	1,156	5.831	39,304	3.240	84	7,056	9.165	592,704	4.380
35	1,225	5.916	42,875	3.271	85	7,225	9.220	614,125	4.397
36	1,296	6.000	46,656	3.302	86	7,396	9.274	636,056	4.414
37	1,369	6.083	50,653	3.332	87	7,569	9.327	658,503	4.431
38	1,444	6.164	54,872	3.362	88	7,744	9.381	681,472	4.448
39	1,521	6.245	59,319	3.391	89	7,921	9.434	704,969	4.465
40	1,600	6.325	64,000	3.420	90	8,100	9.487	729,000	4.481
41	1,681	6.403	68,921	3.448	91	8,281	9.539	753,571	4.498
42	1,764	6.481	74,088	3.476	92	8,464	9.592	778,688	4.514
43	1,849	6.557	79,507	3.503	93	8,649	9.644	804,357	4.531
44	1,936	6.633	85,184	3.530	94	8,836	9.695	830,584	4.547
45	2,025	6.708	91,125	3.557	95	9,025	9.747	857,375	4.563
46	2,116	6.782	97,336	3.583	96	9,216	9.798	884,736	4.579
47	2,209	6.856	103,823	3.609	97	9,409	9.849	912,673	4.595
48	2,304	6.928	110,592	3.634	98	9,604	9.899	941,192	4.610
49	2,401	7.000	117,649	3.659	99	9,801	9.950	970,299	4.626
50	2,500	7.071	125,000	3.684	100	10,000	10.000	1,000,000	4.642

Table B   Base-10 Logarithms

N	0	1	2	3	4	5	6	7	8	9
1.0	.0000	.0043	.0086	.0128	.0170	.0212	.0253	.0294	.0334	.0374
1.1	.0414	.0453	.0492	.0531	.0569	.0607	.0645	.0682	.0719	.0755
1.2	.0792	.0828	.0864	.0899	.0934	.0969	.1004	.1038	.1072	.1106
1.3	.1139	.1173	.1206	.1239	.1271	.1303	.1335	.1367	.1399	.1430
1.4	.1461	.1492	.1523	.1553	.1584	.1614	.1644	.1673	.1703	.1732
1.5	.1761	.1790	.1818	.1847	.1875	.1903	.1931	.1959	.1987	.2014
1.6	.2041	.2068	.2095	.2122	.2148	.2175	.2201	.2227	.2253	.2279
1.7	.2304	.2330	.2355	.2380	.2405	.2430	.2455	.2480	.2504	.2529
1.8	.2553	.2577	.2601	.2625	.2648	.2672	.2695	.2718	.2742	.2765
1.9	.2788	.2810	.2833	.2856	.2878	.2900	.2923	.2945	.2967	.2989
2.0	.3010	.3032	.3054	.3075	.3096	.3118	.3139	.3160	.3181	.3201
2.1	.3222	.3243	.3263	.3284	.3304	.3324	.3345	.3365	.3385	.3404
2.2	.3424	.3444	.3464	.3483	.3502	.3522	.3541	.3560	.3579	.3598
2.3	.3617	.3636	.3655	.3674	.3692	.3711	.3729	.3747	.3766	.3784
2.4	.3802	.3820	.3838	.3856	.3874	.3892	.3909	.3927	.3945	.3962
2.5	.3979	.3997	.4014	.4031	.4048	.4065	.4082	.4099	.4116	.4133
2.6	.4150	.4166	.4183	.4200	.4216	.4232	.4249	.4265	.4281	.4298
2.7	.4314	.4330	.4346	.4362	.4378	.4393	.4409	.4425	.4440	.4456
2.8	.4472	.4487	.4502	.4518	.4533	.4548	.4564	.4579	.4594	.4609
2.9	.4624	.4639	.4654	.4669	.4683	.4698	.4713	.4728	.4742	.4757
3.0	.4771	.4786	.4800	.4814	.4829	.4843	.4857	.4871	.4886	.4900
3.1	.4914	.4928	.4942	.4955	.4969	.4983	.4997	.5011	.5024	.5038
3.2	.5051	.5065	.5079	.5092	.5105	.5119	.5132	.5145	.5159	.5172
3.3	.5185	.5198	.5211	.5224	.5237	.5250	.5263	.5276	.5289	.5302
3.4	.5315	.5328	.5340	.5353	.5366	.5378	.5391	.5403	.5416	.5428
3.5	.5441	.5453	.5465	.5478	.5490	.5502	.5514	.5527	.5539	.5551
3.6	.5563	.5575	.5587	.5599	.5611	.5623	.5635	.5647	.5658	.5670
3.7	.5682	.5694	.5705	.5717	.5729	.5740	.5752	.5763	.5775	.5786
3.8	.5798	.5809	.5821	.5832	.5843	.5855	.5866	.5877	.5888	.5899
3.9	.5911	.5922	.5933	.5944	.5955	.5966	.5977	.5988	.5999	.6010
4.0	.6021	.6031	.6042	.6053	.6064	.6075	.6085	.6096	.6107	.6117
4.1	.6128	.6138	.6149	.6160	.6170	.6180	.6191	.6201	.6212	.6222
4.2	.6232	.6243	.6253	.6263	.6274	.6284	.6294	.6304	.6314	.6325
4.3	.6335	.6345	.6355	.6365	.6375	.6385	.6395	.6405	.6415	.6425
4.4	.6435	.6444	.6454	.6464	.6474	.6484	.6493	.6503	.6513	.6522
4.5	.6532	.6542	.6551	.6561	.6571	.6580	.6590	.6599	.6609	.6618
4.6	.6628	.6637	.6646	.6656	.6665	.6675	.6684	.6693	.6702	.6712
4.7	.6721	.6730	.6739	.6749	.6758	.6767	.6776	.6785	.6794	.6803
4.8	.6812	.6821	.6830	.6839	.6848	.6857	.6866	.6875	.6884	.6893
4.9	.6902	.6911	.6920	.6928	.6937	.6946	.6955	.6964	.6972	.6981
5.0	.6990	.6998	.7007	.7016	.7024	.7033	.7042	.7050	.7059	.7067
5.1	.7076	.7084	.7093	.7101	.7110	.7118	.7126	.7135	.7143	.7152
5.2	.7160	.7168	.7177	.7185	.7193	.7202	.7210	.7218	.7226	.7235
5.3	.7243	.7251	.7259	.7267	.7275	.7284	.7292	.7300	.7308	.7316
5.4	.7324	.7332	.7340	.7348	.7356	.7364	.7372	.7380	.7388	.7396

Table B   (continued)

N	0	1	2	3	4	5	6	7	8	9
5.5	.7404	.7412	.7419	.7427	.7435	.7443	.7451	.7459	.7466	.7474
5.6	.7482	.7490	.7497	.7505	.7513	.7520	.7528	.7536	.7543	.7551
5.7	.7559	.7566	.7574	.7582	.7589	.7597	.7604	.7612	.7619	.7627
5.8	.7634	.7642	.7649	.7657	.7664	.7672	.7679	.7686	.7694	.7701
5.9	.7709	.7716	.7723	.7731	.7738	.7745	.7752	.7760	.7767	.7774
6.0	.7782	.7789	.7796	.7803	.7810	.7818	.7825	.7832	.7839	.7846
6.1	.7853	.7860	.7868	.7875	.7882	.7889	.7896	.7903	.7910	.7917
6.2	.7924	.7931	.7938	.7945	.7952	.7959	.7966	.7973	.7980	.7987
6.3	.7993	.8000	.8007	.8014	.8021	.8028	.8035	.8041	.8048	.8055
6.4	.8062	.8069	.8075	.8082	.8089	.8096	.8102	.8109	.8116	.8122
6.5	.8129	.8136	.8142	.8149	.8156	.8162	.8169	.8176	.8182	.8189
6.6	.8195	.8202	.8209	.8215	.8222	.8228	.8235	.8241	.8248	.8254
6.7	.8261	.8267	.8274	.8280	.8287	.8293	.8299	.8306	.8312	.8319
6.8	.8325	.8331	.8338	.8344	.8351	.8357	.8363	.8370	.8376	.8382
6.9	.8388	.8395	.8401	.8407	.8414	.8420	.8426	.8432	.8439	.8445
7.0	.8451	.8457	.8463	.8470	.8476	.8482	.8488	.8494	.8500	.8506
7.1	.8513	.8519	.8525	.8531	.8537	.8543	.8549	.8555	.8561	.8567
7.2	.8573	.8579	.8585	.8591	.8597	.8603	.8609	.8615	.8621	.8627
7.3	.8633	.8639	.8645	.8651	.8657	.8663	.8669	.8675	.8681	.8686
7.4	.8692	.8698	.8704	.8710	.8716	.8722	.8727	.8733	.8739	.8745
7.5	.8751	.8756	.8762	.8768	.8774	.8779	.8785	.8791	.8797	.8802
7.6	.8808	.8814	.8820	.8825	.8831	.8837	.8842	.8848	.8854	.8859
7.7	.8865	.8871	.8876	.8882	.8887	.8893	.8899	.8904	.8910	.8915
7.8	.8921	.8927	.8932	.8938	.8943	.8949	.8954	.8960	.8965	.8971
7.9	.8976	.8982	.8987	.8993	.8998	.9004	.9009	.9015	.9020	.9025
8.0	.9031	.9036	.9042	.9047	.9053	.9058	.9063	.9069	.9074	.9079
8.1	.9085	.9090	.9096	.9101	.9106	.9112	.9117	.9122	.9128	.9133
8.2	.9138	.9143	.9149	.9154	.9159	.9165	.9170	.9175	.9180	.9186
8.3	.9191	.9196	.9201	.9206	.9212	.9217	.9222	.9227	.9232	.9238
8.4	.9243	.9248	.9253	.9258	.9263	.9269	.9274	.9279	.9284	.9289
8.5	.9294	.9299	.9304	.9309	.9315	.9320	.9325	.9330	.9335	.9340
8.6	.9345	.9350	.9355	.9360	.9365	.9370	.9375	.9380	.9385	.9390
8.7	.9395	.9400	.9405	.9410	.9415	.9420	.9425	.9430	.9435	.9440
8.8	.9445	.9450	.9455	.9460	.9465	.9469	.9474	.9479	.9484	.9489
8.9	.9494	.9499	.9504	.9509	.9513	.9518	.9523	.9528	.9533	.9538
9.0	.9542	.9547	.9552	.9557	.9562	.9566	.9571	.9576	.9581	.9586
9.1	.9590	.9595	.9600	.9605	.9609	.9614	.9619	.9624	.9628	.9633
9.2	.9638	.9643	.9647	.9652	.9657	.9661	.9666	.9671	.9675	.9680
9.3	.9685	.9689	.9694	.9699	.9703	.9708	.9713	.9717	.9722	.9727
9.4	.9731	.9736	.9741	.9745	.9750	.9754	.9759	.9763	.9768	.9773
9.5	.9777	.9782	.9786	.9791	.9795	.9800	.9805	.9809	.9814	.9818
9.6	.9823	.9827	.9832	.9836	.9841	.9845	.9850	.9854	.9859	.9863
9.7	.9868	.9872	.9877	.9881	.9886	.9890	.9894	.9899	.9903	.9908
9.8	.9912	.9917	.9921	.9926	.9930	.9934	.9939	.9943	.9948	.9952
9.9	.9956	.9961	.9965	.9969	.9974	.9978	.9983	.9987	.9991	.9996

Table C Base-e Logarithms

N	0	1	2	3	4	5	6	7	8	9
1.0	.0000	.0100	.0198	.0296	.0392	.0488	.0583	.0677	.0770	.0862
1.1	.0953	.1044	.1133	.1222	.1310	.1398	.1484	.1570	.1655	.1740
1.2	.1823	.1906	.1989	.2070	.2151	.2231	.2311	.2390	.2469	.2546
1.3	.2624	.2700	.2776	.2852	.2927	.3001	.3075	.3148	.3221	.3293
1.4	.3365	.3436	.3507	.3577	.3646	.3716	.3784	.3853	.3920	.3988
1.5	.4055	.4121	.4187	.4253	.4318	.4383	.4447	.4511	.4574	.4637
1.6	.4700	.4762	.4824	.4886	.4947	.5008	.5068	.5128	.5188	.5247
1.7	.5306	.5365	.5423	.5481	.5539	.5596	.5653	.5710	.5766	.5822
1.8	.5878	.5933	.5988	.6043	.6098	.6152	.6206	.6259	.6313	.6366
1.9	.6419	.6471	.6523	.6575	.6627	.6678	.6729	.6780	.6831	.6881
2.0	.6931	.6981	.7031	.7080	.7129	.7178	.7227	.7275	.7324	.7372
2.1	.7419	.7467	.7514	.7561	.7608	.7655	.7701	.7747	.7793	.7839
2.2	.7885	.7930	.7975	.8020	.8065	.8109	.8154	.8198	.8242	.8286
2.3	.8329	.8372	.8416	.8459	.8502	.8544	.8587	.8629	.8671	.8713
2.4	.8755	.8796	.8838	.8879	.8920	.8961	.9002	.9042	.9083	.9123
2.5	.9163	.9203	.9243	.9282	.9322	.9361	.9400	.9439	.9478	.9517
2.6	.9555	.9594	.9632	.9670	.9708	.9746	.9783	.9821	.9858	.9895
2.7	.9933	.9969	1.0006	.0043	.0080	.0116	.0152	.0188	.0225	.0260
2.8	1.0296	.0332	.0367	.0403	.0438	.0473	.0508	.0543	.0578	.0613
2.9	.0647	.0682	.0716	.0750	.0784	.0818	.0852	.0886	.0919	.0953
3.0	1.0986	.1019	.1053	.1086	.1119	.1151	.1184	.1217	.1249	.1282
3.1	.1314	.1346	.1378	.1410	.1442	.1474	.1506	.1537	.1569	.1600
3.2	.1632	.1663	.1694	.1725	.1756	.1787	.1817	.1848	.1878	.1909
3.3	.1939	.1969	.2000	.2030	.2060	.2090	.2119	.2149	.2179	.2208
3.4	.2238	.2267	.2296	.2326	.2355	.2384	.2413	.2442	.2470	.2499
3.5	1.2528	.2556	.2585	.2613	.2641	.2669	.2698	.2726	.2754	.2782
3.6	.2809	.2837	.2865	.2892	.2920	.2947	.2975	.3002	.3029	.3056
3.7	.3083	.3110	.3137	.3164	.3191	.3218	.3244	.3271	.3297	.3324
3.8	.3350	.3376	.3403	.3429	.3455	.3481	.3507	.3533	.3558	.3584
3.9	.3610	.3635	.3661	.3686	.3712	.3737	.3762	.3788	.3813	.3838
4.0	1.3863	.3888	.3913	.3938	.3962	.3987	.4012	.4036	.4061	.4085
4.1	.4110	.4134	.4159	.4183	.4207	.4231	.4255	.4279	.4303	.4327
4.2	.4351	.4375	.4398	.4422	.4446	.4469	.4493	.4516	.4540	.4563
4.3	.4586	.4609	.4633	.4656	.4679	.4702	.4725	.4748	.4770	.4793
4.4	.4816	.4839	.4861	.4884	.4907	.4929	.4951	.4974	.4996	.5019
4.5	1.5041	.5063	.5085	.5107	.5129	.5151	.5173	.5195	.5217	.5239
4.6	.5261	.5282	.5304	.5326	.5347	.5369	.5390	.5412	.5433	.5454
4.7	.5476	.5497	.5518	.5539	.5560	.5581	.5602	.5623	.5644	.5665
4.8	.5686	.5707	.5728	.5748	.5769	.5790	.5810	.5831	.5851	.5872
4.9	.5892	.5913	.5933	.5953	.5974	.5994	.6014	.6034	.6054	.6074
5.0	1.6094	.6114	.6134	.6154	.6174	.6194	.6214	.6233	.6253	.6273
5.1	.6292	.6312	.6332	.6351	.6371	.6390	.6409	.6429	.6448	.6467
5.2	.6487	.6506	.6525	.6544	.6563	.6582	.6601	.6620	.6639	.6658
5.3	.6677	.6696	.6715	.6734	.6752	.6771	.6790	.6808	.6827	.6845
5.4	.6864	.6882	.6901	.6919	.6938	.6956	.6974	.6993	.7011	.7029

Table C (continued)

N	0	1	2	3	4	5	6	7	8	9
5.5	1.7047	.7066	.7084	.7102	.7120	.7138	.7156	.7174	.7192	.7210
5.6	.7228	.7246	.7263	.7281	.7299	.7317	.7334	.7352	.7370	.7387
5.7	.7405	.7422	.7440	.7457	.7475	.7492	.7509	.7527	.7544	.7561
5.8	.7579	.7596	.7613	.7630	.7647	.7664	.7681	.7699	.7716	.7733
5.9	.7750	.7766	.7783	.7800	.7817	.7834	.7851	.7867	.7884	.7901
6.0	1.7918	.7934	.7951	.7967	.7984	.8001	.8017	.8034	.8050	.8066
6.1	.8083	.8099	.8116	.8132	.8148	.8165	.8181	.8197	.8213	.8229
6.2	.8245	.8262	.8278	.8294	.8310	.8326	.8342	.8358	.8374	.8390
6.3	.8405	.8421	.8437	.8453	.8469	.8485	.8500	.8516	.8532	.8547
6.4	.8563	.8579	.8594	.8610	.8625	.8641	.8656	.8672	.8687	.8703
6.5	1.8718	.8733	.8749	.8764	.8779	.8795	.8810	.8825	.8840	.8856
6.6	.8871	.8886	.8901	.8916	.8931	.8946	.8961	.8976	.8991	.9006
6.7	.9021	.9036	.9051	.9066	.9081	.9095	.9110	.9125	.9140	.9155
6.8	.9169	.9184	.9199	.9213	.9228	.9242	.9257	.9272	.9286	.9301
6.9	.9315	.9330	.9344	.9359	.9373	.9387	.9402	.9416	.9430	.9445
7.0	1.9459	.9473	.9488	.9502	.9516	.9530	.9544	.9559	.9573	.9587
7.1	.9601	.9615	.9629	.9643	.9657	.9671	.9685	.9699	.9713	.9727
7.2	.9741	.9755	.9769	.9782	.9796	.9810	.9824	.9838	.9851	.9865
7.3	.9879	.9892	.9906	.9920	.9933	.9947	.9961	.9974	.9988	2.0001
7.4	2.0015	.0028	.0042	.0055	.0069	.0082	.0096	.0109	.0122	.0136
7.5	2.0149	.0162	.0176	.0189	.0202	.0215	.0229	.0242	.0255	.0268
7.6	.0281	.0295	.0308	.0321	.0334	.0347	.0360	.0373	.0386	.0399
7.7	.0412	.0425	.0438	.0451	.0464	.0477	.0490	.0503	.0516	.0528
7.8	.0541	.0554	.0567	.0580	.0592	.0605	.0618	.0631	.0643	.0656
7.9	.0669	.0681	.0694	.0707	.0719	.0732	.0744	.0757	.0769	.0782
8.0	2.0794	.0807	.0819	.0832	.0844	.0857	.0869	.0882	.0894	.0906
8.1	.0919	.0931	.0943	.0956	.0968	.0980	.0992	.1005	.1017	.1029
8.2	.1041	.1054	.1066	.1078	.1090	.1102	.1114	.1126	.1138	.1150
8.3	.1163	.1175	.1187	.1199	.1211	.1223	.1235	.1247	.1258	.1270
8.4	.1282	.1294	.1306	.1318	.1330	.1342	.1353	.1365	.1377	.1389
8.5	2.1401	.1412	.1424	.1436	.1448	.1459	.1471	.1483	.1494	.1506
8.6	.1518	.1529	.1541	.1552	.1564	.1576	.1587	.1599	.1610	.1622
8.7	.1633	.1645	.1656	.1668	.1679	.1691	.1702	.1713	.1725	.1736
8.8	.1748	.1759	.1770	.1782	.1793	.1804	.1815	.1827	.1838	.1849
8.9	.1861	.1872	.1883	.1894	.1905	.1917	.1928	.1939	.1950	.1961
9.0	2.1972	.1983	.1994	.2006	.2017	.2028	.2039	.2050	.2061	.2072
9.1	.2083	.2094	.2105	.2116	.2127	.2138	.2148	.2159	.2170	.2181
9.2	.2192	.2203	.2214	.2225	.2235	.2246	.2257	.2268	.2279	.2289
9.3	.2300	.2311	.2322	.2332	.2343	.2354	.2364	.2375	.2386	.2396
9.4	.2407	.2418	.2428	.2439	.2450	.2460	.2471	.2481	.2492	.2502
9.5	2.2513	.2523	.2534	.2544	.2555	.2565	.2576	.2586	.2597	.2607
9.6	.2618	.2628	.2638	.2649	.2659	.2670	.2680	.2690	.2701	.2711
9.7	.2721	.2732	.2742	.2752	.2762	.2773	.2783	.2793	.2803	.2814
9.8	.2824	.2834	.2844	.2854	.2865	.2875	.2885	.2895	.2905	.2915
9.9	.2925	.2935	.2946	.2956	.2966	.2976	.2986	.2996	.3006	.3016

Use the properties of logarithms and $\ln 10 = 2.3026$ to find logarithms of numbers less than 1 or greater than 10.

Exercise 1.1 (page 12)

1. decimal 3. 2 5. composite 7. decimals 9. negative 11. $x + (y + z)$ 13. $5m + 5 \cdot 2$ 15. interval 17. two
19. positive 21. 1, 2, 6, 7 23. $-5, -4, 0, 1, 2, 6, 7$ 25. $\sqrt{2}$ 27. 6 29. $-5, 1, 7$ 31. (number line: $-1\ 0\ 1\ 2\ 3\ 4\ 5$)
33. (number line: 10 11 12 13 14 15 16 17 18 19 20) 35. (number line: $-5\ -4\ -3\ -2\ -1\ 0\ 1\ 2\ 3\ 4\ 5$)
37. (number line: $-6\ -5\ -4\ -3\ -2\ -1\ 0\ 1\ 2\ 3\ 4$) 39. 41. (at 2) 43. (at 0, 5)
45. (-2, 2) 47. (5) 49. (-5, 0) 51. (-2, 3) 53. (-4; 2, 6)
55. $(-\infty, -2) \cup (2, \infty)$ (-2, 2) 57. $(-\infty, -1] \cup [3, \infty)$ (-1, 3) 59. 13
61. 0 63. -8 65. -32 67. $5 - \pi$ 69. 0 71. $x + 1$ 73. $-(x - 4)$ 75. 5 77. 5 79. natural numbers
81. rational numbers

Exercise 1.2 (page 24)

1. factor 3. 3; $2x$ 5. scientific; integer 7. x^{m+n} 9. $x^n y^n$ 11. 1 13. 169 15. -25 17. $4 \cdot x \cdot x \cdot x$
19. $(-5x)(-5x)(-5x)(-5x)$ 21. $7x^3$ 23. x^2 25. $-27t^3$ 27. $x^3 y^2$ 29. 10.648 31. -0.0625 33. x^5 35. z^6 37. y^{21}
39. z^{26} 41. $27x^3$ 43. $x^6 y^3$ 45. $\dfrac{a^6}{b^3}$ 47. 1 49. 1 51. $\dfrac{1}{z^4}$ 53. $\dfrac{1}{y^5}$ 55. x^2 57. x^4 59. a^4 61. x 63. $\dfrac{m^9}{n^6}$ 65. $\dfrac{1}{a^9}$
67. $\dfrac{a^{12}}{b^4}$ 69. $\dfrac{1}{r^4}$ 71. $\dfrac{x^{32}}{y^{16}}$ 73. $\dfrac{9x^{10}}{25y^2}$ 75. $\dfrac{4y^{12}}{x^{20}}$ 77. $\dfrac{64z^7}{25y^6}$ 79. 4 81. -8 83. 216 85. -12 87. 20 89. $\frac{3}{64}$
91. 3.72×10^5 93. -1.77×10^8 95. 7×10^{-3} 97. -6.93×10^{-7} 99. 1×10^{12} 101. 937,000 103. 0.0000221 105. 3.2
107. -0.0032 109. 1.17×10^4 111. 7×10^4 113. 5.3×10^{19} 115. 1.986×10^4 meters per min 117. 1.67248×10^{-15} g
119. x^{n+2} 121. x^{m-1} 123. x^{m+4} 127. (number line: -2, 4)

Exercise 1.3 (page 38)

1. 0 3. not 5. $a^{1/n}$ 7. $\sqrt[n]{ab}$ 9. $\neq$ 11. 3 13. $\frac{1}{5}$ 15. -3 17. 10 19. -4 21. -4 23. $4|a|$ 25. $2|a|$ 27. $-2a$
29. $-6b^2$ 31. $\dfrac{4a^2}{5|b|}$ 33. $-\dfrac{10x^2}{3y}$ 35. 8 37. -64 39. -100 41. $\frac{1}{8}$ 43. $\frac{1}{512}$ 45. $-\frac{1}{27}$ 47. $\frac{32}{243}$ 49. $\frac{16}{9}$ 51. $10s^2$
53. $\dfrac{1}{2y^2 z}$ 55. $x^6 y^3$ 57. $\dfrac{1}{r^6 s^{12}}$ 59. $\dfrac{4a^4}{25b^6}$ 61. $\dfrac{100s^8}{9r^4}$ 63. a 65. 7 67. 5 69. -3 71. $-\frac{1}{5}$ 73. $6|x|$ 75. $3y^2$ 77. $2y$
79. $\dfrac{|x|y^2}{|z^3|}$ 81. $\sqrt{2}$ 83. $17x\sqrt{2}$ 85. $2y^2\sqrt{3y}$ 87. $12\sqrt[3]{3}$ 89. $6z\sqrt[4]{3z}$ 91. $6x\sqrt{2y}$ 93. 0 95. $\sqrt{3}$ 97. $\sqrt[3]{4}$

99. $\dfrac{2b\sqrt[4]{27a^2}}{3a}$ **101.** $\dfrac{u\sqrt[3]{6uv^2}}{3v}$ **103.** $\dfrac{1}{2\sqrt{5}}$ **105.** $\dfrac{1}{\sqrt[3]{3}}$ **107.** $\dfrac{b}{32a\sqrt[5]{2b^2}}$ **109.** $\dfrac{2\sqrt{3}}{9}$ **111.** $-\dfrac{\sqrt{2x}}{8}$ **113.** $\sqrt{3}$ **115.** $\sqrt[5]{4x^3}$

117. $\sqrt[6]{32}$ **119.** $\dfrac{\sqrt[4]{12}}{2}$ **125.** $(-2, 5]$ **127.** 6.17×10^8

Exercise 1.4 (page 51)

1. monomial; variables **3.** trinomial **5.** one **7.** like **9.** coefficients; variables **11.** yes, trinomial, 2nd degree **13.** no
15. yes, binomial, 3rd degree **17.** yes, monomial, 0th degree **19.** yes, monomial, no defined degree **21.** $6x^3 - 3x^2 - 8x$

23. $4y^3 + 14$ **25.** $-x^2 + 14$ **27.** $-28t + 96$ **29.** $-4y^2 + y$ **31.** $2x^2y - 4xy^2$ **33.** $8x^3y^7$ **35.** $\dfrac{m^4n^4}{2}$ **37.** $-4r^3s - 4rs^3$

39. $12a^2b^2c^2 + 18ab^3c^3 - 24a^2b^4c^2$ **41.** $a^2 + 4a + 4$ **43.** $a^2 - 12a + 36$ **45.** $x^2 - 16$ **47.** $x^2 + 2x - 15$ **49.** $3u^2 + 4u - 4$
51. $10x^2 + 13x - 3$ **53.** $9a^2 - 12ab + 4b^2$ **55.** $9m^2 - 16n^2$ **57.** $6y^2 - 16xy + 8x^2$ **59.** $9x^3 - 27xy - yx^2 + 3y^2$
61. $5z^3 - 5tz - 2tz^2 - 2t^2$ **63.** $27x^3 - 27x^2 + 9x - 1$ **65.** $6x^3 + 14x^2 - 5x - 3$ **67.** $6x^3 - 5x^2y + 6xy^2 + 8y^3$ **69.** $6y^{2n} + 2$
71. $-10x^{4n} - 15y^{2n}$ **73.** $x^{2n} - x^n - 12$ **75.** $6r^{2n} - 25r^n + 14$ **77.** $xy + x^{3/2}y^{1/2}$ **79.** $a - b$ **81.** $\sqrt{3} + 1$ **83.** $x(\sqrt{7} - 2)$
85. $\dfrac{x(x + \sqrt{3})}{x^2 - 3}$ **87.** $\dfrac{(y + \sqrt{2})^2}{y^2 - 2}$ **89.** $\dfrac{\sqrt{3} + 3 - \sqrt{2} - \sqrt{6}}{2}$ **91.** $\dfrac{x - 2\sqrt{xy} + y}{x - y}$ **93.** $\dfrac{1}{2(\sqrt{2} - 1)}$ **95.** $\dfrac{y^2 - 3}{y^2 + 2y\sqrt{3} + 3}$

97. $\dfrac{1}{\sqrt{x + 3} + \sqrt{x}}$ **99.** $\dfrac{2a}{b^3}$ **101.** $-\dfrac{2z^9}{3x^3y^2}$ **103.** $\dfrac{x}{2y} + \dfrac{3xy}{2}$ **105.** $\dfrac{2y^3}{5} - \dfrac{3y}{5x^3} + \dfrac{1}{5x^4y^3}$ **107.** $3x + 2$ **109.** $x - 7 + \dfrac{2}{2x - 5}$

111. $x - 3$ **113.** $x^2 - 2 + \dfrac{-x^2 + 5}{x^3 - 2}$ **115.** $x^4 + 2x^3 + 4x^2 + 8x + 16$ **117.** $6x^2 + x - 12$ **125.** 27 **127.** $\dfrac{125x^3}{8y^6}$
128. $-b\sqrt[3]{2ab}$

Exercise 1.5 (page 62)

1. factor **3.** $x(a + b)$ **5.** $(x + y)(x + y)$ **7.** $(x + y)(x^2 - xy + y^2)$ **9.** $3(x - 2)$ **11.** $4x^2(2 + x)$ **13.** $7x^2y^2(1 + 2x)$
15. $3abc(a + 2b + 3c)$ **17.** $(x + y)(a + b)$ **19.** $(4a + b)(1 - 3a)$ **21.** $(x + 1)(3x^2 - 1)$ **23.** $t(y + c)(2x - 3)$
25. $(a + b)(x + y + z)$ **27.** $(2x + 3)(2x - 3)$ **29.** $(2 + 3r)(2 - 3r)$ **31.** $(x + z + 5)(x + z - 5)$ **33.** prime
35. $(x + y - z)(x - y + z)$ **37.** $-4xy$ **39.** $(x^2 + y^2)(x + y)(x - y)$ **41.** $3(x + 2)(x - 2)$ **43.** $2x(3y + 2)(3y - 2)$
45. $(x + 4)(x + 4)$ **47.** $(b - 5)(b - 5)$ **49.** $(m + 2n)(m + 2n)$ **51.** $(x + 7)(x + 3)$ **53.** $(x - 6)(x + 2)$ **55.** prime
57. $(4x - 3y)(3x + 2y)$ **59.** $(6a + 5)(4a - 3)$ **61.** $(3x + 7y)(2x + 5y)$ **63.** $2(6p - 35q)(p + q)$ **65.** $-(6m - 5n)(m - 7n)$
67. $-x(6x + 7)(x - 5)$ **69.** $x^2(2x - 7)(3x + 5)$ **71.** $(x^2 + 5)(x^2 - 3)$ **73.** $(a^n - 3)(a^n + 1)$ **75.** $(3x^n - 2)(2x^n - 1)$
77. $(2x^n + 3y^n)(2x^n - 3y^n)$ **79.** $(5y^n + 2)(2y^n - 3)$ **81.** $(2z - 3)(4z^2 + 6z + 9)$ **83.** $2(x + 10)(x^2 - 10x + 100)$
85. $(x + y - 4)(x^2 + 2xy + y^2 + 4x + 4y + 16)$ **87.** $(2a + y)(2a - y)(4a^2 - 2ay + y^2)(4a^2 + 2ay + y^2)$
89. $(a - b)(a^2 + ab + b^2 + 1)$ **91.** $(4x^2 + y^2)(16x^4 - 4x^2y^2 + y^4)$ **93.** $(x - 3 + 12y)(x - 3 - 12y)$
95. $(a + b - 5)(a + b + 2)$ **97.** $(x + 2)(x^2 - 2x + 4)(x - 1)(x^2 + x + 1)$ **99.** $(x^2 + x + 2)(x^2 - x + 2)$
101. $(x^2 + x + 4)(x^2 - x + 4)$ **103.** $(2a^2 + a + 1)(2a^2 - a + 1)$ **109.** $2(\tfrac{3}{2}x + 1)$ **111.** $x^{1/2}(x^{1/2} + 1)$ **113.** $ab(b^{1/2} - a^{1/2})$
115. $(x - 2)(x + 3 + y)$ **117.** $(a + 1)(a^3 + a^2 + 1)$ **119.** 1 **121.** x^{20} **123.** 1

Exercise 1.6 (page 72)

1. numerator **3.** $ad = bc$ **5.** $\dfrac{ac}{bd}$ **7.** $\dfrac{a + c}{b}$ **9.** equal **11.** not equal **13.** $\dfrac{a}{3b}$ **15.** $\dfrac{8x}{35a}$ **17.** $\dfrac{16}{3}$ **19.** $\dfrac{z}{c}$ **21.** $\dfrac{2x^2y}{a^2b^3}$ **23.** $\dfrac{2}{x + 2}$

25. $-\dfrac{x - 5}{x + 5}$ **27.** $\dfrac{3x - 4}{2x - 1}$ **29.** $\dfrac{x^2 + 2x + 4}{x + a}$ **31.** $\dfrac{x(x - 1)}{x + 1}$ **33.** $\dfrac{x - 1}{x}$ **35.** $\dfrac{x(x + 1)^2}{x + 2}$ **37.** $\dfrac{1}{2}$ **39.** $\dfrac{z - 4}{(z + 2)(z - 2)}$ **41.** 1

43. 1 **45.** $\dfrac{x(x + 3)}{x + 1}$ **47.** $\dfrac{x + 5}{x + 3}$ **49.** 4 **51.** $\dfrac{-1}{x - 5}$ **53.** $\dfrac{5x - 1}{(x + 1)(x - 1)}$ **55.** $\dfrac{2a - 4}{(a + 4)(a - 4)}$ **57.** $\dfrac{2}{(x + 2)(x - 2)}$

59. $\dfrac{2(x^2 - 3x + 1)}{(x + 1)^2(x - 1)}$ **61.** $\dfrac{3y - 2}{y - 1}$ **63.** $\dfrac{1}{x + 2}$ **65.** $\dfrac{2x - 5}{2x(x - 2)}$ **67.** $\dfrac{2x^2 + 19x + 1}{(x + 4)(x - 4)}$ **69.** 0 **71.** $\dfrac{-x^4 + 3x^3 - 43x^2 - 58x + 697}{(x + 5)(x - 5)(x + 4)(x - 4)}$

73. $\dfrac{b}{2c}$ **75.** $81a$ **77.** -1 **79.** $\dfrac{y+x}{x^2y^2}$ **81.** $\dfrac{y+x}{y-x}$ **83.** $\dfrac{a^2(3x-4ab)}{ax+b}$ **85.** $\dfrac{x-2}{x+2}$ **87.** $\dfrac{3x^2y^2}{xy-1}$ **89.** $\dfrac{3x^2}{x^2+1}$ **91.** $\dfrac{x^2-3x-4}{x^2+5x-3}$

93. $\dfrac{x}{x+1}$ **95.** $\dfrac{5x+1}{x-1}$ **97.** $\dfrac{3x}{3+x}$ **99.** $\dfrac{x+1}{2x+1}$ **105.** 6 **107.** $\dfrac{y^9}{x^{12}}$ **109.** $-\sqrt{5}$

Chapter Summary (page 80)

1. a. $3, 6, 8$ **b.** $0, 3, 6, 8$ **c.** $-6, -3, 0, 3, 6, 8$ **d.** $-6, -3, 0, \frac{1}{2}, 3, 6, 8$ **e.** $\pi, \sqrt{5}$ **f.** $-6, -3, 0, \frac{1}{2}, 3, \pi, \sqrt{5}, 6, 8$
2. a. 3 **b.** $6, 8$ **c.** $-6, 0, 6, 8$ **d.** $-3, 3$ **3. a.** assoc. prop. of add. **b.** comm. prop. of add. **c.** assoc. prop. of mult.
d. distributive prop. **e.** comm. prop. of mult. **f.** comm. prop. of add. **4. a.** (number line: 10 11 12 13 14 15 16 17 18 19 20)

b. (number line: 6 7 8 9 10 11 12 13 14) **5 a.** (number line: -3 to 5) **b.** (number line: -1, 0) **c.** (number line: -2, 4)
d. (number line: -4, 6) **6. a.** 6 **b.** 25 **c.** $\sqrt{2}-1$ **d.** $\sqrt{3}-1$ **7.** 12 **8. a.** $-5aaa$ **b.** $25aa$ **9. a.** $3t^3$
b. $-6b^2$ **10. a.** n^6 **b.** p^6 **c.** $x^{12}y^8$ **d.** $\dfrac{a^{12}}{b^6}$ **e.** $\dfrac{1}{m^6}$ **f.** $\dfrac{q^6}{8p^6}$ **g.** $\dfrac{1}{a^3}$ **h.** $\dfrac{b^6}{a^4}$ **i.** $\dfrac{y^8}{9}$ **j.** $\dfrac{a^8}{b^{10}}$ **k.** $\dfrac{y^4}{9x^4}$ **l.** $-\dfrac{8m^{12}}{n^3}$ **11.** 18

12. a. 6.75×10^3 **b.** 2.3×10^{-4} **13. a.** 480 **b.** 0.00025 **14.** 1.5×10^{14} **15. a.** 11 **b.** $\frac{3}{5}$ **c.** $2x$ **d.** $3|a|$ **e.** $-10x^2$

f. not a real number **g.** $x^6|y|$ **h.** $\dfrac{y^2}{x^6}$ **i.** $-c$ **j.** a^2 **16.** **a.** 16 **b.** $\frac{1}{8}$ **c.** $\frac{8}{27}$ **d.** $\frac{4}{9}$ **e.** $\frac{9}{4}$ **f.** $\frac{125}{8}$ **g.** $36x^2$ **h.** $p^{2a/3}$

17. a. 6 **b.** -7 **c.** $\frac{3}{5}$ **d.** $\frac{3}{5}$ **e.** $|x|y^2$ **f.** x **g.** $\dfrac{m^2|n|}{p^4}$ **h.** $\dfrac{a^3b^2}{c}$ **18. a.** $7\sqrt{2}$ **b.** 0 **c.** $x\sqrt[3]{3x}$ **19. a.** $\dfrac{\sqrt{35}}{5}$ **b.** $2\sqrt{2}$

c. $\dfrac{\sqrt[3]{4}}{2}$ **d.** $\dfrac{2\sqrt[3]{5}}{5}$ **20. a.** $\dfrac{2}{5\sqrt{2}}$ **b.** $\dfrac{1}{\sqrt{5}}$ **c.** $\dfrac{2x}{3\sqrt{2x}}$ **d.** $\dfrac{21x}{2\sqrt[3]{49x^2}}$ **21. a.** 3rd degree, binomial **b.** 2nd degree, trinomial

c. 2nd degree, monomial **d.** 4th degree, trinomial **22.** **a.** $5x-6$ **b.** $2x^3 - 7x^2 - 6x$ **c.** $9x^2 + 12x + 4$ **d.** $6x^2 - 7xy - 3y^2$
e. $8a^2 - 8ab - 6b^2$ **f.** $3z^2 + 10z^2 + 2z - 3$ **g.** $a^{2n} + a^n - 2$ **h.** $2 + 2x\sqrt{2} + x^2$ **i.** $\sqrt{6} + \sqrt{2} + \sqrt{3} + 1$ **j.** -5

23. a. $\sqrt{3} + 1$ **b.** $-2(\sqrt{3} + \sqrt{2})$ **c.** $\dfrac{2x(\sqrt{x}+2)}{x-4}$ **d.** $\dfrac{x - 2\sqrt{xy} + y}{x-y}$ **24. a.** $\dfrac{x-4}{5(\sqrt{x}-2)}$ **b.** $\dfrac{1-a}{a(1+\sqrt{a})}$ **25. a.** $\dfrac{y}{2x}$

b. $2a^2b + 3ab^2$ **c.** $x^2 + 2x + 1$ **d.** $x^3 + 2x - 3 - \dfrac{6}{x^2-1}$ **26. a.** $3t(t+1)(t-1)$ **b.** $5(r-1)(r^2+r+1)$

c. $(3x+8)(2x-3)$ **d.** $(3a+x)(a-1)$ **e.** $(2x-5)(4x^2+10x+25)$ **f.** $2(3x+2)(x-4)$ **g.** $(x+3+t)(x+3-t)$ **h.** prime
i. $(2z+7)(4z^2-14z+49)$ **j.** $(7b+1)^2$ **k.** $(11z-2)^2$ **l.** $8(2y-5)(4y^2+10y+25)$ **m.** $(y-2z)(2x-w)$

n. $(x^2+1+x)(x^2+1-x)(x^4+1-x^2)$ **27. a.** $(x-2)(x+3)$ **b.** $\dfrac{2y-5}{y-2}$ **c.** $\dfrac{t+1}{5}$ **d.** $\dfrac{p(p+4)}{(p^2+8p+4)(p-3)}$

e. $\dfrac{(x-2)(x+3)(x-3)}{(x-1)(x+2)^2}$ **f.** 1 **g.** $\dfrac{3x^2-10x+10}{(x-4)(x+5)}$ **h.** $\dfrac{2(2x^2+3x+6)}{(x+2)(x-2)}$ **i.** $\dfrac{3x^3-12x^2+11x}{(x-1)(x-2)(x-3)}$ **j.** $\dfrac{-5x-6}{(x+1)(x+2)}$

k. $\dfrac{-x^3+x^2-12x-15}{x^2(x+1)}$ **l.** $\dfrac{3x}{x+1}$ **28. a.** $\dfrac{20}{3x}$ **b.** $\dfrac{y}{2}$ **c.** $\dfrac{y+x}{xy(x-y)}$ **d.** $\dfrac{y+x}{x-y}$

Chapter 1 Test (page 86)

1. $-7, 1, 3$ **2.** 3 **3.** comm. prop. of add. **4.** distributive prop. **5.** (number line: -4, 2) **6.** (number line: -3, 6)

7. 17 **8.** $-(x-7)$ **9.** 16 **10.** 8 **11.** x^{11} **12.** $\dfrac{r}{s}$ **13.** a **14.** x^{24} **15.** 4.5×10^5 **16.** 3.45×10^{-4} **17.** 3700 **18.** 0.0012

19. $5a^2$ **20.** $\dfrac{216}{729}$ **21.** $\dfrac{9s^6}{4t^4}$ **22.** $3a^2$ **23.** $5\sqrt{3}$ **24.** $-4x\sqrt[3]{3x}$ **25.** $\dfrac{x(\sqrt{x}+2)}{x-4}$ **26.** $\dfrac{x-y}{x+2\sqrt{xy}+y}$ **27.** $-a^2+7$

28. $-6a^6b^6$ **29.** $6x^2 + 13x - 28$ **30.** $a^{2n} - a^n - 6$ **31.** $x^4 - 16$ **32.** $2x^3 - 5x^2 + 7x - 6$ **33.** $6x + 19 + \frac{34}{x-3}$
34. $x^2 + 2x + 1$ **35.** $3(x + 2y)$ **36.** $(x + 10)(x - 10)$ **37.** $(5t - 2w)(2t - 3w)$ **38.** $3(a - 6)(a^2 + 6a + 36)$
39. $(x + 2)(x - 2)(x^2 + 3)$ **40.** $(3x^2 - 2)(2x^2 + 5)$ **41.** 1 **42.** $\dfrac{-2x}{(x+1)(x-1)}$ **43.** $\dfrac{(x+5)^2}{x+4}$ **44.** $\dfrac{1}{(x+1)(x-2)}$ **45.** $\dfrac{b+a}{a}$

46. $\dfrac{y}{y+x}$

Exercise 2.1 (page 96)

1. root; solution   **3.** no   **5.** linear   **7.** one   **9.** no restrictions   **11.** $x \neq 0$ **13.** $x \geq 0$ **15.** $x \neq 3$ and $x \neq -2$ **17.** $x = 5$;
conditional equation   **19.** no solution   **21.** $x = 7$; conditional equation   **23.** no solution   **25.** identity   **27.** $b = 6$;
conditional equation   **29.** identity   **31.** $x = 1$ **33.** $z = 9$ **35.** $z = 10$ **37.** $x = -3$ **39.** $x = 6$ **41.** $x = -2$ **43.** $x = \frac{5}{2}$
45. $x = 1$ **47.** $x = -\frac{14}{11}$ **49.** identity   **51.** $y = 2$ **53.** $s = 4$ **55.** $x = -4$ **57.** no solution   **59.** $a = 17$ **61.** $x = -\frac{2}{5}$
63. $x = \frac{2}{3}$ **65.** $n = 3$ **67.** $y = 5$ **69.** no solution   **71.** $a = 2$ **73.** $p = \dfrac{k}{2.2}$ **75.** $b_2 = \dfrac{2A}{h} - b_1$ **77.** $r^2 = \dfrac{3V}{\pi h}$
79. $s = \dfrac{f(P_n - L)}{i}$ **81.** $m = \dfrac{r^2F}{Mg}$ **83.** $y = b\left(1 - \dfrac{x}{a}\right)$ **85.** $r = \dfrac{r_1 r_2}{r_1 + r_2}$ **87.** $n = \dfrac{l - a + d}{d}$ **89.** $n = \dfrac{360}{180 - a}$
91. $r_1 = \dfrac{-Rr_2 r_3}{Rr_3 + Rr_2 - r_2 r_3}$ **95.** $5|x|$ **97.** $\dfrac{4y^4}{25x^2}$ **99.** $5|y|$ **101.** $\dfrac{|ab^3|}{z^2}$

Exercise 2.2 (page 105)

1. add   **3.** amount   **5.** rate; time   **7.** 84   **9.** 94   **11.** 7   **13.** $2\frac{1}{2}$ ft   **15.** 20 ft by 8 ft   **17.** 20 ft   **19.** $10,000 at 7%,
$12,000 at 6%   **21.** 327   **23.** $17,500 at 8%, $19,500 at $9\frac{1}{2}$%   **25.** $79.95   **27.** 200 units   **29.** 21   **31.** $\frac{1}{3}$ hr   **33.** $21\frac{1}{0}$ hr
35. 1 liter   **37.** $\frac{1}{15}$ liter   **39.** about 4.5 gal   **41.** 4 liters   **43.** 39 mph going; 65 mph returning   **45.** $2\frac{1}{2}$ hr   **47.** 50 sec
49. 12 mph   **51.** 600 lb barley; 1637 lb oats; 163 lb soybean meal   **53.** about 11.2 mm   **57.** $(x + 7)(x - 9)$
59. $(3x - 5)(3x + 1)$ **61.** $(x + 3)^2$

Exercise 2.3 (page 120)

1. $ax^2 + bx + c = 0$ **3.** $\sqrt{c}$; $-\sqrt{c}$ **5.** equal real numbers   **7.** $3, -2$ **9.** $12, -12$ **11.** $2, -\frac{5}{2}$ **13.** $2, \frac{3}{5}$ **15.** $\frac{3}{5}, -\frac{5}{3}$ **17.** $\frac{3}{2}, \frac{1}{2}$
19. $3, -3$ **21.** $5\sqrt{2}, -5\sqrt{2}$ **23.** $3, -1$ **25.** $2, -4$ **27.** $x^2 + 6x + 9$ **29.** $x^2 - 4x + 4$ **31.** $a^2 + 5a + \frac{25}{4}$
33. $r^2 - 11r + \frac{121}{4}$ **35.** $y^2 + \frac{3}{4}y + \frac{9}{64}$ **37.** $q^2 - \frac{1}{5}q + \frac{1}{100}$ **39.** $5, 3$ **41.** $2, -3$ **43.** $0, 25$ **45.** $\frac{2}{3}, -2$ **47.** $\dfrac{-5 \pm \sqrt{5}}{2}$
49. $\dfrac{-2 \pm \sqrt{7}}{3}$ **51.** $\pm 2\sqrt{3}$ **53.** $3, -\dfrac{5}{2}$ **55.** $2, -\dfrac{1}{5}$ **57.** $1, -2$ **59.** $\dfrac{-5 \pm \sqrt{13}}{6}$ **61.** $\dfrac{-1 \pm \sqrt{61}}{10}$ **63.** rational and equal
65. not real numbers   **67.** rational and unequal   **69.** not real numbers   **71.** yes   **73.** $2, 10$ **75.** $3, -4$ **77.** $\dfrac{3}{2}, -\dfrac{1}{4}$ **79.** $\dfrac{1}{2}, -\dfrac{4}{3}$
81. $\dfrac{5}{6}, -\dfrac{2}{5}$ **83.** $4 \pm 2\sqrt{2}$ **85.** $1, -1$ **87.** $-\dfrac{1}{2}, 5$ **89.** -2 **91.** $3, -\dfrac{8}{11}$ **93.** $t = \pm\sqrt{\dfrac{2h}{g}}$ **95.** $t = \dfrac{8 \pm \sqrt{64 - h}}{4}$
97. $y = \pm\dfrac{b\sqrt{a^2 - x^2}}{a}$ **99.** $a = \pm\dfrac{bx\sqrt{b^2 + y^2}}{b^2 + y^2}$ **101.** $x = \dfrac{-y \pm y\sqrt{5}}{2}$ **109.** $2x^2 - 8x$ **111.** $12m$

Exercise 2.4 (page 126)

1. $A = lw$ **3.** 4 ft by 8 ft   **5.** 9 cm   **7.** 2 in.   **9.** 20 mph going and 10 mph returning   **11.** 7 hr   **13.** 25 sec   **15.** about 9.5 sec
17. 10¢   **19.** 1,440   **21.** 4 hr   **23.** about 9.5 hr   **25.** no   **27.** Matilda at 8%; Maude at 7%   **29.** 10   **31.** 20
33. 10 m and 24 m   **37.** $\frac{x-6}{x(x-3)}$ **39.** $\frac{x+3}{x(x+2)}$

Exercise 2.5 (page 137)

1. imaginary   **3.** imaginary   **5.** $2 - 5i$ **7.** real   **9.** i **11.** -1 **13.** -1 **15.** -1 **17.** $x = 3; y = 5$ **19.** $x = \frac{2}{3}; y = -\frac{2}{9}$
21. $5 - 6i$ **23.** $-2 - 10i$ **25.** $4 + 10i$ **27.** $1 - i$ **29.** $-9 + 19i$ **31.** $-5 + 12i$ **33.** $52 - 56i$ **35.** $-6 + 17i$ **37.** $0 + i$
39. $4 + 0i$ **41.** $\frac{2}{5} - \frac{1}{5}i$ **43.** $\frac{1}{25} + \frac{7}{25}i$ **45.** $\frac{1}{2} + \frac{1}{2}i$ **47.** $-\frac{7}{13} - \frac{22}{13}i$ **49.** $-\frac{12}{17} + \frac{11}{34}i$ **51.** $\dfrac{6 + \sqrt{3}}{10} + \dfrac{3\sqrt{3} - 2}{10}i$ **53.** 5   **55.** $\sqrt{13}$
57. $7\sqrt{2}$ **59.** $\dfrac{\sqrt{2}}{2}$ **61.** 6   **63.** $\sqrt{2}$ **65.** $\dfrac{3\sqrt{5}}{5}$ **67.** 1   **69.** $-1 \pm i$ **71.** $-2 \pm i$ **73.** $1 \pm 2i$ **75.** $\frac{1}{3} \pm \frac{1}{3}i$
77. $(x + 2i)(x - 2i)$ **79.** $(5p + 6qi)(5p - 6pi)$ **81.** $2(y + 2zi)(y - 2zi)$ **83.** $2(5m + ni)(5m - ni)$ **91.** $4x^2\sqrt{2}$
93. $x - 4\sqrt{x + 1} + 5$

Exercise 2.6 (page 145)

1. equal   **3.** extraneous   **5.** $0, -5, -4$ **7.** $0, \frac{4}{3}, -\frac{1}{2}$ **9.** $5, -5, 1, -1$ **11.** $6, -6, 1, -1$ **13.** $3\sqrt{2}, -3\sqrt{2}, \sqrt{5}, -\sqrt{5}$
15. $0, 1$ **17.** $\frac{1}{8}, -8$ **19.** $1, 144$ **21.** $\frac{1}{64}$ **23.** $\frac{1}{9}$ **25.** 27 **27.** 1 **29.** $-\frac{3}{2}, -\frac{5}{2}$ **31.** 9 **33.** 20 **35.** 2 **37.** 2 **39.** $3, 4$
41. $\frac{1}{5}, -1$ **43.** $3, 5$ **45.** $-2, 1$ **47.** $2, -\frac{5}{2}$ **49.** -2 **51.** no solution   **53.** 3 **55.** 1 **57.** 400 ft   **59.** 8 ft

63.

65. $[3, \infty)$

67. $[-2, 1)$

69. $(-\infty, 1) \cup [2, \infty)$

Exercise 2.7 (page 157)

1. right   **3.** $a < c$ **5.** $b - c$ **7.** $>$ **9.** linear   **11.** equivalent   **13.** $(-\infty, 1)$ **15.** $[1, \infty)$

17. $(-\infty, 1)$ **19.** $[1, \infty)$ **21.** $(-\infty, 3]$ **23.** $(-10/3, \infty)$ **25.** $(5, \infty)$

27. $[14, \infty)$ **29.** $(-\infty, 15/4]$ **31.** $(-44/41, \infty)$ **33.** $(6, 9]$ **35.** $(8, 22]$

37. $[-11, 4]$ **39.** $[5, 21]$ **41.** $(-\infty, 0)$ **43.** $(0, \infty)$ **45.** $(3, 14/3)$

47. $(2, \infty)$ **49.** $(-4, 5/6)$ **51.** $[-2, \infty)$ **53.** $(-4, -3)$

55. $(-\infty, 2] \cup [3, \infty)$ **57.** $(-3, -2)$ **59.** $(-\infty, -1/2] \cup [-1/3, \infty)$ **61.** $(1/3, 1/2)$

63. $(-\infty, -3/2] \cup [1, \infty)$ **65.** $(-3, 2)$ **67.** $(-\infty, -1) \cup (-1, 0) \cup (1, \infty)$ **69.** $(-\infty, -3) \cup (-3, -2] \cup (2, \infty)$

71. $(-\infty, -2) \cup (-2, -1/3) \cup (1/2, \infty)$ **73.** $(0, 3/2)$ **75.** $(-\infty, 0) \cup (3/2, \infty)$ **77.** $(-\infty, 2) \cup [13/5, \infty)$

79. $(-\infty, -\sqrt{7}) \cup (-1, 1) \cup (\sqrt{7}, \infty)$ **81.** 17 min   **83.** 12   **85.** 7 hr   **87.** $16\frac{2}{3}$ cm $< s < 20$ cm
89. $40 + 2w < P < 60 + 2w$ **93.** $-2, 2$ **95.** 2   **97.** All are real.

Exercise 2.8 (page 167)

1. x **3.** $x = k$ or $x = -k$ **5.** $-k < x < k$ **7.** $x \le -k$ or $x \ge k$ **9.** 7   **11.** 0   **13.** 2   **15.** $\pi - 2$ **17.** $x - 5$ **19.** x^3 if $x \ge 0$; $-x^3$ if $x < 0$ **21.** $0, -4$ **23.** $2, -\frac{4}{3}$ **25.** $\frac{14}{3}, -2$ **27.** $7, -3$ **29.** no solution   **31.** $\frac{2}{7}, 2$ **33.** $x \ge 0$ **35.** $-\frac{3}{2}$ **37.** $0, -6$

39. 0   **41.** $\frac{3}{5}, 3$ **43.** $-\frac{3}{13}, \frac{9}{5}$

45. $(-3, 9)$ number line, -3 to 9

47. $(-\infty, -9) \cup (3, \infty)$ number line, -9, 3

49. $(-\infty, -7] \cup [3, \infty)$ number line, -7, 3

51. $[-13/3, 1]$ number line, $-13/3$, 1

53. $(-\infty, -3) \cup (-3, \infty)$ number line, -3

55. $(-1, 1/5)$ number line, -1, $1/5$

57. $(-\infty, -7/9) \cup (13/9, \infty)$ number line, $-7/9$, $13/9$

59. $(-5, 7)$ number line, -5, 7

61. $(-2, -1/2) \cup (-1/2, 1)$ number line, -2, $-1/2$, 1

63. $(-7/3, -2/3) \cup (4/3, 3)$ number line, $-7/3$, $-2/3$, $4/3$, 3

65. $(-7, -1) \cup (11, 17)$ number line, -7, -1, 11, 17

67. $(-18, -6) \cup (10, 22)$ number line, -18, -6, 10, 22

69. $(-10, -7] \cup [5, 8)$ number line, -10, -7, 5, 8

71. $\left[-\frac{1}{2}, \infty\right)$ **73.** $(-\infty, 0)$ **75.** $\left(-\infty, -\frac{1}{2}\right)$ **77.** $[0, \infty]$ **79.** $70° \le t \le 86°$ **81.** $|c - 0.6°| \le 0.5°$ **87.** 3.725×10^4 **89.** 523,000

Chapter Summary (page 173)

1. a. no restrictions   **b.** $x \ne 0$ **c.** $x \ge 0$ **d.** $x \ne 2, x \ne 3$ **2. a.** $\frac{16}{27}$; conditional equation   **b.** -14; conditional equation
c. $\frac{16}{5}$; conditional equation   **d.** no solution   **e.** 7; conditional equation   **f.** identity   **g.** 7; conditional equation

h. $\frac{1}{5}$; conditional equation   **3. a.** $F = \frac{9}{5}C + 32$ **b.** $f = \frac{is}{P_n - l}$ **c.** $f_1 = \frac{ff_2}{f_2 - f}$ **d.** $l = \frac{a - S + Sr}{r}$ **4.** 1.5 liters

5. about 3.9 hr   **6.** $5\frac{1}{7}$ hr   **7.** $3\frac{1}{3}$ oz   **8.** \$4500 at 11%; \$5500 at 14%   **9.** 10   **10. a.** $4, -\frac{3}{2}$ **b.** $1, \frac{4}{3}$ **c.** $0, \frac{8}{3}$ **d.** $\frac{2}{3}, \frac{4}{9}$
11. a. $3, 5$ **b.** $-4, -2$ **c.** $\frac{1 \pm \sqrt{21}}{10}$ **d.** $0, \frac{1}{5}$ **12. a.** $2, -7$ **b.** $9, -\frac{2}{3}$ **c.** $\frac{1 \pm \sqrt{21}}{10}$ **d.** $-1 \pm 2l$ **13.** $\frac{1}{3}$ **14.** 10, 2

15. $\frac{8}{5}, 5$ **16.** either 95 by 110 yd or 55 by 190 yd   **17.** 320 mph for prop plane; 440 mph for jet   **18.** 1 sec   **19.** $1\frac{1}{2}$ ft
20. a. $0 + i$ **b.** $0 - i$ **c.** $-2 - i$ **d.** $-2 - 5i$ **e.** $5 - 2i$ **f.** $21 - 9i$ **g.** $0 - 3i$ **h.** $0 - 2i$ **i.** $\frac{3}{2} - \frac{3}{2}i$ **j.** $-\frac{2}{5} + \frac{4}{5}i$ **k.** $\frac{4}{5} + \frac{3}{5}i$
l. $\frac{1}{2} - \frac{5}{2}i$ **m.** $\sqrt{10}$ **n.** 1   **21. a.** $2, -3$ **b.** $4, -2$ **c.** $1, 1, -1, -1$ **d.** $1, -1, 6, -6$ **e.** 9   **f.** $8, -27$ **g.** 5   **h.** 0
i. $4, -4$ **j.** no solution   **22. a.** $(-\infty, 7)$ number line, 7 **b.** $[-1/5, \infty)$ number line, $-1/5$ **c.** $(-\infty, 5/3)$ number line, $5/3$ **d.** $[-3, 5)$ number line, -3, 5

e. $(-\infty, -2) \cup (4, \infty)$ number line, -2, 4 **f.** $(-4, 1)$ number line, -4, 1 **g.** $(-1, 3)$ number line, -1, 3 **h.** $(-\infty, -3/2) \cup (1, \infty)$ number line, $-3/2$, 1

i. $(-\infty, -2] \cup (3, \infty)$ number line, -2, 3 **j.** $(-4, 1]$ number line, -4, 1 **k.** $[-2, 1] \cup (3, \infty)$ number line, -2, 1, 3 **l.** $(-\infty, 0) \cup (5/2, \infty)$ number line, 0, $5/2$

23. **a.** $5, -7$ **b.** 0   **c.** $(-6, 0)$ number line, -6, 0 **d.** $(-\infty, 2] \cup [8/3, \infty)$ number line, 2, $8/3$ **e.** $(-5, 1)$ number line, -5, 1

f. $(-\infty, -29) \cup (35, \infty)$ number line, -29, 35 **g.** $(-7/2, -2) \cup (-1, 1/2)$ number line, $-7/2$, -2, -1, $1/2$ **h.** $(-1, 4/3) \cup (4/3, 11/3)$ number line, -1, $4/3$, $11/3$

Chapter 2 Test (page 178)

1. $x \ne 0, x \ne 1$ **2.** $x \ge 0$ **3.** $\frac{5}{2}$ **4.** 37   **5.** $x = \mu + z\sigma$ **6.** $a = \frac{bc}{c+b}$ **7.** 87.5   **8.** \$14,000   **9.** $\frac{1}{2}, \frac{3}{2}$ **10.** $\frac{3}{2}, -4$

11. $x = \frac{-b \pm \sqrt{b^2 - 4ac}}{2a}$ **12.** $\frac{5 \pm \sqrt{133}}{6}$ **13.** $5, -3$ **14.** 8 sec   **15.** i **16.** 1   **17.** $7 - 12i$ **18.** $-23 - 43i$ **19.** $\frac{4}{5} + \frac{2}{5}i$

20. $0 + i$ **21.** 13   **22.** $\frac{\sqrt{10}}{10}$ **23.** $2, -2, 3, -3$ **24.** $1, -\frac{1}{32}$ **25.** 139   **26.** -1 **27.** $(-\infty, 2]$ number line, 2

28. $(-\infty, 5)$ **29.** $(2, \infty)$ **30.** $[-2, 1)$ **31.** $2, -\frac{10}{3}$ **32.** 0

33. $(-\infty, 3/2) \cup (7/2, \infty)$ **34.** $[-9, 6]$

Cumulative Review Exercises   (page 179)

1. $-2, 0, 2, 6$ **2.** $2, 11$ **3.** $[-4, 7)$ **4.** $(-\infty, 0) \cup [2, \infty)$ **5.** comm. prop. of addition

6. transitive prop.   **7.** $9a^2$ **8.** $81a^2$ **9.** a^6b^4 **10.** $\dfrac{9y^2}{x^4}$ **11.** $\dfrac{x^4}{16y^2}$ **12.** $\dfrac{4y^{10}}{9x^6}$ **13.** a^3b^3 **14.** abc^2 **15.** $\sqrt{3}$ **16.** $\dfrac{\sqrt[3]{2x^2}}{x}$

17. $\dfrac{3(y + \sqrt{3})}{y^2 - 3}$ **18.** $\dfrac{3x(\sqrt{x} + 1)}{x - 1}$ **19.** $5\sqrt{3} - 3\sqrt{5}$ **20.** $3\sqrt{2}$ **21.** $5 - 2\sqrt{6}$ **22.** 4 **23.** $-8x + 8$ **24.** $9x^4 - 4x^3$

25. $6x^2 + 11x - 35$ **26.** $z^3 + z^2 + 4$ **27.** $2x^2 - x + 1$ **28.** $3x^2 + 1 - \dfrac{x}{x^2 + 2}$ **29.** $3t(t - 2)$ **30.** $(3x + 2)(x - 4)$

31. $(x + 1)^2(x - 1)^2(x^2 + 1)^2$ **32.** $(x + 1)(x^2 - x + 1)(x - 1)(x^2 + x + 1)$ **33.** $\dfrac{x - 5}{x + 5}$ **34.** $(2x + 1)(x + 2)$ **35.** $\dfrac{5x^2 + 17x - 6}{(x + 3)(x - 3)}$

36. $\dfrac{-x^2 + x + 7}{(x + 2)(x + 3)}$ **37.** $b + a$ **38.** $-\dfrac{1}{xy}$ **39.** $0, 10$ **40.** 34 **41.** $R = \dfrac{R_1 R_2}{R_1 + R_2}$ **42.** $r = \dfrac{a - S}{l - S}$ or $r = \dfrac{S - a}{S - l}$

43. either 8 ft by 24 ft or 12 ft by 16 ft   **44.** $8000   **45.** $\frac{3}{5} + \frac{4}{5}i$ **46.** $\frac{3}{2} - \frac{1}{2}i$ **47.** 5 **48.** $0 + 10i$ **49.** 3 **50.** $2, -2, 3, -3$

51. $2, 7$ **52.** $64, 729$ **53.**   **54.**   **55.**

56.   **57.**   **58.**

Exercise 3.1   (page 194)

1. quadrants   **3.** to the right   **5.** abscissa   **7.** linear   **9.** x-intercept   **11.** horizontal   **13.** midpoint   **15.** $A(2, 3)$ **17.** $C(-2, -3)$
19. $E(0, 0)$ **21.** $G(-5, -5)$ **31.**   **33.**   **35.**
23. QI   **25.** QIII   **27.** QI
29. positive x-axis

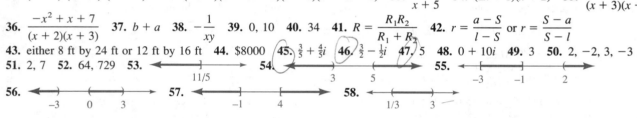

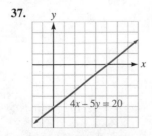

37.

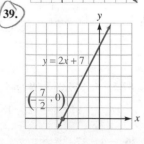

39.

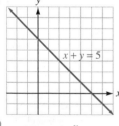

41.

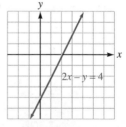

43.

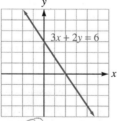

45.
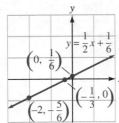

47.
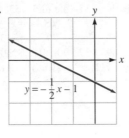

49.
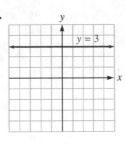

51.
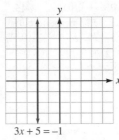

53.
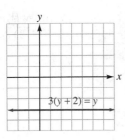

55.
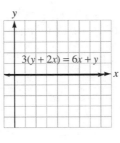

57. 1.22   **59.** 4.67   **61.** $162,500   **63.** 200   **65.** 100 rpm   **67.** 5
69. $\sqrt{13}$ **71.** $\sqrt{2}$ **73.** 2   **75.** 5   **77.** 13   **79.** 25   **81.** $8\sqrt{2}$
83. 7   **85.** (4, 6)   **87.** (0, 1)   **89.** (0, 0)   **91.** $\left(\dfrac{\sqrt{5}}{2}, \dfrac{\sqrt{5}}{2}\right)$
93. (5, 6)   **95.** (5, 15)   **99.** $\sqrt{2}$ units   **103.** approx. 170 mi
107. 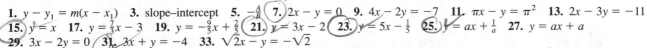   **109.** empty set

Exercise 3.2 (page 207)

1. divided   **3.** run   **5.** the change in   **7.** vertical   **9.** negative   **11.** 5   **13.** $-\frac{7}{4}$ **15.** -2 **17.** undefined   **19.** $\frac{5}{3}$ **21.** -1
23. 3   **25.** $\frac{1}{2}$ **27.** $\frac{2}{3}$ **29.** 0   **31.** negative   **33.** positive   **35.** undefined   **37.** 3.5 students per yr   **39.** $642.86 per year
41. $\frac{\Delta T}{\Delta t}$ is the hourly rate of change in temperature.

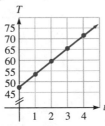

43. The slope is the speed of the plane.

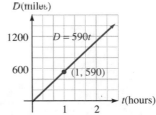

45. perpendicular   **47.** parallel   **49.** perpendicular   **51.** perpendicular   **53.** perpendicular   **55.** parallel   **57.** perpendicular
59. neither   **61.** not on same line   **63.** on same line   **65.** No two are perpendicular.   **67.** PQ and PR are perpendicular.   **69.** PQ
and PR are perpendicular.   **81.** $y = -\frac{3}{7}x + 3$ **83.** $y = -\frac{2}{5}x + 2$

Exercise 3.3 (page 219)

1. $y - y_1 = m(x - x_1)$ **3.** slope–intercept   **5.** $-\frac{A}{B}$ **7.** $2x - y = 0$ **9.** $4x - 2y = -7$ **11.** $\pi x - y = \pi^2$ **13.** $2x - 3y = -11$
15. $y = x$ **17.** $y = \frac{7}{3}x - 3$ **19.** $y = -\frac{9}{5}x + \frac{2}{5}$ **21.** $y = 3x - 2$ **23.** $y = 5x - \frac{1}{5}$ **25.** $y = ax + \frac{1}{a}$ **27.** $y = ax + a$
29. $3x - 2y = 0$ **31.** $3x + y = -4$ **33.** $\sqrt{2}x - y = -\sqrt{2}$
35. 1, (0, −1)

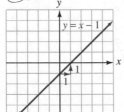

37. $\frac{2}{3}$, (0, 2)

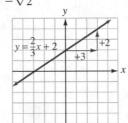

39. $-\frac{2}{3}$, (0, 6)

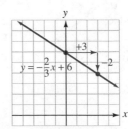

41. $\frac{3}{2}$, $(0, -4)$ **43.** $-\frac{1}{3}$, $\left(0, -\frac{5}{6}\right)$ **45.** $\frac{7}{2}$, $(0, 2)$ **47.** parallel **49.** perpendicular **51.** parallel **53.** perpendicular
55. perpendicular **57.** perpendicular **59.** $y = 4x$ **61.** $y = 4x - 3$ **63.** $y = \frac{4}{5}x - \frac{26}{5}$ **65.** $y = -\frac{1}{4}x$ **67.** $y = -\frac{1}{4}x + \frac{11}{2}$
69. $y = -\frac{5}{4}x + 3$ **71.** $m = -\frac{4}{5}$; $(0, 4)$ **73.** $m = -\frac{2}{3}$; $(0, 4)$ **75.** $x = -2$ **77.** $x = 5$ **79.** $y = -3200x + 24,300$
81. $y = 47,500x + 475,000$ **83.** $y = -\frac{710}{3}x + 1900$ **85.** \$90 **87.** \$890 **89.** \$37,200 **91.** \$230 **93.** about 838
95. $C = \frac{5}{9}(F - 32)$ **97.** $y = 3.75x + 37.5$; \$52$\frac{1}{2}$ **99.** 1655 barrels per day **111.** x^5 **113.** $\frac{125}{729}$ **115.** $-\sqrt{3}$

Exercise 3.4 (page 238)

1. x-intercept **3.** axis of symmetry **5.** x-axis **7.** circle; center **9.** $x^2 + y^2 = r^2$ **11.** $(-2, 0)$, $(2, 0)$; $(0, -4)$ **13.** $(0, 0)$, $\left(\frac{1}{2}, 0\right)$;
$(0, 0)$ **15.** $(-1, 0)$, $(5, 0)$; $(0, -5)$ **17.** $(1, 0)$, $(-2, 0)$; $(0, -2)$ **19.** $(-3, 0)$, $(0, 0)$, $(3, 0)$; $(0, 0)$ **21.** $(-1, 0)$, $(1, 0)$; $(0, -1)$

23.
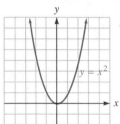

25.
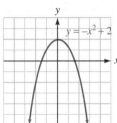

27.
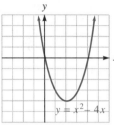

29.
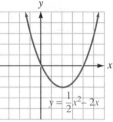

31. about the y-axis **33.** about the x-axis **35.** about the x-axis, the y-axis, and the origin **37.** about the y-axis **39.** none
41. about the x-axis **47.**
43. about the y-axis
45. about the x-axis

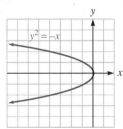

49.
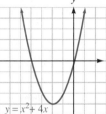

51.
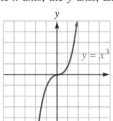

53.
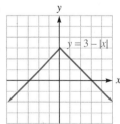

55.
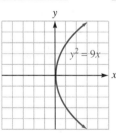

57.
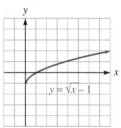

59.
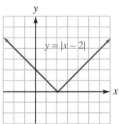

61.
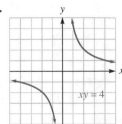

63. $x^2 + y^2 - 1 = 0$ **73.**
65. $x^2 + y^2 - 12x - 16y + 84 = 0$
67. $x^2 + y^2 - 6x + 8y + 23 = 0$
69. $x^2 + y^2 - 6x - 6y - 7 = 0$
71. $x^2 + y^2 + 6x - 8y = 0$

75.

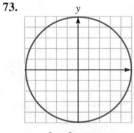

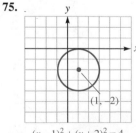

77.

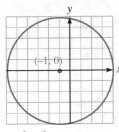

$x^2 + y^2 + 2x - 24 = 0$

79.

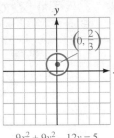

$9x^2 + 9y^2 - 12y = 5$

81.

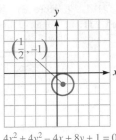

$4x^2 + 4y^2 - 4x + 8y + 1 = 0$

83.

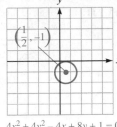

0.25, 0.88

85.

0.50, 7.25

87.

±2.65

89.

1.44

91. 4 sec   **93.**

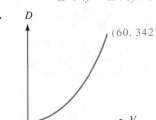

95. $x^2 + y^2 - 14x - 8y + 40 = 0$ **99.** 6
101. -1 **103.** 20 oz

Exercise 3.5   (page 249)

1. quotient   **3.** means   **5.** extremes, means   **7.** inverse   **9.** joint   **11.** 14   **13.** 2, -3 **15.** 18   **17.** $\frac{1}{2}$ **19.** 1000   **21.** 1
23. $\frac{21}{4}$ **25.** -8 **27.** direct variation   **29.** neither   **31.** $14\frac{6}{11}$ ft   **33.** 3 sec   **35.** 20 volts   **37.** 432 hertz   **39.** $\frac{\sqrt{3}}{4}$
47. $\frac{3x+5}{(x+2)(x+1)}$ **49.** $\frac{(x+4)(x+1)}{x-1}$ **51.** $-\frac{1}{x}$

Chapter Summary   (page 254)

1. a. $(2, 0)$ **b.** $(-2, 1)$ **2.**
c. $(0, -1)$ **d.** $(3, -1)$

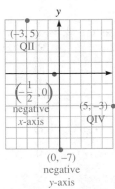

3. a.

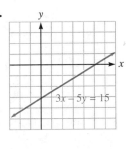

$3x - 5y = 15$

b.

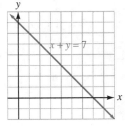

$x + y = 7$

c.

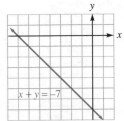

$x + y = -7$

d.

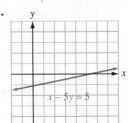

$x - 5y = 5$

4. a.

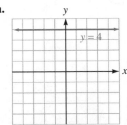

$y = 4$

b.

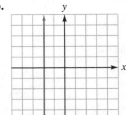

5. $12,150 **6. a.** 10 **b.** 13 **c.** 2 **d.** $2\sqrt{2}|a|$ **7. a.** $(0, 3)$ **b.** $\left(-6, \frac{15}{2}\right)$ **c.** $\left(\sqrt{3}, 8\right)$ **d.** $(0, 0)$ **8. a.** -6 **b.** 2 **c.** -1
d. 1 **9.** 200 ft/min **10. a.** 0 **b.** undefined **11. a.** negative **b.** positive **12.** $y = 7$ **13.** $x = 3$ **14. a.** $y = -\frac{7}{5}x$
b. $y - 1 = -4(x + 2)$ **c.** $y + 5 = -2(x - 7)$ or $y - 1 = -2(x - 4)$ **15. a.** $y = \frac{2}{3}x + 3$ **b.** $y = -\frac{3}{2}x - 5$

16. a.

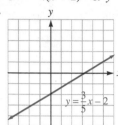

b.

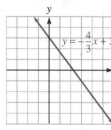

17. a. $y = -7x + 47$ **20. a.**
b. $y = \frac{1}{7}x - 3$
c. $y = \frac{3}{4}x - \frac{3}{2}$
d. $y = 3x + 5$
18. a. slope $= -\frac{5}{2}$; $\left(0, \frac{7}{2}\right)$
b. slope $= \frac{3}{4}$; $\left(0, -\frac{7}{2}\right)$
19. a. $y = 17$
b. $x = -5$

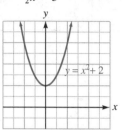

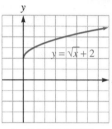

b.

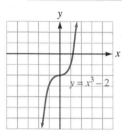

c.

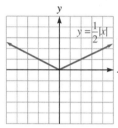

d.

e.

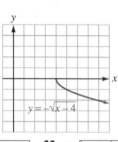

f.

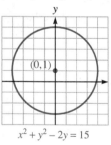
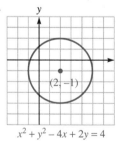

21. a.

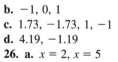

b.

22. a.

b.

23. a. $(x + 3)^2 + (y - 4)^2 = 144$, or $x^2 + y^2 + 6x - 8y - 119 = 0$ **b.** $\left(x + \frac{1}{2}\right)^2 + \left(y - \frac{5}{2}\right)^2 = \frac{121}{2}$, or $x^2 + y^2 + x - 5y - 54 = 0$

24. a.

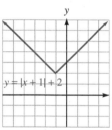

b.

25. a. 3.32, -3.32 **33.** 140 hr
b. $-1, 0, 1$
c. 1.73, -1.73, 1, -1
d. 4.19, -1.19
26. a. $x = 2$, $x = 5$
b. $x = -5$, $x = 5$ **27.** $\frac{9}{5}$ lb
28. $\frac{25}{9}$ **29.** $333\frac{1}{3}$ cc **30.** 1
31. about 117 ohms
32. $385

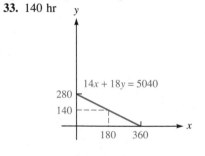

Chapter 3 Test (page 261)

1. QII
2. negative y-axis

3.

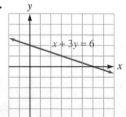

4.

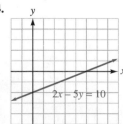

5.

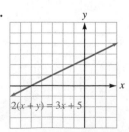

6.

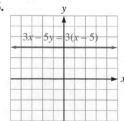

7.

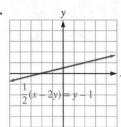

$\frac{1}{2}(x - 2y) = y - 1$

8.

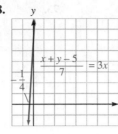

$\frac{x + y - 5}{7} = 3x$

9. $\sqrt{41}$ **10.** approximately 4.44 **11.** $(0, 0)$ **12.** $\left(\sqrt{2}, 2\sqrt{2}\right)$ **13.** $-\frac{5}{4}$
14. $\frac{\sqrt{3}}{3}$ **15.** neither **16.** perpendicular **17.** $y = 2x - 11$
18. $y = 3x + \frac{1}{2}$ **19.** $y = 2x + 5$ **20.** $y = -\frac{1}{2}x + 5$
21. $y = 2x - \frac{11}{2}$ **22.** $x = 3$ **23.** $(-4, 0), (0, 0), (4, 0); (0, 0)$
24. $(4, 0); (0, 4)$ **25.** about the x-axis **26.** about the y-axis

27.

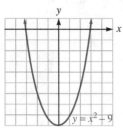

$y = x^2 + 9$

28.

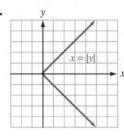

$x = |y|$

29.

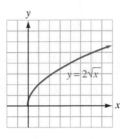

$y = 2\sqrt{x}$

30.

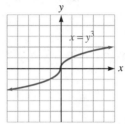

$x = y^3$

31. $(x - 5)^2 + (y - 7)^2 = 64$
32. $(x - 2)^2 + (y - 4)^2 = 32$

33.

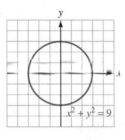

$x^2 + y^2 = 9$

34.

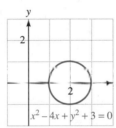

$x^2 - 4x + y^2 + 3 = 0$

35. $y = kz^2$ **36.** $w = krs^2$
37. $P = \frac{35}{2}$ **38.** $x = \frac{27}{32}$
39. $x = 2.65$ **40.** $x = 5.85$

Exercise 4.1 (page 275)

1. function **2.** domain **5.** $y = f(x)$ **7.** x **9.** vertical; once **11.** a function **13.** not a function **15.** a function
17. not a function **19.** not a function **21.** a function **23.** domain: $(-\infty, \infty)$; range: $(-\infty, \infty)$ **25.** domain: $(-\infty, \infty)$;
range: $[0, \infty)$ **27.** domain: $(-\infty, -1) \cup (-1, \infty)$; range: $(-\infty, 0) \cup (0, \infty)$ **29.** domain: $[0, \infty)$; range: $[0, \infty)$
31. domain: $(-\infty, -2) \cup (-2, -1) \cup (-1, \infty)$; range: $(-\infty, 0) \cup (0, \infty)$ **33.** domain: $(-\infty, -3) \cup (-3, \infty)$; range: $(-\infty, 1) \cup (1, \infty)$
35. $4; -11; 3k - 2; 3k^2 - 5$ **37.** $4; \frac{3}{2}; \frac{1}{2}k + 3; \frac{1}{2}k^2 + \frac{5}{2}$ **39.** $4; 9; k^2; k^4 - 2k^2 + 1$ **41.** $\frac{1}{3}; 2; \frac{2}{k + 4}; \frac{2}{k^2 + 3}$

43. $\frac{1}{3}; \frac{1}{8}; \frac{1}{k^2 - 1}; \frac{1}{k^4 - 2k^2}$ **45.** $\sqrt{5}; \sqrt{10}; \sqrt{k^2 + 1}; \sqrt{k^4 - 2k^2 + 2}$ **47.** 3 **49.** $2x + h$ **51.**

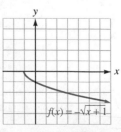

Range

f

Domain

53. a function **59.**

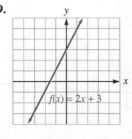

$f(x) = 2x + 3$

55. a function
57. a function

61.

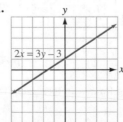

$2x = 3y - 3$

63.

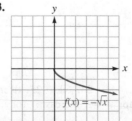

$f(x) = -\sqrt{x}$

65.

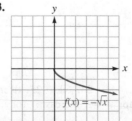

$f(x) = -\sqrt{x + 1}$

67.

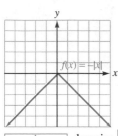

$f(x) = -|x|$

69.

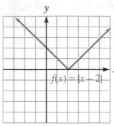

$f(x) = |x - 2|$

71.

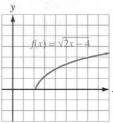

$f(x) = \sqrt{2x - 4}$

73.

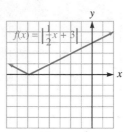

$f(x) = \left|\frac{1}{2}x + 3\right|$

75. domain: $\left[\frac{5}{2}, \infty\right)$; range: $[0, \infty)$ **77.** domain: $(-\infty, \infty)$; range: $(-\infty, \infty)$ **79.** $F = \frac{9}{5}C + 32$

81. $c = \frac{3}{2500}n + \frac{7}{2}$ **87.** 1, 7, 8 **89.** 7 **91.** $(-4, 7]$ **93.**

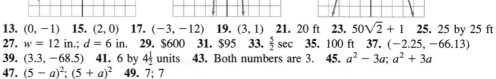
$-3 \qquad 5 \; 6$

Exercise 4.2 (page 285)

1. $y = ax^2 + bx + c$ **3.** $-\frac{b}{2a}$ **5.**

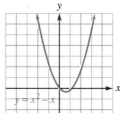

$y = x^2 - x$

7.

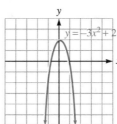

$y = -3x^2 + 2$

9.

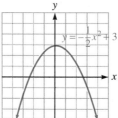

$y = -\frac{1}{2}x^2 + 3$

11.

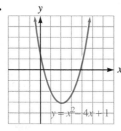

$y = x^2 - 4x + 1$

13. $(0, -1)$ **15.** $(2, 0)$ **17.** $(-3, -12)$ **19.** $(3, 1)$ **21.** 20 ft **23.** $50\sqrt{2} + 1$ **25.** 25 by 25 ft
27. $w = 12$ in.; $d = 6$ in. **29.** \$600 **31.** \$95 **33.** $\frac{5}{2}$ sec **35.** 100 ft **37.** $(-2.25, -66.13)$
39. $(3.3, -68.5)$ **41.** 6 by $4\frac{1}{2}$ units **43.** Both numbers are 3. **45.** $a^2 - 3a; a^2 + 3a$
47. $(5 - a)^2; (5 + a)^2$ **49.** 7; 7

Exercise 4.3 (page 295)

1. 4 **3.** odd
5. piecewise-
defined
7. 3

9.

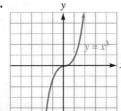

$y = x^3$

11.

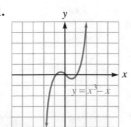

$y = x^3 - x$

13.

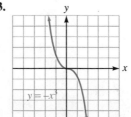

$y = -x^3$

15.

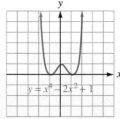

$y = x^4 - 2x^2 + 1$

17. even **19.** neither **21.** odd **23.** odd **25.** decreasing for $x < 0$; increasing for $x > 0$ **27.** increasing for $x < 0$; decreasing
for $x > 0$ **29.** decreasing for $x < 2$; increasing for $x > 2$

31.

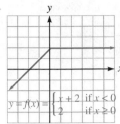

$$y = f(x) = \begin{cases} x+2 & \text{if } x < 0 \\ 2 & \text{if } x \geq 0 \end{cases}$$

33.

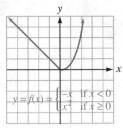

$$y = f(x) = \begin{cases} -x & \text{if } x < 0 \\ x^2 & \text{if } x \geq 0 \end{cases}$$

35.

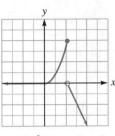

$$y = f(x) = \begin{cases} 0 & \text{if } x < 0 \\ x^2 & \text{if } 0 \leq x \leq 2 \\ 4 - 2x & \text{if } x > 2 \end{cases}$$

37.

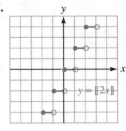

$y = [\![2x]\!]$

39.

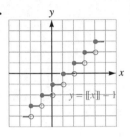

$y = [\![x]\!] - 1$

41. $26

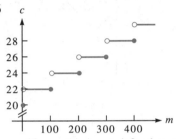

43. $1.60

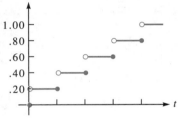

45.

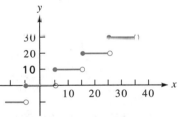

47. no; this is not defined at $x = 0$

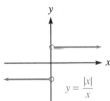

$y = \dfrac{|x|}{x}$

53. $3x + 5$; $3x + 3$

55. $\dfrac{3x-8}{5}$, $\dfrac{3x+1}{5} - 3$ or $\dfrac{3x-14}{5}$

Exercise 4.4 (page 305)

1. upward
3. to the right
5. 2; down
7. y-axis
9. horizontally

11.

$y = x^2 - 2$

13.

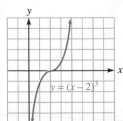

$y = (x + 3)^2$

15.

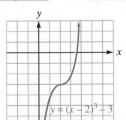

$y = (x + 1)^2 + 2$

17

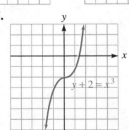

$y = \left(x + \dfrac{1}{2}\right)^2 - \dfrac{1}{2}$

19.

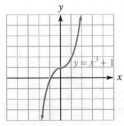

$y = x^3 + 1$

21.

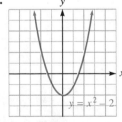

$y = (x - 2)^3$

23.

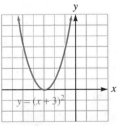

$y = (x - 2)^3 + 3$

25.

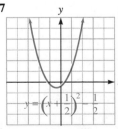

$y + 2 = x^3$

27.

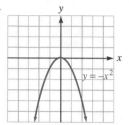

$y = -x^2$

29.

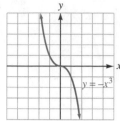

$y = -x^3$

31.

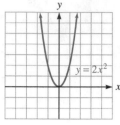

$y = 2x^2$

33.

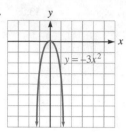

$y = -3x^2$

35.

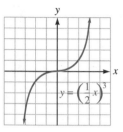

$y = \left(\frac{1}{2}x\right)^3$

37.

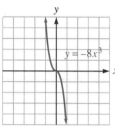

$y = -8x^3$

39.

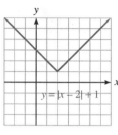

$y = |x - 2| + 1$

41.

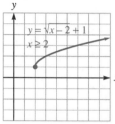

$y = \sqrt{x - 2} + 1$
$x \geq 2$

43.

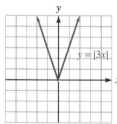

$y = |3x|$

45.

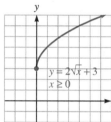

$y = 2\sqrt{x} + 3$
$x \geq 0$

47.

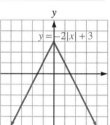

$y = -2|x| + 3$

55. $\frac{x-2}{x+2}$ **57.** all real numbers except 3
59. $x + 2 + \frac{-2}{x+1}$

Exercise 4.5   (page 321)

1. asymptote   **3.** vertical   **5.** x-intercept   **7.** the same   **9.** horizontal   **11.** 20 hr   **13.** 12 hr   **15.** $5555.56   **17.** $50,000
19. $c = f(x) = 1.25x + 700$ **21.** $1325   **23.** $1.95   **25.** $c = f(n) = 0.09n + 7.50$ **27.** $77.25   **29.** 9.75¢
31. $(-\infty, 2) \cup (2, \infty)$ **33.** $(-\infty, -5) \cup (-5, 5) \cup (5, \infty)$ **35.** $(-\infty, -1) \cup (-1, 0) \cup (0, 1) \cup (1, \infty)$ **37.** $(-\infty, \infty)$

39.

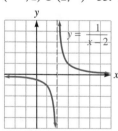

$y = \frac{1}{x-2}$

41.

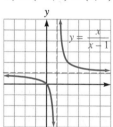

$y = \frac{x}{x-1}$

43.

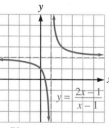

$y = \frac{x+1}{x-2}$

45.

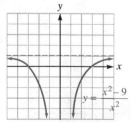

$y = \frac{2x-1}{x-1}$

47.

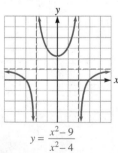

$y = \frac{x^2-9}{x^2-4}$

49.

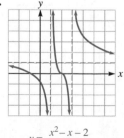

$y = \frac{x^2-x-2}{x^2-4x+3}$

51.

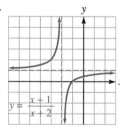

$y = \frac{x^2+2x-3}{x^3-4x}$

53.

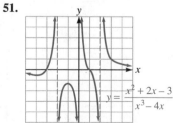

$y = \frac{x^2-9}{x^2}$

55.
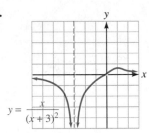
$$y = \frac{x}{(x+3)^2}$$

57.
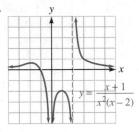
$$y = \frac{x+1}{x^2(x-2)}$$

59.
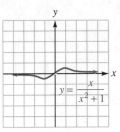
$$y = \frac{x}{x^2+1}$$

61.
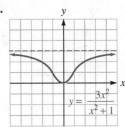
$$y = \frac{3x^2}{x^2+1}$$

63.
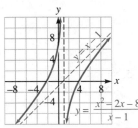
$$y = \frac{x^2-2x-8}{x-1}$$

65.
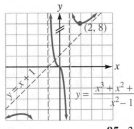
$$y = \frac{x^3+x^2+6x}{x^2-1}$$

67.
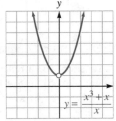
$$y = \frac{x^2}{x}$$

69.
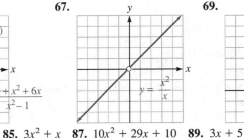
$$y = \frac{x^3+x}{x}$$

71.
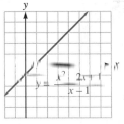
$$y = \frac{x^2 - 2x - 1}{x-1}$$

73.
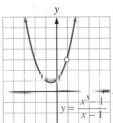
$$y = \frac{x^3-1}{x-1}$$

85. $3x^2 + x$ **87.** $10x^2 + 29x + 10$ **89.** $3x + 5$

Exercise 4.6 (page 334)

1. $f(x) + g(x)$ **3.** $f(x)g(x)$ **5.** $(-\infty, \infty)$.   **7.** $g(f(x))$ **9.** identity   **11.** $(f + g)(x) = 5x - 1;\ (-\infty, \infty)$

13. $(f \cdot g)(x) = 6x^2 - x - 2;\ (-\infty, \infty)$ **15.** $(f - g)(x) = x + 1;\ (-\infty, \infty)$ **17.** $(f/g)(x) = \dfrac{x^2 + x}{x^2 - 1} = \dfrac{x}{x - 1};$

$(-\infty, -1) \cup (-1, 1) \cup (1, \infty)$ **19.** 7   **21.** 1   **23.** 12   **25.** no value   **27.** $f(x) = 3x^2;\ g(x) = 2x$ **29.** $f(x) = 3x^2;\ g(x) = x^2 - 1$
31. $f(x) = 3x^3;\ g(x) = -x$ **33.** $f(x) = x + 9;\ g(x) = x - 2$ **35.** 11   **37.** -17 **39.** 190   **41.** 145   **43.** $(f \circ g)(x) = 3x + 3;$
$(-\infty, \infty)$ **45.** $(f \circ f)(x) = 9x;\ (-\infty, \infty)$ **47.** $(g \circ f)(x) = 2x^2;\ (-\infty, \infty)$ **49.** $(g \circ g)(x) = 4x;\ (-\infty, \infty)$ **51.** $(f \circ g)(x) = \sqrt{x^2 + 1};$
$(-\infty, \infty)$ **53.** $(f \circ f)(x) = \sqrt[4]{x};\ [0, \infty)$ **55.** $(g \circ f)(x) = x;\ [-1, \infty)$ **57.** $(g \circ g)(x) = x^4 - 2x^2;\ (-\infty, \infty)$ **59.** $f(x) = x - 2,$
$g(x) = 3x$ **61.** $f(x) = x - 2, g(x) = x^2$ **63.** $f(x) = x^2, g(x) = x - 2$ **65.** $f(x) = \sqrt{x}, g(x) = x + 2$ **67.** $f(x) = x + 2, g(x) = \sqrt{x}$
69. $f(x) = x, g(x) = x$ **79.** $y = \frac{x+7}{3}$ **81.** $y = \frac{3x}{1-x}$

Exercise 4.7 (page 343)

1. one-to-one   **3.** interchange   **5.** one-to-one   **7.** not one-to-one   **9.** not one-to-one   **11.** not one-to-one   **13.** not one-to-one
15. one-to-one   **17.** one-to-one   **19.** not one-to-one   **25.** $f^{-1}(x) = \frac{1}{3}x$ **27.** $f^{-1}(x) = \frac{x-2}{3}$ **29.** $f^{-1}(x) = \frac{1}{x} - 3$ **31.** $f^{-1}(x) = \frac{1}{2x}$

33.
$f(x) = 5x$
$f^{-1}(x) = \dfrac{1}{5}x$

35.
$f^{-1}(x) = \dfrac{x+4}{2}$
$f(x) = 2x - 4$

37.
$y - x = 2$
$f^{-1}(x) = x + 2$
$x - y = 2$
$f(x) = x - 2$

39.
$2x + y = 4$
$f(x) = 4 - 2x$
$2y + x = 4$
$f^{-1}(x) = \dfrac{4-x}{2}$

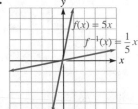
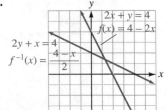

41.

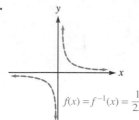

$$f(x) = f^{-1}(x) = \frac{1}{2x}$$

43.

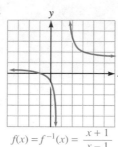

$$f(x) = f^{-1}(x) = \frac{x+1}{x-1}$$

45. $f^{-1}(x) = -\sqrt{x+3}$ $(x \geq -3)$
47. $f^{-1}(x) = \sqrt[4]{x+8}$ $(x \geq -8)$
49. $f^{-1}(x) = \sqrt{4-x^2}$ $(0 \leq x \leq 2)$
51. domain: $(-\infty, 1) \cup (1, \infty)$; range: $(-\infty, 1) \cup (1, \infty)$
53. domain: $(-\infty, 0) \cup (0, \infty)$; range: $(-\infty, -2) \cup (-2, \infty)$
57. $a \geq 0$ **59.** 8   **61.** 4   **63.** $\frac{5}{4}$ **65.** $\frac{1}{7}$

Chapter Summary (page 349)

1. a. a function **b.** a function **c.** not a function **d.** a function   **2. a.** domain: $(-\infty, \infty)$; range: $[-5, \infty)$
b. domain: $(-\infty, 5) \cup (5, \infty)$; range: $(-\infty, 3) \cup (3, \infty)$ **c.** domain: $[1, \infty)$; range: $[0, \infty)$ **d.** domain: $(-\infty, \infty)$; range: $[0, \infty)$
3. a. 8; -17; -2 **4. a.**
b. -2; $-\frac{3}{4}$; $-\frac{6}{5}$
c. 0; 5; 2
d. $\frac{1}{7}$; $\frac{1}{2}$; -1

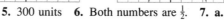

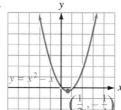

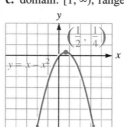

$y = x^2 - x$

b.

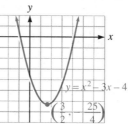

$y = x - x^2$

$\left(\frac{1}{2}, \frac{1}{4}\right)$

c.

$y = x^2 - 3x - 4$

$\left(\frac{3}{2}, -\frac{25}{4}\right)$

d.

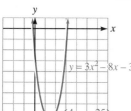

$y = 3x^2 - 8x - 3$

$\left(\frac{4}{3}, -\frac{25}{3}\right)$

5. 300 units   **6.** Both numbers are $\frac{1}{2}$.   **7. a.**

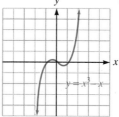

an odd function

$y = x^3 - x$

b.

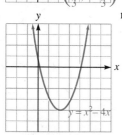

neither even nor odd   **c.**

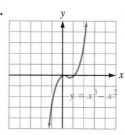

neither even nor odd

$y = x^2 - 4x$

$y = x^3 - x^2$

d.

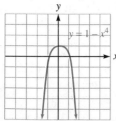

an even function   **8. a.**

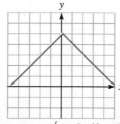

increasing for $x < 0$; decreasing for $x > 0$

$y = 1 - x^4$

$y = f(x) = \begin{cases} x + 5 & \text{if } x \leq 0 \\ 5 - x & \text{if } x > 0 \end{cases}$

b. 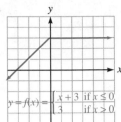   increasing for $x < 0$; constant for $x > 0$ **9. a.**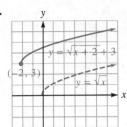

$y = f(x) = \begin{cases} x+3 & \text{if } x \le 0 \\ 3 & \text{if } x > 0 \end{cases}$

b.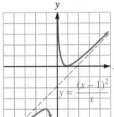

10. a. 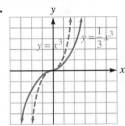   **b.** 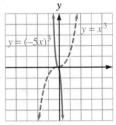   **11. a.** 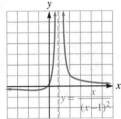   **b.**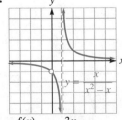

c. 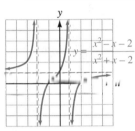   **d.**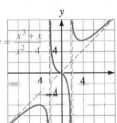

12. a. $(f + g)(x) = x^2 + 2x$ **b.** $(f \cdot g)(x) = 2x^2 + x^2 - 2x - 1$

c. $(f - g)(x) = x^2 - 2x - 2$ **d.** $(f/g)(x) = \dfrac{f(x)}{g(x)} = \dfrac{x^2 - 1}{2x + 1}$

e. $(f \circ g)(x) = f(g(x)) = 4x^2 + 4x$ **f.** $(g \circ f)(x) = g(f(x)) = 2x^2 - 1$

13. a. $f^{-1}(x) = \frac{x+1}{7}$ **b.** $f^{-1}(x) = 2 - \frac{1}{x}$ **c.** $f^{-1}(x) = \frac{x}{x+1}$

d. $f^{-1}(x) = \sqrt[3]{\dfrac{7}{x}}$

Chapter 4 Test (page 354)

1. domain: $(-\infty, 5) \cup (5, \infty)$; range: $(-\infty, 0) \cup (0, \infty)$ **2.** domain: $[-3, \infty)$; range: $[0, \infty)$ **3.** $\frac{1}{2}, 2$ **4.** $\sqrt{6}, 3$ **5.** $(7, -3)$

6. $(1, -4)$ **9.** 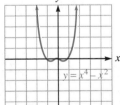   **10.**   **11.** $\frac{25}{8}$ sec   **12.** $\frac{625}{4}$ ft   **13.** 10 ft   **14.** 110 ft

7. $(4, -10)$

8. $(-2, 9)$

15. vertical asymptotes: $x = -3$ and $x = 3$; horizontal asymptote: $y = 0$ **16.** vertical asymptote: $x = 3$; horizontal asymptote: none; slant asymptote: $y = x - 2$

17. 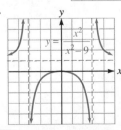   **18** 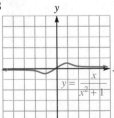   **19.** 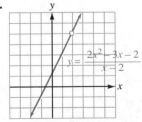   **20.**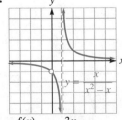

21. $(f + g)(x) = f(x) + g(x) = 3x + x^2 + 2$ **22.** $(g \circ f)(x) = g(f(x)) = 9x^2 + 2$ **23.** $(f/g)(x) = \dfrac{f(x)}{g(x)} = \dfrac{3x}{x^2 + 2}$

24. $(f \circ g)(x) = f(g(x)) = 3x^2 + 6$
25. $f^{-1}(x) = \frac{x+1}{x-1}$
26. $f^{-1}(x) = \sqrt[3]{x+3}$

27.

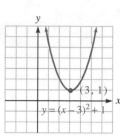

28.

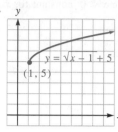

29. range: $(-\infty, -2) \cup (-2, \infty)$
30. range: $(-\infty, 3) \cup (3, \infty)$

Cumulative Review Exercises (page 356)

1.

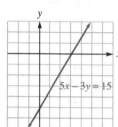

2.

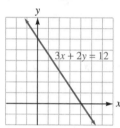

3. $\sqrt{41}$; $\left(\frac{1}{2}, \frac{3}{2}\right)$; $-\frac{4}{5}$ **4.** $2\sqrt{29}$; $(-2, 5)$; $\frac{2}{5}$ **5.** $y = -2x - 1$
6. $y = \frac{7}{2}x - \frac{11}{4}$ **7.** $y = \frac{3}{5}x + 6$ **8.** $y = -4x$

9.

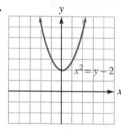

10.

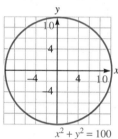

11.

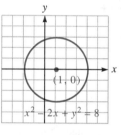

12.

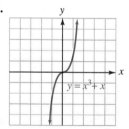

13. $x = 1, x = 10$ **14.** $x = 2, x = 8$ **15.** \$62.50   **16.** $\frac{25}{4}$ **17.** a function   **18.** a function   **19.** a function   **20.** not a function
21. domain: $(-\infty, \infty)$; range: $[5, \infty)$ **22.** domain: $(-\infty, -2) \cup (-2, \infty)$; range: $(-\infty, 0) \cup (0, \infty)$
23. domain: $[2, \infty)$; range: $(-\infty, 0]$ **27.**
24. domain: $[-4, \infty)$; range: $[0, \infty)$
25. $\left(-\frac{5}{2}, -\frac{49}{4}\right)$
26. $\left(\frac{5}{2}, \frac{49}{4}\right)$

27.

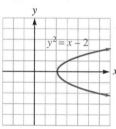

28.

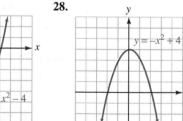

29.

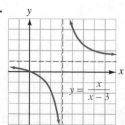

30.

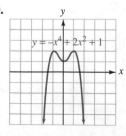

31.

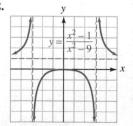

32.

33. $(f + g)(x) = f(x) + g(x) = x^2 + 3x - 3$; domain: $(-\infty, \infty)$ **34.** $(f - g)(x) = f(x) - g(x) = -x^2 + 3x - 5$; domain: $(-\infty, \infty)$

35. $(f \cdot g)(x) = f(x) \cdot g(x) = 3x^3 - 4x^2 + 3x - 4$; domain: $(-\infty, \infty)$ **36.** $(f/g)(x) = \dfrac{f(x)}{g(x)} = \dfrac{3x - 4}{x^2 + 1}$; domain: $(-\infty, \infty)$

37. $(f \circ g)(2) = 11$ **38.** $(g \circ f)(2) = 5$ **39.** $(f \circ g)(x) = 3x^2 - 1$ **40.** $(g \circ f)(x) = 9x^2 - 24x + 17$ **41.** $f^{-1}(x) = \frac{x-2}{3}$

42. $f^{-1}(x) = \frac{1}{x} + 3$ **43.** $f^{-1}(x) = \sqrt{x - 5}$ **44.** $f^{-1}(x) = \frac{x+1}{3}$ **45.** $y = kwz$ **46.** $y = \dfrac{kx}{t^2}$

Exercise 5.1   (page 370)

1. exponential   **3.** $(-\infty, \infty)$, or all real numbers   **5.** $(0, \infty)$, or all positive real numbers   **7.** asymptote   **9.** increasing
11. 11.0357   **13.** 451.8079   **15.** $5^{2\sqrt{2}} = 25^{\sqrt{2}}$ **17.** a^4 **19.** yes   **21.** no   **23.** $b = \frac{1}{2}$ **25.** no value of b **27.** $b = 2$ **29.** $b = e$

31.
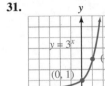
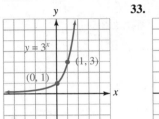

33.
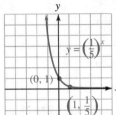

35.
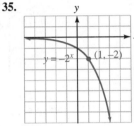

37.
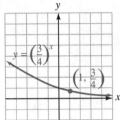

39.
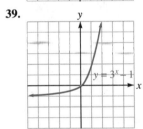

41.
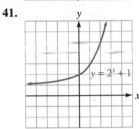

43.
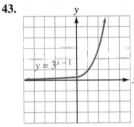

45.
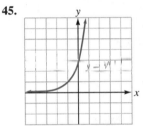

47.
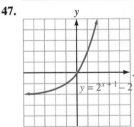

49.
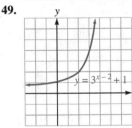

51.
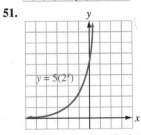

53.
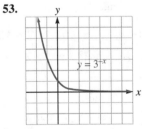

55. 0.1868 g   **57.** about 35.4%   **59.** 0.1575 unit   **61.** \$1104.02   **63.** \$15.78   **65.** \$1425.93   **67.** \$1847.16   **69.** 824
71. 0.933 lumens   **75.** $x^2(1 + 9x^2)$ **77.** $(x + 4)(x - 3)$

Exercise 5.2   (page 380)

1. 2.72   **7.**
3. increasing
5. birth; death

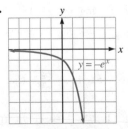

9.

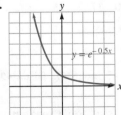

11.
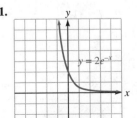

13.
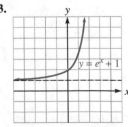

15. yes **17.** no **19.** no **21.** no **23.** $13,375.68 **25.** $6849.16 **27.** $7647.95 from continuous compounding; $7518.28 from annual compounding **29.** 315 **31.** 51 **33.** 10.6 billion **35.** 2.6 **37.** 6.73×10^5 **39.** 0.076 **41.** about 7350 **43.** about 0.07% **45.** 49 mps **47.** about 72.2 years **53.** $k = e^3$ **55.** 8 **57.** 3 **59.** 2 **61.** 3

Exercise 5.3 (page 390)

1. $x = b^y$ **3.** range **5.** inverse **7.** exponent **9.** $(b, 1)$ and $(1, 0)$ **11.** $20 \log \dfrac{E_0}{E_1}$ **13.** $3^4 = 81$ **15.** $\left(\frac{1}{2}\right)^3 = \frac{1}{8}$ **17.** $4^{-3} = \frac{1}{64}$ **19.** $\pi^1 = \pi$ **21.** $\log_8 64 = 2$ **23.** $\log_4 \frac{1}{16} = -2$ **25.** $\log_{1/2} 32 = -5$ **27.** $\log_x z = y$ **29.** 3 **31.** 3 **33.** 3 **35.** $\frac{1}{2}$ **37.** -3 **39.** 64 **41.** 7 **43.** 5 **45.** $\frac{1}{25}$ **47.** $\frac{1}{6}$ **49.** 5 **51.** $\frac{3}{2}$ **53.** 4 **55.** $\frac{2}{3}$ **57.** 5 **59.** 4 **61.** $b = 2$ **63.** $b = 2$

65.
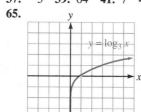

67.
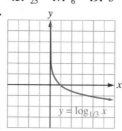

69.
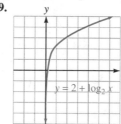

71.
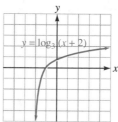

73.
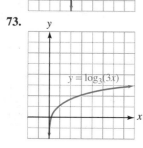

75.
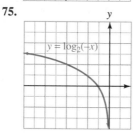

77. 0.5119 **79.** -2.3307 **81.** 25.2522 **83.** 1.9498×10^{-4} **85.** 55 db **87.** 29 **89.** 4.4 **91.** 4 **93.** 3 years old **95.** about 10.8 years **97.** b is larger. **101.** $\left(-\frac{7}{2}, -\frac{37}{4}\right)$ **103.** 425 ft by 850 ft

Exercise 5.4 (page 399)

1. $\log_e x$ **3.** $(-\infty, \infty)$ **5.** 10 **7.** $C = M(1 - e^{-kt})$ **9.** $E = RT \ln \left(\dfrac{V_f}{V_i}\right)$ **11.** 3.5596 **13.** 2.0664 **15.** -0.2752 **17.** undefined **19.** 4.0645 **21.** 69.4079 **23.** 0.0245 **25.** 120.0719 **27.** no **29.** no

31.  **33.**

35. 19.9 hr **37.** about 5.8 yr **39.** about 3654 joules **41.** about 208,000 V **45.** $y = 5x$ **47.** $y = -\dfrac{3}{2}x + \dfrac{13}{2}$ **49.** $x = 2$

51. $\dfrac{1}{2x - 3}$ **53.** $\dfrac{x + 1}{3(x - 2)}$

Exercise 5.5 (page 410)

1. 0 **3.** $M; N$ **5.** $x; y$ **7.** x **9.** $\neq$ **11.** 0 **13.** 7 **15.** 10 **17.** 1 **25.** $\log_b 2 + \log_b x + \log_b y$ **27.** $\log_b 2 + \log_b x - \log_b y$ **29.** $2 \log_b x + 3 \log_b y$ **31.** $\frac{1}{3}(\log_b x + \log_b y)$ **33.** $\log_b x + \frac{1}{2} \log_b z$ **35.** $\frac{1}{3} \log_b x - \frac{1}{3}\log_b y - \frac{1}{3} \log_b z$ **37.** $\log_b \dfrac{x + 1}{x}$

39. $\log_b x^2 \sqrt[3]{y}$ **41.** $\log_b \dfrac{\sqrt{z}}{x^3 y^2}$ **43.** $\log_b \dfrac{\dfrac{x}{z} + x}{\dfrac{y}{z} + y} = \log_b \dfrac{x}{y}$ **45.** true **47.** false **49.** true **51.** false **53.** true **55.** false

57. false **59.** true **61.** true **63.** true **65.** 1.4472 **67.** 0.3521 **69.** 1.1972 **71.** 2.4014 **73.** 2.0493 **75.** 0.4682

77. 1.7712 **79.** 0.9597 **81.** 1.8928 **83.** 2.3219 **85.** 4.77 **87.** from 5.01×10^{-4} to 1.26×10^{-3} **89.** 19 db
91. The original intensity must be raised to the 4th power. **93.** The volume V is squared. **105.** yes **107.** no **109.** $(-\infty, \infty)$
111. $(-\infty, \infty)$

Exercise 5.6 (page 419)

1. exponential **3.** $A_0 2^{-t/h}$ **5.** $x = 1.1610$ **7.** $x = 1.2702$ **9.** $x = 1.7095$ **11.** $x = 0$ **13.** $x = \pm 1.0878$ **15.** $x = 0, x = 1.0566$
17. $x = 3, x = -1$ **19.** $x = -2, x = -2$ **21.** $x = 0$ **23.** $x = 0.2789$ **25.** $x = 1, x = 3$ **27.** $x = 0$ **29.** $x = 7$ **31.** $x = 4$
33. $x = 10, x = -10$ **35.** $x = 50$ **37.** $x = 20$ **39.** $x = 10$ **41.** $x = 3, x = 4$ **43.** $x = 1, x = 100$ **45.** no solution **47.** $x = 6$
49. $x = 9$ **51.** $x = 4$ **53.** $y = 7, y = 1$ **55.** $x = 20$ **57.** $x \approx 1.81$ **59.** about 5.1 yr **61.** about 42.7 days **63.** about 4200 yr
65. about 5.6 yr **67.** about 5.4 yr **69.** because $\ln 2 \approx 0.70$ **71.** about 27 m **73.** $k = \frac{\ln 0.75}{3}$ **75.** $\frac{\ln(2/3)}{5}$ **81.** $y = \frac{x-2}{3}$ **83.** 19
85. $5x^2 - 1$

Chapter Summary (page 425)

1. a. $5^{2\sqrt{2}}$ **2. a.**
b. $2^{\sqrt{10}}$

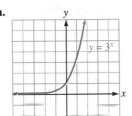

b.

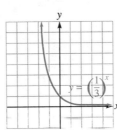

3. $p = 1, q = 6$
4. domain: $(-\infty, \infty)$;
   range: $(0, \infty)$

5. a.

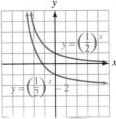

b.

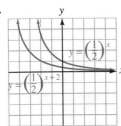

6. about 34.2 yr
7. \$2,189,703.45
8. 0.19 lumens
9. \$2,324,767.37

10. a.

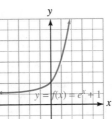

b.

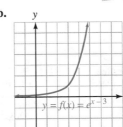

11. 582,000,000
12. $(0, \infty)$; $(-\infty, \infty)$
13. a. 2 **b.** $-\frac{1}{2}$
c. 0 **d.** -2 **e.** $\frac{1}{2}$
f. $\frac{1}{3}$ **14. a.** 32
b. 9 **c.** 8 **d.** -1
e. $\frac{1}{8}$ **f.** 2 **g.** 4
h. 2 **i.** 10 **j.** $\frac{1}{25}$
k. 5 **l.** 3

15. a.

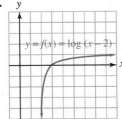

b.

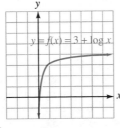

16. a.

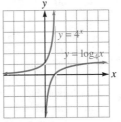

b.

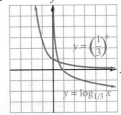

17. 53 db **18.** 4.4
19. a. 6.1137 **b.** -0.1111
20. a. 10.3398 **b.** 2.5715

21. a.

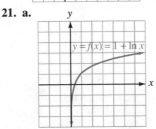

b.

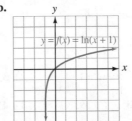

22. 80% **23.** 23 yr **24.** 2017 joules
25. a. 0 **b.** 1 **c.** 3 **d.** 4 **26. a.** 4
b. 0 **c.** 7 **d.** 3 **e.** 4 **f.** 9
27. a. $2 \log_b x + 3 \log_b y - 4 \log_b z$
b. $\frac{1}{2}(\log_b x - \log_b y - 2 \log_b z)$
28. a. $\log_b \dfrac{x^3 z^7}{y^5}$ **b.** $\log_b \dfrac{\sqrt{xy^3}}{z^7}$

29. a. 3.36 **b.** 1.56 **c.** 2.64 **d.** -6.72 **30.** 1.7604 **31.** about 7.94×10^{-4} grams-ions per liter **32.** $k \ln 2$ less

33. a. $\dfrac{\log 7}{\log 3} \approx 1.7712$ **b.** 2 **c.** $\dfrac{\log 3}{\log 3 - \log 2} \approx 2.7095$ **d.** $-1, -3$ **34. a.** $25, 4$ **b.** 4 **c.** 2 **d.** $4, 3$ **e.** 6 **f.** 31

g. $\frac{\ln 9}{\ln 2} \approx 3.1699$ **h.** no solution **i.** $\frac{e}{e-1} \approx 1.5820$ **j.** 1 **35.** about 3300 yr

Chapter 5 Test (page 430)

1.
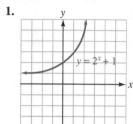
$y = 2^x + 1$

2.
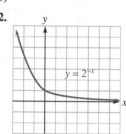
$y = 2^{-x}$

3. $\frac{3}{64}$ g

4. $1060.90

5.
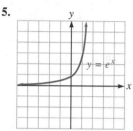
$y = e^x$

6.
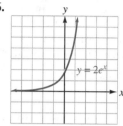
$y = 2e^x$

7. $4451.08 **8.** 3 **13.**
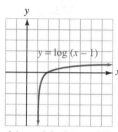
$y = \log(x - 1)$

9. -3 **10.** 17

11. 2 **12.** -3

14.
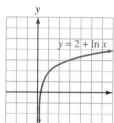
$y = 2 + \ln x$

15. $2 \log a + \log b + 3 \log c$

16. $\frac{1}{2}(\ln a - 2 \ln b - \ln c)$ **17.** $\log \dfrac{b\sqrt{a+2}}{c^2}$

18. $\log \dfrac{\sqrt[3]{\dfrac{a}{b^2}}}{c}$ **19.** 1.3801 **20.** 0.4259

21. $\dfrac{\log 3}{\log 7}$ or $\dfrac{\ln 3}{\ln 7}$ **22.** $\dfrac{\log e}{\log \pi}$ or $\dfrac{\ln e}{\ln \pi}$

23. true **24.** false **25.** false **26.** false **27.** 6.4 **28.** 46 db **29.** $\frac{\log 3}{\log 3 - 2} \approx -0.3133$ **30.** $-1, 3$ **31.** 1 **32.** 10

Exercise 6.1 (page 441)

1. natural **3.** nonreal complex **5.** factor **7.** 4 **9.** 67 **11.** 70,249 **13.** 128.3085938 **15.** true **17.** true **19.** false **21.** true
23. $\{-1, -5, 3\}$ **25.** $\left\{1, 1, \sqrt{3}, -\sqrt{3}\right\}$ **27.** $\{2, 3, i, -i\}$ **29.** $x^2 - 9x + 20$ **31.** $x^3 - 3x^2 + 3x - 1$
33. $x^3 - 11x^2 + 38x - 40$ **35.** $x^4 - 3x^2 + 2$ **37.** $x^3 - \sqrt{2}x^2 + x - \sqrt{2}$ **39.** $x^3 - 2x^2 + 2x$ **41.** $(x - 1)(3x^2 + x - 5) - 9$
43. $(x - 3)(3x^2 + 7x + 15) + 41$ **45.** $(x + 1)(3x^2 - 5x - 1) - 3$ **47.** $(x + 3)(3x^2 - 11x + 27) - 85$ **49.** $x^2 + 2x + 3$
51. $7x^2 - 10x + 5 + \frac{-4}{x+1}$ **53.** $4x^3 + 9x^2 + 27x + 80 + \frac{245}{x-3}$ **55.** $3x^4 + 12x^3 + 48x^2 + 192x$ **57.** 47 **59.** -569 **61.** $\frac{15}{8}$
63. 5 **65.** 0 **67.** 384 **69.** $16 + 2i$ **71.** $40 - 40i$ **73.** 0 **77.** QIV **79.** QI **81.** 10 **83.** 1

Exercise 6.2 (page 450)

1. zero **3.** conjugate **5.** 0 **7.** 2 **9.** lower bound **11.** 10 **13.** 4 **15.** 4 **17.** $x^3 - 3x^2 + x - 3 = 0$
19. $x^3 - 6x^2 + 13x - 10 = 0$ **21.** $x^4 - 5x^3 + 7x^2 - 5x + 6 = 0$ **23.** $x^4 - 2x^3 + 3x^2 - 2x + 2 = 0$ **25.** 0 or 2 positive;
1 negative; 0 or 2 nonreal **27.** 0 positive; 1 or 3 negative; 0 or 2 nonreal **29.** 0 positive; 0 negative; 4 nonreal **31.** 1 positive;
1 negative; 2 nonreal **33.** 0 positive; 0 negative; 10 nonreal **35.** 0 positive; 0 negative; 8 nonreal; yes **37.** 1 positive;
1 negative; 2 nonreal **39.** $-2, 4$ **41.** $-1, 1$ **43.** $-4, 6$ **45.** $-4, 3$ **47.** $-2, 4$ **53.** k units to the right **55.** reflected about
the y-axis **57.** stretched vertically by a factor of a

Exercise 6.3 (page 459)

1. 7 **3.** root **5.** $1, -1, 5$ **7.** $3, -3, 2$ **9.** $1, -1, 2$ **11.** $1, 2, 3, 4$ **13.** $2, -5$ **15.** $1, -1, 2, -2, -3$
17. $0, 2, 2, 2, -2, -2, -2$ **19.** $\frac{2}{3}$ **21.** $-1, 2, 3, \frac{2}{3}$ **23.** $-3, \frac{1}{3}, \frac{1}{2}, \frac{1}{2}$ **25.** $2, 2, -3, -\frac{1}{3}, \frac{1}{2}$ **27.** $\frac{3}{2}, \frac{2}{3}, -\frac{3}{5}$ **29.** $\frac{2}{3}, -\frac{3}{5}, 4$ **31.** $-\frac{1}{2}, \frac{3}{2}, \frac{6}{5}$
33. $3, 1 - i$ **35.** $3, -3, 1 - i$ **37.** $-\frac{2}{3}, 3, -1$ **39.** $\frac{1}{2}, \frac{1}{2}, \frac{1}{2}, 1, 1$ **41.** 10, 20, 60 ohms **45.** $(3, 7)$ or approx. $(1.54, 13.63)$
47. $6ab^2\sqrt{2abc}$ **49.** $8a\sqrt{2b}$ **51.** $\sqrt{3} + 1$

Exercise 6.4   (page 465)

1. $P(a)$ and $P(b)$ **3.** x_l and x_r **5.** $P(-2) = 3$; $P(-1) = -2$ **7.** $P(4) = -40$; $P(5) = 30$ **9.** $P(1) = 8$; $P(2) = -1$
11. $P(2) = -72$; $P(3) = 154$ **13.** $P(0) = 10$; $P(1) = -60$ **15.** 1.7   **17.** -2.2 **19.** 1.7   **21.** -1.2 **23.** $-2.2, 2.2$ **25.** 1, 2;
the bisection method fails to find the solution 2   **27.** $(0.83, 0.56)$, $(-0.83, -0.56)$ **31.** vertical: $x = 2$, $x = -2$; horizontal: $y = 0$;
slant: none   **33.** vertical: $x = 2$; horizontal: none; slant: $y = x + 2$

Chapter Summary   (page 469)

1. a. 1   **b.** 66   **c.** 241   **d.** 34   **2. a.** false   **b.** false   **c.** true   **d.** true   **3. a.** $2x^3 - 5x^2 - x + 6$ **b.** $2x^3 + 3x^2 - 8x + 3$
c. $x^4 + 3x^3 - 9x^2 + 3x - 10$ **d.** $x^4 + x^3 - 5x^2 + x - 6$ **4. a.** $3x^3 + 9x^2 + 29x + 90$ with remainder of 277
b. $2x^3 + 4x^2 + 5x + 13$ with remainder of 25   **c.** $5x^4 - 14x^3 + 31x^2 - 64x + 129$ with remainder of -259
d. $4x^4 - 2x^3 + x^2 + 2x$ with remainder of 1   **5. a.** 6   **b.** 6   **c.** 65   **d.** 1984   **6. a.** $2 - i$ **b.** i **7. a.** 0 or 2 positive;
0 or 2 negative; 0, 2, or 4 nonreal   **b.** 1 or 3 positive; 1 negative; 0 or 2 nonreal   **c.** 1 positive; 0, 2, or 4 negative;
0, 2, or 4 nonreal   **d.** 1 or 3 positive; 0 or 2 negative; 2, 4, or 6 nonreal   **e.** 0 positive; 0 negative; 4 nonreal   **f.** 0 positive;
1 negative; 6 nonreal   **8. a.** $-1, 2$ **b.** $-5, 2$ **9. a.** $-5, -\frac{3}{2}, -2$ **b.** $\frac{1}{3}$ **c.** $2, -2, \frac{3}{2}, -\frac{3}{2}$ **d.** $4, 4, -2, -\frac{1}{2}$
10. a. $P(-1) = -9$; $P(0) = 18$ **b.** $P(1) = -8$; $P(2) = 21$ **11. a.** 4.2   **b.** 2.5   **12. a.** 1.67   **b.** 0.67   **13.** 10 m by 6 m;
12 m by 5 m; 15 m by 4 m   **14.** 17.27 ft

Chapter 6 Test   (page 473)

1. 1   **2.** -30 **3.** $x^3 - 4x^2 - 5x$ **4.** $x^4 - 2x^2 - 3$ **5.** 5   **6.** -28 **7.** $\frac{11}{3}$ **8.** $6 - 3i$ **9.** $(x - 2)(2x^2 + x - 2) - 5$
10. $(x + 1)(2x^2 - 5x + 1) - 2$ **11.** $2x + 3$ **12.** $3x^2 + x$ **13.** $x^3 - 2x^2 + x - 2 = 0$ **14.** $x^3 - 5x^2 + 9x - 5 = 0$
15. 1 or 3 positive; 0 or 2 negative; 0, 2, or 4 nonreal   **16.** 1 positive; 0 or 2 negative; 0 or 2 nonreal   **17.** $-3, 3$ **18.** $-1, 6$
19. 2, 3, $-\frac{1}{2}$ **20.** 23

Cumulative Review Exercises   (page 474)

1.
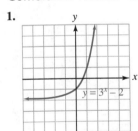
$y = 3^x - 2$

2.
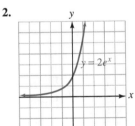
$y = 2e^x$

3.
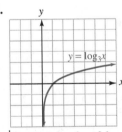
$y = \log_3 x$

4.
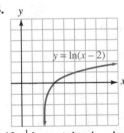
$y = \ln(x - 2)$

5. 6   **6.** -3
7. 3   **8.** 2

9. $\log a + \log b + \log c$ **10.** $2 \log a + \log b - \log c$ **11.** $\frac{1}{2}(\log a + \log b - 3 \log c)$ **12.** $\frac{1}{2} \log a + \log b - \log c$
13. $\log \dfrac{a^3}{b^3}$ **14.** $\log \dfrac{\sqrt{ab^3}}{\sqrt[3]{c^2}}$ **15.** $x = \dfrac{\log 8}{\log 3} - 1$ **16.** $x = -1$ **17.** $x = 500$ **18.** $x = \sqrt{11}$ **19.** 9   **20.** -36 **21.** 4   **22.** $2 - i$
23. a factor   **24.** not a factor   **25.** a factor   **26.** a factor   **27.** 12   **28.** 2000   **29.** 2 or 0 positive; 2 or 0 negative;
4, 2, or 0 nonreal   **30.** 1 positive; 3 or 1 negative; 2 or 0 nonreal   **31.** $-1, -3, 3$ **32.** $-1, 1, 2$

Exercise 7.1   (page 485)

1. system
3. consistent
5. independent
7. consistent
9. dependent
11. is

13.
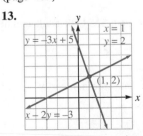
$y = -3x + 5$ $x = 1$ $y = 2$ $(1, 2)$ $x + 2y = -3$

15.
$x = -2$ $y = 4$ $(-2, 4)$ $3x + 2y = 2$ $-2x + 3y = 16$

17. $x = 2.2$, $y = -4.7$ **19.** $x = 1.7$, $y = 0.3$
21. $x = -1$, $y = -2$ **23.** $x = 3$, $y = -2$
25. $x = \frac{1}{2}$, $y = \frac{1}{3}$ **27.** no solution   **29.** $x = 3$, $y = 1$
31. $x = 3$, $y = 2$ **33.** $x = -3$, $y = 0$
35. $x = 1$, $y = -\frac{1}{2}$ **37.** $x = \frac{1}{3}$, $y = \frac{5}{3}$
39. no solution; inconsistent system   **41.** $x = 2$,
$y = -3$ **43.** $x = 9$, $y = -1$ **45.** $x = 1$, $y = 2$, $z = 0$
47. $x = 0$, $y = -\frac{1}{3}$, $z = -\frac{1}{3}$ **49.** $x = 1$, $y = 2$, $z = -1$

51. $x = 1, y = 0, z = 5$ **53.** no solution; inconsistent system **55.** $x = 0, y = 1, z = 0$ **57.** $x = \frac{2}{3}, y = \frac{1}{4}, z = \frac{1}{2}$
59. $x = \frac{1}{2}, y = \frac{1}{2}, z = \frac{1}{2}$ **61.** dependent equations; $x = 2 - y$, $y = $ any number, $z = 1$ **63.** 225 acres of corn; 125 acres of soybeans
65. 210,000 cm² **67.** 40 g and 20 g **69.** 10 ft **71.** $E(x) = 43.53x + 742.72$, $R(x) = 89.95x$; 16 pairs per day

73. 15 hr cooking hamburgers,
   10 hr pumping gas, 5 hr janitorial
75. 1.05 million in 0–14 group,
   1.56 million in 15–49 group,
   0.39 million in 50-and-older group
77. 30°, 50°, 100°

85.

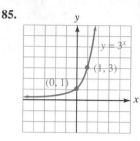

87. $x = 8$ **89.** $\log x - 2 \log y - \log z$ **91.** $\log \dfrac{xy^3}{\sqrt{z}}$

Exercise 7.2 (page 499)

1. matrix **3.** coefficient **5.** equation **7.** row equivalent **9.** interchanged **11.** adding; multiple **13.** $x = \frac{13}{3}, y = \frac{8}{3}$ **15.** $x = 7$,
$y = 6$ **17.** $x = 1, y = 2, z = 3$ **19.** $x = 1, y = 1, z = 3$ **21.** row echelon form **23.** reduced row echelon form **25.** $x = 2$,
$y = -1$ **27.** $x = -2, y = 0$ **29.** $x = 3, y = 1$ **31.** no solution; inconsistent system **33.** $x = 1, y = 0, z = 2$ **35.** $x = 2, y = -2$,
$z = 1$ **37.** $x = 1, y = 1, z = 2$ **39.** $x = -1, y = 3, z = 1$ **41.** $x = 13, y = 3$ **43.** $x = -13, y = 7, z = -2$ **45.** $x = 10, y = 3$
47. $x = 0, y = 0$ **49.** $x = 3, y = 1, z = -2$ **51.** $x = \frac{1}{4}, y = 1, z = \frac{1}{4}$ **53.** $x = 1, y = 2, z = 1, t = 1$ **55.** $x = 1, y = 2, z = 0$,
$t = 1$ **57.** $x = \frac{9}{4}, y = -3, z = \frac{3}{4}$ **59.** $x = 0, y = \frac{20}{3}, z = -\frac{1}{3}$ **61.** $x = 1, y = -3$ **63.** dependent equations; general solution is
$(x, y, z) = \left(\frac{8}{7} + \frac{1}{7}z, \frac{10}{7} - \frac{4}{7}z, z\right)$ **65.** dependent equations; general solution is $(w, x, y, z) = (1 + z, -z, -1 - z, z)$ **67.** no solution;
inconsistent system **69.** 1300 mi **71.** Dictionaries are 4.5 in. wide; atlases are 3.5 in. wide; thesauruses are 4 in. wide.
77. $x = \pm 2, y = \pm 1, z = \pm 3$ **79.** $y = mx + b$ **81.** equal **83.** $y = 2x + 7$ **85.** $x = 2$

Exercise 7.3 (page 513)

1. $i; j$ **3.** every element **5.** additive identity **7.** $x = 2, y = 5$ **9.** $x = 1, y = 2$ **11.** $\begin{bmatrix} 15 & -15 \\ 0 & -10 \end{bmatrix}$

13. $\begin{bmatrix} 25 & 75 & -10 \\ -10 & -25 & 5 \end{bmatrix}$ **15.** $\begin{bmatrix} -1 & 2 & 1 \\ -6 & 0 & 0 \end{bmatrix}$ **17.** $\begin{bmatrix} -6 & 5 & 0 \\ 1 & -1 & 0 \end{bmatrix}$ **19.** $\begin{bmatrix} 18 & -1 & -4 \\ -35 & 0 & -1 \end{bmatrix}$

21. $\begin{bmatrix} -5 & 2 & -7 \\ 5 & 0 & -3 \\ 2 & -3 & 5 \end{bmatrix}$ **23.** $\begin{bmatrix} 2 & -2 \\ 3 & 10 \end{bmatrix}$ **25.** $\begin{bmatrix} -22 & -22 \\ -105 & 126 \end{bmatrix}$ **27.** $\begin{bmatrix} 4 & 2 & 10 \\ 5 & -2 & 4 \\ 2 & -2 & 1 \end{bmatrix}$ **29.** $\begin{bmatrix} 32 \\ 2 \end{bmatrix}$

31. not possible **33.** $\begin{bmatrix} -4 & -19 & 13 \\ -11 & 3 & -11 \\ 4 & 6 & -2 \\ 13 & 13 & -1 \end{bmatrix}$ **35.** $\begin{bmatrix} -36.29 \\ 16.2 \\ -19.26 \end{bmatrix}$ **37.** $\begin{bmatrix} -16.11 & 4.71 & 33.64 \\ -19.6 & 20.35 & 6.4 \\ -100.6 & 72.82 & 62.71 \end{bmatrix}$ **43.** $\begin{bmatrix} 4 & 5 \\ -7 & -1 \end{bmatrix}$

45. not possible **47.** $\begin{bmatrix} 24 & 16 \\ 39 & 26 \end{bmatrix}$ **49.** $\begin{bmatrix} 47 \\ 81 \end{bmatrix}$ **51.** $QP = \begin{bmatrix} 584.50 \\ 709.25 \\ 1036.75 \end{bmatrix}$ Adult males spent \$584.50; adult females spent \$709.25;

children spent \$1036.75. **53.** $\begin{bmatrix} 1 & 1 & 0 \\ 0 & 1 & 1 \\ 1 & 0 & 0 \end{bmatrix}$ **55.** $A^2 = \begin{bmatrix} 5 & 1 & 2 & 2 \\ 1 & 5 & 2 & 2 \\ 2 & 2 & 6 & 0 \\ 2 & 2 & 0 & 4 \end{bmatrix}$ indicates the number of ways two cities can be linked with exactly one intermediate city to relay messages.   **57.** No; if $A = \begin{bmatrix} 1 & 1 \\ 1 & 1 \end{bmatrix}$ and $B = \begin{bmatrix} 1 & 0 \\ 0 & 0 \end{bmatrix}$, then $(AB)^2 \neq A^2B^2$.

59. Let $A = \begin{bmatrix} 1 & 2 \\ 1 & 2 \end{bmatrix}$ and $B = \begin{bmatrix} 2 & 2 \\ -1 & -1 \end{bmatrix}$. Neither is the zero matrix, yet $AB = 0$.   **61.** $6x^2 - 4x - 8$ **63.** $\frac{x+1}{x-1}$

65. $a = \frac{2s}{n} - l$

Exercise 7.4 (page 523)

1. $AB = BA = I$ **3.** $[I \,|\, A^{-1}]$ **5.** $\begin{bmatrix} 3 & 4 \\ 2 & 3 \end{bmatrix}$ **7.** $\begin{bmatrix} 5 & -7 \\ -2 & 3 \end{bmatrix}$ **9.** $\begin{bmatrix} -2 & 3 & -3 \\ -5 & 7 & -6 \\ 1 & -1 & 1 \end{bmatrix}$ **11.** $\begin{bmatrix} 4 & 1 & -3 \\ -5 & -1 & 4 \\ -1 & -1 & 1 \end{bmatrix}$

13. no inverse   **15.** $\begin{bmatrix} 1 & -2 & 1 \\ 0 & 1 & -2 \\ 0 & 0 & 1 \end{bmatrix}$ **17.** no inverse   **19.** $\begin{bmatrix} 1 & -2 & 1 & 0 \\ 0 & 1 & -2 & 1 \\ 0 & 0 & 1 & -2 \\ 0 & 0 & 0 & 1 \end{bmatrix}$ **21.** $\begin{bmatrix} 8 & -2 & -6 \\ -5 & 2 & 4 \\ 2 & 0 & -2 \end{bmatrix}$

23. $\begin{bmatrix} -2.5 & 5 & 3 & 5.5 \\ 5.5 & -8 & -6 & -9.5 \\ -1 & 3 & 1 & 3 \\ -5.5 & 9 & 6 & 10.5 \end{bmatrix}$ **25.** $x = 23, y = 17$ **27.** $x = 0, y = 0$ **29.** $x = 1, y = 2, z = 2$ **31.** $x = 54, y = -37,$

$z = -49$ **33.** $x = 2, y = 1$ **35.** $x = 1, y = 2, z = 1$ **37.** 2 of model A, 3 of model B   **39.** no   **43.** $X = \begin{bmatrix} 0 \\ 0 \\ 0 \end{bmatrix}$ **49.** all reals

except 2 and -2 **51.** all reals   **53.** $y \geq 0$ **55.** all reals

Exercise 7.5 (page 537)

1. $|A|$ or det A **3.** 0   **5.** 0   **7.** 8   **9.** 1   **11.** $\begin{vmatrix} -2 & 3 \\ 8 & 9 \end{vmatrix}$ **13.** $\begin{vmatrix} 1 & -2 \\ 4 & 5 \end{vmatrix}$ **15.** $-\begin{vmatrix} -2 & 3 \\ 8 & 9 \end{vmatrix}$ **17.** $\begin{vmatrix} 1 & -2 \\ 4 & 5 \end{vmatrix}$

19. -54 **21.** -7 **23.** 86   **25.** -2 **27.** 2   **29.** 12   **31.** true   **33.** false   **35.** 3   **37.** 3   **39.** $x = 1, y = 2$ **41.** $x = 3, y = 0$

43. $x = 1, y = 0, z = 1$ **45.** $x = 1, y = -1, z = 2$ **47.** $x = 6, y = 6, z = 12$ **49.** $p = \frac{5}{6}, q = \frac{2}{3}, r = \frac{1}{2}, s = \frac{5}{2}$ **51.** $3x - 2y = 0$

53. $6x + 7y = 9$ **55.** 30 sq. units   **57.** 73 sq. units   **63.** $x = 8$ **65.** $x = -1$ **67.** 10   **69.** 24   **73.** domain: $n \times n$ matrices;

range: reals   **75.** yes   **77.** 21.468   **79.** $(x - 1)(x + 4)$ **81.** $x(3x + 1)(3x - 1)$ **83.** $\frac{4x - 5}{(x - 2)(2x - 1)}$ **85.** $\frac{x^2 + 2x + 1}{x(x^2 + 1)}$

Exercise 7.6 (page 546)

1. first-degree and second-degree   **3.** $\frac{1}{x} + \frac{2}{x - 1}$ **5.** $\frac{5}{x} - \frac{3}{x - 3}$ **7.** $\frac{1}{x + 1} + \frac{2}{x - 1}$ **9.** $\frac{1}{x - 3} - \frac{3}{x + 2}$ **11.** $\frac{8}{x + 3} - \frac{5}{x - 1}$

13. $\frac{5}{2x - 3} + \frac{2}{x - 5}$ **15.** $\frac{2}{x} + \frac{3}{x - 1} - \frac{1}{x + 1}$ **17.** $\frac{1}{x} + \frac{1}{x^2 + 3}$ **19.** $\frac{3}{x + 1} + \frac{2}{x^2 + 2x + 3}$ **21.** $\frac{3}{x} + \frac{2}{x + 1} + \frac{1}{(x + 1)^2}$

23. $\frac{1}{x} + \frac{2}{x^2} - \frac{3}{x - 1}$ **25.** $\frac{2}{x} + \frac{1}{x - 3} + \frac{2}{(x - 3)^2}$ **27.** $\frac{1}{x - 1} - \frac{4}{(x - 1)^3}$ **29.** $\frac{1}{x} + \frac{1}{x^2} + \frac{2}{x^2 + x + 1}$ **31.** $\frac{3}{x} + \frac{4}{x^2} + \frac{x + 1}{x^2 + 1}$

33. $-\frac{1}{x + 1} - \frac{3}{x^2 + 2}$ **35.** $\frac{x + 1}{x^2 + 2} + \frac{2}{x^2 + x + 2}$ **37.** $\frac{1}{x} + \frac{x}{x^2 + 2x + 5} + \frac{x + 2}{(x^2 + 2x + 5)^2}$ **39.** $x - 3 - \frac{1}{x + 1} + \frac{8}{x + 2}$

41. $1 + \frac{2}{3x + 1} + \frac{1}{x^2 + 1}$ **43.** $1 + \frac{1}{x} + \frac{x}{x^2 + x + 1}$ **45.** $2 + \frac{1}{x} + \frac{3}{x - 1} + \frac{2}{x^2 + 1}$ **47.** No; it's the sum of two cubes.

49. $2|a|\sqrt{2ab}$ **51.** $3x^2\sqrt{2x}$ **53.** $x = 9$

Exercise 7.7 (page 554)

1. half-plane; boundary

3. is not

5.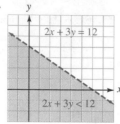

7.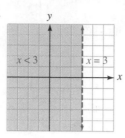

9.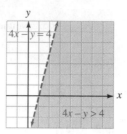

11.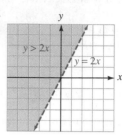

13.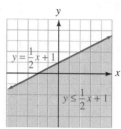

15.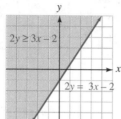

17.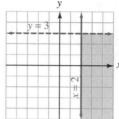

19.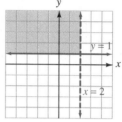

21.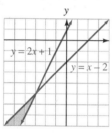

23.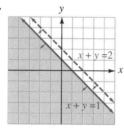

25.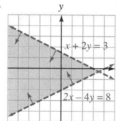

27.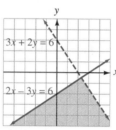

29.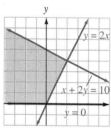

31.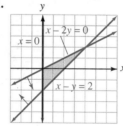

33.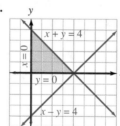

35.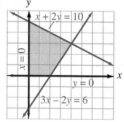

43. one; one   **45.** 0

Exercise 7.8 (page 563)

1. constraints   **3.** objective   **5.** $P = 12$ at $(0, 4)$ **7.** $P = \frac{13}{6}$ at $\left(\frac{5}{3}, \frac{4}{3}\right)$ **9.** $P = \frac{18}{7}$ at $\left(\frac{3}{7}, \frac{12}{7}\right)$ **11.** $P = 3$ at $(1, 0)$ **13.** $P = 0$ at $(0, 0)$

15. $P = 0$ at $(0, 0)$ **17.** $P = 6$ at $(0, 3)$ **19.** $P = -2$ at $(1, 2)$ and $(-1, 0)$ **21.** 3 tables and 12 chairs; $1260

23. 30 IBMs and 30 Macintoshes; $2700   **25.** in batches of 20 oz from A and 40 oz from B; minimum fat content per batch 14 oz

27. 12 cases of chewing gum and 36 cases of bubble gum; $8160   **29.** $150,000 in stocks and $50,000 in bonds; $17,000

31. 2 buses, 2 trucks; $1100   **35.** $\begin{bmatrix} 1 & 0 & 0 \\ 0 & 1 & 0 \\ 0 & 0 & 1 \\ 0 & 0 & 0 \end{bmatrix}$ **37.** $\left(\frac{1}{3} - 3y, y, -\frac{1}{3}\right)$

Chapter Summary (page 569)

1. a.

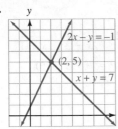

b.

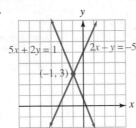

c.

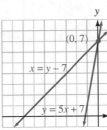

2. a. $x = 2$, $y = -1$ **b.** $x = 0$, $y = -3$ **c.** $x = 1$, $y = 1$
3. a. $x = -3$, $y = 2$ **b.** $x = -2$, $y = 5$ **c.** $x = 2$, $y = -1$
4. a. $x = 1$, $y = 0$, $z = 1$
b. $x = 1$, $y = 1$, $z = -1$ **c.** $x = 0$, $y = 1$, $z = 2$

5. $10,400   **6.** 900 adult tickets, 450 senior tickets, 450 children's tickets   **7. a.** $x = 1$, $y = 1$ **b.** $x = 3$, $y = 1$, $z = -2$

c. $x = -10$, $y = 1$, $z = 10$ **d.** no solution   **8.** $x = -4$, $y = 3$ **9. a.** $\begin{bmatrix} 1 & 3 & 4 \\ 4 & 0 & 2 \end{bmatrix}$ **b.** $\begin{bmatrix} 2 & 5 & 4 \\ -2 & -6 & 6 \\ -4 & 5 & -3 \end{bmatrix}$

c. $\begin{bmatrix} 4 & -1 \\ -7 & -7 \end{bmatrix}$ **d.** $\begin{bmatrix} -17 & 19 \\ 10 & -12 \end{bmatrix}$ **e.** $[5]$ **f.** $\begin{bmatrix} 2 & -1 & 1 & 3 \\ 4 & -2 & 2 & 6 \\ 2 & -1 & 1 & 3 \\ 10 & -5 & 5 & 15 \end{bmatrix}$ **g.** not possible   **h.** $[-24]$ **i.** $\begin{bmatrix} 0 \\ -6 \end{bmatrix}$

10. a. $\begin{bmatrix} 5 & -3 \\ -3 & 2 \end{bmatrix}$ **b.** $\begin{bmatrix} 1 & 0 & 0 \\ -\frac{3}{2} & \frac{1}{2} & \frac{1}{2} \\ 1 & -\frac{1}{2} & 0 \end{bmatrix}$ **c.** $\begin{bmatrix} 9 & 16 & -56 \\ -3 & -5 & 18 \\ -1 & -2 & 7 \end{bmatrix}$ **e.** No inverse exists.

11. a. $x = 1$, $y = 2$, $z = 1$ **b.** $w = 1$, $x = 1$, $y = 0$, $z = -1$ **12. a.** $\ldots$ **b.** $\ldots$ **c.** $\ldots$ **d.** -25 **13. a.** $x = 1$, $y = -2$

b. $x = 1$, $y = 0$, $z = -2$ **c.** $x = 1$, $y = -1$, $z = 3$ **d.** $w = 1$, $x = 0$, $y = -1$, $z = 2$ **14. a.** 21   **b.** 7   **15. a.** $\dfrac{3}{x} + \dfrac{4}{x + 1}$

b. $\dfrac{3}{x} + \dfrac{2}{x^2} + \dfrac{x - 1}{x^2 + 1}$ **16. a.**   **b.**   **17. a.** $P = 6$ at $(3, 0)$
b. $P = 12$ at $(0, -4)$

c. $\dfrac{1}{x} - \dfrac{1}{x^2 + x + 5}$ 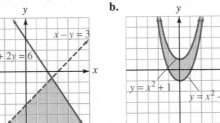   **c.** $P = 2$ at $(1, 1)$ **d.** $P = 3$ at $\left(-\frac{2}{3}, \frac{5}{3}\right)$ **18.** 1000 bags of x, 1400 bags of y

d. $\dfrac{1}{x + 1} - \dfrac{2}{(x + 1)^2} + \dfrac{2}{(x + 1)^3}$

Chapter 7 Test (page 575)

1.

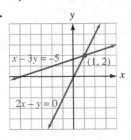

2.

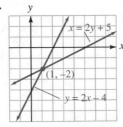

3. $x = 1$, $y = -3$ **4.** $x = 3$, $y = 5$ **5.** 6 liters of 20% solution, 4 liters of 45% solution   **6.** CD World 100 decks, Ace 25 decks, HiFi 50 decks   **7.** $x = 2$, $y = 1$ **8.** $x = 1$, $y = 2$, $z = 1$ **9.** $x = 1$, $y = 0$, $z = -2$ **10.** $x = -\frac{2}{5}y + \frac{7}{5}$, $z = -\frac{8}{5}y - \frac{7}{5}$, $y =$ any number

11. $\begin{bmatrix} 16 & -14 & 20 \\ 0 & -6 & -13 \end{bmatrix}$ **12.** $[-1]$ **13.** $\begin{bmatrix} -\frac{7}{3} & \frac{19}{3} \\ \frac{2}{3} & -\frac{5}{3} \end{bmatrix}$ **14.** $\begin{bmatrix} -13 & -3 & 14 \\ 4 & 1 & -4 \\ 12 & 3 & -13 \end{bmatrix}$ **15.** $x = \frac{17}{3}, y = -\frac{4}{3}$

16. $x = -36, y = 11, z = 34$ **17.** -12 **18.** -24 **19.** $x = -\frac{13}{12}, y = -\frac{5}{4}$ **20.** $x = -\frac{1}{2}, y = 1, z = \frac{3}{2}$ **21.** $\dfrac{3}{2x - 3} + \dfrac{1}{x + 1}$

22. $\dfrac{1}{x} + \dfrac{2x + 1}{x^2 + 2}$ **23.**   **24.**   **25.** $P = 7$ at $(1, 2)$ **26.** $P = -8$ at $(8, 0)$

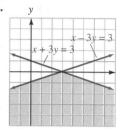

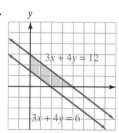

Exercise 8.1   (page 588)

1. $(2, -5); 3$ **3.** $(0, 0); \sqrt{5}$ **5.** to the left   **7.** downward   **9.** directrix; focus   **11.** $x^2 + y^2 = 49$ **13.** $(x - 2)^2 + (y + 2)^2 = 17$
15. $(x - 1)^2 + (y + 2)^2 = 36$ **17.**   **19.**   **21.** yes   **23.** 60 mi
25. $(x - 4)^2 + y^2 = 16$ **27.** $x^2 = 12y$
29. $y^2 = -12x$
31. $(x - 3)^2 = -12(y - 5)$
33. $(x - 3)^2 = -28(y - 5)$
35. $(y - 2)^2 = -2(x - 2)$ or
$(x - 2)^2 = -2(y - 2)$

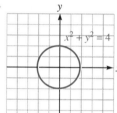

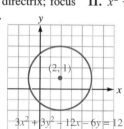

37. $(x + 4)^2 = -\frac{16}{3}(y - 6)$ or $(y - 6)^2 = \frac{9}{4}(x + 4)$ **39.** $(y - 8)^2 = -4(x - 6)$ **41.** $(x - 3)^2 = \frac{1}{2}(y - 1)$
43.   **45.**   **47.**   **49.**

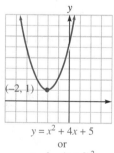

$y = x^2 + 4x + 5$
or
$y - 1 = (x + 2)^2$

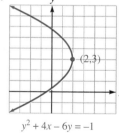

$y^2 + 4x - 6y = -1$
or
$(y - 3)^2 = -4(x - 2)$

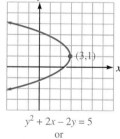

$y^2 + 2x - 2y = 5$
or
$(y - 1)^2 = -2(x - 3)$

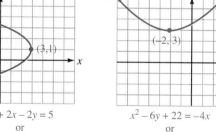

$x^2 - 6y + 22 = -4x$
or
$(x + 2)^2 = 6(y - 3)$

51.

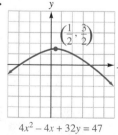

$4x^2 - 4x + 32y = 47$
or
$\left(x - \frac{1}{2}\right)^2 = -8\left(y - \frac{3}{2}\right)$

53. 2 ft   **55.** $x^2 = \frac{-45}{2}y$ **57.** 8   **59.** about 12.6 cm   **61.** about 520 ft
63. $0x^2 + 0xy + y^2 - 8x - 4y + 12 = 0$ **65.** $x^2 + (y - 3)^2 = 25$ **67.** $y = x^2 + 4x + 3$
71. 4   **73.** $\frac{49}{4}$ **75.** $x = 1, -5$ **77.** $x = -2, 9$

Exercise 8.2 (page 602)

1. sum; constant   **3.** vertices   **5.** $(a, 0)$; $(-a, 0)$ **7.** $\dfrac{x^2}{25} + \dfrac{y^2}{16} = 1$ **9.** $\dfrac{9x^2}{16} + \dfrac{9y^2}{25} = 1$ **11.** $\dfrac{x^2}{7} + \dfrac{y^2}{16} = 1$

13. $\dfrac{(x-3)^2}{4} + \dfrac{(y-4)^2}{9} = 1$ **15.** $\dfrac{(x-3)^2}{9} + \dfrac{(y-4)^2}{4} = 1$ **17.** $\dfrac{(x-3)^2}{41} + \dfrac{(y-4)^2}{16} = 1$ **19.** $\dfrac{x^2}{36} + \dfrac{(y-4)^2}{20} = 1$

21. $\dfrac{x^2}{100} + \dfrac{y^2}{64} = 1$ **23.** 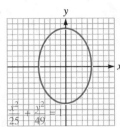   **25.** 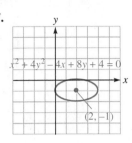   **27.**

29. 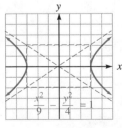   **31.** 199,395 mi   **33.** $\dfrac{x^2}{2500} + \dfrac{y^2}{900} = 1$; 36 m   **35.** about 20.8 in.   **41.** $\dfrac{x^2}{9} + \dfrac{y^2}{8} = 1$

45. $\begin{bmatrix} -1 & 8 \\ 5 & -1 \\ 4 & 7 \end{bmatrix}$ **47.** $\begin{bmatrix} 5 & 8 \\ 12 & -2 \\ -1 & 5 \end{bmatrix}$ **49.** $\begin{bmatrix} -1 & -2 \\ -2 & 0 \\ 1 & -1 \end{bmatrix}$

Exercise 8.3 (page 612)

1. difference; constant   **3.** $(a, 0)$; $(-a, 0)$ **5.** transverse axis   **7.** $\dfrac{x^2}{25} - \dfrac{y^2}{24} = 1$ **9.** $\dfrac{(x-2)^2}{4} - \dfrac{(y-4)^2}{9} = 1$

11. $\dfrac{(y-3)^2}{9} - \dfrac{(x-5)^2}{9} = 1$ **13.** $\dfrac{y^2}{9} - \dfrac{x^2}{16} = 1$ **15.** $\dfrac{(x-1)^2}{4} - \dfrac{(y+3)^2}{16} = 1$ or $\dfrac{(y+3)^2}{4} - \dfrac{(x-1)^2}{16} = 1$ **17.** $\dfrac{x^2}{10} - \dfrac{3y^2}{20} = 1$

19. 24 sq. units   **21.** 12 sq. units   **23.** $\dfrac{(x+2)^2}{4} - \dfrac{4(y+4)^2}{81} = 1$ or $\dfrac{(y+4)^2}{4} - \dfrac{4(x+2)^2}{81} = 1$ **25.** $\dfrac{x^2}{36} - \dfrac{16y^2}{25} = 1$

27. 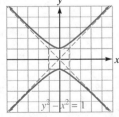   **29.** 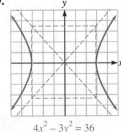   **31.**   **33.**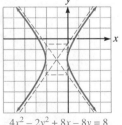

35.
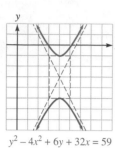
$y^2 - 4x^2 + 6y + 32x = 59$

37.
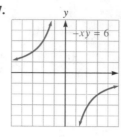
$-xy = 6$

39. $\dfrac{(x-3)^2}{9} - \dfrac{(y-1)^2}{16} = 1$ **41.** $4x^2 - 5y^2 - 60y = 0$

43. $\dfrac{x^2}{100{,}000{,}000^2} - \dfrac{y^2}{200{,}000{,}000^2} = 1$ **45.** hyperbola; $\dfrac{x^2}{144} - \dfrac{y^2}{25} = 1$

51. $f^{-1}(x) = \frac{x+2}{3}$ **53.** $f^{-1}(x) = \frac{2x}{5-x}$ **55.** $f(g(x)) = (x+1)^4 + 1$

57. $f(f(x)) = (x^2+1)^2 + 1$

Exercise 8.4 (page 621)

1. conic **3.**
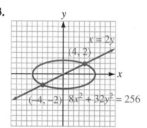
$x = 2y$
$(4, 2)$
$(-4, -2)$ $8x^2 + 32y^2 = 256$

5.
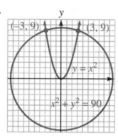
$(-3, 9)$ $(3, 9)$
$y = x^2$
$x^2 + y^2 = 90$

7.
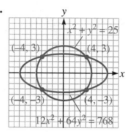
$x^2 + y^2 = 25$
$(-4, 3)$ $(4, 3)$
$(-4, -3)$ $(4, -3)$
$12x^2 + 64y^2 = 768$

9.
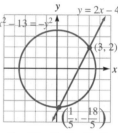
$y = 2x - 4$
$x^2 - 13 = -y^2$
$(3, 2)$
$\left(\frac{1}{5}, -\frac{18}{5}\right)$

11.
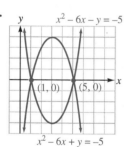
$x^2 - 6x - y = -5$
$(1, 0)$ $(5, 0)$
$x^2 - 6x + y = -5$

13. $(1, 2), (-1, 0)$ **15.** $(1, 0.67), (-1, -0.67)$ **17.** $(3, 0), (0, 5)$ **19.** $(1, 1)$ **21.** $(1, 2), (2, 1)$

23. $(-2, 3), (2, 3)$ **25.** $\left(\sqrt{5}, 5\right)), \left(-\sqrt{5}, 5\right)$ **27.** $(3, 2), (3, -2), (-3, 2), (-3, -2)$

29. $(2, 4), (2, -4), (-2, 4), (-2, -4)$ **31.** $\left(-\sqrt{15}, 5\right), \left(\sqrt{15}, 5\right), (-2, -6), (2, -6)$

33. $(0, -4), (-3, 5), (3, 5)$ **35.** $(-2, 3), (2, 3), (-2, -3), (2, -3)$ **37.** $(3, 3)$

39. $(6, 2), (-6, -2), \left(\sqrt{42}, 0\right), \left(-\sqrt{42}, 0\right)$ **41.** $\left(\frac{1}{2}, \frac{1}{3}\right), \left(\frac{1}{3}, \frac{1}{2}\right)$ **43.** 7 by 9 cm

45. either \$750 at 9% or \$900 at 7.5% **47.** about 23 mi **55.** vertical: $x = 1$; horizontal: $y = 3$

57. vertical: $x = 1, x = -1$; horizontal: $y = 0$ **59.** y-axis **61.** origin

Chapter Summary (page 626)

1. a. $x^2 + y^2 = 16$ **b.** $x^2 + y^2 = 100$ **c.** $(x-3)^2 + (y+2)^2 = 25$ **d.** $(x+2)^2 + (y-4)^2 = 25$ **e.** $(x-5)^2 + (y-10)^2 = 85$

f. $(x-2)^2 + (y-2)^2 = 89$ **2. a.** $(x-3)^2 + (y+2)^2 = 16$ **b.** $(x+2)^2 + (y-5)^2 = 16$ **3. a.** $y^2 = -2x$ **b.** $x^2 = 16y$

4. $(x+2)^2 = -\frac{4}{11}(y-3)$ **5. a.**
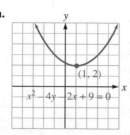
$(1, 2)$
$x^2 - 4y - 2x + 9 = 0$

b.
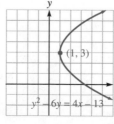
$(1, 3)$
$y^2 - 6y = 4x - 13$

6. $\dfrac{x^2}{36} + \dfrac{y^2}{16} = 1$ **7.** $\dfrac{y^2}{25} + \dfrac{x^2}{4} = 1$

8. $\dfrac{(x+2)^2}{16} + \dfrac{(y-3)^2}{9} = 1$

9. $\dfrac{(x-2)^2}{1} + \dfrac{(y+1)^2}{4} = 1$

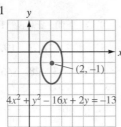

$4x^2 + y^2 - 16x + 2y = -13$

(2, -1)

10. $\dfrac{x^2}{4} - \dfrac{y^2}{12} = 1$

11. $\dfrac{y^2}{9} - \dfrac{x^2}{16} = 1$

12. $\dfrac{x^2}{9} - \dfrac{(y-3)^2}{16} = 1$

13. $\dfrac{y^2}{9} - \dfrac{(x-3)^2}{16} = 1$

14. $y = \pm\dfrac{4}{5}x$

15. $\dfrac{(x-1)^2}{4} - \dfrac{(y+2)^2}{9} = 1$

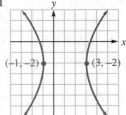

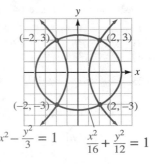

(-1, -2)   (3, -2)

$9x^2 - 4y^2 - 16y - 18x = 43$

16.

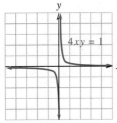

$4xy = 1$

17.

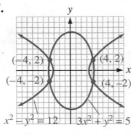

(-4, 2)   (4, 2)
(-4, -2)   (4, -2)

$x^2 - y^2 = 12$ $3x^2 + y^2 = 52$

18.

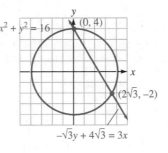

$x^2 + y^2 = 16$ (0, 4)
$(2\sqrt{3}, -2)$
$-\sqrt{3}y + 4\sqrt{3} = 3x$

19.

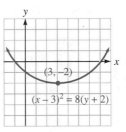

(-2, 3)   (2, 3)
(-2, -3)   (2, -3)

$x^2 - \dfrac{y^2}{3} = 1$ $\dfrac{x^2}{16} + \dfrac{y^2}{12} = 1$

20. $(-4, 2), (-4, -2), (4, 2), (4, -2)$ **21.** $(0, 4), (2\sqrt{3}, -2)$ **22.** $(-2, 3), (-2, -3), (2, 3), (2, -3)$

Chapter 8 Test   (page 632)

1. $(x-2)^2 + (y-3)^2 = 9$
2. $(x-2)^2 + (y-3)^2 = 41$
3. $(x-2)^2 + (y+5)^2 = 169$

4.

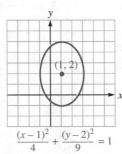

(2, -3)

$(x-2)^2 + (y+3)^2 = 9$

5. $(x-3)^2 = 16(y-2)$
6. $(y+6)^2 = -4(x-4)$
7. $(x-2)^2 = \frac{4}{3}(y+3)$ or $(y+3)^2 = -\frac{9}{2}(x-2)$

8.

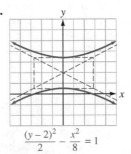

(3, -2)

$(x-3)^2 = 8(y+2)$

9. $\dfrac{x^2}{100} + \dfrac{y^2}{64} = 1$

10. $\dfrac{x^2}{169} + \dfrac{y^2}{144} = 1$

11. $\dfrac{x^2}{4} + \dfrac{y^2}{36} = 1$

12.

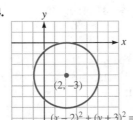

(1, 2)

$\dfrac{(x-1)^2}{4} + \dfrac{(y-2)^2}{9} = 1$

13. $\dfrac{x^2}{25} - \dfrac{y^2}{144} = 1$

14. $\dfrac{x^2}{36} - \dfrac{4y^2}{25} = 1$

15. $\dfrac{(x-2)^2}{64} - \dfrac{(y+1)^2}{36} = 1$

16.

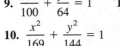

$\dfrac{(y-2)^2}{2} - \dfrac{x^2}{8} = 1$

17. $(\sqrt{7}, 4), (-\sqrt{7}, 4)$ **18.** $(3\sqrt{2}, 3), (-3\sqrt{2}, 3), (3\sqrt{2}, -3), (-3\sqrt{2}, -3)$ **19.** $(y-2)^2 = 6(x+3)$; parabola
20. $\dfrac{(x-1)^2}{3} + \dfrac{(y+2)^2}{2} = 1$; ellipse

Cumulative Review Exercises (page 633)

1. 16 **2.** $\frac{1}{2}$ **3.** y^2 **4.** $x^{17/12}$ **5.** $x^{4/3} - x^{2/3}$ **6.** $\frac{1}{x} + 2 + x$ **7.** $-3x$ **8.** $4t\sqrt{3t}$ **9.** $4x$ **10.** $x + 3$ **11.** $7\sqrt{2}$
12. $-12\sqrt[4]{2} + 10\sqrt[4]{3}$ **13.** $-18\sqrt{6}$ **14.** $\dfrac{5\sqrt[3]{x^2}}{x}$ **15.** $\dfrac{x + 3\sqrt{x} + 2}{x - 1}$ **16.** $\sqrt{xy}$ **17.** 2, 7 **18.** $\frac{1}{4}$ **19.** 1, $\frac{3}{2}$ **20.** $\dfrac{-2 \pm \sqrt{7}}{3}$
21. $7 + 2i$ **22.** $-5 - 7i$ **23.** $13 + 0i$ **24.** $12 - 6i$ **25.** $-12 - 10i$ **26.** $\frac{3}{2} + \frac{1}{2}i$ **27.** $\sqrt{13}$ **28.** $\sqrt{61}$ **29.** -2 **30.** $\left(1, \frac{1}{2}\right)$
31. $(-\infty, -2) \cup (3, \infty)$ **32.** $[-2, 3]$ **33.** 5 **34.** 27 **35.** $12x^2 - 12x + 5$ **36.** $6x^2 + 3$ **37.** $2^y = x$ **38.** $\log_3 a = b$ **39.** 5
40. 3 **41.** $\frac{1}{27}$ **42.** 1 **43.** $y = 2^x$ **44.** x **45.** 1.9912 **46.** 0.301 **47.** 1.6902 **48.** 0.1461 **49.** $\dfrac{2\log 2}{\log 3 - \log 2}$ **50.** 16
51. \$2848.31 **52.** 1.16056 **53.** (2, 1) **54.** (1, 1) **55.** $(-2, -2)$ **56.** (3, 1) **57.** -1 **58.** -1 **59.** $(-1, -1, 3)$ **60.** (1, 1, 1)

Exercise 9.1 (page 642)

1. power **3.** first **5.** $7 \cdot 6 \cdot 5 \cdot 4 \cdot 3 \cdot 2 \cdot 1$ **7.** $(n - 1)!$ **9.** 24 **11.** 4320 **13.** 1440 **15.** $\frac{1}{1320}$ **17.** $\frac{5}{3}$ **19.** 18,564
21. $a^3 + 3a^2b + 3ab^2 + b^3$ **23.** $a^5 - 5a^4b + 10a^3b^2 - 10a^2b^3 + 5ab^4 - b^5$ **25.** $8x^3 + 12x^2y + 6xy^2 + y^3$
27. $x^3 - 6x^2y + 12xy^2 - 8y^3$ **29.** $16x^4 + 96x^3y + 216x^2y^2 + 216xy^3 + 81y^4$ **31.** $x^4 - 8x^3y + 24x^2y^2 - 32xy^3 + 16y^4$
33. $x^5 - 15x^4y + 90x^3y^2 - 270x^2y^3 + 405xy^4 - 243y^5$ **35.** $\dfrac{x^4}{16} + \dfrac{x^3y}{2} + \dfrac{3x^2y^2}{2} + 2xy^3 + y^4$ **37.** $6a^2b^2$ **39.** $35a^3b^4$ **41.** $-b^5$
43. $2380a^{13}b^4$ **45.** $-4\sqrt{2}a^3$ **47.** $1134a^5b^4$ **49.** $\dfrac{3x^2y^2}{2}$ **51.** $\dfrac{-55r^2s^9}{2048}$ **53.** $\dfrac{n!}{3!(n-3)!}a^{n-3}b^3$
55. $\dfrac{n!}{(r-1)!(n-r+1)!}a^{n-r+1}b^{r-1}$ **59.** -252 **67.** $3xyz^2(x^2yz^2 - 2z^3 + 5x)$ **69.** $(a^2 + b^2)(a + b)(a - b)$ **71.** $\frac{3+x}{3-x}$

Exercise 9.2 (page 650)

1. domain **3.** summation notation **5.** 6 **7.** $5c$ **9.** 0, 10, 30, 60, 100, 150 **11.** 21 **13.** $a + 4d$ **15.** 15 **17.** 15 **19.** 15
21. $\frac{242}{243}$ **23.** 35 **25.** 3, 7, 15, 31 **27.** $-4, -2, -1, -\frac{1}{2}$ **29.** k, k^2, k^4, k^8 **31.** $8, \dfrac{16}{k}, \dfrac{32}{k^2}, \dfrac{64}{k^3}$ **33.** an alternating series
35. not an alternating series **37.** 30 **39.** -50 **41.** 40 **43.** 500 **45.** $\frac{7}{12}$ **47.** 160 **49.** 3725 **55.** 6 cm **57.** 26 ft

Exercise 9.3 (page 657)

1. $(n - 1)$ **3.** $l = a + (n - 1)d$ **5.** arithmetic means **7.** 1, 3, 5, 7, 9, 11 **9.** 5, $\frac{7}{2}$, 2, $\frac{1}{2}$, -1, $-\frac{5}{2}$ **11.** 9, $\frac{23}{2}$, 14, $\frac{33}{2}$, 19, $\frac{43}{2}$
13. 318 **15.** 6 **17.** 370 **19.** 44 **21.** $\frac{25}{2}$, 15, $\frac{35}{2}$ **23.** $-\frac{82}{15}$, $-\frac{59}{15}$, $-\frac{36}{15}$, $-\frac{13}{15}$ **25.** 285 **27.** 555 **29.** $157\frac{1}{2}$ **31.** 20,100
33. \$2600 **35.** $\frac{1}{8}$ mi **37.** 304 ft **39.** 210 **45.** 6 **47.** -1 **49.** 1, -1, i, $-i$

Exercise 9.4 (page 665)

1. r^{n-1} **3.** $l = ar^{n-1}$ **5.** geometric means **7.** 10, 20, 40, 80 **9.** $-2, -6, -18, -54$ **11.** 3, $3\sqrt{2}$, 6, $6\sqrt{2}$ **13.** 2, 6, 18, 54
15. 256 **17.** -162 **19.** $10\sqrt[4]{2}$, $10\sqrt{2}$, $10\sqrt[4]{8}$ **21.** 8, 32, 128, 512 **23.** 124 **25.** $-29,524$ **27.** $\frac{1995}{32}$ **29.** 18 **31.** 8 **33.** $\frac{5}{9}$
35. $\frac{25}{99}$ **37.** 23 **39.** \$69.82 **41.** about $\frac{1}{3}c$ **43.** about 8.6×10^{19} **45.** \$180,176.87 **47.** \$2001.60 **49.** \$2013.62
51. \$264,094.58 **53.** 1.8447×10^{19} grains **55.** no **57.** $5 - 3i$ **59.** 10 **61.** $\frac{1+8i}{5}$ **63.** $-i$

Exercise 9.5 (page 672)

1. two **3.** $n = k + 1$ **29.** no **37.**

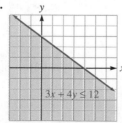

39.

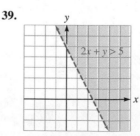

Exercise 9.6 (page 681)

1. 24   **3.** $\frac{n!}{(n-r)!}$ **5.** 1   **7.** $\binom{n}{r}$ **9.** 1   **11.** $\frac{n!}{a! \cdot b! \dots}$ **13.** 840   **15.** 35   **17.** 120   **19.** 5   **21.** 1   **23.** 1200   **25.** 40
27. 2278   **29.** 144   **31.** 8,000,000   **33.** 240   **35.** 6   **37.** 40,320   **39.** 14,400   **41.** 24,360   **43.** 5040   **45.** 48   **47.** 96
49. 210   **51.** 24   **53.** 5040   **55.** 2,721,600   **57.** 1120   **59.** 59,400   **61.** 272   **63.** 28   **65.** 252   **67.** 142,506   **69.** 66   **77.** 2
79. 4   **81.** true   **83.** true

Exercise 9.7 (page 689)

1. experiment   **3.** $\frac{n(E)}{n(S)}$ **5.** {(1, H), (2, H), (3, H), (4, H), (5, H), (6, H), (1, T), (2, T), (3, T), (4, T), (5, T), (6, T)}
7. {a, b, c, d, e, f, g h, i, j, k, l, m, n, o, p, q, r, s, t, u, v, w, x, y, z}   **9.** $\frac{1}{6}$ **11.** $\frac{2}{3}$ **13.** $\frac{19}{42}$ **15.** $\frac{13}{42}$ **17.** $\frac{3}{8}$ **19.** 0   **21.** $\frac{1}{12}$ **23.** $\frac{1}{169}$
25. $\frac{5}{12}$ **27.** about 6.3×10^{-12} **29.** 0   **31.** $\frac{3}{13}$ **33.** $\frac{1}{6}$ **35.** $\frac{1}{8}$ **37.** $\frac{5}{16}$ **39.** (S = survive, F = fail)
{SSSS, SSSF, SSFS, SFSS, FSSS, SSFF, SFSF, FSSF, SFFS, FSFS, FFSS, SFFF, FSFF, FFSF, FFFS, FFFF}   **41.** $\frac{1}{4}$ **43.** $\frac{1}{4}$ **45.** 1
47. $\frac{32}{119}$ **49.** $\frac{1}{3}$ **51.** 0.18   **53.** 0.14   **55.** about 33%   **57.** no   **59.** $-10, 4$ **61.**

Exercise 9.8 (page 695)

1. compound   **3.** mutually exclusive   **5.** $1 - P(A)$ **7.** $P(A) \cdot P(B)$ **9.** $\frac{1}{2}$ **11.** $\frac{7}{13}$ **13.** $\frac{1}{221}$ **15.** $\frac{25}{204}$ **17.** $\frac{11}{36}$ **19.** $\frac{7}{12}$ **21.** $\frac{5}{8}$
23. $\frac{13}{16}$ **25.** $\frac{21}{128}$ **27.** $\frac{3}{20}$ **29.** $\frac{1}{49}$ **31.** 0.973   **33.** $\frac{11}{20}$ **35.** $\frac{1}{8}$ **37.** $\frac{3}{8}$ **39.** $\frac{17}{36}$ **41.** 0.03   **45.** $P = 10$ at (4, 2)

Exercise 9.9 (page 700)

1. odds for an event   **3.** $\frac{1}{5}$ **5.** $\frac{1}{6}$ **7.** 5 to 1   **9.** 1 to 1   **11.** $\frac{5}{36}$ **13.** 31 to 5   **15.** 1 to 1   **17.** 1 to 12   **19.** 3 to 10   **21.** $\frac{5}{7}$
23. 1 to 90   **25.** 15 to 1   **27.** $\frac{1}{9}$ **29.** No; the expected winnings are $\$\frac{9}{11}$ **31.** 7 to 1   **33.** \$1.72   **35.** 6.54   **37.** $\frac{32}{390,625}$
39. less than $\frac{1}{2}$ **41.** $2x + 3y = 5$ **43.** $(x - 3)^2 + (y - 5)^2 = 25$

Chapter Summary (page 704)

1. a. 720   **b.** 30,240   **c.** 8   **d.** $\frac{280}{3}$ **2. a.** $x^3 + 3x^2y + 3xy^2 + y^3$ **b.** $p^4 + 4p^3q + 6p^2q^2 + 4pq^3 + q^4$
c. $a^5 - 5a^4b + 10a^3b^2 - 10a^2b^3 + 5ab^4 - b^5$ **d.** $8a^3 - 12a^2b + 6ab^2 - b^3$ **3. a.** $56a^5b^3$ **b.** $80x^3y^2$ **c.** $84x^3y^6$ **d.** $439,040x^3$
4. a. 63   **b.** 9   **5. a.** 5, 17, 53, 161   **b.** $-2, 8, 128, 32,768$ **6. a.** 90   **b.** 60   **c.** 1718   **d.** -360 **7. a.** 117   **b.** 281
c. -92 **d.** $-\frac{135}{2}$ **8.** $\frac{7}{2}, 5, \frac{13}{2}$ **9.** 25, 40, 55, 70, 85   **10. a.** 3320   **b.** 5780   **c.** -5220 **d.** -1540 **11. a.** $\frac{1}{729}$ **b.** 13,122
c. $\frac{9}{16,384}$ **d.** $\frac{8}{15,625}$ **12.** $2\sqrt{2}, 4, 4\sqrt{2}$ **13.** 4, -8, 16, -32 **14.** 16   **15. a.** $\frac{3280}{27}$ **b.** 6560   **c.** $\frac{2295}{128}$ **d.** $\frac{520,832}{78,125}$ **16.** $\frac{3280}{3}$
17. $16\sqrt{2}$ **18. a.** $\frac{2}{3}$ **b.** $\frac{3}{25}$ **c.** no sum   **d.** 1   **19. a.** $\frac{1}{3}$ **b.** 1   **c.** $\frac{17}{99}$ **d.** $\frac{5}{11}$ **20.** \$4775.81   **21.** 6516, 3134   **22.** \$3486.78
25. a. 6720   **b.** 35   **c.** 1   **d.** 4050   **e.** 564,480   **f.** 840   **g.** 21   **h.** 66   **i.** 120   **j.** $\frac{56}{1287}$ **k.** $\frac{1}{6}$ **l.** $\frac{33}{66,640}$ **26.** 20,160
27. 90,720   **29.** 24   **30.** $\frac{1}{108,290}$ **31.** $\frac{108,289}{108,290}$ **32.** about 6.3×10^{-12} **33.** $\frac{60}{143}$ **34.** $\frac{1}{2}$ **35.** $\frac{7}{13}$ **36.** $\frac{1}{2,598,960}$
37. about $\frac{4}{6.350 \times 10^{11}}$ **38.** $\frac{15}{16}$ **39.** 664   **40.** 1 to 7   **41.** 1 to 15   **42.** 1 to 23   **43.** $\frac{11}{21}$ **44.** \$2.00

Chapter 9 Test (page 709)

1. 144   **2.** 384   **3.** $10x^4y$ **4.** $112a^2b^6$ **5.** 27   **6.** -36 **7.** 155   **8.** -130 **9.** 9, 14, 19   **10.** $-6, -18$ **11.** 255.75
12. about 9   **13.** about \$0.42c   **14.** about \$1.46c   **16.** 800,000   **17.** 42   **18.** 24   **19.** 28   **20.** 1   **21.** 576   **22.** 120   **23.** 60
24. {(H, H, H), (H, H, T), (H, T, H), (H, T, T), (T, H, H), (T, H, T), (T, T, H), (T, T, T)}   **25.** $\frac{1}{6}$ **26.** $\frac{2}{13}$ **27.** $\frac{429}{866,320}$ **28.** $\frac{5}{16}$
29. 40%   **30.** 0.9   **31.** $\frac{8}{13}$ **32.** $\frac{5}{9}$ **33.** $\frac{3}{4}$ **34.** \$1

Exercise 10.1 (page 720)

1. interest   **3.** annual   **5.** future   **7.** interest   **9.** periodic rate   **11.** principal; periodic rate; frequency of compounding;
number of years   **13.** effective   **15.** \$1296   **17.** \$1763.19   **19.** \$1612.70   **21.** \$2840.84   **23.** \$2912.71   **25.** \$2944.91
27. 6.14%   **29.** 9.73%   **31.** \$14,027.60   **33.** \$11,678.47   **35.** \$20,448.10   **37.** \$356,867.13   **39.** \$207,976
41. 8.62 million ft^3 **43.** 7.4%; 7.12%   **45.** \$1267.45   **47.** \$3853.73   **49.** \$50,363.13   **53.** $\frac{x+3}{2x+1}$ **55.** 1

Exercise 10.2 (page 726)

1. annuities   **3.** payments; interest   **5.** principal; periodic rate; frequency of compounding; number of years   **7.** $1318.08
9. $318.36   **11.** $1200.61   **13.** $1556.03   **15.** $5608.49   **17.** $34,603.82   **19.** $1665.81   **21.** $637.52   **23.** $2467.11
25. $86,803,923.58   **27.** $22,177.71   **29.** Bank B   **31.** $114,432.12   **33.** $1466.08   **35.** 12   **37.** 3

Exercise 10.3 (page 733)

1. present value   **3.** promissory note   **5.** amortizing   **7.** $48,116.14   **9.** $300.04   **11.** $27,128.43   **13.** $99,650.84
15. $533.84   **17.** 15-yr: $1560.22; 20-yr: $1341.84   **19.** $936.88   **21.** $146,506.50   **23.** $2\sqrt{6}$ **25.** $13\sqrt{5x}$

Chapter Summary (page 737)

1. $3077.25   **2.** $2405.47   **3.** $6670.80   **4.** BigBank, 6.45%; BestBank, 6.4%; BigBank   **5.** $5622.23   **6.** $8856.49
7. $88,353.06   **8.** $369.89   **9.** $5552.11   **10.** $437,563.50   **11.** $1687.03; $303,665.40   **12.** $1404.89; $505,760.40

Chapter 10 Test (page 740)

1. compound   **2.** periodic   **3.** effective   **4.** annual   **5.** annuities   **6.** sinking fund   **7.** present value   **8.** amortizing
9. $2117.56   **10.** $2141.11   **11.** 5.116%   **12.** $2498.00   **13.** $50,506.10   **14.** $113.20   **15.** $112,652.71   **16.** $910.16

Cumulative Review Exercises (page 740)

1.
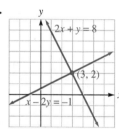

2.
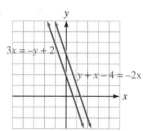
inconsistent system

3. $x = 3, y = 1$ **4.** $x = 3, y = 2, z = 1$ **5.** $x = 1, y = 0, z = 2$

6. $x = 1, y = 2, z = 1, t = 1$ **7.** $\begin{bmatrix} 1 & 3 \\ 3 & 7 \end{bmatrix}$ **8.** $\begin{bmatrix} -3 & 1 \\ 1 & -1 \end{bmatrix}$ **9.** $\begin{bmatrix} 3 & 2 & 0 \\ -2 & 8 & 7 \end{bmatrix}$ **10.** $\begin{bmatrix} 9 & 6 \\ 6 & 21 \end{bmatrix}$

11. $\begin{bmatrix} -1 & \frac{3}{2} \\ \frac{1}{2} & -\frac{1}{2} \end{bmatrix}$ **12.** $\begin{bmatrix} 4 & 5 & -4 \\ -1 & -1 & 1 \\ -4 & -6 & 5 \end{bmatrix}$ **13.** -41 **14.** -1 **15.** $\dfrac{\begin{vmatrix} 11 & 3 \\ 24 & 5 \end{vmatrix}}{\begin{vmatrix} 4 & 3 \\ -2 & 5 \end{vmatrix}}$ **16.** $\dfrac{\begin{vmatrix} 4 & 11 \\ -2 & 24 \end{vmatrix}}{\begin{vmatrix} 4 & 3 \\ -2 & 5 \end{vmatrix}}$

17. $\dfrac{2}{x+1} - \dfrac{3}{x+2}$ **18.** $\dfrac{\frac{1}{2}}{2x-5} - \dfrac{\frac{3}{2}}{(2x-5)^2}$ **19.**
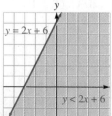

20.
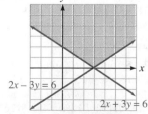

21. $x^2 + y^2 = 16$

22. $(x - 2)^2 + (y + 3)^2 = 121$ **23.**

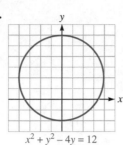

$x^2 + y^2 - 4y = 12$

24.

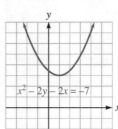

$x^2 - 2y - 2x = -7$

25.

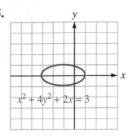

$x^2 + 4y^2 + 2x = 3$

26.

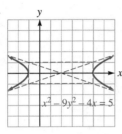

$x^2 - 9y^2 - 4x = 5$

27. $\dfrac{x^2}{36} + \dfrac{y^2}{16} = 1$ **28.** $\dfrac{(x - 2)^2}{21} + \dfrac{(y - 3)^2}{25} = 1$ **29.** $\dfrac{x^2}{4} - \dfrac{y^2}{5} = 1$

30. $\dfrac{(y - 4)^2}{9} - \dfrac{(x - 2)^2}{9} = 1$ **31.** $16x^7y$ **32.** $1792x^3y^5$ **33.** 10   **34.** 65   **35.** 33   **36.** $\frac{364}{9}$

37. 1680   **38.** 1   **39.** 66   **40.** 24   **41.** 17,280   **42.** 495   **43.** $\frac{1}{18}$ **44.** $\frac{506}{19,992} = \frac{253}{9996}$ **45.** 0.48

46. 0.3   **48.** about $8.46   **49.** $3757.02   **50.** $972.08

■ ■ ■ ■ ■ ■ ■ ■ ■ INDEX

5.1 EXPONENTIAL FUNCTIONS

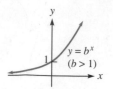

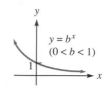

Formula for radioactive decay: $A = A_0 2^{-t/h}$

Formula for compound interest: $A = P\left(1 + \dfrac{r}{k}\right)^{kt}$

5.2 BASE-e EXPONENTIAL FUNCTIONS

$e = 2.718281828 \ldots$

Formula for continuous compound interest: $A = Pe^{rt}$

Formula for population growth: $P = P_0 e^{kt}$

5.3 LOGARITHMIC FUNCTIONS

$y = \log_b x$ is equivalent to $x = b^y$.

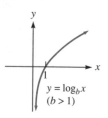

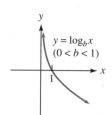

5.5 PROPERTIES OF LOGARITHMS

$\log_b 1 = 0 \quad \log_b b = 1$

$\log_b b^x = x \quad b^{\log_b x} = x$

$\log_b MN = \log_b M + \log_b N$

$\log_b \dfrac{M}{N} = \log_b M - \log_b N$

$\log_b M^p = p \log_b M$

If $\log_b x = \log_b y$, then $x = y$.

The change-of-base formula: $\log_b y = \dfrac{\log_a y}{\log_a b}$

7.5 DETERMINANTS

$\begin{vmatrix} a & b \\ c & d \end{vmatrix} = ad - bc$

8.1 THE CIRCLE AND THE PARABOLA

$(x - h)^2 + (y - k)^2 = r^2$ Circle with center at (h, k) and radius r

$x^2 + y^2 = r^2$ Circle with center at the origin and radius r

$\left.\begin{aligned} (y - k)^2 &= \pm 4p(x - h) \\ (x - h)^2 &= \pm 4p(y - k) \end{aligned}\right\}$ Parabola with vertex at (h, k)

8.2 THE ELLIPSE

$\left.\begin{aligned} \dfrac{(x - h)^2}{a^2} + \dfrac{(y - k)^2}{b^2} &= 1 \\ \dfrac{(y - k)^2}{a^2} + \dfrac{(x - h)^2}{b^2} &= 1 \end{aligned}\right\}$ Ellipse with center at (h, k) $(0 < b \le a)$

8.3 THE HYPERBOLA

$\left.\begin{aligned} \dfrac{(x - h)^2}{a^2} - \dfrac{(y - k)^2}{b^2} &= 1 \\ \dfrac{(y - k)^2}{a^2} - \dfrac{(x - h)^2}{b^2} &= 1 \end{aligned}\right\}$ Hyperbola with center at (h, k)

9.1 THE BINOMIAL THEOREM

$n! = n(n - 1)(n - 2) \cdots \cdot 3 \cdot 2 \cdot 1$

$0! = 1$

$n(n - 1)! = n!$

$(a + b)^n = a^n + \dfrac{n!}{1!(n - 1)!} a^{n-1}b +$

$\dfrac{n!}{2!(n - 2)!} a^{n-2}b^2 + \cdots + b^n$

9.2 SEQUENCES, SERIES, AND SUMMATION NOTATION

If c is a constant, then

$$\sum_{k=1}^{n} c = nc$$

$$\sum_{k=1}^{n} cf(k) = c \sum_{k=1}^{n} f(k)$$

$$\sum_{k=1}^{n} [f(k) + g(k)] = \sum_{k=1}^{n} f(k) + \sum_{k=1}^{n} g(k)$$